Operations Research Proceedings

GOR (Gesellschaft für Operations Research e.V.)

More information about this series at http://www.springer.com/series/722

Karl Franz Doerner · Ivana Ljubic
Georg Pflug · Gernot Tragler
Editors

Operations Research Proceedings 2015

Selected Papers of the International
Conference of the German, Austrian
and Swiss Operations Research Societies
(GOR, ÖGOR, SVOR/ASRO),
University of Vienna, Austria,
September 1–4, 2015

 Springer

Editors

Karl Franz Doerner
Department of Business Administration
University of Vienna
Vienna
Austria

Ivana Ljubic
Decision Sciences and Statistics (IDS)
    Department
ESSEC Business School of Paris
Cergy Pontoise Cedex
France

Georg Pflug
Department of Statistics and Operations
    Research
University of Vienna
Vienna
Austria

Gernot Tragler
Operations Research and Control Systems
TU Wien
Vienna
Austria

ISSN 0721-5924   ISSN 2197-9294   (electronic)
Operations Research Proceedings
ISBN 978-3-319-42901-4   ISBN 978-3-319-42902-1   (eBook)
DOI 10.1007/978-3-319-42902-1

Library of Congress Control Number: 2016947499

Printed on acid-free paper

This Springer imprint is published by Springer Nature
The registered company is Springer International Publishing AG Switzerland
The registered company address is: Gewerbestrasse 11, 6330 Cham, Switzerland

Preface

The operations research conference OR2015, jointly organized by the Austrian Operations Research Society (ÖGOR), the German Operations Research Society (GOR, former DGOR), and the Swiss OR Society (SVOR/ASRO), took place during September 2–4, 2015 at the University of Vienna, Austria.

Nearly 900 participants from 48 countries attended the conference. More than 650 presentations were given in 27 streams, including 2 plenary and 12 semi-plenary talks. The motto "Optimal Decisions and Big Data" was given a special emphasis.

This conference series of DGOR/GOR started in 1971 and the Vienna conference was number 43 in a row. It is now a tradition that every 4 years this conference is jointly organized by the sister societies of the three German speaking countries, the last one of this type was Zürich 2011. All the past conferences were documented by the proceedings volumes edited by Springer. Since 1994 the proceedings have been called OR Proceedings and since 2005 they have appeared in the new yellow/blue design.

Although looking back to a long tradition, this proceedings volume contains timely research of more than 100 selected participants of the conference. All contributions are peer-reviewed and show the broad scope of the OR discipline, ranging from structural and algorithmic issues in optimization and simulation to various applications areas, such as transportation and logistics, cutting and packing, production, supply chains, energy, finance, and more.

The German GOR Society awards excellent Ph.D. theses with a special prize. This year, the award winners were Gregor Hendel, Heide Hoppmann, and Time Berthold. The extended abstracts of their presentations are included in this volume.

Special thanks go to all sponsors and supporters (GOR, ÖGOR, SVOR/ASRO, IBM, Gams, Gurobi, Ampl, Risc, Springer, LocalSolver, Siemens, INFORM, EURO, Universität Wien), the local organizers (W. Gutjahr, R. Hartl, I. Ljubic, M. Rauner, G. Tragler, R. Vetschera), the secretarial help (D. Sundt, G. Kamhuber, K. Traub), and last but not least the staff of the faculty of Business, Economics and Statistics of the University of Vienna (K. Kinast, C. Satzer).

Vienna, Austria   Karl Franz Doerner
Cergy Pontoise Cedex, France   Ivana Ljubic
Vienna, Austria   Georg Pflug
Vienna, Austria   Gernot Tragler
April 2016

Contents

Part X Scheduling, Project Management and Health Services

Part XI Energy

Part I
Award Winners

Exploiting Solving Phases for Mixed-Integer Programs

Gregor Hendel

Abstract Modern MIP solving software incorporates dozens of auxiliary algorithmic components for supporting the branch-and-bound search in finding and improving solutions and in strengthening the relaxation. Intuitively, a dynamic solving strategy with an appropriate emphasis on different solving components and strategies is desirable during the search process. We propose an adaptive solver behavior that dynamically reacts on transitions between the three typical phases of a MIP solving process: The first phase objective is to find a feasible solution. During the second phase, a sequence of incumbent solutions gets constructed until the incumbent is eventually optimal. Proving optimality is the central objective of the remaining third phase. Based on the MIP-solver SCIP, we demonstrate the usefulness of the phase concept both with an exact recognition of the optimality of a solution, and provide heuristic alternatives to make use of the concept in practice.

1 Introduction

The availability of sophisticated solving software technology based on the branch-and-bound approach [8] has made Mixed integer programming (MIP) the modeling tool of choice for many practical optimization problems. One of its main advantages is that after termination, branch-and-bound provides a proof of optimality for the best found solution. In many situations, however, practical limits on the run time and memory consumption prevent the search from completing the proof, although the solution found at termination might already be optimal. During the search process, we typically observe three phases: The first phase until a feasible solution is found, a second phase during which a sequence of improving solutions gets constructed, and a third phase during which the remaining search tree must be fully explored to prove

The work for this article has been conducted within the Research Campus Modal funded by the German Federal Ministry of Education and Research (fund number 05M14ZAM).

G. Hendel (✉)
Konrad Zuse Zentrum für Informationstechnologie, Takustraße 7, 14195 Berlin, Germany
e-mail: hendel@zib.de

© Springer International Publishing Switzerland 2017
K.F. Dœrner et al. (eds.), *Operations Research Proceedings 2015*,
Operations Research Proceedings, DOI 10.1007/978-3-319-42902-1_1

optimality. In [6] we empirically demonstrated that the MIP solver SCIP [1] spends more than 40 % of its average solving time during the third phase.

Since every phase emphasizes a different goal of the solving process, it seems natural to pursue these goals with different search strategies to achieve the phase objective as fast as possible. Research on adaptive solver behavior that reacts on solving phases naturally poses the question how the solver should guess that the current incumbent is optimal prior to termination.

There has been little work on such heuristic criteria for deciding whether a solution can be assumed to be optimal. Such criteria cannot be expected to be exact because the decision problem of proving whether a given solution is optimal is still $\mathcal{NP}$-complete in general, hence the term "heuristic".

A bipartion of the solving process has already been suggested in the literature, see [9] for an overview and further references, where the proposed strategies solely involve the node selection in use. Our suggested three-phase approach gives a more refined control of the solver behaviour.

The remainder of the paper is organized as follows: We formally introduce Mixed-Integer Programs and and the concept of solving phases in Sect. 2. The main novelty of this paper are *heuristic transitions* for deciding when the solver should stop searching for better solutions and concentrate on proving optimality. We present two heuristic transitions that take into account global information of the list of open subproblems in Sect. 3. We conclude with a computational study of the proposed adaptive solvers in Sect. 4.

2   Solving Phases in Mixed Integer Programming

Let $A \in \mathbb{R}^{m \times n}$ a real matrix, $b \in \mathbb{R}^m$, $c \in \mathbb{R}^n$, let $l, u \in \mathbb{R}^n_\infty$ and $\mathcal{I} \subseteq \{1, \dots, n\}$, where $n, m \in \mathbb{N}$. A *mixed-integer program (MIP)* is a minimization problem P of the form

$$c^{\mathrm{opt}} := \inf\{c^t x \ : \ x \in \mathbb{R}^n, \ Ax \leq b, \ l \leq x \leq u, \ x_j \in \mathbb{Z} \quad \forall j \in \mathcal{I}\}.$$

A vector $y \in \mathbb{R}^n$ is called a *solution* for P, if it satisfies all linear constraints, bound requirements, and integrality restrictions of P. We call $\mathcal{I}$ the set of *integer variables* of P. A solution y^{opt} that satisfies $c^t y^{\mathrm{opt}} = c^{\mathrm{opt}}$ is called *optimal*. The *LP-relaxation* of P is defined by dropping the integrality restrictions. By solving the LP-relaxation to optimality, we obtain a *lower bound* δ (also called dual bound) on the optimal objective of P. All commercial and noncommercial general purpose MIP solvers are based on the *branch-and-bound* procedure [8], which they extend by various auxiliary components such as *primal heuristics* [5], *cutting plane routines*, and *node presolving techniques* for improving the primal or dual convergence of the method.

Whenever there is an incumbent solution $\hat{y}$, we measure the relative distance between $\hat{y}$ and the optimal objective value c^{opt} in terms of the *primal gap*

$$\gamma := \begin{cases} 0, & \text{if } c^{\text{opt}} = c^t \hat{y}, \\ 100 * \frac{c^t \hat{y} - c^{\text{opt}}}{\max\{|c^t \hat{y}|, |c^{\text{opt}}|\}}, & \text{if } \text{sig}(c^{\text{opt}}) = \text{sig}(c^t \hat{y}), \\ 100, & \text{otherwise.} \end{cases}$$

A primal gap of 0% means that the incumbent is an optimal solution, although this might not be proven so far because the dual bound for P is less than the optimal objective. Similarly, we use a *dual gap* γ^* to measure the relative distance between c^{opt} and the proven dual bound δ.

In the context of solving phases, elapsed time since the solving process was started plays an important role. All definitions such as the incumbent solution $\hat{y}$ and its objective (the primal bound) $c^t \hat{y}$ or its dual counter parts δ and the corresponding gaps γ and γ^* can be translated into functions of the elapsed time. Let $t_1^* > 0$ denote the point in time when the first solution is found or the *first phase transition*. The *primal gap function* $\gamma : [t_1^*, \infty] \mapsto [0, 100]$ measures the primal gap at every point in time $t \geq t_1^*$ during solving by calculating the primal gap for the best incumbent $\hat{y}(t)$ found until t.

For the solving time $T > 0$ for P, we partition the solving time interval $[0, T]$ into three disjoint *solving phases*:

$$\begin{aligned} \mathscr{P}_1 &:= [0, t_1^*[, & \text{the Feasibility phase,} \\ \mathscr{P}_2 &:= \{t \geq t_1^* : \gamma(t) > 0\}, & \text{the Improvement phase,} \\ \mathscr{P}_3 &:= \{t \geq t_1^* : \gamma(t) = 0, \gamma^*(t) > 0\}, & \text{the Proof phase.} \end{aligned}$$

Every solving phase is named after its main primal objective of finding a first and optimal solution in $\mathscr{P}_1$ and $\mathscr{P}_2$, respectively, and proving optimality during $\mathscr{P}_3$. We presented promising strategies for each phase in [6]; During the Feasibility phase, we search for feasible solutions with a two-stage node selection strategy combining a uct [10] and depth-first strategy with restarts together with an inference branching rule. The Improvement phase is conducted with the default search strategy of SCIP except for the use of uct inside Large Neighborhood Search heuristics. For the Proof phase, we deactivate primal heuristics, and apply cutting planes periodically during a depth-first search traversal of the remaining search tree. Note that a phase-based solver that uses different settings after a heuristic phase transition remains exact; the use of different settings based on the heuristic phase transition might only influence the performance of the solver to finish the solving process.

The desired moment in time when a phase-based solver should switch from an improvement strategy to a proof strategy is given by the *second phase transition*

$$t_2^* := \sup \mathscr{P}_1 \cup \mathscr{P}_2.$$

Because of the practical impossibility to detect t_2^* exactly before the solving process finishes, we dedicate the next section to introduce *heuristic phase transitions* for our phase-based solver.

3   Heuristic Phase Transitions

We propose to use properties of the frontier of open subproblems during the solving process as heuristic phase transitions. Let $\mathscr{Q}$ denote the set of open subproblems. We call $Q \in \mathscr{Q}$ an *active node* and denote by d_Q the *depth* of Q in the search tree. If the solving process has not found an optimal solution yet, there exists an active node $Q \in \mathscr{Q}$ that contains it. We use the *best-estimate* [3] to circumvent the absence of true knowledge about best solutions in the unexplored subtrees. After solving the LP-relaxation of a node P with solution $\tilde{y}_P$, the *best-estimate* defined as

$$\hat{c}_P = c^t \tilde{y}_P + \sum_{j:(\tilde{y}_P)_j \notin \mathbb{Z}} \min\{\Psi_j^- \cdot ((\tilde{y}_P)_j - \lfloor (\tilde{y}_P)_j \rfloor), \Psi_j^+ \cdot (\lceil (\tilde{y}_P)_j \rceil - (\tilde{y}_P)_j)\}$$

is an estimate of the best solution objective attainable from P by adding the minimum *pseudo-costs* [3] to make all variables $j \in \mathscr{I}$ with fractional LP-solution values $(\tilde{y}_P)_j \notin \mathbb{Z}$ integral, where we use average unit gains Ψ_j^-, Ψ_j^+ over all previous branching decisions. For active nodes $Q \in \mathscr{Q}$, an initial estimate can be calculated from the parent estimate and the branching decision to create Q.

Definition 1 (*active-estimate transition*) We define the *active-estimate transition* as the first moment in time t_2^{estim} when the incumbent objective is smaller than the minimum best-estimate amongst all active nodes, i.e.

$$t_2^{\text{estim}} := \min\left\{t \geq t_1^* : c^T \hat{y}(t) \leq \inf\{\hat{c}_Q : Q \in \mathscr{Q}(t)\}\right\}. \tag{1}$$

In practice, the best-estimate may be very inaccurate and over- or underestimate the true objective value obtainable from a node, which may lead to an undesirably early or late active-estimate transition. In order to drop the use of the actual incumbent objective, we introduce another transition that compares all active and already processed nodes only at their individual depths. Let the *rank-1 nodes* be defined as

$$\mathscr{Q}^{\text{rank-1}}(t) := \{Q \in \mathscr{Q}(t) : \hat{c}_Q \leq \inf\{\hat{c}_{Q'} : Q' \text{ processed before } t, d_{Q'} = d_Q\}\}.$$

$\mathscr{Q}^{\text{rank-1}}(t)$ contains all active nodes with very small lower bounds or near-integral solutions with small pseudo-cost contributions compared to already processed nodes at the same depth.

Definition 2 (*rank-1 transition*) The *rank-1 transition* is the moment in time when $\mathscr{Q}^{\text{rank-1}}(t)$ becomes empty for the first time:

$$t_2^{\text{rank-1}} := \min\{t \geq t_1^* : \mathscr{Q}^{\text{rank-1}}(t) = \emptyset\}. \tag{2}$$

The main difference between the rank-1 and the active-estimate transitions is that the former does not compare an incumbent objective with the node estimates. Note that the rank-1 criterion $\mathscr{Q}^{\text{rank-1}} = \emptyset$ is never satisfied as long as there exist active nodes which are deeper in the tree than any previously explored node. The name of this transition is inspired by a node rank definition that requires full knowledge about the entire search tree at completion, see [6] for details.

4 Computational Results

We conducted a computational study to investigate the performance benefits of a phase-based solver that reacts on phase transitions with a change of its search strategy. Apart from the default settings of SCIP we tested an `oracle` that detects the second phase transition exactly, `estim` uses the active-estimate transition (1), and `rank-1` the rank-1 transition (2). For the latter two, we also required that at least 50 branch-and-bound nodes were explored. At the time a criterion is met, we assume that the current incumbent is optimal and let the solver react on this assumption by switching to settings for the Proof phase. We tested with a time limit of 2 h on the 168 instances from three publicly available MIPLIB libraries [2, 4, 7]. We excluded four instances for which no optimal solution value was known by the time of this writing.

In Table 1, we present the shifted geometric means of the measured running times of the different settings with a shift of 10 s. We also show the percentage time compared to `default`, the number of solved instances for every setting, and p-values obtained from a two-sided Wilcoxon signed rank test that takes into account logarithmic shifted quotients, see [6] for details. The `oracle` setting could solve three instances more than the default setting. Over the entire test set, we observe improvements in the shifted geometric mean solving time for every new setting, where the highest improvement of 5.6 % was obtained with the `oracle`-setting. With the `rank-1` setting, we obtain a similar speed-up of 5.4 %. Both are accompanied by small p-values of 0.013 and 0.008. The table also shows the results for two instance groups based on the performance of the slowest of the four tested algorithms. On the 73 easy instances, `oracle` is slower than `default` by almost 5 %, whereas `rank-1` is the fastest amongst the tested settings. The computational overhead of the reactivated separation during the Proof phase seems to outweigh its benefits on this easy group. The p-values, however do not reveal any of the settings to be significantly different from `default`.

Table 1 Shifted geometric mean results for t (s) and number of solved instances

	All instances				Easy (max $t \le 200$)			Hard (max $t > 200$)		
	# solv.	t (s)	%	p	t (s)	%	p	t (s)	%	p
default	127	257.0	100.0		11.7	100.0		1992.8	100.0	
estim	129	245.0	95.3	0.905	11.7	100.7	0.521	1827.1	91.7	0.488
oracle	130	242.7	94.4	0.013	12.2	104.9	0.410	1766.3	88.6	0.000
rank-1	128	243.1	94.6	0.008	11.3	97.0	0.226	1832.9	92.0	0.026

The results on the hard instances show more pronounced improvements with all new settings by up to 11.4 % obtained with the `oracle` setting. The setting `estim` improves the time by 8.2 % but the corresponding p-value of 0.488 does not identify this improvement as significant. A smaller time improvement of 8 % with the `rank-1` setting is indicated as significant by a p-value of less than 5 %. This result indicates a more consistent improvement over the entire test set for `rank-1`, whereas the active-estimate transition could rather improve the performance on a few outliers.

5 Conclusions

In our experiment, the use of a phase-specific solver adaptation could significantly improve the running time, especially on harder instances. Furthermore, we introduced two heuristic phase transitions that yielded performance improvements similar to what can be obtained in principle if we could determine the phase transitions exactly, which is an important first step to make use of such adaptive solver behavior in practice. We attribute the significant improvements with the exact and rank-1 transitions in particular to the judicious reactivation of cutting plane separation locally in the tree at the cost of deactivating primal heuristics. Future work on solving phases could comprise experiments with different heuristic phase transitions, or base the work distribution between primal heuristics and separation on more local properties that are specific to the subtree.

References

1. Achterberg, T.: Constraint integer programming. Ph.D. thesis, Technische Universität Berlin (2007)
2. Achterberg, T., Koch, T., Martin, A.: MIPLIB 2003. Oper. Res. Lett. **34**(4), 1–12 (2006)
3. Bénichou, M., Gauthier, J.M., Girodet, P., Hentges, G., Ribière, G., Vincent, O.: Experiments in mixed-integer programming. Math. Program. **1**, 76–94 (1971)
4. Bixby, R.E., Ceria, S., McZeal, C.M., Savelsbergh, M.W.P.: An updated mixed integer programming library: MIPLIB 3.0. Optima **58**, 12–15 (1998)
5. Fischetti, M., Lodi, A.: Heuristics in mixed integer programming. In: Cochran, J.J., Cox, L.A., Keskinocak, P., Kharoufeh, J.P., Smith, J.C. (eds.) Wiley Encyclopedia of Operations Research and Management Science. Wiley, New York (2010). (Online publication)
6. Hendel, G.: Empirical analysis of solving phases in mixed integer programming. Master thesis, Technische Universität Berlin (2014)
7. Koch, T., Achterberg, T., Andersen, E., Bastert, O., Berthold, T., Bixby, R.E., Danna, E., Gamrath, G., Gleixner, A.M., Heinz, S., Lodi, A., Mittelmann, H., Ralphs, T., Salvagnin, D., Steffy, D.E., Wolter, K.: MIPLIB 2010. Math. Program. Comput. **3**(2), 103–163 (2011)
8. Land, A.H., Doig, A.G.: An automatic method of solving discrete programming problems. Econometrica **28**(3), 497–520 (1960)

9. Linderoth, J.T., Savelsbergh, M.W.P.: A computational study of search strategies for mixed integer programming. INFORMS J. Comput. **11**(2), 173–187 (1999)
10. Sabharwal, A., Samulowitz, H., Reddy, C.: Guiding combinatorial optimization with UCT. In: Beldiceanu, N., Jussien, N., Pinson, E. (eds.) CPAIOR. Lecture Notes in Computer Science, vol. 7298, pp. 356–361. Springer, New York (2012)

An Extended Formulation for the Line Planning Problem

Heide Hoppmann

Abstract In this paper we present a novel extended formulation for the line planning problem that is based on what we call "configurations" of lines and frequencies. Configurations account for all possible options to provide a required transportation capacity on an infrastructure edge. The proposed configuration model is strong in the sense that it implies several facet-defining inequalities for the standard model: set cover, symmetric band, MIR, and multicover inequalities. These theoretical findings can be confirmed in computational results. Further, we show how this concept can be generalized to define configurations for subsets of edges; the generalized model implies additional inequalities from the line planning literature.

1 Introduction

Line planning is an important strategic planning problem in public transport. The task is to find a set of lines and frequencies such that a given demand can be transported. There are usually two main objectives: minimizing the travel times of the passengers and minimizing the line operating costs.

Since the late 90s, the line planning literature has developed a variety of integer programming approaches that capture different aspects, for an overview see Schöbel [10]. Bussieck, Kreuzer, and Zimmermann [5] propose an integer programming model to maximize the number of direct travelers. Operating costs are discussed for instance in the article of Goossens, van Hoesel, and Kroon [7]. Schöbel and Scholl [11] and Borndörfer and Karbstein [1] focus on the number of transfers and the number of direct travelers, respectively, and further integrate line planning and passenger routing in their models. Borndörfer, Grötschel, and Pfetsch [2] also propose an integrated line planning and passenger routing model that allows a dynamic generation of lines.

Supported by the Research Center MATHEON "Mathematics for key technologies".

H. Hoppmann (✉)
Zuse Institute Berlin, Takustr. 7, 14195 Berlin, Germany
e-mail: hoppmann@zib.de

© Springer International Publishing Switzerland 2017
K.F. Dœrner et al. (eds.), *Operations Research Proceedings 2015*,
Operations Research Proceedings, DOI 10.1007/978-3-319-42902-1_2

11

All these models employ some type of *capacity* or *frequency demand constraints* in order to cover a given demand. In this paper we propose a concept to strengthen such constraints by means of a novel extended formulation. The idea is to enumerate the set of possible *configurations* of line frequencies for each capacity constraint. We show that such an extended formulation implies general facet defining inequalities for the standard model.

We consider the following basic *line planning problem*: We are given an undirected graph $G = (V, E)$ representing the transportation network; a *line* is a simple path in G and we denote by $\mathcal{L} = \{\ell_1, \ldots, \ell_n\}$, $n \in \mathbb{N}$, the given set of lines. We denote by $\mathcal{L}(e) := \{\ell \in \mathcal{L} : e \in \ell\}$ the set of lines on edge $e \in E$. Furthermore, we are given an ordered set of *frequencies* $\mathcal{F} = \{f_1, \ldots, f_k\} \subseteq \mathbb{N}$, $k \in \mathbb{N}$, such that $0 < f_1 < \cdots < f_k$, and we define $\mathcal{F}_0 := \mathcal{F} \cup \{0\}$. The cost of operating line $\ell \in \mathcal{L}$ at frequency $f \in \mathcal{F}$ is given by $c_{\ell,f} \in \mathbb{Q}_{\geq 0}$. Finally, each edge in the network bears a positive frequency demand $F(e) \in \mathbb{N}$; it gives the number of line operations that are necessary to cover the demand on this edge.

A *line plan* $(\bar{\mathcal{L}}, \bar{f})$ consists of a subset $\bar{\mathcal{L}} \subseteq \mathcal{L}$ of lines and an assignment $\bar{f} : \bar{\mathcal{L}} \to \mathcal{F}$ of frequencies to these lines. A line plan is *feasible* if the frequencies of the lines satisfy the frequency demand $F(e)$ for each edge $e \in E$, i.e., if

$$\sum_{\ell \in \bar{\mathcal{L}}(e)} \bar{f}(\ell) \geq F(e) \text{ for all } e \in E. \tag{1}$$

We define the cost of a line plan $(\bar{\mathcal{L}}, \bar{f})$ as $c(\bar{\mathcal{L}}, \bar{f}) := \sum_{\ell \in \bar{\mathcal{L}}} c_{\ell, \bar{f}(\ell)}$. The *line planning problem* is to find a feasible line plan of minimal cost.

2   Standard Model and Extended Formulation

A common way to formulate the line planning problem uses binary variables $x_{\ell,f}$ indicating whether line $\ell \in \mathcal{L}$ is operated at frequency $f \in \mathcal{F}$. In our case, this results in the following *standard model*:

$$\text{(SLP)} \quad \min \quad \sum_{\ell \in \mathcal{L}} \sum_{f \in \mathcal{F}} c_{\ell,f}\, x_{\ell,f}$$

$$\text{s.t.} \quad \sum_{\ell \in \mathcal{L}(e)} \sum_{f \in \mathcal{F}} f\, x_{\ell,f} \geq F(e) \qquad \forall e \in E \tag{2}$$

$$\sum_{f \in \mathcal{F}} x_{\ell,f} \leq 1 \qquad \forall \ell \in \mathcal{L} \tag{3}$$

$$x_{\ell,f} \in \{0, 1\} \qquad \forall \ell \in \mathcal{L}, \forall f \in \mathcal{F}. \tag{4}$$

Model (SLP) minimizes the cost of a line plan. The *frequency demand constraints* (2) ensure that the frequency demand is covered while the *assignment constraints* (3) guarantee that every line is operated at at most one frequency.

In the following, we give an extended formulation for (SLP) that aims at tightening the LP-relaxation. Our extended formulation is based on the observation that the frequency demand of an edge can also be expressed by specifying the minimum number of lines that have to be operated at each frequency. We call these frequency combinations *minimal configurations* and a formal description is as follows.

Definition 1 For $e \in E$ define the set of *(feasible) configurations of e* by

$$\bar{\mathcal{Q}}(e) := \left\{ q = (q_{f_1}, \dots, q_{f_k}) \in \mathbb{Z}_{\geq 0}^{\mathcal{F}} : \sum_{f \in \mathcal{F}} q_f \leq |\mathcal{L}(e)|, \sum_{f \in \mathcal{F}} f \, q_f \geq F(e) \right\}$$

and the set of *minimal configurations of e* by

$$\mathcal{Q}(e) := \left\{ q \in \bar{\mathcal{Q}}(e) : (q_{f_1}, \dots, q_{f_i} - 1, \dots, q_{f_k}) \notin \bar{\mathcal{Q}}(e) \quad \forall i = 1, \dots, k \right\}.$$

As an example, consider an edge with frequency demand of 9. Let there be three lines on this edge, that each can be operated at frequency 2 or 8. To cover this demand we need at least two lines with frequency 8 or one line with frequency 2 and one line with frequency 8.

We extend (SLP) using binary variables $y_{e,q}$ that indicate for each edge $e \in E$ which configuration $q \in \mathcal{Q}(e)$ is chosen. This results in the following formulation:

$$(\text{QLP}) \quad \min \quad \sum_{\ell \in \mathcal{L}} \sum_{f \in \mathcal{F}} c_{\ell,f} \, x_{\ell,f}$$

$$\text{s.t.} \quad \sum_{\ell \in \mathcal{L}(e)} x_{\ell,f} \geq \sum_{q \in \mathcal{Q}(e)} q_f \, y_{e,q} \qquad \forall e \in E, \forall f \in \mathcal{F} \qquad (5)$$

$$\sum_{q \in \mathcal{Q}(e)} y_{e,q} = 1 \qquad \forall e \in E \qquad (6)$$

$$\sum_{f \in \mathcal{F}} x_{\ell,f} \leq 1 \qquad \forall \ell \in \mathcal{L} \qquad (7)$$

$$x_{\ell,f} \in \{0, 1\} \qquad \forall \ell \in \mathcal{L}, \forall f \in \mathcal{F} \qquad (8)$$

$$y_{e,q} \in \{0, 1\} \qquad \forall e \in E, \forall q \in \mathcal{Q}(e). \qquad (9)$$

The *(extended) configuration model* (QLP) also minimizes the cost of a line plan. The *configuration assignment constraints* (6) ensure that exactly one configuration is chosen for each edge, while the *coupling constraints* (5) guarantee that a sufficient number of lines is operated at the frequencies w.r.t. the chosen configurations.

2.1 Model Comparison

The configuration model (QLP) provides an extended formulation of (SLP), i.e., the convex hulls of all feasible solutions—projected onto the space of the x variables—

coincide. The same does not hold for the polytopes defined by the fractional solutions, since (QLP) provides a tighter LP relaxation.

Band inequalities were introduced by Stoer and Dahl [12] and can also be applied to the line planning problem. Given an edge $e \in E$, a *band* $f_{\mathcal{B}} : \mathcal{L}(e) \to \mathcal{F}_0$ assigns a frequency to each line containing e and is called *valid band of e* if $\sum_{\ell \in \mathcal{L}(e)} f_{\mathcal{B}}(\ell) < F(e)$. That is, if all lines on the edge are operated at the frequencies of a valid band, then the frequency demand is not covered and at least one line needs to be operated at a higher frequency. Hence, the *band inequality*

$$\sum_{\ell \in \mathcal{L}(e)} \sum_{f \in \mathcal{F}: f > f_{\mathcal{B}}(\ell)} x_{\ell,f} \geq 1 \tag{10}$$

is a valid inequality for (SLP) for all $e \in E$ and each valid band $f_{\mathcal{B}}$ of e. The simplest example is the case $f_{\mathcal{B}}(\ell) \equiv 0$, which states that one must operate at least one line on every edge, i.e., the *set cover* inequality $\sum_{\ell \in \mathcal{L}(e)} \sum_{f \in \mathcal{F}} x_{\ell,f} \geq 1$ is valid for $P_{IP}(SLP)$ for all $e \in E$. We call the band $f_{\mathcal{B}}$ *symmetric* if $f_{\mathcal{B}}(\ell) = f$ for all $\ell \in \mathcal{L}(e)$ and for some $f \in \mathcal{F}$. Note that set cover inequalities are symmetric band inequalities. We call the valid band $f_{\mathcal{B}}$ *maximal* if there is no valid band $f_{\mathcal{B}'}$ with $f_{\mathcal{B}}(\ell) \leq f_{\mathcal{B}'}(\ell)$ for every line $\ell \in \mathcal{L}(e)$ and $f_{\mathcal{B}}(\ell) < f_{\mathcal{B}'}(\ell)$ for at least one line $\ell \in \mathcal{L}(e)$. Maximal band inequalities often define facets of the single edge relaxation of the line planning polytope, see [9]. The symmetric ones are implied by the configuration model.

Theorem 1 ([8]) *The LP relaxation of the configuration model implies all band inequalities (10) that are induced by a valid symmetric band.*

The demand inequalities (2) can be strengthened by the mixed integer rounding (MIR) technique [6]. Let $e \in E$, $\lambda > 0$ and define $r = \lambda F(e) - \lfloor \lambda F(e) \rfloor$ and $r_f = \lambda f - \lfloor \lambda f \rfloor$. The *MIR inequality*

$$\sum_{\ell \in \mathcal{L}(e)} \sum_{f \in \mathcal{F}} (r \lfloor \lambda f \rfloor + \min(r_f, r)) x_{\ell,f} \geq r \lceil \lambda F(e) \rceil \tag{11}$$

is valid for (SLP). These strengthened inequalities are also implied by the configuration model.

Theorem 2 ([8]) *The LP relaxation of the configuration model implies all MIR inequalities (11).*

The configuration model is strong in the sense that it implies several facet-defining inequalities for the standard model. However, the enormous number of configurations can blow up the formulation for large instances. Hence, we propose a mixed model that enriches the standard model by a judiciously chosen subset of configurations; only for edges with a small number of minimal configurations the corresponding variables and constraints are added. This provides a good compromise between model strength and model size. We present computational results for large-scale line

planning problems in [3, 8] that confirm the theoretical findings for the naive configuration model and show the superiority of the proposed mixed model. Our approach
shows its strength in particular on real world instances.

3 Multi-edge Configuration

In this section we generalize the concept of minimal configurations with the goal to
tighten the LP relaxation further. The idea is to define configurations for a subset
of edges. For this purpose we partition the lines according to the edges they pass
in $\tilde{E}$, $\tilde{E} \subseteq E$. Let $E' \subseteq \tilde{E}$, then we denote by $\mathcal{L}(E')|_{\tilde{E}} := \{\ell \in \mathcal{L} : \ell \cap \tilde{E} = E'\}$ the
set of lines such that E' corresponds to the edges they pass in $\tilde{E}$ and define $\mathcal{E}(\tilde{E}) :=$
$\{E' \subseteq \tilde{E} : \mathcal{L}(E')|_{\tilde{E}} \neq \emptyset\}$. A *multi-edge configuration* specifies for each line set in this
partition how many of them are operated at a certain frequency. The formal definition
reads as follows:

Definition 2 For $\tilde{E} \subseteq E$, let $\bar{\mathcal{Q}}(\tilde{E}) \subseteq \mathbb{Z}_{\geq 0}^{\mathcal{E}(\tilde{E}) \times \mathcal{F}}$ be the set of *(feasible) multi-edge*
configuration of $\tilde{E}$ with $Q \in \bar{\mathcal{Q}}(\tilde{E})$ if and only if

$$\sum_{f \in \mathcal{F}} Q_{E',f} \leq \left| \mathcal{L}(E')|_{\tilde{E}} \right| \qquad \forall E' \in \mathcal{E}(\tilde{E}), \qquad (12)$$

$$\sum_{\substack{E' \in \mathcal{E}(\tilde{E}): f \in \mathcal{F} \\ e \in E'}} \sum_{f \in \mathcal{F}} f \cdot Q_{E',f} \geq F(e) \qquad \forall e \in \tilde{E}. \qquad (13)$$

We call a multi-edge configuration $Q \in \bar{\mathcal{Q}}(\tilde{E})$ *minimal* if there is no $\bar{Q} \in \bar{\mathcal{Q}}(\tilde{E})$ such
that $\bar{Q}_{E'',f} \leq Q_{E'',f}$ for all $E'' \in \mathcal{E}(\tilde{E}), f \in \mathcal{F}$ and $\bar{Q}_{E'',f} < Q_{E'',f}$ for some $E'' \in \mathcal{E}(\tilde{E})$,
$f \in \mathcal{F}$. The set of *minimal multi-edge configurations of* $\tilde{E}$ is denoted by $\mathcal{Q}(\tilde{E})$.

Let $\mathcal{E}$ be a cover of E, i.e., $\mathcal{E} \subseteq 2^E$ such that $\bigcup_{E' \in \mathcal{E}} E' = E$. We extend the
standard model (SLP) with binary variables $y_{E',Q}$ indicating for each subset of edges
$E' \in \mathcal{E}$ which minimal multi-edge configuration $Q \in \mathcal{Q}(\tilde{E})$ is chosen. The *multi-*
edge configuration model induced by the edge cover $\mathcal{E}$ is defined as follows:

$$(\mathcal{E}\text{-QLP}) \quad \min \sum_{\ell \in \mathcal{L}} \sum_{f \in \mathcal{F}} c_{\ell,f} x_{\ell,f}$$

$$\sum_{\ell \in \mathcal{L}(E')|_{\tilde{E}}} x_{\ell,f} \geq \sum_{Q \in \mathcal{Q}(\tilde{E})(\tilde{E})} Q_{E',f} \cdot y_{\tilde{E},Q} \qquad \forall \tilde{E} \in \mathcal{E}, \, \forall E' \in \mathcal{E}(\tilde{E}), \, \forall f \in \mathcal{F} \quad (14)$$

$$\sum_{Q \in \mathcal{Q}(\tilde{E})} y_{\tilde{E},Q} = 1 \qquad \forall \tilde{E} \in \mathcal{E} \qquad (15)$$

$$\sum_{f \in \mathcal{F}} x_{\ell,f} \leq 1 \qquad \forall \ell \in \mathcal{L} \qquad (16)$$

$$x_{\ell,f} \in \{0, 1\} \qquad\qquad \forall\, \ell \in \mathcal{L}, \forall f \in \mathcal{F} \qquad\qquad (17)$$

$$y_{\tilde{E},Q} \in \{0, 1\} \qquad\qquad \forall\, \tilde{E} \in \mathcal{E},\ \forall\, Q \in \mathcal{Q}(\tilde{E}). \qquad (18)$$

The multi-edge configuration model ($\mathcal{E}$-QLP) minimizes the cost of a line plan. Since for each edge $e \in E$ in ($\mathcal{E}$-QLP) a minimal multi-edge configuration is chosen for the subset $\tilde{E} \in \mathcal{E}$ containing e, the frequency demand of e is satisfied by every feasible solution of ($\mathcal{E}$-QLP). Model ($\mathcal{E}$-QLP) also provides an extended formulation for (SLP).

Let $\tilde{E} \subseteq E$ such that $\alpha(\tilde{E}) := \max\{|\ell \cap \tilde{E}| : \ell \in \mathcal{L}\} > 0$ and $F(E') = \sum_{e \in \tilde{E}} F(e)$, then the aggregated frequency inequality (Bussieck [4])

$$\sum_{\ell \in \mathcal{L}(E')} \sum_{f \in \mathcal{F}} f\, x_{\ell,f} \geq \left\lceil \frac{1}{\alpha(\tilde{E})} F(E') \right\rceil \qquad (19)$$

and the aggregated cardinality inequality

$$\sum_{\ell \in \mathcal{L}(E')} \sum_{f \in \mathcal{F}} x_{\ell,f} \geq \left\lceil \frac{1}{\alpha(\tilde{E})} |\tilde{E}| \right\rceil \qquad (20)$$

are valid for (SLP). These inequalities are in general not valid for the LP relaxation of the standard model (SLP) and the configuration model (QLP). However, regarding the multi-edge configuration model ($\mathcal{E}$-QLP) we obtain the following result.

Proposition 1 ([8]) *Let $\mathcal{E}$ be a cover of E and $\tilde{E} \in \mathcal{E}$ s.t. $\alpha(\tilde{E}) > 0$. Then the aggregated frequency inequality* (19) *and the aggregated cardinality inequality* (20) *for $\tilde{E}$ are implied by the LP relaxation of the multi-edge configuration model ($\mathcal{E}$-QLP).*

References

1. Borndörfer, R., Karbstein, M.: A direct connection approach to integrated line planning and passenger routing. In: Delling, D., Liberti, L. (eds.) ATMOS 2012 - 12th Workshop on Algorithmic Approaches for Transportation Modelling, Optimization, and Systems, vol. 25, pp. 47–57 (2012)
2. Borndörfer, R., Grötschel, M., Pfetsch, M.E.: A column-generation approach to line planning in public transport. Transp. Sci. **41**(1), 123–132 (2007)
3. Borndörfer, R., Hoppmann, H., Karbstein, M.: A configuration model for the line planning problem. In: Frigioni, D., Stiller, S. (eds.) ATMOS 2013 - 13th Workshop on Algorithmic Approaches for Transportation Modeling, Optimization, and Systems, vol. 33, pp. 68–79 (2013)
4. Bussieck, M.: Optimal lines in public rail transport. Ph.D. thesis, Technische Universität Braunschweig (1998)
5. Bussieck, M.R., Kreuzer, P., Zimmermann, U.T.: Optimal lines for railway systems. Eur. J. Oper. Res. **96**(1), 54–63 (1997)
6. Dash, S., Günlük, O., Lodi, A.: MIR closures of polyhedral sets. Math. Program. **121**(1), 33–60 (2010)

7. Goossens, J.-W., van Hoesel, S., Kroon, L.: A branch-and-cut approach for solving railway line-planning problems. Transp. Sci. **28**(3), 379–393 (2004)
8. Hoppmann, H.: A configuration model for the line planning problem. Master's thesis, Technische Universität Berlin (2014)
9. Karbstein, M.: Line planning and connectivity. Ph.D. thesis, TU Berlin (2013)
10. Schöbel, A.: Line planning in public transportation: models and methods. OR Spectr. 1–20 (2011)
11. Schöbel, A., Scholl, S.: Line planning with minimal traveling time. In: Kroon, L.G., Möhring, R.H. (eds.) Proceedings of 5th Workshop on Algorithmic Methods and Models for Optimization of Railways (2006)
12. Stoer, M., Dahl, G.: A polyhedral approach to multicommodity survivable network design (1994)

Improving the Performance of MIP and MINLP Solvers by Integrated Heuristics

Timo Berthold

Abstract This article provides an overview of the author's dissertation (Berthold, Heuristic algorithms in global MINLP solvers, 2014, [4]). We study heuristic algorithms that are tightly integrated within global MINLP solvers and analyze their impact on the overall solution process. This comprises generalizations of primal heuristics for MIP towards MINLP as well as novel ideas for MINLP primal heuristics and for heuristic algorithms to take branching decisions and to collect global information in MIP.

1 Introduction

The past twenty years have witnessed a substantial progress in the development of MIP (mixed integer programming) software packages, see, e.g., [1, 11]. As a consequence, state-of-the-art MIP solvers, commercial and non-commercial, are nowadays capable of solving a variety of different types of MIP instances arising from real-world applications within reasonable time, and mixed integer linear programming has become a standard technique in Operations Research. For mixed integer non-linear programming (MINLP), today's situation is comparable to that of MIP in the early nineties. Techniques to efficiently solve MINLPs are known in principle. Most of them have been developed in academic proof-of-concept implementations, but are not yet well-tested for a broad range of industrial problems, in particular not for large-scale applications.

On the practical side, there are many real-world applications that are inherently nonlinear and need to be tackled by MINLP. This induces a growing need for MINLP algorithms that are at the same time innovative from a theoretical perspective and efficient in practice.

T. Berthold (✉)
Fair Isaac Germany GmbH, c/o Zuse Institute Berlin, Takustr. 7, 10551 Berlin, Germany
e-mail: timoberthold@fico.com

© Springer International Publishing Switzerland 2017   19
K.F. Dœrner et al. (eds.), *Operations Research Proceedings 2015*,
Operations Research Proceedings, DOI 10.1007/978-3-319-42902-1_3

Towards this goal, we present new ideas for heuristic algorithms that are conceived with a special focus on being employed within a global MINLP solver. In the following, we shortly review the contributions of [4]. Computational experiments were carried out within the MINLP framework SCIP [2, 10]. Significant parts of the thesis have been published in refereed conference proceedings and international journals [3, 5–9].

2   Shift-and-Propagate

Many primal heuristics for MIP and MINLP either solve sequences of linear programs or auxiliary MIPs and MINLPs. Moreover, they rely on an optimal relaxation solution being at hand. Finding this may itself take a significant amount of time.

Shift-and-Propagate [7] is a primal heuristic that does not require a previously found relaxation solution. It alternately shifts variables to promising fixing values in order to make linear constraints feasible and propagates these fixings to get tighter domains for choosing subsequent fixing values. Shift-and-Propagate is specifically designed to be employed as a quick start heuristic inside a global solver.

Our experiments in [4] reveal that this heuristic alone finds solutions on more instances than three rounding heuristics and two combinatorial improvement heuristics together. Combining existing rounding and improvement heuristics with Shift-and-Propagate increases the number of instances on which a feasible solution is found by 60 %. Furthermore, the average primal gap at the end of root node processing can be reduced by 14 %, while the running time increases only by 2 %.

3   An Objective Feasibility Pump for Nonconvex MINLP

The Feasibility Pump is probably the best known primal heuristic for mixed integer programming. The original work by Fischetti, Glover, and Lodi [13] has been succeeded by more than a dozen follow-up publications.

In [4], we present and evaluate three novel ideas for solving nonconvex MINLPs with a Feasibility Pump: the generation of valid cutting planes for nonconvex nonlinearities, using a hierarchy of MIP solving procedures, and applying an objective function for the auxiliary MIPs that incorporates second-order information. For the latter, we introduced the so-called *Hesse-distance*, which measures the distance of two points using the Hessian of the Lagrangian of an NLP as a metric.

In our computational experiments, the dynamic use of various MIP solving strategies shows the favorable behavior to produce more solutions and better solutions in a shorter average running time. A convex combination of the Hesse-distance function and the Manhattan distance likewise improves the number of found solutions and their quality, but at the cost of an increased running time.

4 Large Neighborhood Search: From MIP to MINLP

Large neighborhood search (LNS) is a variant of the local search paradigm that has been widely used in Constraint Programming, Operations Research, and Combinatorial Optimization. LNS heuristics are an important component of modern MIP solvers [12]. To define the neighborhood, the feasible region of the MIP is restricted by additional constraints: most often variable fixings or some very restrictive cardinality constraint.

We discuss a generic and straightforward way of generalizing large neighborhood search heuristics from mixed integer *linear* programming to mixed integer *nonlinear* programming. In this context, we implemented and tested MINLP versions of six LNS heuristics that are known from the literature.

Our computational results show that the generalized LNS heuristics increase the quality of the best feasible solution after root node computation, and further, that they improve the behavior of the solver for the global search. The improvement merely consists of a faster convergence towards the optimal solution in terms of the primal integral [3] while not affecting the time needed to prove optimality.

5 RENS: The Optimal Rounding

Many LNS heuristics, diving and of course all rounding heuristics are based on the idea of fixing some of the variables that take an integral value in a relaxation solution. Therefore, the question of whether a given solution of a relaxation is roundable, i.e., all fractional variables can be shifted to integral values without losing feasibility for the constraint functions, is particularly important for the likelihood of many primal heuristics to succeed.

The RENS algorithm [5] constructs a sub-MINLP of a given MINLP based on an optimal solution of a linear or nonlinear relaxation. RENS is designed to compute an optimal—w.r.t. the original objective function—rounding of a relaxation solution.

In computational experiments, we demonstrate that most MIP, MIQCP, and MINLP instances have roundable LP and NLP optima and that in most cases, the optimal roundings can be computed efficiently. For MINLP, a version of SCIP that applies RENS frequently during search gives an improvement of 8 % in running time.

6 Undercover: The Smallest Sub-MIP

Undercover [6] is a LNS heuristic that explores a mixed integer *linear* subproblem of a given MINLP. This sub-MIP is constructed by fixing a minimal set of variables that suffices to linearize *all* nonlinear constraints. Although general in nature, the heuristic appears most promising for MIQCPs.

Undercover solves a vertex covering problem to identify a smallest set of variables to fix, a so-called *cover*, such that each nonlinear constraint becomes linear in the remaining variables. Subsequently, these variables are fixed to values obtained from a reference point, e.g., an optimal solution of a linear relaxation.

Undercover exploits the fact that small covers correspond to large sub-MIPs. We show that a majority MINLPs from the MINLPLib allow for covers consisting of at most 25 % of the variables. For MIQCPs, Undercover proved to be a fast start heuristic, that often produces feasible solutions of reasonable quality. We further show that for MIQCPs, applying Undercover at the root node improves the overall performance of SCIP by up to 30 %.

7 Rapid Learning: A New CP/MIP Hybrid

Conflict learning is a technique that analyzes infeasible subproblems encountered during a tree search algorithm. Conflict analysis techniques have been developed and furthered by the artificial intelligence research community and, later, the SAT and the CP community.

Rapid Learning [9] is a heuristic algorithm based on the observation that a CP solver can typically perform a partial search on a few hundred or thousand nodes in a fraction of the time that a MIP solver needs for processing the root node. Rapid Learning splits the search in two phases: an incomplete CP search which feeds the global information storage and a subsequent MIP search which starts from scratch, hopefully profiting from the information that the rapid CP search gathered.

Our computational experiments indicate that performing a CP search for valid conflicts once at the root node is particularly beneficial for infeasible instances and IPs for which the primal part of the search is the hard one, improving running times by 25 % on average.

8 Cloud Branching: Exploiting Dual Degeneracy

Branch-and-bound methods for mixed integer linear programming are traditionally based on solving a linear programming relaxation and branching on a variable which takes a fractional value in a (single) relaxation optimum. Cloud Branching [8] is a branching strategy that makes use of multiple relaxation solutions.

Cloud Branching exploits dual degeneracy of the LP relaxation, hence the existence of alternative optimal solutions to filter unpromising branching candidates. Our computational experiments show that a version of Full Strong Branching that uses Cloud Branching ideas is about 30 % faster than default Full Strong Branching on a standard test set with high dual degeneracy.

9 The Impact of Primal Heuristics

In the main computational study of [4], we conducted experiments on three general, publicly available, academic benchmark sets for MIP, MIQCP, and MINLP, namely, the MIPLIB, the GLOMIQO test set, and the MINLPLIB. Further, we included results for real-world applications from projects with industry partners who the author was associated with during his time at ZIB. This comprises MIPs from a a supply chain management application of SAP, MIQCPs from an internal Siemens project on "power management for smart buildings", and nonconvex MINLPs from a nomination validation problem of Open Grid Europe. Those instances are not publicly available since they contain confidential data from our partner's operating procedures.

It turns out that primal heuristics consistently help to significantly improve performance on all six test sets. This holds true for various measures that we used, namely the number of found solutions, the time to find a first solution, the quality of the primal bound at termination, the primal integral [3], and the number of branch-and-bound nodes required to prove optimality, see Fig. 1. These results were consistent over all test sets, on the original formulations as well as on permuted formulations, and pass statistical significance tests. We further observe that primal heuristics have only few impact on easy instances, but are crucial for solving hard optimization problems. In a nutshell: The more difficult the problem, the more important becomes the deployment of primal heuristics within global solvers for MIP and MINLP.

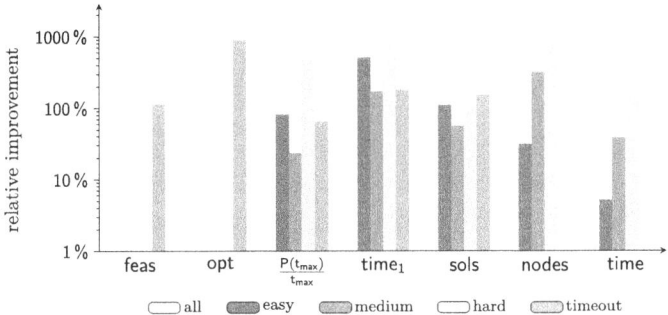

Fig. 1 Bar chart illustrating the relative difference between SCIP running with and without primal heuristics w.r.t. seven performance measures. Scaled such that positive values correspond to improvements by using heuristics. Grouped into easy, medium, hard, and timeout instances by the number of branch-and-bound nodes needed to prove optimality

References

1. Achterberg, T., Wunderling, R.: Mixed integer programming: analyzing 12 years of progress. In: Jünger, M., Reinelt, G. (eds.) Facets of Combinatorial Optimization, pp. 449–481. Springer, Berlin (2013)
2. Achterberg, T., Berthold, T., Koch, T., Wolter, K.: Constraint integer programming: a new approach to integrate CP and MIP. In: Perron, L., Trick, M.A. (eds.) Proceedings of CPAIOR 2008. LNCS, vol. 5015, pp. 6–20. Springer, Berlin (2008)
3. Berthold, T.: Measuring the impact of primal heuristics. Oper. Res. Lett. **41**(6), 611–614 (2013)
4. Berthold, T.: Heuristic algorithms in global MINLP solvers. Ph.D. thesis, Technische Universität Berlin (2014)
5. Berthold, T.: RENS - the optimal rounding. Math. Program. Comput. **6**(1), 33–54 (2014)
6. Berthold, T., Gleixner, A.M.: Undercover: a primal MINLP heuristic exploring a largest sub-MIP. Math. Program. **144**(1–2), 315–346 (2014)
7. Berthold, T., Hendel, G.: Shift-and-propagate. J. Heuristics **21**(1), 73–106 (2015)
8. Berthold, T., Salvagnin, D.: Cloud branching. In: Gomes, C., Sellmann, M. (eds.) Proceedings of the CPAIOR 2013. LNCS, vol. 7874, pp. 28–43. Springer, Berlin (2013)
9. Berthold, T., Feydy, T., Stuckey, P.J.: Rapid learning for binary programs. In: Lodi, A., Milano, M., Toth, P. (eds.) Proceedings of CPAIOR 2010. LNCS, vol. 6140, pp. 51–55. Springer, Berlin (2010)
10. Berthold, T., Heinz, S., Vigerske, S.: Extending a CIP framework to solve MIQCPs. In: Lee, J., Leyffer, S. (eds.) Mixed Integer Nonlinear Programming. The IMA Volumes in Mathematics and its Applications, vol. 154, pp. 427–444. Springer, New York (2011)
11. Bixby, R.E., Fenelon, M., Gu, Z., Rothberg, E., Wunderling, R.: MIP: theory and practice – closing the gap. In: Powell, M.J.D., Scholtes, S. (eds.) Systems Modelling and Optimization: Methods, Theory, and Applications, pp. 19–49. Kluwer Academic Publisher, Boston (2000)
12. Fischetti, M., Lodi, A.: Heuristics in mixed integer programming. In: Cochran, J.J., Cox, L.A., Keskinocak, P., Kharoufeh, J.P., Smith, J.C. (eds.) Wiley Encyclopedia of Operations Research and Management Science. Wiley, New York (2010) (Online publication)
13. Fischetti, M., Glover, F., Lodi, A.: The feasibility pump. Math. Program. **104**(1), 91–104 (2005)

Part II
Discrete Optimization, Integer
Programming, Graphs and Networks

An Experimental Study of Algorithms for Controlling Palletizers

Frank Gurski, Jochen Rethmann and Egon Wanke

Abstract We consider the FIFO STACK-UP problem which arises in delivery industry, where bins have to be stacked-up from conveyor belts onto pallets. Given k sequences $q_1, \ldots, q_k$ of labeled bins and a positive integer p. The goal is to stack-up the bins by iteratively removing the first bin of one of the k sequences and put it onto a pallet located at one of p stack-up places. Each of these pallets has to contain bins of only one label, bins of different labels have to be placed on different pallets. After all bins of one label have been removed from the given sequences, the corresponding stack-up place becomes available for a pallet of bins of another label. The FIFO STACK-UP problem is computational intractable (Gurski et al., Math. Methods Oper. Res., [3], ACM Comput. Res. Repos. (CoRR), 2013, [4]). In this paper we consider two linear programming models for the problem and compare the running times of our models for randomly generated sequences using GLPK and CPLEX solvers. We also draw comparisons with a breadth first search solution for the problem (Gurski et al.,Modelling, Computation and Optimization in Information Systems and Management Sciences, 2015, [7]).

1 Introduction

We consider the combinatorial problem of stacking up bins from conveyor belts onto pallets. This problem originally appears in *stack-up systems* or *palletizing systems* that play an important role in delivery industry and warehouses. Stack-up systems

F. Gurski (✉) · E. Wanke
Institute of Computer Science, University of Düsseldorf, 40225 Düsseldorf, Germany
e-mail: frank.gurski@hhu.de

E. Wanke
e-mail: e.wanke@hhu.de

J. Rethmann
Faculty of Electrical Engineering and Computer Science, Niederrhein University of Applied Sciences, 47805 Krefeld, Germany
e-mail: jochen.rethmann@hs-niederrhein.de

© Springer International Publishing Switzerland 2017
K.F. Dœrner et al. (eds.), *Operations Research Proceedings 2015*,
Operations Research Proceedings, DOI 10.1007/978-3-319-42902-1_4

are often the back end of *order-picking systems*. A detailed description of the applied background of such systems is given in [1, 8].

The bins that have to be stacked-up onto pallets reach the stack-up system on a conveyor belt. At the stack-up system the bins are picked-up by stacker cranes or robotic arms and moved onto pallets. The pallets are located at *stack-up places*. This picking process can be performed in different ways depending on the architecture of the palletizing system. Full pallets are carried away by automated guided vehicles, or by another conveyor system, while new empty pallets are placed at free stack-up places.

We consider so-called *multi-line palletizing systems*, where there are several buffer conveyors from which the bins are picked-up. The robotic arms or stacker cranes and the stack-up places are located at the end of these conveyors. We assume that the assignment of the bins to the conveyors and the order of bins within each conveyor is given. If further each arm can only pick-up the first bin of one of the buffer conveyors, then the system is called a FIFO palletizing system. Such systems can be modeled by several simple queues. Figure 1 shows a sketch of a simplified stack-up system with 2 buffer conveyors and 3 stack-up places.

In the following we describe a stack-up processing using a simple example. For some technical definitions see [3, 5–7]. Given two sequences $q_1 = (b_1, \ldots, b_4) = [a, b, a, b]$ and $q_2 = (b_5, \ldots, b_{10}) = [c, d, c, d, a, b]$. Each bin b_i is labeled with a *pallet symbol* $plt(b_i)$. Every row of Fig. 2 represents a *configuration* $C_Q = (i_1, i_2)$, where value i_j denotes the number of bins which have been removed from sequence q_j, $j = 1, 2$. Already removed bins are shown in grey. The transition from one

Fig. 1 A FIFO stack-up system

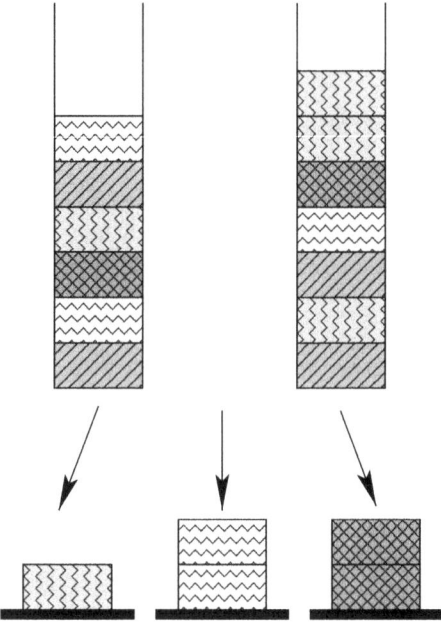

Fig. 2 A processing of two given sequences with 2 stack-up places

i	q_1	q_2	bin to remove	C_Q	$open(C_Q)$
0	$[a,b,a,b]$	$[c,d,c,d,a,b]$	b_5	$(0,0)$	$\emptyset$
1	$[a,b,a,b]$	$[c,\underline{d},c,d,a,b]$	b_6	$(0,1)$	$\{c\}$
2	$[a,b,a,b]$	$[c,d,\underline{c},d,a,b]$	b_7	$(0,2)$	$\{c,d\}$
3	$[a,b,a,b]$	$[c,d,c,\underline{d},a,b]$	b_8	$(0,3)$	$\{d\}$
4	$[\underline{a},b,a,b]$	$[c,d,c,d,a,b]$	b_1	$(0,4)$	$\emptyset$
5	$[a,b,a,b]$	$[c,d,c,d,\underline{a},b]$	b_9	$(1,4)$	$\{a\}$
6	$[a,\underline{b},a,b]$	$[c,d,c,d,a,b]$	b_2	$(1,5)$	$\{a\}$
6	$[a,b,a,b]$	$[c,d,c,d,a,\underline{b}]$	b_{10}	$(2,5)$	$\{a,b\}$
7	$[a,b,\underline{a},b]$	$[c,d,c,d,a,b]$	b_3	$(2,6)$	$\{a,b\}$
8	$[a,b,a,\underline{b}]$	$[c,d,c,d,a,b]$	b_4	$(3,6)$	$\{b\}$
9	$[a,b,a,b]$	$[c,d,c,d,a,b]$	$-$	$(4,6)$	$\emptyset$

row to the next row, i.e. the removal of a bin on position $i_j + 1$ of sequence q_j in configuration $C_Q = (i_1, i_2)$ is called a *transformation step*. A pallet t is called *open* in configuration $C_Q = (i_1, \ldots, i_k)$ during a processing of k sequences, if there is a bin for pallet t at some position less than or equal to i_j in sequence q_j, and if there is another bin for pallet t at some position greater than i_ℓ in sequence q_ℓ. The *set of open pallets* in configuration C_Q is denoted by $open(C_Q)$. A sequence of transformation steps that transforms the list Q of k sequences from the initial configuration $(0, 0, \ldots, 0)$ into the final configuration $(|q_1|, |q_2|, \ldots, |q_k|)$ is called a *processing* of Q. The order in which the bins are removed from the sequences within a processing of Q is called a *bin solution* of Q and the order in which the pallets are opened during the processing of Q is denoted as a *pallet solution*. In Fig. 2 we obtain the bin solution $B = (b_5, b_6, b_7, b_8, b_1, b_9, b_2, b_{10}, b_3, b_4)$, and the pallet solution $T = (c, d, a, b)$.

The FIFO STACK-UP problem is to decide for a given list Q of k sequences and a positive integer p whether there is a processing of Q, such that in each configuration during the processing of Q at most p pallets are open.

The FIFO STACK-UP problem is NP-complete even if the number of bins per pallet is bounded, but can be solved in polynomial time if the number k of sequences or the number p of stack-up places is fixed [3, 4]. A dynamic programming solution for the FIFO STACK-UP problem is shown in [3, 5]. Parameterized algorithms and a linear programming approach for computing a pallet solution for the problem is given in [6]. A breadth first search solution combined with some cutting technique for the problem was presented in [7].

We use the following variables: k denotes the number of sequences, p stands for the number of stack up places, m represents the number of pallets, n denotes the total number of bins, and $N = \max\{|q_1|, \ldots, |q_k|\}$ is the maximum sequence length. In view of the practical background, it holds $p < m$, $k < m$, $m < n$, and $N < n$. For some positive integer n, let $[n] = \{1, \ldots, n\}$ denote the set of all positive integers between 1 and n.

2 Solutions of the FIFO Stack-Up Problem

Linear Programming 1: Computing a bin solution   We have given a list Q of k sequences and n bins $b_1, \ldots, b_n$ which use m pallet symbols. Our aim is to find a bijection $\pi : [n] \to [n]$ for the bins, such that by the removal of the bins from the sequences in the order of π the number of needed stack-up places p is minimized.

To realize a bijection $\pi : [n] \to [n]$ we define n^2 binary variables $x_i^j \in \{0, 1\}$, $i, j \in [n]$, such that x_i^j is equal to 1, if and only if $\pi(i) = j$. In order to ensure π to be surjective and injective, we use the conditions $\sum_{i=1}^n x_i^j = 1$ for every $j \in [n]$ and $\sum_{j=1}^n x_i^j = 1$ for every $i \in [n]$. Further we have to ensure that all variables x_i^j, $i, j \in [n]$, are in $\{0, 1\}$. We will denote the previous $n^2 + 2n$ conditions by Permutation(n, x_i^j).

We have to consider the relative orderings of the bins given by the k sequences $q_1 = (b_1, \ldots, b_{n_1}), \ldots, q_k = (b_{n_{k-1}+1}, \ldots, b_{n_k})$. For every sequence q_s, $1 \le s \le k$, and every bin b_i of this sequence, i.e. $n_{s-1} + 1 \le i \le n_s$, we know that if b_i is placed at position j, i.e. $x_i^j = 1$, then

- every bin $b_{i'}$, $i' > i$ of sequence q_s, i.e. $i < i' \le n_s$, is not placed before b_i, i.e. $x_{i'}^{j'} = 0$ for all $j' < j$, which is ensured by $x_{i'}^{j'} \le 1 - x_i^j$ and
- every bin $b_{i'}$, $i' < i$ of sequence q_s, i.e. $n_{s-1} + 1 \le i' < i$, is not placed after b_i, i.e. $x_{i'}^{j'} = 0$ for all $j' > j$, which is ensured by $x_{i'}^{j'} \le 1 - x_i^j$.

We will denote the previous $\mathcal{O}(k \cdot n^2 \cdot N^2)$ conditions by SequenceOrder(Q, x_i^j).

By an integer valued variable p we count the number of used stack-up places for some given sequence Q as follows.

$$\text{minimize } p \tag{1}$$

subject to

$$\text{Permutation}(n, x_i^j), \text{ and SequenceOrder}(Q, x_i^j) \tag{2}$$

$$\text{and } \sum_{t=1}^m f(t, c) \le p \text{ for every } c \in [n - 1] \tag{3}$$

$$\text{where } f(t, c) = \left(\bigvee_{i \in [n], j \le c, plt(b_i)=t} x_i^j \right) \wedge \left(\bigvee_{i \in [n], j > c, plt(b_i)=t} x_i^j \right) \tag{4}$$

The used logical operations in (4) can be expressed by linear programming constraints as shown in [2]. For the correctness note that subexpression $f(t, c)$ is equal to one if and only if in the considered ordering of the bins there is a bin b' with $plt(b') = t$ opened at a step $\le c$ and there is a bin b'' with $plt(b'') = t$ opened at a step $> c$, if and only if pallet t is open after the cth bin has been removed. Integer program (1)–(4) has $n^2 + 1$ variables and a polynomial number of constraints.

Theorem 1 *For every list Q of sequences the binary integer programm (1)–(4) computes the minimum number of stack-up places p in a processing of Q.*

Linear Programming 2: Computing a pallet solution   A useful relation between an instance of the FIFO STACK-UP problem and the directed pathwidth of a directed graph model was applied in [6]. The *sequence graph* $G_Q = (V, E)$ for an instance $Q = (q_1, \ldots, q_k)$ of the FIFO STACK-UP problem is defined in [3, 6] as follows. The vertex set V is the set of pallet symbols. There is an arc $(u, v) \in E$ if and only if there is a sequence $q_i = (b_{n_{i-1}+1}, \ldots, b_{n_i})$ with two bins b_{j_1}, b_{j_2} such that (1) $j_1 < j_2$, (2) b_{j_1} is destined for pallet u, (3) b_{j_2} is destined for pallet v, and (4) $u \neq v$. Digraph $G_Q = (V, E)$ can be computed in time $\mathcal{O}(n + k \cdot |E|) \subseteq \mathcal{O}(n + k \cdot m^2)$, see [4].

Theorem 2 ([3, 4]) *Let $Q = (q_1, \ldots, q_k)$ then the sequence graph $G_Q = (V, E)$ has directed pathwidth at most $p - 1$ if and only if Q can be processed with at most p stack-up places.*

By Theorem 2 the minimum number of stack-up places can be computed by the directed pathwidth of sequence graph G_Q plus one. Using the fact that the directed pathwidth equals the directed vertex separation number [9], in [6] we have given an integer program on $m^2 + 1$ variables and a polynomial number of constraints that computes the minimum number of stack-up places.

Breadth First Search Solution   In [7] we introduced a breadth first search solution FIND OPTIMAL PALLET SOLUTION (FIFO-FOPS) of running time $\mathcal{O}(n^2 \cdot (m + 2)^k)$. Since for practical instance sizes this running time is too huge for computations, we added some cutting technique to reduce the size of the search space.

3   Experimental Study

Next we evaluate an implementation of algorithm FIFO-FOPS and realizations of our two LP solutions in GLPK and CPLEX.

Creating Instances Since there are no benchmark data sets for the FIFO STACK-UP problem we generated realistic but randomly instances by an algorithm, which allows to give the following parameters.

- p_{max} an upper bound on the number of stack-up places needed to process the generated sequences
- m number of pallets
- k number of sequences
- r_{min} and r_{max} the minimum and maximum number of bins per pallet
- d maximum number of sequences on which the bins of each pallet can be distributed

Algorithm RANDOM INSTANCES

#$open$:= 0; $open$:= $\emptyset$;	▶ Description of Functions and Variables:
for all $i \in [m]$ do no[i]:=random($r_{\min}, r_{\max}$)	▶ random(l, u): choose a value in $[l, u]$ at random
for all $i \in [m]$ and $j \in [d]$ do	▶ no[$1..m$]: int // number bins for pallets $1 \ldots m$
seq[i][j] := random($1, k$)	▶ n: int // total number of bins, $n = \sum_{i=1}^{m} \text{no}[i]$
for all $i \in [k]$ do q_i := ()	▶ seq[$1..m$][$1..d$]: int // for each individual pallet
$unproc$:= $[m]$; i := 0;	▶ there are up to d sequences, on which to
while $i < n$ do	▶ distribute the bins

while $i < n$ do
  $i := i + 1$
  if #$open = p_{\max}$
    $plt :=$ choose some pallet of $open$ at random
  else
    $plt :=$ choose some pallet of $open \cup unproc$ at random
  if $plt \in unproc$ ▷ first bin of pallet plt
    #$open :=$ #$open + 1$; $open := open \cup \{plt\}$; $unproc := unproc - \{plt\}$;
  $x :=$random($1, d$); let $s :=$ seq[plt][x];
  append bin b_i to sequence q_s
  no[plt] := no[plt] - 1
  if no[plt] = 0   ▷ last bin of pallet plt
    #$open$:= #$open - 1$; $open := open - \{plt\}$;

Fig. 3 Construction of random instances for the FIFO STACK-UP problem

Table 1 Running times in seconds for randomly generated instances of the FIFO STACK-UP problem. Running times on more than 1800 s are indicated by a bar (-)

Instance				GLPK		CPLEX	
n	p	m	k	Bin solution	Pallet solution	Bin solution	Pallet solution
15	2	3	2	41.4	0.1	0.1	0.1
20	2	4	2	-	0.1	1.7	0.1
30	4	5	4	-	0.7	66.8	0.2
40	4	6	4	-	2.5	932.6	1.2
50	4	8	5	-	684.6	-	16.3
100	5	10	5	-	-	-	282.3
Instance				BFS and cutting pallet solution			
n	p	m	k				
1000	18	50	12	2.5			
2000	18	100	12	4.2			
5000	20	200	12	7.7			
7500	20	250	12	11.4			
10000	24	300	12	61.9			
20000	24	500	12	89.2			

The idea is to compute a bin solution $B = (b_1, \ldots, b_n)$ with respect to the given parameters and to distribute the bins to the k sequences such that the relative order will be preserved, i.e. if b_i has been placed left of b_j in some sequence, then $i < j$ in B. The algorithm is shown in Fig. 3.

Implementation We have implemented algorithm FIFO-FOPS as a single-threaded program in C++ on a standard Linux PC with 3.30 GHz CPU and 7.7 GiB RAM. GLPK v4.43 and CPLEX 12.6.0.0 have been run on the same machine. GLPK is single-threaded, while CPLEX uses all 4 cores of the CPU.

Evaluation In Table 1 we list some of our chosen parameters. For each assignment we randomly generated and solved 10 instances in order to compute the average time for solving the same instances with the given parameters by our two linear programming models using GLPK and CPLEX and by algorithm FIFO-FOPS.

Our results show that FIFO-FOPS can be used to solve practical instances on several thousand bins of the FIFO STACK-UP problem. Our two linear programming approaches can only be used to handle instances up to 100 bins and less than 10 pallets. As expected, CLPEX can solve the instances much faster than GLPK, and the pallet solution approach is much better than the bin solution approach.

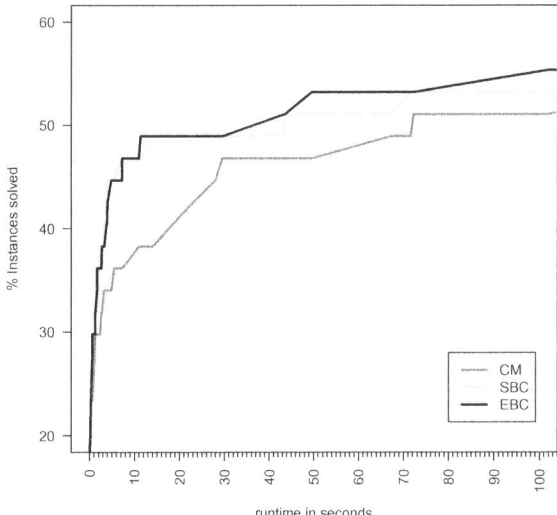

References

1. de Koster, R.: Performance approximation of pick-to-belt orderpicking systems. Eur. J. Oper. Res. **92**, 558–573 (1994)
2. Gurski, F.: Efficient binary linear programming formulations for boolean functions. Stat. Optim. Inf. Comput. **2**(4), 274–279 (2014)
3. Gurski, F., Rethmann, J., Wanke, E.: On the complexity of the FIFO stack-up problem. Math. Methods Oper. Res. (To appear)

4. Gurski, F., Rethmann, J., Wanke, E.: Complexity of the FIFO stack-up problem. ACM Comput. Res. Repos. (CoRR) arXiv:abs/1307.1915(2013)
5. Gurski, F., Rethmann, J., Wanke, E.: Moving bins from conveyor belts onto pallets using FIFO queues. Operations Research Proceedings (OR 2013), pp. 185–191. Springer, New York (2014)
6. Gurski, F., Rethmann, J., Wanke, E.: Algorithms for controlling palletizers. Operations Research Proceedings (OR 2014). Springer, New York (2015) (To appear)
7. Gurski, F., Rethmann, J., Wanke, E.: A practical approach for the FIFO stack-up problem. Modelling, Computation and Optimization in Information Systems and Management Sciences. Advances in Intelligent Systems and Computing, vol. 360, pp. 141–152. Springer, New York (2015)
8. Rethmann, J., Wanke, E.: Storage controlled pile-up systems, theoretical foundations. Eur. J. Oper. Res. **103**(3), 515–530 (1997)
9. Yang, B., Cao, Y.: Digraph searching, directed vertex separation and directed pathwidth. Discret. Appl. Math. **156**(10), 1822–1837 (2008)

Robust Two-Stage Network Problems

Adam Kasperski and Paweł Zieliński

Abstract In this paper a class of network optimization problems is discussed. It is assumed that a partial solution can be formed in the first stage, when the arc costs are precisely known, and completed optimally in the second stage, after a true second-stage cost scenario occurs. The robust min-max criterion is used to compute an optimal solution. Several complexity results for the interval and discrete uncertainty representations are provided.

1 Introduction

Many optimization problems arising in operations research have a two-stage nature. Namely, a partial solution can be computed in the first stage and then completed optimally in the second stage, after a true state of the world reveals. Typically, the second stage costs are uncertain and they are specified as a scenario set. Each particular realization of the second stage costs is called a scenario. If no additional information with the scenario set is provided, then the robust min-max criterion can be applied to choose a solution. In this paper we consider a class of network optimization problems. The input instance is a network with deterministic first stage arc costs and uncertain second stage arc costs. We wish to build an object in this network such as a spanning tree, $s - t$ path, $s - t$ cut, or matching whose maximum total cost in both stages over all scenarios is minimum.

The robust two-stage network models have been discussed in a number of papers. In [8] the two-stage bipartite matching and in [6] the two-stage spanning tree problems have been examined. When the number of second stage cost scenarios is a part

A. Kasperski (✉)
Faculty of Computer Science and Management, Wrocław University of Technology,
Wrocław, Poland
e-mail: adam.kasperski@pwr.edu.pl

P. Zieliński
Faculty of Fundamental Problems of Technology, Wrocław University of Technology,
Wrocław, Poland
e-mail: pawel.zielinski@pwr.edu.pl

© Springer International Publishing Switzerland 2017   35
K.F. Dœrner et al. (eds.), *Operations Research Proceedings 2015*,
Operations Research Proceedings, DOI 10.1007/978-3-319-42902-1_5

of input, then both problems are strongly NP-hard and hard to approximate within $(1 - \varepsilon) \log n$ for any $\varepsilon > 0$. In [6, 8] some positive approximation results for both problems have been shown as well. In [1, 4] the robust two-stage network problems with uncertain demands have been studied. In this paper we provide several complexity results for the problem under study in the case of the interval and discrete uncertainty representations.

2 Problem Formulation

We are given a directed graph $G = (V, A)$. Let Φ be a set of all feasible solutions, where each solution $X \in \Phi$ is formed by a subset of arcs of G. Suppose that for each arc $a \in A$ a nonnegative cost c_a is given. In the deterministic problem $\mathscr{P}$ we seek a solution $X \in \Phi$ minimizing the total cost $\sum_{a \in X} c_a$. In this paper we will consider the following special cases of set Φ: Φ_{ST} contains all spanning trees in G, Φ_{SP} contains all simple $s - t$ paths in G, Φ_{CUT} contains all $s - t$-cuts in G (let $V_1 \cup V_2$ be a partition of V such that $s \in V_1$ and $t \in V_2$; then the cut contains all the arcs starting in V_1 and ending in V_2), Φ_M contains all perfect matchings in G. Given Φ_α, where $\alpha \in \{SP, ST, CUT, M\}$, we will also define Φ'_α as the following relaxation of Φ_α: $X' \in \Phi'_\alpha$ if there is $X \subseteq X'$ such that $X \in \Phi_\alpha$. For example, if $X' \in \Phi'_{SP}$, then graph $G' = (V, X')$ contains a simple path from s to t and, possibly, more arcs. For the deterministic problem $\mathscr{P}$ with nonnegative arc costs we get the same optimal solutions for Φ_α and Φ'_α. However, we will show that the situation is different when a two-stage problem is considered.

Let us now introduce the robust two-stage version of problem $\mathscr{P}$. For each arc $a \in A$ we are given a first stage cost $C_a \geq 0$ and a second stage cost $c_a(S) \geq 0$ under scenario S. All the possible second stage cost scenarios (scenarios for short) are specified as scenario set Γ. Let $X_1 \subseteq A$ be a subset of arc and $S \in \Gamma$. Define

$$f(X_1, S) = \sum_{a \in X_1} C_a + \min_{\{X_2 \subseteq A:\, X_1 \cup X_2 \in \Phi\}} \sum_{a \in X_2} c_a(S).$$

In this paper we consider the following problem:

$$\text{Two-Stage } \mathscr{P} : \min_{X_1 \subseteq A} f_1(X_1) = \min_{X_1 \subseteq A} \max_{S \in \Gamma} f(X_1, S).$$

Consider a sample problem shown in Fig. 1. When $\Phi = \Phi_{SP}$, then the cost of an optimal solution equals $M + 1$, because we can choose either arc e_1 or e_2 in the first stage (but not both). On the other hand, when $\Phi = \Phi'_{SP}$, then we can choose both arcs e_1 and e_2 in the first stage and the cost of an optimal solution equals 3. This example demonstrates that the problems with Φ_α and Φ'_α may be quite different. The problem with Φ_{SP} may appear, for example, if one wants to travel to some point in the first stage and wait in this point for the true second stage scenario. On the other

Fig. 1 A sample problem with sets Φ_{sp} and Φ'_{sp}, where M is a large positive number

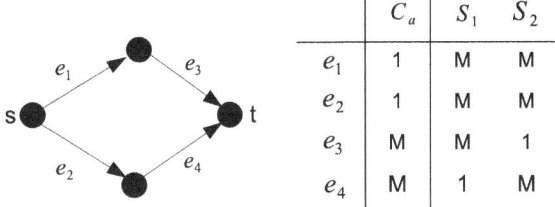

	C_a	S_1	S_2
e_1	1	M	M
e_2	1	M	M
e_3	M	M	1
e_4	M	1	M

hand, Φ'_{SP} is appropriate if one wishes to buy (or rent) some connections in the first stage and connect optimally s and t in the second stage.

3 Interval Uncertainty Representation

Assume that for each arc we have an interval $[c_a, c_a + d_a]$ of all possible second-stage costs, where $c_a \geq 0$ is the nominal second stage cost, and $d_a \geq 0$ is the maximum possible deviation of the second stage cost of a from its nominal value. Then Γ^l, $l \in [0 : |A|]$, is the subset of the Cartesian product $\Pi_{a \in A}[c_a, c_a + d_a]$ such that in any $S \in \Gamma^l$ at most l arcs have the second stage costs greater than their nominal values. The parameter l allows us to model the degree of uncertainty. In particular $l = 0$ means that all the arcs have precise second stage costs, and $l = |A|$ corresponds the the traditional interval uncertainty representation, where $\Gamma = \Pi_{a \in A}[c_a, c_a + d_a]$ (see, e.g. [9]). Scenario set Γ^l was introduced in [3].

Theorem 1 *When l is a part of input, $\Gamma = \Gamma^l$, and $\Phi \in \{\Phi_{ST}, \Phi'_{ST}, \Phi_{SP}, \Phi'_{SP}\}$, then* TWO-STAGE $\mathscr{P}$ *is strongly NP-hard.*

Proof Consider the following MOST VITAL $\mathscr{P}$ problem. Given a network G with arc costs $c_a \geq 0, a \in A$, a set of feasible solutions Φ, and a positive integer k. We seek a set of k arcs whose removal maximizes the increase of the cost of an optimal solution in $\mathscr{P}$. Define this maximum optimal cost as opt_1. Given an instance of MOST VITAL $\mathscr{P}$, focus on the following instance of TWO-STAGE $\mathscr{P}$ with the same network G and Φ. The first stage cost of $a \in A$ is $C_a = M$, and the interval second-stage arc cost of $a \in A$ is $[c_a, M]$, where M is a large constant, say $M = |A| \cdot \max_{a \in A} c_a$. Choose scenario set Γ^l for $l = k$. Since all the first stage costs are very large, no arc is chosen in the first stage. We thus have to evaluate $opt = f_1(\emptyset) = \max_{S \in \Gamma^l} \min_{X \in \Phi} \sum_{a \in X} c_a(S)$. But it is easy to check that $opt = opt_1$. It has been shown in [2, 10] that MOST VITAL $\mathscr{P}$ is strongly NP-hard when $\Phi = \Phi_{ST}$ or $\Phi = \Phi_{SP}$. Hence computing the value of opt is also strongly NP-hard. Observe that opt does not change when we replace Φ_{ST} with Φ'_{ST} or Φ_{SP} with Φ'_{SP} in the instance of TWO-STAGE $\mathscr{P}$. This observation follows from the fact that no arc is chosen in the first stage and $\min_{X \in \Phi} \sum_{a \in X} c_a(S) = \min_{X \in \Phi'} \sum_{a \in X} c_a(S)$. $\square$

Observe that the evaluation of $f_1(\emptyset)$ is an adversarial problem discussed in [11]. In our problem the adversary is the nature which aims to choose the worst cost scenario before a solution is computed. Theorem 1 shows that this problem is NP-hard. Interestingly, the opposite problem in which decision maker first chooses a solution and nature picks then the worst scenario, is polynomially solvable [3]. It is easily seen that TWO-STAGE $\mathscr{P}$ is polynomially solvable for scenario set $\Gamma^{|A|}$. Indeed, let $\hat{c}_a = \min\{C_a, c_a + d_a\}$ for each $a \in A$, and let $\hat{X}$ be an optimal solution to problem $\mathscr{P}$ for the arc costs $\hat{c}_a, a \in A$. An easy verification shows that $\hat{X} = \hat{X}_1 \cup \hat{X}_2$ is an optimal solution to the two-stage problem, where $a \in \hat{X}_1$ if $\hat{c}_a = C_a$ and $a \in \hat{X}_2$ otherwise. The solution $\hat{X}$ can be computed in polynomial time if $\mathscr{P}$ is polynomially solvable.

4 Discrete Uncertainty Representation

In this section we discuss the case when Γ contains K explicitly listed scenarios. Consider first the following min-max version of the representatives selection problem (MIN-MAX RS for short), first studied in [5]. We are given a set T of n tools. This set is partitioned into p disjoint sets $T^1, \ldots, T^p$, where $|T^i| = r_i$ and $n = \sum_{i \in [p]} r_i$. For notation purposes the tools in set T^i will be denoted by $f_1^i, \ldots, f_{r_i}^i$. We wish to choose a subset $Y \subseteq T$ of the tools which contains exactly one tool from each set T^i, i.e. $|Y \cap T^i| = 1$ for each $i \in [p]$. Let $c_j(S)$ be a cost of tool $j \in T$ under scenario $S \in \Gamma$. Let Γ be a finite scenario set, $|\Gamma| = K$. We seek a selection Y whose maximum cost, $\max_{S \in \Gamma} \sum_{j \in Y} c_j(S)$, is minimized. The MIN-MAX RS problem is known to be NP-hard for two scenarios and strongly NP-hard when the number of scenarios is a part of input [5]. In the latter case it is also not approximable within $O(\log^{1-\varepsilon} K)$ for any $\varepsilon > 0$, unless the problems in NP have quasipolynomial algorithms [7].

Theorem 2 *When $\Phi \in \{\Phi_{SP}, \Phi_{CUT}, \Phi_M\}$, then TWO-STAGE $\mathscr{P}$ is NP-hard when $K = 2$. Furthermore, when K is a part of input, then it is strongly NP-hard and not approximable within $O(\log^{1-\varepsilon} K)$ for any $\varepsilon > 0$ unless the problems in NP have quasipolynomial algorithms.*

Proof We will show a polynomial time cost preserving reduction from MIN-MAX RS to TWO-STAGE $\mathscr{P}$ when $\Phi \in \{\Phi_{SP}, \Phi_{CUT}, \Phi_M\}$. Given an instance of MIN-MAX RS with scenario set $\Gamma = \{S_1, \ldots, S_K\}$, we build the corresponding instance of TWO-STAGE $\mathscr{P}$ as follows. Let M be a sufficiently large constant. Network G will contain three types of arcs. Each *tool arc* f_j^i has the first stage cost equal to M and the second stage cost under S_k, $k \in [K]$, equal to the cost of the tool f_j^i under S_k; each *dashed arc* has the first stage cost equal to 0 and the second stage cost equal to M under each scenario; each *bold arc* has the first stage cost equal to M and the second stage cost equal to M under each scenario. The networks for sets Φ_{SP}, Φ_{CUT} and Φ_M are shown in Figs. 2, 3, and 4, respectively.

Consider the problem with Φ_{SP} (see the network shown in Fig. 2). Assume that there is a selection Y of the tools whose min-max cost is c. A solution to the two-stage

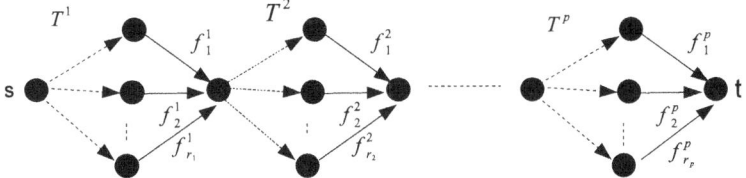

Fig. 2 Network G when $\Phi = \Phi_{SP}$. G contains the tool and the *dashed arcs*

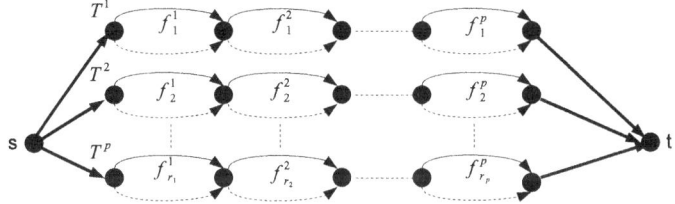

Fig. 3 Network G from the reduction $\Phi = \Phi_{CUT}$. G contains the tool, *dashed*, and *bold arcs*

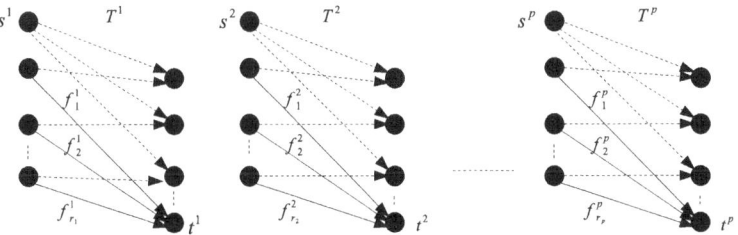

Fig. 4 Network G from the reduction $\Phi = \Phi_M$. G contains the tool and the *dashed arcs*

problem of the same cost can be constructed by considering a path in G of the form $X = X_1 \cup X_2$, where $X_2 = Y$ and X_1 contains the proper dashed arcs. The dashed arcs from X_1 are selected in the first stage. By the construction, we get $f_1(X_1) = c$. Suppose now that there is a subset of arcs X_1 such that $f_1(X_1) = c$. When M is sufficiently large, then we can choose X_1 such that $c < M$. By the construction, X_1 can contain only dashed arcs. Furthermore, since Φ_{SP} contains only simple paths from s to t, the set X_1 contains exactly one dashed arc from each component in G corresponding to T^i. For X_1, there is an unique completion X_2 of X_1 such that $X_1 \cup X_2 \in \Phi_{SP}$. Observe that X_2 corresponds to a feasible tool selection. It is clear that min-max cost of $Y = X_2$ equals c. The reasoning is similar for sets Φ_{CUT} and Φ_M (see Figs. 3 and 4). $\qquad \square$

The reduction in Theorem 2 is not correct when $\Phi \in \{\Phi'_{SP}, \Phi'_{CUT}, \Phi'_M\}$. It follows from the fact that we can choose all the dashed arcs in the first stage, which allows us to obtain a better solution. In the next theorem we show how to modify the reduction to obtain a weaker result.

Theorem 3 When $\Phi \in \{\Phi'_{SP}, \Phi'_{CUT}, \Phi'_M\}$, then TWO-STAGE $\mathscr{P}$ is NP-hard when $K = 2$ and strongly NP-hard when K is a part of input.

Proof The reduction is the same as in the proof of Theorem 2 with the following small modification. We change the first stage cost of all dashed arcs to M and the second stage cost of all dashed arcs under each scenario to $N \gg M$. This ensures that we choose exactly one dashed arc in each component T^i in the first stage and no dashed arc in the second stage. Easy verification shows that there is a selection of the tools of the min-max cost equal to c if and only if there is a subset of arcs X_1 such that $f_1(X_1) \leq pM + c$. A similar reasoning applies to sets Φ'_{CUT} and Φ'_M. Observe that the reduction is not cost preserving, so the negative result is weaker than the corresponding one in Theorem 2.   □

Acknowledgements This work was partially supported by the National Center for Science (Narodowe Centrum Nauki), grant 2013/09/B/ST6/01525.

References

1. Atamturk, A., Zhang, M.: Two-stage robust network flow and design under demand uncertainty. Oper. Res. **55**, 662–673 (2007)
2. Bar-Noy, A., Khuller, A.S., Schieber, B.: The complexity of finding most vital arcs and nodes. Technical report CS-TR-3539, Institute for Advanced Studies, University of Maryland, College Park, MD (1995)
3. Bertsimas, D., Sim, M.: Robust discrete optimization and network flows. Math. Program. **98**, 49–71 (2003)
4. Dhamdhere, K., Goya, V., Ravi, R., Singh, M.: How to pay, come what may: approximation algorithms for demand-robust covering problems. In: Annual IEEE Symposium on Foundations of Computer Science, pp. 367–378 (2005)
5. Dolgui, A., Kovalev, S.: Min-max and min-max (relative) regret approaches to representatives selection problem. 4OR Q. J. Oper. Res. **10**, 181–192 (2012)
6. Kasperski, A., Zieliński, P.: On the approximability of robust spanning problems. Theor. Comput. Sci. **412**, 365–374 (2011)
7. Kasperski, A., Kurpisz, A., Zieliński, P.: Approximability of the robust representatives selection problem. Oper. Res. Lett. **43**, 16–19 (2015)
8. Katriel, I., Kenyon-Mathieu, C., Upfal, E.: Commitment under uncertainty: two-stage matching problems. Theor. Comput. Sci. **408**, 213–223 (2008)
9. Kouvelis, P., Yu, G.: Robust Discrete Optimization and Its Applications. Kluwer Academic Publishers, Boston (1997)
10. Lin, K.C., Chern, M.S.: The most vital edges in the minimum spanning tree problem. Inf. Process. Lett. **45**, 25–31 (1993)
11. Nasrabadi, E., Orlin, J.B.: Robust optimization with incremental recourse. CoRR (2013). arXiv:1312.4075

Composed Min-Max and Min-Sum Radial Approach to the Emergency System Design

Marek Kvet and Jaroslav Janáček

Abstract This paper deals with the emergency service system design, where not only the disutility of an average user is minimized, but also the disutility of the worst situated user must be considered. Optimization of the average user disutility is often related to the weighted p-median problem. To cope with both objectives, we have suggested a composed method. In the first phase, the disutility of the worst situated user is minimized. The second phase is based on the min-sum approach to optimize the average user's disutility. The result of the first phase is used here to reduce the model size. We focus on effective usage of the reduction and explore the possibility of a trade-off between a little loss of optimality and computational time.

1 Introduction

The emergency system design consists in locating a limited number of service centers at positions from a given finite set to optimize the service accessibility to an average user. This way the emergency system design can be tackled as the weighted p-median problem known also as min-sum optimization [1, 3, 4, 10]. Applying of the min-sum objective may lead to such design, where some users are caught in inadmissibly distant locations from any service center, what is considered unfair. The fair designing emerges whenever limited resources are to be fairly distributed among participants. Fairness has been broadly studied in [2, 11]. We focus here on the simple min-max criterion used in the p-center problem. We study a composition of the min-sum and min-max approaches. It is based on the idea that the min-max optimization performed at the first stage of the composition considerably reduces the set of relevant distances considered at the following stage, where the min-sum location problem is solved. The extension of the original composition has been evoked by the finding that the

M. Kvet (✉) · J. Janáček
Faculty of Management Science and Informatics, University of Žilina,
Univerzitná 8215/1, 010 26 Žilina, Slovakia
e-mail: marek.kvet@fri.uniza.sk

J. Janáček
e-mail: jaroslav.janacek@fri.uniza.sk

© Springer International Publishing Switzerland 2017   41
K.F. Dœrner et al. (eds.), *Operations Research Proceedings 2015*,
Operations Research Proceedings, DOI 10.1007/978-3-319-42902-1_6

min-max stage puts on a very strong limit on the second stage of the composition and the price of fairness paid by the average user is too high. That is why the limit given by the min-max solution is relaxed taking some higher distance as the limit of the maximal distance. To be able to make fully use of the reduced set of relevant distances, the radial formulation [3, 5, 6, 8] has been employed for both stages. It was found that this way of problem solving is much more effective than the location-allocation formulation.

2 Composition of Min-Max and Min-Sum Optimization Methods

The emergency system design problem is formulated as a task of location of at most p service centers so that the objective function value derived from distances between center locations and users location is minimal. To describe the problem, we denote by I a set of possible service center locations and by J the set of users locations. The symbol b_j denotes the number of users sharing the location j. The distance between service center location i and users location j is denoted as d_{ij}. The matrix $\{d_{ij}\}$ of all distances from a center location $i \in I$ to a users location $j \in J$ contains a sequence $d^0 < d^1 < \cdots < d^u$ of relevant values, which do not belong to the $p - 1$ highest values in an individual column of the matrix. The sequence plays a key role in the next problem processing, where we define a binary constant a_{ij}^s for each triple $[i, j, s]$, where $i \in I, j \in J$ and $s \in [0 \ldots u]$. The constant a_{ij}^s is equal to 1, if the distance d_{ij} between the user location j and the possible center location i is less than or equal to d^s, otherwise a_{ij}^s is equal to 0. Through all models, we will use binary variables $y_i \in \{0, 1\}$ for $i \in I$ to model the decision on placing a service center at the location i by the value of 1. Further, for the first stage model, we introduce the variables x_j, to indicate, whether the users distance to the nearest service center is greater than d^s. In this case, the variable takes the value of 1, and it takes the value of 0 otherwise. The corresponding model can be formulated according to [9] as follows.

$$Minimize \quad \sum_{j \in J} x_j \tag{1}$$

$$Subject \ to: \quad x_j + \sum_{i \in I} a_{ij}^s y_i \geq 1 \quad for \ \ j \in J \tag{2}$$

$$\sum_{i \in I} y_i \leq p \tag{3}$$

$$x_j \geq 0 \quad for \ \ j \in J \tag{4}$$

$$y_i \in \{0, 1\} \quad for \ \ i \in I \tag{5}$$

The objective function (1) represents the number of user locations, the users distance to the nearest service center of which is greater than d^s. The constraints (2) ensure that the variables x_j are allowed to take the value of 0, if there is at least one center located in radius d^s from the user location j and constraint (3) limits the number of located service centers by p. Solution of the problem (1)–(5) indicates, whether there exists a solution of the emergency system design problem with objective function value less than d^s, what can be easily used by bisection process for determination of the lowest superscript $v + 1$, where the associated value of (1) is zero. The resulting value d^v of the bisection process is used as the highest relevant value for the second stage. In the case, when the superscript v takes its value from the range 20–30, the min-sum stage can be easily solved exactly according to the following model [6, 8], where another series of nonnegative variables x_{js} for $j \in J$ and $s \in [0 \ldots v]$ is introduced. The variable x_{js} indicates by the value of one that no service center is located in the radius d^s from the users location j. The symbol e_s denotes the difference $d^{s+1} - d^s$.

$$Minimize \qquad \sum_{j \in J} b_j \sum_{s=0}^{v} e_s x_{js} \qquad (6)$$

$$Subject \ to: \qquad x_{js} + \sum_{i \in I} a_{ij}^s y_i \geq 1 \qquad for \ \ j \in J \ \ and \ \ s = 0, 1 \ldots v \qquad (7)$$

$$\sum_{i \in I} y_i \leq p \qquad (8)$$

$$\sum_{i \in I} a_{ij}^{v+1} y_i \geq 1 \qquad for \ \ j \in J \qquad (9)$$

$$x_{js} \geq 0 \qquad for \ \ j \in J \ \ and \ \ s = 0, 1 \ldots v \qquad (10)$$

$$y_i \in \{0, 1\} \qquad for \ \ i \in I \qquad (11)$$

In this model, the constraints (7) ensure that the variables x_{js} are allowed to take the value 0, if there is at least one service center located in the distance d^s from the users location j. The constraint (8) puts the limit p on the number of located service centers. The constraints (9) prevent the solution from admitting a distance from a user to the service center higher than d^v. As mentioned in Sect. 1, the limit on the highest relevant distance produced by the first stage may impose a too strong constraint paid by very big loss of the average distance from users to the nearest service center. That is why a designer backs off the demand on fairness to obtain better characteristics of the designed system for the average user. The min-max limit relaxation can be performed by simple setting of the superscript v to some higher value than obtained by the bisection in the first stage. If the new value of superscript v does not exceed the upper bound of the above mentioned range, the model (6)–(11) can be used to obtain the resulting emergency system design. In the case, when the final value of v is too high to process the complete sequence $d^0 < d^1 < \cdots < d^v$, there are two ways to overcome the difficulties brought by big size of the problem (6)–(11). The first way consists in application of a heterogeneous system of radii, where an individual sub-sequence $D_{j0} < D_{j1} < \cdots < D_{jw(j)} = d^v$ is determined for each users location $j \in J$. The sub-sequence contains all distance values from the location j to the possible service center locations, which are not higher than d^v. Then the constants e_{js} and a_{ij}^s must be redefined according to the sub-sequences and the second stage can be solved using the model (6)–(11) with slight changes in coefficients of the objective function (6). The second way makes use of the concept of dividing points [5, 6] and converts the approach to an approximate one. The selection of dividing points is obviously based on so-called "relevance" of a distance d^h, which expresses the strength of our expectation that the distance d^h will be a part of the unknown optimal solution. So far, we have suggested and explored several ways of the relevance estimation and the results are reported in [7, 8].

3 Computational Study

To compare the min-sum, min-max and the composed approaches to the emergency system design, we performed the series of numerical experiments on the pool of benchmarks obtained from the real emergency health care system originally designed for each self-governing region of Slovakia. The instances are organized so that they correspond to the administrative organization of Slovakia. For each self-governing region (Bratislava (BA), Banská Bystrica (BB), Košice (KE), Nitra (NR), Prešov (PO), Trenčín (TN), Trnava (TT) and Žilina (ZA)) all cities and villages with corresponding numbers of inhabitants b_j were taken into account. The coefficients b_j were rounded to hundreds. The set of communities represents both the set J of users locations and the set I of possible center locations. The number p of located centers was derived from the original design and it varies from 9 to 67. To verify the suggested approach on larger instances, we used the benchmark of the whole real emergency system of Slovakia. The cardinality of the set I is 2916 and the value of p was set to 273. To solve the problems, the optimization software FICO Xpress 7.7 (64-bit, release 2014) was used and the experiments were run on a PC equipped with the Intel Core i7 2630 QM 2.0 GHz processor and 8 GB RAM. The obtained results are reported in the Table 1. The computational times are not reported here, but we note that the computational time for any instance solved by radial min-sum approach does not exceed 2 s whereas the min-max solution was completed at most in 0.5 s. As far as the whole road network is concerned, the computational process of the min-sum approach takes 12 s and the min-max approach does not exceed 35 s. For each design, the system objective function was computed by (12).

$$ObjF(y) = \sum_{j \in J} b_j min \left\{ d_{ij} : i \in I, y_i = 1 \right\} \tag{12}$$

Table 1 Comparison of the min-sum, min-max and the composed approaches

	$|I|$	p	MIN-SUM			MIN-MAX			Composed		
			ObjF	MRD	Afl	ObjF	DmM	BD	ObjF	DmM	BD
BA	87	9	20,342	25	2.92	33,018	14	1.86	26,229	14	2.03
BB	515	52	17,289	26	1.48	42,410	13	3.24	21,780	13	1.44
KE	460	46	20,042	23	1.79	39,310	12	2.03	24,117	12	0.85
NR	350	35	22,651	17	1.42	36,810	13	2.73	26,894	13	3.41
PO	664	67	20,025	22	1.75	53,560	12	2.24	24,467	12	1.55
TN	276	28	15,686	30	2.26	33,789	12	2.21	23,476	12	2.41
TT	249	25	18,873	24	1.71	27,823	13	2.84	21,067	13	1.20
ZA	315	32	20,995	26	1.01	39,556	14	4.05	24,424	14	1.52
SR	2916	273	161,448	24	1.66	342,896	13	3.02	193,715	13	1.34

The maximal relevant distance between the users and the nearest located service center is denoted by *MRD*. The column *Afl* is used to express the percentage of the population, whose distance to the nearest service center is greater than the maximal distance *DmM* resulting from the min-max solution. The percentage of users, whose distance to the nearest center equals *DmM*, is given in the columns denoted by *BD*.

The reported results show, that the min-sum, min-max and the composed approaches bring different solutions as concerns the system objective function value. The min-max approach provides a solution with much higher system objective function value, but the maximal relevant distance from the users locations to the nearest service center is approximately half as much smaller. From this point of view, the presented composed method represents a suitable compromise. Since the min-max approach may bring a too strong limit on the maximal distance, we suggested a series of experiments to study the consequences of this limit relaxation. Instead of the maximal distance *DmM* obtained by solving the min-max problem, we took a higher distance d^k, where the superscript k was shifted by $1, 3, 5, \dots, 15$ from the superscript v of the distance d^v obtained by the min-max stage. Then we solved the min-sum problem on the second stage and compared the resulting objective function value to the simple min-sum solution. The average results for self-governing regions are plotted in the Table 2 together with the result for whole Slovakia. The reported values represent the percentual deviation of the composed method solution from the min-sum solution. The obtained results prove that the average deviation of

Table 2 Accuracy evaluation of the composed method with different levels of *DmM* relaxation

Levels of *DmM* relaxation	1	3	5	7	9	11	13	15
Self-governing regions (average)	13.15	4.22	2.19	1.66	1.33	0.25	0.13	0.13
Whole road network of Slovakia	11.24	4.47	2.54	0.95	0.31	0.00	0.00	0.00

the result from the simple min-sum solution decreases with increasing level of *DmM* relaxation.

4 Conclusions

We have suggested a composed approach to the emergency system design. The composition consists of combination of the min-max and min-sum approaches based on radial formulation. It takes into account both the fairness in access to service and overall system effectivity. Concerning the min-max approach, the resulting design prevents about 2 % of population from being exposed to excessively long distance to the nearest service center. But this result is paid by almost twice the system objective function value compared to the min-sum approach. The composed approach considerably reduces this loss of effectiveness and preserves the same maximal distance to the nearest service center as the min-max approach. Since the min-max result may put a very strong limit on the maximal distance allowed in the following min-sum approach, we explored the consequences of this limit relaxation and found that a little turning back of min-max fairness can bring considerable gain of the system effectiveness. Concerning easy implementation on a commercial IP-solver, we have presented a flexible tool for emergency system design.

Acknowledgements This work was supported by the research grants VEGA 1/0518/15 "Resilient rescue systems with uncertain accessibility of service" and APVV-15-0179 "Reliability of emergency systems on infrastructure with uncertain functionality of critical elements".

References

1. Avella, P., Sassano, A., Vasil'ev, I.: Computational study of large scale p-median problems. Math. Program. **109**, 89–114 (2007)
2. Bertsimas, D., Farias, V.F., Trichakis, N.: The price of fairness. Oper. Res. **59**, 1731 (2011)
3. García, S., Labbé, M., Marín, A.: Solving large p-median problems with a radius formulation. INFORMS J. Comput. **23**(4), 546–556 (2011)
4. Ingolfsson, A., Budge, S., Erkut, E.: Optimal ambulance location with random delays and travel times. Health Care Manag. Sci. **11**(3), 262–274 (2008)
5. Janáček, J.: Approximate Covering Models of Location Problems. Lecture Notes in Management Science: ICAOR 08, vol. 1, pp. 53–61. Yerevan, Armenia (2008)
6. Janáček, J., Kvet, M.: Approximate solving of large p-median problems. In: ORP3 Conference 2011, 13–17 Sept, pp. 221–225. Cádiz, Spain. ISBN 978-84-9828-348-8 (2011)
7. Janáček, J., Kvet, M.: Public service system design with disutility relevance estimation. Mathematical Methods in Economics, 11–13 Sept 2013, pp. 332–337. Jihlava (2013)
8. Janáček, J., Kvet, M.: Relevant network distances for approximate approach to large p-median problems. In: Operations Research Proceedings 2012: Selected Papers of the International Conference on Operations Research, 4–7 Sept 2012, pp. 123–128. Springer, Hannover, Germany. ISSN 0721-5924, ISBN 978-3-319-00794-6 (2014)
9. Kvet, M., Janáček, J.: Min-max optimal public service system design. Croat. Oper. Res. Rev. **6**(1), 17–27 (2015)

10. Marianov, V., Serra, D.: Location problems in the public sector. In: Drezner, Z., et al. (eds.) Facility Location: Applications and Theory, pp. 119–150. Springer, Berlin (2002)
11. Ogryczak, W., Sliwinski, T.: On direct methods for lexicographic min-max optimization. In: Gavrilova, M., et al. (eds.) ICCSA 2006. LNCS, vol. 3982, pp. 802–811. Springer, Berlin (2006)

LP-Based Relaxations of the Skiving Stock Problem—Improved Upper Bounds for the Gap

John Martinovic and Guntram Scheithauer

Abstract We consider the one-dimensional skiving stock problem (SSP) which is strongly related to the dual bin-packing problem in literature. In the classical formulation, different (small) item lengths and corresponding availabilities are given. We aim at maximizing the number of objects with a certain minimum length that can be constructed by connecting the items on hand. Such computations are of high interest in many real world application, e.g. in industrial recycling processes, wireless communications and politico-economic questions. For this optimization problem, we give a short introduction by outlining different modelling approaches, particularly the pattern-based standard model, and mentioning their relationships. Since the SSP is known to be NP-hard a common solution approach consists in solving an LP-based relaxation and the application of (appropriate) heuristics. Practical experience and computational simulations have shown that there is only a small difference (called gap) between the optimal objective values of the relaxation and the SSP itself. In this paper, we will present some new results and improved upper bounds for the gap of the SSP that are based on the theory of residual instances of the skiving stock problem.

1 Introduction

We consider the one-dimensional skiving stock problem (SSP) which is strongly related to the dual bin-packing problem (DBBP): find the maximum number of items with minimum length L that can be constructed by connecting a given supply of $m \in \mathbb{N}$ smaller item lengths $l_1, \ldots, l_m$ with availabilities $b_1, \ldots, b_m$. Such objectives are of high interest in many real world application, e.g. in industrial production processes (see [13] for a good overview) or politico-economic questions [2, 6].

J. Martinovic (✉) · G. Scheithauer
Department of Mathematics, Institute of Numerical Mathematics, Technische Universität
Dresden, 01062 Dresden, Germany
e-mail: john.martinovic@tu-dresden.de

G. Scheithauer
e-mail: guntram.scheithauer@tu-dresden.de

© Springer International Publishing Switzerland 2017   49
K.F. Dœrner et al. (eds.), *Operations Research Proceedings 2015*,
Operations Research Proceedings, DOI 10.1007/978-3-319-42902-1_7

Furthermore, also neighboring tasks, such as dual vector packing problems [4] or the maximum cardinality bin-packing problem [3, 10], are often associated or even identified with the dual bin packing problem. These formulations are of practical use as well since they are applied in multiprocessor scheduling problems [1] or surgical case plannings [12].

The considered optimization problem, for the case $b_i = 1$ $(i = 1, \ldots, m)$, was firstly mentioned by Assmann et al. [2] and denoted by dual bin packing problem. Based on practical preliminary thoughts [5], a generalization for larger availabilities b_i $(i = 1, \ldots, m)$ has been considered in [13] also motivating the term of skiving stock problem. In that article, Zak formulates a pattern-oriented model of the SSP with an infinite number of variables and provides first (numerical) results regarding the gap of this optimization problem, i.e., the difference between the optimal values of the continuous relaxation of the SSP and the SSP itself. A finite formulation of Zak's model and other modelling approaches (arcflow model, onestick model, Kontorovich model) have been extensively investigated in [9], whereas first theoretical results concerning the gap of general instances have been dealt with in [7].

In this article, we aim at presenting new and improved upper bounds for the gap which are based on the theory of residual instances. Therefore, the following section briefly introduces the skiving stock problem and summarizes important definitions. The third section deals with the theory of residual instances of the SSP, and Sect. 4 states how this approach can be used to obtain improved bounds for the gap. Note that, due to the space limitation, this paper only deals with one selected topic of the submitted presentation. The proofs of the results in Sects. 3 and 4, and further contributions can be found online in the preprint [8].

2 Preliminaries

Throughout this paper, we will use the abbreviation $E := (m, l, L, b)$ for an instance of the SSP with $l = (l_1, \ldots, l_m)^\top$ and $b = (b_1, \ldots, b_m)^\top$. Without loss of generality, we assume all input-data to be positive integers with $L > l_1 > \cdots > l_m > 0$. Any feasible arrangement of items leading to a final product of minimum length L is called *(packing) pattern* of E. We always represent a pattern by a nonnegative vector $a = (a_1, \ldots, a_m)^\top \in \mathbb{Z}_+^m$ where $a_i \in \mathbb{Z}_+$ denotes the number of items of type $i \in I := \{1, \ldots, m\}$ being contained in the considered pattern. For a given instance E, the set of all patterns is defined by $P(E) := \{a \in \mathbb{Z}_+^m \mid l^\top a \geq L\}$. A pattern $a \in P(E)$ is called *minimal* if there exists no pattern $\tilde{a} \in P(E)$ such that $\tilde{a} \neq a$ and $\tilde{a} \leq a$ hold (componentwise). The set of all minimal patterns is denoted by $P^\star(E)$. Let $x_j \in \mathbb{Z}_+$ denote the number how often the minimal pattern $a^j = (a_{1j}, \ldots, a_{mj})^\top \in \mathbb{Z}_+^m$ $(j \in J^\star)$ of E is used where $J^\star = \{1, \ldots, n\}$ represents an index set of all minimal patterns of E. Then the standard model of the skiving stock problem can be formulated as

$$z^\star(E) = \max \left\{ \sum_{j \in J^\star} x_j \;\middle|\; \sum_{j \in J^\star} a_{ij}x_j \le b_i,\, i \in I,\, x_j \in \mathbb{Z}_+,\, j \in J^\star \right\}. \tag{1}$$

A common (approximate) solution approach consists in considering the continuous relaxation

$$z_c^\star(E) = \max \left\{ \sum_{j \in J^\star} x_j \;\middle|\; \sum_{j \in J^\star} a_{ij}x_j \le b_i,\, i \in I,\, x_j \ge 0,\, j \in J^\star \right\} \tag{2}$$

and the application of appropriate heuristics. Then, the difference

$$\Delta(E) := z_c^\star(E) - z^\star(E) \tag{3}$$

is called *gap* (of E). Note that, similar to the one-dimensional cutting stock problem, there are also other modelling approaches, e.g. the arcflow model and the onestick model, which have been introduced in [9]. Therein, the equivalence of the corresponding continuous relaxations is proved. This implies, in particular, that the gap is independent of the considered modelling framework.

Definition 1 A set $\mathscr{T}$ of instances has the *integer round-down property* (IRDP), if $\Delta(E) < 1$ holds for all $E \in \mathscr{T}$, and it has the *modified integer round-down property* (MIRDP), if $\Delta(E) < 2$ holds for all $E \in \mathscr{T}$. Whenever $\mathscr{T} = \{E\}$ is a singleton, we briefly say that E, instead of $\{E\}$, has the (modified) integer round-down property. An instance E with $\Delta(E) \ge 1$ is called *non-IRDP instance*.

It is conjectured, see [13], that the skiving stock problem possesses the MIRDP. Currently, to our best knowledge, the best upper bound is given by $\Delta(E) < m - 1$ for instances $E = (m, l, L, b)$ with $m \ge 2$ and has been developed in [7] by means of generalizing Theorem 4 of [13].

3 Residual Instances

In this section, we reformulate the calculation of the gap $\Delta(E)$ of an instance $E = (m, l, L, b)$ in some sense. To this end, the instance E is replaced by an instance $\overline{E} = (m, l, L, \overline{b})$ with principally the same input-data, but a (appropriately) reduced vector $\overline{b}$ of availabilities. Hence, due to the smaller total number of objects, the optimization problem becomes more manageable in a certain extent.

Definition 2 Let $E = (m, l, L, b)$ be an instance of the skiving stock problem, and let x^c denote a solution of the corresponding continuous relaxation (2). Then the instance

$$\overline{E} := \overline{E}\left(x^c\right) := \left(m, l, L, b - A \lfloor x^c \rfloor\right)$$

with $\lfloor x^c \rfloor = (\lfloor x^c_1 \rfloor, \ldots, \lfloor x^c_n \rfloor)^\top$ is called *residual instance* of E.

Here $A := A(E)$ denotes the matrix whose columns are equal to the elements of $P^\star(E)$, i.e., to the minimal patterns of the given instance E. Then, we can state the following general result.

Lemma 1 *Let $E = (m, l, L, b)$ be an instance of the skiving stock problem, and let x denote a solution of the continuous relaxation (2) of E. Then $Ax = b$ holds.*

We start with a first result concerning the optimal objective values $z^\star_c(E)$ and $z^\star_c(\overline{E})$ of the continuous relaxations of E and $\overline{E}$, respectively.

Lemma 2 *Let $\overline{E} = \overline{E}(x^c)$ be a residual instance of E, then $z^\star_c(E) - z^\star_c(\overline{E}) \in \mathbb{Z}_+$.*

In a second step, this lemma leads us to a relationship between the gaps of an instance and one of its residual instances.

Lemma 3 *Let $\overline{E} = \overline{E}(x^c)$ be a residual instance of E, then $\Delta(E) \leq \Delta(\overline{E})$.*

Hence, the gap of an instance E can be bounded above by the gap of a corresponding residual instance $\overline{E}$. In particular, any upper bound for the gap $\Delta(\overline{E})$ of a residual instance (of E) is also applicable to the original gap $\Delta(E)$. In particular, the following statements hold.

Corollary 1 *Let $\overline{E} = \overline{E}(x^c)$ be a residual instance of E then:*

1. *If $\overline{E}$ has the IRDP, then E has the IRDP.*
2. *If $\overline{E}$ has the MIRDP, then E has the MIRDP.*

Thus, the consideration of the gaps of residual instances is sufficient, in general.

4   Improved Upper Bounds for the Gap

In order to derive upper bounds for the gap on the basis of residual instances, the following lemma will prove beneficial.

Lemma 4 *Let $\overline{E}$ be a residual instance (of E) and $p, q \in \mathbb{R}$ with $p \geq 1$. Then, the following implication holds:*

$$z^\star_c(\overline{E}) < p \cdot z^\star(\overline{E}) + q \implies \Delta(\overline{E}) < \frac{p-1}{p} \cdot m + \frac{q}{p} = m\left(1 - \frac{1}{p}\right) + \frac{q}{p}. \quad (4)$$

As a first application, the concept of residual instances provides a very simple proof and a slight generalization of [7, Theorem 1].

Theorem 1 *Let $E = (m, l, L, b)$ be an instance of the skiving stock problem, then $\Delta(E) < \max\{1, m - 1\}$.*

A similar, but more powerful, result can be obtained as follows.

Theorem 2 *Let $E = (m, l, L, b)$ be an instance of the skiving stock problem, then $\Delta(E) < \max\{2, m - 2\}$.*

In particular, this observation has an important implication.

Corollary 2 *Let $E = (m, l, L, b)$ be an instance with $m = 4$ item types. Then E has the MIRDP.*

In the previous theorems we did not apply Lemma 4 directly, but only a part of its proof. For almost all values of m, the following theorem improves the upper bound of the gap, given by Theorem 2, significantly by using the sufficient condition in (4).

Theorem 3 *Let $E = (m, l, L, b)$ be an instance of the skiving stock problem, then $\Delta(E) < (m + 1)/2$.*

One of the main ingredients of the previous result is given by the inequality $(l^\top \bar{b})/L < 2 \cdot z^\star(\overline{E})$. Note that this upper bound is (asymptotically) tight and, therefore, cannot be improved in the general setting. The instances $E(L) = (1, L - 1, L, 2)$ with $L \in \mathbb{N}, L \geq 2$ satisfy $z^\star(E(L)) = 1$ and

$$\frac{l^\top \bar{b}}{L} = 2 \cdot \frac{L - 1}{L} \longrightarrow 2 \cdot z^\star(E(L)) \quad \text{as} \quad L \to \infty.$$

Nevertheless, for most instances, this estimate can be improved if we use some more of the problem-specific input-data. To this end, note that the inequaltiy $l^\top a < 2L$ for all $a \in P^\star(E)$ might be very weak in some cases. Instead, it would be better to consider

$$v_{max} := v_{max}(E) := \max\left\{l^\top a \,\middle|\, a \in P^\star(E), a \leq b\right\} \tag{5}$$

Then, due to Lemma 4 with $p = v_{max}/L \in [1, 2)$ and $q = 1$, we can state the following result.

Theorem 4 *Let $E = (m, l, L, b)$ an instance of the skiving stock problem, then*

$$\Delta(E) < m\left(1 - \frac{L}{v_{max}}\right) + \frac{L}{v_{max}}. \tag{6}$$

If the item lengths are sufficiently small, we can state the following special case of the previous proposition.

Corollary 3 *Let $\tau \in \mathbb{N}$ with $\tau \geq 2$ be given. If $l_i \leq (\tau + 1)/\tau^2 \cdot L$ holds for all $i \in I$, then the gap satisfies the inequality*

$$\Delta(E) < \frac{m + \tau}{\tau + 1}.$$

5 Conclusions and Outlook

In this paper, we considered upper bounds for the gap of (general) instances of the one-dimensional skiving stock problem. Therefore, we gave an introduction to the theory of residual instances and investigated the relationship between the gaps of the original and the residual problem. Furthermore, by means of Lemma 4, we provided a very useful sufficient condition to calculate upper bounds for the gap of residual instances. At first, this method has been proven beneficial by reducing the tedious proof of [7, Theorem 1], which was based on the idea of [13], to a small number of lines. Furthermore, we successfully applied Lemma 4 in order to state several new and improved bounds for the gap of general instances of the skiving stock problem.

In the context of 1D cutting, estimates like $\Delta_{CSP}(E) < \max\{2, (m+2)/4\}$ (see [11, p. 61]) are already available. Therefore, we aim at improving the upper bound $(m+1)/2$ introduced in this paper in order to get a bit closer to the MIRDP-conjecture. Additionally, another main objective is given by the consideration of stronger relaxations, such as the proper relaxation.

References

1. Alvim, A.C.F., Ribeiro, C.C., Glover, F., Aloise, D.J.: A hybrid improvement heuristic for the one-dimensional bin packing problem. J. Heuristics **10**(2), 205–229 (2004)
2. Assmann, S.F., Johnson, D.S., Kleitman, D.J., Leung, J.Y.-T.: On a dual version of the one-dimensional bin packing problem. J. Algorithms **5**, 502–525 (1984)
3. Bruno, J.L., Downey, P.J.: Probabilistic bounds for dual bin-packing. Acta Informatica **22**, 333–345 (1985)
4. Csirik, J., Frenk, J.B.G., Galambos, G., Rinnooy Kan, A.H.G.: Probabilistic analysis of algorithms for dual bin packing problems. J. Algorithms **12**, 189–203 (1991)
5. Johnson, M.P., Rennick, C., Zak, E.J.: Skiving addition to the cutting stock problem in the paper industry. SIAM Rev. **39**(3), 472–483 (1997)
6. Labbé, M., Laporte, G., Martello, S.: An exact algorithm for the dual bin packing problem. Oper. Res. Lett. **17**, 9–18 (1995)
7. Martinovic, J., Scheithauer, G.: Integer rounding and modified integer rounding for the skiving stock problem. Discrete Optimiz **21**, 118–130 (2016)
8. Martinovic, J., Scheithauer, G.: Improved upper bounds for the gap of the skiving stock problem. Preprint MATH-NM-03-2015, Technische Universität Dresden (2015)
9. Martinovic, J., Scheithauer, G.: Integer linear programming models for the skiving stock problem. Eur. J. Oper. Res. **251**(2), 356–368 (2016)
10. Peeters, M., Degraeve, Z.: Branch-and-price algorithms for the dual bin packing and maximum cardinality bin packing problem. Eur. J. Oper. Res. **170**(2), 416–439 (2006)
11. Rietz, J.: Untersuchungen zu MIRUP für Vektorpackprobleme. Ph.D. thesis, TU Bergakademie Freiberg (2003)
12. Vijayakumar, B., Parikh, P., Scott, R., Barnes, A., Gallimore, J.: A dual bin-packing approach to scheduling surgical cases at a publicly-funded hospital. Eur. J. Oper. Res. **224**(3), 583–591 (2013)
13. Zak, E.J.: The skiving stock problem as a counterpart of the cutting stock problem. Int. Trans. Oper. Res. **10**, 637–650 (2003)

Creating Worst-Case Instances for Upper and Lower Bounds of the Two-Dimensional Strip Packing Problem

Torsten Buchwald and Guntram Scheithauer

Abstract We present a new approach to create instances with large performance ratio, i.e., worst-case instances, of common heuristics for the two-dimensional Strip Packing Problem. The idea of this new approach is to optimise the width and the height of all items aiming to find an instance with maximal performance ratio with respect to the considered heuristic. Therefore, we model the pattern obtained by the heuristic as a solution of an ILP problem and merge this model with a Padberg-type model of the two-dimensional Strip Packing Problem. In fact, the composed model allows to compute the absolute worst-case performance ratio of the heuristic with respect to a limited number of items. We apply the new approach for the Next-Fit Decreasing-Height, the First-Fit Decreasing-Height, and the Best-Fit Decreasing-Height heuristic. Furthermore, we provide an opportunity to use this idea to create worst-case instances for lower bounds.

1 Introduction

In this paper, we consider the two-dimensional Strip Packing Problem (SPP) with rectangular items. Let a set $I := \{1, \ldots, n\}$ of non-rotatable rectangles R_i (items) of width $w_i \leq 1$ and height $h_i \leq 1$ be given. The items have to be packed into a strip of width 1 and minimal height OPT such that the items do not overlap each other.

A lot of upper and lower bounds are known for this problem, but for most of them the exact absolute worst-case performance ratio, which is the supremum over all instances of the fraction of the bound and the optimal value, is unknown. To reduce the gap between a proven upper bound of the absolute worst-case performance ratio and the performance ratio of an instance having maximal ratio known so far, it is necessary to reduce the theoretical upper bound or to find instances with greater performance ratio.

T. Buchwald (✉) · G. Scheithauer
Institute of Numerical Mathematics, Dresden University of Technology, Dresden, Germany
e-mail: torsten.buchwald@tu-dresden.de

G. Scheithauer
e-mail: guntram.scheithauer@tu-dresden.de

© Springer International Publishing Switzerland 2017
K.F. Dœrner et al. (eds.), *Operations Research Proceedings 2015*,
Operations Research Proceedings, DOI 10.1007/978-3-319-42902-1_8

In this paper, we introduce a new approach to compute such instances. For several heuristics, we show how to model this issue as an optimisation problem which maximises the performance ratio within a subset of SPP instances. In this way, we obtain the absolute worst-case performance ratio of these heuristics for the considered subsets.

2   Modelling Upper Bounds for the SPP

In this section, we consider three heuristics, Next-Fit Decreasing-Height (NFDH), First-Fit Decreasing-Height (FFDH), and Best-Fit Decreasing-Height (BFDH) which place the items sequentially. We show how the optimisation problem addressed above can be modelled. Since we want to maximise the absolute worst-case performance ratio which is a fraction, we linearise this objective function by fixing the optimal value of the SPP and maximising the height of the heuristic. For all heuristics we assume that all items can be packed within a strip height of at most 1, i.e., $OPT \leq 1$. To ensure that this condition is fulfilled, our model contains a Padberg-type model [2] of the two-dimensional SPP. The second part of our model describes the considered heuristic. Hence, we aim to maximise the height of the heuristic solution depending on the widths and heights of the items. According to the three heuristics all items are sorted by non-increasing height and are packed into levels. (For describing the heuristics, we use the notations of [1].)

2.1   Basic Model

First, we introduce a basic model which is common for all three heuristics. Then the optimisation model for a particular heuristic is created by extending this basic formulation. The basic model contains the width w_i and the height h_i of the items as variables as well as their allocation points (lower left corner) (x_i, y_i) in an optimal packing of height 1. Due to the Padberg-type model, the basic model contains also binary variables u_{ij} and v_{ij} for each pair of items i and j to indicate whether i is left to j ($u_{ij} = 1$) or below j ($v_{ij} = 1$). These variables are necessary to model the nonoverlapping of all items. The height of each level defined by the heuristics is denoted by variable $\widetilde{h}_k$. The binary variable s_{ik} equals 1 if and only if item i is a *regular* item which is packed in level k. A regular item r is an item which is packed at the last (highest) opened level at the time when r is packed. Additionally to s_{ik}, the model contains the binary variables $\overline{s}_{ik}$ which equals 1 if and only if item i is a non-regular (fallback) item packed in level k, and the variables t_{ik} which equals 1 if and only if item i is the first (regular) item packed in level k.

For a given maximum number N of items, a maximum number S of levels, a precision ε, and sets $I := \{1, \ldots, N\}$ and $K := \{1, \ldots, S\}$, the basic model is defined as follows:

$$H := \sum_{k=1}^{S} \widetilde{h}_k \rightarrow \max$$

subject to

$$h_{i+1} \leq h_i, \quad i = 1, \ldots, N - 1 \tag{1}$$
$$\varepsilon \leq w_i \leq 1, \quad i \in I \tag{2}$$
$$0 \leq h_i \leq 1, \quad i \in I \tag{3}$$
$$0 \leq x_i \leq 1 - w_i, \quad i \in I \tag{4}$$
$$0 \leq y_i \leq 1 - h_i, \quad i \in I \tag{5}$$
$$x_i + w_i \leq x_j + 1 - u_{ij}, \quad i \in I, \quad j \in I, \quad i \neq j \tag{6}$$
$$y_i + h_i \leq y_j + 1 - v_{ij}, \quad i \in I, \quad j \in I, \quad i \neq j \tag{7}$$
$$u_{ij} + u_{ji} + v_{ij} + v_{ji} = 1, \quad i \in I, \quad j \in I, \quad i < j \tag{8}$$
$$0 \leq \widetilde{h}_k \leq 1, \quad k \in K \tag{9}$$
$$\sum_{k=1}^{S} s_{ik} + \overline{s}_{ik} = 1, \quad i \in I \tag{10}$$
$$t_{ik} \leq s_{ik}, \quad i \in I, \quad k \in K \tag{11}$$
$$\sum_{i=1}^{N} t_{ik} = 1, \quad k \in K \tag{12}$$
$$\sum_{i=1}^{N} t_{ik} h_i = \widetilde{h}_k, \quad k \in K \tag{13}$$
$$\sum_{k=1}^{S-1} s_{ik} \widetilde{h}_{k+1} \leq h_i \leq \sum_{k=1}^{S} s_{ik} \widetilde{h}_k + \sum_{k=1}^{S-1} \overline{s}_{ik} \widetilde{h}_{k+1}, \quad i \in I \tag{14}$$
$$\sum_{i=1}^{N} (s_{ik} + \overline{s}_{ik}) w_i \leq 1, \quad k \in K \tag{15}$$
$$h_i, w_i, x_i, y_i \in \mathbb{R}, \quad i \in I \tag{16}$$
$$\widetilde{h}_k \in \mathbb{R}, \quad k \in K \tag{17}$$
$$u_{ij}, v_{ij} \in \mathbb{B}, \quad i \in I, \quad j \in I, \quad i \neq j \tag{18}$$
$$s_{ik}, \overline{s}_{ik}, t_{ik} \in \mathbb{B}, \quad i \in I, \quad k \in K \tag{19}$$

According to the considered heuristics, condition (1) guaranties the non-increasing (decreasing) item heights. Due to the Padberg-type model, restrictions (2)–(8) ensure that all items can be placed within the maximal (optimal) height 1. The remaining constraints belong to the heuristics. Constraint (10) guarantees that each item is packed into exactly one level, and conditions (12) ensure that at least one item is placed in each level. The relation between item heights and level heights is modelled in (13) and (14). Finally, restrictions (15) ensure that the widths of all items of one level does not exceed the width of the strip.

2.2 Next-Fit Decreasing-Height

The NFDH heuristic places the current item in the highest level which is opened or in a new level if the item does not fit in the current highest level.

Since all items are regular items, we can eliminate the binary variables $\bar{s}_{ik}$ and simplify some constraints. The appropriate model for the NFDH heuristic results from the basic model by substituting conditions (10), (14) and (15) with

$$\sum_{k=1}^{S} s_{ik} = 1, \quad i \in I \tag{10a}$$

$$\sum_{k=1}^{S-1} s_{ik}\widetilde{h}_{k+1} \le h_i \le \sum_{k=1}^{S} s_{ik}\widetilde{h}_{k}, \quad i \in I \tag{14a}$$

$$\sum_{i=1}^{N} s_{ik}w_i \le 1, \quad k \in K \tag{15a}$$

$$\sum_{i=1}^{N} s_{ik}w_i + \sum_{i=1}^{N} t_{i,k+1}w_i \ge 1 + \varepsilon, \quad k = 1, \ldots, S-1 \tag{20}$$

According to the NFDH heuristic, restrictions (20) ensure that the first item of each level does not fit within the previous level.

It is known that the height of the NFDH heuristic is at most 3 if $OPT = 1$ due to [1]. The table in Fig. 1 shows the absolute worst-case performance ratio of the NFDH heuristic for the SPP instances with at most $N \in \{1, \ldots, 10\}$ items in dependence on the number of levels: The maximal performance ratios for a fixed number of items N are marked in bold face. As expected, the absolute worst-case performance ratio increases with the number of items and becomes closer and closer to the value 3. An instance with performance ratio $2\frac{3}{4}$ is illustrated in Fig. 1 and contains the following item sizes:

$w = (\varepsilon; \varepsilon; 1-\varepsilon; 2\varepsilon; \frac{1-\varepsilon}{8}-\varepsilon; \frac{7+\varepsilon}{8}; \frac{1+7\varepsilon}{8}; \frac{1-\varepsilon}{4}; \frac{5+3\varepsilon}{8}; \frac{3+5\varepsilon}{8})$, $\quad h = (1; \frac{1}{2}; \frac{1}{2}; \frac{1}{2}; \frac{1}{4}; \frac{1}{4}; \frac{1}{4}; \frac{1}{8}; \frac{1}{8}; \frac{1}{8})$

$S\backslash N$	1	2	3	4	5	6	7	8	9	10
1	1	1	1	1	1	1	1	1	1	1
2		1	$1\frac{1}{2}$	$1\frac{1}{2}$	$1\frac{1}{2}$	$1\frac{1}{2}$	$1\frac{1}{2}$	$1\frac{1}{2}$	$1\frac{1}{2}$	$1\frac{1}{2}$
3			$1\frac{1}{2}$	2	2	2	2	2	2	2
4				$1\frac{2}{3}$	2	$2\frac{1}{4}$	$2\frac{1}{3}$	$2\frac{1}{3}$	$2\frac{1}{3}$	$2\frac{1}{3}$
5					2	$2\frac{1}{3}$	$2\frac{1}{2}$	$2\frac{1}{2}$	$2\frac{1}{2}$	$2\frac{1}{2}$
6						2	$2\frac{2}{5}$	$2\frac{1}{2}$	$2\frac{5}{8}$	$2\frac{2}{3}$
7							$2\frac{1}{4}$	$2\frac{3}{5}$	$2\frac{2}{3}$	$2\frac{3}{4}$
8								$2\frac{2}{7}$	$2\frac{5}{8}$	$2\frac{7}{10}$
9									$2\frac{3}{7}$	$2\frac{3}{4}$
10										$2\frac{5}{11}$
best	1	1	$1\frac{1}{2}$	2	2	$2\frac{1}{3}$	$2\frac{1}{2}$	$2\frac{3}{5}$	$2\frac{2}{3}$	$2\frac{3}{4}$

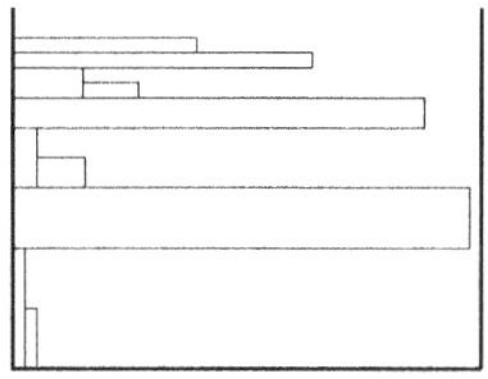

heuristic pattern for $N = 10$

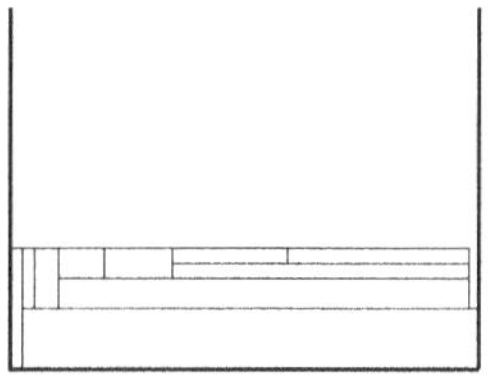

optimal pattern for $N = 10$

Fig. 1 Absolute worst-case performance ratios depending on the number of items

Analysing the worst-case instances obtained for $N \leq 10$, we obtain an asymptotic worst-case example for the NFDH heuristic. Let $f(i)$ denote the Fibonacci numbers, i.e., $f(1) = f(2) = 1, f(i+2) = f(i) + f(i+1)$ for $i \in \mathbb{N}$. For $n = 3m+1$, $m \in \mathbb{N}$ the instance $E(m)$ given by

$$w = (f(1) * \varepsilon, f(1 * 2) * \varepsilon, 1 - (f(1 * 2 + 1) - 1) * \varepsilon, f(1 * 2 + 1) * \varepsilon,$$
$$f(2 * 2) * \varepsilon, 1 - (f(2 * 2 + 1) - 1) * \varepsilon, f(2 * 2 + 1) * \varepsilon,$$
$$\vdots \qquad \vdots \qquad \vdots$$
$$f(m * 2) * \varepsilon, 1 - (f(m * 2 + 1) - 1) * \varepsilon, f(m * 2 + 1) * \varepsilon)$$
$$h = \quad (1, 2^{-1}, 2^{-1}, 2^{-1}, 2^{-2}, 2^{-2}, 2^{-2}, \ldots, 2^{-m}, 2^{-m}, 2^{-m})$$

has asymptotic worst-case performance ratio 3 for $m \to \infty$.

2.3 First-Fit Decreasing-Height

The FFDH heuristic places the current item in the first (lowest) level where the item fits or in a new level if the item does not fit in any opened level. To obtain the appropriate optimisation model for the FFDH heuristic, we only need to extend the basic model by the following terms:

$$\sum_{l=1}^{k-1} s_{il}(1+\varepsilon) + w_i + \sum_{j=1}^{i-1}(s_{jk}+\bar{s}_{jk})w_j \geq (1+\varepsilon)(1-s_{ik}-\bar{s}_{ik}), i \in I, \quad k \in K$$

(21)

These constraints ensure that each item is placed in the first level where it fits.

2.4 Best-Fit Decreasing-Height

The BFDH heuristic places the current item in a level where the item fits and the remaining width after placing the item is minimised, or in a new level if the item does not fit in any opened level. For this heuristic, we introduce binary variables r_{ik} which equals 1 if and only if item i fits in level k. To obtain the appropriate optimisation model for the BFDH heuristic, we extend the basic model by adding variables $r_{ik} \in \mathbb{B}$, $i \in I$, $k \in K$, and the following terms:

$$(2r_{ik}-1)\left(w_i + \sum_{j=1}^{i-1}(s_{jk}+\bar{s}_{jk})w_j\right) \leq 2r_{ik}-1, \quad i \in I, k \in K, \quad (22)$$

$$(2r_{ik}-1)\left(w_i + \sum_{j=1}^{i-1}(s_{jk}+\bar{s}_{jk})w_j\right) \leq (1+\varepsilon)(2r_{ik}-1), \quad i \in I, k \in K, \quad (23)$$

$$\sum_{j=1}^{i-1} w_j \sum_{l=1}^{S}(s_{il}+\bar{s}_{il})(s_{jl}+\bar{s}_{jl}) \geq r_{ik}\sum_{j=1}^{i-1}(s_{jk}+\bar{s}_{jk})w_j, \quad i \in I, k \in K, \quad (24)$$

The constraints (22) and (23) ensure that variables r_{ik} get the correct values. Constraint (24) guarantees that each item is placed in the level where it fits best.

3 Conclusions and Outlook

In this paper, we proposed a new approach to obtain worst-case instances for the two-dimensional Strip Packing Problem when the number of items is limited. We implemented this approach for three heuristics and presented first promising results for one of these heuristics. Moreover, we created an asymptotic worst-case instance for the NFDH heuristic by generalising the structure of the worst-case instances created by our approach. We are optimistic that we obtain similar results for the other two heuristics.

It will be part of our future work to apply this approach to other upper bounds (heuristics). Moreover, we will try to apply this approach also to lower bounds by adding appropriate constraints.

References

1. Coffman Jr., E.G., Garey, M.R., Johnson, D.S., Tarjan, R.E.: Performance bounds for level-oriented two-dimensional packing algorithms. SIAM J. Comput. **9**, 808–826 (1980)
2. Padberg, M.: Packing small boxes into a big box. Math. Meth. OR **1**, 1–21 (2000)

The Maximum Scatter TSP
on a Regular Grid

Isabella Hoffmann, Sascha Kurz and Jörg Rambau

Abstract In the Maximum Scatter Traveling Salesman Problem the objective is to find a tour that maximizes the shortest distance between any two consecutive nodes. This model can be applied to manufacturing processes, particularly laser melting processes. We extend an algorithm by Arkin et al. that yields optimal solutions for nodes on a line to a regular $(m \times n)$-grid. The new algorithm WEAVE(m, n) takes linear time to compute an optimal tour in some cases. It is asymptotically optimal and a $(\frac{\sqrt{10}}{5})$-approximation for the (3×4)-grid, which is the worst case.

1 Introduction

The Maximum Scatter Traveling Salesman Problem (MSTSP) asks for a closed tour in a graph visiting each node exactly once so that its shortest edge is as long as possible. Applications, approximation strategies, and exact algorithms for the geometric version where all vertices and edges are on a line or on a circle were presented by Arkin et al. [1]. Moreover, they proved that the problem is NP-complete in general. Chiang [2] focuses on the max-min m-neighbor TSP (particularly with $m = 2$), which we do not consider here. It is not known whether the geometric MSTSP for points in the plane is NP-hard. In this paper we extend their strategy on the line to the geometric MSTSP in the plane where all points are located on a rectangular equidistant grid with m rows and n columns. More formally, a regular rectangular grid is defined as the complete graph $G(m, n)$ on the vertex set

I. Hoffmann Partially supported by a grant of the Oberfrankenstiftung, project number: 03549.

I. Hoffmann (✉) · S. Kurz · J. Rambau
Universität Bayreuth, Universitätsstr. 30, 95447 Bayreuth, Germany
e-mail: isabella.hoffmann@uni-bayreuth.de

S. Kurz
e-mail: sascha.Kurz@uni-bayreuth.de

J. Rambau
e-mail: joerg.rambau@uni-bayreuth.de

© Springer International Publishing Switzerland 2017
K.F. Dœrner et al. (eds.), *Operations Research Proceedings 2015*,
Operations Research Proceedings, DOI 10.1007/978-3-319-42902-1_9

$V(m, n) = \{\binom{x}{y} \in \mathbb{Z}^2 | 1 \leq x \leq n, \ 1 \leq y \leq m\}$. We call the problem MSTSP(m, n) and our new algorithm WEAVE(m, n).

All current results are summarized in Theorem 1.

Theorem 1 *Let $m \leq n$ be natural numbers, and let $k = \lfloor \frac{n}{2} \rfloor$ and $t = \lfloor \frac{m}{2} \rfloor$. Then there is a linear-time algorithm* WEAVE(m, n) *that specifies a feasible tour for* MSTSP(m, n) *that is*

(i) optimal whenever n is odd, $m = n$, or $m = 2$;

(ii) a $(\sqrt{1 - \frac{2(k-t)}{(k-t)^2 - 2t(k-1)+1}})$-approximation whenever m and n are even;

(iii) a $(\sqrt{1 - \frac{2k-1}{k^2+t^2}})$-approximation whenever n is even and m is odd.

The algorithm is in all cases no worse than a $\frac{\sqrt{10}}{5}$-approximation. More specifically, it will turn out that the additive gap between the objective value of WEAVE(m, n) and the optimal value of MSTSP(m, n) is always strictly smaller than one, the length of a shortest edge in $G(m, n)$.

2　The Algorithm

WEAVE(m, n) is based on an algorithm for a tour for points on a line introduced by Arkin et al. [1]. The resulting order of the points on a line is exactly transferred to the columns in the equidistant grid. That is why we will consider this order here in more detail.

There are two subroutines in the algorithm, one for an odd number of points and one for an even number of points. For an odd number of points $n = 2k + 1 \geq 3$, the distances between subsequent points in the order are either k or $k + 1$, see the left two parts of Fig. 1. So the order of the points in the tour is: $1, k + 2, 2, k + 3, \ldots, n, k + 1$. The tour is completed by returning to the first point. For further use, we call this procedure LINEODD(n). For an even number of points $n = 2k \geq 4$, the distances are chosen to be alternatingly $k - 1$ and $k + 1$. At the points at the end of the line and at the two closest points to the center of the line the distance of k is used as well. So the distance of k is filled in between the alternating distances when reaching the endpoints or the center, see the last two components of Fig. 1. Equally one can say, the end points have distances of k and $k + 1$ and the two points around the center have distances of $k - 1$ and k. This procedure is called LINEEVEN(n) in the following. If k is odd, the order of the points in the tour is $1, k + 2, 3, k + 4, \ldots, n - 1, k, n, k - 1, \ldots, k + 1$. If k is even, the order of the points in the tour is $1, k + 2, 3, k + 4, \ldots, k - 1, n, k, n - 1, \ldots, k + 1$. From Arkin et al. [1] (Sects. 6.1 and 6.2) it follows that the tours generated by LINEODD(n) and LINEEVEN(n) are optimal tours for MSTSP$(1, n)$. Thus, WEAVE$(1, n)$ is set to LINEODD(n) or LINEEVEN(n), respectively.

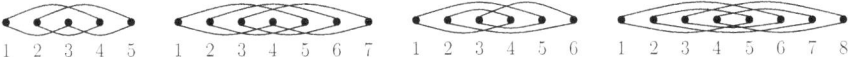

Fig. 1 Order of LINEODD(n) and LINEEVEN(n) for n = 5, 6, 7, 8

In the following we present the new algorithm for $m \geq 2$. The points in the grid are labeled by (i, j), where $i = 1, 2, \ldots, m$ is the row index and $j = 1, \ldots, n$ is the column index. The simple idea to apply WEAVE($1, n$) row by row yields $k - 1$ as the shortest edge length. However, this can be improved by using the freedom in the second dimension. One can increase the length of the shortest edge in WEAVE($1, n$) by using the order of nodes in both dimensions at the same time. However, this yields no closed tour. Thus, we need a new pattern in the second dimension.

To this end, the rows are partitioned in pairs for m even and pairs plus one triple for m odd. The paired rows are at distance t, as are the rows in the triple. The idea is to jump back and forth inside a pair (triple) of rows where the order of columns is given by WEAVE($1, n$). The successor node for a node always lies in the paired (tripled) row to the current node. For example, for odd n this is done until all nodes in the pair have been visited, whereas for even n this is done for half of the nodes. Then we traverse into the next pair. Since switching to the next pair yields a smaller vertical distance, we start in a column such that these pair-switches are accompanied by the largest possible horizontal distances $k + 1$ in WEAVE($1, n$). Thus, WEAVE(m, n) starts each pair in column $k + 2$. The setup is illustrated in Fig. 2.

We first describe WEAVE(m, n) in a way that the construction principle becomes apparent. First, let m be even.

For an odd number of columns the algorithm works in the following way, where in the following the order of column indices in each line is given by WEAVE($1, n$):

$$(1, k + 2), (t + 1, 2), (1, k + 3), (t + 1, 3), \ldots, (1, n), (t + 1, k + 1), (1, 1),$$
$$(t + 1, k + 2), (1, 2), (t + 1, k + 3), (1, 3), \ldots, (t + 1, n), (1, k + 1), (t + 1, 1),$$
$$(2, k + 2), (t + 2, 2), (2, k + 3), (t + 2, 3), \ldots, (2, n), (t + 2, k + 1), (2, 1),$$
$$\ldots$$
$$(t, k + 2), (m, 2), (t, k + 3), (m, 3), \ldots, (t, n), (m, k + 1), (t, 1),$$
$$(m, k + 2), (t, 2), (m, k + 3), (t, 3), \ldots, (m, n), (t, k + 1), (m, 1), (1, k + 2). \quad (1)$$

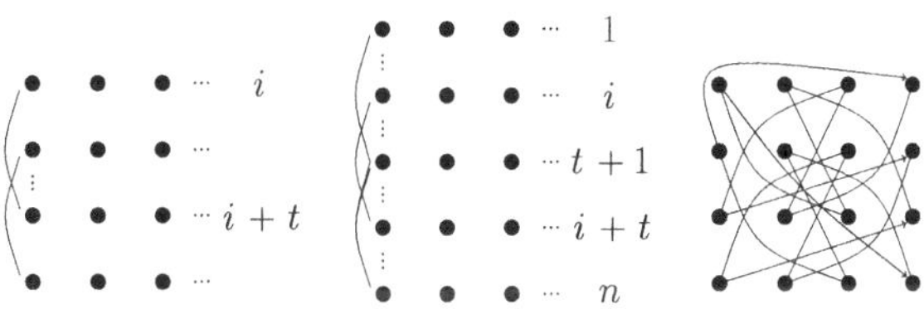

Fig. 2 Pairs of rows for an even number, for an odd number of rows, and an example of a (4×4)-grid

For an even number of columns this procedure fails because it generates a subtour inside a pair of rows. This can be rectified by switching pairs after each completed order of WEAVE$(1, n)$. The resulting order of nodes is as follows (again, the order of columns in each line is given by WEAVE$(1, n)$):

$$(1, k + 2), (t + 1, 3), (1, k + 4), (t + 1, 5), \ldots, (t + 1, 2), (1, k + 1), (t + 1, 1),$$
$$(2, k + 2), (t + 2, 3), (2, k + 4), (t + 2, 5), \ldots, (t + 2, 2), (2, k + 1), (2, 1),$$

$$\ldots,$$

$$(t + 1, k + 2), (1, 3), (t + 1, k + 4), (1, 5), \ldots, (t + 1, 2), (1, k + 1), (t + 1, 1),$$
$$(t + 2, k + 2), (2, 3), (t + 2, k + 4), (2, 5), \ldots, (t + 2, 2), (2, k + 1), (t + 2, 1),$$

$$\ldots,$$

$$(m, k + 2), (t, 3), (m, k + 4), (t, 5), \ldots, (m, 2), (t, k + 1), (m, 1), (1, k + 2). \quad (2)$$

Let now m be odd. First, the triple $(1, t + 1, m)$ of rows is taken. If n is not divisible by 3, then the rows are visited in the order $(1, t + 1, m, 1, t + 1, \ldots, 1)$, $(t + 1, m, 1, t + 1, m, \ldots, t + 1)$, $(m, 1, t + 1, m, 1, \ldots, m)$ or $(1, t + 1, m, 1, t + 1, \ldots, t + 1)$, $(m, 1, t + 1, m, 1, \ldots, 1)$, $(t + 1, m, 1, t + 1, m, \ldots, m)$, depending on n. During this process, the columns are visited in the order of WEAVE$(1, n)$ three times consecutively.

If n is a multiple of 3 the row order pattern is always $(1, t + 1, m, 1, t + 1, \ldots, m)$, $(t + 1, m, 1, t + 1, m, \ldots, 1)$, $(m, 1, t + 1, m, 1, \ldots, t + 1)$, after the row triple has been traversed completely, the procedure is the same as for even m.

We now define for WEAVE(m, n) the successor node $(\text{nextrow}(i, j), \text{nextcol}(i, j))$ of each node (i, j) in every possible case. Together with the starting node $(1, k + 2)$ this yields a complete definition of our algorithm. Recall that $t = m - t$ if m is even and $t = m - 1 - t$ if m is odd.

We start with the successor column of nextcol(i, j) when n is odd:

$$\text{nextcol}(i, j) := \begin{cases} j + k + 1 & \text{if } j \le k \\ j - k & \text{if } j > k. \end{cases}$$

The successor column when n is even and k is even:

$$\text{nextcol}(i, j) := \begin{cases} j + k + 1 & \text{if } j \le k \text{ and } j \text{ odd} \\ j + k - 1 & \text{if } j \le k \text{ and } j \text{ even} \\ j - (k + 1) & \text{if } n > j > k + 1 \text{ and } j \text{ odd} \\ j - (k - 1) & \text{if } n > j > k + 1 \text{ and } j \text{ even} \\ j - k & \text{if } j = k + 1 \text{ or } j = n, \end{cases}$$

The successor column when n is even and k is odd:

$$\text{nextcol}(i, j) := \begin{cases} j + k + 1 & \text{if } j < k \text{ and } j \text{ odd} \\ j + k - 1 & \text{if } j < k \text{ and } j \text{ even} \\ j - (k + 1) & \text{if } j > k + 1 \text{ and } j \text{ even} \\ j - (k - 1) & \text{if } j > k + 1 \text{ and } j \text{ odd} \\ j + k & \text{if } j = k \\ j - k & \text{if } j = k + 1. \end{cases}$$

Let now $m, n > 3$. The successor row $\text{nextrow}(i, j)$ of (i, j) when n is odd [even] and m is even:

$$\text{nextrow}(i, j) := \begin{cases} i + t & \text{if } i \leq t \big[\text{and } j \neq 1\big] \\ i - t & \text{if } i > t \text{ and } j \neq 1 \\ \big[i + (t + 1) & \text{if } i < t \text{ and } j = 1\big] \\ i - (t - 1) & \text{if } m > i > t \text{ and } j = 1 \big[\text{or } (i = m \text{ and } j = 1)\big] \\ \text{TERMINATE} & \text{if } i = m\big[-t\big] \text{ and } j = 1 \end{cases}$$

The successor row when n is odd [even], m is odd, and $n \bmod 3 \neq 0$:

$$\text{nextrow}(i, j) := \begin{cases} i + t & \text{if } i \leq t + 1 \big[\text{and } j \neq 1 \text{ or} \\ & ((i = 1 \text{ or } i = t + 1) \text{ and } j = 1)\big] \\ i - t & \text{if } m > i > t + 1 \text{ and } j \neq 1 \\ 1 & \text{if } i = m \text{ and } j \neq 1 \\ \big[i + (t + 1) & \text{if } 1 < i < t \text{ and } j = 1\big] \\ i - (t - 1) & \text{if } m - 1 > i > t + 1 \text{ and } j = 1 \\ \big[i - (t - 2) & \text{if } i = m - 1 \text{ and } j = 1\big] \\ 2 & \text{if } i = m \text{ and } j = 1 \\ \text{TERMINATE} & \text{if } i = m - 1\big[-t\big] \text{ and } j = 1 \end{cases}$$

The successor row when n is odd [even], m is odd, and $n \bmod 3 = 0$:

$$\text{nextrow}(i, j) := \begin{cases} i + t & \text{if } 1 < i \leq t \big[\text{and } j \neq 1\big] \\ i + t & \text{if } (i = 1 \text{ or } i = t + 1) \text{ and } j \neq 1 \\ i - t & \text{if } m > i > t + 1 \text{ and } j \neq 1 \\ 1 & \text{if } i = m \text{ and } j \neq 1 \\ m & \text{if } i = 1 \text{ and } j = 1 \\ t + 1 & \text{if } i = m \text{ and } j = 1 \\ \big[i + (t + 1) & \text{if } 1 < i < t \text{ and } j = 1\big] \\ i - (t - 1) & \text{if } m - 1 > i \geq t + 1 \text{ and } j = 1 \\ \big[i - (t - 2) & \text{if } i = m - 1 \text{ and } j = 1\big] \\ \text{TERMINATE} & \text{if } i = m - 1\big[-t\big] \text{ and } j = 1. \end{cases}$$

If $m = 3$, then WEAVE(m, n) terminates when for the first time the triple of rows is finished.

3 Optimality and Gaps

We now analyze the edge lengths appearing in the tours of the algorithm, in particular, the shorter ones to get a lower bound for a solution to the MSTSP. After that, we have a look at the upper bounds for the longest possible shortest edge in a tour.

By comparing upper bounds and lower bounds we can prove that WEAVE(m, n) is optimal for any grid with an odd number of columns, for any quadratic grid and for any grid with two rows. In the following, we identify the Euclidean lengths of candidates for the shortest edge in a tour computed by WEAVE(m, n).

For odd n the relevant edge lengths are $\sqrt{t^2 + k^2}$ and $\sqrt{(t-1)^2 + (k+1)^2}$. For n even the lengths are $\sqrt{t^2 + (k-1)^2}$ and $\sqrt{(t-1)^2 + (k+1)^2}$. A straightforward calculation shows: $t^2 + k^2 \leq (t-1)^2 + (k+1)^2$ and $t^2 + (k-1)^2 \leq (t-1)^2 + (k+1)^2$, so the lengths of the shortest edges of WEAVE(m, n) are $\sqrt{t^2 + k^2}$ for n odd and $\sqrt{t^2 + (k-1)^2}$ for n even.

An upper bound for the shortest edge can be derived by considering the distances between the nodes in the center of the grid and the corner nodes. Each node has to be connected to exactly two other nodes in a tour. This holds, in particular, for central nodes, to which the corners of the grid are the furthest. If m and n are odd, there is exactly one central node, and the distance to each corner is $\sqrt{t^2 + k^2}$. If m is even, there are two edges for each of the two middle nodes of length $\sqrt{t^2 + k^2}$. So WEAVE(m, n) is optimal whenever n odd as its shortest edge has length $\sqrt{t^2 + k^2}$. If n is even, there are two respectively four central nodes. For m odd, the longest distance appears twice and is $\sqrt{t^2 + k^2}$. As the longest distance to a corner appears only once if m is even, the second longest distance has to appear in any tour. Thus, $\sqrt{(t-1)^2 + k^2}$ is an upper bound, too. For even n, there is still a gap between the length of the shortest edge of the tour from WEAVE(m, n) and the current upper bound. The absolute value of the gap between lower and upper bound is $\sqrt{t^2 + k^2} - \sqrt{t^2 + (k-1)^2}$ if m is odd and $\sqrt{(t-1)^2 + k^2} - \sqrt{t^2 + (k-1)^2}$ if m is even, respectively. But we can show with Lemma 1 that this gap is smaller than the minimal distance in the grid.

Lemma 1 *For any regular grid with even n, the gap between the solution of* WEAVE(m, n) *and the upper bound of the MSTSP deduced from the distances between corner and central points is always smaller than one.*

Proof As the upper bound for even m is smaller than the upper bound for odd m (recall that $\sqrt{(t-1)^2 + k^2} < \sqrt{t^2 + k^2}$), it suffices to prove $\sqrt{t^2 + k^2} < \sqrt{t^2 + (k-1)^2} + 1$. We can square both sides of the inequality because all the values are greater zero: $t^2 + k^2 < t^2 + (k-1)^2 + 2\sqrt{t^2 + (k-1)^2} + 1 \Leftrightarrow k - 1 < \sqrt{t^2 + (k-1)^2}$. The strict inequality is proved as on the right hand side $t^2 > 0$ for $m > 1$. So the value of the gap between the lower and the upper bound is always strictly smaller than one.

If the grid is quadratic, the number of columns and rows has to be both odd or both even. As $m = n$ and therefore $t = k$, the gap for even m and n vanishes. So WEAVE(m, n) is optimal for quadratic grids.

When the number of rows is two, we claim that WEAVE($2, n$) is optimal for all numbers of columns. As it is always optimal for odd n, it suffices to solely consider even n. Assume that some tour achieves the upper bound, then each central node would have to be connected to both of its farthest corners. But as there are four central nodes, the edges connecting central and corner nodes violate the subtour elimination inequality: a contradiction. Thus, the upper bound cannot be achieved by a feasible tour, and each tour must employ at least one shorter edge incident to a central node. The size of this new bound is $\sqrt{1 + (k - 1)^2}$. Since this equals the lower bound achieved by WEAVE($2, n$), the claim is proved.

Calculating a bound for the approximation factor by the division of the approximate solution and the upper bound for the shortest edge, we can summarize the quality of WEAVE(m, n) in the following proposition.

Proposition 1 WEAVE(m, n) *is optimal whenever n is odd, or $m = 2$, or $m = n$.*

For even n and odd m we have $\text{WEAVE}(m, n) \geq \underbrace{\sqrt{1 - \frac{2k - 1}{t^2 + k^2}}}_{\alpha_{t,k}} \cdot OPT(m,n).$ *For even*

m, n we have $\text{WEAVE}(m, n) \geq \underbrace{\sqrt{1 - \frac{2(k - t)}{(k - t)^2 + 2t(k - 1) + 1}}}_{\alpha_{t,k}} \cdot OPT(m,n).$

Since the approximation guarantee converges to one for increasing m and n, we conclude the following:

Corollary 1 *In all cases, $\alpha \geq \frac{\sqrt{10}}{5}$. Moreover, $\lim_{k \to \infty} \alpha_{t,k} = 1$ for all t and $\lim_{t \to \infty, k \to \infty} \alpha_{t,k} = 1$. In this sense, WEAVE($m, n$) is asymptotically optimal.*

Since WEAVE(m, n) works by a fixed formula for each edge, it is a linear-time algorithm. From this, Theorem 1 follows.

4 Conclusion

We have presented an asymptotically optimal linear-time algorithm WEAVE(m, n) for the MSTSP for the complete graph on the points in the regular $m \times n$-grid in the plane. This is the first polynomial time algorithm that solves an infinite class of two-dimensional MSTSP in the plane to optimality. The complexity of the general MSTSP in the plane remains open.

References

1. Arkin, E., Chiang, Y.-J., Mitchell, J.S.B., Skiena, S.S., Yang, T.-C.: On the maximum scatter TSP. SIAM J. Comput. **29**, 515–544 (1999)
2. Chiang, Y.-J.: New approximation results for the maximum scatter TSP. Algorithmica **41**, 309–341 (2005)

Using Contiguous 2D-Feasible 1D Cutting Patterns for the 2D Strip Packing Problem

Isabel Friedow and Guntram Scheithauer

Abstract We consider the 2D rectangular strip packing problem without rotation. A relaxation of that problem is the 1D horizontal bar relaxation, the LP relaxation of the 1D binary cutting stock problem. To represent a solution of the strip packing problem, a solution of the horizontal bar relaxation has to satisfy, among others, the vertical contiguous condition. To strengthen the bar relaxation with respect to that vertical contiguity, we investigate a cutting plane approach. Furthermore, a solution of the bar relaxation must ensure constant location of items. Because 1D cutting patterns do not provide any information about the location of the items contained, we investigate methods to provide *2D feasibility* of patterns in the column generation and cutting plane process, respectively. Some computational results are also reported.

1 Introduction

Given a set of rectangles $I := \{1, \ldots, n\}$ of width w_i and height h_i, $i \in I$, the objective of the rectangular two-dimensional strip packing problem (SPP) is to pack the set without overlap into a strip of width W and minimal needed height H. The dimensions of the rectangles and the strip are integers and rotation of rectangles is not allowed. The SPP, known to be NP-hard, has various practical applications. Therefore, numerous heuristic algorithms have been proposed in literature. For a survey of some of the most common, see [3]. Exact approaches based on a branch and bound algorithm are proposed in [1, 4]. In Sect. 2, we describe how to formulate inequalities to create sequences of contiguous 1D cutting patterns. To generate only

I. Friedow (✉) · G. Scheithauer
Institute of Numerical Mathematics, Technical University of Dresden, 01069,
Dresden, Germany
e-mail: isabel.friedow@tu-dresden.de

G. Scheithauer
e-mail: guntram.scheithauer@tu-dresden.de

© Springer International Publishing Switzerland 2017
K.F. Dœrner et al. (eds.), *Operations Research Proceedings 2015*,
Operations Research Proceedings, DOI 10.1007/978-3-319-42902-1_10

2D *feasible* sequences, we modify the column generation problems if necessary or identify and forbid *2D infeasible* cutting patterns as explained in Sect. 3. In Sect. 4, we present some computational results and compare them with results of other methods proposed in literature.

2 Contiguous 1D Cutting Patterns

According to the horizontal bar relaxation (HBR) [6], a rectangle $i \in I$ is replaced by h_i items of width w_i and height 1, which can be considered as 1D objects. Then the task is to cut exactly h_i items of item type $i \in I$ by using binary cutting patterns a^j with frequencies λ_j while minimizing the total number of used 1D cutting patterns. Pattern a^j contains item type i if and only if $a_i^j = 1$. It is feasible if $\sum_{i \in I} w_i a_i^j \leq W$. Let J describe the set of feasible cutting patterns, then the horizontal bar relaxation

$$\min \left\{ \sum_{j \in J} \lambda_j : \sum_{j \in J} a_i^j \lambda_j = h_i, i \in I, \lambda_j \geq 0, j \in J \right\} \tag{1}$$

is a relaxation of the SPP. In the following a pattern a^j is represented by index $j \in J$. To create such an ordering of 1D cutting patterns that all items representing one rectangle are located in consecutive patterns, we start at the bottom of the strip. We define a set of rectangles I^B and call a pattern *bottom pattern* if it only contains item types $i \in I^B$. Because we want to minimize the height of the packing it seems reasonable to pack high rectangles as early as possible. Furthermore packing long rectangles in the end may cause a lot of waste. Thus I^B should contain, for example, especially high or long rectangles, but in an exact approach I^B equals I. Let $J^B :=$ $\{j \in J : a_i^j = 0 \, \forall i \in I \setminus I^B\}$ denotes the set of bottom patterns and $u_B := \min\{h_i : i \in I^B\}$. To ensure the usage of bottom patterns, we extend (1) by

$$\sum_{j \in J^B} \lambda_j \geq u_B. \tag{2}$$

In a 2D packing, all rectangles $i \in I(a^j) := \{i \in I : a_i^j = 1\}$ have to be placed at the bottom of the strip if $j \in J^B$. Because of that, we know $y_i = 0$ for all $i \in I(a^j)$, where (x_i, y_i) denotes the allocation point of rectangle i, and $b_{min}^j := \min\{h_i : i \in I(a^j)\}$ is the maximum usage of a^j. Thus, we know that a continuing pattern of a^j has to contain all item types $i \in I_j^+ = \{i \in I(a^j) : h_i > b_{min}^j\}$. Furthermore, all items types $i \in I_j^- := \{i \in I(a^j) : h_i = b_{min}^j\}$ whose demand is covered by usage frequency $\lambda_j = b_{min}^j$, are forbidden. The set of continuing patterns of $j \in J^B$ is denoted by

$$J(a^j) := \{k \in J \setminus J^B : a_i^k = 1 \; \forall \; i \in I_j^+, a_i^k = 0 \; \forall \; i \in I_j^-\}.$$

If a^j is used exactly b_{min}^j-times, then the demand of item types $i \in I_j^+$ is reduced to $\tilde{h}_i := h_i - b_{min}^j$. Thus, the induced overall demand for continuing patterns $k \in J(a^j)$ is $b_{ind}^j := \min\{\tilde{h}_i : i \in I_j^+\}$. To ensure the usage of continuing patterns of a^j we require

$$\sum_{k \in J(a^j)} \lambda_k b_{min}^j - \lambda_j b_{ind}^j \geq 0. \tag{3}$$

Now, we want to show that the method of continuing can be applied to non-bottom patterns because of the knowledge of y-coordinates. For that purpose let $S := \{s_1, \ldots, s_e\} \subseteq J$ be a sequence of continuing patterns with bottom pattern $s_1 \in J^B$ and $s_{j+1} \in J(a^{s_j})$, $j = 1, \ldots, e - 1$. Furthermore let $S(i) := \{s \in S : a_i^s = 1\}$ be the set of patterns that contain $i \in I$. In addition to the types $i \in I_{s_{e-1}}^+$, pattern s_e contains item types $i \in I_{new}(s_e) := I(a^{s_e}) \setminus I_{s_{e-1}}^+$. In a 2D packing the y-positions of that rectangles are known to be $y_i := \sum_{j=s_1}^{s_{e-1}} b_{min}^j$. Let $h^{s_e} := \min\{h_i : i \in I_{new}(s_e)\}$. The maximum usage of pattern s_e is $b_{min}^{s_e} := \min\{h^{s_e}, b_{ind}^{s_{e-1}}\}$, and so the reduced demands are $\tilde{h}_i := h_i - \sum_{s \in S(i)} b_{min}^s$ for all $i \in I(a^{s_e})$. Thus, a continuing pattern of s_e has to contain all $i \in I_{s_e}^+ := \{i \in I(a^{s_e}) : \tilde{h}_i > 0\}$ but items types $i \in I_{s_e}^- := \{i \in I(a^{s_e}) : \tilde{h}_i = 0\} \cup I_{s_{e-1}}^-$ are forbidden. Again the overall usage of continuing patterns $l \in J(a^{s_e}) := \{j \in J \setminus J^B : a_i^j = 1, i \in I_{s_e}^+, a_i^j = 0, i \in I_{s_e}^-\}$ is restricted by $b_{ind}^{s_e} := \min\{\tilde{h}_i : i \in I_{s_e}^+\}$ and we can add a continuing condition (3) for s_e. Assuming that $b_{min}^{s_e} = h^{s_e} < b_{ind}^{s_{e-1}}$, a continuing pattern of s_e automatically continues s_{e-1} and thus for $l \in J(s_e)$ follows $l \in J(s_{e-1})$. Problem (1)–(3) is a relaxation of the 1-contiguous bin packing problem [1] and we call it *HBR with vertical contiguity conditions* (HBRVC).

Our cutting plane approach works as follows. At the beginning the coefficient matrix consists of the identity matrix for (1) and an empty bottom pattern (2). There are no patterns that have to be continued, $C := \emptyset$, and HBRVC does not contain any contiguous conditions (3). Let HBRVC be solved for the current set C by column generation and let $\mathcal{L} := \{j \in J : \lambda_j > 0\}$ be the set of used patterns. For all cutting patterns $j \in Q := \{k \in \mathcal{L} : k \notin C, k \in J^B \vee \exists \, l \in C : a^k \in J(a^l)\}$ the corresponding continuing constraint (3) is added to HBRVC. Furthermore a trivial continuing pattern a^{j_0} with $a_i^{j_0} = 1$ if $i \in I_j^+$ and $a_i^{j_0} = 0$ otherwise is added to the model for all $j \in Q$. After C is updated to $C := C \cup Q$ problem HBRVC is solved again. In the solution process of HBRVC the dual simplex multipliers are obtained by solving the master problem HBRVC for the current set of variables represented by J.

Let $d_1, \ldots, d_n$ be the dual multipliers for the item constraints (1), d_B the multiplier for the bottom constraint (2) and $\tilde{d}_j$, $j \in C$, the dual multipliers for the continuing constraints (3). The slave problem for a new bottom pattern $j \in J^B$ is the following binary problem

$$\sum_{i \in I^B} d_i a_i + d_B - \sum_{j \in J^B} b_{ind}^j \cdot \tilde{d}_j \cdot \mu_j \to \max$$

$$\sum_{i \in I^B} w_i a_i \leq W$$

$$\sum_{i \in I(a^j)} (1 - a_i) + \sum_{i \in I^B \setminus I(a^j)} a_i \leq (1 - \mu_j) \quad j \in J^B$$

$$\mu_j \in \{0, 1\}, \ j \in J^B \qquad a_i \in \{0, 1\}, \ i \in I^B \qquad a_i = 0, \ i \in I \setminus I^B$$

where $\mu_j = 1$ if $a_i = a_i^j$ for all $i \in I$ and $\mu_j = 0$ otherwise. A pattern a equal to a pattern j that is already in the pool $J^B \subset J$ can not improve the objective value of the master problem. Instead of penalizing the equality to an existing pattern $j \in J^B$ by the negative profit $-b_{ind}^j \cdot \tilde{d}_j$, we use the slave problem (P_B)

$$o_B := \max \left\{ \sum_{i \in I^B} d_i a_i + d_B : \sum_{i \in I^B} w_i a_i \leq W, \ \sum_{i \in I(a^r)} (1 - a_i) + \sum_{i \in I \setminus I(a^r)} a_i \leq 1, r \in J^B, \right.$$

$$\left. a_i \in \{0, 1\}, \ i \in I^B, \ a_i = 0, \ i \in I \setminus I^B \right\} \tag{P_B}$$

where such solutions are excluded. Now, we consider a non-bottom pattern a^j, $j \in J \setminus J^B$, with corresponding sequence $S := \{s_1, \ldots, s_{e-1}, j\}$. A continuing pattern of a^j has the fixed profit $p_j := b_{min}^j \tilde{d}_j + \sum_{i \in I_j^+} d_i + p(S)$ where $p(S) := \sum_{t \in F_j} b_{min}^t \tilde{d}_t$ represents the profit caused by implied continuing occurring for $t \in F_j := \{k \in S : b_{ind}^k > \sum_{l \in S: k < l \leq j} b_{min}^l \}$. The slave problem for $l \in J(a^j)$ is problem (P_j)

$$o_j := \max \left\{ \sum_{i \in I} d_i a_i + p_j : \sum_{i \in I} w_i a_i \leq W_j, \ \sum_{i \in I(a^r)} (1 - a_i) + \sum_{i \in I \setminus I(a^r)} a_i \leq 1, r \in J(a^j), \right.$$

$$\left. a_i \in \{0, 1\}, \ i \in I, \ a_i = 1, \ i \in I_j^+, \ a_i = 0, \ i \in I_j^- \right\} \tag{P_j}$$

where $W_j := W - \sum_{i \in I_j^+} w_i$. A pattern that is neither a bottom pattern nor a continuing pattern is obtained by a binary knapsack problem (BKP). Let o_N be the objective value of (BKP). If $o_B > 1$ (resp. $o^j > 1$, $j \in C$, $o_N > 1$) with the solution of (P_B) (resp. (P_j), (BKP)) a new pattern is found and added to the pattern pool J. The master problem for the updated set J is solved and new multipliers are obtained. The process is repeated until $\max\{o_B, \{o_j, j \in J\}, o_N\} \leq 1$ which means HBRVC is solved for C.

3   2D-Feasible 1D Cutting Patterns

A sequence S is called *2D-feasible* if there exist coordinates (x_i, y_i) for all $i \in I_S := \bigcup_{j \in S} I(a^j)$ so that all rectangles $i \in I_S$ can be packed into the strip without overlapping. To determine whether a sequence is 2D-feasible or not, we use a model which is based on the Padberg-type formulation of the SPP [5]. Since the y-coordinates of all rectangles $i \in I_S$ are known, we only have to consider the x-positions and the overlapping in x-direction. A rectangle i only precedes or follows a rectangle $j \neq i$ along the x-axis if a pattern $s \in S$ exists with $a_i^s = 1 \wedge a_j^s = 1$. Thus, with regard to the non-overlapping conditions, it is sufficient to consider the set $I_O := \{(i, j) : i, j \in I_S, i < j, \exists s \in S : a_i^s \cdot a_j^s = 1\}$. We define binary variables u_{ij} with

$$u_{ij} = 1 \quad \text{if } x_j \geq x_i + w_i, \quad u_{ij} = 0 \quad \text{if } x_i \geq x_j + w_j$$

for all $(i, j) \in I_O$ and the following non-overlapping conditions

$$x_i + w_i \leq x_j + W(1 - u_{ij}) \qquad\qquad (i, j) \in I_O \qquad (4)$$

$$x_j + w_j \leq x_i + W u_{ij} \qquad\qquad (i, j) \in I_O \qquad (5)$$

A rectangle $i \in I_S$ is contained in the strip if $0 \leq x_i \leq W - w_i$. With the knowledge of the patterns of S, we can strengthen that constraint. Consider a rectangle i and a pattern $s \in S(i)$. The x-coordinate of rectangle i has to be greater than or equal (resp. less than or equal) to the sum of the widths of all rectangles represented by $j \in I(a^s)$, $j \neq i$, that precede (resp. follow) rectangle i. Thus, we obtain the stronger constraints

$$x_i \leq W - w_i - \sum_{j \in I(a^k), j > i} w_j u_{ij} - \sum_{j \in I(a^k), j < i} w_j (1 - u_{ij}) \quad i \in I_S, k \in S(i) \quad (6)$$

$$x_i \geq \sum_{j \in I(a^k), j > i} w_j (1 - u_{ij}) + \sum_{j \in I(a^k), j < i} w_j u_{ij} \qquad\qquad i \in I_S, k \in S(i) \quad (7)$$

If a solution of problem (4)–(7) exists there is a packing for I_S and thus, S is 2D-feasible. Our intention is to create only 2D-feasible sequences in our cutting plane approach. Let $S := \{s_1, \ldots, s_e\}$ be 2D-feasible where $s_e \in Q$ has to be continued. A pattern k that continues s_e is called 2D-feasible if $\tilde{S} := \{s_1, \ldots, s_e, k\}$ is still 2D-feasible. A 1D cutting pattern does not provide any information about the location of the items contained but in some special cases it is possible to modify the slave problem for continuing patterns of s_e so that 2D-feasibility is guaranteed for $\tilde{S}$. For that purpose, we define $I_{done}(s) := \{i \in I(a^s) \cap I_s^-\}$ for $s \in S$.

If a pattern $s \in S$ does not utilize the whole strip width $n_p := W - \sum_{i \in I(a^s)} w_i$ pseudo items of width 1 and demand b_{min}^s are added to $I_{done}(s)$. It is easy to see that all patterns that continue a bottom pattern are 2D-feasible and all patterns k that continue s_e are 2D-feasible if $|I_{done}(s_e)| = 1$. Now assume $I_{done}(s_e) = \{i_1, i_2\}$. If there is a 2D packing in which rectangle i_1 is located directly next to rectangle i_2 again all continuing patterns are 2D-feasible. Otherwise, only patterns that are solutions of a multiple knapsack problem (MKP) with knapsack capacities w_{i_1} and w_{i_2} are 2D-feasible. If $|I_{done}(s_k)| = 1$ for all $k < e$ it is always possible to place i_1 directly next to i_2. So if $|I_{done}(s_e)| = 2$ and there is a pattern s_k, $k < e$, with $|I_{done}(s_k)| > 1$ problem (4)–(7) with the additional constraint $x_{i_1} + w_{i_1} = x_{i_2}$ is solved with $u_{i_1 i_2} = 1$ if $i_1 < i_2$ and $u_{i_1 i_2} = 0$ otherwise. If there is no solution, the slave problem for s_e is a MKP, otherwise it is (P_{s_e}). Now assume $|I_{done}(s_e)| = 3$ with $I_{done} := \{i_1, i_2, i_3\}$ and w.l.o.g. $w_{i_1} > w_{i_2} > w_{i_3}$. We define test-items $t_1, \ldots, t_4$ of width

$$w_{t_1} := w_{i_1} + w_{i_2} + w_{i_3}, \quad w_{t_2} := w_{i_1} + w_{i_2}, \quad w_{t_3} := w_{i_1} + w_{i_3}, \quad w_{t_4} := w_{i_2} + w_{i_3}.$$

A test-item can only be packed if the corresponding rectangles of $I_{done}(s_e)$ are placed directly side by side. We use again a Padberg-type model to determine which lengths are available in a continuing pattern and how to define the slave problem for continuing patterns of s_e to ensure 2D-feasibility. In case of $|I_{done}(s_e)| \geq 4$, we only consider a reduced set of test-items to reduce the possibility of generating a 2D-infeasible continuing pattern. To ensure the 2D-feasibility of all sequences in that case, we test a pattern $k \in J(s_e) \cap Q$ for 2D-feasibility before we add the continuing condition (3). If problem (4)–(7) is not solvable for $S := \{s_1, \ldots, s_e, k\}$, pattern k is added to the set of forbidden patterns J_{forb} for which we require in HBRVC:

$$\sum_{j \in J_{forb}} \lambda_j \leq 0.$$

4   Numerical Experiments and Conclusions

We tested our approach with $I^B := I$ in an exact algorithm for the SPP and present results for some instances considered in [1, 4]. An initial upper bound for the SPP is obtained by a LP-based constructive algorithm [2] (HLP). To improve the upper bound during the cutting plane process, we compute sequence depending upper bounds. That means, for every sequence S with $V(S) := \sum_{i \in I_S} w_i h_i > 0.2 \cdot \sum_{i \in I} w_i h_i$, we place the remaining rectangles $I \setminus I_S$ into a 2D-packing of I_S by using heuristic HLP. The best value achieved within a time limit of 3600 s is named UB in Table 1. The objective value of HBRVC is the lower bound LB for the SPP. The instances ht1-ht7, ht9, ngcut1-ngcut8, ngcut10-ngcut12, cgcut1, gcut1, gcut3 and beng1-beng10 are optimally solved either in [1] or in [4] and also by our approach. In Table 1 the remaining six instances of these classes are presented. The best upper bound achieved in [1, 4] is UB_{min} and the best lower bound is LB_{max}.

By using contiguous 2D-feasible 1D cutting patterns, we improve the upper bound for four of six instances and optimality could be proven for one of them. For one instance the obtained lower bound is worse than known values. In general the objective value of HBRVC does not provide a sufficiently strong lower bound. With regard

Table 1 HBRVC in an exact algorithm for the SPP

Instance	n	W	LB	t_{LB}	UB	t_{UB}	gap	LB_{max}	UB_{min}	gap
ht8	28	60	30	0.0	31	1.4	3.33	30	31	3.33
cgcut2	23	70	64	82.7	64	5.8	0.00	63	65	3.17
cgcut3	60	70	652	0.0	660	145.9	1.23	652	670	2.76
ngcut9	18	20	49	0.0	50	23.4	2.04	49	50	2.04
gcut2	20	250	1184	0.0	1187	21.2	0.25	1184	1190	0.50
gcut4	50	250	2991	0.0	3000	3184.0	0.30	2995	3003	0.27

to an exact approach, we will focus on new lower bounds and improvement of performance of the algorithm in future work.

References

1. Alvarez-Valdez, R., Parreño, F., Tamarit, J.M.: A branch and bound algorithm for the strip paking problem. OR Spectr. **31**, 431–459 (2009)
2. Friedow, I., Scheithauer, G.: New inequalities for 1D relaxations of the 2D rectangular strip packing problem. In: Operations Research Proceedings 2014. Springer (2016)
3. Lodi, A., Martello, S., Monaci, M.: Two-dimensional packing problems - a survey. Eur. J. Op. Res. **141**, 241–252 (2002)
4. Martello, S., Monaci, M., Vigo, D.: An exact approach to the strip-packing problem. INFORMS J. Comput. **15**(3), 310–319 (2003)
5. Padberg, M.: Packing small boxes into a big box. Math. Methods Op. Res. **52**, 1–21 (2000)
6. Scheithauer, G.: LP-based bounds for the container and multi-container loading problem. Int. Trans. Op. Res. **6**, 199–213 (1999)

Computing Partitions with Applications to Capital Budgeting Problems

Frank Gurski, Jochen Rethmann and Eda Yilmaz

Abstract We consider the following capital budgeting problem. A firm is given a set of investment opportunities $X = \{x_1, \ldots, x_n\}$ and a number m of portfolios. Every investment x_i, $1 \leq i \leq n$, has a return of r_i and a price of p_i. Further for every portfolio j there is capacity c_j. The task is to choose m disjoint portfolios $X'_1, \ldots, X'_m$ from X such that for every $1 \leq j \leq m$ the prices in X'_j do not exceed the capacity c_j and the total return of this selection is maximized. From a computational point of view this problem is intractable, even for $m = 1$ [8]. Since the problem is defined on inputs of various informations, in this paper we consider the fixed-parameter tractability for several parameterized versions of the problem. For a lot of small parameter values we obtain efficient solutions for the partitioning capital budgeting problem. We also consider the connection to pseudo-polynomial algorithms.

1 Introduction

Capital budgeting is a tool for maximizing a company's profit by binary choosing among a given number of investments subject to a budget constraint [2, 12, 13]. Formally there is given a set of investment opportunities $X = \{x_1, \ldots, x_n\}$. Every investment x_i, $1 \leq i \leq n$, has a return of r_i and a price of p_i. Further there is a portfolio of capacity c. The task is to choose a subset X' from X such that the prices in X' do not exceed the capacity c and the total return of this selection is maximized [9]. In [5] we analyzed the generalization where the prices of the investments may change

F. Gurski (✉) · E. Yilmaz
Institute of Computer Science, Algorithmics for Hard Problems Group,
University of Düsseldorf, 40225 Düsseldorf, Germany
e-mail: frank.gurski@hhu.de

E. Yilmaz
e-mail: eda.yilmaz@hhu.de

J. Rethmann
Faculty of Electrical Engineering and Computer Science, Niederrhein University of Applied
Sciences, 47805 Krefeld, Germany
e-mail: jochen.rethmann@hs-niederrhein.de

© Springer International Publishing Switzerland 2017
K.F. Dœrner et al. (eds.), *Operations Research Proceedings 2015*,
Operations Research Proceedings, DOI 10.1007/978-3-319-42902-1_11

during a number of time periods T. In this paper we consider the problem where we have given a number m of different portfolios into which the given investments may be partitioned. For some positive integer n, let $[n] = \{1, \ldots, n\}$ be the set of all positive integers between 1 and n.

Name: MAX PARTITIONING CAPITAL BUDGETING (MAX PCB)

Instance: A set $X = \{x_1, \ldots, x_n\}$ of n investment opportunities, a number m of portfolios, for every $i \in [n]$ investment x_i has a return of r_i and a price of p_i, and for every portfolio $j \in [m]$ there is a capacity c_j.

Task: Find m disjoint (possibly empty) portfolios $X'_1, \ldots, X'_m$ from X such that for every $1 \leq j \leq m$ the prices in portfolio X'_j do not exceed the capacity c_j and the total return of this selection is maximized.

From a computational point of view the partitioning capital budgeting problem is equivalent to the well known multiple 0–1 knapsack problem. This implies that the problem is intractable even for the case $m = 1$ [8] and allows an efficient polynomial-time approximation scheme (EPTAS) [6].

The parameters n, m, r_i, p_i, and c_j are assumed to be positive integers. Further we can assume that $n \geq m$, since otherwise we can eliminate the $m - n$ portfolios of smallest capacity. Let $r_{max} = \max_{1 \leq i \leq n} r_i$, $p_{max} = \max_{1 \leq i \leq n} p_i$, and $c_{max} = \max_{1 \leq j \leq m} c_j$. For some instance I its size $|I|$ can be bounded by

$$|I| \in \mathcal{O}(n + m + \sum_{i=1}^{n} \log_2(r_i) + \sum_{i=1}^{n} \log_2(p_i) + \sum_{j=1}^{m} \log_2(c_j))$$
$$\in \mathcal{O}(n + m + n \cdot \log_2(r_{max}) + n \cdot \log_2(p_{max}) + m \cdot \log_2(c_{max})).$$

The size of the input is important for the analysis of running times.

By choosing a boolean variable $y_{i,j}$ for every investment $x_i \in X$ and every portfolio $1 \leq j \leq m$, indicating whether or not the investment x_i will be put into portfolio j, a binary integer programming (BIP) version of the MAX PCB problem is as follows.

$$\max \sum_{j=1}^{m} \sum_{i=1}^{n} r_i \cdot y_{i,j} \tag{1}$$

$$\text{s.t.} \sum_{i=1}^{n} p_i \cdot y_{i,j} \leq c_j \text{ for } j \in [m] \tag{2}$$

$$\text{and} \sum_{j=1}^{m} y_{i,j} \leq 1 \text{ for } i \in [n] \tag{3}$$

$$\text{and } y_{i,j} \in \{0, 1\} \text{ for } i \in [n], j \in [m] \tag{4}$$

The condition (3) ensures that all portfolios are disjoint, i.e. every investment is contained in at most one portfolio.

A dynamic programming solution for MAX PCB for $m = 2$ portfolios can be found in [11], which can be generalized as follows.

Theorem 1 MAX PCB *can be solved in time* $\mathcal{O}(n \cdot m \cdot \prod_{j=1}^{m} c_j) \subseteq \mathcal{O}(n \cdot m \cdot c_{max}^m)$.

Proof We define $R[k, c_1', \ldots, c_m']$ to be the maximum return of the subproblem where we only may choose a subset from the first k investments $x_1, \ldots, x_k$ and the capacities are $c_1', \ldots, c_m'$. We initialize $R[0, c_1', \ldots, c_m'] = 0$ for all $c_1', \ldots, c_m'$ since when choosing none of the investments, the return is always zero. Further we set $R[k, c_1', \ldots, c_m'] = -\infty$ if at least one of the $c_1', \ldots, c_m'$ is negative in order to represent the case where the price p_k of an investment is too high for packing it into a portfolio of capacity c_j'. The values $R[k, c_1', \ldots, c_m'], 1 \le k \le n$, for every $0 \le c_j' \le c_j$ and every $1 \le j \le m$ can be computed by the following recursion.

$$R[k, c_1', \ldots, c_m'] = \max \begin{cases} R[k-1, c_1', \ldots, c_m'] \\ R[k-1, c_1' - p_k, \ldots, c_m'] + r_k \\ \vdots \\ R[k-1, c_1', \ldots, c_m' - p_k] + r_k \end{cases}$$

All these values define a table with $\mathcal{O}(n \cdot \prod_{j=1}^{m} c_j)$ fields, where each field can be computed in time $\mathcal{O}(m)$. The optimal return is $R[n, c_1, \ldots, c_m]$. Thus we have shown that MAX PCB can be solved in time $\mathcal{O}(n \cdot m \cdot \prod_{j=1}^{m} c_j)$. ☐

In this paper we use standard definitions for pseudo-polynomial algorithms and parameterized algorithms from the textbooks [1] and [3].

2 Pseudo-polynomial Algorithms

The existence of pseudo-polynomial algorithms for MAX PCB depends on the assumption whether the number of portfolios m is given in the input or is fixed.

Theorem 2 MAX PCB *is not pseudo-polynomial.*

Proof We give a reduction from 3- PARTITION which is not pseudo-polynomial by [4]. In this problem we have given $n = 3m$ positive integers $w_1, \ldots, w_n$ such that $\sum_{j=1}^{n} w_j/m = B$ and $B/4 < w_j < B/2$ for every $1 \le j \le n$. The task is to decide whether there is a partition of $N = \{1, \ldots, n\}$ into m sets $N_1, \ldots, N_m$ such that $\sum_{j \in N_i} w_j = B$ for every $1 \le i \le m$.

Let I be an instance for 3- PARTITION. We define an instance I' for MAX PCB by choosing the number of investments as n, the number of portfolios as m, the capacities $c_j = B$ for $1 \le j \le m$, the returns $r_i = 1$ and prices $p_i = w_i$ for $1 \le i \le n$. By this construction every 3- PARTITION solution for I implies a solution with optimal return n for the MAX PCB instance I' and vice versa. ☐

By Theorem 1 we obtain the following result.

Theorem 3 *For every fixed m there is a pseudo-polynomial algorithm that solves* MAX PCB *in time* $\mathcal{O}(n \cdot m \cdot \prod_{j=1}^{m} c_j) \subseteq \mathcal{O}(n \cdot m \cdot c_{max}^m)$.

3 Parameterized Algorithms

A *parameterized problem* is a pair (Π, κ), where Π is a decision problem, $\mathscr{I}$ the set of all instances of Π and $\kappa : \mathscr{I} \to \mathbb{N}$ a so-called *parameterization*. The idea of parameterized algorithms is to restrict the combinatorial explosion to a parameter $\kappa(I)$ that is expected to be small for all inputs $I \in \mathscr{I}$.

An algorithm A is an *fpt-algorithm with respect to* κ, if there is a computable function $f : \mathbb{N} \to \mathbb{N}$ such that for every instance $I \in \mathscr{I}$ the running time of A on I is at most $f(\kappa(I)) \cdot |I|^{\mathcal{O}(1)}$ or equivalently at most $f(\kappa(I)) + |I|^{\mathcal{O}(1)}$. If there is an fpt-algorithm with respect to κ that decides Π then Π is called *fixed-parameter tractable*.

An algorithm A is an *xp-algorithm with respect to* κ, if there are two computable functions $f, g : \mathbb{N} \to \mathbb{N}$ such that for every instance $I \in \mathscr{I}$ the running time of A on I is at most $f(\kappa(I)) \cdot |I|^{g(\kappa(I))}$. If there is an xp-algorithm with respect to κ which decides Π then Π is called *slicewise polynomial*.

For the MAX PCB problem we add a threshold value k for the return to the instance and choose a parameter $\kappa(I)$ from the instance I, in order to obtain the following parameterized problem.

Name: $\kappa(I)$-PARTITIONING CAPITAL BUDGETING ($\kappa(I)$-PCB)
Instance: An instance of MAX PCB and a positive integer k.
Parameter: $\kappa(I)$
Question: Is it possible to choose m disjoint (possibly empty) portfolios $X'_1, \ldots, X'_m$ from X such that for every $1 \le j \le m$ the prices in portfolio X'_j do not exceed the capacity c_j and the total return of this selection is at least k?

For some instance I of the parameterized problem its size $|I|$ can be bounded by $|I| \in \mathcal{O}(n + m + n \cdot \log_2(r_{\max}) + n \cdot \log_2(p_{\max}) + m \cdot \log_2(c_{\max}) + \log_2(k))$.

Parameterization by number of investments n A brute force solution is to consider all $B(n)$ possible partitions of the n investments into disjoint nonempty subsets (portfolios). $B(n)$ is denoted as the nth *Bell number* and asymptotically grows faster than c^n for every constant c but slower than $n!$. For every such partition we check if it consists of at most m sets and after sorting descending w.r.t. the sum of prices in each set X'_j of the partition, we compare this sum with capacity c_j. The capacities are also assumed to be sorted, such that $c_1 \ge c_2 \ge \cdots \ge c_m$. Every of these partitions can be handled in $\mathcal{O}(m \cdot \log(m) + n)$ for sorting the X'_j and summing up the prices.

Theorem 4 *There is an fpt-algorithm that solves n-PCB in time* $\mathcal{O}(B(n) \cdot (m \cdot \log(m) + n))$.

Parameterization by standard parameter k When choosing the threshold value of the return as our parameter, i.e. $\kappa(I) = k$ we obtain the so-called *standard parameterization* of the problem.

Theorem 5 *There is an fpt-algorithm that solves k-PCB in time* $2^{\mathcal{O}(k \cdot \log^4(k))} + n^{\mathcal{O}(1)}$.

Proof The MAX PCB problem is equivalent to the max multiple 0–1 knapsack problem which allows an EPTAS of parameterized running time $2^{\mathscr{O}(1/\varepsilon \cdot \log^4(1/\varepsilon))} + n^{\mathscr{O}(1)}$, see [6]. By Proposition 2 in [10] we can use this EPTAS for $\varepsilon = 1/2k$ in order to obtain an fpt-algorithm that solves the standard parameterization of the corresponding decision problem in time $2^{\mathscr{O}(k \cdot \log^4(k))} + n^{\mathscr{O}(1)}$. $\qquad\square$

To obtain a more simple xp-algorithm we show the following observation. The *size* of a set is the number of its elements and the size of a number of sets is the size of its union.

Lemma 1 (Bounding the size of the solution) *For every instance of* MAX PCB *there is a feasible solution* $X_1', \ldots, X_m'$ *of return at least* k *if and only if there is a feasible solution* $X_1'', \ldots, X_m''$ *of return at least* k, *which has size at most* k.

Proof Let I be an instance of MAX PCB and $X' = X_1' \cup \ldots \cup X_m'$, $X_j' \subseteq X$, be a solution which complies the capacities of every portfolio and the return $\sum_{x_i \in X'} r_i$ is at least k. If there are at least $k + 1$ investments in X' we can remove one of the investments of smallest return r' and obtain a solution X''. The solution X'' still complies the capacities of every portfolio, since X' did and the prices are positive integers. Further, X'' has a return of at least $r' \cdot (k + 1) - r' = r' \cdot k \geq k$, since all returns are positive integers. $\qquad\square$

Thus we can assume that $|X_1' \cup \ldots \cup X_m'| \leq k$ to show the following result.

Theorem 6 *There is an xp-algorithm that solves* k-PCB *in time* $\mathscr{O}(n^k \cdot B(k + 1) \cdot (m \cdot \log(m) + n))$.

Proof By Lemma 1 we can restrict to solutions which have size at most k. The number of families of disjoint subsets $X_1', \ldots, X_m'$ of a set X on n Elements, such that $|X_1' \cup \ldots \cup X_m'| \leq k$ can be bounded as follows.

$$\sum_{i=1}^{k} \binom{n}{i} B(i) = \sum_{i=1}^{k} \binom{k}{i} \frac{(k-i)!n!}{k!(n-i)!} B(i) \overset{i \leq k}{\leq} \frac{n!}{(n-k)!} \sum_{i=1}^{k} \binom{k}{i} B(i) \leq \frac{n!}{(n-k)!} B(k + 1)$$
$$\leq n^k B(k + 1)$$

Every of these families can be handled in time $\mathscr{O}(m \cdot \log(m) + n)$. $\qquad\square$

Parameterization by the sum of capacities $\sum_{j=1}^{m} c_j$ From a practical point of view choosing k as a parameter is not useful, since a large return of the portfolio X' violates the aim that a good parameterization is small for every input. So we suggest it is better to choose other parameters, e.g. $\kappa(I) = \sum_{j=1}^{m} c_j$.

By assuming that we have one portfolio of capacity $\sum_{j=1}^{m} c_j$ using Theorem 1 in [5] we can bound the number of investments w.r.t. the sum of all capacities.

Lemma 2 (Bounding the size of the inut) *Every instance of* MAX PCB *can be transformed into an equivalent one, such that* $n \in \mathscr{O}((\sum_{j=1}^{m} c_j) \cdot \log(\sum_{j=1}^{m} c_j))$.

Thus by Theorem 4 we get the following result.

Theorem 7 *There is an fpt-algorithm that solves the $(\sum_{i=1}^{m} c_j)$-PCB problem in time $\mathscr{O}(B(\mathscr{O}((\sum_{j=1}^{m} c_j) \cdot \log(\sum_{j=1}^{m} c_j))) \cdot (m \cdot \log(m) + n))$.*

Parameterization by number of portfolios m When choosing $\kappa(I) = m$ the parameterized problem does not allow an fpt-algorithm and even no xp-algorithm with respect to m since both would imply a polynomial time algorithm for every fixed m. But even for $m = 1$ the problem is NP-hard [8].

Theorem 8 *There is no xp-algorithm that solves m-PCB, unless $P = NP$.*

Parameterization by combined parameters In the case of hardness with respect to some parameter ℓ a natural question is whether the problem remains hard for *combined* parameters, i.e. parameters (ℓ, ℓ') that consists of two or even more parts of the input.

By Theorem 1 the $(m, c_{\max})$-PCB problem is fixed-parameter tractable with running time $\mathscr{O}(m \cdot n \cdot c_{\max}^m) \subseteq \mathscr{O}(c_{\max}^m \cdot |I|^2)$.

Further we add the parameter $\mathrm{val}(I) =$"length of the binary encoding of maximum number within I". By Theorem 3 we obtain a parameterized running time of $\mathscr{O}(m \cdot n \cdot c_{\max}^m) \subseteq \mathscr{O}(|I|^2 \cdot (2^{\mathrm{val}(I)})^m) \subseteq \mathscr{O}(2^{m \cdot \mathrm{val}(I)} \cdot |I|^2)$. Thus the $(m, \mathrm{val}(I))$-PCB problem is fixed-parameter tractable.

Finally for the parameter (m, n) we can apply the result of [7], which implies that integer linear programing is fixed-parameter tractable for the parameter "number of variables". Thus by BIP (1)–(4) the (m, n)-PCB problem is fixed-parameter tractable with running time $\mathscr{O}((n \cdot m)^{\mathscr{O}(n \cdot m)} \cdot |I|)$.

References

1. Ausiello, G., Crescenzi, P., Gambosi, G., Kann, V., Marchetti-Spaccamela, A., Protasi, M.: Complexity and Approximation: Combinatorial Optimization Problems and Their Approximability Properties. Springer, Berlin (1999)
2. Cornuejols, G., Tütüncü, R.: Optimization Methods in Finance. Cambridge University Press, New York (2013)
3. Downey, R., Fellows, M.: Fundamentals of Parameterized Complexity. Springer, New York (2013)
4. Garey, M., Johnson, D.: Computers and Intractability: A Guide to the Theory of NP-Completeness. W.H. Freeman and Company, San Francisco (1979)
5. Gurski, F., Rethmann, J., Yilmaz, E.: Capital budgeting problems: A parameterized point of view. In: Operations Research Proceedings (OR 2014), Selected Papers. Springer (2015) (To appear)
6. Jansen, K.: A fast approximation scheme for the multiple knapsack problem. In: Proceedings of the Conference on Current Trends in Theory and Practice of Computer Science, vol. 7147, pp. 313–324. Springer, LNCS (2012)
7. Kannan, R.: Minkowski's convex body theorem and integer programming. Math. Op. Res. **12**, 415–440 (1987)
8. Kellerer, H., Pferschy, U., Pisinger, D.: Knapsack Problems. Springer, Berlin (2010)
9. Lorie, J., Savage, L.: Three problems in capital rationing. J. Bus. **28**, 229–239 (1955)

10. Marx, D.: Parameterized complexity and approximation algorithms. Comput. J. **51**(1), 60–78 (2008)
11. Pisinger, D., Toth, P.: Knapsack problems. In: Handbook of Combinatorial Optimization, vol. A, pp. 299–428. Kluwer Academic Publishers (1999)
12. Weingartner, H.: Capital budgeting of interrelated projects: survey and synthesis. Manag. Sci. **12**(7), 485–516 (1966)
13. Weingartner, H., Martin, H.: Mathematical Programming and the Analysis of Capital Budgeting Problems. Prentice Hall Inc, Englewood Cliffs (1963)

Upper Bound for the Capacitated Competitive Facility Location Problem

V.L. Beresnev and A.A. Melnikov

Abstract We consider the capacitated competitive facility location problem (CCFLP) where two competing firms open facilities to maximize their profits obtained from customer service. The decision making process is organized as a Stackelberg game. Both the set of candidate sites where firms may open facilities and the set of customers are finite. The customer demands are known, and the total demand covered by each of the facilities can not exceed its capacity. We propose the upper bound for the leader's objective function based on solving of the estimating MIP.

1 Introduction

Competitive location models form a wide class of optimization problems [2, 6, 7]. In contrast to classic facility location problems, these models consider several competitors acting in a common space and having objective functions, which are usually in conflict with each other. In this paper we deal with competition of two firms, which open their facilities aiming to capture customers and maximize profit from their serving. The decision making process is organized in a Stackelberg game framework [10]. It assumes that competitors are not equal in rights. One of them, called a leader, decides, where to open his facilities first. The second firm, called a follower, knows the leader's decision and uses this information when opens its own facilities at the second step. After the decisions are made, each customer chooses the firm to be served by.

There is a number of customer behavior models depending on the kind of service the facilities provide and many other factors [9]. Similar to the model of the facility location with order [4], we assume that each customer has known preferences. The firm which opens the most preferable facility for the customer captures him.

V.L. Beresnev (✉) · A.A. Melnikov
Sobolev Institute of Mathematics, Koptyuga Avenue, 4, 630090 Novosibirsk, Russia
e-mail: beresnev@math.nsc.ru

A.A. Melnikov
e-mail: melnikov@math.nsc.ru

© Springer International Publishing Switzerland 2017
K.F. Dœrner et al. (eds.), *Operations Research Proceedings 2015*,
Operations Research Proceedings, DOI 10.1007/978-3-319-42902-1_12

After customers have chosen the firm to be served by, each firm assigns facilities to serve the captured customers. In our model both leader and follower use a free choice of suppliers rule [2]. According to this rule, firm can serve the customer only with a facility which is more preferable for him than any of the competitor's facilities. To satisfy the customer's demand, the facility must provide him a known quantity of goods. The capacity of facilities is assumed to be bounded. Since the total demand satisfied by the facility can not exceed facility's capacity, some customers may be ignored. The problem is to find a location and a supply scheme for the leader's facilities maximizing his profit anticipating the follower's rational reaction.

The rest of the paper is organized as follows. In Sect. 2 we give a bilevel integer linear mathematical model of the problem. In Sect. 3 we show that the CCFLP can be represented as a problem of maximization of a pseudo–boolean function, depending only on the leader's facilities location. The upper bound for this function is presented in Sect. 3 as well. Section 4 concludes the paper.

2 Mathematical Model

In the mathematical model of the CCFLP we will use the following entities:

Index sets:

$I = \{1, \ldots, m\}$ set of locations (candidate facility sites);

$J = \{1, \ldots, n\}$ set of customers.

Parameters:

f_i fixed cost of opening a leader's facility $i \in I$;

g_i fixed cost of opening a follower's facility $i \in I$;

p_{ij} profit of the leader's facility $i \in I$ obtained from a customer $j \in J$;

q_{ij} profit of the follower's facility $i \in I$ obtained from a customer $j \in J$;

a_{ij} demand of the customer $j \in J$ for products of the leader's facility $i \in I$;

b_{ij} demand of the customer $j \in J$ for products of the follower's facility $i \in I$;

V_i capacity of the leader's facility $i \in I$;

W_i capacity of the follower's facility $i \in I$.

Variables:

$$x_i = \begin{cases} 1, & \text{if the leader opens facility } i \\ 0, & \text{otherwise} \end{cases} ; \quad z_i = \begin{cases} 1, & \text{if the follower opens facility } i \\ 0, & \text{otherwise} \end{cases} ;$$

$$x_{ij} = \begin{cases} 1, & \text{if the leader's facility } i \text{ serve the customer } j \\ 0, & \text{otherwise} \end{cases} ;$$

$$z_{ij} = \begin{cases} 1, & \text{if the follower's facility } i \text{ serve the customer } j \\ 0, & \text{otherwise} \end{cases} .$$

We assume that the preferences of the customer $j \in J$ are represented with linear order $\succeq_j$ on the set I. The relation $i_1 \succeq_j i_2$ shows that either facility i_1 is more preferable for j than i_2 or $i_1 = i_2$. If $i_1 \neq i_2$ and $i_1 \succeq_j i_2$, we use denotation $i_1 \succ_j i_2$.

Given $j \in J$, we denote the most and the least preferable facilities for j from the nonempty set of facilities K with $\alpha_j(K)$ and $\omega_j(K)$ correspondingly. For a nonzero boolean vector $x = (x_i)$, $i \in I$ we assume that $\alpha_j(x) = \alpha_j(\{i \in I | x_i = 1\})$.

It is assumed that firms use a free choice of suppliers rule for facilities assignment. If we are given with boolean vectors x and z corresponding to the leader's and the follower's facilities locations respectively, then, following this rule, the leader's facility $i \in I$ can serve a customer $j \in J$ iff $i \succ_j \alpha_j(z)$. Similarly, the follower's facility $i \in I$ can serve a customer $j \in J$ iff $i \succ_j \alpha_j(x)$.

Now we can formulate the model of CCFLP:

$$\max_{(x_i),(x_{ij}),(\tilde{z}_i),(\tilde{z}_{ij})} \left\{ -\sum_{i \in I} f_i x_i + \sum_{j \in J} \sum_{i \in I} p_{ij} x_{ij} \right\}, \tag{1}$$

$$\tilde{z}_i + \sum_{k|i \succeq_j k} x_{kj} \leq 1, \quad i \in I, j \in J; \tag{2}$$

$$x_i \geq x_{ij}, \quad i \in I, j \in J; \tag{3}$$

$$\sum_{j \in J} a_{ij} x_{ij} \leq V_i, \quad i \in I; \tag{4}$$

$$x_i, x_{ij} \in \{0, 1\}, \quad i \in I, j \in J; \tag{5}$$

$$(\tilde{z}_i), (\tilde{z}_{ij}) \text{ --- optimal solution of the follower's problem:} \tag{6}$$

$$\max_{(z_i),(z_{ij})} \left\{ -\sum_{i \in I} g_i z_i + \sum_{j \in J} \sum_{i \in I} q_{ij} z_{ij} \right\}, \tag{7}$$

$$x_i + z_i \leq 1, \quad i \in I; \tag{8}$$

$$x_i + \sum_{k|i \succeq_j k} z_{kj} \leq 1, \quad i \in I, j \in J; \tag{9}$$

$$z_i \geq z_{ij}, \quad i \in I, j \in J; \tag{10}$$

$$\sum_{j \in J} b_{ij} z_{ij} \leq W_i, \quad i \in I; \tag{11}$$

$$z_i, z_{ij} \in \{0, 1\}, \quad i \in I, j \in J. \tag{12}$$

We denote the upper level problem (1)–(6) with $\mathcal{L}$ and the lower level problem (7)–(12) with $\mathcal{F}$. For the problem (1)–(12) we use the denotation $(\mathcal{L}, \mathcal{F})$.

The objective function (1) of the problem $\mathcal{L}$ expresses the value of the leader's profit and consists of two components. The first one is the cost of facilities to be opened, and the second summand represents the income collected by them. The problem $\mathcal{F}$ may have a number of optimal solutions for some feasible solution of the problem $\mathscr{L}$. We assume that in this situation the follower cooperates with the leader and chooses the solution, which maximizes (1). Constraints (2) ensure that the leader serve the customer with a facility which is more preferable for the customer than each of the follower's facilities. Also these constraints ensure that the customer is served with not more than one leader's facility. The constraints (3) guarantee that customers are served with open facilities. Inequalities (4) ensure that total demand covered by the facility doesn't exceed its capacity. The objective function and the constraints of the problem $\mathscr{F}$ have similar meaning. Additional constraint (8) ensures that the facility can't be opened by both firms simultaneously.

3   Upper Bound

Note that given boolean vector $x = (x_i), i \in I$, we can obtain the corresponding feasible solution $(X, Z), X = ((x_i), (x_{ij})), Z = ((z_i), (z_{ij}))$ of the problem $(\mathscr{L}, \mathscr{F})$ and calculate the value of its objective function in two steps.

At the first step we solve the problem $\mathscr{F}$ with given $x = (x_i), i \in I$ and get an optimal value of its objective function, F^*. At the second step we solve the auxiliary problem (1)–(5), (8)–(12), with additional constraint

$$-\sum_{i \in I} g_i z_i + \sum_{j \in J} \sum_{i \in I} q_{ij} z_{ij} \geq F^*.$$

Consequently, the problem $(\mathscr{L}, \mathscr{F})$ can be represented as a problem of pseudo–boolean function maximization. To calculate the value of that function $f(x)$ on the boolean vector $x = (x_i), i \in I$ we are to solve the problem $\mathscr{F}$ and the auxiliary problem. This operations provide a corresponding feasible solution of the problem $(\mathscr{L}, \mathscr{F})$ as well.

Consider the problem of calculation of the upper bound for the pseudo–boolean function $f(x), x = (x_i), i \in I$. We call a vector $y = (y_i), i \in I$ which elements take values 0, 1, and uncertain value $*$ by *partial solution*. For the partial solution $y = (y_i)$ we define sets $I^0(y) = \{i \in I \mid y_i = 0\}$ and $I^1(y) = \{i \in I \mid y_i = 1\}$. Partial solution $y = (y_i)$ defines the set of boolean vectors $x = (x_i)$ such, that $x_i = y_i$ for all $i \in I^0(y) \cup I^1(y)$. We denote this set with $P(y)$ and the problem $\mathscr{L}$ with additional constraints $x_i = y_i$ for $i \in I^0(y) \cup I^1(y)$ with $\mathscr{L}(y)$.

To calculate an upper bound for the pseudo–boolean function $f(x)$, where $x \in P(y)$ for some partial boolean vector y, firstly we relax the problem $(\mathscr{L}(y), \mathscr{F})$ by removing the lower level problem $\mathscr{F}$ and its variables. The resulting MIP, which we will address as an *estimating problem*, models the situation where the leader is a monopolist. Its optimal solution is a valid upper bound, which is actually significantly overestimated. We extend the estimating problem with additional inequalities to make it more accurate. These inequalities are satisfied by any feasible solution of the problem $(\mathscr{L}(y), \mathscr{F})$ but reduce the feasible region of the estimating problem.

We adapt the method from [1, 3] to formulate the additional constraints of the estimating problem. Given $j \in J$ we build a set $I_j(y) \subseteq I$ with the following property: if $x \in P(y)$ and $\alpha_j(x) \notin I_j(y)$, then the follower opens some facility which is more preferable for j than $\alpha_j(x)$. Given a partial solution $y = (y_i)$, $i \in I$ and a customer $j \in J$, let us formulate the rule for determining if a facility $i \in I$ belongs to the set $I_j(y)$ or not.

Consider the set $N_j(i) = \{k \in I | k \succ_j i\}$ of facilities which are more preferable for j than i. We suppose that $i \notin I_j(y)$ if $y_i = 0$ or i is less preferable than some leader's facility which is open in partial solution y, i.e. $N_j(i) \cap I^1(y) \neq \emptyset$. If i is the most preferable facility for j, i.e. $N_j(i) = \emptyset$, then $i \in I_j(y)$.

Let $N_j(i) \neq \emptyset$ and $N_j(i) \cap I^1(y) = \emptyset$. Consider the set of customers, which can be captured only with the facility from the set $N_j(i)$:

$$J(i) = \{s \in J | \forall k \in I \text{ if } k \succ_s \alpha_s(I^1(y) \cup \{i\}), \text{ then } k \in N_j(i)\}.$$

Note, that $j \in J(i)$. For each $k \in N_j(i)$ consider the set

$$J(k, i) = \{j \in J(i) | k \succ_j \alpha_j(I^1(y) \cup \{i\})\}$$

of customers from $J(i)$, which the facility k is able to capture. We suppose that $i \notin I_j(y)$ if there exists a subset of customers $S \subseteq J(k, i)$ that makes the facility k profitable for the follower, i.e. $g_k < \sum_{j \in S} q_{kj}$ and $W_k \geq \sum_{j \in S} b_{kj}$. If there is no $k \in N_j(i)$ with a mentioned property, then $i \in I_j(y)$.

Lemma 1 *Let y be an arbitrary partial solution, $(X, \widetilde{Z})$, $X = ((x_i), (x_{ij}))$, $\widetilde{Z} = ((\widetilde{z}_i), (\widetilde{z}_{ij}))$ be a feasible solution of the problem $(\mathscr{L}(y), \mathscr{F})$, and $x = (x_i)$, $\widetilde{z} = (\widetilde{z}_i)$. Then for each $j \in J$ if $\alpha_j(x) \notin I_j(y)$, then $\sum_{k \in N_j(\alpha_j(x))} \widetilde{z}_k > 0$.*

Lemma 1 allows us to formulate additional constraints to strengthen the estimating problem. Consider arbitrary j_1, $j_2 \in J$ and $i_1 \in I$. Note that here j_1 and j_2 are allowed to be equal. Suppose that i_1 is the most preferable leader's facility for the customer j_1 but $i_1 \notin I_{j_1}(y)$. Then from Lemma 1 there is an open facility of the follower in the set $N_{j_1}(i_1)$. Constraints (2) guarantee that in this case $x_{ij_2} = 0$ for any $i \in I$ such that $\omega_{j_2}(N_{j_1}(i_1)) \succeq_{j_2} i$.

The resulting estimating problem is written as follows:

$$\max_{(x_i),(x_{ij}),(t_{ij})} \left\{ -\sum_{i \in I} f_i x_i + \sum_{j \in J} \sum_{i \in I} p_{ij} x_{ij} \right\}, \tag{13}$$

$$x_i \geq x_{ij}, \quad i \in I, j \in J; \tag{14}$$

$$\sum_{j \in J} a_{ij} x_{ij} \leq V_i, \quad i \in I; \tag{15}$$

$$x_i \geq t_{ij}, \quad i \in I, j \in J; \tag{16}$$

$$\sum_{i \in I} t_{ij} = 1, \quad j \in J; \tag{17}$$

$$x_i + \sum_{k | i \succ_j k} t_{kj} \leq 1, \quad i \in I, j \in J; \tag{18}$$

$$1 - t_{i_1 j_1} + \sum_{i \in I_{j_1}(y)} t_{ij_1} \geq \sum_{k | \omega_{j_2}(N_{j_1}(i_1)) \succeq_{j_2} k} x_{kj_2}, \quad i_1 \in I, j_1, j_2 \in J; \tag{19}$$

$$x_i = y_i, \quad i \in I^0(y) \cup I^1(y); \tag{20}$$

$$x_i, x_{ij}, t_{ij} \in \{0, 1\}, \quad i \in I, j \in J. \tag{21}$$

Due to inequalities (16)–(18) the new boolean variable t_{ij}, $i \in I$, $j \in J$ takes the value 1 iff $i = \alpha_j(x)$. Constraints (19) implement the following consequence of Lemma 1: if $i_1 = \alpha_{j_1}(x)$ and $i_1 \notin I_{j_1}(y)$, then $x_{kj_2} = 0$ for each k such that $\omega_{j_2}(N_{j_1}(i_1)) \succeq_{j_2} k$.

Let $X^0 = ((x_i^0), (x_{ij}^0))$, $T^0 = (t_{ij}^0)$ be an optimal solution of the estimation problem (13)–(21) and $B(X^0, T^0)$ be an optimal value of its objective function (13).

Theorem 1 *The inequality* $\max_{x \in P(y)} f(x) \leq B(X^0, T^0)$ *holds.*

4   Conclusion

We have shown, that CCFLP can be represented in the form of maximization of the implicitly given pseudo–boolean function $f(x)$ depending on m boolean variables. It allows to apply methods based on local search ideas [5, 8] for obtaining approximate solutions. For the considered function $f(x)$, where $x \in P(y)$ for some partial solution $y = (y_i)$, $i \in I$, we propose the method of the upper bound calculation. The method consists in formulating and solving of the estimation MIP. Supposed approach of the upper bound calculation may be used in the exact procedure based on the branch–and–bound scheme [1, 3].

Acknowledgements This research is supported by the Russian Foundation for Basic Research (grant No. 15-01-01446)

References

1. Beresnev, V.: Branch-and-bound algorithm for competitive facility location problem. Comput. Op. Res. **40**, 2062–2070 (2013)
2. Beresnev, V.L.: On the competitive facility location problem with a free choice of suppliers. Autom. Remote Control **75**(4), 668–676 (2014)

3. Beresnev, V.L., Mel'nikov, A.A.: The branch–and–bound algorithm for a competitive facility location problem with the prescribed choice of suppliers. J. Appl. Ind. Math. **8**(2), 177–189 (2014)
4. Canovas, L., Garcia, S., Labbe, M., Marin, A.: A strengthened formulation for the simple plant location problem with order. Op. Res. Lett. **35**(2), 141–150 (2007)
5. Davydov, I., Kochetov, Y., Carrizosa, E.: VNS heuristic for the $(r|p)$–centroid problem on the plane. Electron. Notes Discret. Math. **39**, 5–12 (2012)
6. Kress, D., Pesch, E.: Sequential competitive location on networks. Eur. J. Op. Res. **217**, 483–499 (2012)
7. Kucukaydin, H., Aras, N., Kuban Altinel, I.: A leader–follower game in competitive facility location. Comput. Op. Res. **39**, 437–448 (2012)
8. Mel'nikov, A.A.: Randomized local search for the discrete competitive facility location problem. Autom. Remote Control **75**(4), 700–714 (2014)
9. Santos–Penate, D.R., Suares–Vega, R., Dorta–Gonzalez, P.: The leader–follower location model. Netw. Spat. Econ. **7**, 45–61 (2007)
10. Stackelberg, H.: The Theory of the Market Economy. Oxford University Press, Oxford (1952)

The Onset of Congestion in Charging of Electric Vehicles for Proportionally Fair Network Management Protocol

Ľuboš Buzna

1 Introduction

With the expected uptake of electric vehicles in the near future, we are likely to observe overloading in the local distribution networks more frequently. Such development suggests that a congestion management protocol will be a crucial component of future technological innovations in low voltage networks. An important property of a suitable network capacity management protocol is to balance network efficiency and fairness requirements. Assuming a stochastic model, we study the proportional fairness (PF) protocol managing the network capacity in charging of electric vehicles. We explore the onset of congestion by analysing the critical arrival rate, i.e. the largest possible vehicle arrival rate that can still be fully satisfied by the network. We compare the proportionally fair management protocol with the max-flow (MF) management protocol. By numerical simulations on realistic networks, we show that proportional fairness leads not only to more equitable distribution of power allocations, but it can also serve slightly larger arrival rate of vehicles. We consider simplified setup, where the power allocations are dependent on the occupation of network nodes, but they are independent of the exact number of vehicles, and to validate numerical results, we analyse the critical arrival rate on a network with two edges, where the optimal power allocations can be calculated analytically.

2 Optimization Model

We model the electrical distribution network as a directed rooted tree graph composed of the node set $\mathcal{V}$ and edge set $\mathcal{E}$. Only the root node of the tree $r \in \mathcal{V}$ injects the power into the network and electric vehicles can be plugged into all other nodes. By

Ľ. Buzna (✉)
University of Žilina, Univerzitná 8215/1, 01026 žilina, Slovakia
e-mail: lubos.buzna@fri.uniza.sk

© Springer International Publishing Switzerland 2017

K.F. Dœrner et al. (eds.), *Operations Research Proceedings 2015*,
Operations Research Proceedings, DOI 10.1007/978-3-319-42902-1_13

the symbol ⋔ (j) we denote the subtree rooted in the node $j \in \mathcal{V}$. An edge $e_{ij} \in \mathcal{E}$ connects node i to node j, where i is closer to the root than j, and is characterised by the impedance $Z_{ij} = R_{ij} + iX_{ij}$, where R_{ij} is the edge resistance and X_{ij} the edge reactance. The power loss along edge e_{ij} is given by $S_{ij}(t) = P_{ij}(t) + iQ_{ij}(t)$, where $P_{ij}(t)$ is the real power loss, and $Q_{ij}(t)$ the reactive power loss. We model car batteries as elastic loads (i.e. able to absorb any value of power they are allocated). Electric vehicle $l = 1, \ldots, N(t)$ receives only active power $P_l(t)$, where $N(t)$ is the number of vehicles charging at time t. Value $\Delta_{il}(t)$ is one if electric vehicle l is charging on node i and zero otherwise. Vehicle l derives a utility $U_l(P_l(t))$ from the allocated charging power $P_l(t)$. Let $P_{⋔(j)}$ denote the active power, and $Q_{⋔(j)}$ the reactive power consumed by the subtree ⋔ (j) that include power consumed by all vehicles connected to the subtree and power losses dissipated on edges of the subtree. By the symbol $V_i(t)$, we denote the voltage level on the node $i \in \mathcal{V}$. We allocate the power to electric vehicles by maximizing the aggregate utility $U(t)$, while making sure that all nodal voltages are within the interval $((1 - \alpha)V_{\text{nominal}}, (1 + \alpha)V_{\text{nominal}})$, where α is a parameter and V_{nominal} is the nominal voltage level the network is operated on (for more details see Ref. [1]):

$$\underset{W(t)}{\text{maximise}} \quad U(t) = \sum_{l=1}^{N(t)} U_l(P_l(t)) \tag{1}$$

$$\text{subject to} \quad ((1 - \alpha)V_{\text{nominal}})^2 \le W_{ii}(t) \le ((1 + \alpha)V_{\text{nominal}})^2, \quad i \in \mathcal{V}, \tag{2}$$

$$W_{ij}(t) - W_{jj}(t) - P_{⋔(j)}(t)R_{ij} - Q_{⋔(j)}(t)X_{ij} = 0, \quad e_{ij} \in \mathcal{E}, \tag{3}$$

$$\begin{pmatrix} W_{ii}(t) & W_{ij}(t) \\ W_{ji}(t) & W_{jj}(t) \end{pmatrix} \succeq 0, \quad e_{ij} \in \mathcal{E}. \tag{4}$$

With every edge $e_{ij} \in \mathcal{E}$ is associated one decision variable $W_{ij}(t)$ that it is equal to the product of real voltages on edge nodes, i.e. $W_{ij}(t) = V_i(t)V_j(t)$ and similarly with every node $i \in \mathcal{V}$ is associated variable $W_{ii}(t) = V_i(t)^2$. The generalized inequality (4) means that matrices are positive semidefinite. Constraints (2) ensure that all nodal voltages are within the defined limits. Constraints (3)–(4) have been derived in reference [1] and they encode relations between decision variables $W_{ij}(t)$, power allocations $P_l(t)$ and power losses along edges that arise from Kirchhoff's current and voltage laws, where:

$$P_{⋔(k)} = \sum_{i \in \mathcal{V}_{⋔(k)}} \sum_{l=1}^{N(t)} \Delta_{il}(t)P_l(t) + \sum_{i \in \mathcal{V}_{⋔(k)}} \sum_{j:e_{ij} \in \mathcal{E}_{⋔(k)}} P_{ij}(t), \tag{5}$$

and

$$Q_{⋔(k)} = \sum_{i \in \mathcal{V}_{⋔(k)}} \sum_{j:e_{ij} \in \mathcal{E}_{⋔(k)}} Q_{ij}(t), \tag{6}$$

where power losses along edge $e_{ij} \in \mathcal{E}$ can be expressed as:

$$P_{ij}(t) = \left(W_{ii}(t) - 2W_{ij}(t) + W_{jj}(t)\right) \frac{R_{ij}}{R_{ij}^2 + X_{ij}^2}, \tag{7}$$

and

$$Q_{ij}(t) = \left(W_{ii}(t) - 2W_{ij}(t) + W_{jj}(t)\right) \frac{X_{ij}}{R_{ij}^2 + X_{ij}^2}. \tag{8}$$

We consider the *proportional fair* allocation representing a trade-off between network throughput and equality in allocations [2], that maximizes the sum of the logarithm of user rates, i.e. $U(t) = \sum_{l=1}^{N(t)} \log(P_l(t))$. Computationally it is more practical to use the equivalent definition $U(t) = \sum_{i \in \mathcal{V}^+} w_i(t) \log(\mathrm{P}_i(t))$, where $\mathrm{P}_i(t)$ is power allocated to network node i, $\mathcal{V}^+$ is the subset of nodes with at least one charging vehicle, and $w_i(t)$ is a number of vehicles charging at node i at time t, i.e. $w_i(t) = \sum_{l=1}^{N(t)} \Delta_{il}(t)$. Values $P_l(t)$ can be then recovered from $P_l(t) = \frac{\mathrm{P}_i(t)}{w_i(t)}$. As a benchmark representing the efficient network throughput, we consider non-unique *max-flow* allocation given by $U(t) = \sum_{l=1}^{N(t)} P_l(t)$, where we optimise the system whenever the configuration of vehicles changes. Max-flow maximizes the network throughput, however, it can leave some users with zero power, which can be considered as unfair from the user point of view. Both problems are convex, and hence can be solved by general purpose optimization solvers.

To study the behaviour of proportional fairness and max-flow, we implemented a discrete simulator that solves the problem (1)–(4) in discrete time steps. Simulations start with empty network. Vehicles arrive to the network in continuous time, following a Poisson process with rate λ, and choose node to charge randomly with uniform probability. Vehicles have a battery with capacity B that is empty at arrival and leave the network when it is full. The level of battery is given by the time integral of allocated power.

3 Results

3.1 Numerical Experiments

We simulate vehicles charging on the realistic SCE 47-bus network [3] while setting $V_{\mathrm{nominal}} = B = 1.0$ and $\alpha = 0.1$. In order to characterize the behaviour of the network, we adopt the congestion parameter [4]:

$$\eta(\lambda) = \lim_{t \to \infty} \frac{1}{\lambda} \frac{\langle \Delta N(t) \rangle}{\Delta t}, \tag{9}$$

where $\Delta N(t) = N(t + \Delta t) - N(t)$ and $\langle \ldots \rangle$ indicates an average over time window of length Δt. Congestion parameter $\eta(\lambda) = 0$ when all cars leave the network fully charged within a large enough time window, and $\eta(\lambda) > 0$, when some vehicles

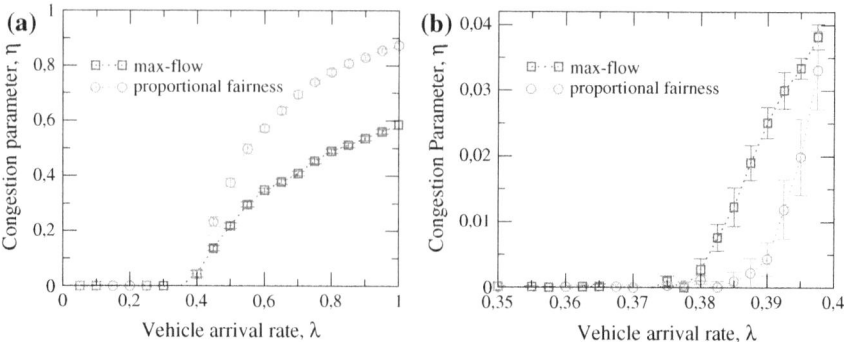

Fig. 1 **a** Congestion parameter η as a function of the vehicle arrival rate for the SCE 47-bus network and for the simulation time horizon of 1.5×10^4. **b** Zoom of the critical region for longer horizon of 10^5 time units. Symbols show average values over an ensemble of 25 independent runs and error bars reflect 95 % confidence intervals

have to wait for increasingly long times to fully charge, i.e. the network is congested. Simulation results in Fig. 1 show that the largest value of the arrival rate λ_c, when all vehicles are still fully charged, is larger for proportional fairness than for max-flow, meaning that proportional fairness is able to charge slightly larger number of vehicles.

3.2 Onset of Congestion in 2-Edge Network

To validate analytically that λ_c can be different in both methods, we analyse a three-node and two-edge network with node 1 (root node), and vehicles arriving at node 2 (the closest node to the root), and at node 3 (the leaf node), respectively, assuming uniform R and X values.

The two congestion control methods lead to different allocations of instantaneous power, with vehicles charging in different order and in different time intervals. The voltage drops with the increasing distance from the root and the lower voltage limit (constraint (2)) is fulfilled at equality for one node. The objective function of proportional fairness guaranties that both nodes (if occupied by vehicles) will receive positive power allocation. Thus, the lower voltage limit constraint is satisfied at equality on the most distant node from the root. In max-flow, however, the maximisation of the aggregate power allocated to vehicles implies also minimising instantaneous power losses, and this is achieved by allocating all power to the closest occupied node from the root node.

Note that optimal max-flow allocation is independent of how many vehicles are charging on each node. To simplify our analysis, we set w_i to value one if at least one vehicle is charging at node i, and zero otherwise for $i \in 2, 3$, and thus proportional fair optimal power allocations will be also independent of the number of vehicles on each node. For this simplified setup we can easily estimate the critical value λ_c analytically.

Under our assumptions, for 2-edge network the problem (1)–(4) can be solved analytically. Optimal power allocation of max-flow at the node $i \in \{2, 3\}$ is:

$$P_i^{MF} = \frac{2\alpha(1 - \alpha)V_{\text{nominal}}^2}{(i - 1)R}.$$

(10)

Optimal proportional fair power allocations are:

$$P_2^{PF} = \frac{2V_{\text{nominal}}^2(3\sqrt{\gamma} - \gamma)}{9R} \text{ and } P_3^{PF} = \frac{(1 - \alpha)V_{\text{nominal}}^2(\sqrt{\gamma} + 3\alpha - 3)}{3R},$$

(11)

where $\gamma = 2\sqrt{\alpha^2 - \alpha + 1}|\alpha - 2| + 2\alpha^2 - 5\alpha + 5$. When deriving the value λ_c, we assume a time interval Δt that is composed of two subintervals t_i, when vehicles situated at node $i \in \{2, 3\}$ are charged. Within this time interval the demand arriving at each of the nodes (i.e. $\frac{B\lambda_c \Delta t}{2}$) has to be the same as the energies $P_2^{MF} t_2$ and $P_3^{MF} t_3$ that max-flow is able to deliver at nodes 2 and 3, respectively. From here we obtain:

$$\lambda_c^{MF} = \frac{P_2^{MF}}{\frac{B}{2}\left(\frac{P_2^{MF}}{P_3^{MF}} + 1\right)}.$$

(12)

Similarly, for proportional fairness we obtain:

$$\lambda_c^{PF} = \frac{P_3^{MF}}{\frac{B}{2}\left(\frac{P_3^{MF}}{P_2^{PF}} - \frac{P_3^{PF}}{P_2^{PF}} + 1\right)}.$$

(13)

We set parameters $R = X = B = V_{\text{nominal}} = 1.0$ and $\alpha = 0.1$, yielding theoretical predictions $\lambda_c^{MF} = 0.12$ and $\lambda_c^{PF} \approx 0.1222$. Thus, our analyses show that proportional fairness may support slightly larger arrival rate, giving support to our numerical simulation on realistic electrical networks. To validate our analyses, we simulated max-flow and proportional fair protocols in 2-edge network with two λ values. When $\lambda < \lambda_c$ number of vehicles is oscillating, while for $\lambda > \lambda_c$ it has a tendency to grow (see Fig. 2). Thus, numerical results are in good agreement with calculated values.

4   Conclusions

The main contribution of this paper is that we showed analytically that PF can accommodate larger arrival rate than MF. This result is surprising, because common expectation is that efficiency of the system comes at the expense of the increased inequality [5]. However, it should be noted that here we optimise the dynamic system over

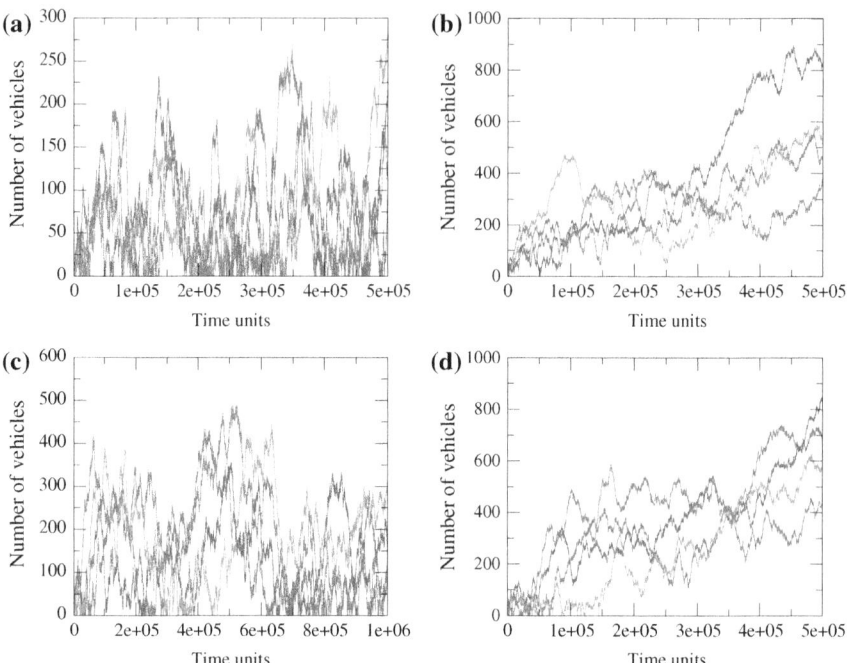

Fig. 2 Representative time series for the 2-edge network. Panel **a** shows that for $\lambda = 0.119$ the max-flow supplies all vehicles, whereas in panel **b**, for $\lambda = 0.121$, it is congested being in the agreement with the calculated value $\lambda_c^{MF} = 0.12$. Panel **c** shows that for $\lambda = 0.122$ the proportional fairness is supplying all vehicles, whereas in panel **d**, for $\lambda = 0.123$, it is congested being in the agreement with the calculated value $\lambda_c^{PF} \approx 0.1222$

a certain time period and our optimisation model is not dynamic, hence, it is only a heuristic.

Acknowledgements This work was supported by VEGA (project 1/0463/16), VEGA (project 1/0339/13) and FP 7 project ERAdiate [621386].

References

1. Carvalho, R., Buzna, L., Gibbens, R., Kelly, F.: Critical behaviour in charging of electric vehicles. New J. Phys. **17**, 095001 (2015)
2. Kelly, F.P., Maulloo, A.K., Tan, D.K.H.: Rate control for communication networks: shadow prices, proportional fairness and stability. J. Oper. Res. Soc. **49**, 237–252 (1998)
3. Gan, L., Li, N., Topcu, U., Low, S.H.: Exact convex relaxation of optimal power flow in radial networks. IEEE Trans. Autom. Control **60**, 72–87 (2015)
4. Arenas, A., Diaz-Guilera, A., Guimera, R.: Communication in networks with hierarchical branching. Phys. Rev. Lett. **86**, 3196–3199 (2001)
5. Bertsimas, D., Farias, V.F., Trichakis, N.: The Price of Fairness. Oper. Res. **59**, 17–31 (2011)

A Comparison of Heuristic Methods for the Prize-Collecting Steiner Tree Problem and Their Application in Genomics

Murodzhon Akhmedov, Ivo Kwee and Roberto Montemanni

Abstract The prize-collecting Steiner tree (PCST) problem is a broadly studied problem in combinatorial optimization. It has been used to model several real world problems related to utility networks. More recently, researchers have started using PCSTs to study biological networks. Biological networks are typically very large in size. This can create a considerable challenge for the available PCST solving methods. Taking this fact into account, we have developed methods for the PCST that efficiently scale up to large biological network instances. Namely, we have devised a heuristic method based on the Minimum Spanning Tree and a matheuristic method composed of a heuristic clustering phase and a solution phase. In this work, we provide a performance comparison for these methods by testing them on large gene interaction networks. Experimental results are reported for the methods, including running times and objective values of the solutions.

1 Introduction

The prize-collecting Steiner tree is a well known problem in combinatorial optimization and graph theory. Within the concept of the PCST, given an undirected network $G = (V, E)$, where nodes are associated with prizes $p_j \geq 0$ and arcs are associated with costs $c_e > 0$, the goal is to construct a sub-graph $G' = (V', E')$ that has a *tree* structure. The researchers have studied different variants of the PCST problem in the literature. One of the broadly studied variant is known as *Goemans–Williamson*

M. Akhmedov (✉) · I. Kwee · R. Montemanni
Dalle Molle Institute for Artificial Intelligence (IDSIA-USI/SUPSI),
Galleria 2, 6928 Manno, Switzerland
e-mail: murodzhon@idsia.ch

I. Kwee
e-mail: ivo.kwee@ior.iosi.ch

R. Montemanni
e-mail: roberto@idsia.ch

M. Akhmedov · I. Kwee
Institute of Oncology Research (IOR), Via Vela 6, 6500 Bellinzona, Switzerland

© Springer International Publishing Switzerland 2017
K.F. Dœrner et al. (eds.), *Operations Research Proceedings 2015*,
Operations Research Proceedings, DOI 10.1007/978-3-319-42902-1_14

Minimization [1], where the objective is to identify a *tree* for a given graph by minimizing the total cost of arcs in a *tree* and minimizing the total prize of nodes excluded from the *tree*. This corresponds to the minimization of the following expression:

$$GW(G') = \sum_{e \in E'} c_e + \sum_{v \notin V'} p_v \tag{1}$$

The PCST has been successfully applied to model several real-world problems in utility networks. Recently, researchers have realized its application to biological networks for discovering the hidden knowledge [2]. Based on this idea, we have applied the PCST to gene interaction networks, where nodes correspond to genes and arcs represent the mutual information between genes. The PCST potentially captures the portion of graphs where genetic aberrations and mutations are highly present. Basically, biological interaction networks are large in size, and this can be remarkable challenge for existing PCST methods. By considering this fact in our previous studies, we have developed methods for the PCST that efficiently scale up to large biological network instances for analyzing the function of genes. In this work, we extensively test previously developed methods on generated gene interaction networks, and compare their performance on large networks.

2   Related Work

The pioneering work was performed by [3] in the PCST literature. The *node weighted Steiner tree problem* was proposed in [4], in which the specific set of nodes have to be covered by output tee. The state-of-the-art exact methods were presented in [5, 6], where the PCST was formulated by means of mixed integer linear programming (MILP) and a branch-and-cut algorithms was employed to solve underlining MILP. Some heuristic and matheuristic algorithms were studied in [1, 7, 8].

There some studies in the literature [2, 9–11] that already applied the PCST for functional analyses of protein interaction networks. As a result of these studies, the authors identified unknown functions of some proteins. They validated their computational findings by biological experiments. This shows the potential of the PCST to generate promising results while analyzing interaction networks.

3   Methodology

Usually, biological interaction networks are complex and huge in size. The PCST belongs to the class of NP-hard problems, where it is time consuming to obtain solutions for large graphs. This was the primary limiting factor for available PCST methods being applied on gene interaction networks. To enable the application of the PCST on biological networks, we have developed a heuristic and a matheuristic solution methods in our previous studies. The methods are shortly outlined in the following subsection.

3.1 The MST-Based Heuristic

This heuristic method is based on the iterative solution of Minimum Spanning Tree (MST) problems. Given an undirected network $G = (V, E)$ and a user-defined para-meter α, the heuristic constructs a complete network $G_1 = (V_1, E_1)$ within the first iteration, where $V_1 : v$ only composed of nodes with $p_v > \alpha$ and $E_1 : (i, j)$ corresponds to the shortest path distance between nodes i and j. The algorithm starts solving the problem by considering the nodes with prize $p(v) \geq \alpha$ at the first iteration. Afterwards, the algorithm solves a MST on G_1 and obtains a tree $T_1 = (V_1', E_1')$ with the cost of $C1$. In the second iteration, the heuristic constructs next complete network $G_2 = (V_2, E_2)$, where $V_2 : v$ formed by all nodes of tree from previous iteration $v \in V_1'$, and $E_2 : (i, j)$ corresponds to the shortest path distance between nodes i and j. Again, the algorithm computes a MST on G_2 and obtains a tree $T_2 = (V_2', E_2')$ with the cost of $C2$. If $C2 \geq C1$, the algorithm terminates. Otherwise, the heuristic continues generating complete graph and solving MST problems until the cost of current tree gets bigger or equal to the cost of the previous tree. Then, the algorithm prunes the leaf nodes of the tree in order to further decrease the cost, and obtains final solution. The interested reader may refer to [12] for further details of the heuristic method.

3.2 The Clustering Matheuristic

The matheuristic algorithm was devised by combination of a heuristic clustering algorithm and an exact PCST solver. The main idea of the matheuristic was to divide the large graph into smaller graph clusters, and to solve each cluster separately using exact solver. The heuristic clustering algorithm clusters the nodes according to the all-pairs shortest path distance. Then, smaller graphs are constructed by inducing the nodes in the same cluster. Every smaller graph is solved by using exact PCST solver. Important to note that any exact solver could be used as inner solver at this stage. We have adapted the method proposed by [5] to our approach, and used it as an exact solver due to its efficiency. In [5], the PCST was formulated by MILP, and a branch-and-cut algorithm was proposed to solve the formulation. The interested reader may refer to [13] for further explanation of the matheuristic method.

4   Experimental Results

In this section, we test the MST-based heuristic and the clustering matheuristic method on large gene interaction network instances, and compare their performance. The benchmark instances are generated based on gene expression profiling data of Diffuse Large B-Cell Lymphoma (DLBCL) cancer patients available online in Gene Expression Omnibus repository.[1] There two subtypes of DLBCL cancer that are: the germinal center B cell (GCB) and an activated B cell (ABC). The goal is to identify a set of genes that are relevant for subtype classification. The networks are generated by using the multiplicative model of ARACNE [14] algorithm, which is a powerful tool for the reconstruction of gene interaction networks. ARACNE uses the mutual information among genes for the network reconstruction. We used two parameters ($eps = 0.01$, $eps = 0.05$) fed into ARACNE in order to generate the test instances. In these networks, every arc represents the interaction between two genes and its weight is labeled as the pairwise correlation of expression values of genes. Each node is labeled with prize $p_v = |E_{ABC} - E_{GCB}|$, where E_{ABC} and E_{GCB} are the mean value of gene expression of ABC and GCB cancer patients for corresponding gene, respectively. All of the nodes have positive prize $p_v > 0$ in generated instances.

The computational experiments have been performed on a machine equipped with an Intel(R) Xeon(R) CPU E5320 1.86 GHz processors and 32 GB of shared memory. A single core was used for the experiments.

Table 1 summarizes the results of both methods for gene interaction network instances generated with the parameter $eps = 0.01$. The first three columns of the table show the names and the sizes of test instances, respectively. From the fourth to the ninth columns we report the objective values and running times of the MST-based heuristic method [12], in which the algorithm employs different values for the parameter α. The tenth and eleventh columns present the objective values and execution times of the clustering matheuristic method [13].

Table 2 delivers the results of the MST heuristic and clustering matheuristic methods for interaction network instances generated with the parameter $eps = 0.05$.

According to the results of the tables, both methods, the MST heuristic and clustering matheuristic, were able to provide solutions in a reasonable time. The solutions obtained by the clustering matheuristic are considerably better than the MST heuristic in terms of solution cost for these instances. The MST heuristic also was able to obtain good quality solutions, and the running times of the instances are improved by decreasing the parameter α. The primary reason for elaborated execution times is the decay in parameter α, where the larger set of nodes are considered in computations during the first iteration. The general pattern of the solution cost is decreased by lowering the α from 0.5 to 0.3 and 0.1, however, lowering the α to 0.0 did not improve the cost further. The main reason for this is the MST heuristic was designed for large networks that have smaller number of positive nodes $p_v > 0$. In contrast, the clustering matheuristic method was developed for large networks where most of the nodes have positive prizes $p_v > 0$. The parameter α can be used to tune a trade

---

[1] http://www.ncbi.nlm.nih.gov/geo/.

Table 1 Results of DLBCL test instances generated with the parameter *eps* = *0.01*

Instance	V	E	MST $\alpha = 0.5$		MST $\alpha = 0.3$		MST $\alpha = 0.1$		MST $\alpha = 0.0$		Clust. MATH	
			OBJ	t (s)	OBJ	t (s)	OBJ	t (s)	OBJ	t (s)	OBJ	t (s)
GSE4732	2407	32503	192.3	1	190.3	4	190.5	32	190.5	143	189.6	120
GSE4475	13211	246942	708.9	24	698.1	96	693.6	855	699.2	6048	692.0	6433
GSE22470	13211	266085	861.1	26	850.2	119	846.3	1158	851.8	6419	842.3	7419
GSE10172	13211	231444	346.3	2	345.7	3	339.0	181	339.8	5861	338.6	5859
GSE19246	21049	82282	1631.8	22	1625.1	102	1622.9	1550	1623.9	6540	1616.9	4686
GSE10846	21049	329576	1556.8	76	1544.7	319	1544.7	1468	1541.7	13908	1538.4	14269
GSE23501	21049	939891	1410.0	27	1406.2	273	1404.8	6593	1405.6	26866	1402.6	33446
GSE31312	21049	1055055	1665.8	212	1662.9	1035	1668.6	8736	1673.9	34988	1662.0	37280

Table 2 Results of DLBCL test instances generated with the parameter *eps* = *0.05*

Instance	V	E	MST $\alpha = 0.5$		MST $\alpha = 0.3$		MST $\alpha = 0.1$		MST $\alpha = 0.0$		Clust. MATH	
			OBJ	t (s)	OBJ	t (s)	OBJ	t (s)	OBJ	t (s)	OBJ	t (s)
GSE4732	2407	40921	192	1	189.4	5	189.8	40	189.8	173	188.8	155
GSE4475	13211	347507	708.3	33	697.0	130	692.5	1134	697.0	8026	690.7	9389
GSE22470	13211	375097	652.5	14	648.3	78	646.7	980	652.9	8570	645.1	9178
GSE10172	13211	325423	364.5	3	363.5	5	354.3	278	355.3	7617	352.6	8336
GSE19246	21049	154469	987.4	4	986.6	34	986.8	643	987.3	8683	985.5	7144
GSE10846	21049	475173	1179.0	38	1173.5	217	1170.7	2341	1170.9	18115	1169.9	19060
GSE23501	21049	1050916	1409.9	30	1405.9	304	1404.6	7250	1405.4	31912	1402.2	36645
GSE31312	21049	1423267	1983.2	510	1979.2	2079	1984.1	14662	1990.4	45707	1977.7	52847

off between the quality and running time for the MST heuristic. The α can be set to a reasonably higher value in order to analyze large interaction networks fast, and also not losing too much from the optimality.

5 Conclusions

In this study, we have compared a MST-based heuristic and a clustering matheuristic methods developed for large prize-collecting Steiner tree problems generated from real biological data describing gene interaction networks. Experimental results support that the performance of the clustering matheuristic is better than the MST heuristic method in terms of solution quality for the interaction network instances, however, MST heuristic also can be used to analyze large interaction networks in a quick manner by tuning the α parameter.

Acknowledgements M. Akhmedov is supported by Swiss National Science Foundation through project 205321-147138/1: "Steiner Trees for Functional Analysis in Cancer System Biology".

References

1. Johnson, D.S., Minkoff, M., Phillips, S.: The prize collecting Steiner tree problem: theory and practice. In: Proceedings of 11th ACM–SIAM Symposium on Discrete Algorithms, pp. 760–769 (2000)
2. Bechet, M.B., Borgs, C., Braunsteinc, A., Chayes, J., Dagkessamanskaia, A., François, J.M., Zecchina, R.: Finding undetected protein associations in cell signalling by belief propagation. PNAS **108**, 882–887 (2010)
3. Bienstock, D., Goemans, M.X., Simchi-Levi, D., Williamson, D.: A note on the prize collecting traveling salesman problem. Math. Progr. **59**, 413–420 (1993)
4. Segev, A.: The node-weighted Steiner tree problem. Networks **17**, 1–17 (1987)
5. Ljubic, I., Weiskircher, R., Pferschy, U., Klau, G.W., Mutzel, P., Fischetti, M.: An algorithmic framework for the exact solution of the prize−collecting Steiner tree problem. Math. Progr. **105**(2), 427–449 (2006)
6. Ljubic, I., Weiskircher, R., Pferschy, U., Klau, G., Mutzel, P., Fischetti, M.: Solving the prize−collecting Steiner tree problem to optimality. Proceedings of ALENEX, Seventh Workshop on Algorithm Engineering and Experiments, pp. 68–76 (2005)
7. Canuto, S.A., Resende, M.G.C., Ribeiro, C.C.: Local search with perturbation for the prize-collecting Steiner tree problem in graphs. Networks **38**, 50–58 (2001)
8. Klau, G.W., Ljubic, I., Moser, A., Mutzel, P., Neuner, P., Pferschy, U., Raidl, G., Weiskircher, R.: Combining a memetic algorithm with integer programming to solve the prize-collecting Steiner tree problem. Genet. Evol. Comput. GECCO **2004**(3102), 1304–1315 (2004)
9. Tuncbag, N., McCallum, S., Huang, S.C., Fraenkel, E.: SteinerNet: a web server for integrating omic data to discover hidden components of response pathways. Nucl. Acids Res. 1–5 (2012)
10. Dittrich, M.T., Klau, G.W., Rosenwald, A., Dandekar, T., Mueller, T.: Identifying functional modules in protein-protein interaction networks: an integrated exact approach. Bioinformatics **26**, 223–231 (2008)
11. Beisser, D., Klau, G.W., Dandekar, T., Mueller, T.: T, and M. Dittrich. BioNet: an R-package for the functional analysis of biological networks. Bioinformatics **26**(8), 1129–1130 (2010)

12. Akhmedov, M., Kwee, I., Montemanni, R.: A fast heuristic for the prize-collecting steiner tree problem. Lect. Notes Manag. Sci. **6**, 207–216 (2014)
13. Akhmedov, M., Kwee, I., Montemanni, R.: A divide and conquer matheuristic algorithm for the prize-collecting Steiner tree problem. Computers and Operation Research (to appear)
14. Margolin, A.A., Nemenman, I., Basso, K., Wiggins, C., Stolovitzky, G., Favera, R.D., Califano, A.: Aracne: an algorithm for the reconstruction of gene regulatory networks in a mammalian cellular context. BMC Bioinform. **7**(Suppl 1), S7 (2006)

A Benders Decomposition Approach for Static Data Segment Location to Servers Connected by a Tree Backbone

Goutam Sen, Mohan Krishnamoorthy, Vishnu Narayanan and Narayan Rangaraj

Abstract We consider the problem of allocating database segments to the locations of a content distribution network (CDN). Many other decisions such as server location, query routing, user assignment and network topology are also addressed simultaneously. We consider the problem with the backbone (server network) being restricted to be a tree. Although it is an extremely hard problem to solve in its original form, prior information on the segment allocation, server location, and the tree backbone, reduces the original problem to a simple assignment problem, and therefore, indicates the suitability of a decomposition approach. We formulate the problem and develop a Benders decomposition approach. Due to the hardness of the master problem, we solve it heuristically to obtain a reasonable upper bound on the original problem in a short period of time. The success of the algorithm is particularly significant in large problems, for which CPLEX struggles to obtain even a feasible integer solution.

1 Introduction

The problem of data location in an information network arises in a distributed architecture, in which multiple data centers store the content and serve the users. The purpose of this architecture is to mitigate load and congestion in the network resulting in faster access to content. However, a service provider would like to minimize the costs associated with the service too. Placement of data could affect the cost of access and transfer to a significant scale. Thus, it is important to consider this problem with the objective of minimizing the total routing costs.

G. Sen (✉)
IITB Monash Research Academy, IIT Bombay, Mumbai 400076, India
e-mail: goutam.sen@iitb.ac.in

M. Krishnamoorthy
Department of Mechanical and Aerospace Engineering, Monash University, Clayton, VIC 3800, Australia

V. Narayanan · N. Rangaraj
Industrial Engineering and Operations Research, IIT Bombay, Mumbai 400076, India

© Springer International Publishing Switzerland 2017   109
K.F. Dœrner et al. (eds.), *Operations Research Proceedings 2015*,
Operations Research Proceedings, DOI 10.1007/978-3-319-42902-1_15

There are three possible units of allocation: files, database segments and the replicas of the entire database (or, mirrors). The optimal solution for the file allocation approach is hard to find due to the presence of a very large number of files. On the other hand, considering mirror allocation leads to significant data consistency cost and data storage cost [1]. So, an intermediate approach could be to partition the database to several segments and locate these segments in appropriate places. This problem is known as Segment Allocation Problem (SAP). If the segments are non-overlapping, the updates are handled locally.

The subproblems associated with the SAP are: (a) segment allocation (b) server location (c) the user assignment to servers (d) query routing, and (e) network topology. All these problems (referred to here as the SAP-tree) are formulated and solved in [6] by using concepts from hub location. The hubs (or data servers) are assumed to be connected by a tree. A file aggregation heuristic [5] is used to pre-compute the segments prior to considering the SAP-tree (a)–(e). A mixed-integer linear programming (MILP) formulation, which is adapted from the well-known "tree of hubs" problem [2], is presented. They also develop a simulated annealing approach for quick heuristic solutions.

In this paper, we analyze a formulation of the SAP-tree in a Benders decomposition framework (BD-tree). As a consequence, we are able to generate upper bounds on the optimal solution within an acceptable computational time. For the variant of SAP with fully connected mesh backbone, upper bounds of the Benders decomposition are reported in [5] along with an empirical performance guarantee (percentage gap from the lower bound). In our current implementation for the variant SAP-tree, such guarantee proves very costly; therefore, we assess the quality of the upper bounds against the CPLEX solutions, if available. Our approach provides mathematical insights into the problem structure and complexity, and the algorithm produces a good feasible solution in cases where CPLEX does not even find the root relaxation within a reasonable time.

2 Formulation of the Problem SAP-Tree

In this section, we formulate the problem SAP-tree by using a 4-subscripted variable based on the formulation given in [3]. For a graph $G = (I, E)$; $i = 1, 2, \ldots, n$; $j = 1, 2, \ldots, n$; $k = 1, 2, \ldots, p$; define: I = set of user nodes in the network G, $i, j \in I$; P = set of database partitions or segments, $k \in P$; d_{ij} = distance between the user nodes i and j; α = discount factor per unit inter-hub transfer cost; q_{ik} = volume of query from user i to segment k. Define the variables: $y_{jk} = 1$ if segment k is located at data hub j, 0 otherwise; $x_{ij} = 1$ if user node i is assigned to the data hub j, 0 otherwise; $h_{jl} = 1$ if (j, l) is a hub arc, 0 otherwise; z_{ijlk} = fraction of query from user i to segment k, that is routed through data hubs located at j and l using tree arc (j, l), in that order. Then, the problem can be formulated as:

$$\min \sum_{i \in I} \sum_{j \in I} \sum_{k \in P} q_{ik} d_{ij} x_{ij} + \sum_{i \in I} \sum_{j \in I} \sum_{l \in I} \sum_{k \in P} \alpha q_{ik} d_{jl} z_{ijlk} \tag{1}$$

subject to

$$\sum_{j \in I} x_{ij} = 1, \quad \forall i \in I, \tag{2}$$

$$\sum_{j \in I} y_{jk} = 1, \quad \forall k \in P, \tag{3}$$

$$\sum_{k \in P} y_{jk} \leq 1, \quad \forall j \in I, \tag{4}$$

$$x_{ij} \leq \sum_{k \in P} y_{jk}, \quad \forall i, j \in I, \tag{5}$$

$$h_{jl} \leq \sum_{k \in P} y_{jk}, \quad \forall j, l \in I; j < l, \tag{6}$$

$$h_{jl} \leq \sum_{k \in P} y_{lk}, \quad \forall j, l \in I; j < l, \tag{7}$$

$$\sum_{j \in I} z_{ijlk} + x_{il} = \sum_{j \in I} z_{iljk} + y_{lk}, \quad i, l \in I, k \in P, \tag{8}$$

$$z_{ijlk} + z_{iljk} \leq h_{jl} \quad \forall i, j, l \in I, j < l, k \in P, \tag{9}$$

$$\sum_{l \in I} z_{ijlk} \leq \sum_{k \in P} y_{jk} \quad \forall i, j \in I, k \in P, \tag{10}$$

$$\sum_{l \in I} z_{iljk} \leq \sum_{k \in P} y_{jk} \quad \forall i, j \in I, k \in P, \tag{11}$$

$$\sum_{j \in I} \sum_{l \in I} h_{jl} = p - 1, \tag{12}$$

$$x_{ij}, y_{jk}, h_{jl} \in \{0, 1\}, \quad \forall i, j, l \in I, k \in P, \tag{13}$$

$$z_{ijlk} \geq 0, \quad \forall i, j, l \in I, k \in P. \tag{14}$$

Constraint (2) represents the single allocation of users to hubs. Constraint (3) ensures that all segments are allocated, and constraint (4) prevents a hub to host more than one segment. Constraint (5) is a link constraint between x_{ij} and y_{jk}. Constraint (8) is the new flow balancing constraint. We have a bunch of new link constraints in this formulation: (6), (7), (9)–(11). Constraints (6) and (7) ensure that the *small tree* arcs connect established data hubs only. Constraint (9) guarantees that the query is routed through the tree edges only. Constraints (10) and (11) ensure that the query is routed through the link (j–l) if these hubs are established.

3   Benders Decomposition Approach

We identify the variables y_{jk} and h_{jl} as complicating variables and obtain the following subproblem and the *initial* master problem to use in the Benders decomposition. The reason for choosing such a decomposition is an expectation that once the tree structure (along with the hub locations) is known, the distances between any two hub nodes can be easily calculated; the remaining subproblem reduces to a simple user assignment problem. In the following mathematical analysis, we prove that this assumption is correct.

The *Primal subproblem (PSP)* consists of the objective (1), subject to constraints (2), (5)–(11), and variables $x_{ij} \in \{0, 1\}, \forall i, j \in I;\ z_{ijlk} \geq 0, \forall i, j, l \in I,$ $k \in P$. The variables y_{jk} and h_{jl} are fixed as $\hat{y}_{jk}$ and $\hat{h}_{jl}$ respectively.

We introduce a new parameter t_{jl} to represent the distance between the data hubs j and l in the small tree (backbone). The computation of t_{jl} for each pair (j, l) is done a-priori. We derive the hub set $H \subset I$ from the values of $\hat{y}_{jk}$. Since there is no user-to-user interaction, we have $x_{ij} = 0, \forall i \in I, j \in I \backslash H$. Therefore, it is sufficient to consider $x_{ij}, \forall i \in I, j \in H$ in the formulation. Again, given the hub set H, the reduced primal subproblem is decomposable for each $i \in I$ and is denoted by *PSP$_i$-R*. The integer program of the *PSP$_i$-R* can be written as follows:

$$\min \sum_{j \in H} \sum_{k \in P} q_{ik}(d_{ij} + \alpha t_{jl}) x_{ij} * \hat{y}_{lk} \tag{15}$$

$$\text{subject to} \qquad \sum_{j \in H} x_{ij} = 1, \tag{16}$$

$$x_{ij} \in \{0, 1\}, \qquad \forall i \in I \backslash H, j \in H, \tag{17}$$

Now, we consider the linearly relaxed *PSP$_i$-R* with $0 \leq x_{ij} \leq 1$. Clearly, the resulting LP is in standard form with only one constraint (16). Hence, the optimal solution to the linearly relaxed *PSP$_i$-R* is always integer.

Since the dual for this type of problems (*PSP*) is generally known to be highly degenerate, we strengthen the values of the dual variables without affecting the value of the objective. In the master problem (MP), we introduce a new set of connectivity constraints (25)–(29) to ensure a spanning tree. A similar strategy is used in [3] by using a dummy source node. We do not use any additional node. We directly adapt the constraints from the *3-subscripted formulation* in [6].

Dual of PSP$_i$ (DPSP$_i$):

$$\max \quad r - \sum_{j \in I}(\sum_{k \in P} \hat{y}_{jk}) s_j + \sum_{l \in I} \sum_{k \in P} \hat{y}_{lk} t_{lk} - \sum_{j \in I} \sum_{l \in I} \sum_{k \in P} \hat{h}_{jl} u_{jlk}$$

$$- \sum_{j \in I} \sum_{k \in P}(\sum_{k \in P} \hat{y}_{jk}) v_{jk} - \sum_{j \in I} \sum_{k \in P}(\sum_{k \in P} \hat{y}_{jk}) w_{jk} \tag{18}$$

$$\text{subject to} \qquad r - s_j + \sum_{k \in P} t_{jk} \le \sum_{k \in P} q_{ik} d_{ij}, \qquad \forall j \in I, \tag{19}$$

$$t_{jk} - t_{lk} - u_{jlk} - v_{jk} - w_{lk} \le \alpha q_{ik} d_{jl}, \qquad \forall j, l \in I, j < l, k \in P \tag{20}$$

$$t_{jk} - t_{lk} - u_{ljk} - v_{jk} - w_{lk} \le \alpha q_{ik} d_{jl}, \qquad \forall l, j \in I, j > l, k \in P \tag{21}$$

$$s_l \ge 0, u_{jlk} \ge 0, v_{jk} \ge 0, w_{jk} \ge 0, r, t_{jk} \in R, \qquad \forall j \in I, l \in I, k \in P. \tag{22}$$

$$\textit{Master Problem (MP):} \qquad \min \sum_{i \in I} \eta_i \tag{23}$$

subject to

$$\eta_i \ge r - \sum_{j \in I}(\sum_{k \in P} y_{jk})s_j + \sum_{l \in I}\sum_{k \in P} y_{lk}t_{lk} - \sum_{j \in I}\sum_{l \in I}\sum_{k \in P} h_{jl}u_{jlk} - \sum_{j \in I}\sum_{k \in P}(\sum_{k \in P} y_{jk})v_{jk}$$

$$- \sum_{j \in I}\sum_{k \in P}(\sum_{k \in P} y_{jk})w_{jk}, \qquad \forall i \in I, (r, s, t, u, v, w) \in \theta \tag{24}$$

subject to (3), (4), (6), (7), (12), and

$$\sum_{j \in I} x_j = 1, \tag{25}$$

$$x_j \le \sum_{k \in P} y_{jk}, \forall j \in I, \tag{26}$$

$$z_{jl} + z_{lj} \le \sum_{k \in P} q_{ik} h_{jl}, j, l \in I; l > j, \tag{27}$$

$$\sum_{i \in I} \eta_i \le UB(1 - \varepsilon), \tag{28}$$

$$\sum_{l \in I} z_{jl} - \sum_{l \in I} z_{lj} = \sum_{k \in P} q_{ik} x_j - \sum_{k \in P} q_{ik} y_{jk}, \forall j \in I, \tag{29}$$

$$x_j \in \{0, 1\}, z_{jl} \ge 0, \eta_i \ge 0, y_{jk} \in \{0, 1\}, h_{jl} \in \{0, 1\}, \forall i, j, l \in I, \text{ and } k \in P. \tag{30}$$

We denote θ as the set of extreme points of the $DPSP_i$, $\forall i \in I$. The MP formulation guarantees a spanning tree only for integer h_{jl}. Since h_{jl} (of the order of n^2) dominates the number of integer variables, the MP proves to be a major bottleneck to the algorithm's performance. Therefore, we develop a heuristic approach to solve the MP. Consequently, MP does not provide a lower bound on the original problem anymore.

The constraint (28) is due to ε-optimal BD algorithm in [4]. MP is solved in CPLEX till the first feasible solution is found. If the program is infeasible, the upper bound (UB) obtained by solving the previous $DPSPs$ is within ε % of the optimal solution, and the algorithm terminates with a proof of the quality of the *UB*. However, this scheme has a drawback. Since the first feasible solution can be weak, it is expected to take a large number of iterations to terminate.

Algorithm 1 Benders Decomposition Algorithm (BD-tree)

1: *Iteration* = 0;
2: Solve the *MP* to feasibility in CPLEX. Let this solution be (y^f_{jk}, h^f_{jl}), for all $j, l \in I, k \in P$. If *MP* is infeasible, terminate and return the UB.
3: Solve the $DPSP_i$ for all $i \in I$ with fixed (y^f_{jk}, h^f_{jl}). If $obj(DPSP) < UB$, $UB = obj(DPSP)$.
4: Generate Benders optimality cuts (24) and add them to *MP*.
5: Apply *RND* to (y^f_{jk}, h^f_{jl}) to obtain an improved solution (y'_{jk}, h'_{jl}), if available.
6: Solve the $DPSP_i$ for all $i \in I$ with fixed (y'_{jk}, h'_{jl}). If $obj(DPSP) < UB$, $UB = obj(DPSP)$.
7: Generate Benders optimality cuts (24) and add them to *MP*.
8: *Iteration* = *iteration* + 1;
9: **if** Iteration=m **then**
10:     return *UB* and terminate;
11: **else**
12:     go to Step 2.

The first feasible solution, as obtained by running MP in CPLEX, is used as an initial solution for a local search heuristic to find a local optimum that is expected to produce a stronger optimality cut. The local search is descent in nature in order to respect constraint (28). We apply a simple random neighborhood descent (RND) algorithm that performs four types of moves by perturbing the location-allocation schema in the present solution. The new solution is accepted in the event of cost improvement. The tree is improved after each move by examining each arc one by one. The moves and the tree improvement heuristic in the RND are exactly the same to those used in the simulated annealing algorithm in [6].

4   Computational Experiments

The *3-subscripted formulation* in [6] for SAP-tree is used for direct CPLEX results. We use CPLEX 12.5 with its default settings for this purpose. The MovieLens dataset [5] contains information about user locations (zipcodes) and movie ratings (used as proxy for queries). The distances between different zipcodes are calculated by the great circle method [5]. The distance between the users in the same zipcode is assumed to be zero, which means that queries from the same zipcode are treated as though they emanate from one user. We consider instances ranging from 10 to 100 nodes(n), and 3 to 5 segments(p). We use $\alpha = \{0.2, 0.4, 0.6, 0.8\}$. The problem instance can be represented by a tuple (n, p, α). The parameter m (number of iterations) in the BD-tree algorithm is fixed to 10 for instances up to (50, 3, 0.8) and set to 5 for larger instances. We run all the experiments on an IBM M3 X3400, Intel Quad core Xeon E5506, 64GB RAM server. The results are listed in the Table 1.

Table 1 Performance of the BD-tree algorithm

n	p	α	Optimal/best-feasible	t_{opt} (s)	$z_{BD\text{-}tree}$ gap (%)	$t_{BD\text{-}tree}$
10	3	0.2	86.04	0.03	0.00	0.65
		0.4	121.40	0.04	0.00	0.46
		0.6	156.76	0.06	0.08	0.45
		0.8	189.40	0.08	0.00	0.42
20	3	0.2	174.60	1.55	0.00	4.91
		0.4	240.03	2.00	3.37	2.82
		0.6	297.51	2.74	4.71	2.78
		0.8	341.77	26.22	0.00	2.82
30	3	0.2	262.18	11.86	0.00	17.69
		0.4	353.76	128.08	9.63	8.16
		0.6	433.21	1386.07	2.76	7.94
		0.8	475.76 (1.11 %)	5000.00	0.00	7.78
30	4	0.2	446.30	1335.41	0.12	75.66
		0.4	637.83	2578.33	3.97	52.36
		0.6	797.50 (2.21 %)	5000.00	5.98	58.22
		0.8	923.35 (2.48 %)	5000.00	6.08	57.26
50	3	0.2	381.17	1423.40	0.00	325.78
		0.4	490.05	4827.79	4.30	178.49
		0.6	559.25 (1.00 %)	5000.00	2.09	188.66
		0.8	598.95 (0.68 %)	5000.00	2.66	189.85
50	5	0.2	1300.88 (3.18 %)	5000.00	9.79	122.17
		0.4	1835.87 (4.14 %)	5000.00	4.59	127.01
		0.6	2321.88 (5.51 %)	5000.00	2.50	137.71
		0.8	2660.86 (5.02 %)	5000.00	**−0.14**	134.09
100	5	0.2	N/A	5000.00	N/A	1295.98
		0.4	N/A	5000.00	N/A	1313.96
		0.6	N/A	5000.00	N/A	1102.78
		0.8	N/A	5000.00	N/A	1134.88

The performance of BD-tree is assessed by the quality of the upper bound (in terms of the percentage *gap* from the optimal solution) and the elapsed CPU time ($t_{BD\text{-}tree}$). We report CPLEX best-feasible solutions obtained after 5000 s along with the ending MIP gap for the cases where the optimality could not be proved. The *gap* is available for instances up to 50 nodes. It is generally within 4–5 % of the optimal solution (or the best-feasible solution) with a few exceptions. We find two instances in which the BD-tree produces weak upper bounds. These instances are (30, 3, 0.4) and (50, 5, 0.2). The gap is reported to be high (about 9 %) in these two cases indicating the need for a more sophisticated local search. For the instance (50, 5, 0.8), BD-tree solution is better (gap of −0.14 %) than the CPLEX best-feasible solution. The primary advantage of the BD-tree is the time saved in reaching a reasonable solution. The BD-tree takes a minuscule portion of the CPLEX solution time to produce an upper bound. For larger instances like (100, 5, 0.6), CPLEX struggles to find even a feasible solution, whereas, BD-tree terminates with a solution in 1102.78 s. This observation is significant keeping in mind that realistic cases deal with hundreds of nodes and direct computation in CPLEX is infeasible. BD-tree may serve as an effective tool that utilizes the easy subproblems to reach an upper bound fairly quickly.

5   Conclusions

The use of the MILP formulation for the SAP-tree is limited to small scale problems in practice. So, we propose a Benders decomposition approach, in which the dual subproblem is solved directly by CPLEX, and the master problem is solved heuristically. We apply the ε-optimal approach [4] to solve the MP to feasibility. This feasible solution is further improved by a local search to facilitate the production of strong cuts. Although we report good upper bounds, our algorithm is still unable to provide an empirical performance guarantee (ε %). This study is important in identifying the complexities of the problem in detail. We plan to extend this work by designing a more efficient cut generation strategy to strengthen the master problem. Although simulated annealing solutions [6] could be used to generate Benders cuts, we feel that a population based heuristic might help to generate multiple cuts based on local solutions thereby accelerating convergence.

References

1. Applegate, D., Archer, A., Gopalakrishnan, V., Lee, S., Ramakrishnan, K.K.: Optimal content placement for a large-scale vod system. In: Proceedings of the 6th International Conference, Co-NEXT '10, pp. 4:1–4:12. ACM, New York, NY, USA (2010)
2. Contreras, I., Fernndez, E., Marin, A.: The tree of hubs location problem. Eur. J. Oper. Res. **202**(2), 390–400 (2010)
3. de Sá, E.M., de Camargo, R.S., de Miranda, G.: An improved benders decomposition algorithm for the tree of hubs location problem. Eur. J. Oper. Res. **226**(2), 185–202 (2013)
4. McDaniel, D., Devine, M.: A modified benders' partitioning algorithm for mixed integer programming. Manag. Sci. **24**(3), 312–319 (1977)
5. Sen, G., Krishnamoorthy, M., Rangaraj, N., Narayanan, V.: Exact approaches for static data segment allocation problem in an information network. Comput. Oper. Res. **62**, 282–295 (2015)
6. Sen, G., Krishnamoorthy, M., Rangaraj, N., Narayanan, V.: Mathematical models and empirical analysis of a simulated annealing approach for two variants of the static data segment allocation problem. Networks (2016, in press). doi:10.1002/net.21675

Mathematical Optimization of a Magnetic Ruler Layout with Rotated Pole Boundaries

Marzena Fügenschuh, Armin Fügenschuh, Marina Ludszuweit,
Aleksandar Mojsic and Joanna Sokół

Abstract Magnetic rulers for measuring systems are either based on incremental or absolute measuring methods. Incremental methods need to initialize a measurement cycle at a reference point. From there, the position is determined by counting increments of a periodic graduation. Absolute methods do not need reference points, since the position can be read directly from the ruler. In the state of the art approach the absolute position on the ruler is encoded using two tracks with different graduation. To use only one track for position encoding in absolute measuring a pattern of trapezoidal magnetic areas is considered instead of the common rectangular ones. We present a mixed integer programming model for an optimal placement of the trapezoidal magnetic areas to obtain the longest possible ruler under constraints conditioned by production techniques, physical limits as well as mathematical approximation of the magnetic field.

1 Introduction

As a matter of fact magnetic rulers with incremental position measurement, see Fig. 1, are easy to produce. They consist of equidistant pole rectangles and the position is determined via counting the switches between the north and south pole. The demand of the market focuses however on absolute position measurement systems. In the state of the art approach the production of such magnetic rulers is cost-intensive mainly due to the need of the two magnetic tracks, see Fig. 2. The one track consists of equidistant pole rectangles. The pole rectangles in the second track are multiples of one rectangle from the first track. In this way the magnetic ruler can be divided into rectangular regions, each one covering a unique pattern and thus encoding a

M. Fügenschuh (✉)
Beuth University of Applied Sciences Berlin, Luxemburger Strae 10, 13353 Berlin, Germany
e-mail: fuegenschuh@beuth-hochschule.de

A. Fügenschuh · M. Ludszuweit · A. Mojsic · J. Sokół
Department of Mechanical Engineering, Helmut Schmidt University of the Federal
Armed Forces Hamburg, Holstenhofweg 85, 22043 Hamburg, Germany

© Springer International Publishing Switzerland 2017
K.F. Dœrner et al. (eds.), *Operations Research Proceedings 2015*,
Operations Research Proceedings, DOI 10.1007/978-3-319-42902-1_16

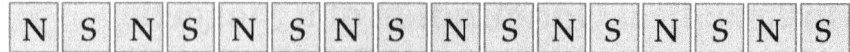

Fig. 1 A magnetic ruler for the incremental position measurement

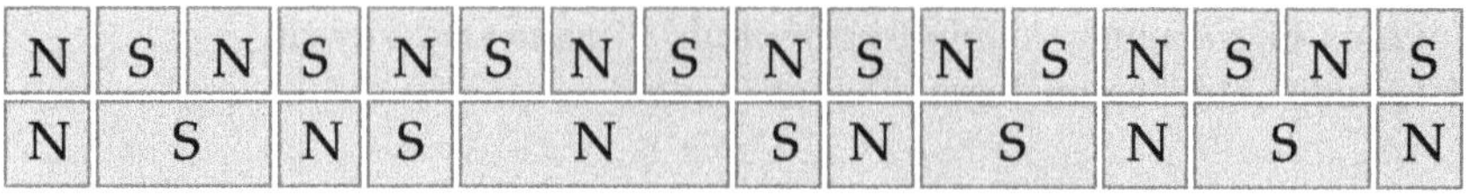

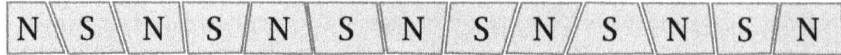

Fig. 2 A magnetic ruler with two tracks for the absolute position measurement

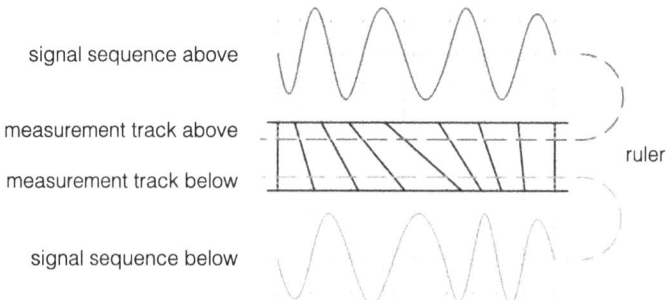

Fig. 3 A magnetic ruler with rotated pole boundaries

unique position [1–4]. The size of such a rectangular region depends on the size of the magnetic reading head.

An approach to reduce the production costs is to elaborate such unique patterns using only one magnetic stripe for the magnetic ruler. This can be achieved, e.g. by the rotation of pole boundaries, see Fig. 3 and [5]. In this way we obtain a magnetic ruler covered by trapezoid shaped poles. The position is encoded via the signal sequence of the magnetic field, see Fig. 4. The different lengths of the upper and lower sides of the trapezoids consecutively placed give rise to the irregularity of the signal curves providing the uniqueness of the encoded values. However, the longer the magnetic ruler the more probable a repetition of a signal value is.

The fundamental task is to find an appropriate placement of the trapezoids supporting a unique signal sequence on an as long as possible magnetic ruler. The placement underlies further conditions due to production techniques, physical limits as well as mathematical approximation of the magnetic field.

signal sequence above

measurement track above

measurement track below

ruler

signal sequence below

Fig. 4 The signal sequence on a magnetic ruler with rotated pole boundaries

In the approach presented in this paper we aim to construct a magnetic ruler with rotated pole boundaries of the maximal length provided only a few production restriction explained in the next section.

2   Mathematical Optimization Model

To obtain a mathematical model we consider the trapezoids as segments, which are uniquely identified via the lengths of its parallel sides. Having the technical specifications as the minimal, l_{min}, as well as the maximal, l_{max}, length of a parallel side as well as the minimum difference between two adjoining sides s, a list of all possible trapezoid side lengths can be determined a priori:

$$L = \{L_1, \ldots, L_n\}, \quad L_1 = l_{min}, \quad \forall_{1 \leq i \leq n-1} \; L_i = L_1 + is, \quad L_n = l_{max},$$

where

$$n = \left\lfloor \frac{l_{max} - l_{min}}{s} \right\rfloor + 1.$$

We index the list with the set $I = \{1, \ldots, n\}$. Each item on the list means a length of a parallel trapezoid's side and can be used for the upper as well as the lower side. We obtain the maximum number of $|I \times I| = n^2$ segments, which can be arranged on one magnetic ruler. One obtains the index set $P = \{1, \ldots, n^2\}$ of positions on magnetic ruler at which a segment can be set. Next, we introduce binary variables $x_{i,j}^p$, for all $i, j \in I$ and $p \in P$. If $x_{i,j}^p$ equals 1 a trapezoid with the upper parallel side L_i and the lower parallel side L_j is placed on position p on the magnetic ruler. Furthermore, we introduce the binary variables $y_{i,j}$ for all $i, j \in I$. A $y_{i,j}$ is set to 1, if the corresponding trapezoid is placed on the magnetic ruler at all.

We seek for a magnetic ruler of a maximal length,

$$\max \sum_{i,j \in I} (L_i + L_j) y_{i,j},$$

provided that each trapezoid is assigned to at most one position,

$$\sum_{i,j \in I} x_{i,j}^p \leq 1, \quad \forall p \in P,$$

can be used at most once,

$$\sum_{p \in P} x_{i,j}^p = y_{i,j}, \quad \forall i, j \in I,$$

and the segments should be placed continuously—without gaps—on the magnetic ruler,

$$\sum_{i,j \in I} x_{i,j}^p \leq \sum_{i,j \in I} x_{i,j}^{p-1}, \qquad \forall\, p \in P \backslash \{1\}.$$

Finally, we formulate the following technical restriction. To magnetize the ruler with rotated pole boundaries, the writing head has to be rotated, but not *too strongly*. The parameter ε determines the maximal possible rotation of the pole boundary. It corresponds to the maximal difference of the cumulated lengths obtained by summing up the upper and the lower parallel sides, respectively, of trapezoids placed upto a given position. We introduce the real variables $\ell_s^p \in \mathbb{R}_+$. Using s we index the sides, $s = 1$ for upper and $s = 2$ for the lower one, and $p \in P$ indicates the position up to which the summation is performed. Thus we obtain the cumulated length on the upper side,

$$\ell_1^1 = \sum_{i,j \in I} L_i x_{i,j}^1,$$

$$\ell_1^p = \ell_1^{p-1} + \sum_{i,j \in I} L_i x_{i,j}^p, \qquad \forall\, p \in P \backslash \{1\},$$

and the cumulated length on the lower side,

$$\ell_2^1 = \sum_{i,j \in I} L_j x_{i,j}^p,$$

$$\ell_2^p = \ell_2^{p-1} + \sum_{i,j \in I} L_j x_{i,j}^p, \qquad \forall\, p \in P \backslash \{1\}.$$

Having that, we can formulate the condition on the difference of both upper and lower total lengths,

$$|\ell_1^p - \ell_2^p| \leq \varepsilon, \qquad \forall\, p \in P.$$

2.1 Preprocessing

The parameter ε gives rise to a sort of a preprocessing condition. If the difference of the upper and lower side of a trapezoid exceeds the value 2ε, it can be removed from considerations. Thus for all $i, j \in I$ with $|L_i - L_j| > 2\varepsilon$ we can immediately set $y_{i,j} = 0$.

3 Computational Results

We apply IBM ILOG CPLEX [6] to solve the model presented above. The computations were performed on a MacOS 10.10 with 1.7 GHz Intel Core i7 processor and 8 GB 1600 MHz DDR3 memory. We summarize the computational results in Table 1. The parameter l_{min}, the minimum length of a trapezoid's side, is generally set to 2 mm, as too small trapezoids would not be recognized by the reading magnetic head. The parameter l_{max}, the maximum length of a trapezoid's side, should not be too big, otherwise breaks in the signal sequence may occur. Small values of the step size s, the minimum difference between two adjoining sides, lead to a bigger number of different trapezoids, but then the uniqueness of the encoded signal is vulnerable as the trapezoids are almost equal. The value ε as already mentioned is specified by the production terms.

Comparing the computation times with and without the preprocessing the advantage of the preprocessing constraint is obvious. In a very short time we obtain layouts of magnetic rulers with lengths of industrial use. As an example we plot in Fig. 5 the magnetic ruler computed due to parameters given in the first line of Table 1. Additionally, by reflecting the ruler on the upright left side and switching the sensitivity of the start pole one obtains the contrary signal sequence and thus a unique signal sequence on a ruler of a double length.

Table 1 The optimal length of magnetic rulers computed due to given production parameters

#	l_{min} (mm)	l_{max} (mm)	s (mm)	ε (mm)	Length of an optimal magnetic ruler (mm)	CPU time without preprocessing (s)	Gap (%)	CPU time with pre-processing (s)
1.	2.0	4.0	0.5	0.5	57.0	1.04		0.17
2.	2.0	4.0	0.5	1.0	75.0	0.40		0.36
3.	2.0	6.5	0.5	0.5	187.0	267.15		0.88
4.	2.0	6.5	0.5	1.0	297.5	1965.47		2.85
5.	2.0	8.0	0.5	2.0	745.0	3600.00[a]	18	11.09
6.	2.0	8.0	0.5	3.0	845.0	32.12		31.35
7.	2.0	8.0	0.4	2.0	1130.0	3600.00[a]	30	56.66
8.	2.0	12.0	0.5	2.0	1809.0	3600.00[a]	575	1234.79
9.	2.0	12.0	0.4	2.0	2819.0	3600.00[a]	∞	3041.26

[a]The computation was interrupted after this time, the optimality gap is given in the next column

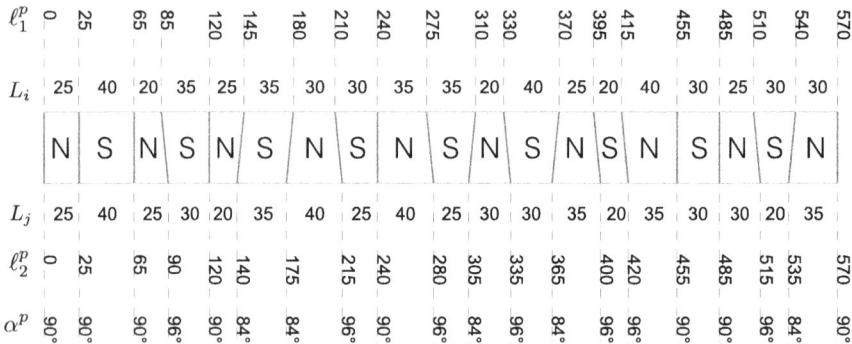

Fig. 5 An optimal magnetic ruler for the parallel side lengths $L := \{20, 25, 30, 35, 40\}$. The unit of the numbers in the figure is $1 \cong 0.1$ mm. The numbers in the *first line* from above are the cumulated *upper sides* of trapezoids up to a given position. The numbers directly above and directly below a trapezoid are the lengths of its *upper* and *lower side*, respectively. In the *next line* the cumulated *lower sides* of trapezoids up to a given position are listed. The *last line* below indicates the measure of the inner *left bottom* angle of each trapezoid

4 Conclusions

After the production of the magnetic ruler layouts we computed by our industrial collaboration partner it turned out that it is preferred to arrange the trapezoids in a gradually growing way with respect to their lengths. Furthermore, it is desired to have the largest parameter for the stepsize when the length of the ruler is fixed. This cannot be formulated directly as a mixed-integer program, because this requires that the stepsize is a real value. Hence we use our approach with a fixed stepsize, and decrease the stepsize in each consecutive run of the model, until the desired length of the ruler is met.

In the approach presented in this paper a trapezoid with the same upper and lower length can be placed at most once and cannot repeat. This assures the uniqueness of the signal sequence in the upper and lower part of the trapezoids. Since the width of the reading head may exceed the width of a single trapezoid, one may consider to cluster two or three adjacent trapezoids, which then cannot repeat on the magnetic ruler. The reformulation of the model in this respect is our ongoing research. We expect to obtain longer rulers with a higher accuracy.

The mathematical optimization model formulated in this paper is a basis to expand, when further production or physical limitations as well as the mathematical approximation by decoding the signal come into play.

Acknowledgements This work is part of a joint project funded by the Central Innovation Program SME supported by the German Federal Ministry for Economic Affairs and Energy. The results were achieved in cooperation with BOGEN Electronic GmbH and the Technical University of Berlin, chair of Electronics and Medical Signal Processing.

References

1. Engelhardt, K., Seitz, P.: High-resolution optical position encoder with large mounting tolerances. Appl. Opt. **36**, 2912–2916 (1997)
2. Perez-Quintin, F., Lutenberg, A., Rebollo, M.A.: Linear displacement measurement with a grating and speckle pattern illumination. Appl. Opt. **45**, 4821–4825 (2006)
3. Nyce, D.S.: Linear Position Sensors: Theory and Application. Wiley, New Jersey (2003)
4. Hans, W.: Position Sensing: Angle and Distance Measurement for Engineers. Butterworth Heinemann (1994)
5. Hoyer, J., Becker, T.: Messvorrichtung und Verfahren zum Messen von Körpern, European Patent no. EP2846126 (2013)
6. IBM ILOG CPLEX Optimization Studio V12.6.0 Documentation. ftp://public.dhe.ibm.com/software/websphere/ilog/docs/optimization/cplex/ps_usrmancplex.pdf

A New Exact Approach to the Space-Free Double Row Layout Problem

Anja Fischer, Frank Fischer and Philipp Hungerländer

Abstract Given a set of departments, two rows with a common left origin and pairwise connectivities between the departments, the Space-Free Double-Row Facility Layout Problem (SF-DRFLP) looks for permutations of the departments in both rows such that the weighted sum of the center-to-center distances between all pairs of departments is minimized. In this paper we present a new mixed-integer linear programming formulation for the (SF-DRFLP) with given assignment of the departments to both rows that combines distance and betweenness variables. Furthermore, we analyze the combinatorial structure of optimal solutions, which allows us to prove a certain balancing condition. We then use this formulation in an enumeration scheme for solving the (SF-DRFLP). Indeed, we test all possible row assignments, where some assignments are excluded by our new combinatorial investigations. This approach allows us to solve (SF-DRFLP) instances with up to 16 departments to optimality for the first time.

1 Introduction

An instance of the *Space-Free Double-Row Facility Layout Problem* (SF-DRFLP) consists of n departments $\{1, \ldots, n\} =: [n]$ with given positive lengths $\ell_i > 0, i \in [n]$, and pairwise non-negative connectivities $w_{ij} \geq 0, i, j \in [n], i < j$. The objective is to find an assignment $r: [n] \to \{1, 2\}$ of the departments to two rows with a

A. Fischer (✉)
Institute for Numerical and Applied Mathematics, University of Goettingen,
Goettingen, Germany
e-mail: anja.fischer@mathematik.uni-goettingen.de

F. Fischer
Department of Mathematics, University of Kassel, Kassel, Germany
e-mail: frank.fischer@uni-kassel.de

P. Hungerländer
Department of Mathematics, University of Klagenfurt, Klagenfurt, Austria
e-mail: philipp.hungerlaender@aau.at

© Springer International Publishing Switzerland 2017
K.F. Dœrner et al. (eds.), *Operations Research Proceedings 2015*,
Operations Research Proceedings, DOI 10.1007/978-3-319-42902-1_17

common left origin and permutations of the departments in both rows minimizing the total weighted sum of the center-to-center distances between all pairs of departments:

$$\min_{\substack{r \in \{1,2\}^n, \\ \pi_1 \in \Pi_1(r), \\ \pi_2 \in \Pi_2(r)}} \sum_{\substack{i,j \in [n] \\ i<j}} w_{ij} \cdot d_{ij}(\pi_1, \pi_2),$$

where $r \in \{1, 2\}^n$ is the row assignment, $\Pi_h(r)$, $h \in \{1, 2\}$, denotes the set of all permutation of the departments assigned to row h and $d_{ij}(\pi_1, \pi_2)$ is the center-to-center distance between departments i and j with respect to $\pi_1 \in \Pi_1(r)$ and $\pi_2 \in \Pi_2(r)$.

A mixed-integer linear programming (MILP) formulation for the (SF-DRFLP) is presented in [3]. With this model instances with up to 13 departments can be solved to optimality. Additionally a semidefinite optimization approach is proposed in [8], which provides high-quality global bounds for instances with up to 15 departments.

In this short paper we adapt the idea of enumerating over all possible row assignments from the semidefinite approach in [8] to an appropriate linear model for the (SF-DRFLP), which we present in Sect. 2. To further improve the efficiency of the enumeration scheme we provide some combinatorial properties of optimal space-free double-row layouts in Sect. 3. In Sect. 4 we provide computational results showing that we are able to solve instances with up to 16 departments to optimality in reasonable time with our new approach. We conclude our paper in Sect. 5. Formal proofs and detailed computational results as well as extensions to related row layout problems are omitted due to space limitations and will be provided in a forthcoming paper.

2 A New Formulation for Space-Free Double-Row Layouts with Fixed Row Assignments

In order to simplify the presentation we introduce two additional dummy departments $D := \{n + 1, n + 2\}$, to be placed at the left and the right boundary of *both* rows, respectively, with $\ell_{n+1} = \ell_{n+2} = 0$ as well as $w_{ij} = 0$ if $i, j \in [n + 2], \{i, j\} \cap D \neq \emptyset$. Note that formally the horizontal position of $n + 2$ is not well-defined (it may be different in both rows), but because of $w_{i(n+2)} = 0$, $i \in \{1, \ldots, n + 1\}$, this is not important. Let $R_h := \{i \in [n]: r(i) = h\}$ be the set of all departments assigned to row $h \in \{1, 2\}$. We define $\tilde{R}_h := R_h \cup D$, $h \in \{1, 2\}$, and set $\ell_{ij} := \frac{1}{2} (\ell_i + \ell_j)$.

Our MILP formulation uses betweenness variables

$$x_{ikj} = x_{jki} = \begin{cases} 1, & \text{if } k \text{ lies between } i \text{ and } j \text{ (in the same row),} \\ 0, & \text{otherwise,} \end{cases}$$

for $h \in \{1, 2\}, i, j, k \in \tilde{R}_h, |\{i, j, k\}| = 3, i < j$, and additionally distance variables $d_{ij} = d_{ji} \geq 0, i, j \in [n + 1], i < j$.

Using these variables we can formulate the (SF-DRFLP) for fixed row assignment r as follows:

$$\min \sum_{\substack{i,j \in [n] \\ i<j}} w_{ij} d_{ij}$$

$$\text{s. t. } x_{ijk} + x_{ikj} + x_{jik} = 1, \qquad h \in \{1,2\}, i,j,k \in \tilde{R}_h, i < j < k, \tag{1}$$

$$x_{(n+1)i(n+2)} = 1, \qquad i \in [n], \tag{2}$$

$$x_{ikj} = 0, \qquad h \in \{1,2\}, i,j \in R_h, k \in D, i < j, \tag{3}$$

$$x_{(n+1)ij} = x_{ij(n+2)}, \qquad h \in \{1,2\}, i,j \in R_h, i \neq j, \tag{4}$$

$$x_{ilj} + x_{ilk} + x_{jlk} \leq 2, \qquad h \in \{1,2\}, i,j,k,l \in \tilde{R}_h, |\{i,j,k,l\}| = 4, \tag{5}$$

$$-x_{ilj} + x_{ilk} + x_{jlk} \geq 0, \qquad h \in \{1,2\}, i,j,k,l \in \tilde{R}_h, |\{i,j,k,l\}| = 4, \tag{6}$$

$$d_{ij} = \sum_{\substack{k \in R_h, \\ k \notin \{i,j\}}} \ell_k x_{ikj} + \ell_{ij}, \qquad h \in \{1,2\}, i,j \in R_h \cup \{n+1\}, i < j, \tag{7}$$

$$d_{ij} \geq d_{i(n+1)} - d_{j(n+1)}, \qquad i,j \in [n], r(i) \neq r(j), \tag{8}$$

$$d_{ij} + d_{jk} \geq d_{ik}, \qquad i,j,k \in [n+1], |\{i,j,k\}| = 3, i < k, \tag{9}$$

$$x_{ijk} \in \{0,1\}, \qquad h \in \{1,2\}, i,j,k \in \tilde{R}_h, |\{i,j,k\}| = 3, \tag{10}$$

$$d_{ij} \geq 0, \qquad i,j \in [n+1], i < j. \tag{11}$$

Equation (1) and inequalities (5)–(6) express that the departments in each row do not overlap. Indeed, these constraints were originally introduced by Amaral in a formulation of the (SRFLP) [2]. If $i,j,k \in [n]$, $|\{i,j,k\}| = 3$, lie in the same row, exactly one of them lies in the middle (see (1)) and then certain transitivity constraints have to be fulfilled. Furthermore we ensure by Eqs. (2)–(4) that in each row all departments lie between the dummy departments $n+1$ (the left border of the layout) and $n+2$ (the right border in each row). The distance of two departments in the same row equals the sum of the length of all departments between them plus half the length of both departments, see (7). Finally the inter-row distances are calculated in (8) and we add triangle inequalities (9) to improve our model.

Remark 1 The model presented above is indeed a formulation for the (SF-DRFLP) with fixed row assignment. Using mainly the betweenness model of [2] for each single row we know that the departments in each row correspond to a feasible ordering. The inner-row distances are calculated by (7) and the absolute value of the difference of the center positions of two departments lying in different rows is modeled via (8). In summary d_{ij}, $i,j \in [n]$, $i < j$, equals the exact distance between departments i and j in all optimal solutions if the corresponding connectivity w_{ij} is positive.

Fig. 1 We consider an instance with $\ell_1 = \ell_2 = 1$, $\ell_3 = k \geq 2$, $w_{12} = 4$, $w_{13} = w_{23} = 1$. In the optimal space-free double-row layout g equals $k = \max_{i \in R_1} \ell_i$

3 Combinatorial Properties for Speeding Up Our Enumeration

Next we incorporate the above models in an enumerative scheme considering all possible row assignments to obtain an exact approach for the $(\mathtt{SF\text{-}DRFLP})$, i.e. we determine the optimal value for each possible row assignment and take a minimal one. Obviously, we can stop the calculation for a row assignment if the lower bound in branch-and-cut exceeds the value of the currently best solution.

In the following we aim to further improve the computational performance of our approach by excluding row assignments that are too unbalanced. We denote the sum of the lengths of the departments in row $h \in \{1, 2\}$ by $L_h := \sum_{i \in R_h} \ell_i$. Obviously we can restrict to all row assignments with $L_1 \geq L_2$ in our enumeration scheme.

Now our goal is to determine the smallest number $g \in \mathbb{R}_+$ such that there always exists an optimal solution of the $(\mathtt{SF\text{-}DRFLP})$, independent of the objective function, where $g \geq L_1 - L_2$ holds for the corresponding row assignment. Note that in general we have $g \geq \ell_{\max,1} = \max_{i \in R_1} \ell_i$, see Fig. 1 for a small instance where this value for g is obtained.

Theorem 2 states that $\ell_{\max,1}$ is also an upper bound to g.

Theorem 2 $\ell_{\max,1}$ *is the smallest g such that there always exists an optimal solution to the $(\mathtt{SF\text{-}DRFLP})$, where $g \geq L_1 - L_2$ holds for the corresponding row assignment r^*.*

4 Computational Experiments

In this section we present computational results for well-known benchmark instances taken from [1, 3–6, 8], which can be downloaded from [7]. All experiments were performed on a $\mathtt{QUAD\text{-}Core\ INTEL\text{-}Core\text{-}I7\text{-}4770}$ (4×3400 MHz) with 32 GB RAM in single processor mode. We used Cplex 12.6.2 [9] as an IP solver and tested two versions. In $\mathtt{Full}$ we included all inequalities at once, and in $\mathtt{Cuts}$ we only used some of the constraints at once and separated the triangle inequalities (9) and inequalities (5)–(6).

Although the number of row assignments grows exponentially and so lots of NP-hard problems have to be considered, we could solve instances with up to 16

Table 1 Running times in seconds and optimal values "optimal" for the (SF-DRFLP) obtained by applying our MILP model for both variants Full and Cuts for each relevant row assignment in the standard variant ($L_1 \geq L_2$) and according to Theorem 2

Name	Source	Standard		Theorem 2		Amaral [3] time (gap)	Optimal
		Full	Cuts	Full	Cuts		
HA5	[8]	0.04	0.02	0.02	0.01	0.02	52.5
HA6	[8]	0.17	0.16	0.09	0.13	0.06	190.5
HA7	[8]	0.39	0.36	0.25	0.29	0.08	166.0
HA8	[8]	1.65	1.51	1.00	1.06	0.57	205.0
HA9	[8]	6.13	6.76	3.79	4.95	4.34	492.5
HA10	[8]	18.04	19.86	11.00	14.57	22.85	838.0
HA11	[8]	30.97	22.88	16.06	16.59	33.89	796.0
HA12	[8]	173.73	164.76	115.03	133.52	331.32	1028.0
HA13	[8]	335.08	227.52	175.57	164.40	832.24	1530.5
HA14	[8]	1868.29	1418.52	1131.58	1043.26	13315.28	1841.0
HA15	[8]	7907.46	7010.24	5449.31	5685.67	TL (2.50)	2643.5
s9	[3, 4]	4.16	3.35	1.47	1.97	9.03	1181.5
s9h	[3, 4]	18.43	28.28	10.99	18.64	55.96	2294.5
s10	[3, 4]	11.78	8.63	4.37	4.39	25.23	1374.5
s11	[3, 4]	48.04	42.65	22.25	24.36	152.59	3439.5
Am12a	[3, 4]	177.61	146.16	110.23	116.41	356.88	1529.0
Am12b	[3, 4]	116.68	84.20	56.44	56.79	421.01	1609.5
Am13a	[3]	446.76	341.49	292.51	282.80	2033.46	2467.5
Am13b	[3]	402.71	267.21	239.40	203.14	1580.31	2870.0
Am14_1	new	2150.47	1645.00	1482.12	1429.62	10058.04	2756.5
Am15_1	[1]	3892.73	1693.54	2077.85	1241.81	TL (5.39)	3195.0
HK15	[6]	3291.68	1132.41	1576.67	764.08	TL (8.73)	16640.0
P16_a	[5]	41468.10	37276.28	29152.76	29800.27	TL (53.48)	7370.0
P16_b	[5]	15619.68	8688.04	8636.34	6466.20	TL (45.08)	5884.5

Additionally, the table shows the running times and the gaps in percent after a time limit (TL) of twelve hours in brackets using the model of Amaral [3]

departments in reasonable time, see Table 1. Note, the previously largest solved instance for the (SF-DRFLP) contained 13 departments. To allow the reader a direct comparison to the approach of Amaral [3] we also tested his model and included the running times as well as the gaps in percent, calculated by ($\frac{\text{optimal}}{\text{lower bound}} - 1) \cdot 100$, for instances not solved within the time limit of twelve hours. With our new approach all instances with $n \leq 15$ could be solved in less than two hours, only the instances with $n = 16$ were costlier, but could be solved within the time limit. This was not possible for five instances using the approach of Amaral. Comparing `Full` and `Cuts` there is no clear winner, although separation often helps if n is increased. In order to show the impact of Theorem 2, we tested a variant using all assignments with $L_1 \geq L_2$, neglecting the ones with $L_1 = L_2$ and $r(1) = 2$ and a variant that restricts to assignments with $\ell_{\max,1} \geq L_1 - L_2$. Reducing the number of row assignments that have to be considered allows to reduce the running times for solving the (`SF-DRFLP`) significantly.

5 Conclusion

In this paper we presented a new formulation for the (`SF-DRFLP`) with fixed row assignment. Using this formulation and enumerating over all relevant row assignments we were able to solve (`SF-DRFLP`) instances with up to 16 departments for the first time.

In a forthcoming paper we will extend our approach to space-free problems with more than two rows. Additionally we will present related models and combinatorial results for general multi-row layouts, where spaces between the departments are allowed.

It remains for future work to further improve our approach. On the one hand the study of the corresponding polyhedron could help deriving stronger relaxations for our formulation and on the other hand it would be interesting to find some criteria for deciding which row assignments should be considered first in order to further speed-up our enumeration scheme.

References

1. Amaral, A.R.S.: On the exact solution of a facility layout problem. Eur. J. Oper. Res. **173**(2), 508–518 (2006)
2. Amaral, A.R.S.: A new lower bound for the single row facility layout problem. Discrete Appl. Math. **157**(1), 183–190 (2009)
3. Amaral, A.R.S.: The corridor allocation problem. Comput. Oper. Res. **39**(12), 3325–3330 (2012)
4. Amaral, A.R.S.: Optimal solutions for the double row layout problem. Optim. Lett. **7**(2), 407–413 (2013)
5. Amaral, A.R.S.: A parallel ordering problem in facilities layout. Comput. Oper. Res. **40**(12), 2930–2939 (2013)
6. Heragu, S.S., Kusiak, A.: Efficient models for the facility layout problem. Eur. J. Oper. Res. **53**(1), 1–13 (1991)
7. Hungerländer, P.: Drflp benchmark problems (2015). http://philipphungerlaender.jimdo.com/benchmark-libraries/
8. Hungerländer, P., Anjos, M.F.: A semidefinite optimization approach to space-free multi-row facility layout. Cahiers du GERAD G-2012-03. GERAD, Montreal, QC, Canada (2012)
9. IBM. IBM ILOG CPLEX V12.6.1 User's Manual for CPLEX (2015)

Part III
Logistics and Transportation

Cost Allocation for Horizontal Carrier Coalitions Based on Approximated Shapley Values

Kristian Schopka and Herbert Kopfer

Abstract To improve competitiveness, small and mid-sized carriers ally in horizontal carrier coalitions for request exchange. A crucial aspect for the long-term viability and stability of coalitions is a fair cost allocation among the agents. Despite of the long computing time, the well-known Shapley value has been used as a scheme for cost allocation. The contribution of this paper lies on the development of a suitable sampling procedure that approximates the Shapley value applied to cost allocations for the collaborative traveling salesman problem with time windows. A computational study identifies the deviation of the values generated by the proposed sampling procedures from the actual Shapley value.

1 Introduction

Rising petrol and labor prices, shorter product life cycles and higher customer expectations reduce the profit margins of carriers [3]. Due to the limited availability of resources and a weak market position, it is difficult for small and mid-sized carriers (SMCs) to generate cost-effective transportation plans. By building horizontal coalitions, SMCs may find a way to overcome their cost disadvantage against large forwarding companies. Thereby, especially auction-based mechanisms for the exchange of transportation orders (referred to as requests) are provided by current studies [8]. Beside the requirement of preserving autonomy of all coalition members, a fair cost allocation permits a long-term and stable coalition. The Shapley value [7] ensures a fair cost allocation in the core if the core is not empty. Several approaches for collaborative transportation planning use the Shapley value to allocate costs (e.g. [4, 5]).

K. Schopka (✉) · H. Kopfer
Chair of Logistics, University of Bremen, Wilhelm-Herbst-Str. 5,
28359 Bremen, Germany
e-mail: schopka@uni-bremen.de
URL: http://www.logistik.uni-bremen.de

H. Kopfer
e-mail: kopfer@uni-bremen.de
URL: http://www.logistik.uni-bremen.de

© Springer International Publishing Switzerland 2017
K.F. Dœrner et al. (eds.), *Operations Research Proceedings 2015*,
Operations Research Proceedings, DOI 10.1007/978-3-319-42902-1_18

Since both the calculation of the Shapley value and the generation of transportation plans are NP-hard problems, strategies that approximate the Shapley value are recommended. The contribution of this paper is the development of a new strategy that approximates the Shapley value and enables a fair cost allocation for coalitions with numerous SMCs. The strategy is compared with an existing sampling procedure for the collaborative traveling salesman problem with time windows (CTSPTW) that is introduced in Sect. 2. Section 3 presents the Shapley value and sampling procedures. The results of a computational study are presented in Sect. 4. Section 5 concludes the paper.

2 The Collaborative TSPTW

The CTSPTW can be formulated as a mixed integer program (Eqs. 1–8). We assume a horizontal coalition with m independent SMCs (referred to as agents) denoted as $K = \{1, 2, \ldots, m\}$. Each agent k has n pick-up requests that are divided into the subsets: private requests P_k and common requests C_k. Whereas the private requests P_k have to be served by the agent k, his common requests C_k can be satisfied by any agent $k' \in K$. Let $C := \cup_{k=1}^{m} C_k$ denote the set of all common requests. To fulfill requests each agent uses one vehicle route $k \in K$ that starts and ends at the individual depot d_k. The transportation network of each agent k is given by the directed graph $G_k := (N_k, A_k)$, where $N_k := \{d_k\} \cup P_k \cup C$ represents the node set and A_k builds the set of connecting arcs. The usage of an arc $(i, j) \in A_k$ on vehicle route k requires travel time g_{ijk}; let one time unit equates one cost unit. The fulfillment of any request $i \in P_k \cup C$ on vehicle route k requires a service time s_{ik} and has to be started within the time window $[a_{ik}, b_{ik}]$. The binary decision variable x_{ijk} is equal to one if vehicle route k uses the arc $(i, j) \in A_k$ and equal to zero otherwise. Let t_{ik} be the service starting time for vehicle tour k and request i.

$$\min \quad c(K) = \sum_{k \in K} \sum_{(i,j) \in A_k} x_{ijk} \cdot g_{ijk} \tag{1}$$

$$\text{subject to:} \quad \sum_{j \in P_k} x_{ijk} = 1, \qquad \forall k \in K, i \in N_k, \tag{2}$$

$$\sum_{k \in K} \sum_{j \in N_k} x_{ijk} = 1, \qquad \forall i \in C, \tag{3}$$

$$\sum_{j \in N_k} (x_{ijk} - x_{jik}) = 0, \qquad \forall k \in K, i \in N_k, \tag{4}$$

$$t_{ik} + s_{ik} + g_{ijk} - M \cdot (1 - x_{ijk}) \leq t_{jk}, \qquad \forall k \in K, i \in N_k, j \in N_k \setminus \{0\}, \tag{5}$$

$$a_i \leq t_i \leq b_i, \qquad \forall k \in K, i \in N_k, \tag{6}$$

$$x_{ijk} \in \{0, 1\}, \qquad \forall (i, j) \in A_k, \tag{7}$$

$$t_{ik} \geq 0, \qquad \forall k \in K, i \in N_k. \tag{8}$$

The mathematical model minimizes the sum of the overall fulfillment costs $c(K)$, while all information is fully transparent and visible for all agents, i.e. aspects of preserving the agents' autonomy are not considered in this paper. Constraints (2)–(4) represent the assignment constraints and ensure that each private request $i \in P_k$ is served exactly once by agent k, respectively each common request $j \in C$ is severed only once by any agent $k' \in K$. Whereat constraint (5) excludes subtours (with M as a large positive number), constraint (6) observes the time windows. Finally, constraints (7)–(8) define the domains of the decision variables.

3 Approximating the Shapley Value

An issue of the CTSPTW is a fair allocation of costs. A popular cost allocation scheme is the single-valued solution concept called Shapley value [7]. Let K be the set of all agents, where any subset (sub-coalition) is given by $S \subset K$. Furthermore, the related costs of any sub-coalition $S \subset K$ are given by $c(S)$ and can in our case be calculated by solving the related CTSPTW including all agents of S. Generally, the Shapley value allocates costs based on the marginal cost contributions of agents and is given by $SV_k(K, c)$ for each agent k as follows:

$$SV_k(K, c) = \sum_{S \subset K \setminus \{k\}} \frac{|S|! \, (|K| - |S| - 1)!}{|K|!} \, (c \, (S \cup \{k\}) - c \, (S))$$

Since the calculation of the Shapley value is NP-hard [1], it is inefficient to determine the actual Shapley value especially for scenarios with numerous agents. However, it seems rather reasonable to approximate the Shapley value by using sampling procedures. To the best of our knowledge, the first sampling procedures were introduced by Mann and Shapley [6], where the *Type-0* sampling has been identified as the best procedure. This procedure repeatedly generates random permutations of all agents until a defined number is reached. For any of those permutations the marginal costs for each agent $k \in K$ are calculated. Therefore the costs $c(S)$ of the sub-coalition including all agents of the viewed permutation before k and the costs $c(S \cup \{k\})$ of this sub-coalition and k are calculated. The marginal costs of k on this permutation are given by the difference of $c(S)$ and $c(S \cup \{k\})$. The calculation of the marginal contribution of all agents on numerous and varying permutations (samples) leads to an approximation of the Shapley value.

The *ApproShapley* algorithm of Castro et al. [2] is a recent variant of the *Type-0* sampling that allows a polynomial calculation of the Shapley value for a large class of theoretical games. Thereby the *ApproShapley* values determine a fair cost allocation for the traveling salesperson game [1]. Indeed by *Type-0* sampling or rather the *ApproShapley* it is possible to reduce the computing time, though by considering only random samples the calculated data (costs) are used ineffective. Furthermore, each iteration of those procedures requires the calculation of costs for a sub-coalition with numerous agents. These two issues still lead to a long and unacceptable computing time for approximating the Shapley value.

A possibility to reduce the computing time and to use the calculated data effective is introduced in this paper. This possibility is implemented in the presented strategic sampling procedure called α-*Sampling*. Against the consideration of samples based on random permutations, α-*Sampling* uses sub-coalitions as samples. By considering only sub-coalitions with a small number of agents, the computing time can be reduced. Thereby, α-*Sampling* calculates the marginal contribution of each agent in every sub-coalition with α agents (α-*coalition*). Let $S^\alpha := \{S^* \subset K : |S^*| = \alpha\}$ store all α-coalition and $S_h^\alpha \in S^\alpha$ represent one concrete α-coalition. Aziz et al. [1] define the marginal contribution of agent k in any sub-coalition, as the cost increase if k joins the coalition or the cost reduction if k leaves the coalition. The marginal contribution $c^k(S_h^\alpha)$ of agent k to a α-*coalition* $S_h^\alpha \in S^\alpha$ is given by:

$$c^k(S_h^\alpha) = \begin{cases} c(S_h^\alpha) - c(S_h^\alpha \setminus \{k\}) & \text{if } k \in S_h^\alpha \\ c(S_h^\alpha \cup \{k\}) - c(S_h^\alpha) & \text{if } k \notin S_h^\alpha \end{cases}$$

After the calculation of the marginal contribution on every α-*coalition* for all agents, the costs of the overall coalition $c(K)$ are allocated among the agents. Therefore, the cost fraction of any agent k is given by the approximated Shapley values $\widetilde{SV_k}$ that the α-*Sampling* procedure calculates as follows:

$$\widetilde{SV_k} = \frac{\sum_{h=1}^{|S^\alpha|} c^k(S_h^\alpha)}{\sum_{l=1}^{|K|} \sum_{q=1}^{|S^\alpha|} c^l(S_q^\alpha)} \cdot c(K) \qquad \forall\, k \in K$$

4 Computational Experiments

In a computational study, the sampling procedures of *ApproShapley* and α-*Sampling* are analyzed in terms of computing time and the deviation to the actual Shapley value for the CTSPTW. Therefore, new instances are generated that differ in the request structure (rs), the number of agents ($|K|$) and collaborative requests ($|C_k|$), and the size of time windows ($|TW|$).[1] For running the experiments of this study, the cost allocation procedures of the Shapley value, the *ApproShapley*, and the α-*Sampling* are implemented in a C++-application on a Windows 7 PC (i7-2600 processor with 3.4 GHz, 16 GB RAM). The mathematical solver CPLEX 12.5.1 was used to calculate the cost for any (sub-)coalition and the computing time for solving any CTSPTW was limited to 600 seconds To reduce the computational effort for all cost allocation procedures the results are stored on a list, so that within one allocation procedure the costs of any (sub-)coalition are calculated only once. Our study includes the *3-*

---

[1]http://www.logistik.uni-bremen.de/english/instances/.

Table 1 Approximation of the Shapley value for instance R2 12 3 80

k	Iso.	Col.	Appro30		Appro10		3-Sampling		4-Sampling	
	IP_k	SV_k	$\widetilde{SV_k}$	$|\Delta_k|$	$\widetilde{SV_k}$	$|\Delta_k|$	$\widetilde{SV_k}$	$|\Delta_k|$	$\widetilde{SV_k}$	$|\Delta_k|$
1	295.85	263.84	273.78	0.0377	269.70	0.0222	248.50	0.0581	254.72	0.0346
2	335.33	299.42	300.76	0.0045	296.74	0.0090	289.54	0.0330	294.96	0.0149
3	314.92	238.74	249.62	0.0456	246.14	0.0310	241.60	0.0120	238.90	0.0007
4	205.70	176.44	169.69	0.0382	171.36	0.0288	184.47	0.0455	184.00	0.0429
5	548.43	406.40	417.02	0.0262	398.81	0.0187	401.22	0.0127	400.49	0.0145
6	384.02	194.66	188.46	0.0318	188.99	0.0291	207.30	0.0650	193.66	0.0052
7	254.77	173.15	174.05	0.0052	150.57	0.1304	179.27	0.0354	173.24	0.0005
8	396.23	361.02	365.84	0.0133	364.72	0.0102	367.22	0.0172	372.92	0.0330
9	280.36	240.77	242.36	0.0066	253.20	0.0517	233.23	0.0313	237.00	0.0157
10	207.07	196.70	195.80	0.0046	200.21	0.0178	193.19	0.0178	198.72	0.0103
11	280.36	252.39	242.88	0.0377	254.41	0.0080	249.89	0.0099	255.23	0.0113
12	311.68	214.43	197.67	0.0782	223.07	0.0403	222.49	0.0376	214.10	0.0016
$\sum$	3814.72	3017.93	MErr	0.0347		0.0460		0.0358		0.0205

Sampling, the 4-Sampling, and two versions of the ApproShapley; namely Appro10 with 10 samples and Appro30 with 30 samples.

Table 1 presents the detailed results for instance R2 12 3 80. The overall costs without request exchange (3814.72) can be reduced by collaborative planning (3017.93). A distribution key for assigning coalition costs to agents is given by the Shapley value SV_k. Thereby SV_k is less than the costs of a planning without request exchange IP_k for all agents $k \in K$. Table 1 includes also the approximated Shapley values $\widetilde{SV_k}$ and the deviation to the actual Shapley value $|\Delta_k|$ for any agent k and all considered sampling procedures. As the mean error (MErr) shows, the results of 4-Sampling are superior to the results of Appro10 and Appro30. The solution quality of 3-Sampling is comparable to the solution quality of Appro30.

The good solution quality of 3-Sampling and 4-Sampling can be confirmed by the results of the overall test cases presented in Table 2. The average mean error for Appro10 is 0.0720 respectively 0.0429 for Appro30. The overall results of 3-Sampling are averaging at a mean error of 0.0395 and are slightly superior to Appro30. The average mean error of 0.0489 for 4-Sampling is inferior to Appro30. Nevertheless, 4-Sampling achieves several best solutions, especially for instances with 10 or 12 agents. Another benefit of 3-Sampling (4-Sampling) refers to the obviously shorter computing time of approximately 10 % (15 %) compared to Appro30. In summary, α-Sampling enables an approximation of the Shapley value for the CTSPTW in obvious shorter computing time that is necessary to calculate the actual Shapley value, where the solution quality is comparable to state-of-the-art sampling procedures.

Table 2 Deviation among sampling procedures and Shapley value

r_s	$\|K\|$	$\|C_k\|$	$\|TW\|$	Shapley	Appro30		Appro10		3-Sampling		4-Sampling	
				Seconds	Seconds	MErr	Seconds	MErr	Seconds	MErr	Seconds	MErr
R1	6	3	60	21	19	**0.0216**	11	0.0279	17	0.0235	21	0.0446
R1	8	3	60	130	73	**0.0109**	31	0.0601	52	0.0261	103	0.0263
R1	10	3	60	854	216	0.0357	88	0.0518	122	0.0274	376	**0.0249**
R1	12	3	60	5894	533	0.0353	216	0.0349	226	0.0305	1099	**0.0288**
R1	6	5	60	37	35	**0.0288**	18	0.0553	27	0.0417	41	0.0673
R1	8	5	60	299	171	**0.0408**	102	0.0672	78	0.0427	226	0.0518
R1	10	5	60	2779	910	0.0477	363	0.0579	210	**0.0280**	885	0.0340
R1	12	5	60	27326	3367	**0.0250**	1516	0.0806	493	0.0382	2627	0.0328
R2	6	3	80	38	35	**0.0379**	26	0.0432	30	0.0618	42	0.1163
R2	8	3	80	339	211	**0.0171**	104	0.0432	104	0.0332	239	0.0516
R2	10	3	80	3017	1006	0.0253	561	0.0549	245	0.0271	834	**0.0213**
R2	12	3	80	20365	3002	0.0347	1803	0.0460	377	0.0358	2156	**0.0205**
R2	6	5	80	107	100	0.1110	74	0.2434	72	**0.0762**	174	0.1804
R2	8	5	80	2472	2090	0.1452	1312	0.1011	459	**0.0802**	1361	0.1799
R2	10	5	80	38930	11296	0.0885	8960	0.0805	1157	**0.0545**	4398	0.0597
R2	12	5	80	180014	23334	0.0481	12641	0.1169	1703	0.0660	9029	**0.0450**
R3	6	3	100	26	21	**0.0280**	14	0.0345	19	0.0284	19	0.0574
R3	8	3	100	172	96	**0.0116**	49	0.0395	66	0.0298	102	0.0296
R3	10	3	100	1230	306	0.0473	145	0.0330	154	0.0328	343	**0.0248**
R3	12	3	100	9983	1078	0.0367	542	0.0578	323	0.0358	882	**0.0239**
R3	6	5	100	47	44	0.0525	28	0.1098	31	**0.0377**	44	0.0691
R3	8	5	100	530	365	0.0479	172	0.1178	120	0.0349	255	**0.0301**

(continued)

Table 2 (continued)

rs	$|K|$	$|C_k|$	$|TW|$	Shapley	Appro30		Appro10		3-Sampling		4-Sampling	
				Seconds	Seconds	MErr	Seconds	MErr	Seconds	MErr	Seconds	MErr
R3	10	5	100	5427	1785	0.0363	785	0.1195	335	0.0378	836	**0.0301**
R3	12	5	100	51036	7346	0.0393	3348	0.0837	752	0.0407	2372	**0.0308**
R3	6	3	140	60	54	0.0447	42	0.0693	44	**0.0337**	50	0.0621
R3	8	3	140	538	353	**0.0299**	193	0.0429	118	0.0308	234	0.0330
R3	10	3	140	4456	1452	0.0275	830	0.0311	419	0.0283	891	**0.0220**
R3	12	3	140	54970	10625	0.0237	4717	0.0258	1130	0.0329	2708	**0.0218**
R3	6	5	140	184	172	0.0648	137	0.1361	95	**0.0482**	174	0.0837
R3	8	5	140	7163	5072	0.0642	3978	0.0742	966	**0.0354**	1927	0.0368
R3	10	5	140	71958	21817	0.0314	10387	0.0749	1810	0.0357	5133	**0.0309**
R3	12	5	140	707124	56569	**0.0345**	23147	0.0895	3849	0.0471	13293	0.0393
Average						0.0429		0.0720		**0.0395**		0.0489

5 Conclusion

In this paper, a new procedure for approximating the Shapley value is introduced called α-*Sampling*. α-*Sampling* considers all sub-coalitions with α agents and calculates the marginal contribution for each agent. By this procedure the computing time can be reduced enormously, particularly when the calculation of the marginal contribution itself is NP-hard (e.g. CTSPTW). Furthermore, a computational study identifies that the solution quality of α-*Sampling* is comparable to state-of-the-art sampling procedures. Hence, α-*Sampling* may ensure a fair cost allocation for horizontal coalitions with numerous agents. Currently, we use α-*Sampling* for the cost allocation within scenarios of dynamic collaborative transportation planning, where typically only less computing time is available.

Acknowledgements The research was supported by the German Research Foundation (DFG) as part of the project "Kooperierende Rundreiseplanung bei rollierender Planung".

References

1. Aziz H., Cahan C., Gretton C., Kilby P., Mattei N., Walsh: A study of Proxies for Shapley Allocations of Transport Costs. arXiv preprint (2014). arXiv:1408.4901
2. Castro, J., Gomez, D., Tejada, J.: Polynomial calculation of the Shapley value based on sampling. Comput. Oper. Res. **36**, 1726–1730 (2009)
3. Cruijssen, F., Cools, M., Dullaert, W.: Horizontal cooperation in logistics: opportunities and impediments. Transp. Res. Part E Log. Transp. Rev. **43**, 129–142 (2007)
4. Frisk, M., Göthe-Lundgren, M., Jörnsten, K., Rönnqvist, M.: Cost allocation in collaborative forest transportation. Eur. J. Oper. Res. **205**, 448–458 (2010)
5. Krajewska, M., Kopfer, H., Laporte, G., Ropke, S., Zaccour, G.: Horizontal cooperation among freight carriers: request allocation and profit sharing. J. Oper. Res. Soc. **59**, 1483–1491 (2008)
6. Mann, I., Shapley, L.: Values of large games IV: Evaluating the electoral college by Monte Carlo techniques. Technical report, The RAND Corporation, Santa Monica, CA, USA (1960)
7. Shapley, L.: A value for n-person games. In: Kuhn, H., Tucker, A. (eds.) Contributions to the Theory of Games II 28, pp. 307–317. Princeton University Press, Princeton (1957)
8. Verdonck, L., Caris, A., Ramaekers, K., Janssens, G.K.: Collaborative logistics from the perspective of road transportation companies. Transp. Rev. **33**, 700–719 (2013)

Collaborative Transportation Planning with Forwarding Limitations

Mario Ziebuhr and Herbert Kopfer

Abstract In collaborative transportation planning, independent forwarders align their transportation plans by exchanging requests within a horizontal coalition. The goal of the coalition members is to increase their profitability and flexibility in competitive markets with high demand fluctuations. In recent publications, it is assumed that each request can be fulfilled by any coalition member. However, in practice some requests are prohibited to be forwarded due to contractual agreements. These requests are known as compulsory requests. The contribution of this paper is to identify the increase of costs caused by compulsory requests of a collaborative pickup and delivery transportation planning problem. To analyze the impact of compulsory requests, an existing column generation-based heuristic with two solution strategies for handling compulsory requests is applied and investigated.

1 Introduction

In competitive transportation markets with high demand fluctuations, forwarders have to reduce their costs and to improve their flexibility by considering different fulfillment modes. As fulfillment modes forwarders use beside their own transportation resources (self-fulfillment), external carriers (subcontracting), and horizontal cooperation (collaborative planning). In collaborative transportation planning (CTP), independent forwarders try to improve their planning situation by reallocating their transportation requests or capacities in a horizontal coalition [6]. The goal of CTP is the identification of a transportation plan where each coalition member reduces his operational costs. In the literature several CTP models are examined. Most of them focus on request exchange where either all requests [1, 6] or just a subset of

M. Ziebuhr (✉) · H. Kopfer
Chair of Logistics, University of Bremen, Wilhelm-Herbst-Str. 5, 28359 Bremen, Germany
e-mail: ziebuhr@uni-bremen.de
URL: http://www.logistik.uni-bremen.de

H. Kopfer
e-mail: kopfer@uni-bremen.de
URL: http://www.logistik.uni-bremen.de

© Springer International Publishing Switzerland 2017   141
K.F. Dœrner et al. (eds.), *Operations Research Proceedings 2015*,
Operations Research Proceedings, DOI 10.1007/978-3-319-42902-1_19

requests are exchanged [2, 5]. In practice, the latter approach seems to be preferable because coalition members do not want to reveal their entire request portfolio in a competitive environment. A common solution approach for these CTP problems is a cherry-picking procedure which identifies profitable requests for self-fulfillment (referred to as reserved requests) and unprofitable requests for outsourcing. In our approach some of these reserved requests are selected due to contractual obligations while the remaining ones are selected due to their profitability. This means that a forwarder has to use a certain fulfillment mode for requests with contractual obligations while the remaining requests can be served by a profitable fulfillment mode. These requests with contractual obligations (e.g. security relevant goods) are known as compulsory requests. The impact of compulsory requests is analyzed in case of transportation planning problems with self-fulfillment and subcontracting as fulfillment modes by Schönberger [4] and Ziebuhr and Kopfer [7]. The contribution of this paper is to identify the increase of costs by considering compulsory requests in a collaborative pickup and delivery transportation problem. To analyze the impact of compulsory requests, a column generation-based heuristic (CGB-heuristic) with different strategies for handling compulsory requests is applied, which were introduced by Ziebuhr and Kopfer [7]. The CGB-heuristic is suitable for linear programming with many variables, where the overall problem is divided in a subproblem for generating vehicle routes and a master problem for selecting vehicle routes. To handle compulsory requests two solution strategies are proposed. One strategy proposes the consideration of the compulsiveness of requests during the subproblem (strict generation procedure) while the other one considers forwarding limitations during the master problem (strict composition procedure). For experiments, the consideration of different request types is proposed, which is motivated by the fact that a forwarder should be able to offer various services to a client and each of these services may differ with respect to its impact on costs. The following request types are considered: standard requests (fulfillment by any fulfillment mode), premium requests (fulfillment by private fleet and coalition members), and premium plus requests (fulfillment by private fleet). In Sect. 2 the CTP problem is presented, while Sect. 3 describes the main solution and Sect. 4 presents the computational experiments.

2   Collaborative Planning with Compulsory Requests

In the pickup and delivery problem with time windows (PDPTW), a transportation plan has to be determined where n less than truckload requests have to be transported from their pickup $P = \{1, \ldots, n\}$ to their delivery location $D = \{n + 1, \ldots, 2n\}$. In the PDPTW, the set of nodes is given by $V = P \cup D \cup \{0\}$ where $\{0\}$ represents the depot. For each edge $(i, j) \in A$ a travel time t_{ij} and a distance d_{ij} is given. At each node $i \in V$ with a demand l_i a service s_i has to be started within a time window $[a_i, b_i]$. For request fulfillment a set of homogeneous vehicles K, each having capacity Q, is assumed. The task is to determine a transportation plan which minimizes the sum of the fixed costs α_k and variable costs β_k by fulfilling routing, time and loading constraints. A transportation plan is defined by the binary decision variable x_{ijk} which is equal to one if a vehicle k travels from node i to j. The PDPTW is solved by isolated planning (IP) assuming a single decision maker and costs defined by the objective function (1). A PDPTW is described by Ropke and Pisinger [3].

$$\min IP = \sum_{k \in K} \alpha_k + \sum_{k \in K} \sum_{(i,j) \in A} \beta_k d_{ij} x_{ijk} \tag{1}$$

This paper looks at the PDPTW from a collaborative perspective, where m forwarders align their transportation plans by exchanging requests. During the collaboration each forwarder i offers his entire request portfolio P_i ($P = \sum_{i=1}^{m} P_i$) for exchange and receives a new portfolio P_i' after the allocation process is completed. Depending on the transportation plans of the coalition members, each member i transfers a set of requests (P_i^-) to the coalition and receives a set of requests (P_i^+) from the coalition. If a member receives more requests than he can fulfill he will employ an external carrier from the spot market. As the coalition members are self-interested they aim to minimize their own costs (local costs) by preserving customer payments and cost structure information. The goal of this CTP problem is to minimize the individual local costs (IP') by considering the updated request portfolio P' and can be modeled as follows:

$$\min CTP = \sum_{i=1}^{m} IP_i' \tag{2}$$

$$P_i' \cap P_j' = \varnothing, \quad \forall i,j = 1, \ldots, m, \; i \neq j, \tag{3}$$

$$\cup_{i=1}^{m} P_i^- = \cup_{i=1}^{m} P_i^+. \tag{4}$$

The described CTP problem involving pickup and delivery nodes assumes that each coalition member i offers his entire request portfolio P_i for exchange. By considering the CTP problem with compulsory requests, it is proposed to separate the request portfolio of a member i into three disjoint sets: standard requests P_i^1, premium requests P_i^2, and premium plus requests P_i^3 with $P_i = P_i^1 \cup P_i^2 \cup P_i^3$. Based on this formulation, the considered CTP problem has to be modified in such a way that premium and premium plus requests cannot be fulfilled by an external carrier and premium plus requests cannot be forwarded to a coalition member.

3 Solution Approach

To solve the CTP problem with forwarding limitations the CGB-heuristic with two strategies for handling compulsory requests is applied and modified. In the following the CGB-heuristic as well as the strategies are briefly described. The idea of the CGB-heuristic is the identification of a transportation plan by using an iterative approach where vehicle routes are generated in a subproblem and promising vehicle routes are selected in a master problem. In a CTP problem, the subproblem is solved by each coalition member separately while the master problem is solved by a neutral software agent. By solving the master problem dual values are generated and forwarded to the subproblem for identifying new vehicle routes which reduce the operational costs. The process is repeated until a specific number of iterations are investigated. The subproblem is solved by an adaptive large neighborhood search (ALNS) and the master problem is solved by the commercial solver ILOG CPLEX. An ALNS is a local search heuristic which uses a simulated annealing acceptance criterion and different removal and insertion heuristics. In CTP it is recommended to execute the CGB-heuristic two times. First, the approach applies to the complete request portfolio P which means that each coalition member is able to bid on all requests of the coalition. Secondly, each coalition member uses this approach for the winning bids of the first application of the CGB-heuristic.

To ensure forwarding limitations for requests, the strict generation procedure is proposed where the compulsiveness of requests is strictly observed by the subproblem. Thereby, only valid vehicle routes are accepted by the simulated annealing during the ALNS. This means that a request is fulfilled by a transportation mode corresponding to the type of service attached to the request. To generate as many valid vehicle routes as possible, three modifications are recommended for the ALNS. First, the external carrier option is penalized by using penalty costs for premium and premium plus requests. Second, the insertion of premium plus requests into vehicle routes from a different coalition member is prohibited by skipping these invalid insertions for the insertion heuristics. Third, the request portfolio of the insertion heuristics is split into one for compulsory requests and one for standard requests so that compulsory requests are preferred for reinsertion. In terms of premium requests, it is observed that each forwarder always generates vehicle routes where all premium requests are served by self-fulfillment in order that similar routes are generated. To solve this issue we eliminate the penalty costs when all requests are visible for each coalition member and reintegrate them after the first application of the CGB-heuristic has assigned a new request portfolio for each coalition member.

A second strategy for handling compulsory requests is the strict composition procedure where forwarding limitations are ignored by the ALNS and considered for the solution of the master problem. This means that many of the routes submitted by the ALNS may contain compulsory requests which are served by an inappropriate fulfillment mode. To ensure that vehicle routes are selected, which are valid for the considered CTP problem, the master problem is extended with new constraints which observe the applied fulfillment modes for compulsory requests. Based on the generated dual values, the ALNS is guided to generate valid vehicle routes in terms of compulsory requests. To ensure feasible solutions by high percentages of compulsory requests in the first round of the column generation, where the ALNS is not guided by dual values, the strict generation procedure is applied.

4   Computational Experiments

In this Sect. 4, the best strategy and the increase of costs in terms of compulsory requests are determined. Therefore, the instances presented by Wang and Kopfer [6] are examined where 2–5 PDPTW instances with the same characteristics (C, R, RC) and size (100 customers) are combined into one instance. The applied fleet sizes are set to the number of used vehicles in the best-known solutions, while the fixed costs are set to zero and the variable costs are set to one per distance unit. In total 24 instances are generated. The instances with five decision makers are skipped due to the computational time. To analyze the CTP problem with compulsory requests the mentioned instances are extended by different ratios of compulsory requests. We analyze ratios of 10, 20, and 30 % compulsory requests, which means e.g. that 20 % of all pickup nodes are compulsory requests and 80 % are standard requests. 15 samples are generated for each ratio and instance. In our study, scenarios with standard requests combined with one type of compulsory requests are analyzed. As parameter setting, the same parameters are used as presented by Ziebuhr and Kopfer [7] except for the extended time limit of the commercial solver (300 s) and reduced iterations during the ALNS (5,000 iterations).

In a benchmark study, the strict generation procedure is compared with the strict composition procedure for all instances with 20% premium or premium plus requests. Thereby, it is identified that on average a forwarder has to charge about 0.27 % (6.13 %) additional costs per 20 % premium request and about 8.25 % (5.60 %) additional costs per 20 % premium plus requests. The results in the brackets are computed by the strict generation procedure, while the remaining ones are computed by the strict composition procedure. As can be seen, the strict composition procedure is preferable for premium requests while the strict generation procedure is preferable for premium plus requests. The strict composition procedure seems to be superior in case that several transportation modes are applicable for request fulfillment which occurs for premium requests and not for premium plus requests.

To investigate the impact of compulsory requests all instances with 2–4 forwarders are solved by the strict generation procedure for premium plus requests and by the strict generation procedure for premium requests. First, the solution quality of the CGB-heuristic is verified by a centralized planning (CP) and collaborative planning approach. The results are compared with the best-known results of Wang and Kopfer [6] which are computed by a CGB-heuristic with a large neighborhood search (LNS) instead of an ALNS. In Table 1, the best-known IP solutions reported by SINTEF, CP and CTP solutions reported by Wang and Kopfer [6], and our solutions are presented. In a centralized planning scenario, the subproblem is solved by a central decision maker who is in charge of all vehicles and request portfolios. In our study, 29 new best-known solutions for the CP and CTP are identified. The best CTP solutions are used to determine the percentaged cost increases per compulsory requests. For the CTP problem with forwarding limitations, it is observed that the impact of premium requests is not significant. However, premium plus requests have a significant impact on the costs where the additional costs depend on the location structure, number of

Table 1 Increase of transportation costs

id	m	n	IP SINTEF	CP LNS [6]	CP ALNS	CTP LNS [6]	CTP ALNS	PP 10%	PP 20%	PP 30%	PR 10%	PR 20%	PR 30%
C101	2	105	1864.29	1699.08	**1669.87**	1691.63	**1669.87**	0.28%	0.20%	0.21%	0.00%	0.01%	0.01%
C102	2	106	1655.38	1539.82	1539.82	1539.82	1539.82	0.36%	0.27%	0.21%	0.00%	0.00%	0.00%
C103	3	159	2658.48	2414.99	**2372.23**	2391.34	**2372.23**	0.33%	0.19%	0.18%	0.00%	0.00%	0.00%
C104	3	159	2517.89	2198.10	**2190.70**	2194.32	**2190.70**	0.49%	0.28%	0.29%	0.00%	0.00%	0.00%
C105	4	212	3518.49	3027.86	**2984.95**	3041.51	**2984.35**	0.20%	0.21%	0.17%	0.00%	0.01%	0.02%
C106	4	211	3519.67	2882.58	**2850.53**	2867.63	**2857.05**	0.34%	0.27%	0.27%	0.01%	0.01%	0.02%
R101	2	104	2452.03	**2263.21**	2276.33	**2269.32**	2276.33	0.29%	0.22%	0.18%	0.08%	0.04%	0.02%
R102	2	104	2363.93	2311.98	**2262.78**	2311.77	**2262.78**	0.07%	0.09%	0.06%	0.01%	0.02%	0.01%
R103	3	160	3600.24	3394.24	**3382.52**	3444.36	**3375.60**	0.08%	0.09%	0.08%	0.01%	0.01%	0.00%
R104	3	154	3239.63	3014.98	**2998.63**	3041.63	**3005.42**	0.14%	0.11%	0.12%	0.01%	0.00%	0.00%
R105	4	208	4434.32	**3906.01**	3907.74	3960.58	**3959.87**	0.13%	0.14%	0.13%	0.02%	0.01%	0.01%
R106	4	215	5624.38	5038.12	**5016.92**	5112.16	**5027.05**	0.13%	0.11%	0.10%	0.01%	0.01%	0.00%
RC101	2	106	2488.89	2386.78	**2378.99**	2378.99	2378.99	0.13%	0.09%	0.08%	0.02%	0.00%	0.00%
RC102	2	107	2867.77	2696.04	**2690.49**	2690.49	2690.49	0.12%	0.14%	0.10%	0.00%	0.00%	0.00%
RC103	3	160	3945.21	3511.99	**3457.16**	3544.73	**3450.44**	0.18%	0.17%	0.16%	0.01%	0.01%	0.01%
RC104	3	161	4190.75	3650.29	**3641.66**	3693.64	**3641.20**	0.25%	0.23%	0.21%	0.01%	0.01%	0.00%
RC105	4	211	5345.12	4539.76	**4489.57**	4680.19	**4489.57**	0.22%	0.18%	0.17%	0.02%	0.01%	0.00%
RC106	4	213	5341.35	4777.83	**4740.34**	4881.84	**4740.34**	0.16%	0.11%	0.12%	0.03%	0.02%	0.02%

(PP = CTP with Premium Plus Requests; PR = CTP with Premium Requests)

Results computed by Intel Core i7-2600 processor (3.4 GHz and 16 GB memory) with C++ and CPLEX 12.51

members, and ratios of premium plus requests. It is identified that random location structures lead to the lowest additional costs, while clustered structures lead to the highest additional costs. This observation can be explained by the distribution of the requests. Thereby, it is more expensive to fulfill premium plus requests in different request clusters instead of considering equally distributed requests. In terms of the ratios, it is observed that the costs per premium plus request decrease by increasing the ratio of premium plus requests. The mentioned behavior can be explained by the first application of the ALNS where it is preferable when several compulsory requests are available for route building. By increasing the number of members m, different results are identified. Thereby, it seems that in random structures additional coalition members reduce the impact of premium plus requests while additional coalition members in clustered structures lead to higher or lower costs depending on the positions of the request clusters.

Acknowledgements This research was supported by the German Research Foundation (DFG) as part of the project "Kooperative Rundreiseplanung bei rollierender Planung".

References

1. Agarwal, R., Ergun, Ö.: Network design and allocation mechanisms for carrier alliances in liner shipping. Oper. Res. **58**(6), 1726–1742 (2010)
2. Özener, O.Ö., Ergun, Ö., Savelsbergh, M.: Lane-exchange mechanisms for truckload carrier collaboration. Transp. Sci. **45**(1), 1–17 (2011)
3. Ropke, S., Pisinger, D.: An adaptive large neighborhood search heuristic for the pickup and delivery problem with time windows. Transp. Sci. **40**(4), 455–472 (2006)
4. Schönberger, J.: Operational Freight Carrier Planning. Springer, Berlin (2005)
5. Schwind, M., Gujo, O., Vykoukal, J.: A combinatorial intra-enterprise exchange for logistics services. Inf. Syst. e-Bus. Manag. **7**(4), 447–471 (2009)
6. Wang, X., Kopfer, H.: Collaborative transportation planning of less-than-truckload freight. OR Spectr. **36**(2), 357–380 (2014)
7. Ziebuhr, M., Kopfer, H.: The integrated operational transportation planning problem with compulsory requests. Proceedings of ICCL 2014. Lecture Notes in Computer Science (LNCS), vol. 8760, pp. 1–15. Springer, Berlin (2014)

A Multi-compartment Vehicle Routing Problem for Livestock Feed Distribution

Levent Kandiller, Deniz Türsel Eliiyi and Bahar Taşar

Abstract In the well-known Vehicle Routing Problem (VRP), customer demands from one or more depots are to be distributed via a fleet of vehicles. Various objectives of the problem are considered in literature, including minimization of the total distance/time traversed by the fleet during distribution, the total cost of vehicle usage, or minimizing the maximum tour length/time. In this study, we consider a multi-compartment VRP with incompatible products for the daily solution of a livestock feed distribution network, where each livestock farm requests one type of feed from a single depot, and the vehicles have several compartments. The objective is to minimize the total cost of distribution. Although VRP is a well-studied problem in literature, multi-compartment VRP is considered only by few authors, and our problem differs from the existing ones due to special operational constraints imposed by the situation on hand. We formulate a basic mathematical model for the problem and present possible extensions. We design a computational experiment for testing the effects of uncontrollable parameters over model performance on a commercial solver and report the results. The proposed model can easily be adapted to other distribution networks such as food and fuel/chemicals.

1 Introduction and Literature Review

Vehicle Routing Problem (VRP) of the logistics field is involved with the distribution of goods to a set of customers via a fleet of vehicles. The problem was proposed by Dantzig and Ramser [2] as a generalization of the well-known Traveling Salesman Problem (TSP), and shown to be NP-hard [6]. In this study, we consider a rich VRP, which involves separated compartments on vehicles that can carry incompatible

L. Kandiller (✉) · D.T. Eliiyi · B. Taşar
Department of Industrial Engineering, Yasar University, Izmir, Turkey
e-mail: levent.kandiller@yasar.edu.tr

D.T. Eliiyi
e-mail: deniz.eliiyi@yasar.edu.tr

B. Taşar
e-mail: bahar.tasar@yasar.edu.tr

© Springer International Publishing Switzerland 2017   149
K.F. Dœrner et al. (eds.), *Operations Research Proceedings 2015*,
Operations Research Proceedings, DOI 10.1007/978-3-319-42902-1_20

products for one or more customers while minimizing total cost. Despite the vast amount of literature on VRP, relatively few studies focused on multi-compartment VRP; Van der Bruggen et al. [8] studied the multi-compartment assignment problem for a large oil company. They considered the number and size of trucks and driver schedules. Avella et al. [1] introduced a multi-period routing problem where each compartment had to be completely full or empty. Suprayogi et al. [7] worked on a multiobjective delivery problem in Indonesia and East Timor. They used sequential insertion and local search algorithms. El Fallahi et al. [4] focused on the distribution of cattle food to farms by deciding the fleet size. They proposed a memetic algorithm and tabu search. Derigs et al. [3] introduced food and petrol distribution examples, a general mathematical formulation for the corresponding VRP, and several heuristic approaches. They assumed that an order should be assigned to exactly one compartment on a vehicle. Henke et al. [5] proposed an mathematical model for the flexible size MCVRP (multi-compartment VRP) and developed neighborhood heuristics for the solution of large size problems.

The next section presents our problem definition and mathematical formulation together with some possible extensions.

2 Problem Definition and Formulation

One of the largest agricultural companies in Turkey seeks a solution for their distribution problem in the Aegean Region. Based on daily demands for different types of feeds, the company delivers to turkey breeding farms in the region. A farm demands only one type of feed, depending on the growth stage of the turkeys at the farm. There is a single depot located in Izmir, Turkey, and all vehicles having several compartments (silos) depart daily from this depot for distribution. The fleet is homogenous and a silo on a vehicle can carry only one type of feed at a time. We assume that all vehicles are available at all times; i.e., there are no breakdowns. The vehicles travel at the same speed. All drivers work in 8 h shifts. Loading and unloading times are included in the travel times. We propose a basic mixed integer programming model for the real-life MCVRP defined above. All decision variables are binary except the one related to time. The vehicle index $k = 1, \ldots, K$ indicates the number of vehicles and the compartment index $l = 1, \ldots, L$ reflects the number of compartments in each vehicle. The customers are indexed by $i, j = 1, \ldots, N$, and the products $p = 1, \ldots, P$.

Parameters

f_k: Fixed cost for vehicle k (TL)
C_{lk}: Capacity of compartment l of vehicle k (tons)
D_{ip}: Demand of customer i for product type p (tons)
s_{ij}: Distance from customer i to customer j (km)
α: Fuel cost (TL/km)

T_k: Time capacity of vehicle k (min)

T_{ij}: Time from customer i to customer j (min)

Decision Variables

$$x_{ki} = \begin{cases} 1, & \text{if vehicle } k \text{ serves farm } i \\ 0, & \text{otherwise.} \end{cases}$$

$$y_k = \begin{cases} 1, & \text{if vehicle } k \text{ is used} \\ 0, & \text{otherwise.} \end{cases}$$

$$u_{ij} = \begin{cases} 1, & \text{if route from customer } i \text{ to customer } j \text{ is used} \\ 0, & \text{otherwise.} \end{cases}$$

$$z_i = \begin{cases} 1, & \text{if route from depot to customer } i \text{ is used} \\ 0, & \text{otherwise.} \end{cases}$$

$$v_i = \begin{cases} 1, & \text{if route from customer } i \text{ to depot is used} \\ 0, & \text{otherwise.} \end{cases}$$

$$a_{lkp} = \begin{cases} 1, & \text{if compartment } l \text{ of vehicle } k \text{ is used for product type } p \\ 0, & \text{otherwise.} \end{cases}$$

$t_{ij}^k = $ The time taken by vehicle k from customer i to customer j

Based on the above definitions, the mathematical model is as follows:

$$\text{Minimize } \sum_{k=1}^{K} f_k y_k + \alpha \sum_{i=1}^{N} (s_{i0} v_i + s_{0i} z_i + \sum_{j=1}^{N} s_{ij} u_{ij}) \tag{1}$$

subject to

$$\sum_{i=1}^{N} x_{ki} \leq N y_k \qquad \forall k \tag{2}$$

$$\sum_{p(i)=1}^{N} D_i x_{ki} \leq \sum_{l=1}^{L} C_{lk} a_{lkp} \qquad \forall k, \forall p \tag{3}$$

$$\sum_{p=1}^{P} a_{lkp} \leq y_k \qquad \forall l, \forall k \tag{4}$$

$$\sum_{k=1}^{K} x_{ki} = 1 \qquad \forall i \tag{5}$$

$$z_j + \sum_{i=1, i \neq j}^{N} u_{ij} = 1 \qquad \forall j \tag{6}$$

$$v_i + \sum_{j=1, i \neq j}^{N} u_{ij} = 1 \qquad \forall i \tag{7}$$

$$\sum_{i=1}^{N} z_i = \sum_{k=1}^{K} y_k \tag{8}$$

$$\sum_{i=1}^{N} v_i = \sum_{k=1}^{K} y_k \tag{9}$$

$$x_{ki} - (1 - u_{ij}) \leq x_{kj} \qquad\qquad \forall k, \ \forall i, j, \ i \neq j \tag{10}$$

$$x_{kj} - (1 - u_{ij}) \leq x_{ki} \qquad\qquad \forall k, \ \forall i, j, \ i \neq j \tag{11}$$

$$t_{ij}^k \geq T_{ij} - T_{ij}(2 - x_{ki} - u_{ij}) \qquad\qquad \forall k, \ \forall i, j, \ i \neq j \tag{12}$$

$$t_{0j}^k \geq T_{0j} - T_{0j}(2 - x_{kj} - z_j) \qquad\qquad \forall k, \ \forall j \tag{13}$$

$$t_{i0}^k \geq T_{i0} - T_{i0}(2 - x_{ki} - v_i) \qquad\qquad \forall k, \ \forall i \tag{14}$$

$$\sum_{k=1}^{K} t_{ij}^k \leq T_{ij} u_{ij} \qquad\qquad \forall i, j, \ i \neq j \tag{15}$$

$$\sum_{k=1}^{K} t_{0j}^k \leq T_{0j} z_j \qquad\qquad \forall j \tag{16}$$

$$\sum_{k=1}^{K} t_{i0}^k \leq T_{i0} v_i \qquad\qquad \forall i \tag{17}$$

$$\sum_{j=1}^{N} t_{0j}^k + \sum_{i=1}^{N} t_{i0}^k + \sum_{i=1}^{N} \sum_{j=1}^{N} t_{ij}^k \leq T_k \qquad\qquad \forall k \tag{18}$$

$$\sum_{i,j \in S} u_{ij} \leq |S| - 1 \qquad\qquad \forall S \subset \{1, \dots, N\} \tag{19}$$

$$x_{ki}, y_k, z_i, v_i, u_{ij} \in 0, 1 \qquad\qquad \forall k, \ \forall i, j, \ i \neq j \tag{20}$$

$$a_{lkp} \in 0, 1 \qquad\qquad \forall k, \ \forall l, \ \forall p \tag{21}$$

$$t_{ij}^k, t_{0j}^k, t_{i0}^k \geq 0 \qquad\qquad \forall k, \ \forall i, j, \ i \neq j \tag{22}$$

The objective function in (1) minimizes the total transportation cost including fixed costs of the vehicles and fuel costs. Constraint set (2) ensures that a vehicle can serve customers only if it is in use. Constraint set (3) imposes capacity restrictions. Constraint set (4) ensures that there can be only one type of product in one silo of a vehicle, while constraint set (5) guarantees that a customer can be served by only one vehicle. Constraint sets (6) and (7) guarantee flow balance. Constraint sets (8) and (9) ensure that the total number of departures (arrivals) from (to) the depot is equal to the total number of vehicles used. Constraint sets (10) and (11) guarantee continuity of the routes. Constraint set (12)–(14) determine travel times. Constraint sets (15)–(17) relate time and route variables. While constraint set (18) imposes time limits for the routes, constraint set (19) enforces sub-tour elimination. Finally, constraint sets (20)–(22) dictate the structures and sign restrictions of the decision variables. As the simple VRP is NP-hard, so is our problem.

For evaluating the possibility of allowing overtime at a higher wage rate, a non-negative continuous decision variable is required as o_k : overtime of vehicle k (min). Constraint (18) should be updated as (18′).

$$\sum_{j=1}^{N} t_{0j}^k + \sum_{i=1}^{N} t_{i0}^k + \sum_{i=1}^{N}\sum_{j=1}^{N} t_{ij}^k \leq T_k + o_k \qquad \forall k \quad (18')$$

Overtime of each vehicle should be penalized in the objective function, replacing (1) with (1′), where c_o is the unit cost of overtime:

$$\text{Minimize} \sum_{k=1}^{K} \left(f_k y_k + c_o o_k \frac{f_k}{T_k} \right) + \alpha \sum_{i=1}^{N} \left(s_{i0} v_i + s_{0i} z_i + \sum_{j=1}^{N} s_{ij} u_{ij} \right) \qquad (1')$$

If multiple trips per vehicle are allowed, an alias vehicle for each possible trip of the same vehicle is created. We define a trip index r for each vehicle where $r = 1, \ldots, R$. With the assumption that a vehicle can perform at most R trips during its shift, the number of vehicles in the model is replaced as $k = 1, \ldots, K, K + 1, \ldots, 2K, 2K + 1, \ldots, RK$. A new constraint should be formed to ensure that, if a vehicle can only perform its next tour if it performs its previous one:

$$y_{kr} \leq y_k \qquad \forall k \quad (23)$$

Finally, to satisfy the time constraint for the same vehicle in different trips, constraint (18) should be updated as (18″):

$$\sum_{j=1}^{N} (t_{0j}^k + t_{0j}^{k+r}) + \sum_{i=1}^{N} (t_{i0}^k + t_{i0}^{k+r}) + \sum_{i=1}^{N}\sum_{j=1}^{N} (t_{ij}^k + t_{ij}^{k+r}) \leq T_k \qquad \forall k \quad (18'')$$

3   Computational Experiment and Results

A pilot experiment is designed to evaluate the performance of the basic model including several factor levels. The number of customers is set as 10 and 15. The locations of the customers are generated uniformly over a 200 by 200 grid, and Euclidean distances are computed between each pair of nodes. The single depot is located at the origin. The demand and product type of each customer is generated from a discrete uniform (DU) distribution, between 1–8 (low variance, low demand), 4–12 (low variance, medium demand) and 1–16 (high variance). After generating the demand for each instance, the minimum number of identical 16-ton vehicles is calculated. The fuel cost is assumed as 1.4 TL/km, while the fixed cost of a vehicle is 200 TL/day and 400 TL/day. Ten instances are generated from each setting and Table 1 shows the summary of the results.

Table 1 Analysis of basic model run results

Setting		Avg. obj.	Max obj.	Min obj.	Avg. CPU	Max CPU	Min CPU	Avg. V.	Max V.	Min V.
$N=10$	$D=1\text{--}16$	$f_k=400$	4202.42	5455.25	3208.92	61.13	108.67	23.17	7	9	5
$N=10$	$D=1\text{--}16$	$f_k=200$	2902.39	3655.25	2208.92	68.56	111.83	30.00	7	9	5
$N=15$	$D=1\text{--}16$	$f_k=400$	5908.12	7328.18	4360.76	67.56	98.43	33.43	9	11	7
$N=10$	$D=1\text{--}16$	$f_k=200$	4068.12	5128.18	2760.76	69.12	109.43	31.29	9	11	7
$N=10$	$D=1\text{--}8$	$f_k=400$	2562.14	2941.97	1923.81	54.55	82.00	19.25	4	4	3
$N=10$	$D=1\text{--}8$	$f_k=200$	1842.27	2141.97	1323.81	45.08	78.25	11.75	4	4	3
$N=15$	$D=1\text{--}8$	$f_k=400$	3330.58	4224.46	2033.31	70.05	109.33	28.67	5	6	4
$N=15$	$D=1\text{--}8$	$f_k=200$	2369.37	3024.46	1233.31	71.73	109.33	29.00	5	6	4
$N=10$	$D=4\text{--}12$	$f_k=400$	4098.74	5227.14	3208.92	70.12	119.17	33.83	6	8	5
$N=10$	$D=4\text{--}12$	$f_k=200$	2838.74	3627.14	2208.92	82.90	123.50	38.83	6	8	5
$N=15$	$D=4\text{--}12$	$f_k=400$	5829.46	7237.26	4818.22	67.98	111.67	23.00	9	11	8
$N=10$	$D=4\text{--}12$	$f_k=200$	3959.70	5037.26	2774.58	75.51	117.00	28.33	9	11	8

In terms of computation time and objective, DU [1, 8] yields the best results for demand, and similar results are expected for the low variance settings computationally. Although DU [1, 8] has low transportation costs, it increases the frequency of orders and there is an economically tradeoff. The combinatorial nature of the problem when the number of customer is increased to 20. Most of the problem instances cannot be solved in reasonable times for this size.

4 Future Work

A multi-compartment vehicle routing problem is considered; a mathematical model and some extensions are proposed. The research is motivated by a real-life distribution problem having several types of incompatible products. As the problem is NP-Hard, only small size instance solutions can be obtained within reasonable time limits even for the basic problem. A heuristic approach is to be developed for obtaining near-optimal solutions in short and practical computation times. For multiple trips per vehicle, split delivery for the customers and overtime considerations, further experimentation will be performed. The developed model can be adapted quite easily to fuel or food delivery.

References

1. Avella, P., Boccia, M., Sforza, A.: Solving a fuel delivery problem by heuristic and exact approaches. Eur. J. Op. Res. **152**(1), 170–179 (2004)
2. Dantzig, G.B., Ramser, J.H.: The truck dispatching problem. Manag. Sci. **6**(1), 80–91 (1959)
3. Derigs, U., Gottlieb, J., Kalkoff, J., Piesche, M., Rothlauf, F., Vogel, U.: Vehicle routing with compartments: applications, modelling and heuristics. OR Spectr. **33**(4), 885–914 (2011)
4. El Fallahi, A., Prins, C., Wolfler, C.: A memetic algorithm and a tabu search for the multi-compartment vehicle routing problem. Comput. Op. Res. **35**(5), 1725–1741 (2008)
5. Henke, T., Speranza, M.G., Wascher, G.: The multi-compartment vehicle routing problem with flexible compartment sizes. Eur. J. Op. Res. **246**, 730–743 (2015)
6. Lenstra, J., Rinnooy Kan, A.: Complexity of vehicle routing and scheduling problems. Networks **11**(2), 221–227 (1981)
7. Suprayogi, Setiawan Komara S., Yamato H.: A local search technique for solving a delivery problem of fuel products. In: Proceedings of the 2nd International Conference on Operations and Supply Chain Management, pp. 18–20 (2007)
8. Van der Bruggen, L., Gruson, R., Salomon, M.: Reconsidering the distribution structure of gasoline products for a large oil company. Eur. J. Op. Res. **81**(3), 460–473 (1995)

Single Vehicle Routing Problem with a Predefined Customer Sequence, Stochastic Demands and Partial Satisfaction of Demands

Epaminondas G. Kyriakidis and Theodosis D. Dimitrakos

Abstract We consider the problem of finding the optimal routing of a single vehicle that starts its route from a depot and delivers a product to N customers that are served according to a particular sequence. The vehicle during its route can return to the depot for restocking. The demands of the customers are random variables with known distributions. The actual demand of each customer is revealed as soon as the vehicle visits the customer's site. It is permissible to satisfy fully or to satisfy partially or not to satisfy the demand of a customer. The cost structure includes travel costs between consecutive customers, travel costs between the customers and the depot and penalty costs if a customer's demand is not satisfied or if it is satisfied partially. A dynamic programming algorithm is developed for the determination of the optimal routing policy. It is shown that the optimal routing policy has a specific threshold-type structure.

1 Introduction

One of the most widely studied problems in combinatorial optimization is the Vehicle Routing Problem (VRP). The objective of the VRP is to minimize the total route cost for one vehicle or for a fleet of identical vehicles that depart from one or several depots and deliver goods to N customers comprising the nodes of a predefined network. The vehicle may also collect expired products from the customers. A first version of the VRP was proposed by Dantzig and Ramser [1].

Much attention has been paid to the capacitated VRP in which the vehicles have limited carrying capacity of the goods that must be delivered (see e.g. Secomandi [11],

E.G. Kyriakidis (✉)
Department of Statistics, Athens University of Economics and Business,
Patission 76, 10434 Athens, Greece
e-mail: ekyriak@aueb.gr

T.D. Dimitrakos
Department of Mathematics, University of the Aegean,
Karlovassi, 83200 Samos, Greece
e-mail: dimitheo@aegean.gr

© Springer International Publishing Switzerland 2017
K.F. Dœrner et al. (eds.), *Operations Research Proceedings 2015*,
Operations Research Proceedings, DOI 10.1007/978-3-319-42902-1_21

Novoa and Storer [7], Goodson et al. [3]). In the last fifteen years some capacitated vehicle routing problems have been studied in which a single vehicle starts its route from a depot and serves N customers according to a predefined sequence. This means that customer 1 is served first, then customer 2 is served, then customer 3 is served and so on. We refer to papers by Yang et al. [12], Kyriakidis and Dimitrakos [4], Minis and Tatarakis [6], Pandelis et al. [8, 9], Dimitrakos and Kyriakidis [2]. Suitable dynamic programming algorithms have been proposed for these problems. It was shown that the structure of the optimal routing strategy is of threshold-type. In all these problems it was assumed that the demands of the customers must be satisfied completely. In the present paper we relax this assumption by allowing the possibility to satisfy partially or not to satisfy the demands of the customers. In the case of partial satisfaction of the demand of a customer a penalty cost is incurred. Specifically, we assume that a vehicle starts its route from a depot loaded with items of a product to its fully capacity and visits N customers according to a predefined sequence $1 \rightarrow 2 \rightarrow \cdots \rightarrow N$. The demands of the customers for the product are random variables with known distributions. The actual demand of a customer becomes known only when the vehicle arrives at his/her site. The vehicle may satisfy fully the demand of a customer or satisfy some part of the demand or not satisfy the demand at all. The vehicle may interrupt its route by going to the depot for replenishment. The total cost consists of (i) travel costs between consecutive customers, (ii) travel costs between customers and the depot and (iii) penalty costs due to unsatisfied demands. A dynamic programming algorithm is constructed for the determination of the optimal routing strategy of the vehicle. It is shown that the optimal routing strategy is characterized for each customer by two critical numbers. Note that the problem studied in the present paper can be considered as a generalization of the problem studied in Kyriakidis and Dimitrakos [5], in which it was assumed that the vehicle visits each customer and satisfies as much demand as possible and then an action is selected that depends on the number of the remaining items of the product carried by the vehicle. A penalty cost was imposed if the demand of the customer was satisfied partially.

The rest of the paper is organized as follows. In Sect. 2 the problem is specified for the case in which the demands of the customers are discrete random variables. A dynamic programming approach is proposed for the determination of the optimal routing strategy. The structure of the optimal routing strategy is given. In Sect. 3 our theoretical results is illustrated by a numerical example. A summary of results and topics for future research are presented in Sect. 4. Note that an extended version of the present paper, in which technical details and many numerical results are included, has been submitted for publication.

2 The Problem When the Demands Are Discrete Random Variables

We assume that a vehicle of capacity Q starts its route from a depot loaded with Q items of a product and visits N customers according to a predefined sequence $1 \rightarrow 2 \rightarrow \cdots \rightarrow N$. The demand of customer $j \in \{1, \ldots, N\}$ for a product is a discrete random variable $\xi_j \in \{0, \ldots, Q\}$ with known distribution. The actual demand of each customer becomes known only when the vehicle visits the customer's site. Let c_{j0} and c_{0j}, $j = 1, \ldots, N$, be the travel cost from customer j to the depot and the travel cost from the depot to customer j, respectively. Let also $c_{j,j+1}$, $j = 1, \ldots, N - 1$, be the travel cost from customer j to customer $j + 1$. These costs can be considered as the costs of the required driver's labor and of the gasoline that the vehicle needs to cover the distances between consecutive customers and the distances between customers and the depot. It is plausible to assume that these costs satisfy the symmetric property and the triangle inequality, i.e. $c_{j0} = c_{0j}$, $j = 1, \ldots, N$ and $c_{j,j+1} \leq c_{j0} + c_{0,j+1}$, $j = 1, \ldots, N - 1$. The road network is depicted in Fig. 1. Suppose that the vehicle arrives at customer's $j \in \{1, \ldots, N - 1\}$ site loaded with $z \in \{0, \ldots, Q\}$ items of the product and suppose that the actual demand of customer is equal to $s \in \{0, \ldots, Q\}$. If $z \geq s$ the possible actions are Action 1_θ, $\theta \in \{0, \ldots, s\}$ and Action 2. Action 1_θ means that the vehicle delivers θ items of the product to customer j and proceeds to customer $j + 1$. Action 2 means that the vehicle delivers s items of the product to customer j, it goes to the depot, restocks with load Q and then visits next customer $j + 1$. If $z < s$ the possible actions are 3_θ ($\theta \in \{0, \ldots, z\}$), $4, 5_\theta$ ($\theta \in \{1, \ldots, s - z\}$) and 6. Action 3_θ means that the vehicle delivers θ items of the product to customer j and then proceeds to customer $j + 1$. Action 4 means that the vehicle delivers z items of the product to customer j, it goes to the depot, restocks with load Q and then visits next customer $j + 1$. Action 5_θ means that the vehicle delivers z items of the product to customer j, it goes to the depot, restocks with load Q, it returns to

Fig. 1 The road network for the problem

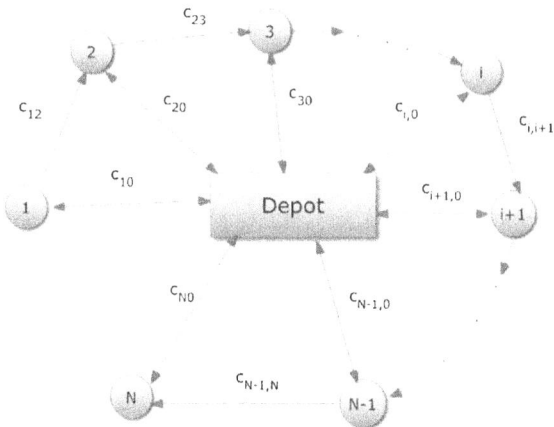

customer j to deliver $\theta \in \{1, \ldots, s - z\}$ owed items and then proceeds to customer $j + 1$. Action 6 means that the vehicle delivers z items of the product to customer j, it goes to the depot to restock with the owed $s - z$ items of the product, it returns to customer j to deliver the owed $s - z$ items of the product, it makes a second trip to the depot to restock with Q items of the product and then goes to next customer $j + 1$. Note that it is assumed that, if Action 5_θ or Action 6 is selected, there is no extra demand when the vehicle returns to customer j, i.e. ξ_j remains unaltered. Suppose that the vehicle arrives at customer's N site and its load is greater or equal to the actual demand of customer N. In this case it satisfies fully the demand and terminates its route by returning to the depot. If the load of the vehicle is less than the actual demand of customer N the possible actions are Action 7 and Action 8. Action 7 means that the vehicle delivers its load to customer N and terminates its route by returning to the depot. Action 8 means that the vehicle delivers its load to customer N, it goes to the depot to restock with the owed quantity, it returns to customer N to deliver the owed quantity and then it terminates its route by returning to the depot. Note that, when Actions 1_θ ($\theta \in \{0, \ldots, s - 1\}$), 3_θ ($\theta \in \{0, \ldots, z\}$), 4, 5_θ ($\theta \in \{1, \ldots, s - z - 1\}$), 7 are selected then some part or the whole of the demand of customer j is not satisfied. In this case a penalty cost is incurred that is equal to π_j per item that is not delivered. The vehicle returns to the depot after servicing (fully or partially) the customers. Our goal is to determine the optimal routing strategy of the vehicle that serves all customers fully or partially. This routing strategy minimizes the expected total cost from the beginning of the route until its end.

Let $f_j(z, s)$, $z, s = 0, \ldots, Q$, denote the minimum expected future cost if the number of items carried by the vehicle when it arrives at customer's $j \in \{1, \ldots, N\}$ site is equal to z and s is the number of items that customer j demands. For $j \in \{1, \ldots, N - 1\}$ this quantity satisfies the following dynamic programming equations (1) and (2) (see Chap. I in Ross [10]). If $z \geq s$ then

$$f_j(z, s) = \min \left\{ H_j(z, s), R_j \right\}, \tag{1}$$

where,

$$H_j(z, s) = c_{j,j+1} + \min_{\theta \in \{0, \ldots, s\}} \left\{ \pi_j(s - \theta) + E f_{j+1}(z - \theta, \xi_{j+1}) \right\},$$

and

$$R_j = c_{0j} + c_{j+1,0} + E f_{j+1}(Q, \xi_{j+1}).$$

If $z < s$ then

$$f_j(z, s) = \min \left\{ A_j(z, s), B_j(z, s), C_j(z, s), D_j \right\}, \tag{2}$$

where,

$$A_j(z, s) = c_{j,j+1} + \min_{\theta \in \{0, \ldots, z\}} \left\{ \pi_j(s - \theta) + E f_{j+1}(z - \theta, \xi_{j+1}) \right\},$$

$$B_j(z, s) = c_{0j} + c_{0,j+1} + \pi_j(s - z) + Ef_{j+1}(Q, \xi_{j+1}),$$

$$C_j(z, s) = 2c_{0j} + c_{j,j+1} + \min_{\theta \in \{1,\dots,s-z\}} \left\{ \pi_j(s - z - \theta) + Ef_{j+1}(Q - \theta, \xi_{j+1}) \right\},$$

$$D_j = 3c_{0j} + c_{0,j+1} + Ef_{j+1}(Q, \xi_{j+1}).$$

The boundary conditions are

$$f_N(z, s) = c_{0N}, \, z \geq s,$$

and

$$f_N(z, s) = c_{0N} + \min \left\{ F_j(z, s), G_j \right\}, z < s, \tag{3}$$

where,

$$F_j(z, s) = \pi_N(s - z),$$

$$G_j = 2c_{N0}.$$

The minimum total expected cost during a visit cycle is equal to

$$f_0 = c_{01} + Ef_1(Q, \xi_1).$$

In the above equations the expected values are taken with respect to the random variables ξ_j, $j = 1, \dots, N$. The terms $H_j(z, s)$ and R_j in the right-hand-side of Eq. (1) correspond to Actions 1_θ ($\theta \in \{0, \dots, s\}$) and to Action 2, respectively. The terms $A_j(z, s)$, $B_j(z, s)$, $C_j(z, s)$, D_j in the right-hand-side of Eq. (2) correspond to Actions 3_θ ($\theta \in \{0, \dots, z\}$), Action 4, Action 5_θ ($\theta \in \{1, \dots, s - z\}$), Action 6, respectively. The terms $F_j(z, s)$ and G_j in the right-hand-side of Eq. (3) correspond to Action 7 and Action 8, respectively. It is possible to prove the following theorem that describes the structure of the optimal routing strategy. The proof is a consequence of the following results: (i) $f_j(z, s)$ is non-increasing in z, (ii) $H_j(z, s)$ is non-decreasing in s and (iii) $A_j(z, s)$, $B_j(z, s)$, $C_j(z, s)$ are non-decreasing in s.

Theorem 1 *For each customer $j \in \{1, \dots, N - 1\}$ and each demand $s \in \{0, \dots, Q\}$ there exist two critical integers $h_1(j, s)$, $h_2(j, s)$ ($h_2(j, s) < s \leq h_1(j, s)$) such that it is optimal to select (i) Action 1_θ for some $\theta \in \{0, \dots, s\}$ if $z \in \{h_1(j, s), \dots, Q\}$ (ii) Action 2 if $z \in \{s, \dots, h_1(j, s) - 1\}$ (iii) Action 3_θ for some $\theta \in \{0, \dots, z\}$ or Action 4 or Action 5_θ for some $\theta \in \{1, \dots, s - z\}$ if $z \in \{h_2(j, s) + 1, \dots, s - 1\}$ and (iv) Action 6 if $z \in \{0, \dots, h_2(j, s)\}$. The critical integers $h_1(j, s)$ and $h_2(j, s)$ are non-decreasing in s.*

Fig. 2 The optimal
decisions for customer 5

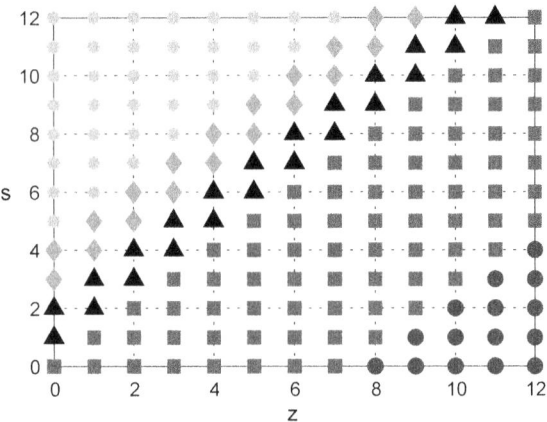

3   Numerical Example

In the following numerical example, we implemented the dynamic programming
algorithm by running the corresponding Matlab program.

Example 1 Suppose that $N = 8$ and $Q = 12$. The travel costs $c_{j,j+1}$ between consecutive customers j and $j + 1$, $j \in \{1, \ldots, 7\}$, are given by: $c_{12} = 5$, $c_{23} = 4$,
$c_{34} = 3$, $c_{45} = 2$, $c_{56} = 5$, $c_{67} = 2$ and $c_{78} = 4$. The travel costs c_{j0} between customers j, $j \in \{1, \ldots, 8\}$ and the depot are given by: $c_{10} = 7$, $c_{20} = 6$, $c_{30} = 5$,
$c_{40} = 4$, $c_{50} = 3$, $c_{60} = 7$, $c_{70} = 6$ and $c_{80} = 5$. Note that these costs satisfy the
triangle inequality. We assume that, for each customer $j \in \{1, \ldots, 8\}$, the penalty
cost π_j is equal to 1.4. We also assume that for each customer $j \in \{1, \ldots, 8\}$, the
demands ξ_j of the customers for the product follow the discrete uniform distribution,
i.e. $Pr(\xi_j = x) = (Q + 1)^{-1}$, $x = 0, \ldots, Q$. In Fig. 2, we present the optimal decisions for customer 5. For $z \geq s$, the action of delivering θ items of the product and
then proceeding directly to next customer (Action 1_θ) is denoted by blue circles and
the action of delivering s items of the product, returning to the depot for restocking
with load Q and then visiting next customer (Action 2) is denoted by red squares.
For $z < s$, we use magenta x-marks for action 3_θ which corresponds to the quantity $A_j(z, s)$, black upper triangles for action 4 which corresponds to the quantity
$B_j(z, s)$, green diamonds for action 5_θ which corresponds to the quantity $C_j(z, s)$ and
cyan stars for action 6 which corresponds to the quantity D_j. It can be seen from Fig. 2
that the structure of the optimal routing strategy is, as expected, of threshold-type
described in Theorem 1. For example, $h_1(5, 1) = 9$ and $h_1(5, 4) = 12$. The value of
the minimum total expected cost f_0 is found approximately equal to 63.82.

4 Summary of Results and Topics for Future Research

In this paper a capacitated vehicle routing problem was studied in which (i) the customers are served according to a predefined sequence, (ii) the demands of the customers are stochastic and each customer's demand is less or equal to the vehicle capacity and (iii) partial satisfaction of demand is allowed. The cost structure included travel costs between consecutive customers, travel costs between customers and the depot and penalty costs due to unsatisfied demands. We selected as decision epochs the epochs at which the vehicle visits for the first time each customer. A dynamic programming approach was proposed for the determination of the optimal routing strategy. It was proved that according to the optimal routing strategy, for a given value of a customer demand, the set of all possible loads carried by the vehicle is divided into four disjoint sets. If the load of the vehicle belongs to the first set, then the optimal action is to satisfy (fully or partially) or not to satisfy the demand of the customer and to proceed to the next customer. If it belongs to the second set, then the optimal action is to satisfy fully the demand of the customer, go to the depot for replenishment and, then proceed to the next customer. If it belongs to the third set, then the optimal action is to satisfy partially the demand of the customer and to proceed (directly or after a trip to the depot for replenishment) to the next customer or to make one trip to the depot for restocking, return to the customer to satisfy fully or partially the demand of the customer and, then proceed to the next customer. If it belongs to the fourth set, then the optimal action is to go to the depot to restock the owed quantity, return to the customer to deliver the owed quantity, make a second trip to the depot for restocking and, then go to the next customer. A possible topic for future research could be the study of a more general problem in which (i) the vehicle delivers new products to the customers, (ii) the vehicle collects expired products from the customers and (iii) partial service is permissible.

References

1. Dantzig, G., Ramser, R.: The truck dispatching problem. Manage Sci **6**, 80–91 (1959)
2. Dimitrakos, T.D., Kyriakidis, E.G.: A single vehicle routing problem with pickups and deliveries, continuous random demands and predefined customer order. Eur J Oper Res **244**, 990–993 (2015)
3. Goodson, J., Ohlmann, J., Thomas, B.: Cyclic-order neighborhoods with application to the vehicle routing problem with stochastic demand. Eur J Oper Res **217**, 312–323 (2012)
4. Kyriakidis, E.G., Dimitrakos, T.D.: Single vehicle routing problem with a predefined customer sequence and stochastic continuous demands. Math Sci **33**, 148–152 (2008)
5. Kyriakidis, E.G., Dimitrakos, T.D. A vehicle routing problem with a predefined customer sequence, stochastic demands and penalties for unsatisfied demands. In: Proceedings of 5th International Conference on Applied Operational Research, *Lecture Notes in Management Science*, 5:10–17, 2013
6. Minis, I., Tatarakis, A.: Stochastic single vehicle routing problem with delivery and pickup and a predefined customer sequence. Eur J Oper Res **213**, 37–51 (2011)

7. Novoa, C., Storer, R.: An approximate dynamic programming approach for the vehicle routing problem with stochastic demands. Eur J Oper Res **196**, 509–515 (2009)
8. Pandelis, D.G., Kyriakidis, E.G., Dimitrakos, T.D.: Single vehicle routing problems with a predefined customer sequence, compartmentalized load and stochastic demands. Eur J Oper Res **217**, 324–332 (2012)
9. Pandelis, D.G., Karamatsoukis, C.C., Kyriakidis, E.G.: Single vehicle routing problems with a predefined customer order, unified load and stochastic discrete demands. Probab Eng Inf Sci **27**(1), 1–23 (2013)
10. Ross, S.M.: An Introduction to Stochastic Dynamic Programming. Academic Press, New York (1983)
11. Secomandi, N.: Comparing neuro-dynamic programming algorithms for the vehicle routing problem with stochastic demands. Comput Oper Res **27**, 1201–1225 (2000)
12. Yang, W.-H., Mathur, K., Ballou, R.H.: Stochastic vehicle routing problem with restocking. Trans Sci **34**, 99–112 (2000)

A Tabu Search Based Heuristic Approach for the Dynamic Container Relocation Problem

Osman Karpuzoğlu, M. Hakan Akyüz and Temel Öncan

Abstract The container relocation problem (CRP) is concerned with clearing out a single yard-bay which contains a fixed number of containers each following a given pickup order so as to minimize the total number of relocations made during their retrieval process. In this work, we consider an extension of the CRP where containers are both received and retrieved at a single yard-bay named Dynamic Container Relocation Problem (DCRP). The arrival and departure sequences of containers are assumed to be known in advance. A tabu search based heuristic approach is proposed to solve the DCRP. Computational experiments are performed on an extensive set of randomly generated test instances from the literature. Our results show that the proposed algorithm is efficient and yields promising outcomes.

1 Introduction

More than 50 % of the world sea borne trade in terms of dollars are carried with containerized cargo [11]. As the emerging technologies increase speeds and sizes of vessels; now, container terminals are forced to transfer larger amounts of containers than before. Therefore, efficient management of container terminals is crucial. The container terminal area can be separated into two parts: quay side area and yard side area. In general, terminal operators give more importance to quay side area operations which include berth allocation, quay crane assignment and scheduling, and vessel storage planning. On the other hand, yard side area operations, which consist of transferring containers from quay side area, yard crane scheduling, and storing and handling of containers at the yard storage area, are equally important in

O. Karpuzoğlu (✉) · M.H. Akyüz · T. Öncan
Department of Industrial Engineering, Galatasaray University,
Çırağan Cad. No: 36, Ortaköy, 34349 Istanbul, Turkey
e-mail: osmankarpuzoglu@gmail.com

M.H. Akyüz
e-mail: mhakyuz@gsu.edu.tr

T. Öncan
e-mail: ytoncan@gsu.edu.tr

© Springer International Publishing Switzerland 2017
K.F. Dœrner et al. (eds.), *Operations Research Proceedings 2015*,
Operations Research Proceedings, DOI 10.1007/978-3-319-42902-1_22

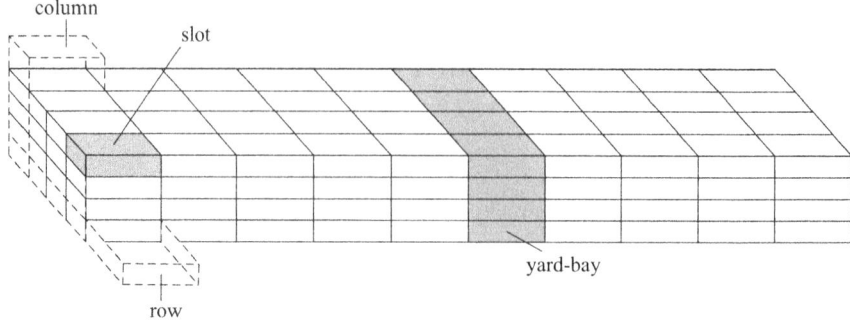

column
slot
yard-bay
row

Fig. 1 A block of containers. *Source* [1]

container terminal management. A yard area includes blocks of containers which is illustrated with Fig. 1.

A yard-bay is served with a yard crane so that containers are received and retrieved at top of the columns. Containers on top of a column are directly accessible for retrieval. However, if a target container (a container that will be retrieved from yard-bay immediately) isn't positioned at top of a column, then all containers above the target container can be relocated to other columns of the yard-bay. Once blocking containers are cleared, target container can be retrieved. These clearing movements are denoted as relocations. Relocations are idle operations for yard cranes. There-fore, yard crane operations should be planned in such a way that the number of relocations are minimized. Given a set of containers in a yard-bay and their retrieval sequence, the Container Relocation Problem (CRP) tries to minimize the total num-ber of relocations made until the yard-bay is empty. A closely related problem is referred to as Dynamic Container Relocation Problem (DCRP) where containers are both received and retrieved at a yard-bay [1]. The DCRP generalizes the CRP which is known to be an NP-hard problem [2] and thus, it is difficult to solve. In this work, we focus on the DCRP and propose an efficient heuristic method to plan operations at a yard-bay. A Tabu Search (TS) based algorithm is suggested for the DCRP. To the best of our knowledge, there are no other work proposing TS algorithms for this problem. Besides, we also present an extensive computational study which shows that the proposed TS algorithm is promising and can be used as an alternative to other heuristics in the literature.

Since the seminal work by Kim and Hong [6], there exist several studies address-ing the CRP. Caserta et al. [2] propose two mathematical formulations and a simple heuristic for the Blocks Relocation Problem (BRP) which is synonymously used to address the CRP. Petering and Hussein [9] propose a new mathematical formulation. They introduce a new look-ahead algorithm that yields better solutions than other algorithms for the CRP. Jovanonic and Voss [5] implement a chain heuristic for BRP where relocation decisions consider properties of the next container to be moved. Recently, Jin et al. [4] developed a greedy look-ahead heuristic and performed exten-sive experiments on existing data sets for the CRP. Unlike the CRP, the DCRP has not received much attention from the researchers. For the first time, Wan et al. [12] introduce the DCRP. In their work, a CRP formulation is suggested and employed to solve both the CRP and DCRP. Akyüz and Lee [1] propose the first mathematical programming formulation for the DCRP and suggest heuristic methods. For an in-depth discussion on container terminal operations and on stacking problems in storage areas, we refer to excellent surveys by Stahlbock and Voss [10] and Lehnfeld and Knust [7], respectively.

The rest of this work consists of the following. In Sect. 2 a formal definition of the DCRP is presented. TS based heuristic is proposed in Sect. 3. Computational results and findings are discussed in Sect. 4. Lastly, Sect. 5 concludes our study.

2   Dynamic Container Relocation Problem

The DCRP is a containers both depart and arrive at the yard-bay. In this work, the following assumptions used by Akyüz and Lee [1] are employed. It is assumed that one yard-crane operates at the yard-bay and can handle one container at a time. The planning horizon of the yard-bay operations are divided into equal time-steps in which only one container arrival or retrieval may occur. The arrival and retrieval time-steps are known a priori. Container relocations occur only at the departure time-step of the outgoing container. In the rest of this paper, we call an arriving (departing) container at the yard-bay as incoming (outgoing) container. Besides a container can be relocated at most once per time-step. No container premarshalling is allowed and all relocations are assumed to be performed within the yard-bay. Containers are assumed to be of the same type and size, i.e., only regular 1 TEU containers. Furthermore, it is assumed that the number of columns is denoted with C and the yard-bay has a height limit given with H. The objective is to minimize the total number of relocations made such that arriving containers are located within the yard-bay and departing containers are retrieved from the yard-bay in the planning horizon. A solution to the DCRP is obtained when all containers are handled.

The difficulty of solving the DCRP can be seen better when a new container arrives at the yard-bay. Observe that, the DCRP reduces to solving a CRP as long as containers depart from the yard-bay until the next container arrival. However, an incoming container is likely to change the retrieval sequence of the existing containers within the yard-bay. Therefore, incoming containers change plans repetitively at each arrival. Now, not only the relocation of the containers but also finding the best location for incoming containers gains importance in order to increase the efficiency of yard cranes in yard-bay planning. For the sake of brevity, the DCRP formulation proposed by Akyüz and Lee [1] is not explicitly given here.

3   Tabu Search Based Heuristic Method

Broadly speaking, TS algorithm moves from one solution to another by changing value of decision variables which represent locations of containers depending on the structure of the problem considered [3]. Unfortunately, changing the value of a decision variable, which represents locations of containers within a yard-bay at a time-step, affects all subsequent container movements to be made in the DCRP. Hence an efficient algorithm is required to restore the feasibility. To this end we employ the Reshuffling Index (RI) heuristic which is tailored for the DCRP. The RI heuristic is presented in the following. Next, we give details of the suggested TS based algorithm, namely TS algorithm (TSA), employed as efficient upper bounding procedure for the DCRP.

RI heuristic is initially used by Murty et al. [8] and adapted for the DCRP by Wan et al. [12]. The RI heuristic is an efficient upper bounding method which is employed in our TS heuristic approach. RI defines the number of relocations that should be made to clear blocking containers above the earliest outgoing container in a column. For each column the RI is calculated and incoming (or relocating) containers are located on top of the column with the lowest RI. Clearly, RI is calculated among the columns which are not full. In case of a tie, the container is placed on top of the highest column. In case of a further tie, the container is randomly assigned to a column. To avoid the randomness of the RI heuristic, the latter tie breaking rule is modified to assigning the container into the leftmost column.

The slot assignment of containers in the yard-bay is denoted as yard-bay configuration. A feasible solution of the DCRP is obtained by keeping track of yard-bay configurations at each time-step. Initially, the RI heuristic is run to obtain a feasible solution. A Tabu List (TL) records the status of columns of the yard-bay for each container. Once a container, denoted by n, is assigned to a column, shown as c, in a feasible solution, then column c is declared as tabu for container n for at least β iterations. Here, β stands for the number of iterations of the tabu duration (tenure). Then, container n can not be positioned at the tabu column c for at least β iterations. The TS heuristic is run for a total number of K iterations. A tabu iteration consists of finding a feasible solution by following the RI heuristic steps described considering the TL which provides diversification of solutions, and hence, the generation of different solutions at each iteration of the TS algorithm. To declare a column c tabu for a container n, α percentage of the incoming containers are randomly selected.

Now, we present the notation used and a generic algorithm for it. Let $TList_{n,c}$ and $\mathcal{C}_n^a$ denote the TL value of container n for column c and the set of available columns on which container n can be placed, respectively. The number of containers that exist in a column c is given by N_c. The total number of time-steps t is shown with T. A formal outline of the TSA is given in Algorithm 1.

4 Computational Experiments

In this section, our computational experiments are presented. The test bed given by Akyüz and Lee [1] is used. There are two groups of instances in the test bed: Group-I and Group-II instances. Each group consists of medium and high density of container traffic at the yard-bay with $C = 6$ columns. The range of height, H, is chosen from the set $\{2, 3, 4, 5, 6\}$ and the number of containers, N, which departs from the yard-bay, is selected from the set $\{5, 50, 100, 200, 400, 800\}$. This makes a total of 60 different combinations for each group of instances. 20 test instances for each combination are randomly generated. Therefore, there are 1200 instances for each group. We refer to the work by Akyüz and Lee [1] for more details on the test bed. In the following, our results obtained by the heuristic method proposed for the DCRP are reported. The experiments are performed on a computer with a Intel Core i7-3630QM CPU 1.40 GHz and 8 GB RAM operating within Microsoft Windows 8 64-bit environment and codes are written in C++.

In Table 1, we summarize the performance of the TSA on Group-I and Group-II instances. The number of TS iterations is set to $K = 10000$. The first column indicates the group and the density of the test instances. The second column gives the size of the test instances so that (C, H) stands for the number of columns and rows (height) in the yard-bay, respectively. The tabu duration parameter β is set to be $\beta = 3$ after our preliminary experiments. The percentage to declare a column as tabu for an incoming container, denoted with parameter α, is calibrated as $\alpha = 2$ and $\alpha = 3$ in

Step 0. *(Initialization):* Set $TList_{n,c} = 0$ for $n = 1, \ldots, N$ and $c = 1, \ldots, C$ and $\mathscr{C}_n^a = \emptyset$ for $n = 1, \ldots, N$. Set tabu iteration number $k = 1$ and time-step $t = 1$.
Step 1. For time-step t if container n is an arriving container then go to Step 2. Otherwise go to Step 6.
Step 2. *(Incoming container):* Check the $TList_{n,c}$ for $c = 1, \ldots, C$ and set $\mathscr{C}_n^a = \{c : TList_{n,c} = 0, N_c < H\}$.
Step 3. If the cardinality of the set $\mathscr{C}_n^a$, $|\mathscr{C}_n^a| > 1$ then use RI heuristic steps to determine the location of container n for columns $c \in \mathscr{C}_n^a$. If $|\mathscr{C}_n^a| = 1$, then place container n to column $c \in \mathscr{C}_n^a$. Otherwise, select the column with lowest $TList_{n,c}$ among the columns satisfying $N_c < H$ to locate container n. Set c^* equals the selected column to place container n.
Step 4. Generate a random number $r \in [0, 100]$. If $r < \alpha$ then $TList_{n,c^*} = \beta + 1$.
Step 5. If $t < T$, set $t = t + 1$ and go to Step 1. Otherwise, go to Step 7.
Step 6. *(Outgoing container):* If there is no container above the outgoing container n then remove container n from the yard-bay configuration and go to Step 5. Otherwise for each container n' above the container n at column c^* repeat the following starting from the highest slot and go to Step 5. Check the $TList_{n',c}$ for all $c = 1, \ldots, C, c \neq c^*$ and set $\mathscr{C}_{n'}^a = \{c : TList_{n',c} = 0, N_c < H, c \neq c^*\}$. If the cardinality of the set $\mathscr{C}_{n'}^a$, $|\mathscr{C}_{n'}^a| > 1$ then use RI heuristic steps to determine the location of container n' for columns $c \in \mathscr{C}_{n'}^a$. If $|\mathscr{C}_{n'}^a| = 1$, then place container n' to column $c \in \mathscr{C}_{n'}^a$. Otherwise, select the column with lowest $TList_{n',c}$ among the columns satisfying $N_c < H$ to locate container n'.
Step 7. If $k < K$ then set $k = k + 1$ a feasible solution is found and update the best upper bound accordingly. Set $TList_{n,c} = TList_{n,c} - 1, t = 1$ and go to Step 1. Otherwise, stop and report the best upper bound.
Algorithm 1: TS Algorithm

Table 1 Summary of the performance of the TSA on Group-I and Group-II instances

Instance group	Size (C, H)	$\alpha = 2, \beta = 3$		$\alpha = 3, \beta = 3$		$\alpha = 2, \beta = 3$	RI	
		UB	CPU	UB	CPU	UB (4 s.)	UB	CPU
Group-I medium	(6, 2)	**0.04**	1.19	**0.04**	0.92	**0.04**	0.11	0.08
	(6, 3)	2.24	1.41	2.3	1.16	**2.2**	3.37	0.08
	(6, 4)	24.68	1.72	24.9	1.66	**24.37**	25.41	0.08
	(6, 5)	63.03	2.22	64.03	2.34	62.33	**57.6**	0.08
	(6, 6)	113.86	2.71	115.16	2.94	113.55	**98.84**	0.09
Group-I high	(6, 2)	25.41	1.2	25.96	0.93	**25.05**	30.57	0.08
	(6, 3)	83.42	1.72	83	1.42	**82.16**	86.3	0.09
	(6, 4)	144.05	2.31	144.15	2.2	143.19	**138.65**	0.09
	(6, 5)	221.84	3.13	222.48	3.09	221.38	**211.95**	0.1
	(6, 6)	293.23	3.68	293.64	3.98	292.51	**276.32**	0.11
Group-II medium	(6, 2)	4.91	0.87	4.92	1.02	**4.85**	5.24	0.07
	(6, 3)	44.67	1.29	44.83	1.36	**44.45**	45.88	0.08
	(6, 4)	92.41	2.12	92.35	1.76	**91.62**	91.63	0.08
	(6, 5)	121.82	2.57	122.24	2.16	122.08	**117.45**	0.09
	(6, 6)	171.92	3.28	171.8	2.7	171.22	**158.92**	0.1
Group-II high	(6, 2)	71.24	1.06	71.68	1.27	**70.81**	76.48	0.08
	(6, 3)	139.43	1.65	139.6	1.8	**139.23**	139.89	0.09
	(6, 4)	200.71	2.5	201.55	2.17	200.16	**188.72**	0.1
	(6, 5)	278.15	3.46	278.81	2.85	278.13	**247.38**	0.11
	(6, 6)	368.43	4.14	369.25	3.67	368.43	**315.57**	0.13

the light of our initial experiments. The columns "UB" and "CPU" indicate the total number of relocations and the CPU times in seconds, respectively. Each cell gives the average of $20 \times 6 = 120$ test instances with different numbers of containers, N. Columns 3 to 6 include the results when $\alpha = 2$, $\beta = 3$ and $\alpha = 3$, $\beta = 3$. In column 7, we remove the limit on number of tabu iterations and impose a time limit of 4 seconds to run the TSA algorithm with the same parameters $\alpha = 2$, $\beta = 3$. The last two columns present the performance of the original RI heuristic whose results are taken from [1] for comparison. The best outcomes are shown with bold characters for each row. Clearly, the performance of TSA increases as the number of TS iterations (or CPU time limit) increases. Observe that the RI heuristic is more efficient than the TSA. It is observed that the suggested TSA performs better than the RI heuristic for $H = 2$ and $H = 3$ on all instances. Moreover, for Group-I and Group-II instances with medium density having a height of $H = 4$ the TSA yields better outcomes than the RI heuristic.

5 Conclusion

In this work, we address the DCRP and suggest a TS based heuristic algorithm. The algorithm, TSA, uses a random selection strategy for tabu declarations for that purpose. To the best of our knowledge, this is the first work which employs TS for the DCRP. According to our computational experiments, it is observed that the proposed TS Algorithm is efficient and yields promising outcomes where it can be used as an alternative to previous heuristic methods. Efforts to design other meta-heuristic algorithms are worthwhile and remain as open research area. The tabu search algorithm to obtain better results worth further research.

Acknowledgements This research is supported by the Turkish Scientific and Technological Research Council (TÜBİTAK) Research Grant no: 214M222, and Galatasaray University Scientific Research Project Grant no: 13.402.010.

References

1. Akyüz, M.H., Lee, C.: A mathematical formulation and efficient heuristics for the dynamic container relocation problem. Nav. Res. Log. **61**, 101–118 (2014)
2. Caserta, M., Schwarze, S., Voss, S.: A mathematical formulation and complexity considerations for the blocks relocation problem. Eur. J. Oper. Res. **219**, 96–104 (2012)
3. Glover, F., Laguna, M.: Tabu search. Kluwer academic publishers, Boston (1997)
4. Jin, B., Zhu, W., Lim, A.: Solving the container relocation problem by an improved greedy look-ahead heuristic. Eur. J. Oper. Res. **240**, 837–847 (2015)
5. Jovanovic, R., Voss, S.: A chain heuristic for the blocks relocation problem. Comput. Ind. Eng. **75**, 79–86 (2014)
6. Kim, K.H., Hong, G.P.: A heuristic rule for relocating blocks. Comput. Oper. Res. **33**, 940–954 (2006)
7. Lehnfeld, J., Knust, S.: Loading, unloading and premarshalling of stacks in storage areas: survey and classification. Eur. J. Oper. Res. **239**, 297–312 (2014)
8. Murty, K., Liu, J., Wan, Y., et al.: A decision support system for operations in a container terminal. Decis. Support. Syst. **39**, 309–332 (2005)
9. Petering, M.E.H., Hussein, M.I.: A new mixed integer program and extended look-ahead heuristic algorithm for the block relocation problem. Eur. J. Oper. Res. **231**, 120–130 (2013)
10. Stahlbock, R., Voss, S.: Operations research at container terminals: a literature update. OR Spectr. **30**, 1–52 (2008)
11. UNCTAD (2014) Review of Maritime Transportation 2014. Paper presented at the United Nations Conference on Trade and Development, New York and Geneva. http://unctad.org/en/PublicationsLibrary/rmt2014_en.pdf. Cited 28 June 2015
12. Wan, Y., Liu, J., Tsai, P.: The assignment of storage locations to containers for a container stack. Nav. Res. Log. **56**, 699–713 (2009)

Optimization of Railway Timetable by Allocation of Extra Time Supplements

Takayuki Shiina, Susumu Morito and Jun Imaizumi

Abstract We consider the allocation of a running time supplement to a railway timetable. Previously, Vekas et al. developed a stochastic programming model. In this paper, their optimization model is improved by adding bound constraints on the supplements. It is shown that the probability of delays decreases when using the proposed model. In addition, an effective L-shaped algorithm is presented.

1 Introduction

In Japan, there is high demand for travel by railway, and this is especially true during weekday rush hours, when the trains become highly congested due to commuters who work in the metropolitan areas. There is a positive correlation between the congestion rate and the times at which travelers embark and disembark at train stations.

On the other hand, it is important to determine an appropriate supplement to the running times between stations and arrival times at each station, in order to create a schedule that is robust against unanticipated delays. In the present study, we develop a way to distribute the running time supplement in such a way that the train can operate according to the schedule, in spite of delays.

Vekas et al. [10] examined the optimal way to allocate the running time supplement for a train. The uncertain disturbances were modeled as a random variable using stochastic programming [2, 3, 6] model. In their model, it was assumed that there was an upper limit to the total supplement, but its allocation was not restricted. In this paper, we suggest an improvement to the previous model and present a new mathematical programming model in which there is a constraint on the running time

T. Shiina (✉) · S. Morito
Waseda University, 3-4-1 Okubo, Shinjuku-ku, Tokyo 169-8555, Japan
e-mail: tshiina@waseda.jp

S. Morito
e-mail: morito@waseda.jp

J. Imaizumi
Toyo University, 5-28-20 Hakusan, Bunkyo-ku, Tokyo 112-8606, Japan
e-mail: jun@toyo.jp

© Springer International Publishing Switzerland 2017   173
K.F. Dœrner et al. (eds.), *Operations Research Proceedings 2015*,
Operations Research Proceedings, DOI 10.1007/978-3-319-42902-1_23

supplement allocated to each trip; this is done in order to minimize the expected delay. In addition, we compare the expected total delay for each model and examine the differences in the delay ratios.

We assume the existence of a railway line with a sequence of $T + 1$ stations, and the stations are assigned numbers $0, 1, \ldots, T$. Thus, we will consider T trips between stations. The expected total delay time of a train traveling from station 0 to station T is set as the objective function that is to be minimized.

To formulate the problem, we make the following assumptions.

1. A stop event in station $t - 1$ and the running event between station $t - 1$ and t are brought together in one event, and this is defined as trip $t (1 \leq t \leq T)$. Each trip corresponds to a stage in the multistage stochastic programming [1, 4, 8].
2. A disturbance that occurs in trip t is defined by the random variable $\tilde{\xi}_t$. It is assumed that $\tilde{\xi}_1, \ldots, \tilde{\xi}_T$ are independent.
3. The total running time supplement is limited, and the amount that can be distributed to each trip has both a lower bound and an upper bound.
4. The actual delay in a trip can be calculated by subtracting the running time supplement from the sum of the delay of the previous trip and the disturbance of the present trip.

2 Formulation of the Problem

In the formulation of the problem, we will use the following notation.

Parameters

T	Number of trips
k_t	Number of realizations of disturbances in trip t
K_t	Number of stage t scenarios $K_t = \prod_{i=1}^{t} k_i$
p_t^s	Probability of stage t scenario s
U_t	Upper bound on the running time supplement allocated to trip t
L_t	Lower bound on the running time supplement allocated to trip t
M	Upper bound on total running time supplement

Random Variables

$\tilde{\xi}_t$	Time of a disturbance in trip t
ξ_t^s	Realization of time of disturbance in trip t for stage t scenario s
$\alpha(t, s)$	Predecessor of stage t scenario s; $\alpha(1, s) = 1$
Ξ_t	Support of $\tilde{\xi}_t$

The decision variables are defined as follows.

First-stage Variable

x_t	Amount of running time supplements exceeding L_t that are allocated to trip t

Stage t Variable

y_t^s Upper bound for actual delay in stage t scenario s

We assume the stochastic elements are defined over a finite discrete probability space $(\Xi, \sigma(\Xi), P)$, where $\Xi = \Xi_1 \times \cdots \times \Xi_T$ is the support of the random data in each stage, with $\Xi_t = \{\xi_t^s, s = 1, \ldots, k_t\}$. The possible sequences of the realization of random variables $(\xi_1, \ldots, \xi_T)$ are called scenarios. The scenarios are often described using a scenario tree, as shown in Fig. 1. In stages $t \leq T$, we have a limited number of possible realizations, which we call the stage t scenarios.

In a scenario tree, the stage t scenario connected to the stage $t - 1$ scenario s is referred to as a successor of the stage $t - 1$ scenario s. The set of all successors of the stage $t - 1$ scenario s is denoted by $D^t(s)$. Similarly, the predecessor of the stage t scenario s is denoted by $\alpha(s, t)$. These relationships are illustrated in Fig. 2.

The stochastic programming problem is formulated as follows.

$$\min \sum_{t=1}^{T} \sum_{s=1}^{K_t} p_t^s Q_t^s (y_{t-1}^{\alpha(t,s)}, x_t, \xi_t^s) \tag{1}$$

Fig. 1 Scenario tree

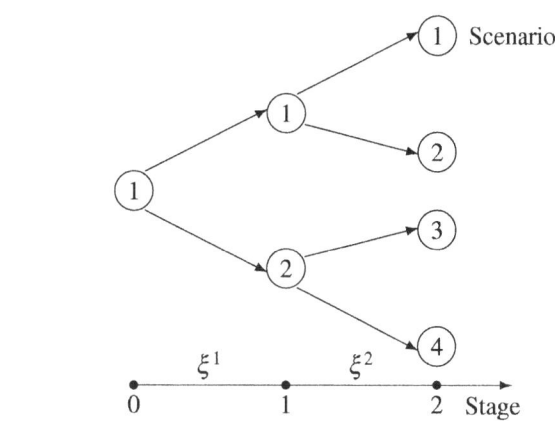

Fig. 2 Successor and predecessor

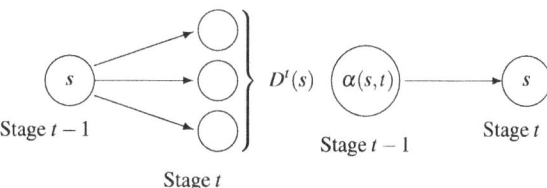

subject to $Q_t^s(y_{t-1}^{\alpha(t,s)}, x_t, \xi_t^s) = \min\{y_t^s \mid y_t^s \geq y_{t-1}^{\alpha(t,s)} + \xi_t^s - (x_t + L_t), y_t^s \geq 0\},$

$$t = 1, \ldots, T, s = 1, \ldots, K_t \tag{2}$$

$$\sum_{t=1}^{T}(x_t + L_t) \leq M \tag{3}$$

$$(x_t + L_t) \leq U_t, t = 1, \ldots, T \tag{4}$$

$$y_0^1 = 0 \tag{5}$$

$$x_t \geq 0, y_t^s \geq 0, t = 1, \ldots, T, s = 1, \ldots, K_t \tag{6}$$

$$\xi_t^s \in \Xi_t, t = 1, \ldots, T, s = 1, \ldots, K_t$$

The objective function (1) minimizes the expectation of the total delay. Constraint (2) represents the definition of the recourse function that denotes the recursion of the actual delay. The actual delay $y_t^s (\geq 0)$ can be calculated by subtracting the running time supplement $x_t + L_t$ from the sum of the delay of the previous trip $y_{t-1}^{\alpha(t,s)}$ and the disturbance of the present trip ξ_t^s. Constraint (3) shows that the total running time supplement is bounded by M. Constraint (4) expresses the upper and lower bounds on the running time supplement for trip t. Constraint (5) indicates the initial delay. Constraints (6) ensure that the variables are nonnegative.

The L-shaped algorithm [7, 9] can be applied to the problem. The Master problem uses y as the upper bound of the expected value for the recourse function.

$$\text{(Master): min} \quad \sum_{t=1}^{T}\sum_{s=1}^{K_t} p_t^s y_t^s \tag{7}$$

$$\text{subject to} \quad \sum_{t=1}^{T}(x_t + L_t) \leq M \tag{8}$$

$$(x_t + L_t) \leq U_t, t = 1, \ldots, T \tag{9}$$

$$x_t \geq 0, y_t^s \geq 0, t = 1, \ldots, T, s = 1, \ldots, K_t \tag{10}$$

The relation between the next y and the recourse function is not included in this problem, and so the optimal cuts are added to a later iteration.

$$y_t^s \geq Q_t^s(y_{t-1}^{\alpha(t,s)}, x_t, \xi_t^s) \tag{11}$$

The Master problem is a linear programing (LP) problem, and various types of algorithms can be used to seek the optimal solution. Here, the Master problem is solved. Next, the solution to the Master problem $(\hat{x}, \hat{y})$ is employed to solve a second-stage problem that defines the recourse function. The dual problem in the second-stage problem is defined as follows.

$$Q_t^s(y_{t-1}^{\alpha(t,s)}, x_t, \xi_t^s) = \min\{y_t^s \mid y_t^s \geq y_{t-1}^{\alpha(t,s)} + \xi_t^s - (x_t + L_t), y_t^s \geq 0\} \quad (12)$$

$$= \max\{\pi_s^t\{y_{t-1}^{\alpha(t,s)} + \xi_t^s - (x_t + L_t)\} \mid 0 \leq \pi_t^s \leq 1\} \quad (13)$$

The second-stage problem has complete recourse, and so this is a feasible problem. In other words, there exist feasible $y_t^s \geq 0$ in an arbitrary ξ_t^s. Also, as long as $\hat{y}$ is not at the upper bound of the recourse function, that is, when $\hat{y} < Q_t^s(\hat{y}_{t-1}^{\alpha(t,s)}, \hat{x}_t, \xi_t^s)$, the dual optimal solution $\hat{\pi}_t^s$ of the second-stage problem can be used to generate the following optimality cut, and this is added to the Master problem.

$$y_t^s \geq \hat{\pi}_s^t\{y_{t-1}^{\alpha(t,s)} + \xi_t^s - (x_t + L_t)\} \quad (14)$$

The Master problem is solved using the L-shaped algorithm, and the cut is added; this is repeated until $\hat{y}_t^s \geq \hat{y}_{t-1}^{\alpha(t,s)} + \xi_t^s - (\hat{x}_t + L_t)$ is satisfied.

When the solution $\hat{x}$, $\hat{y}$ that is obtained in the solution to the Master problem satisfies $\hat{y}_t^s \geq \hat{y}_{t-1}^{\alpha(t,s)} + \xi_t^s - (\hat{x}_t + L_t)$, we can assume that we have obtained an acceptable approximation of the recourse function $Q_t^s(y_{t-1}^{\alpha(t,s)}, x_t, \xi_t^s)$ in the neighborhood of $(\hat{x}, \hat{y})$. Then, we find a first-stage feasible solution, solve a second-stage problem using the first-stage solution, and find the expected value for the recourse function. The L-shaped algorithm for obtaining the optimal solution can be summarized as follows.

L-Shaped Algorithm

Step 1: Solve Master Problem   Solve the Master problem. Let $\hat{x}_t, t = 1, \ldots, T, \hat{y}_t^s$, $s = 1, \ldots, K_t, t = 1, \ldots, T$ be the optimal solution of the Master problem.

Step 2: Solve Recourse Problem   Solve the recourse problem for stage t scenario $s, s = 1, \ldots, K_t, t = 1, \ldots, T$.

Step 3: Add Optimality Cuts   Calculate $Q_t^{s'}(\hat{y}_{t-1}^s, \hat{x}_t, \xi_t^{s'}), \forall s' \in D^t(s), s = 1, \ldots,$ $K_t, t = 0, \ldots, T - 1$. If $\hat{y}_t^{s'} < (1 - \varepsilon)Q_t^{s'}(\hat{y}_{t-1}^s, \hat{x}_t, \xi_t^{s'}), s' \in D^t(s)$, the optimality cut (14) is added to the formulation of Master problem ($\varepsilon > 0$: tolerance). Go to Step 1.

Step 4: Convergence Check   If no optimality cuts are added, then stop.

3   Numerical Examples

The running time supplement was allocated by using the model proposed in the present study. The number of trips in the rail line was set to six. The maximum number of disturbances per trip was set to eight. It was assumed that there was a possibility of either a big or a small disturbance in any trip. The expected values of these big and small disturbance were assumed to be 60 and 20 s, respectively. We assumed three cases: big disturbances occur in each of two of the early trips, big disturbances occur in each of two of the central trips, and big disturbances occur in

Table 1 Result of examples

Position of big disturbance	Forward	Forward	Central	Central	Backward	Backward
Bound constraints	Yes	No	Yes	No	Yes	No
Trip 1 $x_1 + L_1$	100	108	55	33	55	33
Trip 2 $x_2 + L_2$	100	166	55	51	55	55
Trip 3 $x_3 + L_3$	85	55	100	122	55	55
Trip 4 $x_4 + L_4$	55	33	100	156	55	41
Trip 5 $x_5 + L_5$	55	55	85	55	100	166
Trip 6 $x_6 + L_6$	55	33	55	33	100	100
Objective	22.10	14.71	21.89	14.60	19.39	13.42
Probability of delay occurrence (%)	23.44	33.01	23.44	48.71	23.44	33.01
Computing time (s)	42	55	29	57	6	46

each of two of the final trips. The upper bound M on the total running time supplement for the entire rail line was assumed to be 450 s. The bounds for the supplement were set to $L_t = 55$ and $U_t = 100$. The probability of a disturbance in each trip was set to $1/k_t$ for all trips. The disturbances followed a discretized exponential distribution with expected value $1/\lambda_t$.

Experiments were carried out using AMPL [5] CPLEX on a Xeon E5507 2.00 GHz (two processors; memory: 12.0 GB). Table 1 shows the results of the experiment with three disturbance cases. In this table, two types of optimization models which include bound constraints for each trip or not are compared. The models without bound constraints coincide with the previous model [10]. The value of $x_t + L_t$ denotes the planned time for trip t. The total supplement time was equal to the upper bound M, except for one case that had a big disturbance in the final trips and with bound constraints. The probability of delay is the probability of the scenario in which the delay is caused in either of trip in the rail line.

It can be seen that the probability that a delay occurs decreases when upper and lower bounds are added, even though the value of the objective function increases; with upper and lower bounds, the objective function increased by a factor of 1.5. Because the expectation of the total delay is small compared to the upper bound on the total supplement $M = 450$, it has little influence on the schedule. The reason for considering the probability of delay is as follows. In Tokyo area, one of the major feature of the transport by railway is that trains coming from the suburbs, proceed into a different railway lines in the city center, and also go to another direction. The direct operation like this enables passengers to reach wide areas without transfer. But the direct operation may increase the probability of delay which cause a big delay.

In a previous model [10], allocation of the running time supplement was biased because it was not allocated to all trips. We balanced the supplements for each trip by adding upper and lower bounds. The fluctuations of the supplements for each trip

Table 2 Large scale problems

Upper bound U_t	∞	∞	100	100
Lower bound L_t	0	55	0	55
Computing time (s) of deterministic equivalent MIP	43810	8904	17561	1086
Computing time (s) of L-shaped	10846	3539	1018	672

thus became small, and the probability of a delay thus decreased. A certain amount of supplement is necessary for safe operation and as a buffer against delays. An overcrowded schedule can be avoided by setting a lower bound on the supplement for each trip.

Here, we compare the calculation times using the L-shaped algorithm and branch-and-bound for a deterministic equivalent MIP on large-scale problems. The number of scenarios are increased to 12. So the total number of scenarios becomes 12^6 which is approximately three million. Both methods showed calculation times increasing with the problem scale, but the L-shaped algorithm had shorter times (Table 2).

4   Concluding Remarks

In the present study, we showed that our method reduced the probability of delay for a rail line, even though the objective function was slightly larger than that of a previous optimization model. Moreover, it was shown that the problem can be solved effectively by using the L-shaped method.

References

1. Birge, J.R.: Decomposition and partitioning methods for multistage stochastic linear programs. Oper. Res. **33**, 989–1007 (1985)
2. Birge, J.R.: Stochastic programming computation and applications. INFORMS J. Comput. **9**, 111–133 (1997)
3. Birge, J.R., Louveaux, F.V.: Introduction to Stochastic Programming. Springer, Berlin (1997)
4. Birge, J.R., Donohue, C.J., Holmes, D.F., Svintsitski, O.G.: A parallel implementation of the nested decomposition algorithm for multistage stochastic linear programs. Math. Program. **75**, 327–352 (1996)
5. Fourer, R., Gay, D.M., Kernighan, B.W.: AMPL: a modeling language for mathematical programming. Scientific press, South San Francisco (1993)
6. Kall, P., Wallace, S.W.: Stochastic Programming. Wiley, New York (1994)
7. Shiina, T.: L-shaped decomposition method for multi-stage stochastic concentrator location problem. J. Oper. Res. Soc. Jpn. **43**, 317–332 (2000)

8. Shiina, T., Birge, J.R.: Multi-stage stochastic programming model for electric power capacity expansion problem. Jpn. J. Indus. Appl. Math. **20**, 379–397 (2003)
9. Van Slyke, R., Wets, R.J.-B.: L-shaped linear programs with applications to optimal control and stochastic linear programs. SIAM J. Appl. Math. **17**, 638–663 (1969)
10. Vekas, P., van der Vlerk, M.H., Klein Haneveld, W.K. (2012). Optimizing existing railway timetables by means of stochastic programming. Stochastic Programming E-print Series

Train Platforming at a Terminal and Its Adjacent Station to Maximize Throughput

Susumu Morito, Kosuke Hara, Jun Imaizumi and Satoshi Kato

Abstract To cope with growing passenger demand and to provide better services, increasing the number of trains is desired for certain lines. Maximum possible throughput of a line is often limited due to the limited number of platforms at the terminal, which was the case with two bullet-train lines originating from Tokyo. In these lines, there exists an intermediate station in the close vicinity of the terminal. This paper proposes a network-based optimization model to analyze the throughput of a line when turn-backs at the adjacent station are introduced to increase the line capacity.

1 Introduction

Terminal stations in intercity rail transportation are often equipped with many platforms, and trains go into/out of these stations frequently. Reduction of headway time, however, would be limited due to technical reasons of signal equipment as well as safety reasons, and an appropriate methodology will be required to fully utilize the capacity of these terminal stations. These terminal stations are often so-called stub stations, i.e., dead-end stations where trains change their direction. Time for passengers to get off/on a train together with time for clean-up would be needed, but it is not desired for trains to unnecessarily occupy platforms.

We focus on lines in which an intermediate station exists in the close proximity of the dead-end terminal, and propose a mathematical optimization model to quantitatively evaluate the effects of utilizing an intermediate station on the throughput

S. Morito (✉) · K. Hara · S. Kato
Waseda University, 3-4-1 Ohkubo, Shinjuku, Tokyo 169-8555, Japan
e-mail: morito@waseda.jp

J. Imaizumi
Toyo University, 5-28-20 Hakusan, Bunkyo, Tokyo 112-8606, Japan
e-mail: jun@toyo.jp

© Springer International Publishing Switzerland 2017
K.F. Dœrner et al. (eds.), *Operations Research Proceedings 2015*,
Operations Research Proceedings, DOI 10.1007/978-3-319-42902-1_24

of a line. The model gives departure and arrival times of trains as well as platform assignments at the terminal and its adjacent station during one hour horizon assuming a cyclic timetable, so that the number of trains of the line is maximized. The experimental evaluation is performed based on the two bullet-train lines in Tokyo, and effects of such factors as stopping time at stations and the minimum headway time on line throughput are analyzed.

Assigning trains to platforms has been studied as a train platforming problem (TPP) in Europe where train operators are independent from companies which manage infrastructures such as tracks and stations. See, e.g., Billionnet [1].

In Japan, platform assignments are generally separated by in-bound/out-bound and also by lines, and also there exist no need for adjustment of different companies involved. Rather, high-frequency train operations and passenger convenience of transfers between trains are more important. Imaizumi, Kitakoga, and Morito [3] and Wakisaka and Masuyama [2] considered dead-end terminals and studied the relationship between the maximum number of trains that can be run and such factors as the minimum headway time and the minimum turn-back time at the terminal.

This paper, based on Imaizumi, Kitakoga, and Morito [3], considers not only a terminal but also a near-by intermediate station and presents a model that determines arrival and departure times of trains and also platform assignments so that the total number of trains that can be run is maximized. The model thus has both features of time tabling and platform assignments, based on which throughput capacity of the two stations, namely, a terminal and a near-by intermediate station, is evaluated.

2   Throughput Maximization Problem

Below we list basic assumptions of the throughput maximization problem:

1. There exist a terminal and its adjacent station as in Fig. 1a.
2. In Fig. 1, there exist in-bound trains which move from right to left and out-bound trains which move from left to right.
3. At the terminal, all in-bound trains turn back and become out-bound trains.
4. The number of platforms at each station is given, together with track layout (including location of points) around the stations. Platform numbering and track layout of a terminal and of an intermediate station are shown in Fig. 1b and in Fig. 1c, respectively.
5. A part of in-bound trains may turn back at the intermediate station. All other in-bound trains stop at an intermediate station for a fixed amount of time and then head for the terminal. On the other hand, all out-bound trains that leave the

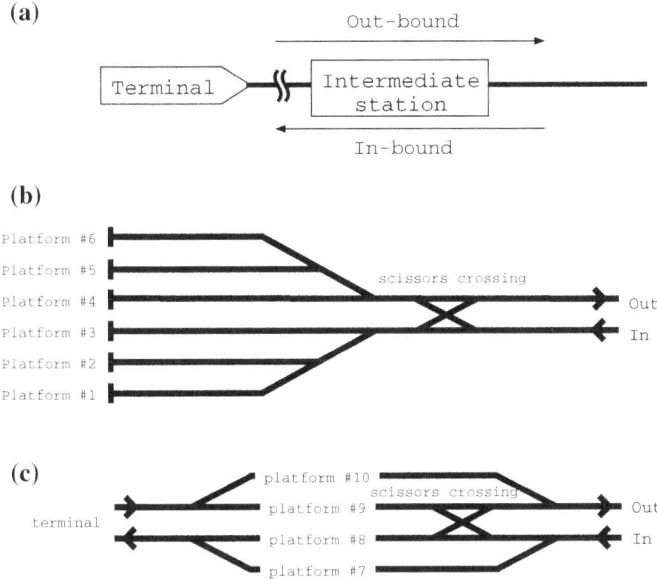

Fig. 1 Terminal, intermediate station, and track layout. **a** A terminal and an intermediate station. **b** Platforms and track layout of the terminal. **c** Platforms and track layout for an intermediate station

terminal stop at the intermediate station for a fixed amount of time. We assume that turn-back at an intermediate station is allowed only at platforms #8 and #9.

6. Turn-back at the terminal or at the intermediate station requires a fixed amount of time. That is, the arrival time of in-bound train and the associated departure time of the out-bound train after turn-back must be separated by at least a fixed amount of time. It is assumed that arriving trains never change parking platform.

7. Time between a platform and the switch is assumed to be 0.

8. Time between the terminal and its adjacent station is a given constant.

9. Any two successive trains of the same direction must be separated by the minimum headway time.

10. Any two trains of the opposite directions passing through crossover structure must be separated by headway limited by crossover structure, which we call the minimum crossover time.

11. We consider a discrete time model.

3 Network Model

State transitions of the system are expressed by taking time as a horizontal axis and platforms as a vertical axis as in Figs. 2 and 3. Each node then corresponds to a status of a platform at the time, and each arc the state transition.

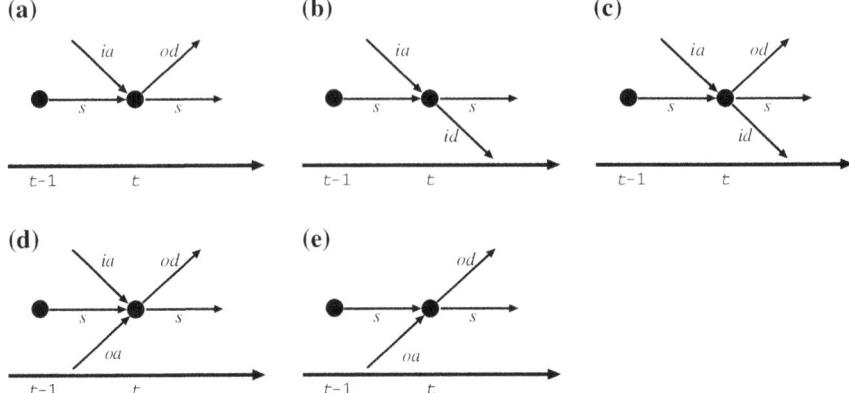

Fig. 2 Arc connections to node. **a** Platform of terminal. **b** Platform #7 of intermediate station. **c** Platform #8 of intermediate station. **d** Platform #9 of intermediate station. **e** Platform #10 of intermediate station

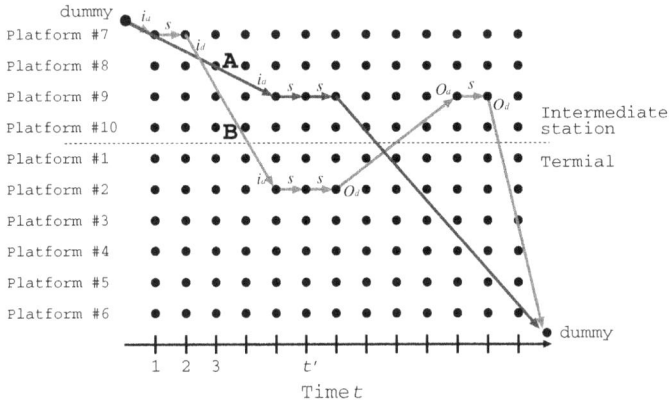

Fig. 3 Network

At a terminal, the following transitions are possible: i_a = arrival of an in-bound train, o_d = departure of an out-bound train, s = stoppage at platform. At an intermediate station, on top of these two transitions of the terminal, two more transitions are possible: i_d = departure of an in-bound train, o_a = arrival of an out-bound train.

A network model will be constructed by (1) considering state transitions from each node and drawing the corresponding directed arcs, and then (2) adding dummy start and end nodes. A path between these dummy nodes indicates the movement of a rolling stock.

For example, path A in Fig. 3 shows movement of a rolling stock which turns back at the intermediate station, whereas path B in Fig. 3 movement of a rolling stock which turns back at the terminal. Note that a rolling stock remains at a station for some time when it is assigned to a different train after turn-back.

We now define notations used in our formulation.

Sets

P_{term}	a set of platforms at the terminal ($P_{term} = \{1 \ldots 6\}$)
P_{int}	a set of platforms at an intermediate station ($P_{int} = \{7 \ldots 10\}$)
T	a set of discrete time ($T = \{0 \ldots t_{max}\}$)

Constants

t_{move}	time between the terminal and an intermediate station
u	minimum turn-back time both at the terminal and an intermediate station
s_{int}	stoppage time at an intermediate station
h	headway time
c	crossover time
α, β	weight for the 1st term and 2nd term in the objective function, respectively

Variables

$x_{p,t}^{o_d}$	departure from platform p of the terminal at time t
$x_{p,t}^{i_a}$	arrival at platform p of the terminal at time t
$x_{p,t}^{term}$	stoppage at platform p of the terminal at time t
$y_{p,t}^{i_a}$	in-bound arrival to platform p of an intermediate station at time t
$y_{p,t}^{i_d}$	in-bound departure from platform p of an intermediate station at time t
$y_{p,t}^{o_a}$	out-bound arrival to platform p of an intermediate station at time t
$y_{p,t}^{i_d}$	out-bound departure from platform p of an intermediate station at time t
$y_{p,t}^{int}$	stoppage at platform p of an intermediate station at time t

The problem then is formulated as a 0–1 integer program. All variables are 0–1 variables and take value 1 if the associated transition occurs, and 0 otherwise. The formulation below only shows typical constraints and similar constraints are omitted to save space.

$$\text{Maximize } \alpha \times \sum_{p \in P_{term}} \sum_{t \in T} x_{t,p}^{i_a} + \beta \times \sum_{p \in P_{int}} \sum_{t \in T} y_{t,p}^{o_d} \tag{1}$$

Subject to

$$\sum_{t \in T} (y_{7,t}^{o_a} + y_{7,t}^{o_d} + y_{8,t}^{o_a} + y_{9,t}^{i_d} + y_{10,t}^{i_a} + y_{10,t}^{i_d}) = 0 \tag{2}$$

$$x_{p,t-1}^{term} + x_{p,t}^{i_a} - x_{p,t}^{o_d} - x_{p,t}^{int} = 0, \, p \in P_{term}, \forall t \in T \tag{3}$$

$$x_{p,t-1}^{term} + x_{p,t}^{i_a} \leq 1, \, p \in P_{int}, \forall t \in T \tag{4}$$

$$y_{p,t'}^{i_d} - y_{p,t'-s_{int}}^{i_a} \leq 0, \, p = 7, 8, \forall t' \in T \tag{5}$$

$$y_{p,t'}^{i_a} + \sum_{t=t'}^{t'+u-1} y_{p,t}^{o_d} \leq 1, \, p = 8, 9, \forall t' \in T \tag{6}$$

$$\sum_{p=7}^{8} y_{p,t'}^{i_d} - \sum_{p \in P_{\text{term}}} x_{p,t'+t_{\text{move}}}^{i_a} = 0, \forall t' \in T \tag{7}$$

$$\sum_{p \in P_{\text{term}}} x_{p,t'}^{o_d} - \sum_{p=9}^{10} y_{p,t'+t_{\text{move}}}^{o_a} = 0, \forall t' \in T \tag{8}$$

$$\sum_{t=t'+1}^{t'+c-1} y_{9,t}^{i_a} \leq M(1 - y_{8,t'}^{o_d}), \forall t' \in T \tag{9}$$

$$\sum_{p \in P_{\text{term}}} \sum_{t=t'}^{t'+h-1} x_{p,t}^{i_a} \leq 1, \forall t' \in T \tag{10}$$

Equation (1) is the objective to maximize throughput. Two alternative measures of performance could be considered for throughput. One is the number of out-bound trains from the intermediate station, which we call the "line throughput," and the other the number of out-bound trains at the terminal, which we call the "terminal throughput." Equation (1) is the weighted sum of the line and terminal throughputs with the corresponding weights of α and β. We currently use 1 and 0.01 for these weights, where the type of throughput with weight 1 becomes a primary objective, whereas the one with weight 0.01 becomes a secondary objective. Which option to choose depends on which throughput measure is more important to the operator.

Equation (2) is a constraint of applicable platforms due to track layout of an intermediate station. Equation (3) is one of the flow balance constraints at platform. Equation (4) is one of two constraints which allow only one train to occupy a platform. Equation (5) is one of the two constraints of stoppage time at an intermediate station. Equation (6) is one of two constraints of turn-back time. Equations (7) and (8) link states of the terminal with those of the intermediate station. There exist three constraints (can be combined into two) like (9) to consider crossover time at an intermediate station and also three (again, can be combined into two) other similar constraints to consider crossover time at the terminal. Here we note that a departure of an out-bound train immediately after an arrival of an in-bound train is permitted in practice. This means that crossover time can be 0 for the sissors crossing of this type, and thus we do not set a constraint for the type of crossover. Equation (10) is one of 6 constraints on headway time.

4 Numerical Experiments

Numerical experiments are performed based on two cases, Case X (Tokyo and Shinagawa of Tokaido Shinkansen) and Case Y (Tokyo and Ueno of Touhoku Shinkansen). We note that in Case Y, the track layout of the intermediate station is slightly different from Case X in that the scissors crossing is located on the out-bound side (that is, the non-terminal side) of the two single switches outside the intermediate station, that is to the right-most position of Fig. 1c. This necessitates an addition of 2 more constraints like (9) to consider scissors crossing.

Parameter setting for $(|P_{term}|, |P_{int}|, t_{move}, u, s_{int}, h, c)$ is (6, 4, 7, 17, 1, 3, 3) for Case X, and (4, 4, 6, 12, 1, 4, 4) for Case Y, unless otherwise specified. These are the parameter settings currently used in practice. The results below are based on $\alpha = 1$ and $\beta = 0.01$, which gives higher priority to the terminal throughput, unless otherwise stated.

Effects of turn-backs at an intermediate station on throughput were studied first assuming the basic parameter settings reflecting the current practice. For Case X, the maximum hourly throughput remains at 18 whether or not turn-backs are allowed at the intermediate station. The results remain the same even if we change the priority in the objective function to $\alpha = 0.01$ and $\beta = 1$, which gives higher priority to the line throughput. However, for Case Y, the throughput of 12 without turn-backs at an intermediate station is increased to 15 when turn-backs are allowed. Out of 15, three utilize intermediate turn-backs. Again, the results remain the same when the priority of the terminal and line throughputs has been switched. The differences of the effects of turn-backs at the two cases are estimated to be due to differences of the number of platforms, together with the differences of the minimum turn-back time.

Some of the parameters might be adjusted with relative ease (than others). The model allows easily to analyze effects of changing parameter values on throughput. Table 1 shows throughputs of Case X when the turn-back time has been changed, which also shows throughputs when turn-backs are allowed/not allowed at the intermediate station. The following observations are in order. First, platforms at the terminal and tracks between the two stations are already fully utilized and one could not expect more throughput originating from the terminal without regard to the values of turn-back time. Second, when turn-back time is increased (which improves passenger service), throughput could be increased by utilizing turn-backs at the intermediate station.

Similar analysis could be made for Case Y which is omitted due to space limitation.

Table 1 Throughputs of Case X for different values of minimum turn-back time

Minimum turn-back time		12	15	16	17	18	19	20
Turn-back only at terminal: # departure from terminal		20	18	18	18	12	12	12
Turn-back possible at both stations	# departures from terminal	20	18	18	18	12	12	12
	# departures from intermediate	0	2	1	0	4	4	4

5 Conclusions

This paper presented a network-based optimization model to analyze throughput when turn-backs at an adjacent station of the terminal are introduced to increase throughput of a given line. The throughput as defined by the two parameters α and β is tried to be maximized by generating a one-hour cyclic timetable which also determines whether a train turns back at the terminal or at the intermediate station, together with assigning platforms to trains at both stations. The model can be used to observe, under the specified throughput measure, not only the effects of allowing intermediate turn-backs on throughput, but also the effects of changing parameter values.

Acknowledgements This work was supported by JSPS Kakenhi Grant No. 26350437.

References

1. Billionnet, A.: Using integer programming to solve the train-platforming problem. Trans. Sci. **37**, 213–222 (2003)
2. Wakisaka, K., Masuyama, S.: Automatic construction of train arival and departure schedules at terminal stations. J. Adv. Mech. Des. Syst. Manuf. **6–5**, 590–599 (2012)
3. Imaizumi, J., Kitakoga, K., Morito, S.: Capacity evaluation of a railway terminal by network flow model. Commun. Oper. Res. Soc. Jpn **56**, 237–243 (2011). (in Japanese)

The Baltic Sea as a Maritime Highway in International Multimodal Transport

Joachim R. Daduna and Gunnar Prause

Abstract The introduction of the *Sulphur Emission Control Areas* is expected to lead to increasing costs in maritime transportation with a high impact on *Short Sea Shipping* in the North Sea and the Baltic Sea. As a consequence a modal shift in multimodal container transport might occur towards road and rail freight transport, which is related to negative ecological effects. Simultaneously, a downside risk of the North Range harbours exist in international container transport due to their competition with the south European harbours. Based on a case study it will be analysed how and to what extent the new frame conditions influence the container business and what consequences can be expected.

1 Introduction

In the North Sea and the Baltic Sea maritime transport, especially multimodal container transport plays an important role as part of the global transport links but also due to the regional transport chains. Since the introduction in January 2015 of the *Sulphur Emission Control Area* (SECA) for reduction of emissions[1] for this region the frame conditions for transport mode selection have been changed. One important question is how and to which extent the changed bunker costs might impact the logistical patterns. But the related economic decisions have to take into account different criteria for modal choice comprising distance, transport time and costs as well as ecological aspects. Literature review reveals that the considered decision variables might differ between the studies [2, 6].

---

[1] See the *International Convention for the Prevention of Pollution from Ships* (MARPOL) with Annex VI in the reviewed version from 2010.

J.R. Daduna (✉)
Berlin School of Economics and Law, Badensche Str. 52, 10825 Berlin, Germany
e-mail: daduna@hwr-berlin.de

G. Prause
Tallinn University of Technology, School of Economics and Business
Administration, Akadeemia Tee 3, 12618 Tallinn, Estonia
e-mail: gunnar.prause@ttu.ee

© Springer International Publishing Switzerland 2017
K.F. Dœrner et al. (eds.), *Operations Research Proceedings 2015*,
Operations Research Proceedings, DOI 10.1007/978-3-319-42902-1_25

Consequently, it has to be analysed in this context if the increase of costs leads to a modal shift from maritime freight transport to road and/or rail freight transport (see e.g. the different results in Notteboom [8] and Holmgren et al. [5]). Additionally, also the related impact of the competition between the North Sea and the Baltic Sea harbours, on one side, and the Eastern Mediterranean and Black Sea harbours, on the other side, have to be discussed with a long-term view of economic efficiency. The future development depends on existing technical alternatives and their related costs.

The paper begins with the consideration of different technical possibilities to comply with the SECA regulations. Afterwards the impact of increased bunker costs on transport system decisions will be discussed in order to assess the possibilities of a modal split in container transport.

2   Technical Solutions for Compliance with SECA Limits

The currently discussed solutions for compliance with SECA limits are presented below. It is important to take into account the related investment and operational costs as well as the necessary realization period for providing a satisfying infrastructure (see e.g. Czermaski et al. [3], 10 pp; Bergqvist et al. [1]).

- Retrofitting of *sulphur scrubbers* together with the use of *heavy fuel oil* (HFO) for a transition period for those ships navigating in SECA areas (see e.g. Jiang et al. [7]).
- Use of nearly sulphur-free *Marine Gas Oil* (MGO).
- Reconstruction of ships with *Liquefied Natural Gas* (LNG) engine which is linked to investments for necessary bunkering and refuelling stations.
- Use of *renewable liquid fuel* (bio oil) with a sulphur share of less than 0.1 %.
- Use of *wind energy* as additional measure to compensate for the increased bunker costs, i.e. the installation of *Flettner Rotors*.

The use of MGO represents only a short-term solution among the five alternatives which can be realized on a larger scale. All others are related to investments in ships and land based infrastructure, i.e., these solutions require long-term realization. Consequently, further research will focus on the use of MGO and its influence on the modal choice in *the multimodal container transport*.

3   Case Study

The assessment of the impact of using MGO will be done by studying a case from Daduna et al. [4] where a container transport between two destinations can be forwarded along different routes and in different combination of the transport modes. This approach allows the analysis of costs as well as of ecological aspects. The starting point of the analysis is the transport of a 40'-ISO-Container from *Leeds* (GB) to

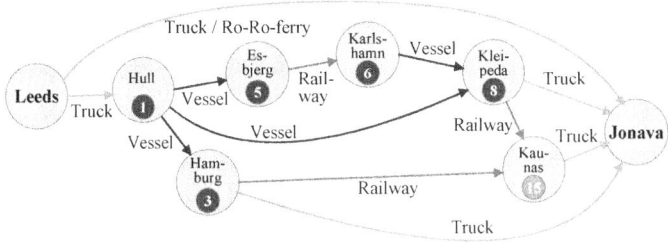

Fig. 1 Connection graph Leeds (GB)—Jonava (LT)

Table 1 Alternative transport routes between Leeds (GB) and Jonava (LT)

(1)	Leeds–(Truck/RoRo-ferry)—Jonava
(2)	Leeds–(Truck)–Hull–(Sea going vessel)–Hamburg–(Truck)—Jonava
(3)	Leeds–(Truck)–Hull–(Sea going vessel)–Hamburg–(Railway)–Kaunas–(Truck)—Jonava
(4)	Leeds–(Truck)–Hull–(Sea going vessel)–Klaipeda–(Truck)—Jonava
(5)	Leeds–(Truck)–Hull–(Sea going vessel)–Klaipeda–(Railway)–Kaunas–(Truck)—Jonava
(6)	Leeds–(Truck)–Hull–(Sea going vessel)–Esbjerg–(Railway)–Karlshamn–(Sea going vessel)–Klaipeda–(Truck)—Jonava
(7)	Leeds–(Truck)–Hull–(Sea going vessel)–Esbjerg–(Railway)–Karlshamn–(Sea going vessel)–Klaipeda–(Railway)–Kaunas–(Truck)—Jonava

Jonava (LT) leading to a connection graph and seven specified transport alternatives which are depicted in Fig. 1 and Table 1 (see e.g. [4]).

The calculations are based on transport data of a 40'-ISO-Container (see e.g. [4], only the data of alternative 1 has been updated due to actual information about RoRo ferry tariffs). The use of exact (route based) data is, apart from distances, only possible to a limited extent. Transport times depend strongly on situational factors like topography, weather conditions or the traffic situation; idle times for modal change have to be added. Cost data are estimations as the pricing results from negotiations depending e.g. on transport volume or transport frequency. The calculation of CO_2 emissions, especially for land based transportation, depends on topography, the technical vehicle standard, infrastructure condition as well as the type of energy generation in case of E-traction.

The starting point of the analysis is a scenario based on the use of HFO before the introduction of the new limits of MARPOL Annex VI. Relating to the initial situation four additional scenarios are considered based on different strong increases of bunker costs. Firstly, increases of bunker costs of 44 and 51 % are analysed, which are in accordance with the forecasts of Bergqvist et al. [1], and after this cost increases of 100 and 200 %. The aim of the research is to analyse tentative modal shifts of maritime transport.

4   Results

The analysis of the considered case leads to the results of Table 2 describing the *length of the routes* (differentiated by transport mode), the *transport time*, the accrued *costs* and the CO_2 emissions.

A closer look at the data of Table 2 reveals some mode-specific tendencies, which are partly contradicting the public perception. In detail it is possible to extract the following results from Table 2 (see Daduna et al. [4]):

- *Transport time*: A high proportion of transportation services by truck leads to a shorter transport time due to the direct links and the high flexibility. A growing share of rail freight transport extends significantly the transport time.
- *Costs*: High shares of road and rail transport cause higher costs. A large proportion of Short Sea Shipping (SSS) yields lower cost constellations.
- *CO_2 emissions:* A high share of road freight transport operations results in significantly increasing emission values whereas rail freight transport and SSS lead to lower emissions.

In case of the price increases of the MGO the question emerges: starting from which price level a modal shift in favour of road and rail transport occurs. The application of the four above mentioned price levels for MBO results are presented in Table 3.

Table 2   Results of the calculations using HFO

	Distances (km)				Dura-tion (hours)	Costs (€)	CO_2 emissions[1]
	Truck	**Rail**	**Vessel**	**Total**			
1	2,270		80	2,350	116	2,700	148.99
2	1,410		710	2,120	137	1,880	104.43
3	130	1,320	710	2,160	205	2,200	51.59
4	340		1,710	2,050	119	1,130	52.88
5	130	320	1,710	2,160	136	1,250	46.59
6	340	500	1,020	1,860	168	1,870	51.96
7	130	820	1,020	1,970	185	1,990	45.67
[1] kg per t/km							

Table 3   Comparison of cost level for different scenarios

	0 %		**44 %**		**51 %**		**100 %**		**200 %**	
1	2,700	7	2,720	7	2,725	7	2,740	7	2,780	7
2	1,880	4	1,960	3	1,980	3	2,060	3	2,240	3
3	2,200	6	2,280	6	2,300	6	2,380	6	2,560	6
4	1,130	1	1,320	1	1,350	1	1,560	1	1,990	1
5	1,250	2	1,440	2	1,470	2	1,680	2	2,110	2
6	1,870	3	1,990	4	2,005	4	2,140	4	2.410	4
7	1,990	5	2,110	5	2,125	5	2,260	5	2,530	5

Table 4 Multi-criterial comparison using the Savage-Niehans rule

	Duration (hours)	Costs (€)	CO_2 emissions	Rating
1	0.000	1.389	2.262	3.651
2	0.181	0.664	1.287	2.132
3	0.767	0.947	0.130	1.844
4	0.026	0.000	0.158	0.184
5	0.172	0.106	0.020	0.298
6	0.448	0.655	0.138	1.241
7	0.595	0.761	0.000	1.356

The calculations reveal that even significant price increases of MBO do not generate substantial changes in the ranking of the seven alternatives. In contrast, the maximum cost differences of the alternatives decrease from 139 to 40 %, but this does not deliver an economic reason for a modal shift from sea freight to road and/or rail freight transport.

So the analysis produces a unique solution concerning the *cost criterion*. By taking additionally into account the *transport time* and the CO_2 *emissions* it turns out that the available data are not generating a unique solution due to missing comparability of the different criteria. Consequently, standardization is necessary for achieving a multi-criterial comparison, which can be realized by applying the *Savage Niehans* rule (also: *Minimax regret rule*), which sets all values of the different alternatives in relation to each best value. The result of this calculation for the starting situation is shown in Table 4.

The application of the *Savage-Niehans* rule to all alternatives identifies the alternatives 4 and 5 as the best solutions. The disadvantage of alternative 5 compared to alternative 4 consists of the longer transport time, which can be compensated by an earlier departure time, i.e., if only costs and CO_2 emissions are included into the decision then the ranking will change. The calculation of the ratings for all five cases is shown in Table 5.

The assessment of the calculations reveals that even significantly higher bunker costs for the use of MGO have no impact on the mode selection. Additionally, a further rise or fall of the oil prices will also have no consequences since all price

Table 5 Comparison of the scenarios considering costs and emissions

	0 %		44 %		51 %		100 %		200 %	
1	3.651	7	3.323	7	3.281	7	3.018	7	2.659	7
2	1.951	6	1.472	6	1,857	6	1.608	6	1.413	6
3	1.077	5	0.857	5	0.730	5	0.656	5	0.416	5
4	0.158	2	0.158	2	0.158	2	0.158	2	0.158	2
5	0.126	1	0.111	1	0.109	1	0.099	1	0.080	1
6	0.793	4	0.653	4	0.623	4	0.510	4	0.349	4
7	0.761	3	0.598	3	0.574	3	0.449	3	0.271	3

changes will influence all transport modes equally. In the long run advantages for rail transport may appear in case of using electrical traction under the precondition of cheap and sufficiently available renewable energy.

Proceeding in this way, it is possible to determine alternative decisions, which enable the decision-maker to include problem-specific preferences (e.g. costs, emissions) in the sense of generating a Pareto-optimal solution. Based on invited offers from logistics service providers a detailed calculation of the possible alternatives must take place to allow for the making of the final decision regarding the preferred multimodal transportation route.

5 Conclusions

After the introduction of SECA regulations in the areas of the North Sea and the Baltic Sea, the Baltic Sea Region will keep its importance as a maritime highway. Modal shifts are not expected with the exception of the very short route sections where they might appear. These conclusions are generally valid for all sea freight transport in the frame of multimodal transport chains *within* the considered area. More research has to be done on the regional integration within intercontinental shipping routes as well as on the assessment of unequal competition conditions due to different emission regulations in the North Sea and the Baltic Sea, on one side, and in the Mediterranean and the Black Sea, on the other side. The research question in this context will be if and to what extent changes will appear in the transport demand of the container traffic.

References

1. Bergqvist, R., Turesson, M., Weddmark, A.: Sulphur emission control areas and transport strategies-The case of Sweden and the forest industry. Eur. Trans. Res. Rev. **7**(2), 10 (2015)
2. Cullinane, K., Toy, N.: Identifying influential attributes in freight route/mode choice decisions: a content analysis. Trans. Res. Part E **36**(1), 41–53 (2000)
3. Czermański, E., Droździecki, S., Matczak, M., Spangenberg, E., Wiśnicki, B.: Sulphur regulations-Technology solutions and economic consequences for the Baltic Sea Region shipping market. Institute of Maritime Transport and Seaborne Trade, University of Gdańsk, Technical report (2014)
4. Daduna, J., Hunke, K., Prause, G.: Analysis of short sea shipping-based logistics corridors in the Baltic Sea region. J. Shipp. Ocean Eng. **2**(5), 304–319 (2012)
5. Holmgren, J., Nikopoulou, Z., Ramstedt, L., Woxenius, J.: Modelling modal choice effects of regulation on low-sulphur marine fuels in Northern Europe. Trans. Res. Part D **28**, 62–73 (2014)
6. Hunke, K., Prause, G.: Sustainable supply chain management in German automotive industry-Experiences and success factors. J. Secur. Sustain. Issues **3**(3), 15–22 (2014)
7. Jiang, L., Kronbak, J., Christensen, L.P.: The costs and benefits of sulphur reduction measures: Sulphur scrubbers versus marine gas oil. Trans. Res. Part D **28**, 19–27 (2014)
8. Notteboom, T.: The impact of low sulphur fuel requirements in shipping on the competitiveness of roro shipping in Northern Europe. WMU J. Marit. Aff. **10**(1), 63–95 (2011)

Computing Indicators for Differences and Similarities Among Sets of Vehicle Routes

Jörn Schönberger

Abstract The quantification of the differences of two sets of vehicle routes is addressed. We present procedures to compute indicators quantifying differences in the tour composition among both sets. Furthermore, we propose two different ideas to compare sequencing decisions related to the routing sub-problem in vehicle routing. Finally, we combine clustering and sequencing decision comparison procedures in order to quantify differences of the vehicle route sets.

1 Introduction

The solving of vehicle routing problems plays an important role in both logistics process planning as well as in the evaluation of planning algorithms. In both applications the comparison of two given sets of vehicle route is necessary in the analysis of impacts of changing underlying decision models or solving procedures. Clustering (tour composition) decisions as well as the sequencing (routing) decisions must be evaluated regarding preserved similarities and resulting differences.

Permutations representing routes can be compared using Hamming-distances [3]. Fagerholt et al. [1] introduce a distance indicator for comparing clustering decisions made in two sets of routes. Løkketangen et al. [4] append attribute vectors to the components of route sets and compare the values of the attribute vectors found in two route sets.

This article first discusses the comparison of clustering decisions (Sect. 2). Similarities of sequencing decisions are evaluated among pairs assigned to the same route (Sect. 3) or to different routes (Sect. 4). Two combined clustering and sequencing decision comparison procedures are proposed in Sect. 5.

J. Schönberger (✉)
Chair of Transport Services and Logistics, Technische Universtät Dresden,
01062 Dresden, Germany
e-mail: joern.schoenberger@tu-dresden.de

© Springer International Publishing Switzerland 2017
K.F. Dœrner et al. (eds.), *Operations Research Proceedings 2015*,
Operations Research Proceedings, DOI 10.1007/978-3-319-42902-1_26

2 Comparison Based on Clustering

Two sets P_1 and P_2 of M vehicle routes (some may be empty) are given and the set O of operations must be covered by both route sets. The ith **tour** contains the subset of O served by the ith route r_i^P in $P \in \{P_1; P_2\}$. The **route** r_i^P is a permutation of the ith tour in P. All operations $o \in O$ are uniquely indexed. The fleet is homogeneous and we define $O^* := (O \times O) \setminus \{(i; i) \mid i \in O\}$.

Figure 1 shows two example sets of three routes P_1 as well as P_2 covering seven operations $1, \ldots, 7$. By shifting customer 1 from route 1 to the beginning of route 3 we transform P_1 into P_2. In P_2, route 1 is empty, route 2 is extended and route 3 remains unchanged.

Shifting of a complete route between two identical vehicles does not lead to an effective route set variation. From this perspective the donating vehicle as well as the receiving vehicle are equivalent. Vice versa, a shift of a route between two equivalent vehicles must not change the quantified distance value between two compared route sets. Fagerholt et al. [1] claim that Hamming-distances are not precise enough to detect these kinds of non-effective shifts.

Two route sets can be compared by analyzing the extent to which the clustering as well as the sequencing decisions associated with the two route sets match. The general idea is to count identical clustering decisions as well as to count identical sequencing decisions in the two route sets P_1 and P_2. In order to overcome the challenge of equivalent routes, we propose the check only if two pairwise different operations i and j from O are clustered in a common tour in P_1 as well as in a common tour in P_2. Following this way to compare the clustering decisions in the two route sets it is not necessary to check if i and j are contained in the tour with the same number in both route sets. Given route set P and two distinct operations $i, j \in O$ we define the binary indicator C_{ij}^P to be 1 if and only if these two operations are contained in the same route. Otherwise, we set $C_{ij}^P := 0$.

$$C(P_1; P_2) := \begin{cases} \frac{c^{used\_both}(P_1;P_2)}{c^{used}(P_1;P_2)}, & \text{if } c^{used}(P_1; P_2) > 0 \\ 1, & \text{if } c^{used}(P_1; P_2) = 0 \end{cases} \tag{1}$$

Fig. 1 Example route sets P_1 and P_2 to be compared

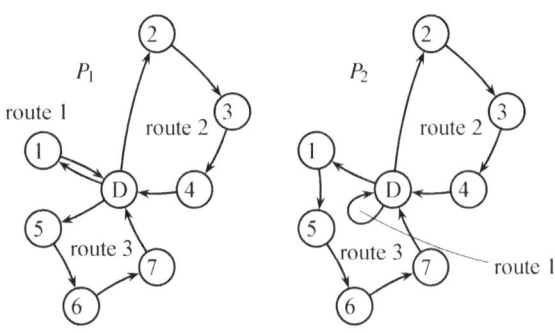

Let $c^{used}(P_1; P_2) := \left\| \left\{ (i; j) \in O^* \mid C_{ij}^{P_1} = 1 \vee C_{ij}^{P_2} = 1 \right\} \right\|$ be the number of pairs of distinct operations that are assigned to a common route either in route set P_1 and/or in route set P_2 and $c^{used\_both}(P_1; P_2) \left\| \left\{ (i; j) \in O^* \mid C_{ij}^{P_1} = 1 \wedge C_{ij}^{P_2} = 1 \right\} \right\|$ carries the number of pairs $(i; j) \in O^*$ that are clustered together in both route sets. We have $0 \leq \frac{c^{used\_both}(P_1;P_2)}{c^{used}(P_1;P_2)} \leq 1$ if $c^{used}(P_1; P_2) > 0$. In case that $c^{used}(P_1; P_2) = 0$ all routes contain at most one customer location. Here, we do not find a pair of operations assigned to the same route neither in P_1 nor in P_2. However, we also do not find a pair of operations that is assigned to different routes in P_1 as well as P_2. Therefore, we define the percentage $C(P_1; P_2)$ of pairs of operations that are assigned to the same route in both considered route sets P_1 as well as P_2 as shown in (1).

3 Comparison of Processing Sequences Inside a Tour

If two distinct operations i and j are found in the same tour in a given route set P ($C_{ij}^P = 1$) we check next if i is processed before j ($S_{ij}^P = 1$) or not ($S_{ij}^P = 0$). We define $s^{used}(P_1; P_2) := \left\| \left\{ (i; j) \in O^* \mid S_{ij}^{P_1} = 1 \vee S_{ij}^{P_2} = 1 \right\} \right\|$ to be the number of pairwise precedencies detected in at least one of the two considered route sets P_1 and P_2. Further, $s^{used\_both}(P_1; P_2) := \left\| \left\{ (i; j) \in O^* \mid S_{ij}^{P_1} = 1 \wedge S_{ij}^{P_2} = 1 \right\} \right\|$ carries the number of pairs of distinct operations that are clustered in a tour in both route sets and that are processed in the same order.

$$S(P_1; P_2) := \begin{cases} \frac{s^{used\_both}(P_1;P_2)}{s^{used}(P_1;P_2)}, & \text{if } s^{used}(P_1; P_2) > 0 \\ 1, & \text{if } s^{used}(P_1; P_2) = 0 \end{cases} \qquad (2)$$

The indicator $S(P_1; P_2)$ determines the percentage of the same processing orders of two distinct operations that are clustered together in a route in both route sets (2). In case that $s^{used}(P_1; P_2) = 0$ both route sets have the common property that no pair $(i; j) \in O^*$ is contained in a tour. No different processing orders are observed and therefore we define $S(P_1; P_2)$ to be 1 in case that $s^{used}(P_1; P_2) = 0$.

4 Sequencing Comparison Based on Giant-Routes

An obvious shortcoming of the intra-tour sequence comparison is the small number of available pairs of operations contained in a tour. In situations where tours are small, even few processing order variations among the rare pairs of operations in a route let $S(P_1; P_2)$ vary significantly. In order to reduce the sensitivity of the distance measure with respect to the sequencing decisions, we propose another way to quantify differences as well as similarities in pairwise precedence relations among two operations.

The basic concept used for increasing the number of comparable pairs of operations is to establish an inter-route comparison of processing sequences. For this, we build a so-called giant-route from the determined routes [2]. A giant-route R is an M-tuple of the available M routes.

The route set comparison starts with the calculation of $C(P_1; P_2)$. In order to prepare the determination of the giant-route we first sort the routes contained in P_1. Each route is evaluated with the smallest unique operation index observed for the operations it contains. Empty routes are evaluated with ∞. Afterwards, the routes are sorted by increasing evaluation values. Finally, the giant-route $R^{P_1} := (r_{i_1}, \ldots, r_{i_M})$ is compiled. In the same way, a giant-route R^{P_2} is setup for P_2.

The binary parameter T_{ij}^P is set to 1 if and only if operation i is processed prior to operation j in the giant-route R^P. The number of pairs of distinct operations that can be compared with respect to their precedence relation is significantly larger compared to the number of pairs considered for the specification of S_{ij}^P (2) since it is not necessary any more that both operations i as well as j are contained in a common route.

$$T(P_1; P_2) := \frac{t^{used\_both}(P_1; P_2)}{t^{used}(P_1; P_2)} \tag{3}$$

Let $t^{used}(P_1; P_2) := \left\| \left\{ (i; j) \in O^* \mid T_{ij}^{P_1} = 1 \vee T_{ij}^{P_2} = 1 \right\} \right\|$ be the number of found processing orders in at least one of the compared giant-routes. The number of matching processing orders of two distinct operations that are observed in both giant-routes is stored in $t^{used\_both}(P_1; P_2) := \left\| \left\{ (i; j) \in O^* \mid T_{ij}^{P_1} = 1 \wedge T_{ij}^{P_2} = 1 \right\} \right\|$. We calculate the percentage of observed processing orders of two distinct operations that are observed in both giant-routes and store it in $T(P_1; P_2)$ (3).

5 Combined Comparison of Clustering and Routing Decisions

$$H^1(P_1; P_2) = 1 - \frac{1}{2} \cdot (C(P_1; P_2) + S(P_1; P_2)) \tag{4}$$

Using the percentage of consistent operation pairings $C(P_1; P_2)$ as well as the percentage of consistent operation processing sequences $S(P_1; P_2)$ we are able to define the first difference indicator $H^1(P_1; P_2)$ for the similarity of two route sets P_1 and P_2 (4). We calculate H^1 for the example scenario from Fig. 1. The left part of Fig. 2 shows the intermediate results for the calculation of $C(P_1; P_2)$. In total $c^{used}(P_1; P_2) = 18$ pairs of clustered operations are found either in P_1 or in P_2 but only $c^{used}(P_1; P_2) = 12$ pairs are detected in both plans P_1 as well as P_2. Therefore, we get $C(P_1; P_2) := \frac{12}{18} = \frac{2}{3}$.

Intermediate results from the calculation of $S(P_1; P_2)$ are shown in the right part of Fig. 2. Overall, $s^{used}(P_1; P_2) = 9$ precedence relations are detected in P_1 or in P_2. Among them, $s^{used\_both}(P_1; P_2) = 6$ are found in both route sets, so that $S(P_1; P_2) := \frac{6}{9} = \frac{2}{3}$. This leads to the $H^1(P_1; P_2)$-distance of $1 - \frac{1}{2} \cdot \left(\frac{2}{3} + \frac{2}{3}\right) = \frac{1}{3}$. From a naive perspective, this H^1-value can be interpreted in the way that the shifting of operation 1 from route 1 to route 3 varies one third of the properties of P_1.

$$H^2(P_1; P_2) = 1 - \frac{1}{2} \cdot (C(P_1; P_2) + T(P_1; P_2)) \tag{5}$$

If we replace $S(P_1; P_2)$ by $T(P_1; P_2)$ in the definition of H^1 we get a second distance indicator H^2 (5) for the comparison of two vehicle route sets.

In the considered example both giant-routes (left part of Fig. 3) provide together $t^{used}(P_1; P_2) = 30$ precedence relations from which $t^{used\_both}(P_1; P_2) = 12$

	1	2	3	4	5	6	7
1					2	2	2
2			1/2	1/2			
3	1/2			1/2			
4		1/2	1/2				
5	2					1/2	1/2
6	2				1/2		1/2
7	2				1/2	1/2	

	1	2	3	4	5	6	7
1					2	2	2
2			1/2	1/2			
3				1/2			
4							
5						1/2	1/2
6							1/2
7							

$C_{ij}^{P_1}$-and $C_{ij}^{P_2}$-values $S_{ij}^{P_1}$-and $S_{ij}^{P_2}$-values

Fig. 2 The *left* tabular carries the clustering information used in the calculation of $C(P_1; P_2)$ (1 means both operations are contained in the same route in P_1 and 2 indicates that the two operations are clustered together in P_2). The *right* tabular contains the sequencing information related to the calculation of $S(P_1; P_2)$

$\left.\begin{array}{l} r_1 = (1) \\ r_2 = (2;3;4) \\ r_3 = (5;6;7) \end{array}\right\} \Rightarrow R^{P_1} = \underbrace{(1;2;3;4;5;6;7)}_{\text{giant-route of } P_1}$

$\left.\begin{array}{l} r_1 = () \\ r_2 = (2;3;4) \\ r_3 = (1;5;6;7) \end{array}\right\} \Rightarrow R^{P_2} = \underbrace{(1;5;6;7;2;3;4)}_{\text{giant-route of } P_2}$

	1	2	3	4	5	6	7
1		1/2	1/2	1/2	1/2	1/2	1/2
2			1/2	1/2	1	1	1
3				1/2	1	1	1
4					1	1	1
5	2	2	2			1/2	1/2
6	2	2	2				1/2
7	2	2	2				

Fig. 3 Determination of $T(P_1; P_2)$: In the *right* tabular all pairs of operations $(i; j) \in O^*$ in which operation i precedes operation j in P_1 (P_2) are indicated by 1 (2)

precedence relations are detected in both route sets. Therefore, we have $T(P_1; P_2) = \frac{12}{30} = 0.4$. The H^2-distance between P_1 as well as P_2 is $H^2(P_1; P_2) = 1 - \frac{1}{2}\left(\frac{2}{3} + \frac{4}{10}\right) \approx 0.47$. From a naive perspective and with regard to the observed H^2-value 47 % of the properties of P_1 are changed if operation 1 is moved from route 1 to route 3.

6   Summary and Conclusions

We have proposed two indicators to quantify structural differences among two given vehicle route sets. Both indicators shown different distance values. The next research steps comprise the evaluation of both indicators in several VRP-like scenarios including larger sets of operations. Furthermore, we will specify additional comparison approaches in order to take care of additional challenges in route set comparison like the processing independence of routes in the CVRP.

References

1. Fagerholt, K., Korsvik, J.E., Løkketangen,A.: Ship routing and scheduling with persistence and distance objectives. In: Bertazzi, L., Grazia Speranza, M., van Nunen, J.A.E.E. (eds.), Innovation in Distribution Logistics. pp. 98–107. Springer, Berlin, Heidelberg (2009)
2. Funke, B., Grünert, T., Irnich, S.: Local search for vehicle routing and scheduling problems: review and conceptual integration. J. Heuristics **11**, 267–306 (2005)
3. Hamming, R.W.: Error-detecting and error-correcting codes. Bell Syst. Tech. J. **29**(2), 147–160 (1950)
4. Løkketangen, A., Oppen, J., Oyola, J., Woodruff, D.L.: An attribute based similarity function for vrp decision support. Decis Mak. Manuf. Serv. **6**(2), 65–83 (2012)

Optimal Transportation Mode Decision for Deterministic Transportation Times, Stochastic Container Arrivals and an Intermodal Option

Klaus Altendorfer and Stefan Minner

Abstract For container transport from an overseas port to the final customer destination, the decision whether the transport should be by direct truck or by an intermodal option, including train and truck, is rather relevant. An optimization problem minimizing transportation costs and delay costs is developed. For unknown container arrival times, deterministic transportation times and a predefined container delivery date, the optimal transportation mode is discussed. For a predefined train departure time, the relationship between unplanned truck delivery if arrival time is after train departure and expected delay for later train departure times is shown. For the intermodal option, an optimization problem minimizing transportation and expected delay costs is solved. Results show that there is a certain delivery date threshold above which the intermodal option becomes optimal. A counter-intuitive finding from the numerical results indicates that better information quality does not necessarily favour the intermodal option. Especially for low delivery date values, the additional cost to be paid for the intermodal option increases with better information quality.

1 Introduction and Literature Review

A problem often faced by logistics providers as well as international supply chains is the transportation mode decision. If containers arrive with an overseas ship at one of the major European ports, several transportation options can be applied. We compare either the option to ship a container directly by truck to its final destination, which is fast but may be expensive, or the option to ship a container initially by

K. Altendorfer (✉)
Department for Operations Management, Upper Austria University of Applied Sciences,
4400 Steyr, Austria
e-mail: klaus.altendorfer@fh-steyr.at

K. Altendorfer
Department for Production and Logistics Management, Johannes Kepler University Linz,
4040 Linz, Austria

S. Minner
TUM School of Management, Technische Universität München, 80333 Munich, Germany

© Springer International Publishing Switzerland 2017   201
K.F. Dœrner et al. (eds.), *Operations Research Proceedings 2015*,
Operations Research Proceedings, DOI 10.1007/978-3-319-42902-1_27

train to a certain inland hub and cover only the on-carriage by truck, which is usually slower but cheaper. Recent literature on intermodal transport focusses on the cost calculation for both modes and evaluates the environmental impact of this decision (e.g. [1, 2]) or technical and infrastructure prerequisites (e.g. [3, 6]). In a survey of intermodal transport decision problems, [4] identify that for the decision itself, the time related measures expected delay and on-time probability are among the most important decision factors. Related to the timing of information, [5] discuss the impact of advance load information on transportation costs. Reference [7] discuss the effect of information uncertainty on the on-time probability and the transportation costs for different transportation modes. Even though they do not specifically cover the intermodal transportation option, they provide some insights on the value of information and its influence on the transportation mode decision.

We extend available literature by analytically evaluating the influence of a stochastic arrival process on the expected delay. Furthermore, a single objective optimization problem is stated to identify the optimal transportation mode decision and an optimal train departure time is identified.

2   Model Development

A set of containers with stochastic arrival time A from the port has to be transported to a final destination either by direct truck transport or by a combination of train and truck. This decision has to be taken before the stochastic arrival occurs. As the arrival time depends on the unloading of the containers from the overseas ship and no further information is available, it is assumed to be uniformly distributed between A_{min} and A_{max}. All containers have the same deterministic delivery date T when they are expected at their final destination and delay costs are c_d per container and period of time. We assume that $T \geq A_{max}$. Costs are c_t for the direct truck transport and c_i for the intermodal option. The train leaves the port at departure time R and whenever $A > R$ occurs, the container is transported via an unplanned truck that has to be provided on a short term basis with costs c_u. The transportation times are deterministic with L_t, L_u, L_i for direct truck, unplanned truck and intermodal option, respectively. The following relationship between the transportation times and costs is assumed: $L_t \leq L_u \leq L_i$ and $c_i \leq c_t \leq c_u$ (see also [7] for similar assumptions). The random delay a container faces is D_t and D_i for the direct truck or intermodal option, respectively. For the direct truck option, the truck waits for the container and the deterministic transportation time applies after container arrival, therefore the expected delay (with $f_A()$ and $F_A()$ being the pdf (probability density function) and cdf (cumulative distribution function) of the arrival time, respectively) is:

$$E[D_t] = \int\limits_{T-L_t}^{A_{max}} (L_t - (T-\kappa)) f_A(\kappa)\, d\kappa = \int\limits_{T-L_t}^{A_{max}} 1 - F_A(\kappa)\, d\kappa \tag{1}$$

The probability that a container arrives after the scheduled train departure time, which means it takes an unplanned truck to be delivered to its final destination, is:

$$\alpha(R) = 1 - F_A(R) = (A_{\max} - R) / (A_{\max} - A_{\min}) \tag{2}$$

The delay of a container that takes the intermodal option for $A \leq R$ is $[L_i - (T - R)]^+$ (which might be zero) and whenever $A > R$ (and therefore the unplanned truck option is realized) it is the random value $[L_u - (T - A)]^+$. The expected delay can therefore be constructed as:

$$E[D_i] = \int_R^{A_{\max}} [(L_u - (T - \kappa))]^+ f_A(\kappa)\, d\kappa + [R - (T - L_i)]^+ F_A(R) \tag{3}$$

Based on $L_t \leq L_u \leq L_i$, several cases can be distinguished with respect to R:

$$E\left[D_i \mid R > T - L_u\right] = \int_{T-L_u}^{A_{\max}} 1 - F_A(\kappa)\, d\kappa + (L_i - L_u)\, F_A(R) + \int_{T-L_u}^{R} F_A(\kappa)\, d\kappa$$

$$E\left[D_i \mid {R > T - L_i \atop R \leq T - L_u}\right] = \int_{T-L_u}^{A_{\max}} 1 - F_A(\kappa)\, d\kappa + (L_i - (T - R))\, F_A(R) \tag{4}$$

$$E\left[D_i \mid R \leq T - L_i\right] = \int_{T-L_u}^{A_{\max}} 1 - F_A(\kappa)\, d\kappa$$

For a very late train departure with $R > T - L_u$, the container is late if it catches the train $A \leq R$ and it is also late if it arrives after train departure. For a medium train departure time $T - L_u \geq R > T - L_i$, the container is late if it catches the train but it might be on time if it uses the unplanned truck, which is more expensive. For a low departure time $R \leq T - L_i$, which includes the case that the train leaves to deliver exactly on time ($R = T - L_i$), containers are only late if they arrive after $T - L_u$ and therefore the unplanned truck delivery, too, does not reach the planned delivery date. Equation (4) shows that an increase in train departure time leads to an increase in expected delay time and Eq. (2) shows that this increase leads to a lower probability of unplanned truck transport. If the train departure time is a decision variable, i.e. a set of containers coming from one ship has to be delivered and the mode has to be decided, these results indicate that $R = T - L_i$ might be a good solution. This train departure time minimizes the expected delay while also minimizing the unplanned truck delivery.

The overall costs for intermodal and direct truck transportation, O_i and O_t, respectively, are:

$$O_i = F_A(R)\, c_i + (1 - F_A(R))\, c_u + E[D_i]\, c_d \tag{5}$$

$$O_t = c_t + E[D_t]\, c_d \tag{6}$$

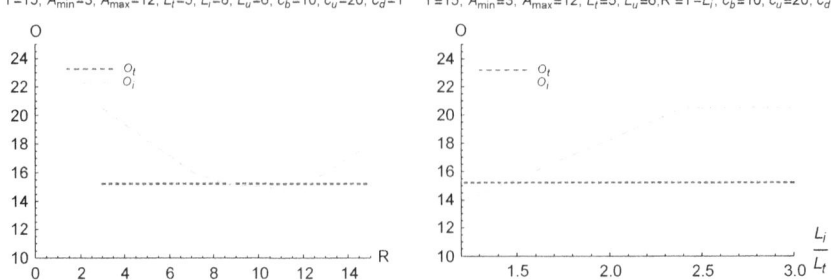

Fig. 1 Cost comparison with respect to predefined train departure

Figure 1 shows a comparison of costs for different transportation modes with respect to the train departure time on the left hand side. It identifies that the minimum costs are not reached with $R = T - L_i$ (which would be $R = 7$) but with a rather high train departure time $R = 11$. This happens because the low delay costs are outweighed by the larger cost difference between the modes of transportation. For $R = 11$, the container has a rather low probability of being transported by the expensive unplanned truck. If the train departure time can also be decided, this shows that the aforementioned trade-off cannot intuitively be solved but has to be optimized. On the right hand side of Fig. 1, the overall costs for intermodal transportation are presented with respect to L_i/L_t showing its relative transportation time in comparison to direct truck transportation. Note that here the intuitive train departure time $R = T - L_i$ is applied, which has already been shown to be not optimal.

From the cost function (5) it is obvious that a train departure time below the intermodal transportation time, i.e. $R < T - L_i$, is never optimal as this leads to unnecessary unplanned truck transport. However, the trade-off between accepting a certain delay and increasing the probability of shipping the container by the cheaper intermodal option has to be further investigated. Therefore, two relevant cases from Eq. (4) can be identified with Case 1 being $R > T - L_u$ and Case 2 being $T - L_u \geq R \geq T - L_i$. For Case 1, the overall costs are:

$$O_i = F_A(R)(c_i - c_u + (L_i - L_u)c_d) + c_u + \left(L_u - T + A_{max} - \int_R^{A_{max}} F_A(\kappa)d\kappa\right)c_d \quad (7)$$

Setting the first derivative with respect to R to zero leads to:

$$R_1^* = \max\left[T - L_u, \frac{c_u - c_i}{c_d} + A_{min} + L_u - L_i\right] \quad (8)$$

The optimal train departure time R_1^* is independent of the required delivery date that only defines the lower boundary from Case 1 definition. The overall costs for Case 2 are:

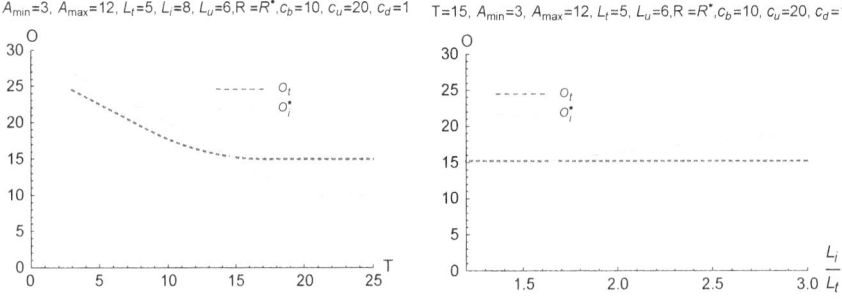

Fig. 2 Optimal cost comparison

$$O_i = F_A(R)(c_i + c_u + (R - T + L_i)c_d) + c_u + \int_{T-L_u}^{A_{max}} 1 - F_A(\kappa)\,d\kappa\,c_d \quad (9)$$

and again setting the first derivative with respect to R to zero leads to:

$$R_2^* = \min\left[T - L_u, \max\left[T - L_i, \left(\frac{c_u - c_i}{c_d} + A_{min} + T - L_i\right)\bigg/2\right]\right] \quad (10)$$

To identify the optimal solution for train departure time, $O_i(R_1^*)$ and $O_i(R_2^*)$ have to be compared. Figure 2 shows the optimal costs for direct truck and intermodal transportation either with respect to the delivery date (left hand side) or with respect to the relative intermodal transportation time (right hand side).

The results with respect to T show that, up until a certain threshold value, an increase in delivery date leads to a similar cost decrease for both modes. With respect to intermodal transportation time, the cost difference is far less than in Fig. 1 with predefined train departure time $R = T - L_i$ where all containers that catch the train are shipped on time. This is an interesting finding as it shows that it can be optimal to accept a certain delay in order to exploit the cheaper intermodal transportation option even for deterministic transportation times with a stochastic arrival process.

3 Numerical Example

To identify the influence of the different model parameters on overall costs, Fig. 3 shows the relative cost difference $(O_i - O_t)/O_t$ as average value over all parameter combinations. For each model parameter (i.e. c_i, c_u, c_t, L_u and L_i), three different values near the values of the examples in Figs. 1 and 2 are tested. For the arrival time information uncertainty, $A_{max} \in \{9, 12, 15\}$ and $A_{min} = 15 - A_{max}$ are applied, which leads to $E[A] = 7.5$. Whenever the values in Fig. 3 are negative, the

L_i	T						
	10	12.5	15	17.5	20	22.5	Avg
6	0.17	0.12	-0.11	-0.23	-0.29	-0.32	-0.11
8	0.20	0.17	-0.02	-0.14	-0.23	-0.29	-0.05
11	0.24	0.22	0.07	-0.04	-0.13	-0.23	0.02
Avg	0.20	0.17	-0.02	-0.14	-0.22	-0.28	-0.05

A_{max} A_{min}	T						
	10	12.5	15	17.5	20	22.5	Avg
3	0.28	0.25	-0.11	-0.20	-0.22	-0.26	-0.04
9	0.17	0.13	0.01	-0.15	-0.25	-0.30	-0.07
15	0.16	0.12	0.04	-0.06	-0.18	-0.28	-0.03
Avg	0.20	0.17	-0.02	-0.14	-0.22	-0.28	-0.05

Fig. 3 Numerical example results

intermodal option leads to lower average costs. Detailed results from the numerical study (not presented) confirm the intuitive results that higher cost for intermodal and unplanned truck transportation and lower costs for direct truck transportation reduce the probability of the intermodal option to be optimal. However, the results for an increasing T value show that the threshold above which the intermodal option is optimal depends very much on the parameter combination. An increase in intermodal delivery time of 80 % leads to a cost delta of only 7 to 9 %, which shows that this parameter's effect is less pronounced than intuitively conjectured. The results concerning $A_{max} - A_{min}$ provide the counter-intuitive finding that better information does not necessarily increase the probability that the intermodal option becomes optimal. For low T values, the additional costs for the intermodal option decrease with higher information uncertainty. Also for high T values, a better information quality does not consistently increase the cost benefits of the intermodal option.

4 Conclusion

In this paper a simplified decision problem for either direct truck or intermodal transportation of containers between a port and their inland destination is solved. Numerical results show that, even in this setting with deterministic transportation times, it is not (always) optimal to schedule train departure such that a container transported by train is on time, because delay costs and transportation costs have to be balanced. Furthermore, a counter-intuitive finding is that better information quality does not necessarily increase the probability that the intermodal option becomes optimal. These results are the basis for further research covering more stochastic influences and richer problem structures, such as train capacity reservation.

References

1. Janic, M.: Modelling the full costs of an intermodal and road freight transport network. Transp. Res. Part D **12**, 33–44 (2007)
2. Macharis, C., van Hoeck, E., Pekin, E., van Lier, T.: A decision analysis framework for intermodal transport: comparing fuel price increases and the internalisation of external costs. Transp. Res. Part A **44**, 550–561 (2010)
3. Roso, V., Woxenius, J., Lumsden, K.: The dry port concept: connecting container seaports with the hinterland. J. Transp. Geogr. **17**, 338–345 (2009)

4. Sommar, R., Woxenius, J.: Time perspectives on intermodal transport of consolidated cargo. Eur. J. Transp. Infrastruct. Res. **7**, 163–182 (2007)
5. Tjokroamidjojo, D., Kutanoglu, E., Taylor, D.: Quantifying the value of advance load information in truckload trucking. Transp. Res. Part E **42**, 340–357 (2006)
6. Trip, J., Bontekoning, Y.: Integration of small freight flows in the intermodal transport system. J. Transp. Geogr. **10**, 221–229 (2002)
7. Zuidwijk, R., Veenstra, A.: The value of information in container transport. Transp. Sci. **49**(3), 675–685 (2014). doi:10.1287/trsc.2014.0518

Optimization Models for Decision Support at Motorail Terminals

Pascal Lutter

Abstract Motorail transportation covers the loading of various types of vehicles onto transportation wagons. The detailed treatment of realistic technical and legal constraints by integer linear programming models is challenging especially in terms of computation times. This paper considers decision support for the loading process at motorail terminals with the goal of speeding up the entire loading process while guaranteeing the feasibility of the proposed loading plan at all times. Specially tailored integer programming formulations are proposed and their suitability for real-world use is evaluated by means of a case study.

1 Introduction

The problem under consideration deals with the loading of cars and motorcycles onto motorail wagons under realistic technical and legal constraints. The load planning problem for motorail trains, introduced as motorail transportation problem (MTP) [7, 8], aims at assigning a given set of vehicles to transportation wagons. Decision support for order acceptance management has already been developed in [8] and is currently being used by DB Fernverkehr AG. This paper focuses on the loading process at motorail terminals. Vehicles are pre-booked and the train length, e.g. the number of transportation wagons, is known before departure. Due to differing and unknown arrival times of vehicles, pre-calculated loading plans can hardly be implemented in practice. Once loaded, vehicles are not allowed to change their position anymore. Thus, loading plans need to be consecutively revised in accordance with the current terminal situation and the already loaded vehicles. We propose and evaluate three integer programming formulations for decision support at motorail terminals.

Related problems arise at container terminals. Ambrosino et al. [1] address the loading of import containers at seaport terminals. In reference to a case study at an

P. Lutter (✉)
Chair of Operations Research and Accounting, Faculty of Management and Economics,
Ruhr University Bochum, Bochum, Germany
e-mail: pascal.lutter@rub.de

© Springer International Publishing Switzerland 2017
K.F. Dœrner et al. (eds.), *Operations Research Proceedings 2015*,
Operations Research Proceedings, DOI 10.1007/978-3-319-42902-1_28

Italian harbor, they develop an integer linear optimization model for decision support at seaport terminals. The goal is to fulfill the physical loading requirements while minimizing rehandling operations as well as penalty costs. As the original integer programming formulation is hard to solve with standard optimization software, the authors develop two heuristics to generate solutions of better quality within reasonable time. Corry and Kozan [4, 5] as well as Bruns et al. [2, 3] consider the load planning problem at intermodal container terminals. While Corry and Kozan explicitly focus on terminal operations management, Bruns et al. emphasize the detailed treatment of physical loading constraints and the integration of uncertainties. In contrast to recent contributions to container terminal optimization, this paper focuses on developing exact approaches to solving the MTP.

The remainder of this paper is structured as follows. Section 2 describes technical details of motorail transportation and processes at motorail terminals. This is followed by different integer linear programming formulations for terminal optimization in Sect. 3. Section 4 presents a case study for evaluating the proposed models. Section 5 concludes with some final remarks and perspectives for future research.

2 Problem Description

The problem under consideration focuses on the optimization of load planning at motorail terminals: pre-booked vehicles consecutively arrive at the terminal and are to be loaded on transportation wagons soon. The goal is to select subsets of vehicles from the waiting queue to be loaded onto the same loading deck, such that all currently remaining vehicles can be loaded as well. The order of the arriving vehicles is unknown due to uncertain vehicle arrival times. Parking spaces are commonly limited, preventing the usage of pre-determined loading plans. Decision support at motorail terminals can be achieved by embedding the MTP in a rolling horizon framework. The goal is to consecutively construct a loading plan for the entire train. A crucial aspect is to guarantee the feasibility of the loading plan at all times. This makes it necessary to partly revise the loading plan for currently not loaded vehicles in each optimization run. Hence, it is required to generate optimal solutions very quickly in order to support the loading process at motorail terminals.

Let $\bar{V}$ denote the number of vehicles and let $V := \{1, \ldots, \bar{V}\}$ denote the set of vehicles. A vehicle $v \in V$ is described by five different characteristics, namely, weight W_v, height H_v, roof width category R_v, position capacity requirement C_v, and type T_v. The entire motorail train consists of $\bar{I}$ identical vehicle transportation wagons i which are summarized in the set $I := \{1, \ldots, \bar{I}\}$. Vehicle transportation wagons offer two loading decks $e \in E := \{1, 2\}$ each providing $\bar{P}$ different positions $p \in P := \{1, \ldots, \bar{P}\}$. The total weight limit for a single loading deck equals $\overline{W}$. The upper loading deck is identified by $e = 1$ while the lower loading deck is given by $e = 2$. Length limits are included into the position capacity requirement parameter C_v and hence incorporated by position capacity constraints. Each vehicle v is associated with a position capacity demand parameter C_v which is determined by its type coding T_v. In

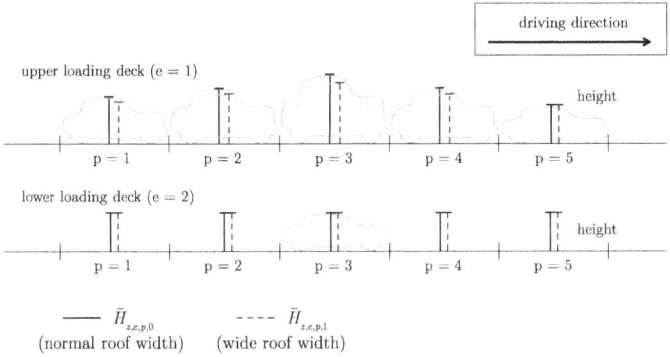

Fig. 1 Illustration of both loading decks of a wagon i (taken from [8])

general, each loading deck provides the same length and contains five positions which are numbered in ascending order corresponding to the driving direction as illustrated by Fig. 1 Maximal heights differ with respect to the loading deck under consideration: upper loading decks have different maximal heights for each position which are further dependent on the roof width of the assigned vehicle. Solid lines indicate maximal heights for vehicles with normal roof width (≤ 135 cm) while dashed lines show maximal heights for vehicles with large roof width (>135 cm). On the lower loading deck, all positions feature the same, but smaller maximum height and no distinction is made between different vehicle roof widths. The maximum height of each position p on loading deck e is given by the parameter $\overline{H}_{e,p,R_v}$ which is dependent on vehicle width R_v. The next section shows the integration of optimization based decision support in order to help terminal staff to select the right vehicles from the queue within a short period of computation time.

3 Optimization Models for Motorail Terminal Optimization

The original four-index formulation of the MTP (MTP-4Idx) was introduced in [8]. Here, a modified version for motorail terminal optimization is presented. The decision variables x_{viep} indicate the assignment of a vehicle v to a position on the train and equal 1 if and only if vehicle v is assigned to position p on loading deck e on wagon i. Let V denote the set of currently not loaded vehicles and let the subset $\mathcal{T} \subset V$ denote vehicles waiting at the terminal. The goal is to select vehicles $v \in \mathcal{T}$, indicated by the decision variables x_{viep}, maximizing the terminal specific objective function (1). In contrast to the original problem formulation, only a single loading deck $(i^\star, e^\star)$ is considered in the objective function. Loading deck selection is done by the terminal staff. Terminal staff is responsible for selecting the right loading deck in each loading iteration. It is important to notice that a loading deck can only be loaded, if all assigned vehicles are available for loading, e.g. they have already arrived at the terminal. Due to the rolling horizon framework, the sets $\mathscr{T}$, V as well as the available transportation wagons change in each iteration. In analogy with [8], the MTP-4Idx for optimizing a fixed loading deck $(i^\star, e^\star)$ reads as

$$\max \quad \sum_{v \in \mathscr{T}} \sum_{p \in P} \left(W_v x_{v i^\star e^\star p} - \mathrm{M} \cdot \sum_{v \in \complement V} x_{v i^\star e^\star p} \right) \tag{1}$$

$$\text{s. t.} \quad \sum_{i \in I} \sum_{e \in E} \sum_{p \in P} x_{viep} = 1 \qquad\qquad \forall v \in V \tag{2}$$

$$H_v x_{viep} \le \overline{H}_{z,e,p,R_v} \qquad\qquad \forall v \in V, i \in I, e \in E, p \in P \tag{3}$$

$$\sum_{v \in V} \sum_{p \in P} W_v x_{viep} \le \overline{W} \qquad\qquad \forall i \in I, e \in E \tag{4}$$

$$\sum_{v \in V} C_v x_{viep} \le 1 \qquad\qquad \forall i \in I, e \in E, p \in P \tag{5}$$

$$x_{viep} \in \{0, 1\} \qquad\qquad \forall v \in V, i \in I, e \in E, p \in P. \tag{6}$$

The objective function (1) maximizes the total weight of vehicles loaded onto the current loading deck $(i^\star, e^\star)$. As only vehicles waiting at the terminal can be loaded, the first sum is restricted to the subset $\mathscr{T}$. The additional term $\mathrm{M} \cdot \sum_{v \in \complement V} x_{v i^\star e^\star p}$ with $\complement V := V \setminus \mathscr{T}$ assures that the loading deck under consideration is preferably loaded with currently available vehicles. If M is sufficiently large, a negative objective value indicates that the current vehicle selection is not realizable. Thus, at least one more vehicle needs to arrive until the particular loading deck can be loaded. Alternatively, this part of the objective function could be included in the constraint set to adjust the domain of decision variables corresponding to currently unavailable vehicles. The inclusion of such constraints changes the structure of the coefficient matrix. As a consequence, an additional constraint appears in the master problem of the column generation approach (discussed at the end of this section). Constraints (2) ensure that each vehicle is assigned to one position on the train. Constraints (4) guarantee a maximal weight of $\overline{W}$ for each loading deck. Maximum heights for each position are considered by constraints (3) and the position capacity is modeled by constraints (5). Finally, the domain of the decision variables is defined by (6).

A crucial component of the MTP-4Idx concerns the formulation of the height constraints (3) as argued by Lutter [6]. By grouping all vehicles which can be placed onto the same positions, a different representation of the problem is derived. The entire height spectrum of a transportation wagon is grouped into four different categories in ascending order of maximal heights. Given these height categories, the set of vehicles

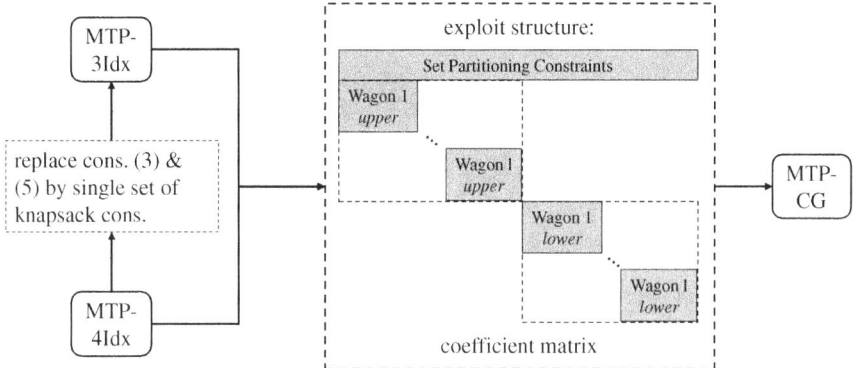

Fig. 2 Three formulations of the MTP. Reformulating height and capacity constraints and dropping the position index from the decision variables in the MTP-4Idx leads to the MTP-3Idx. Both formulations inherit a bordered-block diagonal structure allowing the application of Dantzig-Wolfe (DW) decomposition. Finally, the MTP-CG formulation results from the application of DW decomposition to one of the compact formulations

V is partitioned into four disjoint subsets $\mathcal{H}_1, \ldots, \mathcal{H}_4$, each representing vehicles of a given height category. Let a_h denote the number of feasible positions on a single loading deck for vehicles belonging to height category h. A three-index formulation of the problem (MTP-3Idx) is obtained by removing the position index p from the decision variables. Enforcing height (3) and capacity (5) constraints in a single set of combined knapsack constraints $\sum_{v \in \bigcup_{h'=h}^{4} \mathcal{H}_{h'}} C_v x_{vi1} \leq a_h \quad (h = 1, \ldots, 4)$ gives a proper representation of the MTP. The MTP-3Idx yields a tighter LP-relaxation than the original formulation, but preserves the bordered-block diagonal structure of the coefficient matrix [6].

The third formulation emerges from exploiting the special structure. The problem is split into a master problem and into two sub-problems by applying Dantzig-Wolfe decomposition (MTP-CG). In the following, a subset of vehicles forming a feasible solution for a single loading deck is referred to as a loading pattern. While the previous formulations of the MTP treat loading patterns as decision variables, the third formulation considers loading patterns as input parameters and decides about their use. In order to guarantee that each vehicle is loaded once, the master problem consists of set packing constraints (2). In addition, the master problem assures that no more than $\overline{I}$ loading patterns are selected for upper and lower loading decks. In case of decomposing the MTP-3Idx problem, resulting sub-problems are multi-dimensional knapsack problems ensuring that only feasible solutions, e.g. loading patterns, are generated. Figure 2 summarizes the three proposed model formulations and shows their interdependencies. The next section analyses the computational performance of the developed models on the basis of a real-world case study.

Fig. 3 Runtimes of the
MTP-4Idx, MTP-3Idx and
MTP-CG formulations

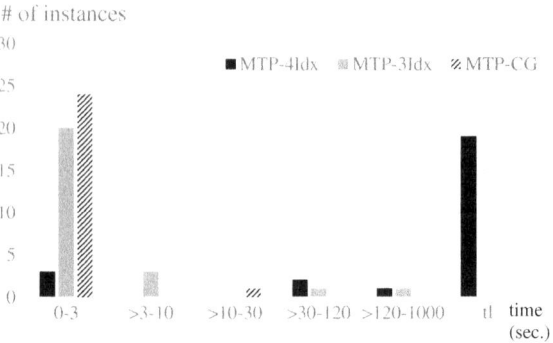

4 Case Study and Results

To illustrate the described problem and to evaluate the computational performance of the three formulation, the following situation is considered: 29 vehicles consecutively arrive at the terminal. The motorail train consists of three wagons. The sum of vehicle weights is 44,829 kg while total train capacity is 45,000 kg. It is assumed, that the loading process starts after 12 vehicles have arrived. Loading starts, if the optimal solution to the MTP indicates a positive objective value. In case of a negative objective, loading is currently not possible and it has to be waited until the next vehicle arrives. In the considered case study, the entire train is loaded after 25 iterations. Figure 3 summarizes running times for all models and all instances. It turns out that the MTP-3Idx and MTP-CG formulations were able to prove optimality in all instances very quickly. The MTP-4Idx formulation performed considerably worse: an optimal solution was only found in 15 instances while optimality was proven in only 6 of 25 test instances. In consequence, it also had the longest runtimes with an average of 1,156 s. The MTP-3idx formulation showed an average runtime of 34 s (due to one outlier) while the MTP-CG formulation exhibited the shortest runtime with only 2.1 s on average and a maximum runtime of 29.5 s.

5 Conclusion and Outlook

In this paper, three formulations of the motorail transportation problem specially tailored for decision support at motorail terminals were described and analyzed on the basis of a real-world case study. As indicated by the computational experiments, the MTP-3Idx and the MTP-CG show the best overall performance and qualify for real-world use at motorail terminals. Further improvements should focus on the inclusion of vehicle shunting operations at the terminal. As these extensions require the introduction of further binary variables which may slow down the solution process, specific solution approaches have to be developed.

References

1. Ambrosino, D., Silvia, S.: Comparison of solution approaches for the train load planning problem in seaport terminals. Transp. Res. Part E Logistics Transp. Rev. **79**, 65–82 (2015)
2. Bruns, F., Goerigk, M., Knust, S., Schöbel, A.: Robust load planning of trains in intermodal transportation. OR Spectr. **36**(3), 631–668 (2013)
3. Bruns, F., Knust, S.: Optimized load planning of trains in intermodal transportation. OR Spectr. **34**(3), 511–533 (2012)
4. Corry, P., Kozan, E.: An assignment model for dynamic load planning of intermodal trains. Comput. Oper. Res. **33**(1), 1–17 (2006)
5. Corry, P., Kozan, E.: Optimised loading patterns for intermodal trains. OR Spectr. **30**(4), 721–750 (2008)
6. Lutter, P.: Optimized load planning for motorail transporation. Comput. Oper. Res. **68**, 63–74 (2016)
7. Lutter, P., Werners, B.: Optimierung der Autozugverladung. In: Zukunftsperspektiven des Operations Research-Erfolgreicher Einsatz und Potenziale, Festschrift zum 80. Geburtstag von Hans-Jürgen Zimmermann, pp. 99–115. Springer, Heidelberg (2014)
8. Lutter, P., Werners, B.: Order acceptance for motorail transportation with uncertain parameters. OR Spectr. **37**(2), 431–456 (2015)

Continuity Between Planning Periods in the Home Health Care Problem

Daniela Lüers and Leena Suhl

Abstract Home health care providers face a complex routing and scheduling task to plan their services because their clients stay at their own homes. As the solution of this task may be inefficient or infeasible for a subsequent planning period, a new optimization is inevitable at the end of each period. The consideration of continuity in multi-period planning by avoiding extensive changes between periods is essential to ensure client and nurse satisfaction. To address this issue, we consider the home health care problem in a rolling planning horizon. Our heuristic solution method determines a new plan while preserving the continuity between periods. Since there are many possibilities to quantify continuity, we compare different measures and show their impact on the solutions.

1 Introduction

Home health care is a growing sector in public health. In some cases it provides an alternative to stationary institutions like hospitals and elderly homes. Due to the spread locations of their clients, home health care providers have to perform a weekly route planning, which respects the working regulations of their employees. In home health care, amongst economic goals, the preferences of clients and nurses are important. One major criterion for satisfaction is continuity [6].

Currently, two types of continuity are considered in previous work on the home health care problem (HHCP): Assigning only a small number of nurses to a client is referred to as *continuity of care* [2]. Whereas keeping similar appointment times, e.g.

D. Lüers (✉) · L. Suhl
Decision Support and Operations Research Lab, University of Paderborn,
Warburger Str. 100, 33098 Paderborn, Germany
e-mail: lueers@dsor.de

L. Suhl
e-mail: suhl@dsor.de

D. Lüers
International Graduate School for Dynamic Intelligent Systems,
University of Paderborn, Warburger Str. 100, 33098 Paderborn, Germany

© Springer International Publishing Switzerland 2017
K.F. Dœrner et al. (eds.), *Operations Research Proceedings 2015*,
Operations Research Proceedings, DOI 10.1007/978-3-319-42902-1_29

in Nickel et al. [7], can be referred to as *continuity of time*. Especially the continuity of care is considered in most of the solution approaches for static planning horizons. However, both types are important to ensure client satisfaction. The demands and capacities of a home health care provider are dynamic. Clients and nurses constantly enter or leave the services or change their demands. Therefore, it is crucial to consider these changes and at the same time assure continuity between periods.

Previous work on rolling horizon planning for the home health care problem followed different approaches. One is the decomposition in two stages, where the first stage assigns new clients to a reference nurse and the scheduling is performed for each nurse and day separately in the second stage [2, 4, 10]. To ensure continuity of care, either the nurse assignments are fixed [10], reassignments are minimized [2] or both [4]. Bowers et al. [3] model continuity of care by maximizing the preferences of clients. Another approach is the construction of master schedules, which are a template for future periods and therefore influence the continuity of time [5, 7]. Nickel et al. [7] also consider the continuity of time by minimizing the sum of changes in job start times, while inserting new clients. Bennett and Erera [1] ensure continuity of time by setting fixed appointment times for new clients for the length of their care period. Nowak et al. [8] model the HHCP as a consistent vehicle routing problem to assure continuity of care.

In this paper, we propose a rolling horizon approach to the HHCP which considers the continuity between planning periods. Because the duty scheduling and routing component depend on each other, we integrate both. Our approach extends previous work by respecting changes not only for clients, but also on the level of nurses and jobs. We allow entering or leaving the services, as well as changing demands or capacity. Continuity is modeled with respect to care and time by using different measures. We also present an additional type for securing the satisfaction of nurses by implementing *continuity of duty schedules*. We compare the different measures and show their impact on the solution. The numerical results are computed by our implementation of an adaptive large neighborhood search (ALNS) [9].

2   Problem Description

The basic HHCP is an integration of the vehicle routing problem with time windows (VRPTW) and the nurse rostering problem. We define the HHCP on the set of clients C, jobs J, nurses N and shift types S for a planning horizon of T days (here $T = 7$). Each client may request multiple jobs, which have a required qualification and duration. The day of a job is known and each job has a fixed time window.

Each nurse n has a set of qualifications. To model work contracts, the set of shift types S is given. Each shift type is defined by its earliest start time and latest end time. To adhere to legal and contract requirements the daily and weekly working time as well as the number of weekly workdays is limited for each nurse n individually. This allows different work contracts, e.g. part-time employments. Further constraints are rest times, the assignment of both days on weekends and the limitation of consecutive workdays. Additionally, we model the assignment of breaks in routes. Finally, the working regulations have to be considered also across periods.

The HHCP assigns each nurse n to a shift type s on each day t and determines the sequence of jobs in this shift, simultaneously. Thus, the routing is restricted by the shift assignments and determines the working times. The objective function minimizes the route lengths of all nurses and a penalty term for unassigned jobs.

Every week alterations in demands or capacities are possible: Clients enter (C^+) or leave (C^-) the services. Known clients can change their job demand concerning durations $(\hat{J}^D)$, time windows $(\hat{J}^T)$ or number (J^+, J^-). Nurses may start (N^+) or terminate their employment (N^-) as well as change their work contract $(\hat{N})$. The unchanged sets of nurses, clients and jobs are denoted by $\bar{N}$, $\bar{C}$ and $\bar{J}$, respectively.

3   Continuity Measures in a Rolling Horizon Approach

In this section we present the implemented measures for continuity of time, care and duty schedules in the HHCP using the previously defined sets $\bar{N}$, $\bar{C}$ and $\bar{J}$.

3.1   Continuity of Time

Let z_j be the start time in minutes of job j in the current and z'_j in the previous period. Nickel et al. [7] model the continuity of time (CoT) as the sum of deviations in start times (in minutes):

$$CoT^{Sum} = \sum_{j \in \bar{J}} |z_j - z'_j|. \tag{1}$$

Another possibility is to minimize the maximum of all time deviations:

$$CoT^{Max} = \max_{j \in \bar{J}} \{|z_j - z'_j|\}. \tag{2}$$

From a practical point of view, a small deviation in times may be tolerated, as they also occur due to uncertainties in driving times. Therefore, we introduce another measure, that considers only deviations of more than θ minutes:

$$CoT^t = \sum_{j \in \bar{J}} \delta_j, \quad \text{with } \delta_j = \begin{cases} 0, & \text{if } |z_j - z'_j| \leq \theta \\ |z_j - z'_j|, & \text{if } |z_j - z'_j| > \theta. \end{cases} \tag{3}$$

For heuristics, we further investigate the influence of using the quadratic sum:

$$CoT^{Quad} = \sum_{j \in \bar{J}} \delta_j^2. \tag{4}$$

3.2 Continuity of Care

Let N_c be the set of nurses assigned in the current period and N'_c the set of assigned nurses in the previous period. The continuity of care (CoC) can be modeled by minimizing the number of new nurses assigned to a client:

$$CoC^{Client} = \sum_{c \in \bar{C}} |\{N_c\} \setminus \{N'_c\}|. \tag{5}$$

In comparison to the client-based calculation, a job-based calculation is also possible:

$$CoC^{Job} = \sum_{j \in \bar{J}} \gamma_j, \qquad \text{with } \gamma_j = \begin{cases} 1, & \text{if } n_j \neq n'_j \\ 0, & \text{otherwise,} \end{cases} \tag{6}$$

where n_j is the currently and n'_j the previously assigned nurse.

3.3 Continuity of Duty Schedules

The first measure for continuity of duty schedules (CoD) is based on shift types. For each nurse and day, we determine if there is a difference in the assigned shift type:

$$CoD^{Type} = \sum_{t \in T} \sum_{n \in \bar{N}} \beta_{n,t}, \qquad \text{with } \beta_{n,t} = \begin{cases} 1, & \text{if } s_{n,t} \neq s'_{n,t} \\ 0, & \text{otherwise,} \end{cases} \tag{7}$$

where $s_{n,t}$ contains the assigned shift type of nurse n on day t in the current period and $s'_{n,t}$ in the previous period. To further consider the start and end of the shifts, a measure similar to the continuity of time is used as an alternative. We calculate the time deviations in start $(a_{n,t}, a'_{n,t})$ and end times $(e_{n,t}, e'_{n,t})$ for each nurse n:

$$CoD^{Time} = \sum_{t \in T} \sum_{n \in \bar{N}} \left(|e_{n,t} - e'_{n,t}| + |a_{n,t} - a'_{n,t}| \right). \tag{8}$$

3.4 Heuristic Rolling Horizon Approach

The implemented heuristic method is based on the ALNS metaheuristic introduced by Ropke and Pisinger [9]. Our destroy operators are job-, client-, nurse- or assignment-based. The set of jobs to remove is determined by the chosen operator, different sorting options and the degree of destruction. The repair operators are based on construction and insertion heuristics for the VRPTW. In the VRPTW no incorporation

Fig. 1 Continuity of time

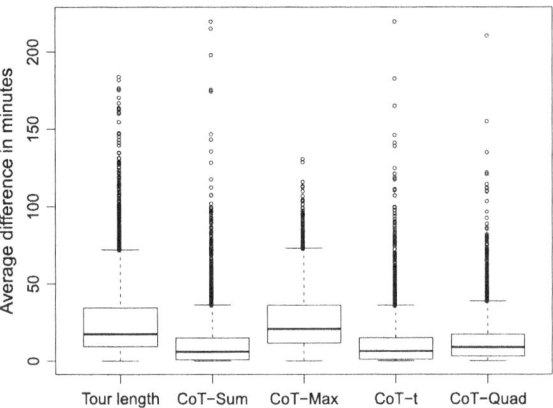

of qualifications and working regulations is required, we therefore use different greedy strategies for shift assignments if a new route is necessary. The objective represents the continuity between the previous and current week.

We use the solution σ' of the previous period as an initial solution. Therefore, we have to remove all jobs in J^- and of the clients in C^-. The new jobs J^+ and jobs assigned to the nurses in N^- and $\hat{N}$ are moved to the set of unassigned jobs. New nurses in N^+ stay unassigned. Thus, we already have a partial solution to start our ALNS with a repair operator and not with a destruction by a destroy operator.

4 Discussion

In this section we compare the results of the ALNS regarding continuity measures.[1] The results are computed for 15 instances that are generated based on previous literature, labor laws and data from the German Federal Statistical Office. We simulate the changes of a subsequent week by a Poisson process and calculate the changes using the arrival and departure times. The arrival rates depend on the type of alteration. The solution for the previous week is also computed by the ALNS. The results are calculated with the noted continuity measure (and penalty cost) as objective.

Figure 1 shows the results for the four CoT measures in comparison to the original objective in a Boxplot diagram. Each data point is a job in an instance. Three of the measures reduce deviations in starting times. Minimizing the maximum deviation is not beneficial in this case (2). On our instances considering the sum (1) and the sum with threshold $\theta = 10$ (3) performed best. We achieved only small deviations in start times for about 75 % of the jobs with these measures.

---

[1] All calculations were performed on resources provided by the Paderborn Center for Parallel Computing (Intel Xeon E5 processor 4 × 2.6 GHz CPUs and 2 GB RAM per run). The results are averaged over five runs with a computation time of five minutes.

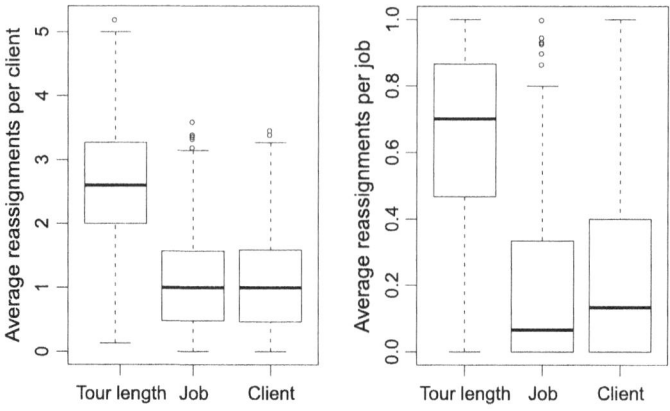

Fig. 2 Continuity of care: results for clients (*left*) and jobs (*right*)

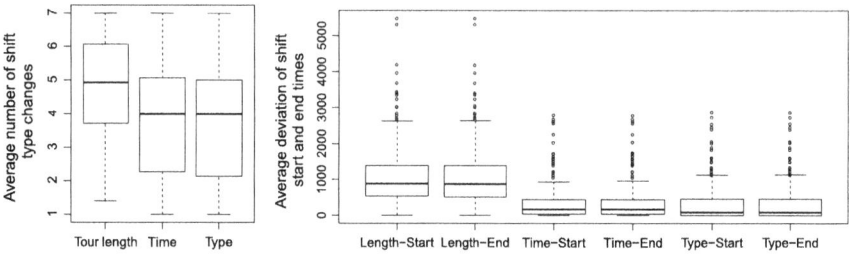

Fig. 3 Continuity of duty schedules analyzed by shift type (*left*) and start and end times (*right*)

Figure 2 compares the CoC measures. Both measures achieve a reduction in the number of newly assigned nurses. Analyzing the results on a client basis we get similar results. On a job-level, the difference of the measures becomes apparent. The job-based measure (6) explicitly reduces the reassignments for each job. In contrast, the client-based measure (5) allows reassignments to familiar nurses.

The results for CoD measures are shown in Fig. 3. Both measures reduce the number of shift changes. Nonetheless, the median is rather high (four reassignments in seven days). The reason for this are the working regulations. Especially, the limitation due to consecutive workdays and rest times influences the possible shift assignments. Looking at the sum of deviations in start and end times of shifts per week, the time-based measure (8) leads to a more balanced distribution.

5 Conclusion

In summary, we presented different continuity measures for the HHCP considering the categories care, time and duty schedules and integrated them into an ALNS heuristic. Nearly all the evaluated measures achieve improvements regarding the target category. In future work, we want to analyze the trade-off between tour lengths and continuity and compare the results of our heuristic to an exact approach.

References

1. Bennett, A.R., Erera, A.L.: Dynamic periodic fixed appointment scheduling for home health. IIE Trans. Healthc. Syst. Eng. **1**(1), 6–19 (2011)
2. Borsani, V., Matta, A., Beschi, G., Sommaruga, F.: A home care scheduling model for human resources. In: Proceedings of International Conference on Service Systems and Service Management, pp. 449–454. IEEE (2006)
3. Bowers, J., Cheyne, H., Mould, G., Page, M.: Continuity of care in community midwifery. Health Care Manag. Sci. **18**(2), 195–204 (2014)
4. Carello, G., Lanzarone, E.: A cardinality-constrained robust model for the assignment problem in home care services. Eur. J. Oper. Res. **236**(2), 748–762 (2014)
5. Gamst, M., Sejr Jensen, T.: Long-term home care scheduling. Technical report 12.2011, Technical University of Denmark, DTU Management Engineering (2011)
6. Haggerty, J.L., Reid, R., Freeman, G.K., Starfield, B.H., Adair, C., McKendry, R.: Continuity of care: a multidisciplinary review. Brit. Med. J. **327**(7425), 1219–1221 (2003)
7. Nickel, S., Schröder, M., Steeg, J.: Mid-term and short-term planning support for home health care services. Eur. J. Oper. Res. **219**(3), 574–587 (2012)
8. Nowak, M., Hewitt, M., Nataraj, N.: Planning strategies for home health care delivery. Technical report, Loyola University Chicago–Information Systems and Operations Management: Faculty Publications & Other Works. Paper 4
9. Ropke, S., Pisinger, D.: An adaptive large neighborhood search heuristic for the pickup and delivery problem with time windows. Transp. Sci. **40**(4), 455–472 (2006)
10. Yalcindag, S., Matta, A., Sahin, E., Shanthikumar, J.: A two-stage approach for solving assignment and routing problems in home health care services. In: Matta, A., Li, J., Sahin, E., Lanzarone, E., Fowler, J. (eds.) Proceedings of the International Conference on Health Care Systems Engineering, Springer Proceedings in Mathematics & Statistics, vol 61, pp. 47–60. Springer, Heidelberg (2013)

Map Partitioning for Accelerated Routing: Measuring Relation Between Tiled and Routing Partitions

Maximilian Adam, Natalia Kliewer and Felix König

Abstract In this paper we propose key figures to compare two different partitions of street maps. The first partition is used in navigation and reduce the cutting edges in the partition. The second is dependent on a new standard for the transmission of navigational data and has a rectangular shape. We build and analyze the relationship of the partitions with real map data and present first results.

1 Introduction

Onboard maps and routing function usually on a device in the car and do not crucially rely on cloud infrastructure. Navigation is an essential basic function of cars that is typically required to work even when not online. Precise and quick onboard routing requires preprocessing, while onboard maps are becoming increasingly modular to facilitate partial map downloads and updates for dynamic routing. Routing preprocessing typically relies on partitioning the street network while minimizing the number of roads crossing the partition. Whereas, modular maps are typically organized in rectangular tiles. When updating one part of a map for routing, preprocessed partitions, which overlap rectangular tiled data clusters have to be transmitted.

Considering this, it is likely that some of the data sent to the devices in tiled data clusters are irrelevant because they are not covered by the routing area as illustrated in Fig. 1. The dark area represents a routing area and the grey tiles represent the tiled data clusters. The dashed grey areas represent the tiled data clusters covering the same routing area. Therefore, the whole data from the dashed grey areas has to be sent when an update of a routing area is needed.

M. Adam (✉) · N. Kliewer
Freie Universität Berlin, Garystr. 21, 14195 Berlin, Germany
e-mail: max.adam@fu-berlin.de

N. Kliewer
e-mail: natalia.kliewer@fu-berlin.de

F. König
TomTom International B.V, An den Treptowers 1, 12435 Berlin, Germany
e-mail: Felix.Koenig@tomtom.com

© Springer International Publishing Switzerland 2017
K.F. Dœrner et al. (eds.), *Operations Research Proceedings 2015*,
Operations Research Proceedings, DOI 10.1007/978-3-319-42902-1_30

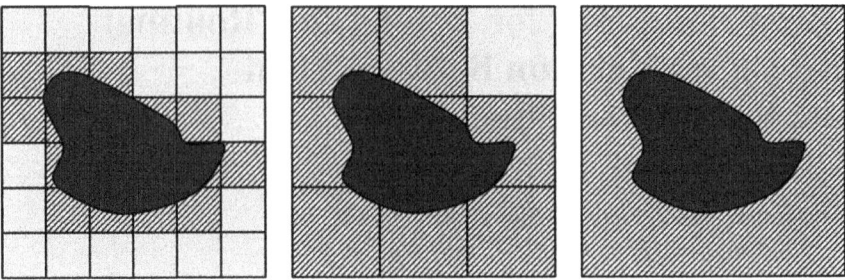

Fig. 1 Tiled data clusters overlapping a routing partition for modular maps with different granularity

Recent literature contains a vast amount of work on graph partitioning for routing preprocessing. The most promising approach of coarsening multilevel scheme has lately been applied by multiple clustering libraries and algorithms such as SCOTCH [5], METIS [3, 4], JOSTLE [6], KaPPa [2] or PUNCH [5]. All of them focus on minimizing the sum of cutting edges between areas and getting good overall balanced partitions with good connectivity. Sending large amounts of data through communication networks remains a major challenge for online routing, even though the existing literature has not yet covered this aspect.

In this paper, the map data structure is represented as a tiled partition of the street graph. We investigate the question of how to measure the relationship between tiled and routing partitions as well as what influences this relationship. To make the relationship of the two partitions measurable we propose performance indicators in Sect. 2 and evaluate these indicators in Sect. 3 by applying them on real street graphs sourced from Open Street Maps. An algorithm based on PUNCH [1] is used for the generation of balanced routing partitions. The results of the indicators are presented for the map group parameter combinations. Chapter 4 gives a brief summary and discusses further research possibilities.

2   Evaluation of Corresponding Partitions

For the definition of the measurements the following notation according to (Delling et al. [1]) is used in this paper. A partition $P = \{V_1, V_2, \ldots V_k\}$ of a street graph $G = (V, E)$ is a set of nonempty disjoint subsets such that $\bigcup_{i=1}^{k}(V_i) = V$. Given $V_i \in P$ is additionally defined, the function $eo(V_i) = E_i = \{e_1, \ldots, e_l\}$ with $\forall e \in E_i, e = u, v : u \in V_i \vee v \in V_i$. Following [5] the size $s_V(v)$ of a vertice $v \in V$ is defined as the number of points in the original graph represented by v and $s_E(e)$ gives back the number of edges represented in G by edge e. By extension, the vertice size $s_V(V_i) = \sum_{v \in V_i}(s_V(v))$ and the edge size $s_E(E_i) = \sum_{e \in E_i}(s_E(e))$ are used. The edges $e = \{u, v\}$ with $u, v \in V$ where $u \in V_i, v \notin V_i$ are denoted as $\delta(V_i)$, the number of cutting edges of the area V_i. By extension $\delta(P)$ is the count of all cutting edges between all areas of the partition P. A routing partition is denoted as P^R and a tiled partition as P^T. Likewise routing areas are named V^R and tiled areas V^T. The value ε controls the tolerated imbalance of routing areas when generating partitions.

For measuring the relationship between tiled and routing areas, we propose the following alternative performance indicators. Note that each performance indicator is computed for each area of a certain partition. In order to analyze a whole partition, these values have to be summarized with statistics, such as the arithmetic mean.

1. ToR (Tiled areas of Routing areas):
   This straightforward performance indicator measures how many tiled areas are sharing vertices with each individual routing area. This performance indicator enables an assessment of the unnecessary sent data as well as insight into the relationship.

$$ToR(V^R) = |\forall V^T \in P^T : \exists v \in V^T \wedge v \in V^R| \tag{1}$$

2. RoT (Routing areas of Tiled areas):
   For each area of the tiled partition, the number of routing areas that share at least one vertice with this tiled area are computed. As the complimentary indicator to ToR, RoT also gives insight into the relationship of the partitions.

$$RoT(V^T) = |\forall V^R \in P^R : (\exists v \in V^R \wedge v \in V^T)| \tag{2}$$

3. ERT (Edge size of Routing areas per edge size of associated Tiled areas):
   Taking into account that the dynamic data is mostly bounded to streets which themselves are represented as edges in the street graphs, this performance indicator shows the percentage of data required (edges of the routing area) of all sent data (all edges of the overlapping tiled areas with this routing area). Edges that are in two tiled areas are counted only once.

$$ERT(V^R) = \frac{s_E(eo(V^R))}{\sum_{V^T \in P^T : (\exists v \in V^T \wedge v \in V^R)} (s_E(eo(V^T)))} \tag{3}$$

4. VRT (Vertice size of Routing areas per vertice size of associated Tiled areas):
   The VRT computes the ratio of necessary to unneccesary data for each routing area. The difference to the ERT is that the assessment of data is obtained by counting the vertices of the areas.

$$VRT(V^R) = \frac{s_V(V^R)}{\sum_{V^T \in P^T : (\exists v \in V^T \wedge v \in V^R)} (s_V(V^T))}. \tag{4}$$

3 Experiments and Results

Our algorithm is based on Graph partitioning using Natural Cuts [1]. However, in our algorithm, some steps of the original PUNCH algorithm are modified and some are excluded. We were able to validate that our generated routing partitions show approximately the same vlaues for apparent key figures compared with the results of PUNCH.

Our algorithm consists of three phases to achieve a balanced partition. The goal is to find a balanced partition $P = \{V_1, V_2, \ldots, V_k\}$ of a graph $G = (V, E)$ where $\forall V_i \in P : s_v(V_i) = \lfloor (1 + \varepsilon)\lceil n/k \rceil \rfloor$ with $n = s_v(V)$ and ε as the balance parameter. Differing from PUNCH, in this paper, k is chosen automatically in dependence of the upper bound for the number of vertices of each routing area U. Therefore, the upper bound U is applied as $\forall V_i \in P : s_v(V_i) \leq U$. However, simplified balancing heuristics of PUNCH are applied resulting in a partition where the areas are relatively balanced. As in the original PUNCH, the balanced partition minimizes the sum of the cutting edges between the areas by applying a filtering step and an assembly step. For the values of parameter U we chose 10,000 and 20,000 in order to generate differing but comparable routing areas. The remaining parameters correspond to PUNCH.

We applied the proposed indicators on nine real street maps divided into three map groups depending on the average vertices per square kilometer:

- Urban Maps with the graphs of Berlin, Hamburg and Bremen (75 to 90 vertices per km^2),
- Regional Maps containing graphs of Thringen, Sachsen Anhalt and the country Luxembourg (vertices per km^2 of 9 to 14) as well as
- Sparse Maps with the maps of Iceland and the US states Wyoming and Alaska (0.8 to 0.03 vertices per km^2).

The indicators vary among the map groups and different U values. RoT have values of four for Urban Maps and 250 for Sparse Maps. This is because routing areas are mostly dependent on the chosen U whereas the tiled areas are dependent on the depicted geographical surface. On average, doubling of routing area size led to only slightly larger values for RoT. ToR had values of nearly one for Regional Maps and Sparse Maps. This shows that the routing partition have a good structure. Lot of tiled areas having only one corresponding routing area means retrospective that the routing partition is already optimal for this tiled areas. It can be stated that RoT and ToR are good for giving an intuitive approach to the analysis how good the partitions are.

Figure 2 visualizes the results of the experiments for these measurements for different groups of maps and two different values of U. The red bars show the average values of the VRT and the green bars display the average values of the ERT. It is notable that the method of measuring area size by the number of vertices or by the number of edges has an insignificant effect on the comparison between the routing area size and the size of all associated tiled areas for map groups with high geographic densities. For Sparse Maps we observe up to 4 % higher VRT-values. We suggest using solely the VRT for further measurements as it is more common for partitioning algorithms to measure the size of areas with the number of vertices and not with the number of edges.

The results of all indicators support the assumption regarding influence of vertices per depicted space-ratio on the relation of the corresponding partitions. More vertices per km^2 leads to smaller routing areas in comparison with tiled areas. It follows that

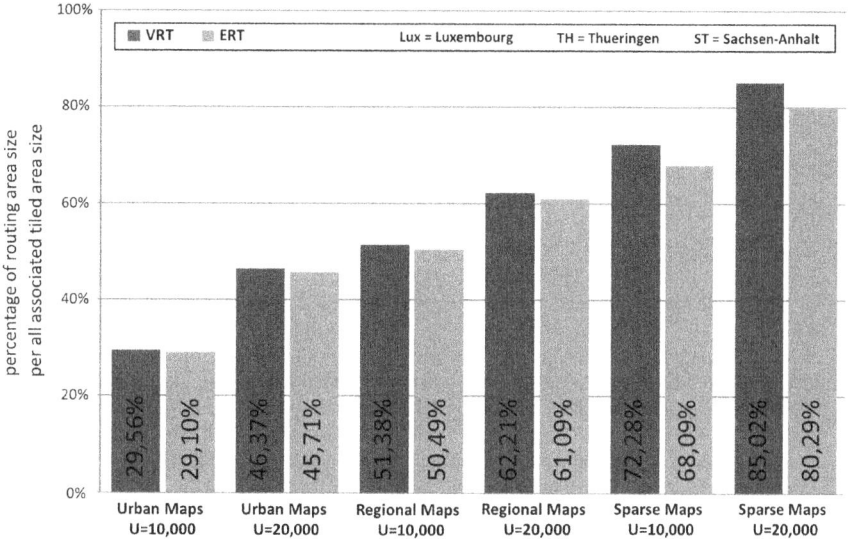

Fig. 2 Average values for the indicators *VRT* and *ERT* for different map groups and different values of upper bound *U*

on average a higher percentage of vertices and edges is not included in a certain routing area but in the associated tiled areas of this routing area. Vice versa, we observe that less vertices per km^2 leads to significantly better relations for the ratio of required and sent data amount reflected also in *ToR* values of 1.02 and *RoT* values up to 258 for Sparse Maps.

4   Summary and Outlook

The proposed performance indicator *VRT* enables a good estimation of the ratio of required to sent data, even when assuming that the data is dependent on the streets. With the indicators *ToR* and *RoT*, it is possible to assess and evaluate the relationship of the routing partitions with tiled data clusters. We observed that street graphs with less vertices per km^2 have clearly better results for the ratio of required and sent data than street graphs with more vertices per km^2.

Further work is required for constructing indicators that are more specific to improve the mapping of the applied algorithms in the automotive navigation industry. Additionally, there is a need to investigate whether existing algorithms such as PUNCH can be modified in such a way that they consider a given tiled partitioning. This modifications is necessary to produce routing partitions which offer a good trade-off between routing benefits and compatibility with the tiled partition to yield a better ratio of required to sent data.

References

1. Delling, D., Goldberg, A.V., Razenshteyn, I., Werneck, R.F.: Graph partitioning with natural cuts. In: 2011 IEEE International Parallel and Distributed Processing Symposium (IPDPS), pp. 1135–1146. IEEE (2011)
2. Holtgrewe, M., Sanders, P., Schulz, C.: Engineering a scalable high quality graph partitioner. In: 2010 IEEE International Symposium on Parallel and Distributed Processing (IPDPS), pp. 1–12. IEEE (2010)
3. Karypis, G., Kumar, V.: Multilevelk-way partitioning scheme for irregular graphs. J. Parallel Distrib. Comput. **48**(1), 96–129 (1998)
4. Karypis, G., Kumar, V.: Parallel multilevel series k-way partitioning scheme for irregular graphs. SIAM Rev. **41**(2), 278–300 (1999)
5. Pellegrini, F., Roman, J.: Scotch: a software package for static mapping by dual recursive bipartitioning of process and architecture graphs. In: High-performance Computing and Networking, pp. 493–498. Springer, Heidelberg (1996)
6. Walshaw, C., Cross, M.: JOSTLE: parallel multilevel graph-partitioning software an overview. Mesh partitioning techniques and domain decomposition techniques, pp. 27–58 (2007)

OD Matrix Estimation Using Smart Card Transactions Data and Its Usage for the Tariff Zones Determination in the Public Transport

Michal Koháni

Abstract OD matrix is an important input parameter for a large number of optimization problems especially in the public transportation. Traditional approaches of obtaining OD matrix, such as surveys, could not enable us to obtain comprehensive and complex data on passengers and their journeys. In cases where passengers in transportation use smart cards, we can obtain more accurate data about the passengers journeys even in cases where these data are incomplete. In this contribution we present a trip-chaining method to obtain passenger journeys from smart card transactions data. Using these transactions data with the combination of data from other sources such as street maps, timetable and bus line routes, we are able to obtain origins and destinations of passenger journeys and also information about the changes between the lines on the passengers journey. Designed approach is verified on the case of the Zilina municipality with a data set with real passengers smart card transactions for a period of one week. Obtained OD matrix is later used as the input for the solving of tariff zones partitioning problem in the Zilina municipality area and these results are also presented in this paper.

1 Introduction

Public transport system planners deal with various types of optimisation problems such as routes design, timetable design, tariff design etc. In almost all problems, one of the major input is the origin-destination (OD) matrix, which describes passenger flows between stops, stations and municipalities in transportation network. Obtaining the OD matrix is often very difficult task. Traditional approaches, such as surveys, could not enable us to obtain comprehensive and complex data on passengers and their journeys and are also quite expensive. At this time, there are more possibilities, how to get input data for OD matrix. Planners can use data from various types of Automated Data Collection Systems, such as the Automated Fare Collection system

M. Koháni (✉)
Department of Mathematical Methods and Operations Research,
University of Žilina, Univerzitna 8215/1, Žilina, Slovakia
e-mail: michal.kohani@fri.uniza.sk

© Springer International Publishing Switzerland 2017
K.F. Dœrner et al. (eds.), *Operations Research Proceedings 2015*,
Operations Research Proceedings, DOI 10.1007/978-3-319-42902-1_31

(AFC), the Automated Vehicle Location system (AVL), and the Automatic Passenger Counting system (APC) [6]. In this paper we propose a trip-chaining algorithm for OD matrix calculation based on municipality smart card transaction data. Proposed approach is verified on data from the Zilina municipality, where we have a data set with passengers smart card transactions for the period of one week. Obtained OD matrix is used as the input for the solving of the tariff zones partitioning problem and these results are also presented in this paper.

2   OD Matrix Calculation

In cases where passengers in transportation use the smart cards, we can obtain more accurate data about the passengers journeys even in cases where these data are incomplete. In Slovakia and surrounding countries, smart cards for public transport become more and more popular. Due to the distance tariff, in regional transport we can get information about boarding and destination stop of all passengers, so the calculation of the OD matrix is not a problem. In our case we are dealing with the municipal public transportation, where due to the unit tariff we know information only about boarding stop, so the destination stop must be calculated or estimated. We can get following information from AFC system:

- serial number of passenger's smart card, name and ID number of boarding stop,
- date and time of boarding, line and connection (trip) number and route variant.

Based on these data, we propose a trip-chaining algorithm to obtain OD matrix from smart card transactions. Similar approaches were mentioned in [1, 3, 6]. Basic idea of the algorithm assumes that the destination stop of the journey can be the boarding stop of passenger's immediately following journey. We need to sort all transaction data by unique serial card number, day and time. Then the algorithm can be described by following steps:

Trip-chaining Algorithm

Step 0: Sort all transaction data (record) by unique serial card number and then by day and time in ascending order.

Step 1: Select next card number. If all serial card numbers were processed, terminate.

Step 2: Select next day of selected card number. If all days were processed or selected day contains only one record, go to the Step 1.

Step 3: Select next record of selected card in selected day. If all records of selected card in selected day were processed, go to the Step 2.

Step 4: Evaluate selected record and go to the Step 3.

Evaluation of records in the Step 4 and journey construction consists of following rules:

- If selected record is the first record of the day, boarding stop of the journey is the stop in this record.

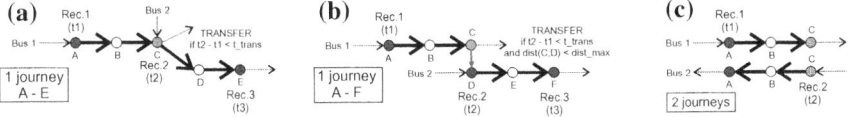

Fig. 1 Transfer evaluation in the algorithm

- If the time difference between validation time of two consecutive records 1 and 2 is less than t_{trans} minutes, evaluate transfer, see Fig. 1.
- A transfer is in the situation where the routes of two evaluated consecutive records are not in the opposite direction or if routes of two consecutive evaluated records do not intersect and the distance between the boarding stop of record 2 and the nearest stop on route from record 1 is at most $dist_{max}$ meters.
- If the record is not considered as transfer, set the destination stop of the journey to boarding stop of record 2, if the boarding stop of record 2 is on the route of record 1. Otherwise find the nearest stop on the route of record 1 to the boarding stop of record 1.

3 Tariff Zones Partitioning Problem

Let all stops in the public transport network constitute the set of nodes I. Stops i and j from set I are connected by the edge $(i, j) \in E$, if there is a direct connection by public transport line between these two stops. Number of passengers between stations i and j is b_{ij} (OD matrix). To describe paths of passenger flows we introduce parameter a_{ij}^{rs}. Value of parameter a_{ij}^{rs} is equal to 1 if the edge (r, s) is used for travelling from stop i to stop j and 0 otherwise. For each pair of stops i and j is c_{ij} the current or fair price of travelling between these two stations.

We introduce binary variables y_i, which represent a fictional centre of the zone and is equal to 1 if there is a centre of the zone in node i and 0 otherwise. For each pair of stops i and j we introduce variable z_{ij} which is equal to 1 if the station j is assigned to the zone with centre in the node i and 0 otherwise. We expect to create at most p_{max} tariff zones.

There are more possibilities how to set new price in zone tariff. Schöbel in [5] proposed solution of fare problem with fixed zones to obtain new fares for trips with various number of travelled zones. In [2] a unit prices f_1 and f_2 were proposed. Parameter f_1 is the price for travelling in the first zone and unit price f_2 for travelling in each additional zone. To calculate new price of the trip between stops i and j, we need to know number of zones crossed on this trip. This can be replaced by the calculation of crossed zone borders accordingly to [5]. We assume that stop can be assigned only to one zone, so the border between zones is on the edge. We introduce binary variable w_{rs} for each existing edge $(r, s) \in E$, which is equal to 1 if stops r and s are in the different zones and is equal to 0 otherwise. New price n_{ij} determined by the number of crossed zones will be calculated as follows:

$$n_{ij} = f_1 + \sum_{(r,s) \in E} f_2 \cdot a_{ij}^{rs} \cdot w_{rs} \tag{1}$$

As the objective function in the model we use the average deviation between current and new price for all passengers, according to the advices of experts in [5]. Mathematical model of zone partitioning with fixed prices and number of zones can be written as follows (2)–(8):

$$minimize \ dev_{avg} = \frac{\sum_{i \in I} \sum_{j \in I} |c_{ij} - n_{ij}| b_{ij}}{\sum_{i \in I} \sum_{j \in I} b_{ij}} \tag{2}$$

$$subject \ to : \sum_{i \in I} z_{ij} = 1, \ for \ j \in I \tag{3}$$

$$z_{ij} \leq y_i, \ for \ i, j \in I \tag{4}$$

$$\sum_{i \in I} y_i \leq p_{max} \tag{5}$$

$$z_{ij} - z_{ik} \leq w_{jk}, \ for \ i \in I, (j, k) \in E \tag{6}$$

$$y_i \in \{0, 1\}, z_{ij} \in \{0, 1\}, \ for \ i, j \in I \tag{7}$$

$$w_{ij} \in \{0, 1\}, \ for \ (i, j) \in E \tag{8}$$

Conditions (3) ensure that each stop is assigned to one zone only. Conditions (4) ensure that each stop is assigned only to the existing zone centre. Condition (5) ensures that at most p_{max} tariff zones is created. Conditions (6) are coupling between variables for allocation of the station to the zone and variables for determining the zone border on the edge (j, k).

Objective function (2) in the model is not a linear function. In [2] the linearisation of the model was proposed and the model was solved using IP solver. To determine the optimal values of parameters in the model, a two-phase procedure was used. In the first phase the optimal number of zones was determined and the model with different settings of parameters f_1, f_2 and p_{max} was solved in the second phase.

4 Numerical Experiments—Case Study

To verify proposed algorithm and subsequent solution of tariff zones partition problem, we make the case study based on smart card transaction data provided by Žilina Municipality transport operator DPMŽ. Data set consist from 111293 records of

Table 1 Percentage of the successfully processed records from the data set DPMŽ

Transfer time t_{trans} (min)	$dist_{max} = 300$ m (%)	$dist_{max} = 600$ m (%)	$dist_{max} = 900$ m (%)
30	71.73	79.64	82.19
60	70.27	78.51	81.76

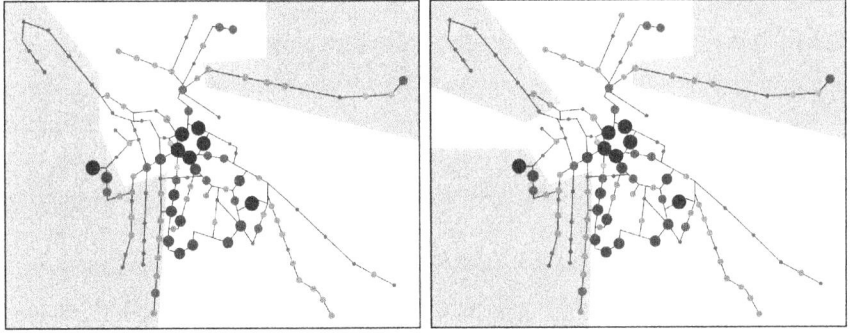

Fig. 2 Solution of zone partition problem with 3 and 4 zones

11300 smart cards used in the time period between October 6 to October 12, 2014. To process the dataset we use software tool developed in [4].

We evaluate various setting of parameters t_{trans} and $dist_{max}$ in trip-chaining algorithm to get OD matrix with the highest percentage of the successfully processed records from the data set. We use values 30 and 60 min as the value of parameter t_{trans} and values 300, 600 and 900 m as the value of parameter $dist_{max}$ (Table 1).

We tried to evaluate obtained results in terms of accuracy. Unfortunately, there is no relevant data source to compare obtained results, so the evaluation was made by manual check of OD matrix and tracking data. We found that results with $dist_{max} = 900$ m contain wrongly assessed journeys. Due to this fact we select as the best OD matrix with parameters $dist_{max} = 600$ m and $t_{trans} = 30$ min.

In the next step we use obtained OD matrix and tracking data to solve tariff zones partitioning problem. Transport network of Žilina Municipality consists of 120 stops. Current prices of travelling depend on the distance only partly. For all journeys up to 5 stops without transfer passenger pays 0.55 Eur, for all journeys for more than 5 stops without transfer 0.65 Eur and for journey with transfer 0.80 Eur.

In our case study we solve the problem for parameter $p_{max} = \{3, 4, 5, 6\}$. Regarding to the current prices, we use for parameter f_1 values 0.50 and 0.55 Eur and for parameter f_2 values 0.05 and 0.1 Eur. In all cases the best results are obtained for values $f_1 = 0.55$ and $f_2 = 0.1$. Results are shown on the scheme of transport network in the Figs. 2 and 3. The diameter of the circles represent approximate number of passengers using the stop for starting or ending their journeys.

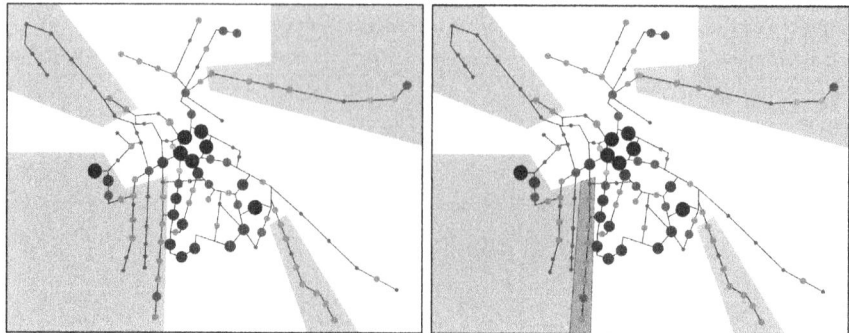

Fig. 3 Solution of zone partition problem with 5 and 6 zones

5 Conclusion

In this paper we described the algorithm for obtaining OD matrix from incomplete passengers' smart card data and resulted OD matrix is then used for the calculation of tariff zones partition problem. We performed the case study with real smart card data. From the results it is obvious that it is necessary to carry out several experiments in setting parameters of the algorithm, as in certain values the result appeared erroneous data. From the results of zone partitioning problem it is obvious, that created input data are relevant. In newly created zones are mostly suburban zones with smaller demand for travelling and higher distances to the centre, as we expected. Results were discussed with local authorities and to be able to apply results in practice it is necessary to test robustness of the design e.g. by fuzzy approach.

Acknowledgements This work was supported by the research grants VEGA 1/0463/16 - "Economically efficient charging infrastructure deployment for electric vehicles in smart cities and communities" and APVV-0760-11 "Designing Fair Service Systems on Transportation Networks". We would like to thank the transport operator DPMŽ for providing necessary data for this research.

References

1. Jánošíková, Ľ., Slavík, J., Koháni, M.: Estimation of a route choice model for urban public transport using smart card data. Transp. Plann. Technol. **37**, 638–648 (2014)
2. Koháni, M.: Zone partitioning problem with given prices and number of zones in counting zones tariff system, In: SOR 2013: Proceedings of the 12th International Symposium on Operational Research: Dolenjske Toplice, Slovenia, pp. 75–80 (2013)
3. Majer, T.: Modelovanie prúdov cestujúcich z neúplných údajov, In: Využitie kvantitatívnych metód vo vedeckovýskumnej praxi, Bratislava: EKONÓM, pp. 62–66 (2009) (in Slovak)
4. Múčka, J.: Aplikácia na analýzu dát o dopravných prúdoch cestujúcich vo verejnej doprave, Bechelor thesis, University of Žilina (2015) (in Slovak)
5. Schöbel, A.: Optimization in Public Transportation. Stop Location, Delay Management and Tariff Zone Design in a Public Transportation Network. Springer, Heidelberg (2006)
6. Zhao, J.: The Planning and Analysis Implications of Automated Data Collection Systems: Rail Transit OD Matrix Inference and Path Choice Modeling Examples: Diploma thesis, Tongji University (2001)

Location Planning of Charging Stations for Electric City Buses

Kilian Berthold, Peter Förster and Brita Rohrbeck

Abstract Fuel prices on the rise and ambitious goals in environment protection make it increasingly necessary to change for modern and sustainable powertrain technologies. This trend also affects the transportation sector, and electric buses with stationary charging technology grow in popularity. Their launch however is still costly, and an optimal choice of the charging stations locations is crucial. In our paper we present a mixed integer model that determines an optimal solution concerning the investment costs for a single bus line. It is mainly constrained by an energy balance. Hence, energy consumption on the driven paths and of auxiliary consumers have to be considered as well as holding times at bus stops and thus the potentially recharged amount of electric energy. Additionally, we take account of service life preservation of the batteries as well as beneficial existing infrastructure and constructional restrictions. We give an overview of our results obtained from real world data of the bus network of Mannheim. In our tests we consider different scenarios regarding passenger volume, traffic density and further factors.

1 Introduction

Electric buses can be configured and charged by different technical approaches [7]. Charging strategies could be exclusively over-night-charging in the bus depot, battery swap [6] or opportunity charging during the daily service at defined charging points. The advantage of opportunity charging is the possibility of using smaller batteries with lower vehicle costs, less weight and less technical ageing-effects. On the

K. Berthold (✉)
Karlsruhe Institute of Technology, Kaiserstrasse 12, 76131 Karlsruhe, Germany
e-mail: kilian.berthold@kit.edu
B. Rohrbeck
e-mail: brita.rohrbeck@kit.edu

P. Förster
Westernacher Business Management Consulting AG, Im Schuhmachergewann 6,
69123 Heidelberg, Germany
e-mail: peter.foerster@westernacher.com

© Springer International Publishing Switzerland 2017
K.F. Dœrner et al. (eds.), *Operations Research Proceedings 2015*,
Operations Research Proceedings, DOI 10.1007/978-3-319-42902-1_32

downside, a charging infrastructure along the bus lines is needed. A city bus usually follows a route with a lot of physically identical stops that have to be connected logically. This makes planning for charging stations of city buses more complex than other location problems, and few literature exists. We developed a mixed integer model for one bus line. Contrary to positioning charging stations for individual electric vehicles, this predetermined routine has to be taken into account. References [3–5] suggest different approaches, the latter with a more technical background. To determine optimal charging locations for private electric vehicles [2] developed a model basing on car park sites.

In the next section we explain our model in detail. Section 3 focusses on computational tests and evaluation. Finally, we give an outlook on future research.

2 Problem Formulation

The main challenge in modelling the problem of locating stationary charging stations is to connect the stops of each tour with each other, i.e. if at location i of the first tour a charging station is built, then this charging station also exists in every other tour at location i. Therefore, we introduce the set $\mathcal{T} = \{0, 1, \ldots, T, T + 1\}$ of tours. Here, $t = 1, \ldots, T$ stand for the actual tours the bus drives during one day, whereas $t = 0$ stands for the outbound trip from the depot to the first bus stop and $t = T + 1$ for the inbound path from the terminus back to the depot after the last tour. Let $\mathcal{N}$ subsume all potential charging stations, comprising the respective T replications of the bus stops of the tour route ($\mathcal{N}^{tour}$) as well as the nodes $\mathcal{N}^{out}$ and $\mathcal{N}^{in}$ from and back to the depot. Hence, $\mathcal{N} = \mathcal{N}^{out} \cup \mathcal{N}^{tour} \cup \mathcal{N}^{in}$, see Fig. 1.

To model this location problem several decisions have to be taken. First of all we need an indicator y_i, $i \in \mathcal{N}$ whether a charging station is built in a node i ($y_i = 1$) or not ($y_i = 0$). For the ease of understanding every potential charging location is called bus stop as well, no matter whether it is actually a bus stop or not. If a bus

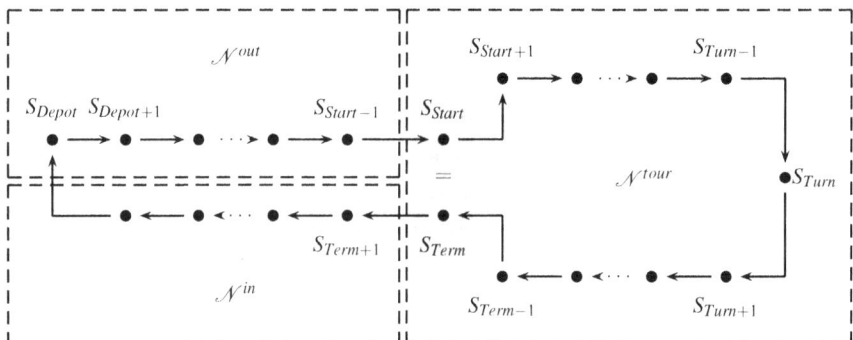

Fig. 1 The route of a bus as graph

stops at a station with a charging infrastructure, i.e. $y_i = 1$, it will load a certain amount of electricity. x_{it} designates the energy charged at a node i, $i \in \mathcal{N}$, during tour t, $t \in \mathcal{T} = \{0, 1, \ldots, T, T+1\}$. While driving, the bus consumes and charges energy. Let e_{it}, $i \in \mathcal{N}$, $t \in \mathcal{T}$ be the amount of energy that the battery of the bus contains when leaving node i in tour t.

The objective of our model is to minimise the overall costs. These costs consist of the installation costs for the charging stations as well as of the acquisition costs for the buses. The installation costs sum up to $\sum_{i \in \mathcal{N}} \frac{1}{a_i} f_i y_i$: If a charging station is installed in a node $i \in \mathcal{N}$, i.e. $y_i = 1$, fixed installation costs f_i arise. In the case of multiple use of a potential charging station in node i, the costs have to be distributed equally among the physically identical nodes. Therefore, a parameter $a_i \in \mathbb{N}, i \in \mathcal{N}$ is required that indicates the number of nodes located at the same physical location as node i. The costs for the buses amount to $f^b \cdot b$ with f^b being the fixed costs per bus and b being the number of buses circulating on the regarded bus line per day. Note that the number of buses is a parameter in the basic model. If different types of buses are considered or restructuring the timetable is possible, we will gain further variability, and we will leave this figure open for decision, too.

Placing the charging stations has to follow numerous constraints. It is reasonable that a charging station is installed in the depot where the bus can charge over night. We hence set $y_{S_{Depot}} = 1$ (1). As a result, the bus starts from the depot with the maximal battery level $\overline{E}$: $e_{S_{Depot},0} = \overline{E}$ (2). This level must not be exceeded throughout all the trip. Therefore, we need to set $e_{it} \leq \overline{E}$ (3) for all segments of the trip. The first segment concerns all stations on the outbound tour, i.e. $i \in \mathcal{N}^{out}$ during tour $t = 0$. The second segment contains all stations $i \in \mathcal{N}^{tour}$ during all actual tours $t \in \{1, \ldots, T\}$, whereas the inbound segment of the trip contains all stations $i \in \mathcal{N}^{in}$ during the return marked by $t = T + 1$. Let $\mathcal{N}_1$ denote the union of these three sets.

On its trip the energy level of the bus must not fall to zero or rather below a safety value $\underline{E} > 0$. Leaving a station i in tour t, a bus has the energy level e_{it}. Driving to the next station it consumes c_{it}. Hence, arriving at the following station the energy level is $e_{it} - c_{it}$, thus we need $e_{it} - c_{it} \geq \underline{E} \ \forall \ (i, t) \in \mathcal{N}_2 := (\mathcal{N}^{out} \times \{0\}) \cup (\mathcal{N}^{tour} \setminus \{S_{Term}\} \times \{1, \ldots, T\}) \cup (\mathcal{N}^{in} \cup \{S_{Term}\} \times \{T + 1\})$ (4). At a certain station i while on tour t the bus charges the amount of energy $x_{i,t}$. Leaving the previous station $i - 1$ it has the energy level $e_{i-1,t}$ and consumes on the connecting edge $c_{i-1,t}$. The energy balance can thus be calculated by $e_{it} = e_{i-1,t} - c_{i-1,t} + x_{it} \ \forall \ (i, t) \in \mathcal{N}_3 := (\mathcal{N}^{out} \setminus \{S_{Depot}\} \times \{0\}) \cup (\mathcal{N}^{tour} \setminus \{S_{Start}\} \times \{1, \ldots, T\}) \cup (\mathcal{N}^{in} \times \{T + 1\})$ for the three parts of the day trip being outbound path, loop and inbound path (5). In a similar way we model the transition from the outbound path to the first tour, depicting the energy level of the first station of the first tour: $e_{S_{Start},1} = e_{S_{Start-1},0} - c_{S_{Start-1},0} + x_{S_{Start},1}$ (6). The transition from one tour to the next can be modelled in an easier way since the terminus node of the preceding journey corresponds to the starting node of the following tour. Thus, their energy levels are identical: $e_{S_{Start},t+1} = e_{S_{Term},t} \ \forall \ t \in \{1, \ldots, T\}$ (7). Analogously, the energy level at the first node of the inbound path is equal to the energy level of the last node during the last tour: $e_{S_{Term},T+1} = e_{S_{Term},T}$ (8). Moreover, the charged energy before the first stop in a tour t, $t \in \{2, \ldots T\}$, equals the amount of energy charged at the terminus of the previous tour: $x_{S_{Start},t} = x_{S_{Term},t-1}$ (9). Apart from modelling the energy development some further conditions must be fulfilled. Energy can only be charged at a station equipped with the charging infrastructure. Additionally, the amount that can be charged at a station depends on the power of the charging station and the time spent at that station. $l_{it} \in \mathbb{R}$, $i \in \mathcal{N}$, $t \in \mathcal{T}$ limits the amount of energy that can be charged on tour t at the potential charging station i. Hence, $x_{it} \le l_{it} y_i \ \forall \ (i, t) \in \mathcal{N}_1$ (10). Due to structural conditions it might not be possible to install a charging station in some nodes. The set of these restricted nodes shall be denoted by $\mathcal{R}$. $y_i = 0 \ \forall i \in \mathcal{R}$ (11) averts the implementation of a charging station in these nodes. The last constraint is a technical one. We need it to link the nodes of the different tours, but also to associate other potential charging stations that are located in the same point. This is for instance the case for S_{Start} and S_{Term}, therefore $(S_{Start}, S_{Term}) \in L$. We introduce the parameter l_{i_1,i_2},

$$l_{i_1,i_2} = \begin{cases} 1 & i_1 \text{ and } i_2 \text{ are situated in the same physical location} \\ 0 & \text{otherwise} \end{cases}$$

The set $L := \{(i_1|i_2): l_{i_1,i_2} = 1, i_2 > i_1\}$ contains all tuples of nodes that are situated in the same physical location. If a charging station is built in any node, then it is also built in its equivalents, so for all tuples $(i_1|i_2) \in L$ we need $y_{i_1} = y_{i_2}$ (12). Below, we summarise the whole model:

$$\min \sum_{i \in \mathcal{N}} \frac{1}{a_i} \cdot f_i \cdot y_i + f^b \cdot b$$

$$\text{s.t.} \quad y_{S_{Depot}} = 1 \tag{1}$$

$$e_{S_{Depot},0} = \overline{E} \tag{2}$$

$$e_{it} \le \overline{E} \qquad\qquad \forall \ (i, t) \in \mathcal{N}_1 \tag{3}$$

$$e_{it} - c_{it} \ge \underline{E} \qquad\qquad \forall \ (i, t) \in \mathcal{N}_2 \tag{4}$$

$$e_{it} = e_{i-1,t} - c_{i-1,t} + x_{it} \qquad\qquad \forall \ (i, t) \in \mathcal{N}_3 \tag{5}$$

$$e_{S_{Start},1} = e_{S_{Start-1},0} - c_{S_{Start-1},0} + x_{S_{Start},1} \tag{6}$$

$$e_{S_{Start},t+1} = e_{S_{Term},t} \qquad\qquad \forall \ t \in \{1, \dots, T\} \tag{7}$$

$$e_{S_{Term},T+1} = e_{S_{Term},T} \tag{8}$$

$$x_{S_{Start},t} = x_{S_{Term},t-1} \qquad\qquad \forall \ t \in \{2, \dots T\} \tag{9}$$

$$x_{it} \le l_{it} y_i \qquad\qquad \forall \ (i, t) \in \mathcal{N}_1 \tag{10}$$

$$y_i = 0 \qquad\qquad \forall i \in \mathcal{R} \tag{11}$$

$$y_{i_1} = y_{i_2} \qquad\qquad \forall \ (i_1|i_2) \in L \tag{12}$$

$$y_i \in \{0, 1\} \qquad\qquad \forall i \in \mathcal{N} \tag{13}$$

$$x_{it}, e_{it} \ge 0 \qquad\qquad \forall \ (i, t) \in \mathcal{N}_1 \tag{14}$$

3   Computational Results

We compared our model using the bus line 63 in the city of Mannheim, Germany with the work of the PRIMOVE Mannheim Project. The tour of line 63 is 9 km long, contains 23 bus stops (way and return) and takes 40 min for the whole circuit. Two buses do service at 20 min intervals.

For a charging station we assumed €250 K being the average costs of a Bombardier PRIMOVE 200 charging station as used in Mannheim. We deduced this value from the public project description [8] and the fact that in Mannheim six charging stations have been installed. We assume these costs to be less by €10 K if the existing bus stop is not situated on the traffic lane and by further €25 K if an overhead wire already exists at that station. For our experiments we measured the exact duration of a stop and the number of people boarding and deboarding over a couple of days, in particular at different times of day. From these data and from the technical data of the buses [1] as well as from the traffic properties of the buses' routing we could derive the energy amounts c_{it} needed for every part of the trip. Alike, we determined l_{it}, the limits of the amount of energy that can be charged maximally on tour t at the potential charging station i. Where data was uncertain we decided for a conservative value.

The lifespan of the buses' batteries extends tremendously if the State of Charge (SOC) does not drop to very low values or rises close to its maximum. We hence restrict the energy level to fluctuate between 20 and 80 % of the battery's 60 kWh capacity and chose these values for $\underline{E}$ and $\overline{E}$.

We implemented our formulation in OPL using the optimization software IBM ILOG CPLEX Optimization Studio, version 12.5.1 on a four kernel computer with 2.4 GHz and 8 GB RAM. We tested different instances regarding traffic, external factors like temperature and driving behaviour. Solving the different instances never took more than 10 s. According to our model for an average parameter setting eight charging stations would be optimal with a cost of €1.78 M. In practice, the PRIMOVE Mannheim only envisages six, partially different charging stations. With our data this would be realised at a cost of €1.39 M which lies 22 % below our solution but would not be feasible. Indeed, also in practice the current configuration of PRIMOVE Mannheim turned out to be insufficient: At present, in addition to the electric buses regular diesel buses are deployed occasionally over the course of a day to give the electric buses time to recharge. In fact, the system in Mannheim is still in a development phase, and experience shall be gathered on line 63 before electrifying more lines.

In a worst case scenario we start from the premise that at the terminus of line 63 a major event takes place. Consequently, passenger volume and traffic density are very high. Additionally, we assume very low temperature values and an aggressive driving behaviour. This leads to a 33 % higher energy consumption and a solution with charging stations at every stop except three and costs of €5.1 M. This scenario might be greatly pessimistic. Yet, it shows the sensitivity of the solution to the underlying values which motivates and justifies a carefully considered planning. This also includes a more in-depth evaluation of passenger flows and energy consumption of auxiliary consumers, like air conditioning, in the case of exceptional circumstances.

It might be more efficient to neglect some extreme situations and in the case they happen to fall back on alternative or additional vehicles.

In addition to several tests with the current timetable of line 63 we also analysed the sensitivity of the solution with regard to small changes in waiting times. By dropping one bus stop and thus saving time or by prolonging line 63's frequency by one minute we gain a minute of charging time at the turning stations. In this case the number of charging stations reduced to four. Costs decreased to €0.92 M.

Our results emphasise that the electrification of a city bus line under the given circumstances is a challenging task. Further practical tests will have to show to what extend an exclusively electric bus operation is viable.

4 Conclusion and Outlook

Our current model depicts the route and energy course of a single bus line and gives in very short time an optimal solution. Since city buses are usually part of a whole bus network, it is reasonable that modelling this is the next step to take. Connecting logically the same station of different circulations to one already gives the basis for integrating more bus lines to a network. The technical progress also prompts the consideration of different battery sizes, ageing-effects, vehicle configurations as well as charging technologies.

For the future work we also plan to scrutinise the current routes and timetables or incorporate municipal vehicles like street cleaning vehicles or rubbish collection.

References

1. Bombardier Transportation GmbH. Reference sheet PRIMOVE Mannheim (2013). http://primove.bombardier.com. Accessed 17 Oct 2014
2. Chen, D., Kockelman, K., Khan, M.: The electric vehicle charging station location problem: a parking-based assignment method for seattle. Transp. Res. Board 92nd Annu. Meet. **340**, 13–1254 (2013)
3. Frade, I., Ribeiro, A., Antunes, A.P., Goncalves, G.A.: An optimization model for locating electric vehicle charging stations in central urban area. Transp. Res. Rec. J. Transp. Res. Board **3582**, 1–19 (2011)
4. Kley, F.: Ladeinfrastrukturen für Elektrofahrzeuge: Entwicklung und Bewertung einer Ausbaustrategie auf Basis des Fahrverhaltens. Fraunhofer Verlag, Stuttgart (2011)
5. Kunith, A., Göhlich, D., Mendelevitch, R.: Planning and optimization of a fast charging infrastructure for electric urban bus systems. In: Proceedings of the 2nd International Conference on Traffic and Transport Engineering, pp. 43–51 (2014)
6. Mak, A.-H., Rong, Y., Shen, Z.-J.: Infrastructure planning for electric vehicles with battery swapping. Manag. Sci. **59**(7), 1557–1575 (2013)
7. Müller-Hellmann, A.: Überlegungen zu Ladeverfahren für Batteriebusse im ÖPNV. Der Nahverkehr **2014**(7–8), 40–43 (2014)
8. Nationale Organisation Wasserstoff- und Brennstoffzellentechnologie (2012). http://www.now-gmbh.de. Accessed 31 July 2015

A Hybrid Solution Approach for Railway Crew Scheduling Problems with Attendance Rates

Kirsten Hoffmann

Abstract This paper presents a model for railway crew scheduling problems dealing with attendance rates for conductors. Afterwards we discuss a hybrid solution approach for these kind of problems. This approach consists of a column generation framework using genetic algorithm to solve the pricing problem. Based on a real-world instance, we compare our hybrid solution approach with the enumeration approach with respect to resulting total costs and computation time.

1 Introduction

Apart from energy costs (fuel, electricity), crew costs are the largest cost factor in rail passenger transport. Therefore, the efficient assignment of crews is becoming increasingly more important. Previous crew scheduling models and solution approaches mostly deal with covering all trips of the given train timetable. The diminishing importance of operational tasks and increasing cost pressure, however, force responsible authorities in Germany to reduce the deployment of conductors. Therefore, transportation contracts defining all frame conditions for different transportation networks determine one or more percentage rates of trains or kilometres that have to be attended by conductors.

In regional rail transport, crew members are train drivers (operator of a train) and conductors (tasks: ticket collection and other customer services). We focus on the latter, as variable attendance rates cannot be applied to train drivers, obviously. Nevertheless, train drivers could be included with attendance rates of 100 %.

There is a wide range of models and algorithms concerning transport crew scheduling and rostering, respectively. For a recent review on passenger railway optimization, see [1]. Due to the size of crew scheduling problems (up to several millions of possible duties), metaheuristics are increasingly gaining in importance. Reference [8] introduce a tabu search algorithm for bus and train drivers. Reference [3] present a mathematical model for railway crew scheduling solved by simulated annealing.

K. Hoffmann (✉)
Faculty of Business and Economics, TU Dresden, 01062 Dresden, Germany
e-mail: kirsten.hoffmann@tu-dresden.de

© Springer International Publishing Switzerland 2017
K.F. Dœrner et al. (eds.), *Operations Research Proceedings 2015*,
Operations Research Proceedings, DOI 10.1007/978-3-319-42902-1_33

Genetic algorithms can be applied in two different ways: After generating feasible duties, genetic algorithms are used to find the optimal shift schedule. On the other hand, the pricing problem (generation of new duties) can be solved with genetic algorithms [5, 7].

To the best of our knowledge, there are no appropriate models or algorithms dealing with attendance rates. Therefore, we define a new model with attendance rates in Sect. 2. Section 3 presents the hybrid solution approach, containing a column generation framework with a genetic algorithm to solve the pricing problem. Section 4 reports the results of our computational experiments.

2 Crew Scheduling Problem with Attendance Rates

In public transport, especially railway traffic, there are several requirements that have to be satisfied. For operational and legal requirements of the German railways see [4]. Additionally, the transportation contract regulates attendance rates, possibly distinguished between product types, lines, train numbers, track sections, and time windows. Moreover, responsible authorities determine penalization if real attendance rates are too low. The railway crew scheduling problem with attendance rates is to find a minimum cost shift schedule satisfying operating conditions and legal requirements. The schedule should cover a subset of all trips such that the different attendance rates specified in the transportation contract are met.

The most common model for crew scheduling problems is the set covering problem, which is known to be NP-hard. To modify this set covering model for our purposes, we need a set M of all trips assigned to the considered transportation network and a set N of all feasible duties. Each duty $j \in N$ is represented by one column in matrix $A \in \{0, 1\}^{|M| \times |N|}$ with $a_{ij} = 1$ if duty $j \in N$ covers trip $i \in M$, 0 otherwise. Parameter c_j displays the costs of duty $j \in N$. Let x_j be the binary decision variables such that $x_j = 1$, if duty j is part of the solution schedule, 0 otherwise. Let $G \subset (0, 1]$ be a set of all attendance rates defined in the transportation contract. Let d_{ig} be the distance of trip $i \in M$ with attendance rate $g \in G$ and y_i the decision variable at which $y_i = 1$ if trip $i \in M$ is covered in the solution schedule. The railway crew scheduling problem with attendance rates for a single day (CSPAR) is

$$\min \sum_{j \in N} c_j x_j \tag{1}$$

$$\text{s.t.} \sum_{i \in M} d_{ig} y_i \geq g \sum_{i \in M} d_{ig} \qquad \forall g \in G \tag{2}$$

$$\sum_{j \in N} a_{ij} x_j \geq y_i \qquad \forall i \in M \tag{3}$$

$$y_i \geq a_{ij} x_j \qquad \forall i \in M, j \in N \tag{4}$$

$$x_j \in \{0, 1\} \qquad \forall j \in N \qquad\qquad (5)$$

$$y_i \in \{0, 1\} \qquad \forall i \in M. \qquad\qquad (6)$$

The objective function (1) minimizes the total costs over all chosen duties. Constraints (2) guarantee that the accumulated distance of the covered trips in the solution schedule is greater than or equal to the requested percentage of the total distance assigned to the special attendance rate. Constraints (3) and (4) are linking constraints for the x_j and y_i variables.

We can relax constraints (4) without changing the value of the objective function. However, variables y_i can no longer be interpreted as existence of trip i in the solution schedule. In practice, we have some additional constraints, e.g. the average working time of the shift schedule or the personnel capacity of each crew base. This model can easily be transferred to multiple days which is, due to the increasing complexity of sets, variables, and indices, not part of the paper.

3   Hybrid Solution Approach

A simple solution approach for the CSPAR consists of two phases: First, generate all possible duties N, e.g. with depth-first search, than solve model (1)–(6) with suitable algorithms or optimisation solver. For practical application, where several million feasible duties are possible, this approach is too time consuming and the memory usage is high.

To get shift schedules for few days or weeks within reasonable time, we have to speed up the solution process. To achieve this goal, column generation is an useful instrument. The problem (1)–(6) is decomposed into master and pricing problem. In the restricted master problem (RMP) the linear programming relaxation of (1)–(6) is solved to optimality with respect to a small subset of duties. To generate new duties that lower the objective value dual values are used. Let π_i, $i \in M$, be the dual values of constraints (3) then $\bar{c}_j = c_j - \sum_{i \in M} \pi_i a_{ij}$ specifies the reduced costs of duty j. To lower the objective value add duties with negative values (3) to the restricted master problem. The question is how to generate feasible duties with negative reduced costs (pricing problem, PP). The most common approach uses the shortest path problem with resource constraints to solve the pricing problem [2]. To reduce the problem size and accelerate the pricing problem, we apply a genetic algorithm for the generation of new duties.

Genetic algorithms adopt techniques derived from natural evolution to search for solutions of optimisation problems. For further information we refer to [6]. The individuals of the population represent the duties. We implement a trip based representation, where the trips are sorted by increasing departure time. The basic operators and properties are discussed below.

Initial schedule (overall) Since we have to start with a feasible schedule to generate dual values, the initial schedule has to suffice (2)–(6). In order to determine such set of duties, we use a block generation approach with depth-first search. For each trip starting at a crew base, duty blocks with given minimum and maximum duration and maximum transition time are formed. The resulting highly productive blocks are randomly matched to feasible duties with required breaks.

Fitness function The fitness function equates the objective function of the pricing problem, i.e. the reduced costs.

Initial population (pricing problem) In each iteration of the column generation approach, the set of all duties is available for the pricing problem, whether from the initial schedule or generated in earlier iterations. We select the best `popSize` individuals of the initial population starting the genetic algorithm procedure in each column generation iteration.

Crossover We implement a special kind of one-point crossover. This operator is applied in four stages: First, select an individual of the current population randomly. Then, choose a different individual starting at the same crew base. After random choice of a cut point of the first parent, we search for a proper cut point of the second parent. On the one hand, the stations at the cut points must match. On the other hand, arrival and departure times have to be compatible. All trips located right of the cut points swap places.

Mutation The mutation operator replaces the randomly chosen genome (trip) with a suitable one. The new trip has to fit into the resulting gap with respect to departure and arrival time/station. The operator is configured in such way that either the resulting duties are always feasible or the mutation will not be executed.

Selection for replacement The population of the next generation contains all generated duties without duplicates and the best old ones not included in the former set.

Figure 1 summarizes the introduced hybrid solution approach. Compared to [5], the operators are highly adapted to the crew scheduling problem, so that just a few generated duties are infeasible concerning the break times. The whole framework exclusively deals with feasible duties. Reference [7] solve the pricing problem in an exact way if the genetic algorithm fails to generate new duties. For instances with up to 1,500 trips per day and a planning period up to 14 days, this approach is too time consuming, especially considering the fact that these duties reduce the value of the linear relaxation, not necessarily of the integer problem. Furthermore, some practical requirements can hardly or not be modelled mathematically. For this reasons, we decided to use a completely metaheuristic pricing step.

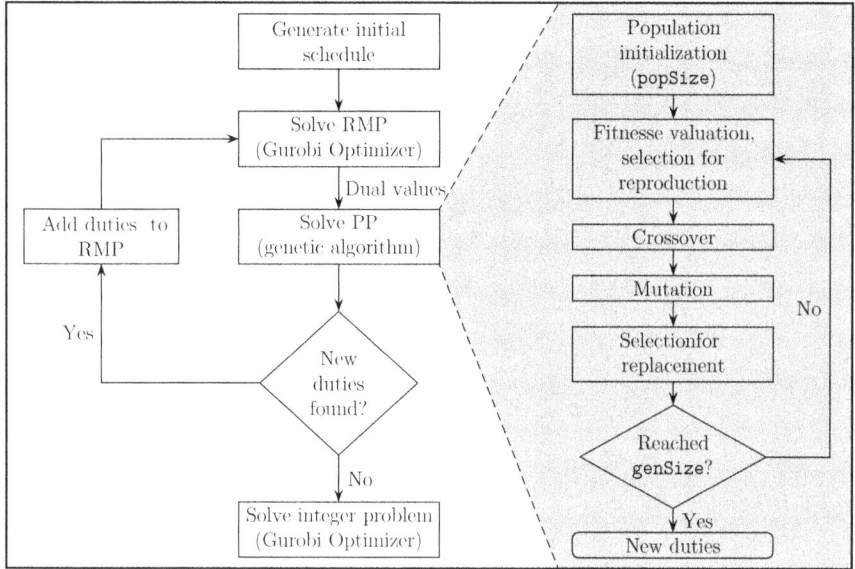

Fig. 1 Hybrid solution approach

4   Computational Results

All computational tests were performed on a real-world test instance with 713 trips, 18 relief points, 10 crew bases and 5 break rooms. Further requirements are: maximum duty time (640 min), maximum working time (600 min), minimum paid time (300 min), and average paid time ($\in [418, 512]$ min). The attendance rates are predefined to 30% of the total distance in the transportation network during the day and 90% from 7pm. Observed cost factors are fixed costs of 2,000 per duty and 50 per working minute. This values are prescribed by our project partners and influence the duration of the duties in relation to the number. We use a Intel(R) Xenon(R) CPU E5-2630 v2 @ 2.6 GHz server with 12 cores and 384 GB RAM and Gurobi 6.0.

The results of the two-phase approach (depth-first search, solve model with the Gurobi Optimizer) are summarized in Table 1. Parameter sets `set1`, `set2`, `set3`,

Table 1 Computational results of the two-phase approach

Parameter set	# Duties	Time generator (s)	Time solver (s)	Total time (s)	Costs
`set1`	121,892	50	31	81	407,150
`set2`	3,821,534	796	578	1,374	397,850
`set3`	2,807,746	16,224	322	16,546	396,400
`set4`	7,512,654	45,532	1,351	46,883	394,450

Table 2 Computational results of the hybrid approach (selection)

Pop Size	Gen Size	Iter	# Duties	Time (s)				Costs			
				PP	LP	IP	Total	Mean	SD	Min	Max
500	50	60	2,099	104	14	231	384	406,061	4.222	394.501	412.200
500	100	50	2,568	142	15	298	489	403,464	5.536	394.100	412.200
1,000	10	105	1,228	138	25	163	362	406,992	4.813	395.250	412.201
1,000	30	83	2,206	198	19	153	403	401,459	5.792	391.950	412.200
1,000	50	61	2,863	212	15	178	440	399,824	5.374	391.450	408.650
1,000	75	50	3,341	245	13	113	406	401,959	4.673	392.150	412.200
1,000	100	42	4,076	265	13	212	524	400,139	5.460	392.250	412.200
2,000	10	119	1,314	276	27	286	623	406,517	3.977	399.850	412.550
2,000	30	84	2,303	473	20	74	601	396,694	5.102	390.650	408.150
2,000	50	60	3,688	557	16	162	769	398,299	5.213	391.200	410.650
2,000	75	55	5,135	755	15	55	859	396,965	4.327	390.900	406.450
2,000	100	45	6,107	828	14	71	948	396,604	4.546	390.800	405.100
4,000	50	70	4,659	2,072	18	34	2,158	394,672	3.902	390.050	403.600
4,000	100	49	9,768	2,922	16	51	3,023	393,889	3.955	389.350	404.200

`set4` define various values of the maximum transition time, productivity of the duties, and minimum duration of connected trips. Both phases run parallel with 24 threads. Considering the limited memory capacity of our system, we receive the best solution within 13 h, whereby 7.5 million duties are generated in the first phase.

For the computational tests of the hybrid solution approach we perform 30 runs per parameter configuration to measure the random component. The crossover rate is fixed at 0.9, the mutation rate at 0.1, and the initial schedule contains 52,900–53,000 duties. The Gurobi Optimizer runs parallel on 24 threads, the pricing problem on one thread. The values stated in Table 2 are mean values, and all real numbers are rounded up to the next integer number.

As shown in the table, an increasing generations size (`genSize`) results in a decreasing number of Iterations (Iter) and an increasing number of duties generated in the pricing problem (# Duties). This implies a rising computation time of the pricing problem (PP). Increasing the population size (`popSize`), the pricing problem lasts longer, too. Furthermore, the mean (Mean), standard deviation (SD) and minimum/maximum value (Min/Max) of the costs are lowered. Reference [7] recommend that the population size has to be equal to three times the number of trips. The proportion of the computation time the Gurobi Optimizer uses for solving the RMP (LP) is insignificant, whereas the time for the integer solution (IP) is fluctuating.

5   Conclusions and Further Research

In this paper we presented a new model for crew scheduling problems with special focus on variable attendance rates for conductors. Although this work is still at an early stage, the introduced solution approach composed of column generation and genetic algorithm delivers good results within reasonable time. In comparison with the simple two-phase approach we obtain better solutions in less time in many cases.

The results show that, in many cases, the time for solving the integer problem with the Gurobi Optimizer exceeds the time for the column generation process. Therefore, we have to check heuristic solutions for the integer problem to reduce the computing time, such as variable fixing. Furthermore, various methods for generating the initial schedule have to be tested. In the pricing problem, we could modify the way of selection or diversify the crossover and mutation operator. In implementation, multi-threading has to be applied to the pricing procedure. Moreover, computational tests for instances covering multiple days have to be analysed with respect to computing time and memory usage.

References

1. Caprara, A., Kroon, L., Monaci, M., Peeters, M., Toth, P.: Passenger railway optimization. In: Barnhart, C., Laporte, G. (eds.), Transportation, of Handbooks in Operations Research and Management Science, Chap. 3, vol. 14, pp. 129–187. Elsevier, Amsterdam (2007)
2. Desrochers, M., Gilbert, J., Sauve, M., Soumis, F.: CREW-OPT: subproblem modeling in a column generation approach to urban crew scheduling. In: Desrochers, M., Rousseau, J.-M. (eds.) Computer-Aided Transit Scheduling. Lecture Notes in Economics and Mathematical Systems, vol. 386, pp. 395–406. Springer, Heidelberg (1992)
3. Hanafi, R., Kozan, E.: A hybrid constructive heuristic and simulated annealing for railway crew scheduling. Comput. Ind. Eng. **70**, 11–19 (2014)
4. Jütte, S., Albers, M., Thonemann, U.W., Haase, K.: Optimizing railway crew scheduling at DB Schenker. Interfaces **41**, 109–122 (2011)
5. Liu, M., Haghani, A., Toobaie, S.: Genetic algorithm-based column generation approach to passenger rail crew scheduling. Transp. Res. Rec. J. Transp. Res. Board **2159**, 36–43 (2010)
6. Michalewicz, Z.: Genetic Algorithms + Data Structures = Evolution Programs. Springer, Heidelberg (1996)
7. Santos, A.G., Mateus, G.R.: General hybrid column generation algorithm for crew scheduling problems using genetic algorithm. In: IEEE Congress on Evolutionary Computation, CEC 2009, pp. 1799–1806 (2009)
8. Shen, Y., Kwan, R.S.: Tabu search for driver scheduling. In: Voß, S., Daduna, J.R. (eds.) Computer-Aided Scheduling of Public Transport. Lecture Notes in Economics and Mathematical Systems, vol. 505, pp. 121–135. Springer, Heidelberg (2001)

A Comparison of Hybrid Electric Vehicles with Plug-In Hybrid Electric Vehicles for End Customer Deliveries

Christian Doppstadt

Abstract In this paper we present a method, which is able to compare the use of pure combustion vehicles with hybrid electric vehicles and plug-in hybrid electric vehicles for end customer deliveries. Benchmark instances representing typical delivery areas for small package shipping companies are introduced. For small instance sizes, we are able to generate exact solutions with standard mixed-integer program solver software. In contrast to the exact approach, our heuristic allows us to solve practical instance sizes.

1 Introduction

Reducing exhaust gases by electric vehicles (EVs) instead of vehicles with internal combustion engine (ICEs) is one of the most important challenges for the future. However, EVs have some disadvantages: their range is limited and additional infrastructure for recharging is required. These limitations can be eliminated by hybrid electric vehicles (HEVs) or plug-in hybrid electric vehicles (PHEVs), which combine combustion and electric engine. In contrast to HEVs, the battery of PHEVs is not only charged while driving, but also by connecting the vehicle to the electric supply network. HEVs and PHEVs are widely used as passenger cars, but uncommon as delivery vehicles. We extend the well-known Traveling Salesman Problem (TSP) by integrating different modes of operation. The aim is to determine which arcs in which mode to use, to minimize the total costs. To the best of our knowledge, this problem setting has not been researched before. For an overview on the classical TSP, existing problem variants, and different solution methods we refer to [1, 5]. The newly introduced operation modes have different costs and travel times for each arc. Usually HEVs and PHEVs distinguish four different modes: pure combustion, pure electric, charging the battery while driving in combustion mode, and a boost mode, which combines combustion and electric engine. We also implemented these four modes. For a review on the technical aspects of electric, hybrid, and fuel

C. Doppstadt (✉)
Goethe University, Frankfurt, Germany
e-mail: doppstadt@wiwi.uni-frankfurt.de

© Springer International Publishing Switzerland 2017
K.F. Dœrner et al. (eds.), *Operations Research Proceedings 2015*,
Operations Research Proceedings, DOI 10.1007/978-3-319-42902-1_34

cell vehicles we refer to [3]. Having four modes of operation extends the graph of a TSP to a multigraph with four arcs between all vertices. The four arcs between each pair of vertices dramatically increases the number of possible round tours for the already NP-hard TSP. Therefore, we implemented a heuristic solution method to be able to solve problems with a practical number of customers. In addition, we modeled the problem as mixed-integer program (MIP) and are able to solve problems with a small number of customers to optimality.

2 Problem Description

We define the Problem on a directed network $G = (V, A)$, with $V = \{0, 1, \ldots, n, n + 1\}$ as set of vertices, including the starting (0) and ending depot $(n + 1)$, with the ending depot as a copy of the starting depot. The set of arcs A has different costs and travel times for each mode of operation indicated as c_c and t_c for the pure combustion mode, c_e and t_e for the electric mode, c_{cc} and t_{cc} for the charging mode, and c_b and t_b for the boost mode. We set a maximum battery capacity l_{max}, a charging rate r_c, and a discharging rate r_d. The objective is to minimize the cost of the arcs travelled in each mode of operation:

$$min \sum_{i=0}^{n} \sum_{j=1:j\neq i}^{n+1} c_{ij}^{cc} x_{ij}^{cc} + c_{ij}^{b} x_{ij}^{b} + c_{ij}^{c} x_{ij}^{c} + c_{ij}^{e} x_{ij}^{e}$$

To ensure the feasibility of a solution, a set of constraints has to be fulfilled. First, the decision variables for each mode are aggregated to a single one, which is used afterwards within the model (1). Additionally, it is guaranteed that each node is visited exactly once (2) and (3). Now, the time flow has to be defined. Therefore, the time of arriving at a node is defined by the departure time at the previous node plus the service time s_i and the travel time in a specific mode of operation (4)–(7). In addition to the time flow, it is also required to define a flow for the battery charging. For the charging mode the battery charging increase according to the time driven on an arc (8). While driving in electric or boost mode, the charging decreases (9) and (10), and using the combustion mode does not change the charging (11). Additionally, the charging has to be positive (12) and a maximum tour duration B is set (13). Finally, the start time is set to 0 (14) and the decision variables are set binary (15).

$$x_{ij}^{cc} + x_{ij}^{b} + x_{ij}^{c} + x_{ij}^{e} = x_{ij} \quad \forall i \in \{0, \ldots, n\}; j \in \{1, \ldots, n + 1\}; i \neq j \tag{1}$$

$$\sum_{i=0}^{n} x_{ij} = \sum_{j=1}^{n+1} x_{ij} = 1 \quad \forall j = 1, \ldots, n; i \neq j \tag{2}$$

$$\sum_{j=1}^{n} x_{0j} = \sum_{i=1}^{n} x_{i(n+1)} = 1 \tag{3}$$

$$x_{ij}^{cc}(w_i + s_i + t_{ij}^{cc} - w_j) = 0 \quad \forall i \in \{0, \ldots, n\}; j \in \{1, \ldots, n+1\}; i \neq j \tag{4}$$

$$x_{ij}^{b}(w_i + s_i + t_{ij}^{b} - w_j) = 0 \quad \forall i \in \{0, \ldots, n\}; j \in \{1, \ldots, n+1\}; i \neq j \tag{5}$$

$$x_{ij}^{c}(w_i + s_i + t_{ij}^{c} - w_j) = 0 \quad \forall i \in \{0, \ldots, n\}; j \in \{1, \ldots, n+1\}; i \neq j \tag{6}$$

$$x_{ij}^{e}(w_i + s_i + t_{ij}^{e} - w_j) = 0 \quad \forall i \in \{0, \ldots, n\}; j \in \{1, \ldots, n+1\}; i \neq j \tag{7}$$

$$x_{ij}^{cc}(min\{l_i + t_{ij}^{cc}r^c, l_{max}\} - l_j) = 0 \quad \forall i \in \{0, \ldots, n\}; j \in \{1, \ldots, n+1\}; i \neq j \tag{8}$$

$$x_{ij}^{b}(l_i - t_{ij}^{b}r^d - l_j) = 0 \quad \forall i \in \{0, \ldots, n\}; j \in \{1, \ldots, n+1\}; i \neq j \tag{9}$$

$$x_{ij}^{e}(l_i - t_{ij}^{e}r^d - l_j) = 0 \quad \forall i \in \{0, \ldots, n\}; j \in \{1, \ldots, n+1\}; i \neq j \tag{10}$$

$$x_{ij}^{c}(l_i - l_j) = 0 \quad \forall i \in \{0, \ldots, n\}; j \in \{1, \ldots, n+1\}; i \neq j \tag{11}$$

$$l_i \geq 0 \quad \forall i = 0, \ldots, n+1 \tag{12}$$

$$w_{n+1} + s_{n+1} \leq B \tag{13}$$

$$w_0 = 0 \tag{14}$$

$$x_{ij}, x_{ij}^{cc}, x_{ij}^{b}, x_{ij}^{c}, x_{ij}^{e} \in \{0, 1\} \tag{15}$$

The difference between HEV and PHEV is the initial charging of the battery. For the HEV we set it 0, as the battery cannot be charged overnight. In contrast, we assume a fully charged battery for the PHEV.

3 Solution Approach

In this section we describe how our solution methods work. The chapter is split into two parts: One for the exact and one for the heuristic solution approach.

3.1 Exact Solution Methods

First, the instances allowing only the pure combustion mode are solved. This provides a measure to calculate the savings gained with a HEV or PHEV. Therefore, our instances are transformed into TSP instances and solve them to optimality with the Concorde solver software using the costs for the combustion mode [4].

Second, we developed a MIP formulation of the HEV and PHEV case. We used the CPLEX solver to solve the benchmark instances to optimality. As the runtime for this approach is huge, we were only able to solve instances with 10 customers within a reasonable runtime.

3.2 Heuristic Solution Method

To be able to solve larger instances, we develop a heuristic solution approach. Again, we create a solution with the Concorde method in pure combustion mode. Based on that, we developed the χ-Mode Change (χ-MC) to change the mode of operation on a given number of arcs. Preliminary studies showed that it is most profitable to change the mode on three (3-MC), four (4-MC), or five (5-MC) arcs. We run all χ-MC moves with a hill-climbing strategy, repeating the move until no further improvement is found and therefore, a local optimum is reached.

4 Numerical Studies

In this section, we describe our numerical results. First, the generation of the benchmark instances is described. Afterwards, we give the results for the exact and heuristic solutions and compare the results of the HEV and PHEV.

4.1 Benchmark Instances

We generated a set of benchmark instances, based on real world delivery tours. The instances differ in number of customers and the location of the depot. In the first set, the depot is adjoining the delivery area. In the second set the depot is 28 km and in the third set 57 km apart from the delivery area. For each customer and depot location characteristic we have three different instances, varying service times and customer locations. Costs for each arc based on Euclidean distances are calculated and different factors for each mode are used to generate a realistic structure. We set the cost for the pure combustion mode to 100% of the distance, for the electric mode to 20%, for the charging mode to 110%, and for the boost mode to 120%. Travel times are calculated based on different speed values, which depend on the arc length and the mode of operation. These speed values were deduced from real world delivery tours. The full instance set can be requested from the author. We assume a battery capacity of 16.8 kWh, a charging rate of 12.0 kW/h, and a discharging rate of 48.0 kW/h. These values are sound with a presumed electric engine with a performance of 60.0 kW as used in [2]. The maximal tour duration is set to 8 h.

4.2 Results of Exact Solution Approach

We ran all our numerical studies on a cloud server with up to 8 Opteron cores. In Table 1 we present the CPLEX solutions for the 10 customer instances. On the left

Table 1 CPLEX solutions for the HEV and PHEV instances

Instance	Runtime	Upper bound	Savings	Upper bound	Runtime	Instance
HEVTSP_1_10_1	85,092	1798.94	−27.16	1310.29	5478	PHEVTSP_1_10_1
HEVTSP_1_10_2	340,079	1601.73	−34.52	1048.74	65,550	PHEVTSP_1_10_2
HEVTSP_1_10_3	255,839	1513.31	−35.49	976.18	129,130	PHEVTSP_1_10_3
HEVTSP_2_10_1	303,680	7308.15	−7.39	6768.08	4895	PHEVTSP_2_10_1
HEVTSP_2_10_2	52,417	7362.93	−7.61	6802.55	804	PHEVTSP_2_10_2
HEVTSP_2_10_3	21,274	7290.82	−7.01	6779.89	243	PHEVTSP_2_10_3
HEVTSP_3_10_1	322,010	12747.22	−5.38	12061.38	1009	PHEVTSP_3_10_1
HEVTSP_3_10_2	104,211	12772.44	−4.92	12143.54	1549	PHEVTSP_3_10_2
HEVTSP_3_10_3	226,579	12935.48	−4.43	12362.03	1298	PHEVTSP_3_10_3

side we report the runtime in seconds and the best CPLEX solution (upper bound) for the HEV instances, on the right side we report the same values for the corresponding PHEV instances, and the middle column denotes the savings of the PHEV compared to the HEV vehicle.

4.3 Comparison of Combustion, Hybrid Electric, and Plug-In Hybrid Electric Vehicle

For the comparison of the three vehicles types we aggregate their values in Table 2. Beside the solution for the pure combustion vehicle in the middle column, we give

Table 2 Comparison of combustion vehicles with hybrid electric vehicles and plug-in hybrid electrical vehicles

HEV instance	Costs	Savings (%)	Combustion	Savings (%)	Costs	PHEV instance
HEVTSP_1_10_1	1798.94	−7.59	1946.71	−32.69	1310.29	PHEVTSP_1_10_1
HEVTSP_1_10_2	1601.73	−4.59	1678.77	−37.53	1048.74	PHEVTSP_1_10_2
HEVTSP_1_10_3	1513.31	−5.81	1606.63	−39.24	976.18	PHEVTSP_1_10_3
HEVTSP_2_10_1	7308.15	−1.39	7410.99	−8.68	6768.08	PHEVTSP_2_10_1
HEVTSP_2_10_2	7362.93	−1.07	7442.58	−8.60	6802.55	PHEVTSP_2_10_2
HEVTSP_2_10_3	7290.82	−1.51	7402.67	−8.41	6779.89	PHEVTSP_2_10_3
HEVTSP_3_10_1	12747.22	−0.56	12818.75	−5.91	12061.38	PHEVTSP_3_10_1
HEVTSP_3_10_2	12738.38	−0.63	12819.10	−5.27	12143.54	PHEVTSP_3_10_2
HEVTSP_3_10_3	12935.48	−0.59	13012.25	−5.00	12362.03	PHEVTSP_3_10_3

(continued)

Table 2 (continued)

HEV instance	Costs	Savings (%)	Combustion	Savings (%)	Costs	PHEV instance
HEVTSP_1_20_1	2005.89	−8.41	2190.03	−30.99	1511.42	PHEVTSP_1_20_1
HEVTSP_1_20_2	1969.86	−8.63	2155.98	−31.73	1471.78	PHEVTSP_1_20_2
HEVTSP_1_20_3	1610.30	−9.66	1782.54	−38.99	1087.46	PHEVTSP_1_20_3
HEVTSP_2_20_1	7807.56	−2.18	7981.50	−8.79	7279.92	PHEVTSP_2_20_1
HEVTSP_2_20_2	7671.69	−1.99	7827.29	−8.90	7130.70	PHEVTSP_2_20_2
HEVTSP_2_20_3	7717.25	−2.09	7882.14	−8.68	7198.23	PHEVTSP_2_20_3
HEVTSP_3_20_1	13343.32	−1.22	13508.19	−5.58	12754.35	PHEVTSP_3_20_1
HEVTSP_3_20_2	13289.94	−1.21	13452.72	−5.50	12713.10	PHEVTSP_3_20_2
HEVTSP_3_20_3	13275.98	−1.07	13419.8	−5.30	12707.99	PHEVTSP_3_20_3
HEVTSP_1_50_1	2716.76	−13.02	3123.30	−28.24	2241.15	PHEVTSP_1_50_1
HEVTSP_1_50_2	2494.49	−12.46	2849.41	−28.99	2023.41	PHEVTSP_1_50_2
HEVTSP_1_50_3	2505.77	−11.21	2822.16	−25.75	2095.46	PHEVTSP_1_50_3
HEVTSP_2_50_1	8338.12	−4.40	8721.75	−9.53	7890.30	PHEVTSP_2_50_1
HEVTSP_2_50_2	8427.86	−3.80	8760.68	−8.87	7983.52	PHEVTSP_2_50_2
HEVTSP_2_50_3	8436.68	−3.90	8779.23	−9.08	7982.18	PHEVTSP_2_50_3
HEVTSP_3_50_1	13853.20	−2.61	14224.77	−5.73	13409.95	PHEVTSP_3_50_1
HEVTSP_3_50_2	13985.94	−2.14	14292.47	−5.78	13466.67	PHEVTSP_3_50_2
HEVTSP_3_50_3	13805.96	−2.35	14138.87	−5.95	13297.19	PHEVTSP_3_50_3

the best values found by either CPLEX or our heuristic for both HEV on the left and for the PHEV on the right side of the table. Based on this, we calculated the potential savings for both types of vehicles compared to the combustion one. The PHEV is able to achieve at least twice as high savings as the HEV on all of our test instances.

5   Conclusion and Outlook

We presented a comparison of hybrid electric vehicles with plug-in hybrid electric vehicles for end customer delivers and showed that using these types of vehicles has high potential savings compared to pure combustion engines. We introduced benchmark instances and solved small instances with an exact approach. Moreover, we developed a heuristic solution approach, which is able to solve large problems. It turned out, that for all test cases the initial charging of the battery for the plug-in vehicles could almost completely be used to drive in electric mode. This gained additional savings compared to the hybrid electrical vehicle.

References

1. Applegate, D.L., Bixby, R.E., Chvatal, V., Cook, W.J.: The Traveling Salesman Problem: A Computational Study. Princeton Series in Applied Mathematics. Princeton University Press, Princeton (2007)
2. ARADEX AG: Retrofit electric drive kit for diesel delivery vehicles (2014). http://www.aradex.de/en/electric-mobility/electric-drive-for-hybrid-hev-and-electric-vehicles-ev/elektroantrieb-nachruestsatz-fuer-diesel-lieferwagen/
3. Chan, C.C.: The state of the art of electric, hybrid, and fuel cell vehicles. Proc. IEEE **95**(4), 704–718 (2007)
4. Concorde: the concorde TSP solver (2014). http://www.math.uwaterloo.ca/tsp/concorde/
5. Gutin, G., Punnen, A.P. (eds.) The Traveling Salesman Problem and Its Variations. Combinatorial Optimization. Kluwer Academic, Dordrecht (2002)

Robust Efficiency in Public Bus Transport and Airline Resource Scheduling

Bastian Amberg, Lucian Ionescu and Natalia Kliewer

Abstract In this work we address the concept of robust efficiency of resource schedules in public bus transport and airline industry, dealing with the competing objectives of cost-efficiency and robustness. Generalizing the findings from two research projects we provide techniques that lead to an improvement of the pareto-front between robustness and cost-efficiency. These techniques include the improvement of scheduling and optimization approaches as well as a refinement of delay prediction models enabling a robustness evaluation closer to reality. Additionally, problem characteristics in public transport and airline network topologies and their influence on the degree of freedom for robust resource scheduling and dispatching strategies are examined.

1 Introduction—Robust Efficiency in Resource Scheduling

In this paper, we aim at the generalization of findings of two closely related research projects, dealing with crew and aircraft or vehicle scheduling in the airline industry and public bus transport, respectively.

In both problem domains, there is a necessity of coping with contrary objectives in scheduling and operations. The planning stage primarily deals with the cost-efficient use of resources like crews, vehicles or aircraft, desirably leading to tight schedules with low idle and buffer times. However, during operations unforeseen

B. Amberg (✉) · L. Ionescu · N. Kliewer
Department of Information Systems, Freie Universität Berlin,
Garystr. 21, 14195 Berlin, Germany
e-mail: bastian.amberg@fu-berlin.de

L. Ionescu
e-mail: lucian.ionescu@fu-berlin.de

N. Kliewer
e-mail: natalia.kliewer@fu-berlin.de

© Springer International Publishing Switzerland 2017
K.F. Dœrner et al. (eds.), *Operations Research Proceedings 2015*,
Operations Research Proceedings, DOI 10.1007/978-3-319-42902-1_35

disruptions occur frequently and may lead to delayed departure and arrival of flights or service trips. Therefore, additional reactionary costs for delay compensation may occur. In this regard, a main goal during operations is punctuality and reliability, demanding for buffer times and possibilities to adapt resource schedules to the current circumstances.

Consequently, the real costs for resource usage are determined by the sum of planned costs and reactionary costs. Considering planned cost minimization as the only objective leads to an underestimation of the real costs. Therefore, the concept of *robust efficiency in resource scheduling* can be applied, aiming at the minimization of real costs by already taking into account probable costs of delays during the scheduling stage. Hence, there is a trade-off between cost-efficiency and robustness. Resource schedules are called *robust* if delays are less likely to be propagated to succeeding tasks. In this context, a distinction can be drawn between exogenous and propagated delays. The former is caused directly by exogenous disruptions and cannot be prevented in scheduling. In contrast, the latter is induced by delayed arrival of previous tasks using the same resources and therefore can be influenced by scheduling decisions.

Considering this, there are two eligible properties of resource schedules, namely *stability* and *flexibility*. Stability describes the capability of schedules to absorb delays by buffer times between tasks so that no or less delay propagation occurs. A high degree of *flexibility* provides simple possibilities to react to delay occurrences on a cost-neutral basis, e.g. by swaps of resources. If delays occur in operations, resources can be swapped in order to prevent delay propagation. Note that both properties aim at the same target of avoiding delay propagation by absorption. Stability aims at carrying out resource schedules exactly as planned. In contrast, the intention of flexibility is to increase the ability to react to varying circumstances during operations by restricted schedule modifications. For detailed literature reviews on robust resource scheduling strategies please refer to [1–4].

For a graphical explanation of the trade-off between cost-efficiency and robustness, see Fig. 1. Line I illustrates the case of cost-efficient scheduling as the only

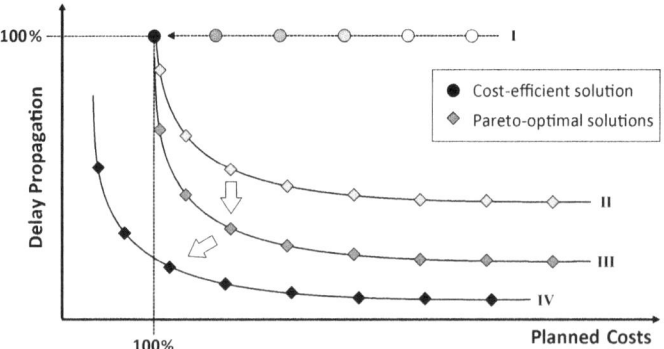

Fig. 1 Pareto-optimal solutions for the robust resource scheduling problem

objective. By additionally taking into account the robustness as a second objective, pareto-optimal solutions can be generated, see line II.

Based on this, the remainder of this paper considers techniques on incorporating robust efficiency into resource schedules. Therefore, a generic framework for generating robust and cost-efficient solutions is addressed in Sect. 2. Furthermore, techniques for improving the pareto-front are presented. Section 3 examines different problem characteristics arising from differing network topologies and their influence on the degree of freedom for scheduling and dispatching strategies in simulation. Finally, the findings are summarized in Sect. 4.

2   Improving Stability and Flexibility—A Generic Robust Scheduling Framework

In this section we present a generic framework for generating pareto-optimal resource schedules following the concept of robust efficiency. Afterwards, techniques for further improving the pareto-front are provided.

For given flight or service trip schedules, column generation approaches are used with a subsequent IP phase in order to generate resource schedules. The pricing algorithms are modeled as resource constrained shortest path problems and solved by dynamic programming. Both the column generation master problem and the final IP phase are usually solved by standard solvers, e.g. CPLEX. For a simultaneous consideration of different scheduling stages, e.g. integrated vehicle and crew scheduling, column generation is combined with Lagrangian relaxation. This enables us to relax constraints and decompose integrated models that are naturally hard to solve. The column generation approaches are extended by delay propagation mechanisms in order to incorporate stability and flexibility in the resulting schedules, see [1, 3] for further details.

Important robustness measures are the overall punctuality of flights/service trips and the total amount of propagated delays. Therefore, the robustness of resulting crew and aircraft or vehicle schedules is evaluated by simulating delay propagation in an event-based simulation. As an input factor, scenarios for exogenous delays are generated based on delay prediction models during simulation.

Evaluating the impact of stability on the robustness demands for propagation only during simulation as delay management and dispatching strategies may overestimate the robustness. In contrast, for the evaluation of flexibility dispatching strategies must be applied during simulation, e.g. by enabling swapping opportunities.

The framework is used to evaluate several techniques that may lead to an improvement of the pareto-front for robust efficiency. On the one hand, prediction models for exogenous delays can be improved. The desired result of more realistic delay prediction is that buffer times or swap opportunities are considered only for tasks that imply a considerable delay propagation risk. As a consequence, the pareto-front is moved downwards, i.e. the robustness is further improved without an additional increase of planned cost, see Fig. 1 (line III). Reference [5] examine the potential of data-driven delay prediction models for airline resource scheduling. Based on this study, [6] generate delay prediction models with different degrees of prediction accuracy and compare their effects on robust crew and aircraft scheduling.

On the other hand, both delay propagation and planned costs of resource schedules can be further reduced by an improvement of scheduling approaches, resulting in additional degrees of freedom for scheduling, see Fig. 1 (line IV). In this context, [3] examines the potential of different scheduling approaches in public bus transport resource scheduling. In particular, sequential, partially integrated, and integrated vehicle and crew scheduling schemes are investigated. The results show that sequential scheduling allows the computation of robust schedules, but it is (far) away from obtaining the cost-to-robustness ratio attained by integrated scheduling. Due to the combined consideration of delay propagation between tasks, and the mutual dependencies between vehicle and crew schedules during optimization, the proposed integrated scheduling scheme is able to build both robust and cost-efficient resource schedules. In addition, it is shown that a further integration of planning stages results in further improvement of the pareto-front. Therefore, integrated scheduling is combined with the ability to shift service trips within time windows in order to modify the given timetable.

3   Examining the Degrees of Freedom in both Problem Domains

The ability of generating resource schedules with high degrees of cost-efficiency and robustness is closely related to the degree of freedom determined by decisions taken in previous planning stages. In particular, this includes the timetabling and network design which show substantial differences in the airline industry and public transport. In this section the emerging consequences are examined with regard to potential differences in the degree of freedom that limit decisions during the resource scheduling stage (Study A) as well as for dispatching strategies during simulation (Study B).

Study A: Figure 2 shows the number of different successor tasks in all crew duties generated in the pricing steps during column generation. With 202 flights for the airline case and 210 service trips for the public transport case, the underlying problems are comparable in size. The airline hub-and-spoke network implies fixed successor flights for about one fourth of all flights. This unexceptionally appears at spoke airports as a result of the out-and-back principle in hub-and-spoke structures. In these cases, scheduling decisions cannot be made, irrespective whether the connection is likely to propagate delay or not. On average there are 5.1 possibilities to connect flights during pricing. Note that deadheading for crews or ferry-flights for aircraft are not allowed.

In contrast, the considered public transport network provides a significantly higher degree of freedom for resource scheduling. This may depend on the consideration of vehicle deadheads, and transfer/walking of drivers allowing different locations for

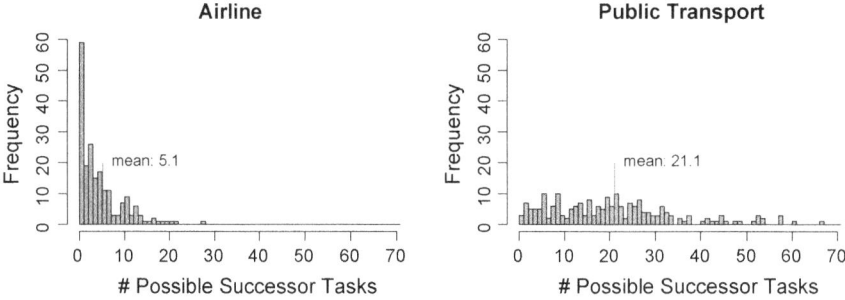

Fig. 2 Different degrees of freedom for the generation of exemplary airline and public transport crew schedules

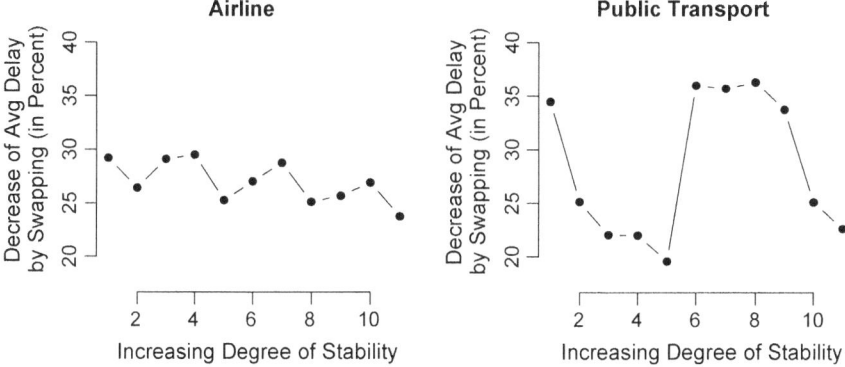

Fig. 3 Impact of swapping during simulation for schedules with different stability degrees

starting or finishing work or having a break. On average there are up to four times as many connection possibilities for every service trip. For only 1.5 % of all service trips the succeeding service trip is fixed.

Study B: Figure 3 presents the reduction of average propagated delay for each task during simulation for exemplary crew and aircraft or vehicle schedules with an increasing degree of stability (from left to right). The overall benefit of allowing resource swaps in case of delays is obvious. On average 26.7 and 28.5 % of propagated delay can be prevented in the airline and public transport resource schedules, respectively. Furthermore, the different degrees of freedom resulting from the network topologies is not apparent here as the amount of delay absorbed due to swapping crews and aircraft or vehicles is in the same order for both problem domains.

Referring to the assumption that stable schedules do not tend to imply a high necessity of dispatching, it is also important to see that an increasing degree of stability does not necessarily decrease the impact of dispatching strategies. Instead, this effect is superimposed by the fact that stable schedules often contain more resources than cost-efficient schedules, offering additional possibilities for dispatching.

This behavior can be observed in the public transport example. In contrast to the less stable solutions (1)–(5), the solutions (6)–(11) contain additional resources that imply more swap possibilities. However, if stability is increased further by incorporating additional buffer times, less swaps are necessary. This leads to the recurrent decrease between solutions (6) and (11).

4 Summary and Conclusions

In this paper, we addressed the concept of robust efficiency in resource scheduling for both airlines and public bus transport. In both problem domains there is a trade-off between cost-efficiency and robustness. We presented two fundamental techniques to improve the resulting pareto-fronts. Firstly, the prediction accuracy can be improved in order to place buffer times only at connections between tasks with a certain delay propagation risk. Secondly, the degree of freedom for resource scheduling can be improved by partial or full integration of planning stages that previously have been solved separately. Sophisticated modeling and optimization techniques are necessary to cope with the resulting increase in complexity.

Comparing the network topologies, they imply different degrees of freedom for resource scheduling. Airline hub-and-spoke networks contain a certain number of fixed connections between flights, mainly at spoke airports. In contrast, public bus transport networks provide a higher degree of freedom for scheduling decisions. However, it becomes apparent that resulting resource schedules provide a considerable degree of flexibility in both domains.

Acknowledgements This research was supported by a grant from the German Research Foundation (DFG, Grant No. KL2152/3-1).

References

1. Dück, V., Ionescu, L., Kliewer, N., Suhl, L.: Increasing stability of crew and aircraft schedules. Transp. Res. Part C: Emerg. Technol. **20**(1), 47–61 (2012)
2. Froyland, G., Maher, S.J., Wu, C.L.: The recoverable robust tail assignment problem. Transp. Sci. **48**(3), 351–372 (2013)
3. Amberg, B.: Robuste Effizienz des Ressourceneinsatzes im ÖPNV – Minimierung von Verspätungspropagation in kosteneffizienten Ressourceneinsatzplänen. Ph.D. thesis, Freie Universität Berlin (2016, to appear) (in German)
4. Ibarra-Rojas, O.J., Delgado, F., Giesen, R., Munoz, J.C.: Planning, operation, and control of bus transport systems: a literature review. Transp. Res. Part B: Methodol. **77**, 38–75 (2015)
5. Ionescu, L., Gwiggner, C., Kliewer, N.: Data analysis of delays in airline resource networks. Bus. Inf. Syst. Eng. **2015**, 1–15 (2015)
6. Ionescu, L., Kliewer, N.: Delay prediction models for robust airline resource scheduling. In: 12th Workshop on Models and Algorithms for Planning and Scheduling Problems (MAPSP) (2015)

Route Minimization Heuristic for the Vehicle Routing Problem with Multiple Pauses

Alexey Khmelev

Abstract In this work we introduce the vehicle routing problem with multiple pauses, where the fleet is heterogeneous in terms of capacity and drivers availability. Each shift has a time interval when the driver is available and a set of breaks that needs to be scheduled in the route during this shift. The objective is to minimize the number of vehicles and the travel distance. To tackle large instances, we develop a three-phase local search algorithm taking multiple breaks into account by introducing an ejection pool and randomized variable neighborhood descent as local improvement procedure. For effective break scheduling, we develop a special dynamic programming routine. Computational experiments are done on the data set provided by a delivery company situated in Novosibirsk, Russia. The instances contain 1000 customers and 30 vehicles. Experiments show effectiveness of our algorithm. It substantially reduces the fleet and travel distance.

1 Introduction

This paper addresses the vehicle routing problem with multiple pauses (VRPMP). We are given a set of customers with known demands. For each customer we know a time window when she has to be served and we know the service time. We are given a heterogeneous fleet of vehicles in terms of capacity and work shifts. In the VRPMP each vehicle has a driver assigned to it. Driver is available for a given shift. Moreover, according to the labor union restrictions, the driver has to schedule a number of short breaks during the shift. Each break pause has a time window and some duration. The routes should be designed in such a way that each customer is visited exactly once by exactly one vehicle. All routes start and end at the depot and all capacity and time constraints should be satisfied. The primary objective is to minimize the total number of routes (vehicles). The secondary objective is to minimize the total distance traveled.

A. Khmelev (✉)
Novosibirsk State University, 2 Pirogova Str., Novosibirsk 630090, Russian Federation
e-mail: avhmel@gmail.com

© Springer International Publishing Switzerland 2017   265
K.F. Dœrner et al. (eds.), *Operations Research Proceedings 2015*,
Operations Research Proceedings, DOI 10.1007/978-3-319-42902-1_36

The VRPMP is a generalization of the well-known vehicle routing problem with time windows (VRPTW). It also generalizes the single-shift single-break vehicle routing problem (VRPP) [1].

The most effective route minimization heuristic was proposed in Nagata and Bräysy [2]. The linear programming model and a heuristic was provided for VRPP by Gagliardi et al. [1]. Another heuristic for VRPP was Unified Hybrid Genetic Search proposed by Vidal et al. [3]. More detailed overview on routing problems with driver's availability restrictions are discussed in Gagliardi et al. [1].

In this paper we present an efficient heuristic reducing the number of routes/vehicles in VRPMP based on the idea of the ejection pool [2]. The key optimization module of this heuristic is Randomized Variable Neighborhood Descent (RVND). It is integrated in a special procedure, which tries to remove a route and allocate unassigned customers to the other routes. For effective break scheduling, we developed a dynamic programming algorithm (DP).

The remainder of this paper is organized as follows. First, the notations are shortly described in Sect. 2. Search space, neighborhoods and break scheduling method described in Sect. 3. The framework algorithm is presented is Sect. 4. The computational experiments and the outcomes are discussed in Sect. 5.

2 Formulation

The VRPMP is defined on directed graph $G = (V, A)$, where $V = \{0, \ldots, n\}$ is the vertex set and A is the arc set. Vertex 0 corresponds to the depot, while the remaining vertices of $V' = \{1, \ldots, n\}$ represent the customers. Each arc $(i, j) \in A$ is associated with travel time t_{ij} and travel distance d_{ij}. Each customer i is associated with service time τ_i and demand q_i. The service of a customer i has to start within a time window $[e_i, l_i]$.

In the beginning of the time, heterogeneous fleet of K vehicles is located at the depot. Let a vehicle k has capacity Q_k. Furthermore, a vehicle/driver is available during a shift $u \in S = \{1, \ldots, m\}$. If vehicle is available within shift $u \in S$, its route must begin and end within a time window $[E_u, L_u]$ and p_u breaks $P_u = \{P_u^1, \ldots, P_u^{p_u}\}$ have to be scheduled in the route. Each break P_u^j of length τ_u^j should be scheduled within $[e_u^j, l_u^j]$. Time windows of breaks do not overlap, i.e., if $i, j \in \{1, \ldots, p_u\}$ such that $i > j$ then $e_u^i \geq l_u^j$. Having two breaks next to each other without customer inbetween is not allowed.

The problem is to find feasible routes in such a way that each customer is visited exactly once; all time windows, vehicle/driver availability and capacity constraints are satisfied; the total number of routes is minimized; the total distance traveled in routes is minimized.

3   Search Space and Neighborhoods

Let r be a route with n_r customers. Let $\sigma = (\sigma_0^r, \sigma_1^r, \ldots, \sigma_{n_r+1}^r)$ be a sequence of visits in route r where the first and the last elements correspond to depot: $\sigma_0^r = \sigma_{n_r+1}^r = 0$. For a given sequence we can evaluate the load of a vehicle $Q(r) = \sum_{i=1,\ldots,n_r} q_{\sigma_i^r}$ and the travel distance $C(r) = \sum_{i=0,\ldots,n_r} d_{\sigma_i^r \sigma_{i+1}^r}$.

To evaluate the duration of a route we use time warps, proposed by Nagata et al. [4]. It is related to the time windows. If vehicle k arrives to customer σ_i^r and this happens earlier than the time window starts ($s_{ik}^r < e_{\sigma_i^r}$), then we have an idle of length $e_{\sigma_i^r} - s_{ik}^r$. Otherwise, if the vehicle is late to a customer ($s_{ik}^r > l_{\sigma_i^r}$), we suggest that the vehicle arrives at time ($s_{ik}^r = l_{\sigma_i^r}$) but pays for a time warp $tw_{i-1,i} = \max\{s_{i-1,k}^r + \tau_{\sigma_{i-1}^r} + t_{\sigma_{i-1}^r, \sigma_i^r} - s_{ik}^r, 0\}$. The total time warp for the route r is $TW(r) = \sum_{i=1,\ldots,n_r} tw_{i-1,i}$. Then, the duration of the route is $D(r) = s_{n_r+1,k}^r - s_{0k}^r + TW(r)$.

The penalty $F_{penalty}$ of route R with vehicle k and shift u is defined as the weighted sum of its excess duration, load, and time-warp: $F_{penalty}(r, k, u) = c_d \max\{0, D(r) - L_u + E_u\} + c_Q \max\{0, Q(r) - Q_k\} + c_{tw} TW(r)$. Finally, the penalty $F_{penalty}(s)$ of solution s, involving a set of routes $R(s)$, is given by the sum of the penalties of all its routes.

The local improvement procedure, called randomized variable neighborhood descent (RVND) [5], uses well-known VRP neighborhoods, 2-opt, 2-opt*, Swap and Relocate [6].

We also exclude unpromising neighboring solutions using mechanism based on the idea of customer correlation [7].

Observed that such neighborhood moves can be viewed as a separation of routes into subsequences, which are then concatenated into new routes. Let $\sigma = (\sigma_i, \ldots, \sigma_j)$ and $\sigma' = (\sigma_{i'}', \ldots, \sigma_{j'}')$ be two subsequences with known D, TW, E and L values, where E and L are the earliest and latest visits to the first vertex allowing a schedule with minimum duration and minimum time-warp use. For concatenation of σ and σ', denoted by $\sigma \oplus \sigma'$ we evaluate this values in amortized constant time [7]. Thus any of neighborhood moves can be computed in amortized constant time.

Insertion of Breaks

To insert break in sequence $\sigma = (\sigma_i, \ldots, \sigma_j)$ we need to find the best division of σ into two subsequences σ^1 and σ^2 ($\sigma^1 \oplus \sigma^2 = \sigma$) to insert the break in. For such insertion, we should know D, TW, E and L values for σ^1 and σ^2. Thus, we have to calculate an optimal travel time from the last element in σ^1 to the break (Fig. 1).

Theorem 1 *Optimal travel time* $t_{\sigma^1 v_p}$ *from the last element in* σ^1 *to the break* v_p *equals* $\min\{\max\{e_{v_p} - D(\sigma^1) + TW(\sigma^1) - L(\sigma^1), 0\}, t_{\sigma^1 \sigma^2}\}$, *where* $t_{\sigma^1 \sigma^2}$ *is the travel time from the last element in* σ^1 *to the first element in* σ^2.

Let $\overline{D}(\sigma, u, i, j), \overline{TW}(\sigma, u, i, j), \overline{E}(\sigma, u, i, j)$ and $\overline{L}(\sigma, u, i, j)$ be the values of σ when the breaks are $\{P_u^i, \ldots, P_u^j\}, 1 \le i \le j \le p_u$ for a shift u. We denote such

(a)   **(b)**   **(c)**

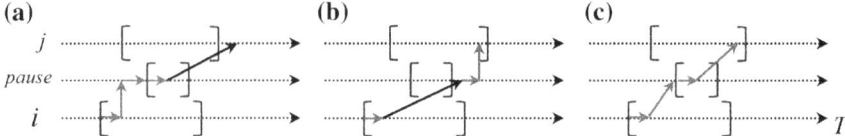

Fig. 1 Pause insertion cases

schedule for σ as σ_{uij}. The break scheduling for concatenation $(\sigma^1 \oplus \sigma^2)_{uij}$ is chosen from the set:

$$\Sigma_{uij}^{\sigma^1\sigma^2} = \{\sigma^1 \oplus \sigma_{uij}^2\} \cup \{\sigma_{uik}^1 \oplus \sigma_{u.k+1,j}^2 | i \le k < j\} \cup \{\sigma_{uij}^1 \oplus \sigma^2\} \cup$$

$$\{\sigma^1 \oplus P_u^i \oplus \sigma_{u,i+1,j}^2\} \cup \{\sigma_{u,i,k-1}^1 \oplus P_u^k \oplus \sigma_{u,k+1,j}^2 | i \le k \le j\} \cup \{\sigma_{u,i,j-1}^1 \oplus P_u^j \oplus \sigma^2\}.$$

Thus, $(\sigma \oplus \sigma')_{uij} = \arg\min_{\sigma \in \Sigma_{uij}^{\sigma^1\sigma^2}} D(\sigma)$.

For a route r with sequence σ_r and shift u we have to find the best division of σ_r into two subsequences σ_r^1 and σ_r^2 minimizing duration D of $(\sigma_r^1 \oplus \sigma_r^2)_{u1p_u}$.

Theorem 2 *We can find* $(\sigma_r)_{u1p_u}$ *using DP in amortized* $O(|r|^2 p_u^3)$ *time.*

4 Route Minimization Heuristic

The proposed heuristic has three phases. In the first phase we try to find a feasible solution. If a feasible solution is found, we minimize the number of routes in the second phase. In the third phase, we minimize the total distance traveled.

First two phases of our algorithm use customer allocation procedure. This procedure tries to allocate a given set of customers C to a partial solution s without creating any new routes. This idea was first introduced by Nagata et al. [2] for VRPTW. We adjust this procedure for VRPMP. The pseudocode of the adapted procedure is presented in Fig. 2.

If there are many feasible insertion positions for v_{in}, we choose a random one (line 6). Otherwise, we minimize the penalty function using RVND (line 8). If the obtained solution is still infeasible after local improvement, we use special ejection procedure proposed by Nagata et al. [2] (line 10). Then, we add ejected customers at the end of the ejection pool (line 11) and perturbate the solution with random neighborhood moves (line 12).

To find a feasible solution we use the constructive procedure presented in Penna et al. [5] with Cheapest Feasible Insertion Criterion and Sequential Insertion Strategy. Upon completion of this procedure, we obtain a partial solution and a set of unserved customers C. Afterwards, we call the customer allocation procedure

```
1   Procedure AllocateCustomers(C, s, T_max);
2   begin
3       EP ← C; // Initialize ejection pool
4       while EP ≠ ∅ ∧ time < T_max do
5           Remove first customer v_in from EP;
6           Insert v_in in s minimizing F_penalty(s);
7           if s is not feasible then
8               RVND(s);
9           if s is still not feasible then
10              Eject a set of customers N_ej from s making it feasible;
11              Add N_ej at the end of EP;
12              Perturb solution s;
13          end
14          if EP ≠ ∅ then
15              Restore s to the input state;
16          Return s;
17  end
```

Fig. 2 Customer allocation procedure

with parameters C, s and limited running time. If the allocation was not finished in time we stop our algorithm.

On each iteration of the route minimization phase we choose a random route r. Then we remove it from the solution s. After that, customers of the route r become unassigned. We call customer allocation procedure trying to insert these customers in the partial solution $s \setminus r$ without adding a route/vehicle.

In the third phase of our heuristic we use education and repair procedure proposed by Vidal et al. [3] to minimize the total travel distance for obtained set of routes.

5 Computational Experiments

The algorithm is implemented in Java and was executed on an Intel Xeon 3.07 GHz. We use two real-life instances with 1000 customers and 30 vehicles. In the first instance, customers are located randomly in the plane with uniform distribution, and in the second the customers are located in clusters. Each vehicle works in one of the three shifts: 7:00–6:00; 13:00–23:00; 7:00–23:00. We solve instances for 0–3 breaks per shift.

In the first experiment, we studied the performance of local search depending on the number of breaks per shift (N_p). The second column in Table 1 shows the average time of the Relocate neighborhood evaluation T_{avg}^{Rel} in milliseconds.

In the second experiment, we investigate the possibility of reducing the vehicle fleet and on how the number of breaks influences this process. Each instance was solved 30 times for different numbers of breaks. Time limits per 10, 60 and 50 min were set for the first, second and the third phases, respectively. The time to allocate one route in the second phase was limited by 3 min.

Table 1 Algorithm results

N/N_p	T_{avg}^{Rel}	N_r	Distance					
		$min/$ max	D_{min}^0	D_{max}^0	D_{min}	D_{max}	D_{avg}	σ (%)
1/0	202	25/25	1,007,427	1,137,817	373,722	385,801	376,743	3766(1, 00)
1/1	480	25/25	988,574	1,088,719	388,511	467,417	408,824	23,208(5, 68)
1/2	794	25/25	1,014,496	1,117,109	412,426	552,966	498,162	43,194(8, 67)
1/3	1015	25/25	1,020,285	1,113,830	482,812	630,404	542,472	46,021(8, 48)
2/0	192	17/18	1,138,816	1,316,069	876,867	1,114,978	1,008,252	81,300(8, 06)
2/1	299	18/19	1,167,424	1,318,621	886,321	1,160,035	1,006,542	98,025(9, 74)
2/2	543	20/21	1,324,955	1,514,604	927,951	1,169,061	1,044,429	93,869(8, 99)
2/3	694	20/21	1,169,061	1,531,777	980,749	1,234,636	1,093,872	104,534(9, 56)

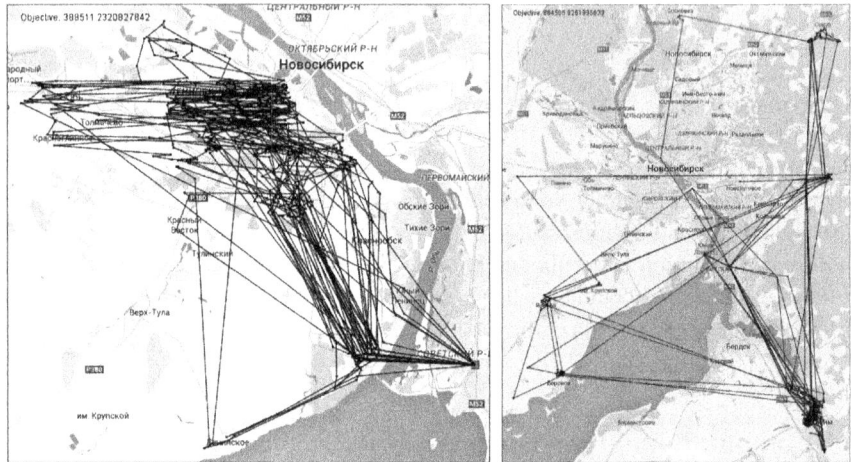

Fig. 3 Obtained solutions for two instances with reference to map

The third column in Table 1 shows the number of routes N_r obtained in the second phase of the algorithm. The fourth and the fifth columns show the total travel distance D^0 after the second phase. The next three columns display the values for the travel distance D after the third phase. Let σ be the standard deviation of value D in percentage of the average. Figure 3 shows the examples of solutions obtained by the algorithm for both instances with reference to the map.

Having breaks significantly changes the structure of the solutions and leads to substantial increase in transportation costs. In general, the main purpose of this work, namely the optimization of the fleet, has been reached. The developed algorithm have shown its efficiency in reducing the fleet by at least 20%.

Acknowledgements This research was partially supported by the RFBR grand 15-07-01141.

References

1. Gagliardi, J.-P., Renaud, J., Ruiz, A., Coehlo, C. L.: The Vehicle Routing Problem with Pauses. Technical report 2014-22, CIRRELT, Montreal, QC, Canada
2. Nagata, Y., Bräysy, O.: A powerful route minimization heuristic for the vehicle routing problem with time windows. Oper. Res. Lett. **37**, 333–338 (2009)
3. Vidal, T., Crainic, T.G., Gendreau, M., Prins, C.: A unified solution framework for multi-attribute vehicle routing problems. Eur. J. Oper. Res. **234**(3), 658–673 (2014)
4. Nagata, Y., Bräysy, O., Dullaert, W.: A penalty-based edge assembly memetic algorithm for the vehicle routing problem with time windows. Comput. Oper. Res. **37**, 724–737 (2010)
5. Penna, P.H.V., Subramanian, A., Ochi, L.S.: An iterated local search heuristic for the heterogeneous fleet vehicle routing problem. J. Heuristics **19**, 202–232 (2013)
6. Khmelev, A., Kochetov, Y.: A hybrid local search for the split delivery vehicle routing problem. Int. J. Artif. Intell. **13**(1), 147–169 (2015)
7. Vidal, T., Crainic, T.G., Gendreau, M., Prins, C.: A hybrid genetic algorithm with adaptive diversity management for a large class of vehicle routing problems with time-windows. Comput. Oper. Res. **400**, 475–489 (2013)

Combining NLP and MILP in Vertical Flight Planning

Liana Amaya Moreno, Zhi Yuan, Armin Fügenschuh, Anton Kaier
and Swen Schlobach

Abstract Vertical flight planning of commercial aircrafts can be formulated as a Mixed-Integer Linear Programming (MILP) problem and solved with branch-and-cut based solvers. For fuel-optimal profiles, speed and altitude must be assigned to the corresponding segments in such a way that the fuel consumed throughout the flight is minimized. Information about the fuel consumption of an aircraft is normally given by the aircraft manufacturers as a black box function, where data is only available on a grid points depending on speed, altitude and weight. Hence, some interpolation technique must be used to adequate this data to the model. Using piecewise linear interpolants for this purpose is suitable for the MILP approach but computationally expensive, since it introduces a significant amount of binary variables. The aim of this work is to investigate reductions of the computation times by using locally optimal solutions as initial solutions for a MILP model which is, thereafter solved to global optimality. Numerical results on test instances are presented.

1 Introduction

Fuel-efficient flight trajectory planning has been the focus of researchers over the years with the objective of helping on one hand, the Air Traffic Management (ATM) to cope with the constantly growing air traffic flow and on the other hand, the aviation

L. Amaya Moreno (✉) · Z. Yuan · A. Fügenschuh
Professorship of Applied Mathematics, Department of Mechanical Engineering,
Helmut Schmidt University, Hamburg, Germany
e-mail: lamayamo@hsu-hh.de

Z. Yuan
e-mail: yuan@hsu-hh.de

A. Fügenschuh
e-mail: fuegeschuh@hsu-hh.de

A. Kaier · S. Schlobach
Lufthansa Systems AG, Kelsterbach, Germany
e-mail: anton.kaier@lhsystems.com

S. Schlobach
e-mail: swen.schlobach@lhsystems.com

© Springer International Publishing Switzerland 2017
K.F. Dœrner et al. (eds.), *Operations Research Proceedings 2015*,
Operations Research Proceedings, DOI 10.1007/978-3-319-42902-1_37

273

industry to deal with skyrocketing fuel costs. The computation of such a trajectory includes calculating a horizontal (2D) route on the earth's surface connecting a departure and an arrival point, assigning to this route an admissible altitude level (1D) and all together guaranteeing that the arrival point is reached within a strict time window (1D). Due to the computational difficulty of this 4D problem, it is normally solved by dividing it into two phases: a horizontal and a vertical one [1]. Moreover, information about fuel consumption is required to optimally assign speed and altitude to the corresponding trajectory-composing segments in order to minimize the amount of fuel. This information is commonly given as a black box function where data is only available on a grid of points depending in general on discrete speed, altitude and weight levels. To handle this type of data different interpolation techniques are used, their selection depends heavily on the solution approach. Furthermore, these interpolation techniques drastically determine the computation times of the solution. This study concentrates on the vertical flight planning of commercial aircrafts using a MILP approach. More precisely, we focus on how to reduce computation times of the proposed MILP model in [6] by providing the MILP solver with good feasible initial solutions. The key for quickly obtaining such initial solutions is to use nonlinear global interpolants for fitting the fuel consumption data instead of locally piecewise linear ones.

2   Mathematical Model

A model for the vertical flight planning of a commercial aircraft is presented in [6] where unit fuel consumption data is available for discrete levels of speed, altitude and weight. In the absence of wind, the optimal altitude can be computed given the speed and weight optimal profiles by enumerating all the possible altitudes and choosing the best one. Hence its calculation does need to be included in the model. The objective of the model is then to assign to each of the n segments composing the flight's trajectory $S = \{1 \ldots n\}$, a speed v_i for $i \in S$ and a weight w_i for $i \in S \cup \{0\}$ at the corresponding time t_i for $i \in S \cup \{0\}$, with Δt_i for $i \in S$ being the time spent on each segment. Furthermore the middle weight w_i^{mid} for $i \in S \cup \{0\}$ of the aircraft along each segment is also computed to finally calculate the fuel consumption f_i for $i \in S$. Among the parameters we have the segment length L_i for $i \in S$, the minimum and maximum duration of the trip, $\underline{T}$ and $\overline{T}$ respectively; W^{dry} denotes the weight of the loaded aircraft including the contingency fuel. The mathematical model is as follows:

$$\min \quad w_0 - w_n \tag{1}$$

$$\text{s.t.} \quad t_0 = 0, \quad \underline{T} \le t_n \le \overline{T} \tag{2}$$

$$\forall i \in S : \quad \Delta t_i = t_i - t_{i-1} \tag{3}$$

$$\forall i \in S : \quad L_i = v_i \cdot \Delta t_i \tag{4}$$

$$w_n = W^{dry} \tag{5}$$

$$\forall i \in S: \quad w_{i-1} = w_i + f_i \tag{6}$$

$$\forall i \in S: \quad w_{i-1} + w_i = 2 \cdot w_i^{mid} \tag{7}$$

$$\forall i \in S: \quad f_i = L_i \cdot \widehat{F}(v_i, w_i^{mid}) \tag{8}$$

The objective function (1) minimizes the fuel consumed throughout the trip calculated as the difference between the starting weight w_0, and the arrival weight w_n. The starting time t_0, and the duration between the given bounds is enforced by constraint (2). The time consistency is preserved by (3) and the equation of motion is given by (4). This quadratic constraint can be reformulated as a second-order cone constraint if Eq. (2) requires to speed up form the unconstrained fuel-optimal travel time. Moreover the resulting constraint can further be reformulated using a system of linear inequalities as introduced by Ben-Tal and Nemirovwski [2] and refine by Glineur [4] (see [6] for more details). As for the weight, Eq. (5) ensures that all the fuel is consumed during the flight. The weight consistency is preserved by (6). The middle weight is computed in (7) which is further used to calculate the fuel consumption per segment in (8), where $\widehat{F}(v_i, w_i^{mid})$ is the function that interpolates all continue values of v and w within the $V \times W$ grid.

Here we focus on how to reduce computation times of fuel-optimal vertical profiles. For this, we first consider the (1)–(8) MILP model presented in [6] where $\widehat{F}$ is a piecewise linear function denoted by $\widehat{F}_L$, obtained using the *Lambda Method* [3, 5]. In order to reduce the computation times of this model, we provide it with a feasible initial solution so the solver does't have to spend time searching for one, which takes, according to our experiments, a large portion of the total computation time. This solution besides being feasible is in most cases very close to the optimal one, given its calculation as follows: we let $\widehat{F}$ be a smooth nonlinear globally interpolated function, $\widehat{F}_N$, obtained with conventional surface fitting techniques, assuming the fuel consumption data is drawn from a nonlinear function. This way we not only avoid the binary variables introduced by $\widehat{F}_L$ that contribute, to some extent, to large computation times but also are able to formulate a Nonlinear Programming (NLP) model that can be quickly solved to local optimality with common nonlinear solvers. This locally optimal solution, would only be a partial initial solution to the MILP model in the sense that all variables are provided except the binary associated to $\widehat{F}_L$. Nevertheless building a complete initial solution from that is not a difficult task, we simply feed this partial solution the initial MINLP model a let solver calculate the binary variables. Finally we proceed to solve once more the MILP model starting from the created initial solution. Figure 1 depicts the above mentioned interpolant $\widehat{F}_L$ and the residuals for the interpolant $\widehat{F}_N$. On the left side we see the 2D piecewise linear interpolant, on the right side we see the residuals of the 5° polynomial interpolant, whose Adjusted R-square statistic is approximately 1 (i.e. the model is able to explain almost 100 % of the variation of the data around the mean) and Root Mean Squared Error (RMSE) is 0.4 which accounts for the standard deviation of the random component present in the data. Note that these residuals differ at most on a 0.5 % from the original data.

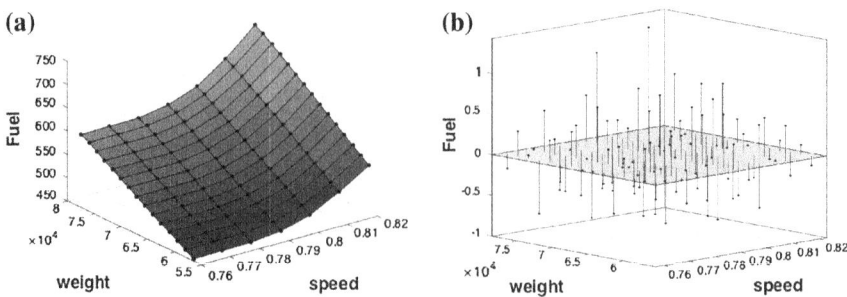

Fig. 1 Unit fuel consumption (kg per nautical mile) for the Airbus 320. On the x-axis the aircrafts speed (mach number from optimal speed to maximal speed) and on the y-axis the weight (kg). **a** Piecewise linear interpolant. **b** Residuals of the nonlinear interpolant

3 Numerical Results

The models were written using AMPL as modelling language and solved afterwards with Gurobi 6.0.0 and SNOPT 7.2-5 for the MILP and the NLP respectively. We adopted similar test instances as in [6], that is, the aircrafts Airbus 320, 380, Boeing 737 and 772 for two speed up factors from the unconstrained optimal. For each aircraft several travel distances were tested ranging form 800 Nautical Miles (NM) for the B737 to 7500 NM for A380 and B772. Each flight is divided into equidistant segments of 100 NM. Table 1 summarizes the features of the test instances. All instances were solved on a 4-core Intel Core i7 at 2.6 GHz and 16 GB RAM computing machine. In the NLP column of Table 2 the computation times to solve this model are listed. As it can be seen they are extremely fast, it takes SNOPT no longer than 1.6 s to reach local optimality. Another NLP solver, CONOPT 3.15C, was also used to solve the instances and the differences between the solvers were minimal, with respect to both the objective function and the computation times. Nevertheless SNOPT was chosen for the further calculation because it performs better with bigger models (see Fig. 2). A polynomial of degree 2 (whose coefficients may physically be easier to interpret) was also proposed for $\widehat{F}_N$ resulting in lower computation times for the NLP model. However the time differences with respect to the 5° $\widehat{F}_N$ were not significant while

Table 1 For each instance we listed the optimal and maximal speed (in Mach number), the dry weight and maximal weight (in kg), the maximal distance (in NM), the number of speed levels $|V|$ (between optimal and maximal speed) and weight levels $|W|$

Type	Opt. speed	Max. speed	Dry weight	Max. weight	Max. distance	$|V|$	$|W|$
A320	0.76	0.82	56,614	76,990	3500	7	15
A380	0.83	0.89	349,750	569,000	7500	7	24
B737	0.70	0.76	43,190	54,000	1800	7	12
B772	0.82	0.89	183,240	294,835	7500	8	16

Table 2 Computation times (s) for the test instances with the two different speed-up factors, 2.5 or 5 % and different number of segments |S|

Aircraft	|S|	2.5 % speed up			5 % speed up		
		NLP	MILP	MILP-In.sol	NLP	MILP	MILP-In.sol
A320	15	0.018	3.521	1.954	0.016	5.115	2.463
	20	0.025	13.750	4.577	0.023	18.306	4.231
	25	0.042	29.195	7.667	0.076	22.774	7.186
	30	0.030	43.318	14.543	0.033	31.031	18.982
	35	0.039	54.525	42.734	0.048	50.409	30.142
A380	30	0.123	14.515	18.265	0.148	17.606	14.623
	40	0.208	138.755	30.841	0.212	97.241	34.455
	50	0.343	136.468	65.821	0.469	168.055	211.726
	60	0.636	311.382	175.656	0.707	273.739	187.716
	70	1.122	341.211	218.609	1.033	596.243	458.675
	75	1.198	413.155	358.756	1.548	348.766	861.872
B737	8	0.012	0.473	0.847	0.012	0.510	0.689
	12	0.018	0.846	1.238	0.018	1.188	1.558
	15	0.027	2.542	2.539	0.024	3.146	3.369
	18	0.034	7.963	7.561	0.038	5.689	6.072
B772	25	0.325	30.769	7.120	0.274	30.582	11.695
	35	0.166	49.582	24.406	0.165	76.427	33.243
	45	0.350	145.996	41.735	0.310	128.898	51.452
	55	0.611	257.124	50.161	0.482	306.332	135.605
	65	1.023	492.730	85.266	0.874	641.873	187.851
	75	1.175	322.360	192.436	1.202	2586.521	831.517

Each segment represents 100 NM

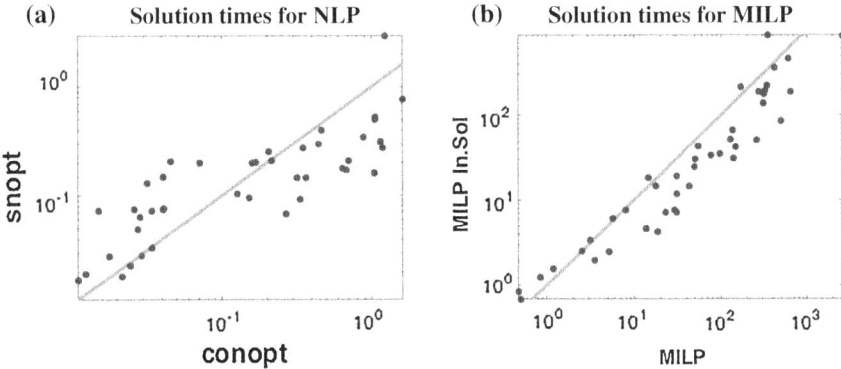

(a) Solution times for NLP

(b) Solution times for MILP

Fig. 2 a SNOPT versus CONOPT: The points under the *green line* on the upper tight side of the plot, represent lower computation times for the solver SNOPT compared to CONOPT. **b** MILP-In.sol versus MILP: The majority of the points are below the *green light* indicating that in fact the computation times with the initial solution are lower than without one

the goodness of fit was very low (not enough evidence to support the least-squares fitting assumptions). It is important to remark that, when comparing the objectives from the NLP and the MILP model they differ at most in a 0.25 %. For the latter model, computation times range from around 0.5–50 s for smaller aircrafts, for the bigger from 17 to 2586 s without providing an initial solution (the results are listed in the MILP column). But the more interesting results are those for the MILP model with the created initial solution (these are listed in the MILP-In.Sol column). They correspond to the time used for computing the initial solution plus the time used to solve the MILP model with the starting solution to optimality or near-optimality (less than 0.05 % gap). These results suggest that providing the MILP model with an initial solution is definitely beneficiary for bigger instances ($|S| \times |V| \times |W| > 3000$), in average the solver requires only 49 % of the time than without an initial solution, see Fig. 2a. For the instances with $1500 < |S| \times |V| \times |W| < 3000$ on the other hand, the improvement average is more (about 35 %). For small instances ($|S| \times |V| \times |W| < 1500$), providing an initial solution does not help reducing the computation times, on the contrary it increases them in average (122 %). Hence the overhead for handling data clearly dominated the solution time.

Reducing the computation times for this model could have a great impact on solving extended versions of this model (more realistic ones). For example, the model presented in [7] takes into account the influence of the wind in both the vertical and the horizontal phase, that is, areas with strong tail-wind are pursued and the ones with strong head-wind are avoided. Similarly to fuel consumption data, wind data is commonly given as a black function and once more suitable interpolation techniques must be used. It is natural then to expect that adding this feature to the model increases not only its size but also its complexity and is precisely in this cases where potential reductions in computation times are of great value.

Acknowledgements This work is supported by BMBF Verbundprojekt E-Motion.

References

1. Altus, S.: Flight Planning – The Forgotten Field in Airline Operations (2007). http://www. agifors.org/studygrp/opsctl/2007/, presented at AGIFORS Airline Operations 2007
2. Ben-Tal, A., Nemirovski, A.: On Polyhedral Approximations of the Second-Order Cone. Operational Research (2001)
3. Dantzig, G.B.: Linear Programming and Extensions. Princeton University Press, Princeton (1963)
4. Glineur, F., De Houdain, R., Mons, B.: Computational Experiments with a Linear Approximation of Second-Order Cone Optimization (2000)
5. Wilson, D.: Polyhedral methods for piecewise-linear functions. Ph.D. thesis, TUniversity of Kentucky (1998)
6. Yuan, Z., Fügenschuh, A., Kaier, A., Schlobach, S.: Variable speed in vertical flight planning. In: Operations Research Proceedings. Springer (2014)
7. Yuan, Z., Amaya Moreno, L., Fügenschuh, A., Kaier, A., Mollaysa, A., Schlobach, S.: Mixed integer second-order cone programming for the horizontal and vertical free-flight planning problem. Technical Report AMOS # 21, Helmut Schmidt University (2015)

European Air Traffic Flow Management with Strategic Deconfliction

Jan Berling, Alexander Lau and Volker Gollnick

Abstract To guarantee a safe journey for each aircraft, air traffic controllers make sure that separation minima are maintained. To prevent controller overburdening, each controller team is responsible for one confined sector. Furthermore, sectors limit the number of flights entering each hour. Compliance to all sector and airport capacity constraints in daily business is ensured by EUROCONTROL's Network Management. This function balances the flights' demand of airspace with available capacity by re-allocating departure timeslots. However, when minimum separation between two aircraft may become compromised, a conflict occurs. Conflicts are solved by controllers who provide pilots with instructions to maintain separation. Hence, conflicts increase controller workload and thus tighten sector capacity. The aim of this work is a prevention of actual conflicts by strategic deconfliction. Strategic conflicts refer to planned trajectories which violate separation minima in any future point in space and time. Mimicking a future Network Management, departure times are re-allocated to reduce the number of strategic conflicts while satisfying both sector and airport capacity constraints. As a basis for the deconfliction, actual datasets of planned trajectories, sector bounds and airport features are aggregated to model the European Air Traffic Flow Management. The allocation problem of departure timeslots is formulated as a Quadratic Binary Problem with linear delay costs, quadratic conflict costs and linear constraints. In an optimal solution, all strategic conflicts are solved. Finally, a trade-off between conflict reduction and delay is performed.

J. Berling (✉) · V. Gollnick
Institute of Air Transportation Systems, Hamburg University of Technology (TUHH),
Blohmstraße 20, 21079 Hamburg, Germany
e-mail: jan.berling@tuhh.de

A. Lau
German Aerospace Center (DLR), Institute of Air Transportation Systems,
Blohmstraße 20, 21079 Hamburg, Germany

© Springer International Publishing Switzerland 2017
K.F. Dœrner et al. (eds.), *Operations Research Proceedings 2015*,
Operations Research Proceedings, DOI 10.1007/978-3-319-42902-1_38

1   Introduction

One of the key performance areas of the air transportation system is safety. In particular, high velocities of aircraft and a dense packaging of airspaces require a cautious and fail-safe cooperation between all stakeholders. As a safety net, aircraft must always maintain a minimum separation standard. When two aircraft are in a situation that may lead to compromised separation minima, a conflict occurs [8]. In day-to-day operations, these conflicts are solved by air traffic controllers. Each controller team is responsible for one specific sector. Sectors are capacity-constrained, so that controllers are able to solve all conflicts. Compliance of aircraft to all sector and airport capacities in the European air transportation system is supervised by EU-ROCONTROL's Network Management. To improve safety, they seek to reduce the number of conflicts [6]. Because conflicts are time-dependent, improved separation of aircraft in time reduces conflict potential. For strategic deconfliction, the Network Management could allocate alternative departure timeslots.

The vast body of literature on Air Traffic Flow Management performing demand-capacity-balancing, e.g. [2, 13, 15], does not include deconfliction. Conversely, research about strategic deconfliction, like [3, 5, 10, 14], does not include demand-capacity-balancing. Hence, there is research potential for the combination of strategic deconfliction with demand-capacity-balancing. Consequently, this work develops an Air Traffic Flow Management approach that combines strategic deconfliction with adherence to system capacities. To achieve both objectives, alternative departure timeslots are allocated. A model for the European air transportation system, the Network Flow Environment, is used. From that, a Quadratic Binary Problem is created and solved. We expect that conflict reduction by departure timeslot allocation increases delay. Accordingly, a trade-off between delay and conflicts is evaluated here.

2   European ATFM Model: Network Flow
Environment (NFE)

European Air Traffic Flow Management (ATFM) is modelled by the Network Flow Environment (NFE), which contains actual sector boundaries with calculated sector capacities, actual airports with approximated capacities as well as planned trajectories on the route network (see [11, 12] for additional information on the model). Airport arrival and departure capacities are based on the runway systems. Sector capacities are calculated by the method of Monitor Alert Parameter (MAP) [7]. Planned trajectories are gathered from Network Managements' flight plan data. The flight data consist of departure aerodrome, intermediate navaids and the arrival aerodrome. Each data-point includes a timestamp, coordinates and flight level. Trajectory points are interpolated with a sampling time of one minute and mapped to the air traffic control sectors (Fig. 1).

Fig. 1 Capacity constrained sectors and trajectories that were planned for departure on the 07.06.2012 between 18:00z and 22:00z

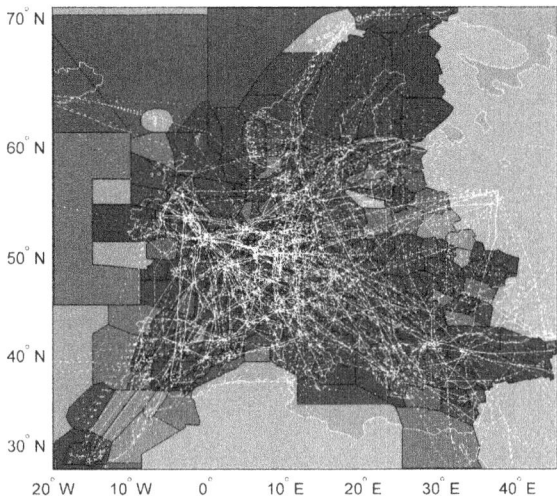

Detection of Strategic Conflicts

When two aircraft converge in space and time so that they may compromise the minimum separation, they are in conflict (see Fig. 2a). The separation standard determines a distance of five nautical miles lateral and thousand feet vertical between all aircraft in the upper airspace [8]. When two aircraft violate the separation minima, a loss of separation occurs (see Fig. 2b).

On the other hand, a *strategic* conflict exists, when two *planned* trajectories violate the separation minima in any points in the future. To enable strategic deconfliction, every planned flight path point is checked against points of other flights for violations of the separation minima.

When a flight is delayed, the whole trajectory shifts in time. Hence, delay allocation can result in new conflicts between trajectories that were separated in time beforehand. Conflicts between delayed flights are only possible, if they are planned to arrive at the critical points with a time difference which is a multiple of the time-granularity. For example, if flight F1 crosses a point 15 min after flight F2, there is no conflict (see Fig. 2c). Conversely, if flight F2 is delayed 15 min, there is a strategic conflict (see Fig. 2d).

A check for separation of every trajectory point against each point of all other trajectories is computationally expensive. That's because the number of possible conflict combinations grows quadratically with the number of flights and points. Therefore, checks for conflicts are only conducted between points in close vicinity, as e.g. in [9]. For that purpose, trajectory points are mapped to a grid which spans the European airspace (see Fig. 3). All trajectory points of each grid element are checked for conflict potential with the points in all nearby grid elements. Finally, the potential conflicts are determined by geometric methods. Moreover, only conflicts in the en-route segments above 10 000 feet are considered.

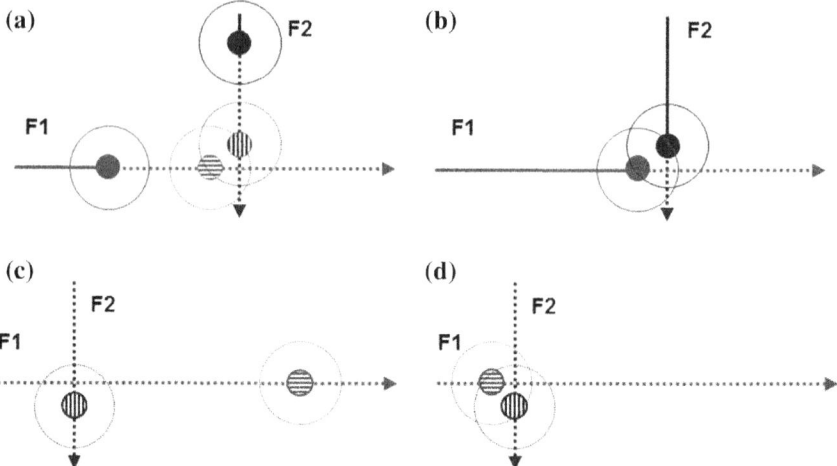

Fig. 2 **a** Conflict between two aircraft converging to a violation of the minimum separation (indicated by *surrounding circles*). **b** Loss of separation between two aircraft because of a violation of the separation minima at that point in time. **c** Two planned trajectories that are not in conflict. They are sufficiently separated because they cross with a time difference. **d** Two planned trajectories that are in conflict because there is a future point in time where they violate the separation minima

Fig. 3 Surrounding area of Hamburg partitioned to grid elements for detection of strategic conflicts between trajectories

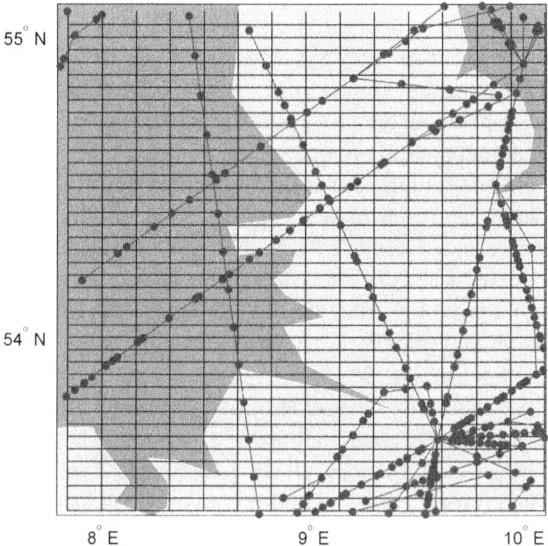

Binary Quadratic Problem: ATFM with Strategic Deconfliction

The goal of this work is to deconflict planned trajectories while adhering to capacity constraints. The ATFM problem is formulated as a Quadratic Binary Problem with binary decision variables, a quadratic cost function and linear constraints. Binary decision variables $x_{f,d}$ represent the allocation of a departure timeslot d to a flight f (4). Each departure timeslot has delay cost $w_{f,d}$ which corresponds to the delay (1). Every strategic conflict between two flights i and j with departure slots $x_{fi,di}$ and $x_{fj,dj}$ has conflict cost $k_{fi,di,fj,dj} = k$. The coefficient k in delay minutes is identical for all strategic conflicts. The conflict cost term depends on variables of two flights and therefore forms a quadratic objective. The objective function consists of both delay and conflict cost and is to be minimized. Set partitioning constraints assign one departure slot to each flight (2). Knapsack constraints maintain that all sector and aerodrome capacity limits are respected (3). The capacity $c_{s,t}$ of every entity s exists for each time-window t individually for the entire model-time. By setting the coefficient k to different delay costs, a trade-off between delay and strategic conflits is performed.

$$\min \mathbf{w}^\mathsf{T}\mathbf{x} + \mathbf{x}^\mathsf{T}\mathbf{K}\mathbf{x} \tag{1}$$

$$\text{so that} \quad \mathbf{F}\mathbf{x} = \mathbf{e} \tag{2}$$

$$\mathbf{A}\mathbf{x} \leq \mathbf{c} \tag{3}$$

$$\forall x_{f,d} \in [0, 1] \tag{4}$$

Due to the fact that the conflict cost term is quadratic, the problem is nonlinear. Therefore, the search for an optimal solution is complicated. For implementation, the quadratic objectives are surrogated by a combination of quadratic and linear constraints.

The NFE and conflict detection together with the function of demand-capacity-balancing are implemented in Matlab. Data are passed to a Matlab-Executable, which generates a Quadratic Binary Problem with linear constraints. The executable is partially based on the OPTI Toolbox [4] and uses the framework SCIP [1] to call the solver Soplex [16].

3   Optimization Results

A trade-off between strategic conflicts and departure delay is performed for a scenario with 3 238 flights. The scenario contains 632 sectors and 489 aerodromes and all flights scheduled for departure between 18:00z and 22:00z. The number of flights entering each sector or aerodrome is capacity-constrained in time-windows of 15 min. each. To avoid cancellations, maximum allocated atfm-delay per flight is 2:30h. With a departure timeslot granularity of 15 min., the maximum delay equals 10 departure slots. All possible conflicts which can arise from all possible departure time combinations were detected. Since departure granularity is 15 min., only conflicts

Table 1 Optimal results of ATFM with strategic deconfliction, depending on conflict cost k

Cost coefficient k (min)	0	5	10	15	20	45
Conflicts (#)	288	255	151	136	9	0
Delay (min)	795	840	1560	1785	3690	3930
Computation time (s)	10.3	73.2	101.4	119.3	136.0	230.2
Saturated capacities (#)	40	40	39	40	36	39

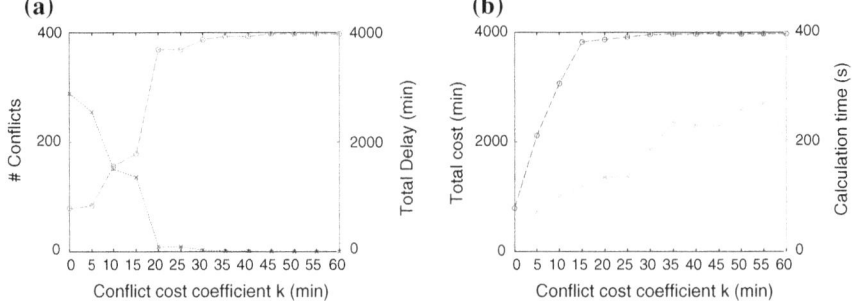

Fig. 4 a Number of conflicts *(dotted line)* and total delay in minutes *(dashed line)* by conflict cost coefficient k. **b** Total cost *(dashed line)* and calculation time *(dotted line)* by conflict cost coefficient k

which occur when two flights depart with a time difference of a multiple of 15 min. are considered. The time difference of the potential conflicts is between zero and the maximum allocated delay. Under these conditions, there are 3 180 potential conflicts between flight-pairs. The number of possible strategic conflicts between all possible departure slot combinations is 23 768. Calculation time to detect all potential conflicts between flight pairs is 129 s.[1]

Optimal results of the problem with different conflict cost coefficients are shown in Table 1. With a conflict cost of zero there are 288 conflicts. The total delay is 795 min. while 40 sector and aerodrome capacities are saturated. With a conflict cost of $k = 20$, the number of conflicts is 9, accompanied by 3 690 min. delay. With a conflict cost of $k = 45$, there are no unsolved conflicts and total delay is 3 930 min. For all conflict cost coefficients, the total number of saturated capacity constraints is between 36 and 40. Calculation time with conflict cost of zero is 10.3 s. At conflict cost of $k = 20$, calculation time is 136 s. With conflict cost of $k = 45$, calculation time is 230.2 s. With a rising conflict-cost, delay increases whereas conflict-count decreases (see Fig. 4). Likewise, with a rising conflict-cost, calculation time and total cost of the optimal solution increases (see Fig. 4).

---

[1] All computations took place on an AMD FX(tm)-4100 processor (3.60 GHz) with 16GB RAM.

4 Conclusion and Outlook

A Quadratic Binary Integer Problem that reduces strategic conflicts between air-craft while adhering to the capacity constraints of the air transportation system was solved to optimality. We showed that the allocation of alternative departure slots can simultaneously reduce the number of strategic conflicts and ensure compliance to capacity constraints. Calculation times remained below five minutes for a Euro-pean scenario with 3 238 flights, which makes this approach viable for short-term planning. A trade-off between delay and conflicts was performed by introducing a conflict cost coefficient weighted in minutes. Conflict costs just above the smallest delay lead to a reduction of strategic conflicts by 97 %. However, the reduction in conflicts increased delay by 2 895 min. An increased cost coefficient leads to higher calculation times because of a rising share of the quadratic term in the objective.

We expect that a smaller granularity for departure timeslots would achieve com-parable conflict counts with less delay. The system efficiency could be improved by additional pre-departure rerouting. Moreover, consideration of stochastic trajec-tory and capacity deviations is desirable. However, the resulting growth in problem size leads to increased computation times. Solving these extended models requires more powerful optimization methods. For complete day scenarios, we envision a rolling-horizon method that sequences optimization through the day and benefits from information gain over time.

References

1. Achterberg, T.: SCIP: solving constraint integer programs. Math. Progr. Comput. **1**(1), 1–41 (2009)
2. Bertsimas, D., Lulli, G., Odoni, A.: An integer optimization approach to large-scale air traffic flow management. Oper. Res. **59**(1), 211–227 (2011)
3. Chaimatanan, S., Delahaye, D., Mongeau, M.: Strategic deconfliction of aircraft trajectories. In: ISIATM, 2nd International Conference on Interdisciplinary Science for Innovative Air Traffic Management (2013)
4. Currie, J., Wilson, D.I.: Opti: lowering the barrier between open source optimizers and the industrial MATLAB user. In: Sahinidis, N., Pinto, J. (eds.) Foundations of Computer-Aided Process Operations. Savannah, Georgia (2012)
5. Durand, N., Allignol, C., Barnier, N.: A ground holding model for aircraft deconfliction. In: 29 Digital Avionics Systems Conference, DASC, Salt Lake City (2010)
6. EUROCONTROL: SESAR Concept of Operations Step 1 (2012)
7. FAA: Order JO 7210. 3V Facility Operation and Administration (2008)
8. ICAO: Doc 4444 Air Traffic Management - Procedures for Air Navigation Services (PANS-ATM) Ed 15 (2007)
9. Koeners, J., Vries, M.D.: conflict resolution support for air traffic control based on so-lution spaces: Design and implementation. In: 27th Digital Avionics Systems Conference IEEE/AIAA, IEEE, St. Paul, MN, pp. 1–9 (2008). doi:10.1109/DASC.2008.4702808
10. Kuenz, A., Schwoch, G.: Global time-based conflict solution: towards the overall optimum. In: In: 31 Digital Avionics Systems Conference, DASC, Williamsburg (2012)

11. Lau, A., Budde, R., Berling, J., Gollnick, V.: The network flow environment: slot allocation model evaluation with convective nowcasting. In: 29 International Council of Aeronautical Sciences, ICAS, St. Petersburg (2014)
12. Lau, A., Berling, J., Linke, F., Gollnick, V., Nachtigall, K.: Large-scale network slot allocation with dynamic time horizons. In: Eleventh USA/Europe Air Traffic Management Research and Development Seminar, Lisboa (2015)
13. Lulli, G., Odoni, A.: The European air traffic flow management problem. Transp. Sci. **41**(4), 431–443 (2007)
14. Ruiz, S., Piera, M.A., Nosedal, J., Ranieri, A.: Strategic de-confliction in the presence of a large number of 4D trajectories using a causal modeling approach. Transp. Res. Part C Emerg. Technol. **39**, 129–147 (2014)
15. van den Akker, J., Nachtigall, K.: Slot Allocation by Column Generation (1999)
16. Wunderling, R.: SoPlex Paralleler und Objektorientierter Simplex. Ph.D. thesis (1996)

On Optimally Allocating Tracks in Complex Railway Stations

Reyk Weiß, Michael Kümmling and Jens Opitz

Abstract Timetabling and capacity planning of railway transport faces ever-growing challenges. Due to the high number of different influences on capacity, timetable optimization in the railway network cannot be efficiently handled by manual effort. The software system TAKT is a state-of-the-art realization, which allows to compute automatically strictly synchronized and conflict-free timetables for very large railway networks. The complexity increases significantly in the consideration of single tracks and highly frequented main railway stations which may also have a extensive track layout. This work shows how the complexity of the timetables process can be reduced by ignoring selected minimum headway constraints. As a result, timetables with possible conflicts in those covered regions will be computed. Consequently, there is the need for efficient algorithms and its corresponding conjunction to solve the remaining conflicts by detecting alternative stopping positions and routes within a main railway station and the optimized selection.

1 Introduction

In large and intermeshed networks, timetabling is a protracted process, which, despite computer aided methods, comes along with a high manual effort today.[1] The reason is based on the huge amount of technical, operative and economical requirements and restrictions and their dependencies among each other. Due to the large amount of constraints and the operator's obligatory needed knowledge about geographic

---

[1]More than 800 employees work in the timetabling department of DB Netz, the largest railway infrastructure manager in Germany.

R. Weiß (✉) · M. Kümmling · J. Opitz
Chair of Traffic Flow Science, TU Dresden, Dresden, Germany
e-mail: Reyk.Weiss@tu-dresden.de

M. Kümmling
e-mail: Michael.Kuemmling@tu-dresden.de

J. Opitz
e-mail: Jens.Opitz@tu-dresden.de

© Springer International Publishing Switzerland 2017
K.F. Dœrner et al. (eds.), *Operations Research Proceedings 2015*,
Operations Research Proceedings, DOI 10.1007/978-3-319-42902-1_39

circumstances as well as the infrastructure, the manual editing is only possible for small subnetworks. For example, DB Netz AG's timetabling is operated decentralized and at some predefined points in the network the train paths are coordinated to fit together. Due to this process the overall optimization of a train path is hard and inefficient. Particularly for the strategic long-term view and developing the infrastructure based on long-term scenarios, timetable optimization poses an outstanding challenge.

In recent years, the team of the Traffic Flow Science chair at TU Dresden in close collaboration with DB Netz AG has successfully developed a prototype software system to support the decision makers with their strategic timetable planning tasks. The software system TAKT automatically calculates and optimizes periodic train paths for complex railway networks by an innovative approach in strategic passenger and freight train's timetabling [3, 4]. All information about model trains like minimum headway times as well as additional constraints like connections and symmetry properties will be used to generate a periodic event scheduling problem (PESP) [6]. Minimum headway constrains are essential to ensure that only one train will use the same part of the infrastructure at the same time, which is the main property of a conflict-free timetable. Subsequently, the PESP will be encoded as a Boolean Satisfiability Problem (SAT) [1] and be solved by a state-of-the-art SAT solver.

Even that step made it possible to calculate timetables for the whole German railway network at once, the complexity is still high and needs to be simplified. Especially highly frequented railway stations as well as single tracks represent a challenge for the automated timetable generation process. This work shows how the complexity can be reduced for the timetable calculation and how the resulting conflicts can be solved by an approach for an optimized and conflict free selection of alternative routes in railway stations as a post-process.

2   The Approach

The software system TAKT uses the built-in routing algorithm [5] to detect the best route with respect to running time and route priority. As a result, the routing algorithm will try to generate routes using the main track. The route priority is given by the infrastructure information. For the main track the highest possible value 100 is used. For siding values between 60 and 90 and for sidings which are crossing the main track of the opposite direction values less then 40 are usual. The generated trains are the base for the timetable. Due to the fact that the route of a train cannot be changed during PESP-based timetable computation, a situation like shown in Fig. 1 will never be computed. Figure 1 shows exemplarily two trains using the same track for both directions. One train starts in station A at 9:57 and arrived station B at 10:05. After a stopping time of about six minutes the train will depart at 10:11 until it finally reaches station C at 10:20. The other train starts at 10:00 and is passing station B at 10:08. This results in a conflict, because both trains are using the main track at the same time. This conflict could be easily prevented, if an alternative stopping position on the siding of station B would be used. The stopping time could be automatically

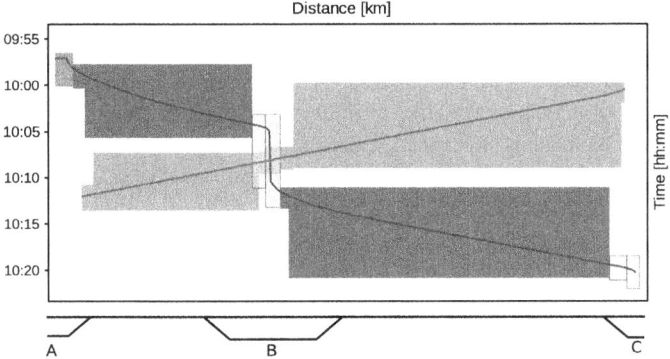

Fig. 1 Illustration of two trains causing a conflict because both are using the same infrastructure in station B at the same time (Screenshot: TAKT)

extended, until the other train passed by. However, the necessary use of an alternative stopping position depends on the resulting timetable. Creating all possible routing variants would increase the complexity of the timetabling problem dramatically. To avoid this problem the knowledge about alternative stopping positions can be used to modify the restrictions of the PESP in that way, that the blocking time on a stop will be ignored if an alternative stopping position exists. Figure 1 illustrates the ignored blocking time in station B by the faded rectangles. Additionally, the use of static train routes cause significant problems in very complex railway stations e.g., Dortmund main station. In order to reduce the complexity in that case it is possible to ignore the minimum headways for all trains in user-defined railway stations.

In fact, by using this approach the resulting timetable will probably contain trains with intersecting blocking times which need to be resolved in a post-process.

2.1 Alternative Routes Detection

After a feasible timetable with respect to the PESP is calculated, all railway stations with conflicts need to be detected. For each train with a stop in those stations a set of alternative variants will be generated. Therefore, it is necessary to detect all available alternative stopping positions and their routes in relation to the train and the current stopping position. The alternative route needs to enter the station at the same entry and also needs to depart from the station at the same exit like the origin route with the exception of the first and the last station of a train. Furthermore, all requirements for the route like power supply for electrical trains and needs for the stopping position like minimum length have to be fulfilled. For each new alternative stopping position a new train will be generated which will start in the last stopping station before the current one and will end as well at the next stopping station. Choosing an alternative route in the current station that will probably change the running time of the current train due to different velocities which will have effects on the blocking time. Compensating longer running times is possible by the reduction of the stopping time, if it was increased during timetable computation. It is very important to consider the influences of those changes not just in the current railway station, especially because it can also effect other trains of the timetable. Finally, for all alternative variants the train dynamics as well as the blocking times based on the new route will be calculated. This set of variants are the base for the optimized alternative route selection as described in Sect. 2.2.

2.2 Optimized Alternative Route Selection

Finding optimal routes in complex stations is a well known problem. Former researches [2, 7] already point to promising approaches which will be adapted for the current requirements. If A describes a set of all trains of the current timetable, for each railway station in which conflicts between trains exists, the set of trains $T \subset A$ operating in this station will be determined. T can be separated in a set of stopping trains T_s and a set of passing trains T_p, with $T_s \cup T_p = T$ and $T_s \cap T_p = \emptyset$. The method described in Sect. 2.1 will be used to generate a set of different variants for each train $t \in T_s$. Those variants, which cause a conflict with at least one train of the current timetable, except trains of T_s and corresponding variants, will be discarded. The result is a set of variants V which are conflict-free to $A \backslash T_s$. The function $O(t)$ returns a set of variants for the train $t \in T_s$. In contrast to [7] the optimization model have to find a feasible solution for all trains and considers the train category, the priority of the alternative route and the runtime difference and minimizes the product between the weighted train category value and the sum of the weighted priority and the weighted difference such that

$$\sum_{i \in T_s} \sum_{j \in O(i)} x_{i,j} \cdot w_i \cdot \left(w_\alpha \cdot \left(100 - \alpha_j \right) + w_\delta \cdot \delta_{i,j} \right) \rightarrow \min \tag{1}$$

subject to

$$\sum_{j \in O(i)} x_{i,j} = 1, \quad \forall i \in T_s, \tag{2}$$

$$\forall i, a \in T_s; j \in O(i); b \in O(a);$$
$$x_{i,j} + x_{a,b} \leq 1, \qquad e_{i,j,a,b}, \tag{3}$$

with $x_{i,j} \in \{0, 1\}$ indicating whether variant j of train i is active or not. $w_i \neq 0$ can be used to prioritize certain train categories for example $w_{ICE} = 1.0$ and $w_{RB} = 0.7$. The global constant variables w_α and w_δ can be used to scale the priority value

$\alpha_j \in \{40 \ldots 100\}$ of the alternative path j proportional to the time difference between the alternative route j and the origin route of train i. The constraint (2) ensures that exactly one variant j of train i will be chosen. Additionally only variants should be chosen which do not have conflicts. $e_{i,j,a,b} \in \{0, 1\}$ indicates whether variant $j \in O(i)$ ($i \in T_s$) and variant $b \in O(a)$ ($a \in T_s$) with $i \neq a$ is in conflict or not. For all variant pairs with $e_{i,j,a,b} = 1$ an allocation at the same time must be excluded as in (3). Since only $x_{i,j}$ are variables, the objective function in (1) is linear.

3   Results

This approach was tested in a long-term scenario of the DB Netz AG in order to determine a feasible and optimal track allocation for selected railway stations. Figure 2 shows an examplifying result for the optimal, conflict-free track allocation for Dortmund main station. The calculation took about 312 s whereas the main time consumption was used for generating all possible train variants and conflict detection. For now the variables were set to $w_\alpha = 1$, $w_\delta = 2$. The w_i for long distance trains was set to 1, for short distance trains to 0.7 and for freight transport trains to 0.5. The routes for 26 trains were changed to achieve a conflict-free and optimal solution with respect to the objective function (1). For some railway stations like Hannover main station this approach was not able to detect a feasible solution, due to the high runtime offsets which have an impact to the blocking times and the influences to other trains in the timetable. For those cases the approach described in [2, 7] is also implemented and can be used to detect the maximum number of possible trains within a railway station.

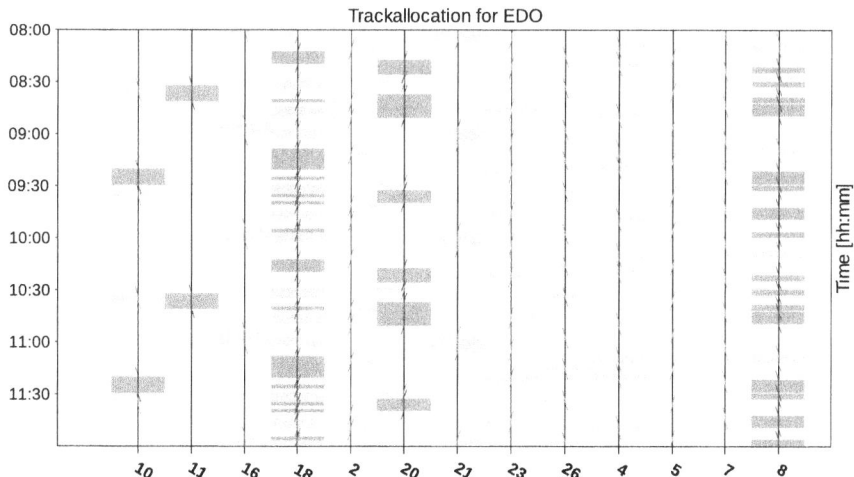

Fig. 2 Feasible and optimal track allocation for Dortmund main station by changing 26 train routes and the use of 13 different tracks (Screenshot: TAKT)

4   Conclusion

This work shows an approach to reduce the complexity of the timetabling problem by ignoring selected minimum headway constraints. This can be done by ignoring the blocking time of train stops if valid alternative stopping positions exist. In that case the current train can be used as a representative for all alternative trains. Furthermore, it could be necessary to ignore the minimum headway constraints of all trains in very complex and highly frequented railway stations to compute a valid timetable. In both cases it is necessary to determine an optimized track allocation in a post-process. It turned out that this approach is very useful and fast. In some cases the changes within a station had so much influences to other trains of the timetable, that it was not possible to calculate a feasible solution. The generation of alternative train paths with the built-in routing algorithm generates only trains with maximum allowed velocity. In some cases, a slower velocity seems to be useful. Additional stops, e.g., on signals are not used in this approach either, but are heavily used during manual timetabling.

References

1. Großmann, P., Hölldobler, S., Manthey, N., Nachtigall, K., Opitz, J., Steinke, P.: Solving periodic event scheduling problems with SAT. In: IEA/AIE. LNAI, vol. 7345, pp. 166–175. Springer (2012)
2. Kroon, L.G., Romeijn, H.E., Zwaneveld, P.J.: Routing trains through railway stations: complexity issues. Eur. J. Oper. Res. **98**, 485–498 (1997)
3. Nachtigall, K.: Periodic Network Optimization and Fixed Interval Timetable. Habilitation thesis, University Hildesheim (1998)
4. Opitz, J.: Automatische Erzeugung und Optimierung von Taktfahrplänen in Schienen-verkehrsnetzen. Mobilität und Verkehr, Gabler Verlag, Logistik (2009)
5. Pöhle D., Weiß R., Opitz J., Großmann P., Nachtigall K.: Neuartiges automatisches Routing in realen Eisenbahnnetzen mit dem Programmsystem TAKT. ETR Band 61 (2012)
6. Serafini, P., Ukovich, W.: A mathematical model for periodic scheduling problems. SIAM J. Discrete Math. **2**(4), 550–581 (1989)
7. Zwaneveld, P.J., Kroon, L.G., Romeijn, H.E., Solomon, M., Dauzre-Prs, S., Van Hoesel, S.P.M., Ambergen, H.W.: Routing trains through railway stations: model formulation and algorithms. Transp. Sci. **30**, 181–194 (1996)

Part IV
Metaheuristics and Multiple Criteria Decision Making

Optimal and Near-Optimal Strategies in Discrete Stochastic Multiobjective Quasi-hierarchical Dynamic Problems

Maciej Nowak and Tadeusz Trzaskalik

Abstract Multi-stage, multi-criteria discrete decision processes under risk are considered and Bellmans principle of optimality is applied. Quasi-hierarchy of multi-period criteria is determined by the decision maker. The aim of the paper is to propose an algorithm to solve quasi-hierarchical problem according to decision maker's requirements. The process of obtaining the final solution is interactive.

1 Introduction

Many decision problems are dynamic by their very nature. In such cases the decision is not made once, but many times. Partial choices are mutually related, since earlier decisions influence which decisions can be considered in the consecutive stages of the process.

The consequences of decisions become apparent in the near or remote future, which is uncertain by its very nature. In such situations we can apply methods using discrete stochastic dynamic programming approach based on Bellman's optimality principle [1]. For these processes it is characteristic that at the beginning of each stage, the decision process is in a certain state. In each state, a set of feasible decisions is available. The process is discrete when all sets of states and decisions are finite. These processes are stochastic which means that the probability of achieving the final state for the given stage is known when at the beginning of this stage the process was in one of the admissible state and when an feasible decision has been made.

We will consider additive multi-criteria processes. At each stage we use stage criteria. The sum of the stage criteria gives the value of the multi-stage criterion. In the classical approach, the task consists in obtaining a strategy for which the expected value of the given criterion is optimal.

M. Nowak (✉) · T. Trzaskalik
Faculty of Informatics and Communication, Department of Operations Research,
University of Economics in Katowice, 1 Maja 50, 40-287 Katowice, Poland
e-mail: maciej.nowak@ue.katowice.pl

T. Trzaskalik
e-mail: tadeusz.trzaskalik@ue.katowice.pl

© Springer International Publishing Switzerland 2017
K.F. Dœrner et al. (eds.), *Operations Research Proceedings 2015*,
Operations Research Proceedings, DOI 10.1007/978-3-319-42902-1_40

Multi-criteria problems can be regarded as hierarchical problems. This means that the decision maker is able to formulate a hierarchy of criteria so that the most important criterion is assigned the number 1; the number 2 is reserved for the second-most important criterion, and so on. We assume that all criteria considered in the problem can be numbered in this way.

In this paper we assume that the decision maker considers the problem as a quasi-hierarchical one. This means that he/she accepts that the most important criterion can reach near-optimal value in the final solution, which usually makes possible to improve less important criteria values.

The problem of identifying near-optimal solution in a discrete deterministic dynamic programming problem was considered in [7]. The proposed algorithm was a generalization of the method developed by Elmaghraby in [2]. This approach enables to generate subsequent paths in the network.

In this paper we consider a quasi-hierarchical stochastic problem. We use the algorithm for identifying near-optimal strategies, described in detail in [5]. Another technique for generating such solutions, initially proposed in [3], was also described in [4].

2   Single-Criterion Stochastic Dynamic Programming

We will use the following notation [6, 7]:

T—number of stages of the decision process under consideration,

y_t—state of the process at the beginning of stage t ($t \in \overline{1, T}$),

$\mathbf{Y}_t$—finite set of process states at stage t,

x_t—feasible decision at stage t,

$\mathbf{X}_t(y_t)$—finite set of decisions feasible at stage t, when the process was in state $y_t \in \mathbf{Y}_t$ at the beginning of this stage,

$F_t(y_{t+1}|y_t, x_t)$—value of stage criterion at stage t for the transition from state y_t to state y_{t+1}, when the decision taken was $x_t \in \mathbf{X}_t(y_t)$,

$P_t(y_{t+1}|y_t, x_t)$—probability of the transition at stage t from state y_t to state y_{t+1}, when the decision taken was $x_t \in \mathbf{X}_t(y_t)$. The following holds:

$$\forall_{t \in \overline{1,T}} \forall_{y_t \in \mathbf{Y}_t} \forall_{x_t \in \mathbf{X}_t(y_v)} \sum_{y_{t+1} \in \mathbf{Y}_{t+1}} P_t(y_{t+1}|y_t, x_t) = 1, \tag{1}$$

$\{x\}$—strategy—a function assigning to each state $y_t \in \mathbf{Y}_t$ exactly one decision $x_t \in \mathbf{X}_t(y_t)$,

$\{\mathbf{X}\}$—the set of all strategies of the process under consideration,

$\{x_{\overline{t,T}}\}$—shortened strategy, encompassing stages from t to T.

We assume that a single state is considered at the first stage. Let us consider a strategy $\{\overline{x}\} \in \{\mathbf{X}\}$. The expected value for the strategy $\{\overline{x}\}$ is calculated as follows:

Algorithm 1

1. For each $y_T \in \mathbf{Y}_T$ we calculate:

$$G_T(y_T, \{\overline{x_{T,T}}\}) = \sum_{y_{T+1} \in \mathbf{Y}_{T+1}} F_T(y_{T+1}|y_T, \overline{x}_T) P_T(y_{T+1}|y_T, \overline{x}_T), \qquad (2)$$

2. For each $y_t \in \mathbf{Y}_t$, $t \in \overline{T-1,1}$, we calculate:

$$G_t(y_t, \{\overline{x_{t,T}}\}) = \sum_{y_{t+1} \in \mathbf{Y}_{t+1}} (F_t(y_{t+1}|y_t, \overline{x}_t) + G_{t+1}(y_{t+1}, \{\overline{x_{t+1,T}}\})) P_t(y_{t+1}|y_t, \overline{x}_t),$$

$$(3)$$

3. As we assume that a single state is considered at the beginning of the first stage, the expected value of the strategy $\{\overline{x}\} \in \{\mathbf{X}\}$ is calculated from the formula:

$$G\{\overline{x}\} = G_1(y_1, \{\overline{x}\}). \qquad (4)$$

Using Bellmans optimality principle [1], we determine the optimal expected value for the process and the optimal strategy.

Algorithm 2

1. For each $y_T \in \mathbf{Y}_T$ we calculate the optimal value:

$$G_T^*(y_T) = \max_{x_T \in \mathbf{X}_t(y_t)} \sum_{y_{T+1} \in \mathbf{Y}_{T+1}} F_T(y_{T+1}|y_T, x_T) P_T(y_{T+1}|y_T, x_T), \qquad (5)$$

and find the decision $x_t^*(y_t)$, for which this maximum is attained. This decision forms a part of the optimal strategy being constructed.

2. For each $y_t \in \mathbf{Y}_t$, $t \in \overline{T-1,1}$, we calculate the optimal value:

$$G_t^*(y_t) = \max_{x_t \in \mathbf{X}_t(y_t)} \sum_{y_{t+1} \in \mathbf{Y}_{t+1}} (F_t(y_{t+1}|y_t, x_t) + G_{t+1}^*(y_{t+1})) P_t(y_{t+1}|y_t, x_t),$$

$$(6)$$

and find the decision $x_t^*(y_t)$, for which this maximum is attained. This decision forms a part of the optimal strategy being constructed.

3. The optimal expected value for the process is calculated from the formula:

$$G^*\{y_1\} = G_1^*(y_1). \qquad (7)$$

The finite set of all strategies $\{\mathbf{X}\}$ can be sorted into M classes in such a way that:

$$\{\mathbf{X}\} = \{\mathbf{X}^1\} \cup \{\mathbf{X}^2\} \cup \cdots \cup \{\mathbf{X}^M\}. \qquad (8)$$

where:

1. $\{\mathbf{X}^i\} \cap \{\mathbf{X}^j\} = \varnothing$ for $i \neq j$
2. $\forall_{i=1,\ldots,M} \forall_{\{x^k\},\{x^l\}\in\{\mathbf{X}^i\}} G\{x^k\} = G\{x^l\}$
3. $\forall_{i<j} \forall_{\{x^i\}\in\{\mathbf{X}^i\}} \forall_{\{x^j\}\in\{\mathbf{X}^j\}} G\{x^i\} > G\{x^j\}$

Let $G\{\mathbf{X}\} = \{G\{x^1\}, \ldots, G\{x^M\}\}$ where $\{x^1\} \in \{\mathbf{X}^1\},\ldots, \{x^M\} \in \{\mathbf{X}^M\}$. We define ith optimal expected value for strategies from $\{\mathbf{X}\}$ denoted as $\max_i G\{\mathbf{X}\}$ in the following way:

$$\max_i G\{\mathbf{X}\} = G\{x^i\}. \tag{9}$$

The way of determining ith optimal expected value and ith optimal strategy is described below.

Algorithm 3

1. For each $y_T \in \mathbf{Y}_T$ we calculate ith optimal value:

$$G_T^i(y_T) = \max_{x_T \in \mathbf{X}_T(y_T)}^{i} \sum_{y_{T+1}\in\mathbf{Y}_{T+1}} F_T(y_{T+1}|y_T, x_T) P_T(y_{T+1}|y_T, x_T), \tag{10}$$

and find the decision $x_T^i(y_T)$, for which this value is attained. This decision forms a part of ith optimal strategy being constructed.
2. For each $y_t \in \mathbf{Y}_t, t \in \overline{T-1, 1}$ we calculate ith optimal value:

$$G_t^i(y_t) = \max_{\substack{x_T \in \mathbf{X}_T(y_T) \\ j\in\overline{1,i}}}^{i} \left\{ \sum_{y_{t+1}\in\mathbf{Y}_{t+1}} \left[F_t(y_{t+1}|y_t, x_t) + G_{t+1}^j(y_{t+1}) \right] P_t(y_{t+1}|y_t, x_t) \right\}, \tag{11}$$

and find the decision $x_t^i(y_t)$, for which this value is attained. This decision forms a part of ith optimal strategy being constructed.

3   Determination of Near-Optimal Strategies

The strategy $\{x^m\}$ is called near-optimal if its expected value differs from the expected value of the optimal strategy $\{x^*\}$ by at most the given value z, that is:

$$G\{x^*\} - G\{x^m\} \leq z. \tag{12}$$

Let **LS** be the set of near-optimal strategies. In order to determine **LS** we can use the following algorithm.

Algorithm 4

1. Set $i := 1$, **LS** $:= \varnothing$.
2. Use algorithm 3 to determine the set $\{\mathbf{X}^i\}$.
3. If for $\{x^i\} \in \{\mathbf{X}^i\}$, $G\{x^i\} \geq G\{x^*\} - z$, then: **LS** $:=$ **LS** $\cup \{\mathbf{X}^i\}$, else go to 5.
4. Set $i := i + 1$, go to 2.
5. End of the procedure.

4 Application of the Quasi-hierarchical Approach to the Solution of the Multi-criteria Problem

Let us assume that the solution of the dynamic problem is evaluated with respect to K multi-stage criteria, each of which is the sum of T stage criteria.

The evaluation of each strategy with respect to each criterion is based on the expected value. We assume that the decision maker ordered the criteria starting with the one he or she regards as the most important. We assume therefore that he is first of all interested in the optimisation of the criterion number 1, then of the criterion number 2, etc. The determination of the solution by means of the quasi-hierarchical approach is performed as follows:

Algorithm 5

1. Determine the optimal solutions of the problem with respect to each criterion.
2. Present the optimal values of each criterion to the decision maker.
3. Ask the decision maker to determine the aspiration thresholds, that is the values which should be attained by each criterion in the final solution.
4. For each criterion determine the set $\mathbf{LS}_k$ of strategies satisfying the requirements determined by the decision maker.
5. Set $J := K$.
6. Determine the set $\mathbf{LS} := \bigcap_{k=\overline{1,J}} \mathbf{LS}_k$.
7. If $\mathbf{LS} \neq \varnothing$, go to step 9.
8. Set $J := J - 1$. Go to step 6.
9. From among the solutions in the set **LS** select those for which the first criterion attains the highest value. If there are more than one such solutions, then in your selection take into account the values of the next criteria in the order determined by the hierarchy formulated by the decision maker.
10. If $\mathbf{LS} := \bigcap_{k=\overline{1,J}} \mathbf{LS}_k = \varnothing$, check if the strategy obtained by this procedure satisfies the decision maker. If not, return to step 3.
11. End of procedure.

In the procedure we determine the set of strategies which satisfy all the requirements determined by the decision maker. In many cases it may turn out that such solutions do not exist. In such cases we try to determine solutions satisfying the requirements formulated for those criteria, which the decision maker regards as the most important ones. Gradually, we therefore omit the requirements formulated for the least important criteria, until the set **LS** containing the solutions satisfying the requirements of the decision maker ceases to be empty. From among the solutions contained in this set we select the one for which the first criterion attains the highest value. If there are more than one such solutions, then in our selection we take into account the values of the next criteria according to the hierarchy defined by the decision maker.

5 Final remarks

Usually we solve the hierarchical problem sequentially. First we find the set of solutions which are optimal with respect to the most important criterion. Out of this set, we select the subset of solutions which are optimal with respect to the criterion number 2. We continue this procedure until we determine the subset of solutions which are optimal with respect to the least important criterion.

The method described in this paper uses a different approach for solving a quasi-hierarchical problem. Using the information provided by the decision maker, we construct the set of optimal and near optimal strategies for each criterion. Next we identify the intersection of the sets identified for all criteria. If it is not empty, we use a quasi-hierarchical approach to select the final solution. When the intersection is empty, the decision maker can choose between two options: to continue the procedure, ignoring the least important criterion or to start the procedure once more, assuming new thresholds for criteria values.

Acknowledgements This research was supported by National Science Centre, decision no DEC-2013/11/B/HS4/01471.

References

1. Bellman, R.: Dynamic Programming. Princenton University Press, Princenton (1957)
2. Elmaghraby, S.E.: The theory of networks and management science. Part 1. Manag. Sci. **18**, 1–34 (1970)
3. Nowak, M.: Quasi-hierarchical approach in multiobjective decision tree. In: Kopanska-Brodka, D. (ed.) Analysis and Decision Aid. pp. 59–73, University of Economics in Katowice, Katowice (2014) (in Polish)
4. Nowak, M., Trzaskalik, T.: Quasi-hierarchical approach to discrete multiobjective stochastic dynamic programming. J. Adv. Comput. Intell. Intell. Inform. (in press)
5. Trzaskalik, T.: MCDM Applications of near optimal solutions in dynamic programming. Multi. Criteria Decis. Mak. **10** (2015)
6. Trzaskalik, T.: Multiobjective analysis in dynamic environmnent. The Karol Adamiecki University of Economics. Katowice Press, Katowice (1998)
7. Trzaskalik, T.: Multiobjective discrete dynamic programming, theory and applications in economics. The Karol Adamiecki University of Economics. Katowice Press, Katowice (1990). (in Polish)

Multicriterial Design of a Hydrostatic Transmission System Via Mixed-Integer Programming

**Lena C. Altherr, Thorsten Ederer, Lucas S. Farnetane,
Philipp Pöttgen, Angela Vergé and Peter F. Pelz**

Abstract In times of planned obsolescence the demand for sustainability keeps growing. Ideally, a technical system is highly reliable, without failures and down times due to fast wear of single components. At the same time, maintenance should preferably be limited to pre-defined time intervals. Dispersion of load between multiple components can increase a system's reliability and thus its availability inbetween maintenance points. However, this also results in higher investment costs and additional efforts due to higher complexity. Given a specific load profile and resulting wear of components, it is often unclear which system structure is the optimal one. Technical Operations Research (TOR) finds an optimal structure balancing availability and effort. We present our approach by designing a hydrostatic transmission system.

1 Motivation and Technical Application

Technical systems have become more and more complex. In recent years, mathematical methods have been applied to find not only optimal operation strategies [3], but also the optimal layout of a technical system [6]. Often, one has to consider

L.C. Altherr (✉) · T. Ederer · L.S. Farnetane · P. Pöttgen · A. Vergé · P.F. Pelz
Chair of Fluid Systems, TU Darmstadt, Darmstadt, Germany
e-mail: lena.altherr@fst.tu-darmstadt.de

T. Ederer
e-mail: thorsten.ederer@fst.tu-darmstadt.de

L.S. Farnetane
e-mail: lucas.farnetane@fst.tu-darmstadt.de

P. Pöttgen
e-mail: philipp.poettgen@fst.tu-darmstadt.de

A. Vergé
e-mail: angela.verge@fst.tu-darmstadt.de

P.F. Pelz
e-mail: peter.pelz@fst.tu-darmstadt.de

© Springer International Publishing Switzerland 2017
K.F. Dœrner et al. (eds.), *Operations Research Proceedings 2015*,
Operations Research Proceedings, DOI 10.1007/978-3-319-42902-1_41

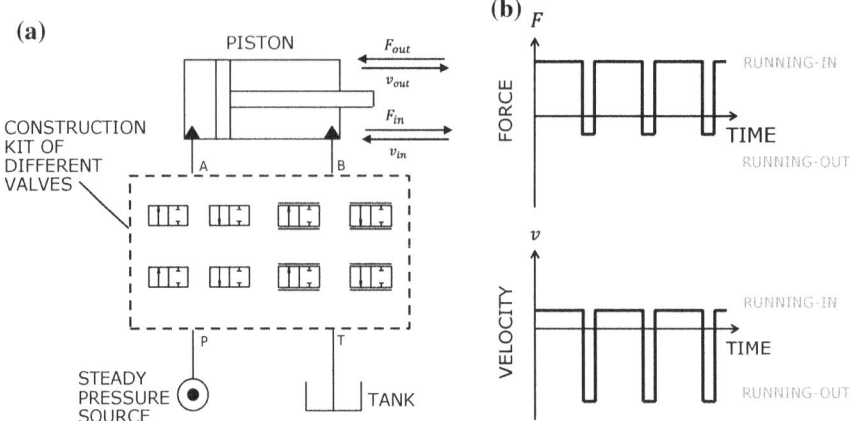

Fig. 1 Details on the hydrostatic transmission system. **a** Design problem: the construction kit consists of four proportional and four shift valves. Which combination results in the best system? **b** Load profile: the piston has to move outwards and inwards with a certain velocity against a specified force

conflicting objectives, like investment costs, energy consumption and maintenance effort. We present an approach that finds an optimal design for a hydrostatic transmission system in a multicriterial setting, cf. Fig. 1. The purpose of the system is to fulfill a predefined load cycle: A piston has to move outwards and inwards with a certain velocity against a specified force. The required power is provided by a steady pressure source and transmitted hydraulically via oil. To do so, volume flow and pressure are controlled by proportional and shift valves.

The valve system will be designed from scratch. Its components can be chosen from a construction kit consisting of four proportional and four shift valves. Serial and parallel connections are possible. We aim to find the system structure that is optimal regarding the conflictive objectives of low complexity and high availability, i.e., a tradeoff between a smaller system (favorable in planning) and fewer replacements due to wear (favorable in operation).

2   Mixed Integer Linear Program

We model the optimization problem as a Mixed Integer Linear Program (MILP) consisting of two stages: First, find an investment decision in the set of components. Secondly, find operating settings for the selected valves that fulfill the given load profile while taking into account the wear of each proportional valve. Both stages are coupled since a larger system may result in a suitable load dispersion between the individual components and therefore reduced wear.

All possible designs for the hydrostatic transmission system are modeled by a complete graph $G = (N, E)$. Its edges E correspond to optional proportional and shift valves and to pipes between them. Furthermore, one edge represents the piston of the transmission system. The nodes N of the graph correspond to connection points between the components. Two additional vertices represent the steady pressure source and the tank of the system. A binary variable $K_{i,j}$ for each optional component $(i, j) \in E$ indicates whether it is purchased ($K_{i,j} = 1$) or not ($K_{i,j} = 0$).

The hydrostatic transmission system has to fulfill two different load scenarios: One for moving the piston outwards and one for moving it inwards. We copy the system graph once for each of these load cases $l \in L = \{1, 2\}$. Since we also want to model that worn out components can only be replaced at predefined points in time, we further expand the index set by adding an index $w \in W = \{0, 1, 2, 3, \ldots\}$ that represents the operating intervals. For each of these intervals both graphs are copied once again. Inbetween these intervals, worn out components may be replaced. We denote the fully expanded graph by $\tilde{G} = (\tilde{N}, \tilde{E})$. Binary variables $\alpha_{w,l,i,j}$ for each valve edge of the expanded graph allow to deactivate purchased components during operation – pipe edges cannot be deactivated, so in the following we presume that $\forall (w, l, i, n) \in \text{pipes}(\tilde{E}) : \alpha_{w,l,i,j} = 1$ by convention.

The conservation of the volume flow $Q_{w,l,i,j}$ is given by

$$\forall (w, l, n) \in \tilde{N} : \sum_{(w,l,i,n) \in \tilde{E}} Q_{w,l,i,n} = \sum_{(w,l,n,j) \in \tilde{E}} Q_{w,l,n,j}. \tag{1}$$

A positive value of $Q_{w,l,i,j}$ represents flow from node i to node j, while a negative value represents flow in the opposite direction. An on/off-constraint makes sure that only active components contribute to the volume flow conservation:

$$\forall (w, l, i, j) \in \tilde{E} : \quad Q_{w,l,i,j} \leq Q^{\max} \cdot \alpha_{w,l,i,j}, \tag{2}$$

$$Q_{w,l,i,j} \geq -Q^{\max} \cdot \alpha_{w,l,i,j}. \tag{3}$$

Another physical on/off-constraint is the pressure propagation

$$\forall (w, l, i, j) \in \tilde{E} : \quad p_{w,l,i} - p_{w,l,j} \leq \Delta p + p^{\max} \cdot (1 - \alpha_{w,l,i,j}), \tag{4}$$

$$p_{w,l,i} - p_{w,l,j} \geq \Delta p - p^{\max} \cdot (1 - \alpha_{w,l,i,j}), \tag{5}$$

which has to be fulfilled along each edge, if the component is active. Regarding open shift valves, we assume $\Delta p = 0$. Regarding proportional valves, the resulting pressure decrease depends on the valve lift $u_{w,l,i,j}$ and on the volume flow:

$$\forall (w, l, i, j) \in \tilde{E} : \quad Q_{w,l,i,j} = \sqrt{\frac{2 \cdot \Delta p}{\rho}} \cdot \zeta \cdot d \cdot u_{w,l,i,j}, \tag{6}$$

where ρ is the oil density, ζ its pressure loss coefficient and d is the valve's diameter.

Inbetween maintenance points, the system has to fulfill the load defined by each loading case $l \in L$. The volume flow and the pressure decrease over the edge (P_i, P_j) representing the piston of the transmission system have to satisfy

$$\forall\, w \in W,\ l \in L: \qquad Q_{w,l,P_i,P_j} = v_l \cdot A_{\text{piston}}, \tag{7}$$

$$p_{w,l,P_i} - p_{w,l,P_j} = \frac{F_l}{A_{\text{piston}}}, \tag{8}$$

where v_l is the required velocity of the piston in loading case l, A_{piston} is the area of the piston, and F_l is the required force in loading case l.

According to [2] the wear $\Delta u_{w,i,j}$ of the control edge of a proportional valve (i,j) in interval w can be calculated by

$$\Delta u_{w,i,j} = \sum_{l \in L} u_{w,l,i,j} \cdot \left(\exp\left[-\frac{1}{3} \cdot \mathscr{K}_{w,l,i,j} \cdot t_l \right] - 1 \right). \tag{9}$$

where $\mathscr{K}_{w,l,i,j} = \mathscr{K}(\Delta p_{w,l,i,j}, u_{w,l,i,j}, u^{\max}, d, \rho, c_v)$ depends on operating parameters such as the pressure loss Δp, the valve lift u, the maximum valve lift $u^{\max}$, the valve's diameter d, the oil density ρ and the volume fraction of suspended solids c_v. With this dependency, we can calculate the wear of each bought valve for a specific loading case l with length t_l. To do so within the MILP, all of the above mentioned nonlinear equations were approximated by piecewise linear functions [7].

If the wear of a control edge exceeds a maximum value $\Delta u^{\max}$ in operating period $w \in W$, the valve is worn out:

$$\Delta u_{w,i,j} \leq \Delta u^{\max} + \delta \cdot D_{w+1,i,j}, \tag{10}$$

where δ is a small value, and $D_{w+1,i,j}$ is a binary indicator. A worn out valve with $D_{w+1,i,j} = 1$ cannot be used in interval $w+1$ if it is not replaced by a new one. A binary indicator $T_{w,i,j}$ represents if the valve has been replaced at the beginning of interval w ($T_{w,i,j} = 1$) or not ($T_{w,i,j} = 0$). For the flow through a valve (i,j) holds:

$$Q_{w,l,i,j} \leq\ (K_{i,j} - D_{w,i,j} + T_{w,i,j}) \cdot Q^{\max}, \tag{11}$$

$$Q_{w,l,i,j} \geq\ -(K_{i,j} - D_{w,i,j} + T_{w,i,j}) \cdot Q^{\max}. \tag{12}$$

$T_{w,i,j}$ is coupled with variable $K_{i,j}$ that represents the purchase of component (i,j):

$$T_{w,i,j} \leq K_{i,j}. \tag{13}$$

Furthermore, each valve's wear in one operating interval has to be coupled with consecutive intervals, cf. Fig. 2:

Fig. 2 The wear of a valve's control edge saturates with time, given constant operating parameters. If the wear reaches the upper limit, the valve is replaced. We assume a constant pressure difference due to a closed-loop control

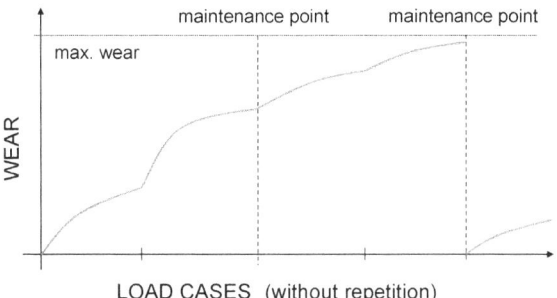

LOAD CASES (without repetition)

$$\forall w_1 \in W, \ w_2 \in W \setminus \{1, \ldots, w_1\}, \ (i, j) \in \text{valves}(E):$$

$$\sum_{w=w_2-w_1-1}^{w_2} u_{w,i,j} \leq \left(K_{i,j} + \sum_{w=w_2-w_1}^{w_2} T_{w,i,j} \right) \cdot (\Delta u^{\max} + \delta). \tag{14}$$

We want to find an optimal system structure for the hydrostatic transmission system that balances two diverse objectives: An increase in availability by introducing redundancy implies a decreased maintenance effort due to wear. However, the system becomes more complex and thus more costly. In order to find an optimal tradeoff, we use the following objective function which calculates the overall lifetime costs consisting of the initial investment and costs due to replacement of worn components at maintenance points:

$$\min \sum_{(i,j)\in\text{shift}(E)} K_{i,j} + 2 \cdot \sum_{(i,j)\in\text{proportional}(E)} \left(K_{i,j} + \sum_{w\in W} T_{w,i,j} \right). \tag{15}$$

In this example we make the assumption that proportional valves are twice as expensive as shift valves and that shift valves do not wear out.

3 Optimal System Structure

Our aim is to find an optimal design of a hydrostatic transmission system with a given maintenance plan. We solve the MILP with the commercial solver *cplex*. If we assume 200,000 load cycles in each operating interval w and change the number of maintenance points considered, we get the result depicted in Fig. 3.

If we consider only one operating interval, i.e., no opportunity for replacements at all, the smallest possible system consisting of four components is chosen: Two shift and two proportional valves. The system meets the required function during the interval and the individual components' wear does not reach the upper limit.

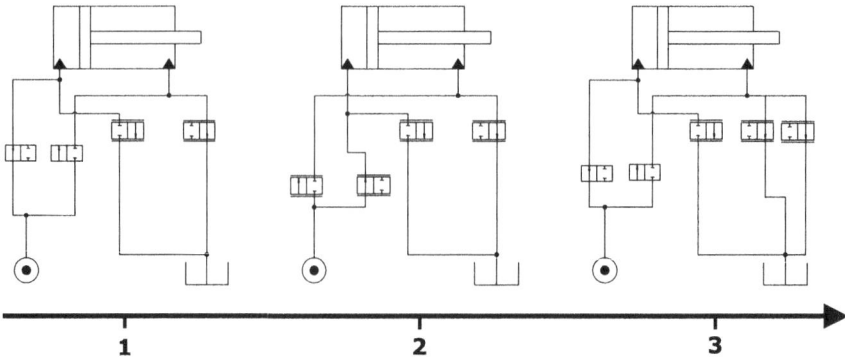

Fig. 3 Depending on the number of operating intervals that are considered in the optimization model, the optimal system structure changes

For an extended period with two operating intervals, the two shift valves are replaced by proportional valves, although we assumed them to be twice as expensive as the shift valves. By doing so, the wear is shared by two proportional valves in each load case. Therefore, no replacement happens at the maintenance point.

When considering a third operation interval, the solution contains one additional valve compared to the smallest possible system consisting of two shift and two proportional valves. During the outwards-movement of the piston, the throttling is performed via a parallel connection of two proportional valves instead of one. This hot spare distributes the wear over both components, making their life cycle more compatible with the time between maintenance points. In this solution, three replacements occur: The proportional valve on the left is replaced two times since it throttles the inward-movement alone and suffers from a high wear. One of the other two proportional valves has to be replaced one time. There was no combination of valves in the construction kit which can sustain the wear for three operating periods.

4 Conclusion and Outlook

The presented optimization model finds an optimal system structure for a specific technical problem with a maintenance plan, given deterministic wear of the components. It can be applied to other technical systems, since boundary conditions, like the number of loading cycles per interval, can be adapted. For each further maintenance point, new second stage variables have to be added to the model, and the run time increases exponentially. While we focused on modeling aspects in this paper, we plan to reduce run time by techniques like dual decomposition in future work [4]. For a large number of time periods it might be necessary to discount the cash flows. Having increased computation speed, interesting extensions of our model are to include reliability aspects [1] and to determine the pareto frontier [5].

Acknowledgements The authors thank the German Research Foundation DFG for funding this research within the Collaborative Research Center 805 "Control of Uncertainties in Load-Carrying Structures in Mechanical Engineering".

References

1. Altherr, L.C., Ederer, T., Pöttgen, P., Lorenz, U., Pelz, P.F.: Multicriterial optimization of technical systems considering multiple load and availability scenarios. In: Applied Mechanics and Materials, vol. 807, pp. 247–256. Trans Tech Publ (2015)
2. Altherr, L.C., Ederer, T., Vergé, A., Pelz, P.F.: Lebensdauer als Optimierungsziel - Algorithmische Struktursynthese am Beispiel eines hydrostatischen Getriebes. VDI-Berichte: Elektrik, Mechanik, Fluidtechnik in der Anwendung, vol. 2264 (2015)
3. Bischi, A., Taccari, L., Martelli, E., Amaldi, E., Manzolini, G., Silva, P., Campanari, S., Macchi, E.: A detailed milp optimization model for combined cooling, heat and power system operation planning. Energy **74**, 12–26 (2014)
4. CarøE, C.C., Schultz, R.: Dual decomposition in stochastic integer programming. Oper. Res. Lett. **24**(1), 37–45 (1999)
5. Ehrgott, M.: Multicriteria Optimization. Springer Science and Business Media, Berlin (2006)
6. Humpola, J.: Gas network optimization by MINLP. Ph.D. thesis, TU Berlin (2014)
7. Vielma, J.P., Ahmed, S., Nemhauser, G.: Mixed-integer models for nonseparable piecewise-linear optimization: Unifying framework and extensions. Oper. Res. **58**(2), 303–315 (2010)

Optimal Pulse-Doppler Waveform Design for VHF Solid-State Air Surveillance Radar

Miloš Jevtić, Nikola Zogović and Stevica Graovac

Abstract VHF radars are suitable in some air surveillance applications, due to their cost-effectiveness and the fact that radar cross section of an aircraft is larger at VHF band than at higher frequencies, making detection easier. To ensure coverage of all ranges and velocities of interest, contemporary VHF radars utilize a complex waveform. We formulate the design of this waveform as a multiobjective optimization problem, with signal-to-noise ratio (SNR), Doppler visibility and Doppler resolution as objectives, which should be maximized. We show that the objectives are in conflict and use a particular example to explore the Pareto frontier (PF) for the problem. We find that reasonable tradeoff can be made between SNR and Doppler visibility, leading to an idea of multiple modes of operation, selectable at run time. We conclude that this subject is worth of further investigation, and that finding an efficient method for determining the PF would facilitate further research.

1 Introduction

VHF radars are desirable in some air surveillance applications, due to their cost-effectiveness and the fact that radar cross section (RCS) of an aircraft is larger at VHF band than at higher frequencies, making detection easier. The use of solid-state transmitter (SST) is almost a norm in modern VHF radars. Low peak power (on the order of 10 kW) inherent to SST requires the use of long pulses to achieve acceptable signal-to-noise ratio (SNR) for long distance targets, which causes significant range eclipsing, since receiver must be turned off while transmitting. Common approach for overcoming this problem is alternating pulses of different durations: long pulses, which enable detection on long ranges only, and short pulses, which enable detection on short ranges only. To provide target detection in environments dominated by

M. Jevtić (✉) · S. Graovac
School of Electrical Engineering, University of Belgrade, Bulevar Kralja Aleksandra 73,
11020 Belgrade, Serbia
e-mail: milos.jevtic@pupin.rs

M. Jevtić · N. Zogović
Institute "Mihailo Pupin", University of Belgrade, Volgina 15, 11060 Belgrade, Serbia

© Springer International Publishing Switzerland 2017
K.F. Dœrner et al. (eds.), *Operations Research Proceedings 2015*,
Operations Research Proceedings, DOI 10.1007/978-3-319-42902-1_42

clutter, recently developed VHF radars employ a processing approach called moving target detection (MTD) [1, 2], which is essentially a combination of clutter mapping and pulse-Doppler (PD) processing [2]. PD processing requires that n_P elementary waveforms (consisting of one short and one long pulse) are transmitted using the same pulse repetition frequency (PRF). Time interval in between successive elementary waveforms beginnings (equal to inverse of PRF) is termed pulse repetition interval (PRI). Time interval consisting of n_P equal PRIs is a coherent processing interval (CPI). For each time sequence consisting of n_P samples taken at the same range during a CPI, discrete Fourier transform (DFT) is applied. Then, spectral samples near zero, corresponding to unwanted stationary or slowly moving objects are discarded, leaving targets of interest to compete only with noise in subsequent detection process. However, in a low PRF (LPRF) regime, typically used in air surveillance radars, target's Doppler shift can be greater than PRF/2, which causes target's energy to fold into [−PRF/2, PRF/2] interval. If target's energy folds into notched out region (near zeroth DFT bin), target will go undetected. In such case, it is said that target's radial velocity (or speed) is blind for that PRF [2]. To reduce effects of blind speeds, several different PRFs are used in consecutive CPIs [2]. Complex waveform which enables coverage of all ranges and velocities of interest, is shown in Fig. 1.

Standard textbooks on radar, such as [1, 2] give basic recommendations on how to choose the PRFs. In [3–5], PRF set selection for a medium PRF (MPRF) airborne radar using evolutionary algorithms was studied. Same topic was formulated as a multiobjective optimization (MOO) problem in [6]. In [7], we formulated the optimal LPRF PD waveform design as a MOO problem, with pulses durations and PRIs as design variables, and SNR maximization and blind speed mitigation efficiency as objectives. We illustrated how this problem can be solved using an approach with a posteriori articulation of preferences [8]. In this paper, we extend the research from [7] by introducing an additional objective—Doppler resolution maximization, and an improved metric for blind speed mitigation efficiency, based on clutter model and blind zones extents calculation.

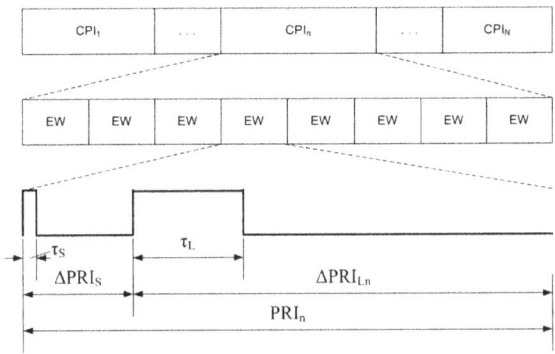

Fig. 1 Complex waveform transmitted during a dwell. Long and short pulses durations, τ_S and τ_L, and the number of elementary waveforms (EW) within a CPI, n_P, are the same in each of N CPIs

2   System Model and Problem Formulation

Let SNR_L denote received SNR when target at maximal range of interest (instrumented range) R_I is illuminated with long pulse. Increasing SNR_L is equivalent to increasing the maximal detection range. For calculating SNR_L, we use the same model as in [7].

We define a blind zone as a region of Doppler spectrum in which target's radial speed is blind for more than one PRF. In a blind zone, target can be hard or impossible to detect. We define Doppler visibility percentage (*DVP*) as percentage of Doppler spectrum of interest which is not covered by blind zones. Increasing *DVP* is equivalent to reducing the extents of blind zones. To calculate *DVP*, we first need to determine the extents of notched out region of spectrum for each PRF. We assume that clutter has Gaussian power spectral density, modeling combined effects of windblown trees and scanning modulation, with standard deviation determined following [1, pp. 2.11–2.16], and assuming windy conditions. We approximate the effects of discarding spectral samples with an ideal highpass filter, with cutoff frequency restricted to odd multiples of half the DFT bin width. Then we seek for a smallest cutoff frequency f_C for which the clutter attenuation (CA) is at least CA_{min}, using [2, Eq. 17.20] to calculate CA. Notched out zones of width $2f_C$ centered at integer multiples of the PRF are combined for all PRFs, and *DVP* is calculated.

Doppler resolution is commonly determined with the Rayleigh width of the DFT mainlobe [2, p. 509], which is inversely proportional to observation time. Multiplying the Doppler resolution with half the wavelength, we obtain velocity resolution (VR). Capability to resolve targets on the basis of their velocities is improved if VR is decreased. The worst resolution VR_{max} is achieved in a CPI with the smallest PRI.

As in [7], we assume that PRIs are chosen according to scheme recommended in [1, p. 2.41, p. 2.91], so that the PRI set is fully determined with only two design variables - the smallest PRI in a set, PRI_{min}, and the parameter r^*, which determines the ratios between PRIs, and in turn how closely spaced they are. As noted in [7], short pulse duration τ_S is not considered a design variable, but it influences the problem through an inequality constraint.

Design variables x_1, x_2 and x_3, design variable vector **x** and feasible design space **X** are defined in (1), with functions $g_j(\mathbf{x})$ same as in [7, Eq. 15]. Objective functions $F_1(\mathbf{x})$, $F_2(\mathbf{x})$ and $F_3(\mathbf{x})$, objective function vector **F(x)** and feasible objective space **Z** are defined in (2), where P_t is peak transmitted power, G is antenna gain, λ is carrier wavelength, σ is target's RCS, k is Boltzmann's constant, T_0 is standard temperature, F is receiver noise factor, L_S is system loss, θ_3 is 3dB antenna beamwidth, and ω is antenna rotation speed. *DVP* can not be expressed in closed form. The problem is formulated in (3).

$$x_1 \equiv r^*, \ x_2 \equiv \tau_L, \ x_3 \equiv PRI_{min}, \quad \mathbf{x} = \begin{bmatrix} x_1 \ x_2 \ x_3 \end{bmatrix}^T$$
$$\mathbf{X} = \left\{ \mathbf{x} \mid g_j(\mathbf{x}) \geq 0, \ j = 1, 2, \ldots, 5 \right\} \tag{1}$$

$$F_1(\mathbf{x}) \equiv SNR_L = \underbrace{\frac{P_t G^2 \lambda^2 \sigma}{(4\pi)^3 R_t^2 kT_0 FL_S}}_{C_{SNR}} \cdot \underbrace{\left\lfloor \frac{\theta_3}{\omega \cdot PRI_{avg} \cdot N} \right\rfloor}_{n_P} \cdot \tau_L$$

$$= C_{SNR} \cdot \underbrace{\left\lfloor \frac{\theta_3}{\omega} \cdot \frac{r^* + \triangle r_{min}(N)}{r^* + \triangle r_{max}(N)} \cdot \frac{1}{PRI_{min} \cdot N} \right\rfloor}_{n_P} \cdot \tau_L$$

$$F_2(\mathbf{x}) \equiv DVP = DVP(r^*, PRI_{min})$$

$$F_3(\mathbf{x}) \equiv -VR_{max} = -\frac{1}{PRI_{min} \cdot n_P} \cdot \frac{\lambda}{2}$$

$$\mathbf{F}(\mathbf{x}) = \left[F_1(\mathbf{x}) \ F_2(\mathbf{x}) \ F_3(\mathbf{x}) \right]^T, \quad \mathbf{Z} = \left\{ \mathbf{F}(\mathbf{x}) \mid \mathbf{x} \in \mathbf{X} \right\} \tag{2}$$

$$\mathbf{x}^* = \arg \max_{\mathbf{x}} \mathbf{F}(\mathbf{x}) \ \mathbf{s.t.} \ g_j(\mathbf{x}) \geq 0, \ j = 1, 2, \ldots, 5 \tag{3}$$

3 Analysis, Optimization Example and Discussion

Table 1 shows our expectations on how increasing design variables affects the objectives. These expectations are partly a result of analyzing formulas for F_1 and F_3 in (2). Discontinuities (stairs, saw teeth) are expected, due to presence of rounding down operator. As for F_2, increasing both x_1 and x_3 leads to more closely spaced PRFs, which increases the likelihood that the aliases of notched out regions will overlap, in turn decreasing F_2. Although we expect decreasing trend for F_2, there may be local extrema. Conflicts between the objectives are expected.

We propose the same optimization approach as in [7], i.e. finding the Pareto frontier (PF) and then imposing preferences. To illustrate the approach, we use the same example as in [7], corresponding to hypothetical modern VHF radar. We assume $N_{DFT} = 64$ and $CA_{min} = 40$ dB. Since design space has continuous parts, we sample it uniformly, calculate values of objective functions for each sampled point, and perform an exhaustive search for Pareto optimal (PO) points in the objective space. The results are shown in Fig. 2.

PO points marked with A, B and C in Fig. 2 correspond to maximal attainable values of objectives. Points between A and C (including C) are not acceptable,

Table 1 Expected influence of design variables on objective functions

	$x_1 \equiv r^* \uparrow$	$x_2 \equiv \tau_L \uparrow$	$x_3 \equiv PRI_{min} \uparrow$
$F_1 \equiv SNR_L$	$\uparrow$ (stair)	$\uparrow$	$\downarrow$ (stair)
$F_2 \equiv DVP$	$\downarrow$ (trend)	–	$\downarrow$ (trend)
$F_3 \equiv -VR_{max}$	$\uparrow$ (stair)	–	$\updownarrow$ (saw tooth)

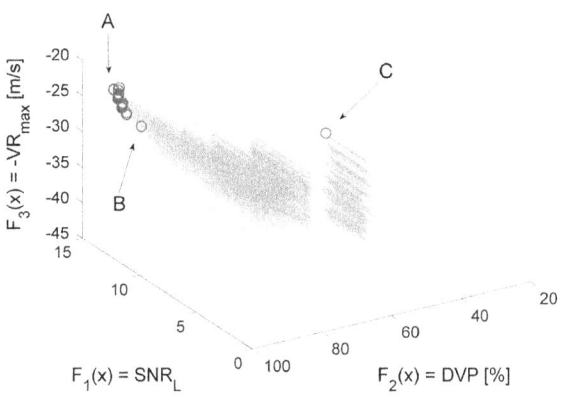

Fig. 2 Objective space (*gray*) with PO points (*red circles*). PO points which correspond to maximal attainable values of objectives are A (**12.81**, 98.26, −22.05), B (10.13, **99.06**, −24.37) and C (3.388, 67.83, **−22.03**). PO points between A and B (including A and B) are the acceptable solutions (color figure online)

because they provide negligible increase in F_3 at the expense of decreasing F_1 and/or F_2. However, moving from B to A, reasonable trade-off is made, i.e. F_1 is increased for a factor of 1.26 (26%) and F_3 is increased for 2.32 m/s (9.5% improvement), at the expense of decreasing F_2 for 0.8%. Thus, A and B and PO points between them are the acceptable solutions. We propose postponing the ultimate choice of a single PO point until run time. For example, operator could choose between two modes of operation, one which favors *DVP* (point B), and the other one which favors improved SNR and VR (point A). The results and suggestions given here, are in general agreement with, but more accurate than those of [7].

Since exhaustive search provided useful insights into achievable tradeoffs, an efficient solution method was not needed in this phase of the research. However, we had to put restrictions on design variables to limit the total number of points, because exhaustive search does not scale well. An upper bound of 35 was put on x_1, while x_2 and x_3 were discretized with resolution of 10 μs, as in [7]. For application in real radar system it would be interesting to increase the resolution and the upper bound, which would require a more efficient solution method. Such a method could even enable real time adaptation to a change of a system parameter (e.g. antenna rotation speed). Since there are three objectives, it seems that a mature evolutionary MOO algorithm such as NSGA-II [9] could be used as the solution method.

4 Conclusion

In this paper, we treat the VHF MTD solid state air surveillance radar PD waveform design as a MOO problem, taking SNR, Doppler visibility and Doppler resolution as objectives. Our findings are as follows:

- Since Doppler visibility is in conflict with SNR and Doppler resolution, there is no single optimal solution for PD waveform design problem, but there are multiple PO solutions among which a compromise solution should be chosen.
- Particular example that we used showed that only a part of PF contains the acceptable solutions, in the sense that compromise achieved by choosing among these solutions is reasonable from the practical perspective. Moving between two PO points at the opposite ends of the acceptable part of the PF, 0.8 % loss in Doppler visibility can be traded for 9.5 % improvement in Doppler resolution and 26 % increase of SNR. The significance of 26 % SNR increase can be illustrated by comparison. To achieve the same effect through increasing the peak transmitted power, P_t, 26 % more power amplifier modules in the SST would be required, which would increase SST size, weight, power consumption and price.
- The ultimate choice of a single solution among the acceptable PO points could be performed by an operator, at run time. In the simplest case, there could be two modes of operation, one which maximizes SNR and Doppler resolution, an the other one which maximizes Doppler visibility. Active mode could be selected as the operator seems fit, according to the current situation.

Based on these results, we believe that the subject of PD waveform design using MOO deserves further research. One line of future work could involve increasing resolution and extents of the design space to allow for a more accurate assessment of the PF. As discussed in the third section, such research would require finding an efficient method for determining the PF.

Acknowledgements Ministry of Education, Science, and Technological Development of Republic of Serbia provided financial support for this work, grant TR32051.

Authors also wish to thank Mr. Vladimir Simeunović, Head of Institute Mihailo Pupin, Computer Systems Dept., for providing additional financial support.

References

1. Skolnik, M.I.: Radar Handbook, 3rd edn. McGraw-Hill, New York (2008)
2. Richards, M.A., Scheer, J.A., Holm, W.A.: Principles of Modern Radar, Vol. I: Basic Principles. SciTech Publishing (2010)
3. Davies, P.G., Hughes, E.J.: Medium PRF set selection using evolutionary algorithms. IEEE Trans. Aerosp. Electron. Syst. (AES) **38**(3), 933–939 (2002)
4. Alabaster, C.M., Hughes, E.J., Matthew, J.H.: Medium PRF radar PRF selection using evolutionary algorithms. IEEE Trans. AES **39**(3), 990–1001 (2003)
5. Wiley, D., Parry, S., Alabaster, C., Hughes, E.: Performance comparison of PRF schedules for medium PRF radar. IEEE Trans. AES **42**(2), 601–611 (2006)
6. Hughes, E.J.: Radar waveform optimisation as a many-objective application benchmark. In: Proceedings of 4th International Conference, EMO 2007, pp. 700–714 (2007)
7. Jevtić, M., Zogović, N., Graovac, S.: Multiobjective approach to optimal waveform design for solid-state VHF pulse-doppler air surveillance radar. In: Proceedings of 2nd International Conference on Electrical, Electronic and Computing Engineering IcETRAN 2015, pp. AUI2.3.1–6 (2015)

8. Marler, R.T., Arora, J.S.: Survey of multi-objective optimization methods for engineering. Struct. Multidiscip. Optim. **26**(6), 369–395 (2004)
9. Deb, K., Pratap, A., Agarwal, S., Meyarivan, T.: A fast and elitist multiobjective genetic algorithm: NSGA-II. IEEE Trans. Evol. Comput. **6**(2), 182–197 (2002)

A Hybrid Approach of Optimization and Sampling for Robust Portfolio Selection

Omar Rifki and Hirotaka Ono

Abstract Dealing with ill-defined optimization problems, where the actual values of the input parameters are unknown or not directly measurable, is generally not an easy task. In order to enhance the robustness of the final solutions, we propose in the current paper a hybrid metaheuristic approach that incorporates a sampling-based simulation module. Empirical application to the classical mean-variance portfolio optimization problem, which is known to be extremely sensitive to noises in asset means, is provided through a genetic algorithm solver. Results of the proposed approach are compared with that specified by the baseline worst-case scenario and the two approaches of stochastic programming and robust optimization.

1 Introduction

Dealing with ill-defined optimization problems, where the actual values of the input parameters are unknown or not directly measurable, is generally not an easy task. For example, Markowitz-type portfolio optimization problems are formulated with many probabilistic coefficients, but the coefficients are actually impossible to measure. Instead, they are approximately calculated from past histories of returns on assets based on some assumptions, but this often raises the problem that the best portfolio on the approximate model could be very bad for the real scenario. This might be due to an over-fitting to the approximate model, and we can say that the computed portfolio is not "robust" for the difference between the calculated coefficients and the "true" coefficients. This kind of problem is frequently seen in the fields where true parameters are difficult or impossible to obtain, but still hard to resolve in the ordinary optimization framework. In this paper, we propose a method to obtain a robust solution for the problem with approximate parameters. The main idea is to combine a metaheuristic (MH) algorithm and simulation techniques. Such

O. Rifki (✉) · H. Ono
Department of Economic Engineering, Kyushu University, Fukuoka 812-8581, Japan
e-mail: rifki@kyudai.jp

H. Ono
e-mail: hirotaka@econ.kyushu-u.ac.jp

© Springer International Publishing Switzerland 2017
K.F. Dœrner et al. (eds.), *Operations Research Proceedings 2015*,
Operations Research Proceedings, DOI 10.1007/978-3-319-42902-1_43

combination has been already covered [9, 10]. The basic role of simulation up to now in these combinations is the estimation of the solution performance, or in other terms the computation of the objective function, which is usually done by taking expectations over simulation's scenarios. Some approaches extend this role to ranking and selection procedures at the end of MH search, which includes screening out inferior solutions and conducting additional simulations [3, 10]. However, our use of simulation is rather different. It is intended to detect robust solutions among several good candidate solutions, with a main focus on robustness assessment. For this purpose, two simple robustness criteria are proposed, such that robustness of each solution depends not only on its performance over scenarios, but on the other solutions performances as well. The simulation module is run once at the end of the MH search, with a synergic exchange between both parts. For the application on portfolio optimization, the hybrid MH is compared to two models of optimization under uncertainty, one of robust optimization and the other of stochastic programming. In a second experiment, the contrast between the worst-case baseline solutions and that of the hybrid approach is discussed. The overall hybrid method is described in the next section, while the application is provided in Sect. 3. Section 4 concludes.

2   The Hybrid Approach

Simulation Module. Consider the following constrained optimization problem,

$$\max_{x} f(\tilde{\theta}, x) \quad \text{subject to} \quad x \in \mathscr{C} \subseteq \mathbb{R}^n, \ \tilde{\theta} \in \Delta \subseteq \mathbb{R}^m, \tag{1}$$

where the uncertainty solely affects the objective function $f(\tilde{\theta}, x) : \Delta \times \mathscr{C} \to \mathbb{R}$, x denotes a vector of design variables constrained to a set $\mathscr{C} \subseteq \mathbb{R}^n$, and $\tilde{\theta} \in \Delta \subseteq \mathbb{R}^m$ is a vector of random variables representing uncertain parameters. We can make the problem (1) well-defined by assuming that $\tilde{\theta}$ varies within an uncertainty set $\Theta \equiv \{\theta \in \Delta : ||\theta - \theta_0||_p \le \xi\} = B_\xi(\theta_0) \subseteq \mathbb{R}^m$, centered at θ_0 which we shall call the *nominal value* of $\tilde{\theta}$. $||.||_p$ denotes the l_p norm in $\mathbb{R}^m$, e.g. $p = 1, 2$. The focus of the simulation is on detecting design variables x that have desirable robustness properties within Θ. To this purpose, given a set $\mathscr{X} = \{x_1, x_2, \ldots, x_K\}$ from which robust solutions will be chosen, we proceed by studying the x_i's behavior in terms of performance across a number of randomly generated samples $\tilde{\theta}$ from Θ, that we shall call *scenarios*. It is important to mention that the simulation module does not involve any optimization operation, but rather a number of evaluations. Suppose that N independent and identically distributed scenarios of $\tilde{\theta}$ are drawn, say $\theta_1, \theta_2, \ldots, \theta_N \in \Theta$. To measure robustness of an instance $x \in \mathscr{X}$, two measures $R_1^{\mathscr{X}}(x)$ and $R_2^{\mathscr{X}}(x)$ are introduced. The first one reports the ratio across scenarios of being *top-ranked solution*, which is defined to be having the best function value f compared with the other solutions of $\mathscr{X}$. Using the following indicator function,

$$I_{\mathscr{X}}(\theta, x) = \begin{cases} 1, & \text{if } f(\theta, x) = \max_{y \in \mathscr{X}} f(\theta, y) \\ 0, & \text{otherwise} \end{cases} \qquad \forall \theta \in \{\theta_1, \ldots, \theta_N\}, \quad \forall x \in \mathscr{X},$$

we can obtain an estimate of $R_1^{\mathscr{X}}(x)$ as, $\qquad \widehat{R_1^{\mathscr{X}}}(x) = \dfrac{1}{N} \sum_{i=1}^{N} I_{\mathscr{X}}(\theta_i, x) \quad \forall x \in \mathscr{X}.$

The second measure $R_2^{\mathscr{X}}(x)$ starts by computing for each scenario θ_i the performance ratio $f(\theta_i, x)/f_{max}^{\mathscr{X}}(\theta_i)$, where $f_{max}^{\mathscr{X}}(\theta_i) = \max_{y \in \mathscr{X}} f(\theta_i, y)$ is the maximum function value reached for the scenario θ_i. These ratios are averaged afterwards. An estimate of $R_2^{\mathscr{X}}(x)$ can be written as,

$$\widehat{R_2^{\mathscr{X}}}(x) = \frac{1}{N} \sum_{i=1}^{N} \frac{f(\theta_i, x)}{f_{max}^{\mathscr{X}}(\theta_i)}.$$

The reason of using the first measure is intuitive. It relates the robustness of a solution to the number of times it is optimum (within $\mathscr{X}$) for a number of samples of $\tilde{\theta}$. Whereas the advantage of the second measure is to take into account the relative span of each solution to the top-ranked solutions across scenarios. Indeed, this latter ratio is desirable to know if a solution is rarely ranked best, but frequently exhibits small gaps to top-ranked solutions. Table 1 illustrates the simulation module. Three elements need to be set: the solution set $\mathscr{X}$, the sample size N, and the distribution of the N random draws of $\tilde{\theta} \in \Theta$, which includes the distribution shape, e.g. Gaussian, uniform and the noise magnitude parameter ξ of $B_\xi(\theta_0) = \Theta$.

Hybridization. The combination between the MH and the simulation is designed to be of high level, allowing each part to retain its own identity. Theoretically any MH independently of its structure can be used within the approach. Before running the simulation, the MH will be performed several times, denoted by V, for several instances of the uncertain parameters. In Fig. 1, the MH is run $V = 3$ times for 3 randomly generated variables $\theta_1', \theta_2', \theta_3' \in \Theta$ with θ_0 the nominal value. The overall solution pool outputted by the MH runs are grouped to constitute the input of the simulation module $\mathscr{X}$. The size of the final output is equal to the MH population size

Table 1 Description of the simulation module

	x_1	x_2	...	x_K	
(I_1) scenario θ_1	$f(\theta_1, x_1)$	$f(\theta_1, x_2)$	...	$f(\theta_1, x_K)$	$\rightarrow f_{max}^{\mathscr{X}}(\theta_1)$
(I_2) scenario θ_2	$f(\theta_2, x_1)$	$f(\theta_2, x_2)$	...	$f(\theta_2, x_K)$	$\rightarrow f_{max}^{\mathscr{X}}(\theta_2)$
...	...	...	...	...	
(I_N) scenario θ_N	$f(\theta_N, x_1)$	$f(\theta_N, x_2)$	...	$f(\theta_N, x_K)$	$\rightarrow f_{max}^{\mathscr{X}}(\theta_N)$
$\widehat{R_1^{\mathscr{X}}}(x)/\widehat{R_2^{\mathscr{X}}}(x)$	-/-	-/-	...	-/-	

The table is headed by a brace labelled $\mathscr{X}$.

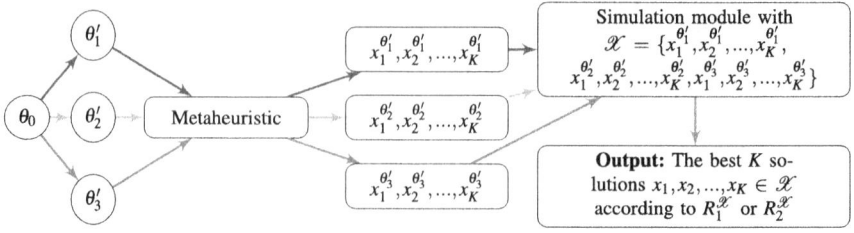

Fig. 1 General scheme of the proposed hybrid algorithm

$K \geq 1$. Each scenario adds a cost of $V \times K$ evaluations, thus finding a mechanism that reduces the sample size N is of high interest.

3   Application to Portfolio Optimization Problem

Methodology. The portfolio selection model considered in this study is the standard single-period mean-variance without short-selling, stated for n risky assets as [11],

$$\max_{w} \{EU(w)\} = \{\mu_p(w) - \lambda \sigma_p^2(w)\} \text{ subject to } \sum_{i=1}^{n} w_i = 1, \quad w_i \geq 0 \quad \forall i \in ||1, n|| \qquad (2)$$

where w_i is the fraction held in the ith asset. The expected return of the portfolio is equal to $\mu_p(w) = \sum_{i=1}^{N} w_i \mu_i$, such that μ_i is the ith asset mean. The variance of the portfolio corresponds to $\sigma_p^2(w) = \sum_{i=1}^{N} \sum_{j=1}^{N} w_i w_j \sigma_{ij}$, such that σ_{ij} is the covariance between the returns of the ith and jth assets. The parameter λ indicates the degree of the investor's risk aversion. This problem is famously known to be extremely sensitive to perturbations in the inputs, especially to assets means μ_i [4, 6]. The uncertainty set wherein the vector $\mu = (\mu_i)$ is considered to take value is the ball $B_{\xi|\mu_0|}(\mu_0) = \{\mu \in \mathbb{R}^n : |\mu_i - \mu_{0i}| \leq \xi |\mu_{0i}|, \forall i \in [|1, n|]\}$ centered at the nominal asset means vector μ_0. Nominal data values are taken from one of the four data instances of the OR-library [1] illustrated in Table 2. In the first experiment, our hybrid method is empirically compared to two robust formulations of the problem (2). To keep those compared models computationally tractable, hard constraints are not

Table 2 Benchmark instances

No.	Index	Assets	Γ_{bound}	No.	Index	Assets	Γ_{bound}
1	Hang Seng	31	13.95	3	FTSE 100	89	22.94
2	DAX 100	85	22.44	4	S&P 100	98	24.03

added to (2). The first formulation is a robust optimization (RO) based on Bertsimas and Sim approach [2]. It leads to the following quadratic constrained program,

$$\max \ z \quad \text{s.t.} \quad \sum_{i=1}^{n} w_i \mu_{0i} - \lambda \sum_{i=1}^{n} \sum_{j=1}^{n} w_i w_j \sigma_{ij} \geq z + p\Gamma + \sum_{i=1}^{n} q_i \quad \text{and} \quad \sum_{i=1}^{n} w_i = 1$$

$$\text{and} \quad -v_i \leq \mu_{0i} \leq v_i, \quad p + q_i \geq \xi v_i w_i, \quad w_i, q_i, v_i, p \geq 0 \quad \forall i \in ||1, n|| \tag{3}$$

where Γ is a real valued parameter within $[0, n]$ used to balance the problem robustness against the level of conservatism of solutions. $\Gamma = 0$ leads to the nominal formulation (2). We set Γ at the approximate bound Γ_{bound} introduced in [2]. The solution of (3) will be noted w^{ro}. A stochastic programming (SP) based model is also considered. According to *mean-risk stochastic models* from Fabozzi et al. [8], achieving robustness can be done through expectation of our objective function $EU(w)$ and subtraction of a penalty function expressed as a risk variance. By disposing of a number of S equally-weighted scenarios μ_i^s, $s \in [|1, S|]$, supposed to be drawn from a distribution similar to that of the hybrid MH scenarios in order to make the analysis easier, we obtain the following robust counterpart of problem (1),

$$\max_{w} \sum_{i=1}^{n} w_i \mu_{0i} - \lambda \sum_{i=1}^{n} \sum_{j=1}^{n} w_i w_j \sigma_{ij} - \kappa \left[\sum_{s=1}^{S} \frac{1}{S} \left(\sum_{i=1}^{n} \mu_i^s w_i \right)^2 - \left(\sum_{s=1}^{S} \sum_{i=1}^{n} \frac{\mu_i^s w_i}{S} \right)^2 \right]$$

$$\text{s.t.} \quad \sum_{i=1}^{n} w_i = 1, \quad \text{and} \quad w_i \geq 0 \quad \forall i \in ||1, n|| \tag{4}$$

where κ is a robustness averseness parameter. We fix $\kappa = 1$ and $S = 50$. The solution of (4) shall be noted w^{sp}. Both w^{ro} and w^{sp} are optimally solved using CPLEX 12.5, while the hybrid method is run in a Java environment. For the metaheuristic, Genetic algorithm (GA) is applied. The choice of GA, which is a well-known population-based MH grounded on the darwinian natural selection principle—*survival of the fittest*, is due to its high performance reported for solving mean-variance models when compared with other MHs [5, 12]. We use a straightforward application of

Table 3 Values of the GA parameters initially obtained through a tuning experiment

GA Parameter	Value	GA Parameter	Value
Representation	Real-valued	Population size (=K)	300
Selection	Tournament	Max generations	800
Crossover	SBX [7]	Crossover probability	0.25
Mutation	Polynomial [7]	Mutation probability	0.01

Fig. 2 First experiment
description

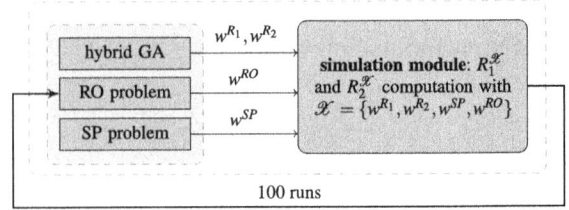

GA with Table 3 considerations. In the first experiment, w^{R_1} and w^{R_2}, respectively the top-ranked portfolios of the hybrid approach according to $R_1^{\mathscr{X}}$ and to $R_2^{\mathscr{X}}$, are compared to w^{ro} and w^{sp}. After each solving run, a new simulation module is performed as indicated by Fig. 2. The final results are averaged over 100 runs. To construct the solution pool of the hybrid approach, 10 MH runs are performed. The remaining parameters are set as follows, the simulation sample size $N = 1.2 \times 10^5$, $\lambda = 2$, plus two noise magnitudes are considered, a low $\xi = 0.01$ and a large one $\xi = 0.3$. The sampling is assumed to be uniform. The second experiment compares the two final lists of solutions of the hybrid approach (of size K) ranked according to $R_1^{\mathscr{X}}$ and $R_2^{\mathscr{X}}$ to that induced by μ_{wc} the baseline worst-case scenario (wc). To spot $\mu_{wc} \in \{\mu_1, \mu_2, \ldots, \mu_N\}$, we seek μ_i with a lower value of the summation $\sum_{w \in \mathscr{X}} EU(w, \mu_i)$, $\forall i \in [|1, N|]$.

Results. The results of the first experiment are shown in Table 4. The dark background box indicates the best value obtained for each row. All the percentages of the $R_2^{\mathscr{X}}$ measure are of very high value, which suggests good performance for the four portfolios. For low noise magnitude, all portfolios have remarkably close $R_2^{\mathscr{X}}$ values, while in high noise magnitude the finer $R_2^{\mathscr{X}}$ ratios are for w^{R_2} and w^{SP}. On the other hand, $R_1^{\mathscr{X}}$ ratios reveal a couple of variations between the portfolios, though in general w^{R_1} has the best value, then followed either by w^{R_2} or w^{R0} and finally by w^{SP}. For the second experiment, the GA population size is reduced to $K = 50$ to prevent memory overflow. Figure 3 shows the composition of the three considered populations averaged over 100 runs. The μ_j in the figure's legend represent the instances of asset means used in the portfolio generation. The three considered populations (of size K) differ in their repartition. The ones according to wc and $R_2^{\mathscr{X}}$ allow more space for *nominal portfolios*, which are generated using μ_0, while the reparation of $R_1^{\mathscr{X}}$ population is more fair between asset means instances, for around 10 % for each μ_j. $R_1^{\mathscr{X}}$ and $R_2^{\mathscr{X}}$ populations are ranked respectively according to $R_1^{\mathscr{X}}$ and $R_2^{\mathscr{X}}$ measures. The rank of the portfolios induced the wc scenario is set according to $EU(w, \mu_{wc})$. For top ranked portfolio of each population, i.e. rank=1, we examine its analogous ranking in the other wc, $R_1^{\mathscr{X}}$ and $R_2^{\mathscr{X}}$ populations. This is done 100 times in order to construct the box plots of Fig. 4. We observe inter alia that the best portfolios of $R_1^{\mathscr{X}}$ population, have spread ranks in wc and $R_2^{\mathscr{X}}$ populations (the gray boxplots), while portfolios ranked best in wc and $R_2^{\mathscr{X}}$, still hold high ranking in $R_1^{\mathscr{X}}$ population. Similar observations can also be made for $\xi = 0.3$ case.

Table 4 First experimental results: the obtained $R_1^{\mathcal{X}}$ and $R_2^{\mathcal{X}}$ ratios for the 4 datasets

	measure	w^{R_1}	w^{R_2}	w^{RO}	w^{SP}	
$\xi = 0.01$	$R_1^{\mathcal{X}}$ (%)	38.67613	37.52136	20.33198	03.47052	No.1 dataset
	$R_2^{\mathcal{X}}$ (%)	99.99999	99.99999	99.99999	99.99999	(31 assets)
$\xi = 0.3$	$R_1^{\mathcal{X}}$ (%)	25.28785	33.32587	29.96074	11.42552	
	$R_2^{\mathcal{X}}$ (%)	99.99830	99.99937	99.99860	99.99936	
$\xi = 0.01$	$R_1^{\mathcal{X}}$ (%)	30.33653	27.12493	20.26699	22.27154	No.2 dataset
	$R_2^{\mathcal{X}}$ (%)	99.99999	99.99999	99.99999	99.99999	(85 assets)
$\xi = 0.3$	$R_1^{\mathcal{X}}$ (%)	34.09839	22.80066	30.17639	12.92455	
	$R_2^{\mathcal{X}}$ (%)	99.99815	99.99933	99.99893	99.99933	
$\xi = 0.01$	$R_1^{\mathcal{X}}$ (%)	38.34715	15.23001	27.24987	19.17295	No.3 dataset
	$R_2^{\mathcal{X}}$ (%)	99.99999	99.99999	99.99999	99.99999	(89 assets)
$\xi = 0.3$	$R_1^{\mathcal{X}}$ (%)	30.16063	24.40341	33.37719	12.05875	
	$R_2^{\mathcal{X}}$ (%)	99.99796	99.99906	99.99851	99.99905	
$\xi = 0.01$	$R_1^{\mathcal{X}}$ (%)	38.57140	18.30873	36.42604	06.69381	No.4 dataset
	$R_2^{\mathcal{X}}$ (%)	99.99999	99.99999	99.99999	99.99999	(98 assets)
$\xi = 0.3$	$R_1^{\mathcal{X}}$ (%)	30.89806	23.09983	24.35261	21.64948	
	$R_2^{\mathcal{X}}$ (%)	99.99820	99.99935	99.99830	99.99935	

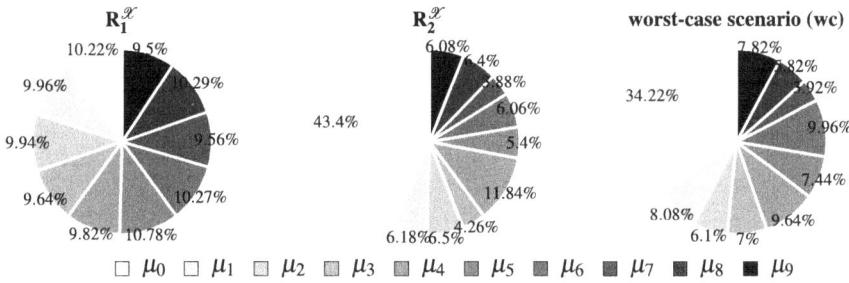

Fig. 3 Population repartitions in terms of asset means instances for $\xi = 0.01$ (for dataset No. 1)

Fig. 4 Distribution of analogous rankings of top portfolios by wc, $R_1^{\mathcal{X}}$ and $R_2^{\mathcal{X}}$ for $\xi = 0.01$ (for dataset No. 1)

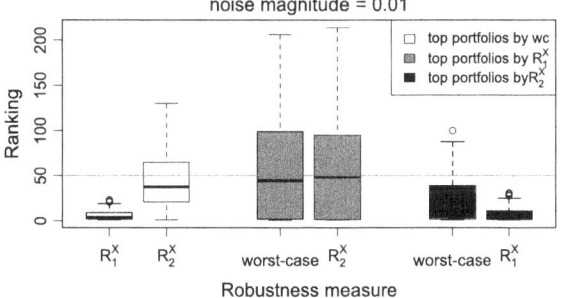

4 Conclusion

In order to find less sensitive solutions to perturbations affecting input parameters of ill-defined problems, we proposed a hybrid metaheuristic approach incorporating a simulation module. Two measures of robustness are used ($R_1^{\mathscr{X}}$ and $R_2^{\mathscr{X}}$). Empirical experiments conducted on the problem of portfolio optimization against noises in assets means show that for low and high noise magnitudes, hybrid GA exhibits better robustness in terms of the $R_1^{\mathscr{X}}$ measure, not far followed by RO based portfolio and then lastly by the SP based portfolio, while $R_2^{\mathscr{X}}$ ratios are quite close for all portfolios. Results of comparing hybrid GA results to the baseline worst-case scenario (wc), i.e. the most less favorable scenario among the generated ones, favors $R_1^{\mathscr{X}}$ measure when it comes to ranking top portfolios according to $R_2^{\mathscr{X}}$ or wc.

References

1. Beasley, J.E.: OR-Library: distributing test problems by electronic mail. J. Oper. Res. Soc. **41**(11), 1069–1072 (1990)
2. Bertsimas, D., Sim, M.: The price of robustness. Oper. Res. **52**(1), 35–53 (2004)
3. Boesel, J., Nelson, B.L., Kim, S.-H.: Using ranking and selection to "clean up" after simulation optimization. Oper. Res. **51**(5), 814–825 (2003)
4. Broadie, M.: Computing efficient frontiers using estimated parameters. Ann. Oper. Res. **45**(1), 21–58 (1993)
5. Chang, T.J., Meade, N., Beasley, J.E., Sharaiha, Y.M.: Heuristics for cardinality constrained portfolio optimisation. Comput. Oper. Res. **27**(13), 1271–1302 (2000)
6. Chopra, V.K., Ziemba, W.T.: The effect of errors in means, variances, and covariances on optimal portfolio choice. J. Portf. Manag. **19**(2), 6–11 (1993)
7. Deb, K., Agrawal, R.B.: Simulated binary crossover for continuous search space. Complex Syst. **9**(2), 115–148 (1995)
8. Fabozzi, F.J., Kolm, P.N., Pachamanova, D., Focardi, S.M.: Robust portfolio optimization and management. Wiley, Hoboken (2007)
9. Glover, F., Kelly, J.P., Laguna, M.: New advances and applications of combining simulation and optimization. In: Proceedings of the 28th Conference on Winter Simulation, pp. 144–152. IEEE Computer Society (1996)
10. Juan, A.A., Faulin, J., Grasman, S.E., Rabe, M., Figueira, G.: A review of Simheuristics: extending metaheuristics to deal with stochastic combinatorial optimization problems. Oper. Res. Perspect. **2**, 62–72 (2015)
11. Markowitz, H.: Portfolio selection. J. Financ. **7**, 77–99 (1952)
12. Woodside-Oriakhi, M., Lucas, C., Beasley, J.E.: Heuristic algorithms for the cardinality constrained efficient frontier. Eur. J. Oper. Res. **213**(3), 538–550 (2011)

Tabu Search Heuristic for Competitive Base Station Location Problem

Marceau Coupechoux, Ivan Davydov and Stefano Iellamo

1 System Model

We consider the following location problem. New 5G networks are deployed by two competitive operators (called resp. O_1 and O_2), which we refer to as a leader and a follower due to their sequential entering the market. They compete to serve clients by installing and configuring base stations (BS). We assume that the leader had already made a decision and is operating a 5G network. Follower arrives at the market knowing the decision of the leader and set it's own 5G network. Follower is able to set up his BSs on all the available sites. It is also possible to share the site with the leader. In the latter case follower pays leader an additional sharing price. Each client chooses the network considering the average quality of the service provided. The aim of the follower is to choose locations for his BS in order to maximize his profit.

1.1 Network and Propagation Model

Let $\mathscr{S}$ be the set of all sites, where base stations can be installed. This set is made of three subsets: $\mathscr{S} = \mathscr{S}_f \cup \mathscr{S}_1^o \cup \mathscr{S}_2^o$, where $\mathscr{S}_i^o$ is the set of sites having a 4G base station installed by $O_i, i \in \{1, 2\}$, and $\mathscr{S}_f$ is a set of free sites for potential new installations.

M. Coupechoux (✉) · S. Iellamo
Télécom ParisTech and CNRS LTCI, 6, rue Barrault, Paris, France
e-mail: marceau.coupechoux@telecom-paristech.fr

S. Iellamo
e-mail: iellamo@enst.fr

I. Davydov
Sobolev Institute of Mathematics, Av.Koptyuga 4, Novosibirsk, Russia
e-mail: vann.davydov@gmail.com

© Springer International Publishing Switzerland 2017
K.F. Dœrner et al. (eds.), *Operations Research Proceedings 2015*,
Operations Research Proceedings, DOI 10.1007/978-3-319-42902-1_44

O_1 have a subset $\mathscr{S}_1^n$ of new 5G base stations installed. His new BSs are installed among the free sites and the sites having old BSs of O_1: $\mathscr{S}_1^n \subset \mathscr{S}_f \cup \mathscr{S}_1^o$. The rent price is set by O_1 for every site with its BS, i.e. in the set $\mathscr{S}_1^n \cup \mathscr{S}_1^o$. Now O_2 is deploying its 5G network by choosing a set $\mathscr{S}_2^n$ for its 5G BSs. He has a choice among all the sites in $\mathscr{S}$, i.e. $\mathscr{S}_2^n \subset \mathscr{S}$. If a 5G BS of O_2 is placed on a site in $\mathscr{S}_1^n$, he will have to pay the sharing price fixed by O_1. Otherwise, he will have to pay for maintenance. At the end of this phase, some users leave O_1 and take a subscription with O_2.

Now let consider a user located at x. Let define the channel gain between location x and BS b as $g_b(x)$ and let assume that the transmit power of b is P_b. Signal to interference and noise ratio (SINR) of the considered user in x with respect to b is given by:

$$\gamma_b(x) = \frac{P_b g_b(x)}{\sum_{i \neq b} P_i g_i(x) + N}, \tag{1}$$

where N is the thermal noise power in the band. User in x is said to be *covered* by b if $\gamma_b(x) \geq \gamma_{min}$ for some threshold γ_{min}. User in x is said to be *served* by b if he is covered and $P_b g_b(x) \geq P_i g_i(x)$ for all $i \neq b$. Note that at every location, users can be served by at most one BS from each operator.

For a user located in x and served by station b, the physical data rate achievable by this user is denoted by $c_b(x)$, which is an increasing nonlinear function of $\gamma_b(x)$ with $c_b(x) = 0$ if $\gamma_b(x) < \gamma_{min}$.

1.2   Traffic Model

We assume there is a constant traffic demand in the network that operators will potentially serve. In every location x, there is a demand $\lambda(x)/\mu(x)$, where $\lambda(x)$ is the arrival rate and $1/\mu(x)$ is the average file size. Note that this demand in x is statistical and can be shared by O_1 and O_2 or not served at all. Let assume that x is covered by O_1 and a proportion $p_1(x)$ of the demand is served by BS b from O_1. A proportion $p_2 = 1 - p_1(x)$ of the demand is served by O_2. We focus in this paper on a specific case for p_1: If location x is not covered by O_1, $p_1(x) = 0$. Otherwise, p_1 does not depend on the location and depends only on the overall relative quality of service in the network O_1 compared to O_2. The idea behind this assumption is that users are mobile and they choose their operator not only with respect to the quality of service at a particular location but rather to the average experienced quality.

Then, the load created by x on b in the network of operator O_i is $p_i(x)\rho_{ib}(x)$, where $\rho_{ib}(x) = \frac{\lambda(x)}{\mu(x)c_{ib}(x)}$, where $c_{ib}(x)$ is the physical data rate in x and is an increasing function of $\gamma_{ib}(x)$. The index i is here to recall that the SINR, so the physical data rate, and the load are computed in the network of O_i. This is an important point because in the rest of the paper, station b is likely to be shared by both operators. We can now define the *load* of station b as: $p_i\rho_{ib}$, where $\rho_{ib} = \sum_{\mathscr{A}_{ib}} \rho_{ib}(x)$, where

$\mathscr{A}_{ib}$ is the serving area of b, i.e., the set of locations served by b, in network O_i. BS b is stable if $p_i \rho_{ib} < 1$ and we will consider only scenarios where this condition is fulfilled.

Let us define by Λ, $\Lambda = \sum_{x \in \mathscr{A}} \lambda(x)$, the total arrival rate in the network. The average throughput obtained by a random user from operator O_1 is given by:

$$t_i = \frac{1}{\Lambda} \sum_{b \in \mathscr{S}_i^n} \sum_{\mathscr{A}_{ib}} p_i \lambda(x)(1 - p_i \rho_{ib}) c_{ib}(x). \tag{2}$$

We assume that users are the players of an evolutionary game. In this framework, the choice of a single user does not influence the average throughput of an operator. An equilibrium is reached when both average throughputs are the same. In this case, we equalize t_1 and t_2, which leads to the following quadratic equation:

$$f(p_1) = p_1^2 \Big(\sum_{b \in \mathscr{S}_2^n} \rho_{2b} \sum_{\mathscr{A}_{2b}} \lambda(y) c_{2b}(y) - \sum_{b \in \mathscr{S}_1^n} \rho_{1b} \sum_{\mathscr{A}_{1b}} \lambda(x) c_{1b}(x) \Big) +$$

$$+ p_1 \Big(\sum_{b \in \mathscr{S}_1^n} \sum_{\mathscr{A}_{1b}} \lambda(x) c_{1b}(x) - 2 \sum_{b \in \mathscr{S}_2^n} \rho_{2b} \sum_{\mathscr{A}_{2b}} \lambda(y) c_{2b}(y) + \sum_{b \in \mathscr{S}_2^n} \sum_{\mathscr{A}_{2b}} \lambda(y) c_{2b}(y) \Big) +$$

$$+ \sum_{b \in \mathscr{S}_2^n} \rho_{2b} \sum_{\mathscr{A}_{2b}} \lambda(y) c_{2b}(y) - \sum_{b \in \mathscr{S}_2^n} \sum_{\mathscr{A}_{2b}} \lambda(y) c_{2b}(y) = 0.$$

Let p_1^* be the operator 1 market share at equilibrium. Several cases arise:

- If $f(p_1) > 0$ for all $p_1 \in [0; 1]$, then operator 1 is always preferred to operator 2, and $p_1^* = 1$.
- If $f(p_1) < 0$ for all $p_1 \in [0; 1]$, then $p_1^* = 0$.
- if $f(p_1) = 0$ for some $p_1 \in [0; 1]$, then there are one or several equilibrium points. In this case, we set $p_1^* = \max\{p_1 \in [0; 1] : f(p_1) = 0\}$. The assumption behind this choice is that operator 1 has come first on the market. The dynamics of p_1 thus starts from 1 and decreases to the first encountered equilibrium point.

1.3 Pricing Model and Objective Function

There are operational costs that have to be paid regularly. These operational costs include traditional costs like electricity, maintenance, site renting, and possibly a sharing price. The sharing price is paid by O_2 to O_1 for every site where BSs are shared. Let λ be the traditional operational cost per unit of time for a single operator BS. Let $(1 + \alpha)\lambda$ with $0 < \alpha < 1$ be the traditional operation cost for a shared BS. Let s_b the sharing price set by O_1 for its BS b. We assume that the revenues of an operator are proportional to the market share, i.e. $P_1 = p_1 * C$, where C is the total capacity of the market. The objective function (i.e., the profits) is revenues minus operational costs.

2 Problem Formulation

As mentioned in the introduction, we are interested in the problem of base station placement where one provider (follower) enters the market competing for the clients with already existing network (leader). Follower deploys his base stations on possible candidate sites so as to maximize his profit. In this section we present the mathematical model of this optimization problem. Let us introduce the decision variables: $y_j = 1$, if the follower installs a BS on a site $j \in \mathscr{S}$, $y_j = 0$ otherwise. $y_{ij} = 1$, if the location i is served from station j, $y_{ij} = 0$ otherwise.

Now the competitive location problem can be written as a following mixed integer programming model:

$$\max_y ((1 - p_1)C - \sum_{j \in \mathscr{S}_1^o} s_j y_j - \sum_{j \in \mathscr{S}_f} s_j x_j y_j - \sum_{j \in \mathscr{S}_f \cup \mathscr{S}_2^o} \lambda y_j (1 - x_j)) \quad (3)$$

subject to

$$P_b g_{ib} y_b \geq \gamma_{min} \sum_{j \in \mathscr{S}, j \neq b} P_j g_{ij} y_j + \gamma_{min} N - \Gamma (1 - y_{ib}) \quad \forall b \in \mathscr{S}, i \in I \quad (4)$$

$$P_b g_{ib} y_b \geq P_j g_{ij} y_j - \Gamma (1 - y_{ib}) \quad \forall b, j \in \mathscr{S}, i \in I \quad (5)$$

$$y_{ij} \leq y_j \quad \forall i \in I, j \in \mathscr{S} \quad (6)$$

The objective function (3) can be understood as the total profit obtained by the follower, computed as the difference between the expected revenue from clients served and the operational costs and sharing prices paid for the stations installed. The sharing payment gives additional profit to the leader, and reduces the gain of the follower. Constraints (4) are the SINR conditions for a location to be covered. When $y_{ib} = 1$, the expression boils down to the SINR condition with respect to the SINR threshold γ_{min}. Whenever $y_{ib} = 0$ then the condition is always fulfilled because of the large value of Γ. Constraints (5) combined with (4) state that the location satisfying the minimal SINR constraint is served by a BS providing the most powerful signal. Constraints (6) state that a service is possible only if a station is installed.

3 Tabu Search Approach

Although, the constraints of the problem are linear, due to realistic model of clients behavior it is not the case for the goal function. Latter fact makes it hard to apply a broad variety of approaches, which works well with linear integer programming problems. In order to tackle the follower problem we propose a tabu search heuristic framework, which performs well on similar problems [1].

The tabu search method has been proposed by Fred Glover. It is a so called trajectory metaheuristic and has been widely used to solve hard combinatorial optimization problems [2]. The method is based on the original local search scheme that lets one "travel" from one local optimum to another looking for a global one, avoiding local optimum traps. The main mechanism that allows it to get out of local optima is a tabu list, which contains a list of solutions from previous iterations which are prohibited to be visited on the subsequent steps. We use well-known Flip and Swap neighborhoods to explore the search space within y variables. Together with the tabu list, we exploit the idea of randomized neighborhoods. This feature allows to avoid looping, significantly reduces the time per iteration, and improves search efficiency. We denote by $Swap_q$ the part q of the $Swap$ neighborhood chosen at random. $Flip_q$ neighborhood is defined in the same way, but with the different value of parameter q.

Scheme of STS method:

1. Build an initial solution Y, define the randomization parameter q, initialize an empty tabu list.

2. Repeat until the stopping criterion is satisfied:

2.1. Construct neighborhoods $Flip_q(Y)$ and $Swap_q(Y)$ and remove forbidden elements from them.

2.2. If the neighborhood is empty, return to step 2.1, else find an adjacent solution Y' with the largest value of the follower's profit.

2.3. Let $Y = Y'$, update tabu list and if $Y' > Y^*$ update the record.

3. Show the best solution found Y^*.

The initial solution is chosen at random. The randomization parameters q for the neighborhoods are set to be sufficiently small. As the tabu list, we use an ordered list of units or pairs of the follower's facilities that have been closed and opened over the last few iterations. The length of the tabu list changes in a given interval during local search. If the best found solution begins to repeat itself, we increase the tabu list length by one; otherwise, we reduce it by one. The method stops after a given number of iterations or after a certain amount of computation time.

4   Experimental Studies

The proposed approach have been implemented in C++ environment and tested on the randomly generated and real data instances. We generated 10 sets of instances with different number of client locations (20, 40, ..., 200). All locations are chosen with the uniform distribution over the square area. The number of sites was 1/4 of number of clients locations. Leader occupies exactly half of the sites at random. All the other data was also generated at random. The aim of the experiment was to study the behavior and convergence of the approach. We run the algorithm on all the 100 instances, 10 runs per instance. Time limit was set to 5 s. for each run. The algorithm has demonstrated strong convergence. Among all the instances there was only 3 examples, with different results on different runs. All of the examples was rather

Table 1 Profit and market share of the follower

Sharing price	Leader share (p_1)	Follower profit	Leader profit	N shared sites	N opened sites
200	0.327	1755	1314	2	7
220	0.327	1715	1354	2	7
250	0.371	1633	1436	2	7
280	0.393	1589	1420	1	6
310	0.416	1574	1338	0	6

big (with 180, 200 and 200) locations. And the relative difference between outputs was less then 1 %. The second experimental study concerns the real data. We use the client locations and base station cites of the part of 13th arrondissement of Paris [3]. The geometric centers of the blocks are assumed to be client locations. We also use the coordinates of existing base stations in this area. We have run a number of experiments in order to testify the believability of the proposed model. The following table contains the results of the dependance of followers behavior from the sharing price, proposed by the leader. It can be seen from the table that high sharing price is not always the optimal one for the leader (Table 1).

5 Conclusions

We have considered new competitive base stations location problem with sharing. We have proposed a mathematical model for this problem and tabu search based heuristic for obtaining good solutions rather fast. Computational results shows the believability of the model and allows to expand both the model and method on the bilevel problem.

Acknowledgements The second author's work is supported by Russian Foundation for Basic Research (grant 15-07-01141 A).

References

1. Iellamo, S., Alekseeva, E., Chen, L., Coupechoux, M., Kochetov, Y.: Competitive location in cognitive radio networks. 4OR **13**(1), 81–110 (2015)
2. Glover, F., Laguna, M.: Tabu Search. Kluwer Academic Publishers, Dordrecht (1997)
3. http://www.cartoradio.fr/cartoradio/web/

Part V
OR for Security, Policy Modeling and Public Sector OR

Robust Optimization of IT Security Safeguards Using Standard Security Data

Andreas Schilling

Abstract Finding an appropriate IT security strategy by implementing the right security safeguards is a challenging task. Many organizations try to address this problem by obtaining an IT security certificate from a recognized standards organization. However, in many cases the requirements of a standard are too extensive to be implemented, particularly by smaller organizations. But the knowledge contained in a security standard may still be used to improve security. Organizations that have an interest in security but not in a certificate, face the challenge of utilizing this knowledge and selecting appropriate safeguards from the given standard. To solve this problem, a new robust optimization model to determine an optimal safeguard configuration is proposed. By incorporating multiple threat scenarios, obtained solutions are robust against uncertain security threats.

1 Introduction

Existing models addressing the problem of establishing an effective IT security strategy require organizations to provide a lot of exact input data like threat probabilities or asset valuations. However, these values are difficult to obtain in practice and hard to verify but nonetheless critical to the quality of solutions. To address this issue, Schilling and Werners [5] proposed an approach that solves the safeguard selection problem on a higher level of abstraction than previous models. For this purpose, a comprehensive IT security knowledge base, the IT baseline protection catalogues published by the German Federal Office for Information Technology [1], is utilized. In this paper, we present a robust extension of this model. By means of simulation, we show in a realistic example that a robust solution stochastically dominates a deterministic one.

A. Schilling (✉)
Chair of Operations Research and Accounting, Faculty of Management and Economics,
Ruhr University Bochum, Bochum, Germany
e-mail: andreas.schilling@rub.de

© Springer International Publishing Switzerland 2017   333
K.F. Dœrner et al. (eds.), *Operations Research Proceedings 2015*,
Operations Research Proceedings, DOI 10.1007/978-3-319-42902-1_45

2 Model Formulation

The IT baseline protection catalogues divide an IT system into several components $p \in \mathscr{P}$. Each component encapsulates knowledge of a specific part of the overall system. This includes a list of threats $i \in \mathscr{I}$ that are endangering the component's security. To counteract these threats, a wide range of safeguard alternatives $k \in \mathscr{K}$ is available. By deploying a safeguard, the criticality of one or more threats is reduced. It is important to note that safeguards are different in their effectiveness and they affect each other when they relate to the same threat. To express how effective a safeguard is, the effectiveness coefficient $\sigma_k \in [0, 1]$ is introduced. If more than one safeguard applies to the same threat, the effectiveness coefficients are multiplied. As a result, the effectiveness of safeguards is increasing with a decreasing slope if more safeguards are selected [4]. The product of these coefficients is then multiplied by the criticality coefficient $\gamma \in \mathbb{R}_{>0}$ of a threat to reduce it. To address the inherent uncertainty of threats, we introduce a number of scenarios for the threats' criticality. By selecting safeguards, the criticality of threat i in scenario $\omega \in \Omega$ is reduced:

$$t_{\omega,i} = \gamma_{\omega,i} \cdot \prod_{k \in \mathscr{K}} \sigma_k^{s_k \cdot T_{k,i}} \quad \forall \omega \in \Omega, i \in \mathscr{I}. \tag{1}$$

Equation (1) yields variable criticality index $t_{\omega,i}$ of threat i in scenario ω for a given safeguard configuration s_k. Binary matrix $T = (T_{k,i})$ is given as an input to the model and defines the relation between safeguards and threats. The exponent of σ_k states that a safeguard only causes a reduction of t_i if and only if it is implemented ($s_k = 1$) and relates to the particular threat ($T_{k,i} = 1$). If $s_k = 0$ or $T_{k,i} = 0$, the corresponding coefficient of the product defaults to 1 and the threat is unaffected by the safeguard.

The definition of $t_{\omega,i}$ leads to a nonlinear problem which has some drawbacks compared to linear models [5]. Due to the structure of the problem, it can be linearized by taking the natural logarithm of $t_{\omega,i}$. This transformation is feasible because we only need to distinguish between threats' relative criticality indexes $t_{\omega,i}$ which remains possible after transformation. When the problem is solved, all logarithmic values can be reconverted.

To define component security, variable $c_{\omega,p}$ is introduced. It is defined as the maximum of criticality indexes of threats associated with the component. Whether threat i is associated with component p is defined by matrix $C = (C_{i,p})$. The definition of component criticality is based on the idea that the security of a component depends on the most critical of its threats, i.e., the weakest link in the security chain:

$$c_{\omega,p} = \max_{i \in \mathscr{I}} \left\{ t_{\omega,i} | C_{i,p} = 1 \right\} \quad \forall \omega \in \Omega, p \in \mathscr{P}. \tag{2}$$

Figure 1 illustrates how component criticality is defined and how security is maximized by reducing the criticality of components. The objective is to select a feasible subset from the pool of safeguards that maximizes security. This is achieved by select-

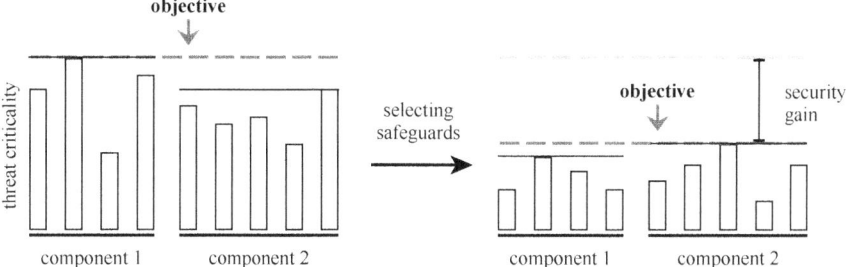

Fig. 1 Safeguard configuration minimizes maximum threat criticality over all components

ing safeguards such that their combined impact on the reduction of threat criticality is maximized.

With these definitions, we formulate a MILP to obtain a robust safeguard config-uration. To ensure robustness against uncertain threat realizations, we minimize the maximal regret in each scenario [6]. By considering multiple scenarios instead of one, the selected safeguard configuration performs better in an uncertain environment. To obtain the robust configuration, we first determine scenario optimal solutions $\ln(Z_\omega^*)$ by solving the problem for all $\omega \in \Omega$:

$$\min \quad \ln(Z_\omega) = \max_{p \in \mathscr{P}} \ln(c_{\omega,p}) \tag{3}$$

$$\text{s.t.} \quad \ln(c_{\omega,p}) = \max_{i \in \mathscr{I}} \left\{ \ln(t_{\omega,i}) | C_{i,p} = 1 \right\} \qquad \forall p \in \mathscr{P} \tag{4}$$

$$\ln(t_{\omega,i}) = \ln(\gamma_{\omega,i}) + \sum_{k \in \mathscr{K}} s_k \cdot T_{k,i} \cdot \ln(\sigma_k) \qquad \forall i \in \mathscr{I} \tag{5}$$

$$\sum_{k \in \mathscr{K}} s_k \leq \overline{N} \tag{6}$$

$$s_k \in \{0, 1\} \qquad \forall k \in \mathscr{K}. \tag{7}$$

The objective is to minimize the logarithmic system criticality $\ln(Z_\omega)$ which is de-fined as the highest logarithmic component criticality index (3). Note that constraint (5) is obtained by taking the natural logarithm of Eq. (1). Constraint (4) defines $\ln(c_{\omega,p})$ as the maximum of associated $\ln(t_{\omega,i})$ values. The total number of safe-guards is limited to $\overline{N}$ in constraint (6) and s_k is set to be binary (7). In the next steps, a robust solution is obtained by considering all scenarios in a single model. First, we minimize the maximal regret:

$$\min \quad \xi = \max_{\omega \in \Omega} \ln(Z_\omega) - \ln(Z_\omega^*) \tag{8}$$

$$\text{s.t.} \quad \ln(Z_\omega) = \max_{p \in \mathscr{P}} \ln(c_{\omega,p}) \qquad \forall \omega \in \Omega \tag{9}$$

$$\text{and } (4)-(7).$$

In constraint (9), the scenario dependent objective is defined. Using the scenario optimal solution $\ln(Z_\omega^*)$, the regret $\ln(Z_\omega) - \ln(Z_\omega^*)$ is calculated and the maximum of these values is minimized (8). The result is the optimal regret ξ^* which serves as an upper bound for the distance of the robust solution to all scenario optimal ones. In the following step, this value is used to guarantee that the robust solution performs well in all considered scenarios. Note that we are using a slightly altered formulation for calculating the regret in (8) and (11) since $\ln(Z_\omega)$ and $\ln(Z_\omega^*)$ are logarithmic values. The used formulation is feasible since it leaves the solution space intact.

Finally, in a third step, we maximize security by minimizing maximum component criticality indexes over all scenarios. In order to adhere to the maximal regret ξ^* in each scenario, constraint (11) is introduced. The relevant output is the robust safeguard configuration $s_k^{rob^*}$:

$$\min \quad \ln(Z^{rob}) = \max_{\omega \in \Omega} \ln(Z_\omega) \tag{10}$$

$$\text{s.t.} \quad \ln(Z_\omega) - \ln(Z_\omega^*) \leq \xi^* \qquad \forall \omega \in \Omega \tag{11}$$

$$(4), (5) \qquad \forall \omega \in \Omega$$

$$\text{and } (6), (7), (9).$$

3 Computational Results

To illustrate the application of the model, an exemplary IT system with 12 components is analyzed. The problem comprises 151 threats and 254 safeguard alternatives. To estimate threat criticality γ and safeguard effectiveness σ, we use the method proposed by [5]: they determine a deterministic single point estimate for γ by considering the number and qualification of safeguards associated with each threat. In the following this value is called γ-base. The safeguard effectiveness σ is estimated based on the qualification level defined in [1].

Schilling and Werners [5] demonstrate how security is increased by applying a deterministic version of the presented model. There an optimal safeguard configuration is obtained by solving the model with a single criticality estimate γ-base per threat. They show that the number of safeguards can be reduced significantly compared to official implementation recommendations published by the BSI while achieving a greater or equal overall security level. The question is how such a solution performs in an uncertain environment and how it compares to a robust one. To answer these questions, we generate a set of scenarios to solve the robust model and a second set to evaluate and compare the obtained solutions.

The goal when generating scenario values $\gamma_{\omega,i}$ is to represent uncertainty in a sensible way. Since security threats are inherently uncertain, we approximate variations of threat criticality coefficients by a normal distribution with expected value γ-base. The standard deviation is set to a value of 5 which results in reasonable variability in relation to γ-base. We initially generate a pool of 250 scenario values for

γ which give a good indication on how the criticality of threats may vary. For computational reasons, the number of scenarios has to be reduced significantly before optimization. To utilize the obtained data most effectively, we employ the scenario reduction approach proposed by [3]. Here scenarios are selected that represent the initial 250 ones as good as possible. This is achieved by iteratively selecting one scenario at a time that represents all remaining scenarios as closely as possible and that is not already represented by a previously selected scenario. In each iteration, the scenario with the minimal cumulated Euclidean distance to all other scenarios is selected and then removed from the initial scenario pool [3]. After 10 iterations, we obtain 10 scenarios which are used as input to the optimization. The scenario generation process is visualized in Fig. 2. The model is implemented and solved with the standard optimization software Xpress Optimization Suite [2].

The goal of this analysis is to determine how the robust solution performs compared to the deterministic γ-base solution in an uncertain environment. For this purpose, we simulate how the system criticality (objective value) of our model varies in different threat scenarios. We generate 1000 additional scenarios using the same procedure as before. In each iteration of the simulation, we compute the objective value with one threat criticality scenario and a fixed safeguard configuration. The two configurations that are of interest are the deterministically determined configuration $s_k^{\gamma-\text{base}^*}$ and the robust configuration $s_k^{rob^*}$. In Fig. 3, the cumulative distribution function of both solutions is plotted. The figure clearly visualizes that the robust solution outperforms the deterministic solution. As a matter of fact, the robust solution is first-order stochastically dominant over the deterministic one. In other words, the objective value of the model is very likely to be smaller throughout various scenarios when applying the robust safeguard configuration $s_k^{rob^*}$ compared to the deterministic configuration $s_k^{\gamma-\text{base}^*}$. This is a much desired result since a lower objective value implies higher security.

The reason for this dominance lies within the structure of the problem and is a combination of two factors: first, we have a lot of safeguards to choose from but only choose a relatively small subset. In addition, safeguards affect different subsets of threats and, by combining them, their impact on different threats changes. As a result, if we incorporate multiple scenarios into the model, the resulting safeguard configuration performs better in each and every one. And since these scenarios are

Fig. 2 Visualization of threat criticality scenario generation with subsequent scenario reduction. Initially, 250 scenarios are generated on the basis of γ-base and then reduced to 10 scenarios that are input to the model

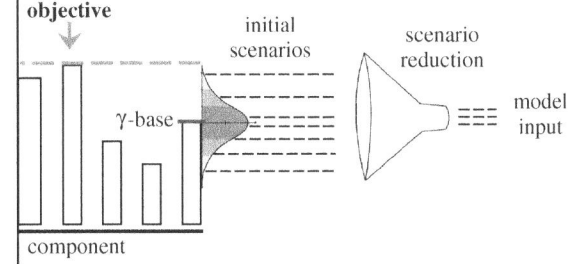

Fig. 3 Plotted cumulative
distribution functions
(CDFs) of objective value of
robust and deterministic
solutions (1000 iterations).
The CDFs indicate the
probability that the objective
value is less than or equal to
a value on the x-axis.
Therefore, robust safeguard
configuration $s_k^{rob^*}$ (*upper
curve*) is more likely to yield
smaller objective values than
deterministic configuration
$s_k^{\gamma-base^*}$ (*lower curve*)

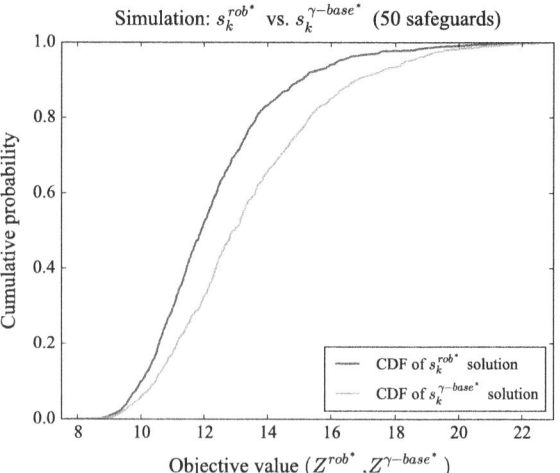

selected to be good representatives of reality, we obtain this strong property of first-order stochastic dominance. This is a significant observation because it implies that even if threat criticality is not fully known, the obtained robust solution will be very effective in protecting an IT system in an uncertain environment.

4 Conclusion

In this paper, we presented a robust extension of the model developed by [5]. The model solves the safeguard selection problem to effectively protect IT systems. By means of simulation, we demonstrated that a robust safeguard configuration performs significantly better under the assumption of unknown threat criticality. Future research may extend the scope of the model further, including a multi-period formulation to support mid- and long-term decision making.

Acknowledgements This work was partially supported by the Horst Görtz Foundation.

References

1. Federal Office for Information Security. IT-Grundschutz-Catalogues: 13th Version (2013)
2. FICO. FICO Xpress Optimization Suite (2015). http://www.fico.com/en/products/fico-xpress-optimization-suite
3. Heitsch, H., Römisch, W.: Scenario tree modeling for multistage stochastic programs. Math. Program. **118**(2), 371–406 (2009)
4. Schilling, A., Werners, B.: Optimizing information security investments with limited budget. Operations Research Proceedings. Springer, New York (2014)

5. Schilling, A., Werners, B.: Optimal selection of IT security safeguards from an existing knowledge base. Eur. J. Oper. Res. **248**(1), 318–327 (2016)
6. Werners, B., Wülfing, T.: Robust optimization of internal transports at a parcel sorting center operated by Deutsche Post World Net. Eur. J. Oper. Res. **201**(2), 419–426 (2010)

Cologne Mass Casualty Incident Exercise 2015—Using Linked Databases to Improve Risk and Crisis Management in Critical Infrastructure Protection

Florian Brauner, Andreas Lotter, Ompe Aime Mudimu
and Alex Lechleuthner

Abstract Critical Infrastructure Protection (CIP) is a challenging operation for all involved organisations: Authorities, critical infrastructure providers and even policy makers. Integrated risk management is required to keep risks as low as possible and well-developed crisis management helps to mitigate the effects of events that have occurred. Achieving the right balance is difficult especially for anthropogenic threats such as terrorist threats, which are difficult to assess with normative risk management approaches. In May 2015, the TH Köln executed two exercises to address risk and crisis management in case of terrorist threats. The exercises were embedded in the research project RiKoV, which was being funded by the German Federal Ministry of Research and Education.

1 Introduction

Critical Infrastructure Protection (CIP) is a major challenge especially when normative risk analysis methodologies fail through (a) lack of data and/or (b) low probabilities of events. Terrorism, as an anthropogenic threat and systemic risk, is an example for an event that has a low probability, but high consequences. New approaches of risk management are needed to analyse such threats, to find the most appropriate security measure to reduce the consequences or prevent the event [1].

F. Brauner (✉) · A. Lotter
Department of Public Safety and Emergency Management, Bergische Universität Wuppertal,
Gaußstr. 20, 42119 Wuppertal, Germany
e-mail: brauner@uni-wuppertal.de

A. Lotter
e-mail: lotter@uni-wuppertal.de

O.A. Mudimu · A. Lechleuthner
TH Köln, Institute for Rescue Engineering and Civil Protection, Betzdorfer Straße 2,
50679 Cologne, Germany
e-mail: ompe_aime.mudimu@th-koeln.de

A. Lechleuthner
e-mail: alex.lechleuthner@th-koeln.de

© Springer International Publishing Switzerland 2017
K.F. Dœrner et al. (eds.), *Operations Research Proceedings 2015*,
Operations Research Proceedings, DOI 10.1007/978-3-319-42902-1_46

The project RiKoV developed such a risk management approach for terrorist threats in public transportation systems. However, even during the next step in our research, a new problem arose: How can such an approach be evaluated? The TH Köln decided to execute two exercises.

In a first exercise, a new approach to determine vulnerability was validated by verifying the determined values with reallife values in a scenario. In a second exercise, the crisis management personnel were trained to improve the efficiency of the involved forces. To collect the necessary data, the Institute of Rescue Engineering and Civil Protection at the TH Köln used a methodical framework they had developed, which consists of technical support systems. For example, it uses a Mass Casualty Incident Benchmark that rates the patient care according to the individual satisfaction of basic needs in the incident with a mobile tele-dialog system. A local positioning system (RTLS) additionally collects the locations and times of forces or victims and gains special events. So, it is possible to evaluate which limited resources came into action and when and where.

All the data are combined in a complex database to understand the processes of prevention and mitigation of terrorist attacks in a critical infrastructure (CI).

2   Methodology

To evaluate methodology, data are needed that can be processed to obtain a result. Nevertheless, the structure and type of the methodology itself, has to be considered in the evaluation process as well. The TH Köln methodology is a scenario-based approach that consists of a generic process model combined with an expert interview guideline [2], which is considered in the evaluation process as well. To evaluate the methodology, a three step evaluation process was executed, including a table-top exercise, a multi-agent-simulation and real-life exercise. This article focuses on the real-life exercise.

In addition, the exercise identifies possible links between risk and crisis management. What information flows from risk management into the on-scene handling of a crisis and also what information flows back later on into risk management cycles [3]?

3   Experiment

On the 25th and 26th of May in 2015, two real-life exercises were executed in a subway station of the city of Cologne. The first day addressed risk management while the second day addressed especially the crisis management of the cities authorities. Especially linking the scenarios for both days to capture information flows from risk into crisis management processes was important.

We chose a terrorist bomb scenario with one attacker (no suicide) to set up two exercises.

1st Exercise "Risk Management"

The environment was a Cologne subway station including four different entrances, two train platforms and two additional intermediate levels each having stairs and escalators to both platforms. Besides already existing security measures such as video surveillance and service personnel, additional security measures were implemented on the intermediate level such as different kinds of metal detectors, explosive detectors and face recognition (smart) cameras. To give the environment a realistic scenario, 100 passengers were played by TH Köln students plus one fictitious attacker. Like a normal station, some of the passengers were carrying luggage including liquids and electronic devices. The attacker also had luggage and was hiding an explosive device in his suitcase.

In ten experiments, the security measures had to find the attacker or the explosives by using different security settings. The overall goal was to find an appropriate setting that allowed detecting the attacker/explosives as well as dealing the passenger flows over time without having increased waiting times for passengers. Therefore, a local positioning system tracked every movement of passengers, attacker or service personnel in the station and stored them in a 3D visualization model in an integrated database.

2nd Exercise "Crisis Management"

In an additional experiment, we went one step further and assumed the scenario could not be stopped by the security measures. Now, the 100 passengers were casualties according to an explosive analysis with 15 very badly injured, 15 moderately injured and 70 lightly injured casualties. The student actors were dressed up and wore make-up according their specific injury role. The train driver started the exercise by sending an alarm by to the local authorities (911 hotline). Over 140 members of the local fire brigade, emergency medical services and police joined the exercise to handle the situation. While our observers and cameras documented and recorded the information flows, the individual care of patients was recorded by a mobile television dialogue system based on APP/Tablet version connected via WLAN.

4 Technique

We used different methods for evaluating the ongoing processes and the influence of the newly implemented security measures. The first method was installing observers at the defined key positions, such as the entrances to the station, the metal detectors and so on. These observers monitored the time it takes to pass a specific security measure or the time the passengers had to stay in the line before entering the station. They also noticed if passengers with dangerous goods were detected. Using observers to evaluate exercises is a well-known method that has been implemented for long-time.

Furthermore we used technical support to obtain results from the exercise, such as video surveillance to record the paths of the passengers. Therefore we installed four cameras in the station: Two in the incoming intermediate level and two on the platform level. In combination with the results of the observers, the recorded videos gave us the possibility to find reasons for increased waiting time.

In addition, we used a Real-Time Locating System (RTLS), which was developed by ubisense. Every passenger and also the service personnel were tagged with a small transmitter. This system gives us the possibility to track the paths of the passengers and to track the duration of stay in different zones. These zones were defined in the planning process and divided the station into the following parts:

- Entrance and waiting line → incoming intermediate level
- Zone of the security measures (where the passengers were checked) → incoming intermediate level
- Zone of luggage return (if the passengers had luggage) → incoming intermediate level
- Platform → platform level
- Exit → outgoing intermediate level

Afterwards, the RTLS even gave us the possibility to see when the service personnel and the attacker met. Together with the observers and the video surveillance we could determine, why the attacker was recognized or why he was not.

For the second exercise we used observers again. Every observer focussed on a special part of the exercise and was given only a few central questions to focus on. In addition to the observers, we also used the Mass Casualty Incident (MCI)-Benchmark, which had been previously developed by the TH Köln [4].

In the planning process, every passengers injuries had been described and then afterwards the recommended methods of treatment were figured out. During the exercise, every passenger received a tablet or smartphone and then logged into a wireless network and to start a web-based application, where all the recommended methods of treatment were on buttons. If the passenger received one of the defined methods of treatment, the passenger was to press the appropriate button and the application defined a time-stamp. All the gained time-stamps were put in the benchmark system to calculate point value. The reachable point value is based on the complexity of the recommended methods of treatment.

5   Integrated Database

The results were captured in an integrated database for evaluation purposes. The database pools the data of the observers, camera surveillance, RTLS, security measures, as well as scenario details such as passenger/terrorist description and luggage/weapon description. The architecture was developed according the preference to analyse the passenger flows for the points of interests (Fig. 1).

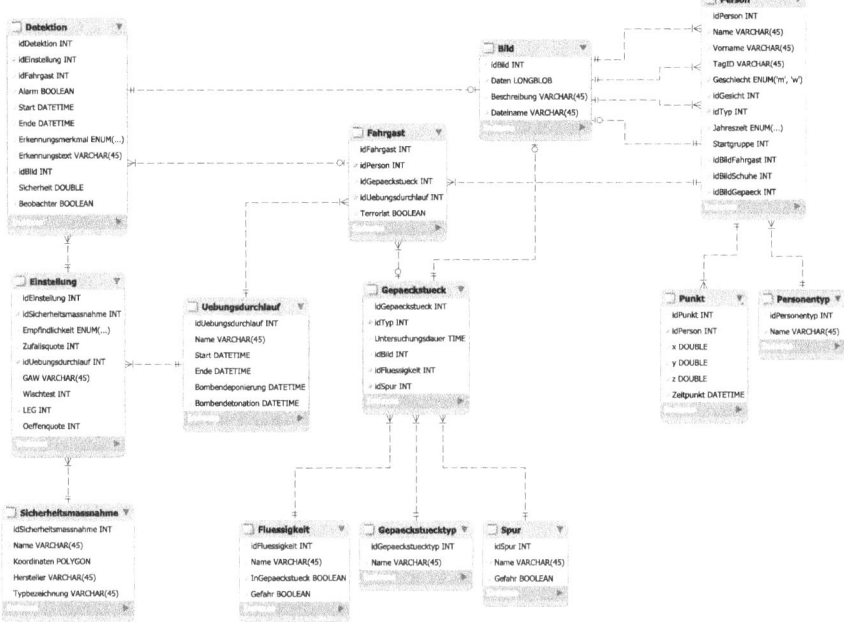

Fig. 1 Database architecture (*source* TH Köln)

6 Results

The results can be displayed in a 3D model visualization of the environment. By using the integrated database, the different data sources can be compared by using a common network time protocol (ntp). This enhances the possibility to look at a same situation with a different point of view.

3D modelling supports the evaluation of the pathways and can identify specific regions of crowding or occasions. Especially, the effects of security measures on the passenger flows can be easily compared according the security level (setting

(a)   **(b)**   **(c)**

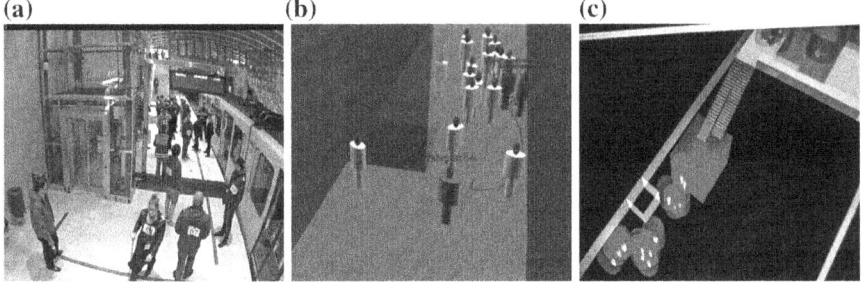

Fig. 2 Exemplary snapshot of real-life exercise and person tracking (*source* TH Köln). **a** Real-life exercise. **b** RTLS. **c** Event-interaction

of measures). The data of the experiments provided first realistic results on the effectiveness of specific data, such as service personnel in context of preventive terrorist or asset detection (Fig. 2).

7 Conclusion

Real-life exercises provide a good possibility to capture data for the evaluation of methodologies in case of unavailable other data or lack of methodologies. The effort needed to prepare and perform such an exercise is high. However, the results can also be used to answer further research questions due to the integrated database. The collection of data needs special techniques to capture the quantitative data as well as qualitative data by observers. The technical framework using a RTLS, video surveillance, MCI-Benchmark and observers worked to obtain the data in both exercises.

In the future, we are going to focus on the social patterns of customers to improve the settings of multi-agent simulations (MAS) in this topic [5]. The optimization of MAS will help to elaborate data, e.g., data farming [6] and to analyse the effectiveness of security measures in a more diverse variety of scenarios [7]. Especially the interconnectedness and the enhancing effects of different security measures will be a major part of this research towards integrated security concepts that cover variety of different scenarios. Also, MAS will allow the investigation of future measures, which have not been implemented in such dynamical systems yet, in order to simulate a proof-of-concept and find new beneficial application possibilities.

Acknowledgements The presented work in progress is a result from our research within the project 'RiKoV'. The research presented in this paper is supported by the German Federal Ministry of Education and Research (BMBF) and we would like to thank the BMBF for their support and funding.

References

1. Brauner, F., Baumgarten, C., Schmitz, W., Neubecker, K.A., Mudimu, O.A., Lechleuthner, A.: RiKoV risk analysis of terrorist threats to rail-bound public transportation: development of an integrated planning solution for efficient economic and organisational measures. In: 10th World Congress on Railway Research 2013, Sydney/Australia; Paper ID 112 Conference Proceedings (2013)
2. Brauner, F., Baumgarten, C., Kornmayer, T., Bentler, C., Mudimu, O.A., Lechleuthner, A.: A methodology for a vulnerability analysis of public transportation systems in context of terrorist attacks. In: Thoma, K., Hring, I., Leismann, T. (eds.) 9th Future Security, Security Research Conference 2014, Berlin/Germany; Fraunhofer Verlag, Stuttgart, pp. 271–277 (2014)
3. Bentler, C., Baumgarten, C., Brauner, F., Kornmayer, T., Mudimu, O.A., Lechleuthner, A.: An integrated risk and crisis management approach for terrorist attacks in public transport networks. Short abstract published in: Ammann, W.J. (ed.) Global Risk Forum GRF Davos, 5th International Disaster and Risk Conference - IDRC 2014, Davos/Switzerland, pp. 89–92 (2014)

4. Stiehl, M., Brauner, F., Lechleuthner, A.: Evaluation of mass casualty incident exercises (MCI) - MCI-benchmark, a scientific evaluation method for comparison of performances in mass casualty incident exercises. Notarzt (2014). doi:10.1055/s-0034-1370094
5. Fonseca, S.P., Griss, M.L., Letsinger, R.: Agent behavior architectures - A MAS framework comparison. In: Proceedings of AAMAS (Bologna, Italy). ACM Press, New York (2002)
6. Horne, G., Meyer, T.: Data farming: discovering surprise. In: Ingalls, R., Rossetti, M.D., Smith, J.S.. Peters, B.A. (eds.) Proceedings of the 2004 Winter Simulation Conference. Piscataway, New Jersey: Institute of Electrical and Electronics Engineers, Inc., pp. 171–180 (2004)
7. Brauner, F., Maertens, J., Bracker, H., Mudimu, O.A., Lechleuthner, A.: Determination of the effectiveness of security measures for low probability but high consequence events: a comparison of multi-agent-simulation and process modelling by experts. In: Palen, L., Buscher, M., Comes, T., Hughes, A. (eds.) The 12th International Conference on Information Systems for Crisis Response and Management (2015)

Simulation-Based Analyses for Critical Infrastructure Protection: Identifying Risks by Using Data Farming

Silja Meyer-Nieberg, Martin Zsifkovits, Dominik Hauschild and Stefan Luther

Abstract Critical infrastructure protection represents one of the main challenges for decision makers today. This paper focuses on rail-based public transport and on the interaction of the station layout with passenger flows. Recurring patterns and accumulation points with high passenger densities are of great importance for an analysis since they represent e.g. critical areas for surveillance and tracking and further security implementations. An agent-based model is developed for crowd behavior in railway stations. For the analysis, we apply the methodology of data farming, an iterative, data-driven analysis process similar to the design of simulation experiments. It uses experimental designs to scan the parameter space of the model and analyses the data of the simulation runs with methods stemming from statistics and data mining. With its help, critical parameter constellations can be identified and investigated in detail.

1 Introduction

Critical infrastructures as for example energy, transportation, telecommunication, water, or health services are essential for the functioning of a modern state [3]. The protection of critical infrastructures is therefore of high importance. The different sectors are also highly interconnected which may lead to cascading effects if disturbances as e.g. blackouts arise. This paper reports first results from an ongoing study in the context of a joint research project [4, 13]. The so-called RiKoV project

S. Meyer-Nieberg (✉) · M. Zsifkovits · D. Hauschild · S. Luther
Fakultät für Informatik, Universität der Bundeswehr München,
Werner-Heisenberg-Weg 39, 85577 Neubiberg, Germany
e-mail: silja.meyer-nieberg@unibw.de

M. Zsifkovits
e-mail: martin.zsifkovits@unibw.de

D. Hauschild
e-mail: dominik.hauschild@unibw.de

S. Luther
e-mail: stefan.luther@unibw.de

© Springer International Publishing Switzerland 2017
K.F. Dœrner et al. (eds.), *Operations Research Proceedings 2015*,
Operations Research Proceedings, DOI 10.1007/978-3-319-42902-1_47

focuses on transportation systems, specifically on rail-bound public transportation. From an abstract viewpoint, two main components from the railway network have to be taken into account: the network connections themselves and the nodes—in other words—the stations. To safeguard the public against e.g. terroristic attacks, the implementation of various security measures is the focus of recent research [14] and also part of RiKoV [4, 13]. Here, the emphasis lies on the protection of train stations as one of the most vulnerable parts of the network. The effectiveness of security measures depends, however, strongly on the location where they are placed. While the problem of optimal sensor placement is often considered in literature, see e.g. [16], we argue that additional factors have to be taken into account. It is standard to consider the station layout to provide an optimal sensor coverage, see e.g. [15]. However, the influence of crowds on the effectiveness of the security measures or surveillance sensors themselves is only seldom investigated. We could only identify some approaches in the area of crowd estimation [8]. In addition, implementing security measures may change the passenger flows and the usage of the station. Installing security technologies as e.g. checkpoints may lead to bottlenecks and may impair the primary function of the station. Analyzing the traffic patterns inside a particular station and investigating the interactions between flows, structure, and security installations appears therefore of great importance.

This paper explores the use of agent-based simulation to gain more insights into the everyday interactions of the station layout with passenger flows. The goal is to identify recurring patterns and accumulation points with high passenger densities. They are of great importance for an analysis since they represent among others critical areas for surveillance and tracking and further security implementations. The paper reports first results of the ongoing analysis focussing on the methodology. We suggest to use data farming, a relatively new approach stemming from military operations research. It represents an interactive, iterative method which appears to be well suited to the task at hand.

The paper is structured as follows. Section 2 introduces agent-based modelling and the pedestrian flow model developed for the analysis. Afterwards, the analysis methodology, data farming, is introduced, before it is applied to the model in Sect. 3. Section 4 summarizes the main findings of the paper and presents an outlook regarding further research topics.

2   An Agent-Based Model for Pedestrian Flows

Agent-based models (ABMs) are concerned with the behavior of autonomous agents, depicting the actions and the interactions between them and the environment [9]. They are used to analyze the collective, emerging behavior of the modelled system. Agent-based models follow a ground-up or bottom-up approach by focussing on the individual and the interactions of the individual. Several type of agents exist with agents having either simple or complex behavioral routines [1]. Since this paper is concerned with individual passenger behaviour, an agent-based model appears as an appropriate choice. It should be noted that the analysis of agent-based models is not easy due to stochastic influences and to the interactions on the individual level. This necessitates an experimental analysis in nearly all cases. The paper focuses on the Munich main station, a terminal station located in the city center. At present, this station represents the second largest train station in Germany. Approximately 350,000 passengers are served by 240 long distance trains and over 500 short distance trains per day, see [10]. Furthermore, concerning the city itself, the central station is an essential traffic hub since it provides connections to suburban trains and underground trains in the station itself as well as to many bus and tram lines on the outside. The terminal station with over 30 tracks consists of three distinct parts: the central platforms dedicated mainly to the long distance travel and two separate stations for regional lines. All platforms are accessible from the main hall where shops and restaurants are located. Currently, the model considers three agent types: tourists, commuters, and business travelers. In each case, arriving (via train) or departing pedestrians (other station entrances) are distinguished. The model can be easily augmented to include further types. The differentiation was introduced to analyze how groups with different basic behavioral patterns influence the passenger flows. At the moment, these pattern representatives serve as demonstrative examples. The value ranges of their control parameters are based on the findings of first data analyses [11]. Future work will consider behavioral types constructed based on an extensive data collection. The model is developed for the purpose to analyse the flows that are caused by the individual paths through the station. The main behavioral routines of the agents reflect this: While the agent group determines the probabilities for choosing particular entry or exit points and influences their movement through the station, each agent makes a stochastic decision causing sufficient diversity of the agent behavior.

The model distinguishes between regional and long distance trains. The assignment is performed randomly based on the agent class. Trains are an important factor of the model. They provide entry and exit points for agents. Their arrival and departure is assumed to be stochastic with the time between the departure of a train and the arrival of a new train following a truncated normal distribution. The time a train stays in the station is uniformly distributed.

Furthermore, the walking behavior is decisive. Pedestrians chose their global path with the A* path finding algorithm. Locally, the path calculated can be varied in order to avoid collisions. The model was implemented using NetLogo which is released under the GNU General Public License. While it has originally be aimed at students, it has emerged as a serious research tool, see e.g. [12].

3 Experimental Analysis: Data Farming

The experiments follow the general *data farming* methodology [2]. Data farming aims to provide insights into the potential system responses and into the underlying causes, that is, into the interaction and effects of the variables. The input variables are typically referred to as factors. Data farming is a interdisciplinary approach that is usually performed iteratively, since information gained from the experiments may lead towards new research questions. Many agent-based models include stochastic effects. In order to arrive at statistically sound conclusions with respect to e.g. the effect of a certain factor, the model has to be run repeatedly. If additionally, the number of agents in the simulation is large and/or the simulation requires a large number of iterations, the time for a single run may become quite long. Thus, computer clusters are often used to perform the analyses. The more influence factors shall be analyzed, the more experiments have to be performed. Furthermore, the parameter space of the influence factors has to be covered in detail. Finally, potential interactions between the factors need to be accounted for. Therefore, modern statistics has devoted a huge amount of attention on finding good experimental designs. In order to analyze the experiments, methods from statistics (descriptive and inferential) and data mining are applied. Commonly used approaches include among others polynomials (typically of first and second order) classification and regression trees, Kriging models, artificial neural networks, support vector regression and many more [6].

Two main scenarios: monday morning and a week-day afternoon are considered in this paper. The scenarios differ in the proportion and rate of tourists, commuters and business travelers entering the station. The morning scenario has among others a higher ration of commuters, whereas tourists represent the dominant group in the afternoon. People choose their specific entrance or exit corresponding based on the findings of a data analysis [11]. Summarizing, the scenarios assume that the primary goal of a commuter is to reach either his train or further connections outside the station. The group of business travelers is assumed to favor long distance trains. Intermediate stops are more probable than for commuters in departure, since we assume that they make purchases with respect to their longer travels. The third group of travelers are tourists, whose behavior strongly differs with respect to whether they just arrived or wish to depart. When they arrive, we assume that they leave the station soon since they want to reach their destination. On departure, their behavior is similar to that of departing business travelers with a preference for more intermediate stops. Since tourists often explore the city itself and since the main station is also one of the main connecting hubs concerning the inner city transportation, the probability of choosing a train as an exit is lower as in the other cases. The available empiric data of Munich central station [11] is used infer waiting times and length of stays at the intermediate stops.

The focus of our current series of experiments lies on the distribution of the pedestrian in the different parts of the stations. We are interested in identifying on the one hand areas with high passenger densities and on the other hand areas with long lengths of stays. Due to time constraints it was not possible to vary all parameters of the model in the data farming process. Therefore, only the parameters of the dominant groups in the respective scenario were taken into account. A nearly orthogonal latin hypercube design was chosen as experimental design [5, 7]. The design resulted in 33 data points. For each data point, 30 simulation runs were conducted, resulting in 990 experiments. Due to space restriction, we only show the main findings as heatmaps in Figs. 1 and 2. This heatmaps are based on the average values of the two

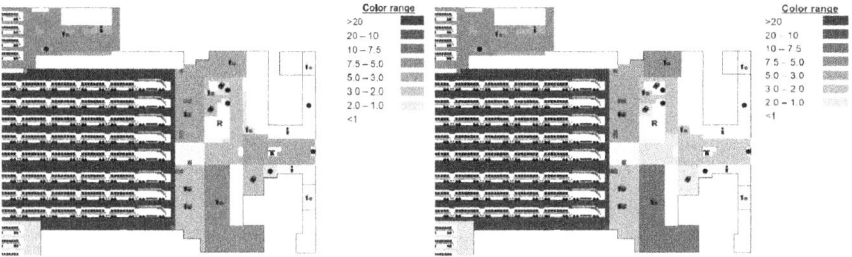

Fig. 1 Heatmap for the performance measures agent distribution

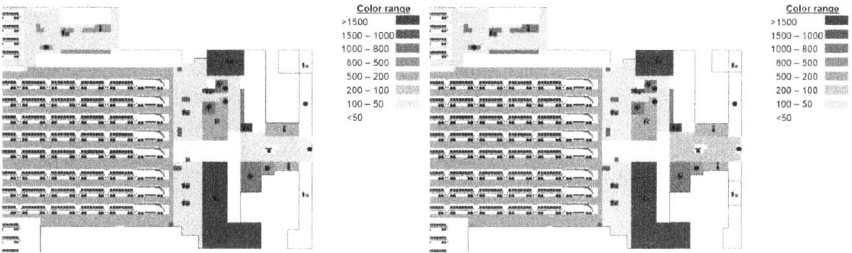

Fig. 2 Heatmap for the performance measure length of stay

performance measures and scaled in representative groups which can be seen in the color range at the top right. As the figures show, the platforms represent the areas with the highest passenger densities whereas people stay longest in the restaurant areas.

4 Conclusions

The protection of critical infrastructures is of considerable importance, where transportation systems represent a vital part. Therefore, research often addresses the implementation, effects, and costs of safeguarding the infrastructure and the population against threats posed for example by terrorist attacks—especially in train stations. The influence and the interactions of security and surveillance installations on the traffic patterns is however seldom considered. Therefore, this paper focuses on the experimental analysis of the interactions of structure, pedestrian traffic patterns, and security measures reporting first results of the ongoing research project. The analysis shows that in the observed station the highest passenger densities can be found on the platforms, wheras people spend most time in the restaurant areas. One has to note that these results might differ according to different stations architectures. The analysis will be extended in several ways in future work. Among other, we will focus on the development of new prototypical behavioral models that will be used as the foundation of a new series of data farming experiments. Furthermore, we are planning to optimize the combination and placement of security installations in different stations and trying to identify resulting changes in the crowds' movement.

Acknowledgements The support from the German Federal Ministry of Education and Research (BMBF) (project RiKoV, Grant No.13N12304) is gratefully acknowledgement.

References

1. Billari, F.C., Fent, T., Prskawetz, A., Scheffran, J. (eds.) Agent-Based Computational Modelling. Physica-Verlag, Heidelberg (2006)
2. Brandstein, A.G., Horne, G.E.: Data farming: a meta-technique for research in the 21st century. Maneuver warfare science 1988, US Marine Corps Combat Development Command Publication (1998)
3. Dehmer, M., Meyer-Nieberg, S., Mihelcic, G., Pickl, S., Zsifkovits, M.: Collaborative risk management for national security and strategic foresight. EURO J. Decision Process. **3**, 1–33 (2015)
4. Dehmer, M., Nistor, M., Schmitz, W., Neubecker, K.: Aspects of quantitative analysis of transportation networks. Future Security (2015)
5. Gu, L., Yang, J.-F.: Construction of nearly orthogonal latin hypercube designs. Metrika **76**(6), 819–830 (2013)
6. Hastie, T., Tibshirani, R., Friedman, J.: The Elements of Statistical Learning. Springer Series in Statistics, 2nd edn. Springer, New York (2009)
7. Hernandez, A.S., Lucas, T.W., Carlyle, M.: Constructing nearly orthogonal latin hypercubes for any nonsaturated run-variable combination. ACM Trans. Model. Comput. Simul. **22**(4), 1–17 (2012)
8. Liebig, T., Xu, Z., May, M.: Incorporating mobility patterns in pedestrian quantity estimation and sensor placement. In: Nin, J., Villatoro, D. (eds.) Citizen in Sensor Networks. Lecture Notes in Computer Science, vol. 7685, pp. 67–80. Springer, Berlin (2013)
9. Macal, C.M., North, M.J.: Tutorial on agent-based modelling and simulation. J. Simul. **4**(3), 151–162 (2010)
10. Meyer-Nieberg, S., Zsifkovits, M., Pickl, S., Brauner, F.: Assessing passenger flows and security measure implementations in public transportation systems. Future Security (2015)
11. Neutert, J.-P.: Empirische Studie über das Personenverhalten am Münchener Hauptbahnhof. Praktikumsbericht. Universität der Bundeswehr München (2015)
12. North, M., Macal, C.: Foundations of and recent advances in artificial life modeling with repast 3 and repast simphony. Artificial Life Models in Software, pp. 37–60. Springer, New York (2009)
13. Pickl, S., Lechleuthner, A., Mudimu, O., Dehmer, M.: Exploring data analysis techniques for threat estimation. Future Security (2015)
14. Räty, T.: Survey on contemporary remote surveillance systems for public safety. IEEE Trans. Syst. Man Cybern. Part C: Appl. Rev. **40**(5), 493–515 (2010)
15. Sforza, A., Starita, S., Sterle, C.: Optimal location of security devices. In: Setola, R., Sforza, A., Vittorini, V., Pragliola, C. (eds.) Railway Infrastructure Security, Topics in Safety, Risk, Reliability and Quality, vol. 27, pp. 171–196. Springer International Publishing, Cham (2015)
16. Valera, M., Velastin, S.A.: Intelligent distributed surveillance systems: a review. IEE Proc. Vis. Image Signal Process. **152**, 192–204. IET (2005)

Time-Based Estimation of Vulnerable Points in the Munich Subway Network

Marian Sorin Nistor, Doina Bein, Wolfgang Bein,
Matthias Dehmer and Stefan Pickl

Abstract In this paper, the frequency of trains in the Munich subway network is analyzed. Using influence diagrams the stations and edges in the network that are most vulnerable to catastrophic attacks are determined. Upon obtaining the number of trains in each station at a certain moment in time, the most vulnerable stations will be automatically identified. This process is discrete in time, and various existing train schedules available to the general public are considered. Considering each schedule, the gain and the cost of destroying a station is calculated. Based on utility values for each station representing the difference between the gain and the cost, an influence diagram decides which stations are most vulnerable to attacks.

M.S. Nistor (✉) · S. Pickl
Universität der Bundeswehr München, Werner-Heisenberg-Weg 39,
85577 Neubiberg, Germany
e-mail: sorin.nistor@unibw.de

S. Pickl
e-mail: stefan.pickl@unibw.de

D. Bein
California State University, Fullerton, CA 92831, USA
e-mail: dbein@fullerton.edu

W. Bein
University of Nevada, Las Vegas, NV 89154, USA
e-mail: wolfgang.bein@unlv.edu

M. Dehmer
The Health and Life Sciences University, Eduard Wallnöfer-Zentrum 1,
6060 Hall in Tirol, Austria
e-mail: matthias.dehmer@umit.at

© Springer International Publishing Switzerland 2017
K.F. Dœrner et al. (eds.), *Operations Research Proceedings 2015*,
Operations Research Proceedings, DOI 10.1007/978-3-319-42902-1_48

1   Introduction

The Munich subway network is the most populous public transportation network in the city of Munich region, carrying over one million passengers [1] each day. There are 100 stations in the network,[1] and eight subway lines totaling 95 km of routes [1]. Each station has a certain level of protection against man-made attacks and natural disasters. Security cameras, security personnel, building infrastructure, are used to proactively keep a station safe. Since there is no absolute protection from destruction we consider each station to have a percentage chance of being destroyed or rendered useless by an attack. These values are calculated using intelligence data which represents known information. Such information has either been collected by analyzing data trends over a substantial period of history or has been obtained through human intelligence gathering. Thus intelligence data impacts the probabilities of events.

For a certain time period each station hosts a number of trains which are passing through and are stopping briefly on their way to their destination (or returning back in case this is the end of the line). The maximum number of trains at a station is bounded by the known total number of (operational) lines in a station. The actual number depends on the time of day, and can be ascertained by calculating the number of trains temporarily stopping at that station at a certain moment of time, day of the week, and type of day in the year, namely whether it is a public holiday or not.

A week-interval during school time, the subway schedule is divided in four different schedules: Monday-Thursday (one schedule), Friday (one schedule), Saturday (one schedule), Sunday and holidays (one schedule) [2]. The exception schedules with reduced traffic or for special events days are not considered in this project. This is due to the smaller impact of an attack, or because the schedule is not fixed and is subject to change by operator.

If the intelligence data is relatively stable (i.e. the values change very slowly as the time passes) the frequency of trains changes abruptly from peak times, to shoulder times, and to quiet times. If a large number of trains are in a station A at time t, then station A could be a vulnerable target. But if the station has a low percentage of being attacked from intelligence data, then another station may be a better candidate for attack. The decision regarding which station is most vulnerable thus depends on numerical values which change over time, and identifying a pattern of time in which a certain station is the most vulnerable is the goal of this paper.

2   Existing Work: Decision Trees and Influence Diagrams

In decision making, a selection is made from a set of alternatives or options so as to minimize or maximize an objective. In the rational model of decision making [3], all aspects of the process such as the set of choices, results and decision criteria are

---

[1]There are 100 stations when four stations are doubly counted: these stations are not physically in the same location, but in close proximity.

known from the outset. A decision tree is a decision analysis tool in which a problem is represented as a directed acyclic graph with nodes and arcs [4]. A node is an element in which either a decision is evaluated or an uncertainty is calculated. The alternative or branch which gives the best overall value is the best path within the tree. A decision tree is useful in graphically displaying the decision alternatives of a problem that has a relatively small number of decision alternatives. But if there are too many variables—decision alternatives and uncertainties—the graph can quickly become large and complex. In such cases, we can represent the same problem with influence diagrams since such diagrams focus on relationship among various elements and less on statistical values.

An influence diagram (ID) is a graphical representation of a decision situation. The graph consists of nodes representing various values and arcs which represent relationships between the nodes [5]. An ID is a directed acyclic graph. It describes a decision problem on three levels: relational, functional and numerical. The relational level is the actual graphical representation of a problem using nodes and arcs that connect one node to another. The arcs define relationships or dependencies among different nodes. The functional level specifies the actual function that describes the dependencies indicated by the graphical structure at the relational level. The numerical level specifies the numerical values related to probability and utility functions. The Maximum Expected Utility Principle (MEU) principle [6] says that a rational agent should choose an action that maximizes its expected utility given the current state of knowledge. Based on the concepts from influence diagrams, we propose an algorithm which solves a specific decision problem. The specific decision problem is to determine nodes and edges in the graph that are most vulnerable to attacks.

3   Data Collection

The data is public, and available via the Munich subway operator website at http://www.mvv-muenchen.de/. As mentioned in Sect. 1, the subway network consists of eight lines (U1–U8) with 100 stations [1, 7]. Four stations are counted twice due to the different physical locations of the subway lines platforms: *Hauptbahnhof*, *Odeonsplatz*, *Olympia-Einkaufszentrum*, and *Sendlinger Tor* [2]. To be able to differentiate between these stations, their names incorporate the according subway lines. The subway lines week-interval schedule is not uniform, but the most common schedule, available for five from eight lines, can be considered. Thus, four distinct schedules can be deemed: Monday-Thursday (one schedule), Friday (one schedule), Saturday (one schedule), Sunday and Holidays (one schedule) [2].

For some of the subway lines, the schedules of working days from Monday to Friday are also divided in two different schedules: normal, and reduced for non-school days. The last schedule will not be considered in this work due to the reduced number of days (84) [2], and available only for five subway lines. The same holds for the few schedule exceptions during Oktoberfest, Carnivals, Christmas, and New Year Eve [2].

4 Algorithm for Time-Based Estimation of Vulnerable Stations

We propose and implement an algorithm which identifies the most vulnerable stations within continuous time intervals in a 24-h time period and a week-interval. The algorithm uses the train frequency data and the available intelligence data to calculate the utility values of each station. We apply the algorithm to simulated data, obtained by manually computing the train frequency data for different timestamps from text data files. Each text file represents the adjacency matrix of the subway network corresponding to a timestamp. The adjacency matrix contains the number of trains passing between each pair of stations of the subway network for that timestamp.

These values are compounded with intelligence data, obtained by estimating the probabilities of completely destroying stations. Our data is simulated and stored in another data text file, and is obtained by assigning a percentage value for each of the 100 stations. A station given a 0 % percentage indicates that there is no likelihood of its destruction, while a value of 100 % means that the station is highly suscepti- ble to attacks. Intelligence data can be collected by visually inspecting stations or following the media (including Web reports), or for that matter any other type of public knowledge. For this paper, the intelligence data indicates the general pattern of security enforced at train stations over a long period of time, generally a month or even longer. For each station, two functions are computed: the gain and the cost.

Definition 1 The gain function for station A represents the gain of completely destroying the station A, measured in the number of trains that cannot go through the station A. The domain of the function is a partition of a time interval, such as a 24-h time interval or a week-interval.

Definition 2 The cost function for station A represents the maximum gain of not destroying that station but completely destroying another station. The domain of the function is a partition of a time interval, such as a 24-h time interval or a week-interval.

The cost function can be computed using the gain functions for all the stations that are available for the same time interval. If some gain functions cannot be computed or need more time to be computed, then these values are ignored, and the cost is computed using only the ones that are available.

The following steps are then performed for a particular timestamp. For each station, the gain function is computed using the frequency train data. Using the available gain functions, the cost function is computed for each station as a result of compounding the gain function with the intelligence data. For each station, a utility value is computed where $utility = gain - cost$. We apply the principle of Maximizing Expected Utility (MEU) to choose one or more stations whose utility values are above a certain threshold.

5   Software Implementation and Simulation Results

We define the Munich subway network as a direct graph G with the pair $G = (V, E)$ where V is a set of subway stations called vertices (nodes); E is a set of connections between stations called edges (links), where each edge is a pair of vertices; W is a set of trains passing between two stations called edge weights, where each edge has a weight. In this work, we consider the train frequency in a week-interval to be analyzed. As mentioned in the previous section, four different schedules are relevant to identify the vulnerable stations. Thus, four different directed graphs can be computed, where the weights of each edges differ from one graph to another. Based on the adjacency matrix of the four computed graphs, the proposed algorithm is applied in order to calculate the utility function of each station for the four proposed schedules to analyze.

Due to the large number of stations, a selection criteria is proposed. In Fig. 1 are represented the top ten vulnerable stations according to the proposed utility function for each schedule of a week-interval. We can observe from Fig. 1 that the most vulnerable day of a week is Friday. Then the utility value reaches its maximum

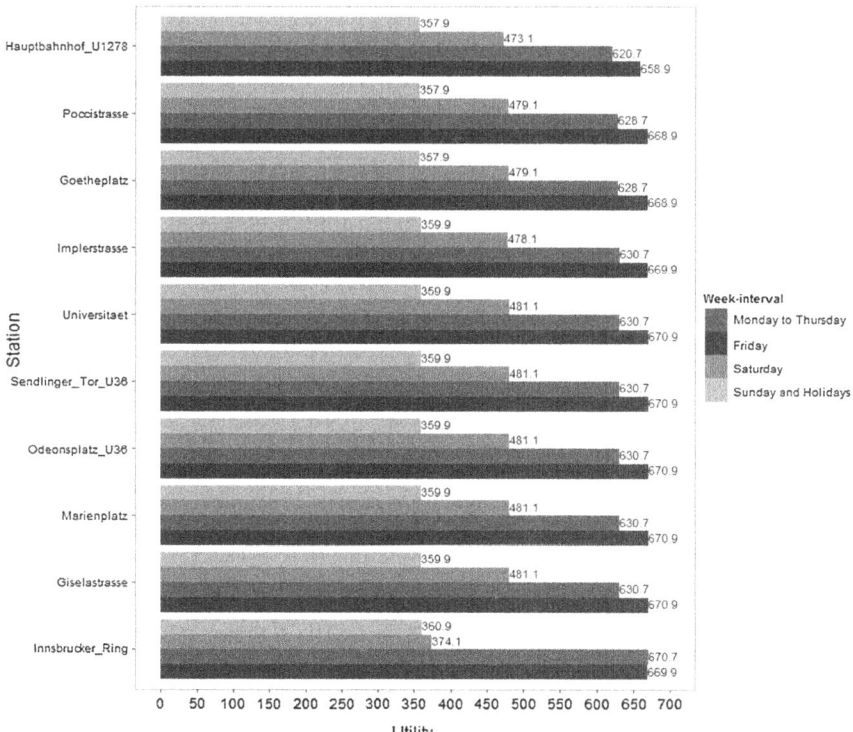

Fig. 1   A graphic representation of the top ten vulnerable stations in a week-interval in an ascending order

value in this network, with one exception, *Insbrucker Ring* station, where the utility value is almost the same from Monday to Friday. The utility value of Friday, compared with the less vulnerable day in a week-interval, it is almost double. Another observation based on these top ten vulnerable stations is regarding the train stations relevance. These are located more closely to the city center, or represent important connections/transfers between two or more subway lines.

The statistical analysis has been performed using the programing language R [8] (Release version 3.1.3).

6 Conclusions and Future Work

We utilize the concept of influence diagram to build an algorithm which solves a specific decision problem. The specific decision problem is to determine the station and edges in the network which are most vulnerable to catastrophic attacks, using readily available data. By utilizing an algorithm, the decision making becomes less prone to human error and hence more efficient in terms of human resources.

The algorithm can be extended to estimate the vulnerable tracks by considering the tracks between stations and calculating the gain, cost and utility functions.

We have considered the data for an entire day, but in practice there are significant differences between the number of trains travelling between stations during peak time, shoulder time, and off time. A more accurate analysis needs to be done by considering a 24-h interval instead of a week interval and identify the most vulnerable tracks at various moments of time.

Acknowledgements Research of author Marian Sorin Nistor, was funded by the People Programme (Marie Curie Actions) of the European Union's Seventh Framework Programme FP7/2007-2013/ under REA Grant Agreement Number 317382.

References

1. Muenchner Verkehrsgesellschaft mbH (MVG) Muenchen, MVG in figures. http://www.mvg. de/dam/en/mvg/ueber/unternehmensprofil/mvg-in-figures-s. Accessed 13 Aug 2015
2. Muenchner Verkehrs- und Tarifverbund GmbH, Alle Informationen zu den Bahnhoefen im MVV (2015). http://www.mvv-muenchen.de/de/netz-bahnhoefe/bahnhofsinformation/index. html. Accessed 13 Aug 2015
3. Eisenfuehr, F., Langer, T., Weber, M.: Rational Decision Making. Springer, New York (2010)
4. Eriksen, S., Keller, L.R.: Decision Trees. Kluwer Academic Publishers, Boston (2001)
5. Shachter, R.D.: Probabilistic inference and influence diagrams. Oper. Res. **36**, 589–604 (1988)
6. Von Neumann, J., Morgenstern, O.: Theory of Games and Economic Behavior. Princeton University Press, Princeton (1953)
7. Muenchner Verkehrs- und Tarifverbund GmbH, Netzplaene (2015). http://www.mvv-muenchen. de/de/netz-bahnhoefe/netzplaene/index.html. Accessed 13 Aug 2015
8. R Development Core Team, R: A Language and Environment for Statistical Computing, Vienna, Austria (2008). http://www.R-project.org. Accessed 13 Aug 2015

Two Approaches to Cooperative Covering Location Problem and Their Application to Ambulance Deployment

Hozumi Morohosi and Takehiro Furuta

Abstract This study proposes two approximation methods to define the coverage probability in ambulance location problems based on the model of cooperative covering proposed by Berman et al. (IEE Trans. 40, 232–245, 2010, [1]) as an extension to classical covering problems. A key ingredient of the model is the estimation of the coverage probability by multiple facilities. We introduce a simple parametric model for the travel time of ambulances and propose two methods to calculate the coverage probability approximately. We report and discuss two solutions obtained from computations using actual data.

1 Introduction

This paper considers a practical implementation of models to deal with ambulance location problems based on a cooperative covering model [1, 2] that has recently been developed in facility location planning to generalize classical covering problems. In classical covering problems, every demand point can take one of two states, covered or not covered by a single facility, e.g. [4]. A cooperative covering problem extends the model so that each demand point can be covered with some probability by not only a single facility but by multiple facilities as well. In order to implement a model to deal with a cooperative covering problem, the key is to estimate the coverage probability of the demand point by multiple facilities.

We are mostly interested in modeling an ambulance location problem in a cooperative covering setting, where the coverage probability is defined as the probability that a demand point is accessed by any ambulance within a prescribed time. As an illustrative example, suppose there are two ambulances A and B for a demand point, and that A and B can cover the point with probability p_A and p_B, respectively. Then

H. Morohosi (✉)
National Graduate Institute for Policy Studies, Tokyo, Japan
e-mail: morohosi@grips.ac.jp

T. Furuta
Nara University of Education, Nara, Japan
e-mail: takef@fw.ipsj.or.jp

© Springer International Publishing Switzerland 2017
K.F. Dœrner et al. (eds.), *Operations Research Proceedings 2015*,
Operations Research Proceedings, DOI 10.1007/978-3-319-42902-1_49

the probability of the point being covered by at least one of the ambulances is calculated as $1 - (1 - p_A)(1 - p_B)$, if we assume that A and B operate independently. We extend this calculation to cope with the actual extensive ambulance service system.

First, we attempt to provide an estimation of the coverage probability with a single ambulance by using the conditional probability distribution of the travel time when the travel distance is given. We propose a parametric model of the travel time distribution and give the estimate of the parameters in the model using the actual dispatch data of ambulances. The proposed model can include the uncertainty of the travel time resulting from congestion or from other kinds of time loss in ambulance service. By virtue of the function, we can estimate the coverage probability within a prescribed time from the distance between the ambulance station and the demand point.

Combining the individual covering probabilities, we propose two methods to approximate the cooperative coverage probability. One is to calculate the covering probability directly in a heuristic way, while the other depends on the use of the uncovered probability. We compare the two approaches in several settings.

In Sect. 2 we introduce our model of the covering probability of ambulances and provide an estimation using actual data. After presenting our models of cooperative covering in Sect. 3, we report some numerical examples in Sect. 4.

2 Covering and Uncovering Probability

To build the model of cooperative covering, we first try to estimate the probability distribution of an ambulance's travel time given a travel distance. The covering probability as well as the uncovered probability is composed of the ambulance's probability of being busy and the probability that an ambulance can reach the scene after the arrival of the ambulance call, which is often referred to as the response time, within a prescribed time. Since the majority of the response time is attributed to travel time, the covering probability can be estimated from the distribution of the travel time. Recent research on travel time estimation can be found in [3, 5], while in this paper we use a simple approach to the problem based on the observations of data.

We assume that the mean traveling time $\bar{t}$ is well approximated by a linear function of distance l: $\bar{t} = rl + a$, and the variance σ^2 is proportional to l: $\sigma^2 = vl$, where, r, a, and v are the parameters to be estimated. Their meanings would be apparent from the expressions. As a simple parametric model, we attempt a normal distribution with these parameters for the conditional cumulative distribution function: $F(t | l) \sim N(rl + a, vl)$. The parameters can be estimated by the maximum likelihood method. The distribution function can be used to calculate the probability that the travel time to the scene at a distance l is shorter than a prescribed time t_c, which is given by $F(t_c | l)$.

Figure 1 displays the relation between the travel distance and the probability. The left panel is for $t_c = 7$ and the right panel is for $t_c = 8$. These values are used in our

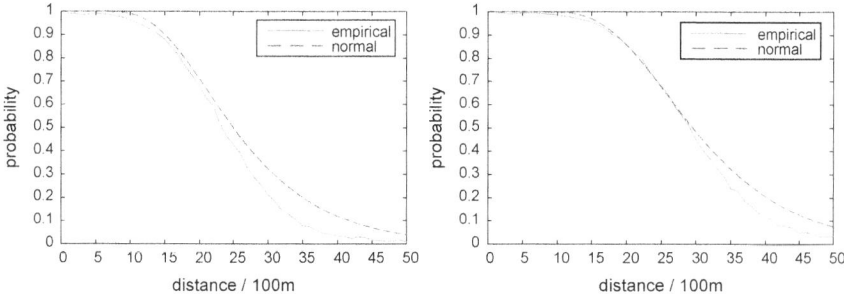

Fig. 1 Travel distance and coverage probability. *Left panel* is for $t_c = 7$. *Right* for $t_c = 8$

numerical experiments. They show good agreement of the model to the data up to 3 km, which is sufficient for our analysis.

Let h be the average probability of an ambulance being busy. Using h and $F(t|l)$, we define the uncovering probability q_{ij} of the demand point i by ambulance j as

$$q_{ij} = h + (1 - h)(1 - F(t_c|l_{ij})), \tag{1}$$

where l_{ij} is the distance between i and j. We introduce a relevant set J_i of ambulances for demand point i, which is considered as the potential sites covering i. One possible definition of J_i could be $J_i = \{j | F(t_c|l_{ij}) > \varepsilon\}$ for some $\varepsilon > 0$, that is ambulances in J_i can respond to the call from i with probability more than ε. Assuming every ambulance operates independently, the probability that demand point i is covered is

$$1 - \prod_{j \in J_i} q_{ij}. \tag{2}$$

Another approach to estimate the covered probability is straightforward. Let J_i be the relevant set for i, and the element of J_i is ordered according to the covering probability $F(t_c|l_{ij})$, namely $J_i = (j_1, \ldots, j_k)$ so that

$$F(t_c|l_{ij_1}) \geq F(t_c|l_{ij_2}) \geq \cdots \geq F(t_c|l_{ij_k}). \tag{3}$$

The probability that ambulance j_1 is dispatched to point i is given by

$$p_{ij_1} = (1 - h)F(t_c|l_{ij_1}). \tag{4}$$

The second ambulance j_2 would be dispatched if the first ambulance j_1 is busy and j_2 is idle, and that probability is

$$p_{ij_2} = h(1 - h)F(t_c|l_{ij_2}). \tag{5}$$

Eventually the kth ambulance would be dispatched with probability

$$p_{ij_k} = h^{k-1}(1-h)F(t_c|l_{ij_k}) \tag{6}$$

Summing these up leads to the total covering probability for i:

$$\sum_{j \in J_i} p_{ij}. \tag{7}$$

Expressions (2) and (7) are utilized to build the cooperative covering models in the next section.

Assuming independence among ambulances could involve some controversy. However, since our focus is on comparing two models, this assumption is made as a conventional approximation.

3   Cooperative Covering Models

Cooperative covering models introduce a novel concept of group coverage into the ambulance location problem. They allow for the probabilistic covering of a demand point by multiple ambulances. In the model, each ambulance has a covering probability. Even if the probability is small, the group of ambulances could cover the point with high probability. With the help of the estimated covering and uncovering probability, we propose two models of cooperative covering. Let I and J be the set of demand points and potential site of ambulances, respectively. We assume K ambulances to be located. The demand, i.e., number of calls, at $i \in I$ is denoted by a_i. If the point i has a total probability larger than the required covering probability level α, it is thought of as "covered." The decision variables are $x_j \in \mathbb{Z}_+$, $j \in J$, which is equal to the number of ambulances located at j, and $y_i \in \{0, 1\}$, $i \in I$, which is equal to 1, if i is covered, and 0, otherwise.

Our first model is based on the uncovered probability (2). The condition on the uncovered probability can be linearized by taking the logarithm.

$$\begin{aligned}
\text{maximize} \quad & \sum_{i \in I} a_i y_i \\
\text{subject to} \quad & \sum_{j \in J_i} x_j \log q_{ij} \le y_i \log(1-\alpha) \quad \forall i \in I \tag{8} \\
& \sum_{j \in J} x_j = K \\
& x_j \in \mathbb{Z}_+, \ y_i \in \{0, 1\}
\end{aligned}$$

The second model uses the covered probability (7).

$$\text{maximize} \quad \sum_{i \in I} a_i y_i$$

$$\text{subject to} \quad \sum_{j \in J_i} p_{ij} x_j \geq \alpha y_i \qquad \forall i \in I \tag{9}$$

$$\sum_{j \in J} x_j = K$$

$$x_j \in \mathbb{Z}_+, \ y_i \in \{0, 1\}$$

It should be noted that the constraint (9) in the second model does not necessarily give the exact covering probability in some cases. For example, if two ambulances are located at the same place j, the probability p_{ij} is doubled and would be overestimated. On the other hand, if no ambulance is located at j, the covered probability would be underestimated. These examples tell us that this model merely gives an approximate heuristic solution. Nevertheless, it often gives a solution similar to the first model. The solutions of the two models are explored in the next section.

4 Numerical Experiments

This section presents a numerical example of the two models using real data from the Tokyo metropolitan area in 2007. The number of demand points is $|I| = 3110$, and the total demand, i.e., sum of calls is 458,903. We conducted the numerical experiments in a conservative manner. In Fig. 2, the dots represent the actual sites of ambulance stations, and the total number of sites is $|J| = 155$. There are two conditions: each site should have at least one ambulance, and the total number of ambulances is $K = 163$. Hence the problem is limited to finding the positions of the eight stations (i.e., $163 - 155$) that have a second ambulance at the existing station. The reason for the partial modification of the model is, firstly to make the problem tractable by reducing the number of decision variables, and secondly, to have a clear comparison between the actual location and the optimized one.

The problem parameters in the model are set to as follows, coverage probability: $\alpha = 0.7$, traveling time: $t_c = 8$. These parameter values are chosen to provide a similar condition to the actual system. Figure 2 shows the optimal location of the eight stations that each have an additional ambulance (circled) for the uncovered model (8) (left panel) and the covered model (9) (right panel). They look similar to each other in this parameter setting; namely, the peripheral area is more likely to require ambulances. On the other hand, the objective function values, the amount of covered demand, have considerable differences between the two models. The percentage of covered demand in the covered model is 96.3 %, while that of the uncovered model is 77.5 %. In our experiments, the covered model always has a greater covered amount than the uncovered model. As discussed in Sect. 3, overestimation is unavoidable in this case because every station is always equipped with at least one ambulance.

Fig. 2 Optimal solution for the uncovered and covered probability models. The uncovered model is on the *left*, and the covered model on the *right*. The parameters are set to $t_c = 6$, $\alpha = 0.7$

Nevertheless, if we can choose the model parameters appropriately, we might use the two models for mutual reference.

5 Summary and Concluding Remarks

We have proposed two practical models for the cooperative covering location problem. Introducing a conditional probability of the travel time readily enables us to compute the covering probability and leads to easy implementation of the cooperative covering model. We demonstrated the solution of the models using the data of actual ambulance service to compare the two models. They show similarities and seem to be complementary to each other in the numerical experiments, although further investigations on the properties of the models is warranted.

References

1. Berman, O., Drezner, Z., Krass, D.: Cooperative cover location problems: the planar case. IEE Trans. **40**, 232–245 (2010)
2. Berman, O., Drezner, Z., Krass, D.: Discrete cooperative covering problems. J. Oper. Res. Soc. **62**, 2001–2012 (2011)
3. Budge, S., Ingolfsson, A., Zerom, D.: Empirical analysis of ambulance travel times: the case of calgary emergency medical services. Manag. Sci. **56**, 716–723 (2010)
4. ReVelle, C., Hogan, K.: The maximum availability location problem. Transp. Sci. **23**, 192–200 (1989)
5. Westgate, B.S., Woodard, D.B., Matteson, D.S., Henderson, S.G.: Travel time estimation for ambulances using bayesian data augmentation. Ann. Appl. Stat. **7**, 1139–1161 (2013)

Part VI
Production, Operations Management, Supply Chains, Stochastic Models and Simulation

Change Point Detection in Piecewise Stationary Time Series for Farm Animal Behavior Analysis

Sandra Breitenberger, Dmitry Efrosinin, Wolfgang Auer,
Andreas Deininger and Ralf Waßmuth

Abstract Detection of abrupt changes in time series data structure is very useful in modeling and prediction in many application areas, where time series pattern recognition must be implemented. Despite of the wide amount of research in this area, the proposed methods require usually a long execution time and do not provide the possibility to estimate the real changes in variance and autocorrelation at certain points. Hence they cannot be efficiently applied to the large time series where only the change points with constraints must be detected. In the framework of the present paper we provide heuristic methods based on the moving variance ratio and moving median difference for identification of change points. The methods were applied for behavior analysis of farm animals using the data sets of accelerations obtained by means of the radio frequency identification (RFID).

1 Introduction

The detection of structure changes is a very common task in time series analysis performed in many application areas like finance, biometrics, climatology and telecommunication systems. The application of automated monitoring systems in agriculture, e.g. for indication of animal welfare, has generated similar problems connected with a pattern recognition and anomaly detection in recorded time series.

S. Breitenberger (✉)
Linz Center of Mechatronics GmbH (LCM), Altenberger Str. 69, 4040 Linz, Austria
e-mail: Sandra.Breitenberger@lcm.at

D. Efrosinin
Institute for Stochastics, Johannes Kepler University Linz, Altenberger Str. 69,
4040 Linz, Austria

W. Auer
Smartbow GmbH, Jutogasse 3, 4675 Weibern, Austria

A. Deininger
Urban GmbH & Co. KG, Auf der Striepe 9, 27798 Hude-Wüsting, Germany

R. Waßmuth
Hochschule Osnabrück, Postfach 1940, 49090 Osnabrück, Germany

© Springer International Publishing Switzerland 2017
K.F. Dœrner et al. (eds.), *Operations Research Proceedings 2015*,
Operations Research Proceedings, DOI 10.1007/978-3-319-42902-1_50

The majority of the change point detection methods can be classified into two main groups, namely, parametric and nonparametric. The first group requires the knowledge about distribution of the data which is incorporated into the detection scheme, while the second one makes no such distributional assumptions regarding the data. The change point problems from both groups have been intensively studied before by various authors. The authors in [1] have studied the multiple change point problem in a correlated series based on Schwarz Information Criterion (SIC). In [5] it was proposed a procedure to detect variance changes based on an iterated cumulative sums of squares (ICSS) algorithm. An exhaustive literature overview and a large number of algorithms were proposed in [2]. Different parametric and nonparametric methods were discussed in [6]. In [8], the focus lied on non-parametric methods. The paper [10] instead were devoted to the change point detection based on a Bayesian approach.

Despite the wide amount of research in the area of change point detection, the detection of structural changes in the variance and autocorrelation with given constraints, what we have exactly in our real data sets, is a quite new research topic. To get the data sets a herd of dairy calves were equipped with wireless ear tag sensors equipped with accelerometers generating in online regime the data sequences of three dimensional acceleration. Such information can assist in understanding of the animal behavior, managing of the farm infrastructure and automatic detection of the animal welfare. In [7] the authors have analyzed the data obtained by means of radio frequency identification (RFID) to measure different behavior patterns like mean daily frequency of visits to the milk feeder of dairy calves, mean daily time spent feeding at each day before and after calving, mean daily duration of the time spent lying down and so on. In paper [4] the authors have combined the data obtained by the GPS collars and satellite images in a wireless sensor networks to monitor behavioral preferences and social behavior of cattle. The work [9] presents two more experiments carried out to investigate the possibility of behavioral classification using GPS data.

The aim of the proposed analysis consists in the following: based only on the given acceleration sample data to obtain the moments of time when the milk feeding of the calves starts. This information can be used then for optimal managing of the diet plan, measuring of the individual volume of the milk feed and have potential to recognize sick animals in a group as discussed in [3]. Furthermore, it is possible to identify calves at the calf feeder without the need of RFID-Chip-Recognition as the system itself knows which calf is drinking. Mastering this method, simplified calf feeders for restricted feeding of group-housed calves could be realized. In this paper, the sample data of the absolute values of the three dimensional acceleration is used. Before the discussion of the change point detection in real data sets will be proposed, the change point model must be defined.

2   The Change Point Model

Let $\mathbf{x}_{1:n} = \{x_1, x_2, \ldots, x_n\}$ be a given series data with length n, where x_i is a realization of the random variable X_i. The problem of the single change point detection of the variance can be formally rewritten as hypothesis testing.

The general approach consists in calculation of some test statistic $S(t_0, n)$, where two-samples (ahead and behind the potential change point t_0) and subsequent maximization of this function for all t_0 are required, i.e.

$$S(max, n) = |\max_{t_0} S(t_0, n)|, \ 1 < t_0 < n.$$

The hypothesis H_0 is rejected if $S(max, n) > c_n$ for some pre-specified threshold c_n which can be estimated from the data sets.

Based on the given sample data we take two sliding time windows corresponding to the moment t_0: The test window $W_1(t_0) = \mathbf{x}_{t_0:t_0+d-1}$ and the reference window $W_2(t_0) = \mathbf{x}_{k:t_0-1}$. The window $W_1(t_0)$ has always a fixed size d, whereas $W_2(t_0)$ can also be of a fixed size d, when $k = t_0 - d$, or it can be growing with the size $t_0 - k$. To check whether t_0 is a change point two phases of the data sample study are used. The first phase consists in a retrospective detection, where n observations are required. This phase is responsible for the parameter estimation and calibration of the threshold levels. The second phase includes the online detection where new observations are received over time. In this case the length of the available sequence of data is not more a constant value and at least $t_0 + d - 1$ observations are needed. In this case the problem of the multiple change point detection can be solved by simple restarting the detection algorithm after the last recognized change point.

Here we define a new statistic which provides better results for the real data, although it is not optimal in case of simulated sample data. Divide the windows $W_1(t_0)$ and $W_2(t_0)$ with the fixed size d into h equal intervals of the length $d/h \in \mathbb{N}$. The next heuristic statistic is derived from the *Mann–Whitney statistic* for medians and is based on the difference of median of variances,

$$H(t_0, n) = |\tilde{y}_{W_1(t_0)} - \tilde{y}_{W_1(t_0)}|, \tag{1}$$

or the normalized version $\tilde{H}(t_0, n) = H(t_0, n)/|\tilde{y}_{W_1(t_0) \cup W_2(t_0)}|$. Where $\tilde{y}_{W_1(t_0)}$ and $\tilde{y}_{W_2(t_0)}$ are the medians obtained respectively for the samples $\mathbf{y}_{t_0:t_0+d/h-1}$ and $\mathbf{y}_{t_0-d/h:t_0-1}$ which consist of empirical variances constructed by the formula

$$y_k = \frac{1}{d/h - 1} \sum_{t=t_0+(k-t_0)d/h}^{t_0+(k-t_0+1)d/h-1} (x_t - m_{x_{t_0+(k-t_0)d/h:t_0+(k-t_0+1)d/h-1}})^2.$$

For the change point in correlation there are a number of statistics. We present a new statistic which turned out to be quite appropriate for the real data samples studied in the framework of this paper,

$$K(t_0, n) = \Big| \max_{2 \leq \tau \leq m} \{\rho_{W_1(t_0)}(\tau)\} - \min_{2 \leq \tau \leq m} \{\rho_{W_1(t_0)}(\tau)\} \tag{2}$$
$$- \max_{2 \leq \tau \leq m} \{\rho_{W_2(t_0)}(\tau)\} + \min_{2 \leq \tau \leq m} \{\rho_{W_2(t_0)}(\tau)\} \Big|,$$

where $\rho_{W_1(t_0)}(\tau)$ and $\rho_{W_2(t_0)}(\tau)$ are empirical autocorrelation functions and m denotes the maximal lag.

The change point detection algorithm consists of the following main steps that are carried out for each sliding time window of size $n \geq 2d$:

Step 1. For the given sample $\mathbf{x}_{1:n}$ evaluate the statistic $S(t_0, n)$.
Step 2. Calculate the extrema $S(max, n)$ for the variance and correlation hypothesis testing.
Step 3. Compare the value $S(max, n)$ with a critical threshold c_n. If $S(max, n) < c_n$, there is no significant variance change. Otherwise, if $S(max, n) \geq c_n$, a potential variance change at point max is detected.
Step 4. Check if the change point from Step 3 satisfies the conditions,

$$S^2_{W_1(t_0)} \in [a_1, b_1], \quad \rho_{W_1(t_0)}(\tau) \in [a_{2,\tau}, b_{2,\tau}],$$

where $[a_1, b_1]$ and $[a_{2,\tau}, b_{2,\tau}]$ are intervals obtained from the real data sets.
Step 5. To evaluate the next change point the change detector can be resetted to the data after the time $t_0 + 1$. To use the previous observations the shift in variance at point t_0 can be eliminated by the following transformation (F denotes the Fisher ratio statistic):

$$x_t^* = \begin{cases} \frac{1}{t_0-1} \sum_{t=1}^{t_0-1} x_t + \sqrt{F(t_0, n)}\left(x_t - \frac{1}{t_0-1} \sum_{t=1}^{t_0-1} x_t\right) & t < t_0, \\ x_t, & t \geq t_0. \end{cases}$$

Go to Step 1 and treat $x_{t_0-d+1:t_0-d+1+n}^*$ as a new sample data.

If n is set to $2d$, no information of the past is used. If so, Step 2 can be ignored (computation of the maximum of one element).

3   Application to the Real Observations

The ear tag records ten values of 3D accelerations per second. For the analysis we evaluate the absolute values of acceleration. The autocorrelation analysis of the data inside and outside the feeding interval has shown that the data $W_1(t_0)$ exhibits the periodic oscillations during the feeding process whereas the data $W_2(t_0)$ can be treated as a realization of the White-Noise processes. Therefore the change in correlation must be also taken into account. The parameter τ was set to 30 during our analysis.

The proposed change point detection algorithm has been carried out for different statistics, for different values of d ($d = 60, 200$) and for different values of n

$(n = 2d, 1000)$. For each possible combination a table was generated that shows for different thresholds of variance change and correlation change respectively how many data sets out of 50 data sets have found the correct change point at the beginning of the drinking period. Table 1 shows for $H(t_0, n), d = 200$ and $n = 400$ such a table where $c1$ corresponds to the threshold of variance change and $c2$ to the change in correlation.

For the comparison of different methods, it is also important to take a look at the false detected change points. For the statistic $H(t_0, n), d = 200$ and $n = 400$ Table 2

Table 1 Amount of time series with a change point at the beginning of the drinking phase

Amount of data sets with correct change point	$c2 = 0.01$	$c2 = 0.04$	$c2 = 0.05$	$c2 = 0.06$	$c2 = 0.08$	$c2 = 0.10$	$c2 = 0.12$	$c2 = 0.15$	$c2 = 0.20$
$c1 = 3$	48	47	46	45	44	41	38	37	25
$c1 = 4$	46	45	45	42	41	38	35	34	25
$c1 = 5$	44	43	42	39	39	35	33	33	25
$c1 = 6$	42	41	39	36	35	32	30	30	24
$c1 = 7$	38	36	35	33	32	29	27	27	23
$c1 = 8$	36	35	35	33	31	29	27	26	22
$c1 = 9$	34	32	32	30	29	29	26	25	20
$c1 = 10$	30	29	28	26	25	25	22	22	18
$c1 = 12$	24	24	23	21	20	18	17	17	15

Table 2 Mean amount of false detected drinking starts between two drinking periods

Mean amount of false change points	$c2 = 0.01$	$c2 = 0.04$	$c2 = 0.05$	$c2 = 0.06$	$c2 = 0.08$	$c2 = 0.10$	$c2 = 0.12$	$c2 = 0.15$	$c2 = 0.20$
$c1 = 3$	13.76	11.94	11.64	11.3	11.12	10.88	9.92	8.06	5.30
$c1 = 4$	12.54	10.72	10.62	10.38	10.18	9.74	8.80	6.88	4.70
$c1 = 5$	11.76	10.20	9.94	9.72	8.92	8.32	7.76	6.06	4.04
$c1 = 6$	10.96	9.44	8.94	8.50	7.82	7.28	6.68	5.30	3.54
$c1 = 7$	9.92	8.40	7.86	7.50	6.86	6.30	5.84	4.62	3.12
$c1 = 8$	8.86	7.38	6.98	6.68	6.10	5.68	5.28	4.14	2.88
$c1 = 9$	7.94	6.72	6.38	6.18	5.72	5.28	4.92	3.80	2.58
$c1 = 10$	7.32	6.28	5.92	5.78	5.20	4.88	4.60	3.54	2.36
$c1 = 12$	6.38	5.40	5.06	4.82	4.44	4.10	3.80	3.04	2.04

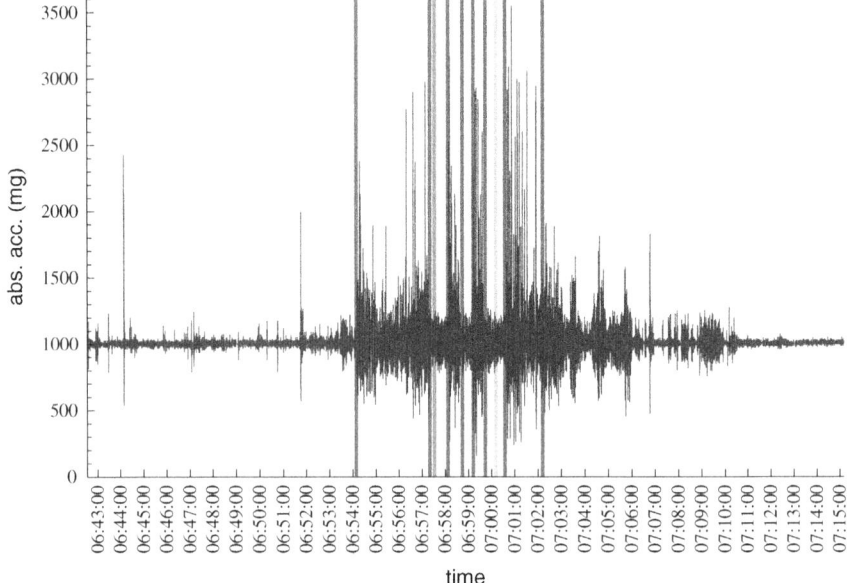

Fig. 1 Absolute acceleration of an ear tag in the course of time with change points shown as *red lines* (color figure online)

shows the mean of false detected change points between two drinking periods for 50 data sets.

The comparison of these tables for all possible combinations of the statistics and parameters d and n showed that actually this method with $H(t_0, n)$, $d = 200$ and $n = 400$ produced the best results. In Fig. 1 values of absolute acceleration in the course of time of a special calf are shown. In the middle of the time series a drinking period took place. The green vertical line denotes the beginning of the drinking period, the cyan vertical line marks the end of the drinking period and the detected change points are shown as red lines (here for the thresholds $c1 = 15$, $c2 = 0.20$). In this figure, the start of the drinking period was accepted as correct. In analysis we put a short time interval around the correct change point ($+/-20$ s) and all detected change points that lay in that interval were considered as correct.

The results show that more data, e.g. localization data, will be needed to get an completely reliable animal identification system at the feeding station. As calves can also suck on other objects in the box as on the plastic teat of the calf feeder, an additional knowledge about the approximate positions of the calves would be advantageous.

Acknowledgements This work has been supported by the Linz Center of Mechatronics (LCM) in the framework of the Austrian COMET-K2 programme and by Smartbow GmbH, which provided the real data sets.

References

1. Al Ibrahim, A., Ahmed, M., BuHamra, S.: Focus on applied statistics. Chapter testing for multiple change-point in an autoregressive model using SIC criterion, pp. 37-51. Nova Publishers, New York (2003)
2. Badagián, A.L.: Time series segmentation procedures to detect, locate and estimate change points. Ph.D. thesis, Universidad Carlos III de Madrid, Spain (2013)
3. Breitenberger, S., Efrosinin, D., Auer, W., Deininger, A., Waßmuth R.: Automated detection of amount and period of drinking for calves equipped with eartags producing acceleration data (in German). 12. Tagung: Bau, Technik und Umwelt (2015)
4. Handcock, R.N., Swain, D.L., Bishop-Hurley, G.J., Patison, K.P., Wark, T., Valencia, P., Corke, P., O'Neill, C.J.: Monitoring animal behaviour and environmental interactions using wireless sensor network, GPS collars and satellite remote sensing. Sensors **9**, 3586–3603 (2009)
5. Inclán, C., Tiao, G.: Use of cumulative sums of squares for retrospective detection of changes of variance. J. Am. Stat. Assoc. **427**, 913–923 (1994)
6. Ross, G.J.: Parametric and nonparametric sequential change detection in R: the cpm Package. J. Stat. Softw. To appear (2013)
7. Rushen, J., Chapinal, N., de Pasillé, A.M.: Automated monitoring of behavioural-based animal welfare indicators. Anim. Welf. **21**, 339–350 (2012)
8. Sharkey, P., Killick, R.: Nonparametric methods for online change point detection. In: STOR601 Research Topic II (2014)
9. Spink, A., Cresswell, B., Közsch, A., van Langevelde, F., Neefjes, M., Noldus, L.P.J.J., van Oeveren, H., Prins, H., van der Wal, T., de Weerd, N., Frederik der Boer, W.: Animal behaviour analysis with GPS and 3D accelerometers. In: Precision Livestock Farming. Contribution in Proceedings, pp. 229–239. Leoven, Belgium (2013)
10. Wei-xing, S., Yun-long, Z., Fang, L., Kun-yuan, H.: On-line outlier and change point detection for time series. J. Cent. South Univ. **20**, 114–122 (2013)

Reliable Order Promising with Multidimensional Anticipation of Customer Response

Ralf Gössinger and Sonja Kalkowski

Abstract Reliable order promising is a key competitive factor for MTO companies. Several measures to maintain reliability in an uncertain production system are proposed in literature, but so far a multidimensional anticipation of customer response to modified order specifications has not been adequately taken into account. The purpose of this paper is to demonstrate the potential of this measure by modeling and numerically analyzing the impacts on reliability and profit.

1 Problem

During the order promising process potential customers and the company decide interactively on placing resp. accepting orders as well as on order terms. Thereby, customers' decisions to place orders are not only determined by quoted prices but also by non-monetary factors [1, 2]. In particular, reliable delivery dates are of strategic importance since tardy deliveries can induce losses of market shares in the long-term [1]. Hence, in an uncertain business environment most likely realizable delivery dates have to be determined and supply-related measures to compensate deviations need to be identified in advance. Among these measures are e.g. capacity nesting [3] or safety capacity [4]. Additionally, interacting with customers by proposing order specifications that deviate from customers' inquiries in multiple dimensions can influence demand. This can be favorable if the acceptable extent of deviations in different dimensions as well as the interactions between the different dimensions can be estimated precisely enough. To get first insights onto the impacts of this measure we focus on the dimensions price and delivery time.

The aim of the present paper is to extent an existing capable-to-promise (CTP) approach [5] that determines reliable delivery date promises while anticipating time-

R. Gössinger (✉) · S. Kalkowski
Department of Business Administration, Production Management and Logistics,
University of Dortmund, Martin-Schmeißer-Weg 12, 44227 Dortmund, Germany
e-mail: ralf.goessinger@udo.edu

S. Kalkowski
e-mail: sonja.kalkowski@tu-dortmund.de

© Springer International Publishing Switzerland 2017
K.F. Dœrner et al. (eds.), *Operations Research Proceedings 2015*,
Operations Research Proceedings, DOI 10.1007/978-3-319-42902-1_51

oriented customer behavior (one-dimensional response function [1]). The extension refers to a dynamic time-related price differentiation, that is prices for the same product are adapted to changing production situations [6]. For this purpose a discount is granted in case of a deviation between the preferred and the offered delivery date in order to increase the acceptance probability of the offer. Thereby customer behavior is anticipated by a response function with the dimensions discount and delivery date deviation. In the paper essential characteristics of this response function and resulting modifications of the original decision model are derived. Based on this, a numerical analysis is performed to investigate the impacts of adjusting delivery dates and prices on profit and reliability of delivery dates. Essential findings and implications for future research are summarized in the final section.

2   Modeling Delivery Date-Dependent Discounts

Offering delivery date-dependent discounts is motivated by two observations: (1) A dynamic time-related price differentiation can be considered as to be unfair if differentiation criteria are not customary or transparent [7]. (2) In MTO production the significance level of the estimated willingness to pay can be very low due to a high degree of individualization [8] so that the risk of generating losses by an inappropriate price differentiation is not acceptable. Thus, granting discounts for expected deviations from customary delivery dates seems preferable. To take the impacts of discounts independently from the level of market prices into account relative discounts are considered.

The purpose of granting delivery date-dependent discounts is to offer deviating conditions (delivery date, price) for order inquiries $i \in \bar{A}$ which would need to be rejected at the desired conditions (price ρ_i, quantity q_i, delivery time interval $[t_i^e, t_i^l]$). In the current planning period t_a the company decides on the delivery date $D_{i,t} \in \{0, 1\}$ ($t \in [t_a, T]$) (and thus on the extent of deviation V_i^{des} from the requested delivery time interval), on the production quantities $P_{i,t}$ as well as on the discount Φ_i. These decisions influence the fulfillment of previously accepted orders $i \in \hat{A}$. A deviation from production quantities and contractually fixed delivery dates is possible for these orders, but induces penalty costs γ_i^e, γ_i^l for premature/tardy delivery.

To anticipate customer response to deviating order conditions a two-dimensional response function $\beta(V_i^{des}, \Phi_i)$ which specifies the acceptance probability in dependence of the delivery time deviation V_i^{des} and the discount Φ_i forms a central element of the planning model. For the delivery date deviation negative or positive (premature or tardy delivery) values may occur (1). To reduce computational effort the value range is limited to a maximum deviation V^{max} (2). Discounts are granted as positive relative reductions of customary prices up to a maximum value ψ which depends on the relation between manufacturing costs and customary price (3).

Fig. 1 Continuous
two-dimensional response
function: $\alpha = 0$, $\beta = 0$,
$\gamma = 0.02$, $\delta = 20$,
$V^{max} = 30$, $\psi = 0.1$

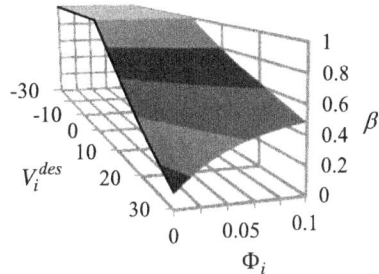

$$V_i^{des} = min\left(\sum\nolimits_{t=t_a}^{T} D_{i,t} \cdot t - t_i^e, 0\right) + max\left(\sum\nolimits_{t=t_a}^{T} D_{i,t} \cdot t - t_i^l, 0\right) \qquad \forall i \in \bar{A} \quad (1)$$

$$- V^{max} \le V_i^{des} \le V^{max} \qquad \forall i \in \bar{A} \quad (2)$$

$$0 \le \Phi_i \le \psi < 1 \qquad \forall i \in \bar{A} \quad (3)$$

A continuous response function $\beta(V_i^{des}, \Phi_i)$ possesses the characteristics (a)–(d):
(a) non-decreasing in Φ_i: the acceptance probability does not decrease with increasing discount; (b) non-increasing in $V_i^{des} > 0$: with increasing positive deviation the
acceptance probability does not increase; (c) $\beta(0, \Phi_i)$: certain acceptance if requested
and offered conditions match; (d) non-decreasing in $V_i^{des} < 0$: the acceptance probability does not increase with increasing negative deviation. An example of a two-
dimensional response function possessing these characteristics is given in (4):

$$\beta(V_i^{des}, \Phi_i) = \begin{cases} e^{+\alpha \cdot V_i^{des}} - e^{+\beta \cdot \Phi_i} + e^{+\alpha \cdot V_i^{des} + \beta \cdot \Phi_i} & : V_i^{des} \in]-V^{max}, 0]; \; \Phi_i \in [0, \psi] \\ e^{-\gamma \cdot V_i^{des}} - e^{-\delta \cdot \Phi_i} + e^{-\gamma \cdot V_i^{des} - \delta \cdot \Phi_i} & : V_i^{des} \in [0, V^{max}]; \quad \Phi_i \in [0, \psi] \end{cases}$$
$$(4)$$

Graphically a hypersurface of the acceptance probability depending on delivery date
deviation and discount results (see Fig. 1).

Compared to the original decision model the relevant stochastic part of the objective function is adjusted to consider the modified response function as well as discounts (5). The remaining terms of the objective function do not change substantially.

$$\sum\nolimits_{t=t_a}^{T} \sum\nolimits_{i \in \bar{A}} \beta(V_i^{des}, \Phi_i) \cdot (q_i \cdot (1 - \Phi_i) \cdot \rho_i \cdot D_{i,t}) \qquad (5)$$

3   Numerical Analysis

This analysis is performed to systematically investigate the impacts of "proposing
deviating order conditions" (no proposal (A0); proposal of deviating delivery dates
(A1) and proposal of deviating delivery dates and discounts (A2)) in the context of
order- and resource-related uncertainty. By means of descriptive statistics interactions with "capacity nesting" (B) and "providing safety capacity" (C) implemented

in the original model are to be analyzed. For this purpose generated profits and the reliability from the customers' point of view (planning robustness) which is measured by penalty costs for deviations from contractually fixed delivery dates [9] are considered. The following suppositions can be justified for these indicators: (S1) A1 and A2 extend the degrees of freedom in planning so that a higher amount of accepted orders and therefore an enhancement of profits are to be expected. Since A2 offers higher degrees of freedom than A1 higher profits will probably result. (S2) Between A, B and C interactions exist with regard to planning robustness since the measures have contrary effects on the amount of accepted orders. As the probability to meet delivery dates increases with decreasing workload A probably influences planning robustness in a negative and B resp. C in a positive way.

Real order and capacity data as well as systematically generated data form the data basis. Order-related uncertainty is present for each product configuration with respect to order quantity and interarrival time. On this basis five order streams are generated by means of truncated normal distributions that only permit positive values. All orders have $t_i^l - t_i^e = 10$, $\gamma_i^l = 3\%$, $\gamma_i^e = 0\%$, $k^{Tr} = 59$ and $k_i^{L.FP} = 0.25\%$ per tied up capital in common. To take resource-related uncertainty into account symmetric triangular distributions are estimated for capacity situations with a low/high level of uncertainty (III and IV) and certain capacity situations (I and II) are taken as references (Table 1). The parameters of the response function (4) were empirically estimated based on the collected data: $\alpha = 0$, $\beta = 0$, $\gamma = 0.02$, $\delta = 20$, $V^{max} = 30$, $\psi = 0.1$. To identify interactions of A and B resp. C test runs with (A1, A2) and without (A0) this measure are performed. Thereby a normal (Ca: $1 \cdot \sigma^c$) and a high level of safety capacity (Cb: $2 \cdot \sigma^c$) are assumed. Considering the observed capacity situations (I, II, III, IV) all test constellations can be assigned to 18 scenarios within which the B-parameters are varied (costs for utilizing premium capacity (PC): 500, 1000, 1500, 2000, 2500, 3000; shares of PC: 1/3, 1/2, 2/3). In total 7920 test constellations result (99.67% solved to optimality). For the analysis two scenario-specific values are determined: While B^{equ} indicates which observation value is to be expected for a uniform choice of B-parameters, B^{opt} results from an optimal choice.

To evaluate the monetary impacts of A1 and A2 the percentage profit changes generated compared to A0 are considered for low/high uncertainty (Fig. 2). A horizontal comparison provides information on the effect of switching from A1 to A2. Comparisons along the depth axis illustrate the influence of C-parameters on A. Furthermore, comparing B^{equ} and B^{opt} reflects the influence of B-parameters on A. Several tendencies can be observed: Measure A leads to profit enhancements whereby the extent of increases is influenced by A-, B- and C-parameters as well as capacity uncertainty: A2 generates stronger relative impacts than A1. The effects are stronger for a change from A0 to A1 compared to switching from A1 to A2. For low uncertainty the relative impacts of A1 and A2 are strengthened by an increasing C-parameter, whereas they are first strengthened and then damped for high uncertainty (exception: A2 for B^{opt}). By optimizing B-parameters the relative impacts of

Table 1 Order and capacity data

Product configuration c	1	2	3	4	5	6	7
Order quantity $\mu_c^q\|\sigma_c^q$	4.7|3.3	4.9|2.5	7.8|3.9	4.7|6.5	7.7|0.6	4.8|3.3	6.4|2.6
Interarrival time $\mu_c^t\|\sigma_c^t$	13.0|5.9	11.4|12.7	9.1|4.9	22.8|16.8	30.3|22.5	22.8|31.5	11.4|7.4
Price ρ_c	5750	3999	3999	3299	2990	2699	2599
Manufacturing costs k_c^M	570.93	384.93	416.50	281.19	289.65	251.72	462.44
Capacity scenario	μ^{cap}	σ_1^{cap}	σ_2^{cap}	σ_3^{cap}	σ_4^{cap}	σ_5^{cap}	$\sigma_{6...T}^{cap}$
I|II	2.75|2.5	0|0	0|0	0|0	0|0	0|0	0|0
III	2.75	0	0.0204	0.0408	0.0612	0.0816	0.1021
IV	2.5	0	0.0408	0.0816	0.1225	0.1633	0.2041

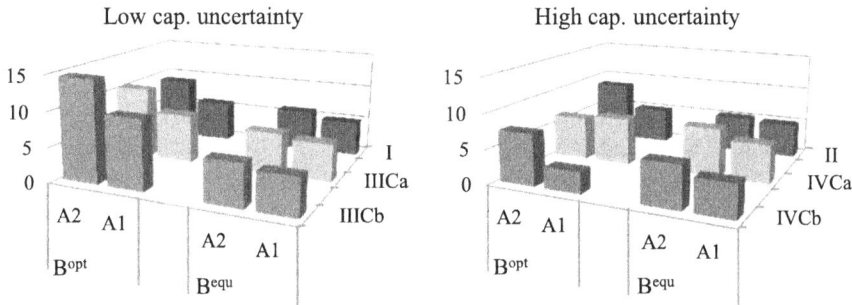

Fig. 2 Percentage profit changes compared to A0

A1 and A2 are strengthened in case of low uncertainty. For high uncertainty heterogeneous effects result. Additionally, the relative impacts of A1 and A2 are higher for low than for high uncertainty. In total (S1) is not disproved by the test results.

Analyzing the impacts of A on reliability of promised delivery dates (planning robustness) several tendencies can be identified by evaluating penalty cost changes (Fig. 3): Depending on A-, B- and C-parameters as well as capacity uncertainty A has robustness enhancing or reducing effects. A2 unfolds more beneficial impacts (lower increases/ higher reductions) than A1. The positive relative effects induced by A1 and A2 at first tend to be strongly weakened by increasing C-parameters before being less influenced. The more advantageous impacts of A1 are intensified by optimizing B-parameters whereas a heterogeneous picture emerges for A2. With regard to the disadvantageous relative impacts heterogeneous results are achieved for A1 and A2. For low uncertainty the relative effects of A1 and A2 are lower than for high uncertainty. In total (S2) is not disproved by the test results.

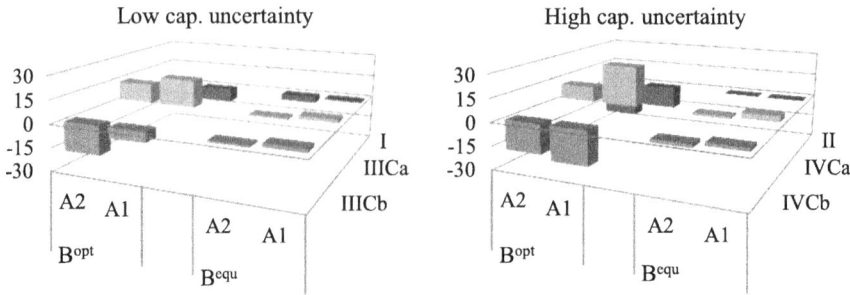

Fig. 3 Percentage penalty cost changes compared to A0

4 Conclusions

In the present paper an existing CTP approach is extended by dynamic time-related price differentiation to generate reliable order promises which are economically acceptable for customers and the company. For this purpose order conditions that deviate from customer requests are proposed while customer response is anticipated by means of a response function with the dimensions deviation from delivery date and discount. The extended planning model is numerically analyzed to investigate the impacts of this measure on economic success and reliability. The following main impacts of proposing deviating delivery dates have to be highlighted: (1) Generated profits can be increased by proposing deviating delivery dates. This effect is strengthened by additionally granting discounts. (2) The impact direction on reliability from customers' point of view depends on the chosen parameter settings; both enhancing and reducing effects, are observed. In an overall view proposing deviating order specifications mainly induces economic advantages. Adverse constellations indicate that this measure should not be applied regardless of other supply-related measures but needs to be coordinated. Since the results relate to the underlying data basis they have to be verified in future research using expanded data constellations. Due to the constant level of order-related uncertainty in the tests the behavior of the planning approach needs to be investigated for different levels of order-related uncertainty.

References

1. Kingsman, B.[G.], Worden, L., Hendry, L., Mercer, A., Willson, E.: Integrating marketing and production planning in make-to-order companies. Int. J. Prod. Econ. **30/31**, 53–66 (1993)
2. Stevenson, M., Hendry, L.C., Kingsman, B.G.: A review of production planning and control: the applicability of key concepts to the make-to-order industry. Int. J. Prod. Res. **43**, 869–898 (2005)
3. Harris, F.H.deB., Pinder, J.P.: A revenue management approach to demand management and order booking in assemble-to-order manufacturing. J. Oper. Manag. **13**, 299–309 (1995)

4. Pibernik, R., Yadav, P.: Dynamic capacity reservation and due date quoting in a make-to-order system. Naval Research Logistics **55**, 593–611 (2008)
5. Gössinger, R., Kalkowski, S.: Robust order promising with anticipated customer response. Int. J. Prod. Econ. **170**, 529–542 (2015)
6. Talluri, K.T., van Ryzin, G.J.: The Theory and Practice of Revenue Management. Springer, New York (2005)
7. Shen, Z.-J.M., Su, X.: Customer behavior modeling in revenue management and auctions: a review and new research opportunities. Prod. Oper. Manag. **16**, 713–728 (2007)
8. Wertenbroch, K., Skiera, B.: Measuring customers' willingness to pay at the point of purchase. J. Market. Res. **39**, 228–241 (2002)
9. Sridharan, S.V., Berry, W.L., Udayabhanu, V.: Measuring master production schedule stability under rolling planning horizons. Decision Science **19**, 147–166 (1988)

A Scalable Approach for the *K*-Staged Two-Dimensional Cutting Stock Problem

Frederico Dusberger and Günther R. Raidl

Abstract This work focuses on the *K*-staged two-dimensional cutting stock problem with variable sheet size. High-quality solutions are computed by an efficient beam-search algorithm that exploits the congruency of subpatterns and takes informed decisions on which of the available sheet types to use for the solutions. We extend this algorithm by embedding it in a sequential value-correction framework that runs the algorithm multiple times while adapting element type values in each iteration and thus constitutes a guided diversification process for computing a solution. Experiments demonstrate the effectiveness of the approach and that the sequential value-correction further increases the overall quality of the constructed solutions.

1 Introduction

We consider the *K-staged two-dimensional cutting stock problem with variable sheet size* (*K*2CSV) in which we are given a set of n_E rectangular *element types* $E = \{1, \ldots, n_E\}$, each $i \in E$ specified by a height $h_i \in \mathbb{N}^+$, a width $w_i \in \mathbb{N}^+$, and a demand $d_i \in \mathbb{N}^+$. Furthermore, we have a set of n_T *stock sheet types* $T = \{1, \ldots, n_T\}$, each $t \in T$ specified by a height $H_t \in \mathbb{N}^+$, a width $W_t \in \mathbb{N}^+$, an available quantity $q_t \in \mathbb{N}^+$, and a cost factor $c_t \in \mathbb{N}^+$. Both elements and sheets can be rotated by 90°. A feasible solution is a set of *cutting patterns* $\mathscr{P} = \{P_1, \ldots, P_n\}$, i.e. an arrangement of the elements specified by E on the stock sheets specified by T without overlap and using guillotine cuts up to depth K. Each pattern $P_j, j = 1, \ldots, n$, has an associated stock sheet type t_j and a quantity a_j specifying how often the pattern is to be applied, i.e. how many sheets of type t_j are cut following pattern P_j.

We thank LodeStar Technology for their support and collaboration in this project.

F. Dusberger (✉) · G.R. Raidl
Institute of Computer Graphics and Algorithms, Tu Wien, Vienna, Austria
e-mail: dusberger@ac.tuwien.ac.at

G.R. Raidl
e-mail: raidl@ac.tuwien.ac.at

© Springer International Publishing Switzerland 2017
K.F. Dœrner et al. (eds.), *Operations Research Proceedings 2015*,
Operations Research Proceedings, DOI 10.1007/978-3-319-42902-1_52

The $K2$CSV occurs in many industrial applications and we are thus considering here in particular large-scale instances from industry. For these instances, typically the number of different element and sheet types is moderate but the demands of the element types are rather high. Nonetheless, reasonable solutions need to be found within moderate runtimes. The objective is to find a feasible set of cutting patterns $\mathscr{P}$ minimizing the number of used sheets $c(\mathscr{P})$ weighted by their cost factors, i.e.

$$\min\ c(\mathscr{P}) = \sum_{t \in T} c_t \sigma_t(\mathscr{P}) \tag{1}$$

where $\sigma_t(\mathscr{P})$ is the number of used sheets of type $t \in T$ in the set $\mathscr{P}$.

Each cutting pattern $P_j \in \mathscr{P}$ is represented by a (cutting) tree structure where the leaf nodes correspond to individual (possibly rotated) elements and the internal nodes are either *horizontal* or *vertical compounds* containing at least one subpattern. Vertical compounds always only appear at odd stages (levels), starting from stage one, and represent parts separated by horizontal cuts of the respective stage. Horizontal compounds always only appear at even stages and represent parts separated by vertical cuts. Each node thus corresponds to a rectangle of a certain size (h, w), which is in case of compound nodes the bounding box of the respectively aligned subpatterns. A pattern's root node always has a size that is not larger than the respective sheet size, i.e. $h \leq H_{t_j}$ and $w \leq W_{t_j}$. Analogously to a_j denoting the quantity of sheets cut according to pattern P_j, compound nodes store congruent subpatterns only by one subtree and maintain an additional quantity. In this tree structure, residual (waste) rectangles are never explicitly stored, but can be derived considering a compound node's embedding in its parent compound or sheet.

Each pattern $P_j \in \mathscr{P}$ can be transformed into a *normal form* of equal objective value, hence it is sufficient to consider patterns in normal form only. In normal form, subpatterns of vertical (horizontal) compounds are arranged from top to bottom (left to right), ordered by nonincreasing width (height), and aligned at their left (top) edges, i.e. in case the subpatterns have different widths (heights), remaining space appears to their right (at their bottom). Figure 1 shows a 3-staged cutting tree and the corresponding pattern.

2 Related Work

Many state of the art approaches for cutting and packing problems employ column generation, dynamic programming, or a combination thereof [3, 10]. However, in the light of large-scale instances, exact approaches cannot compute solutions within reasonable time. Instead, heuristics and metaheuristics are far more promising in

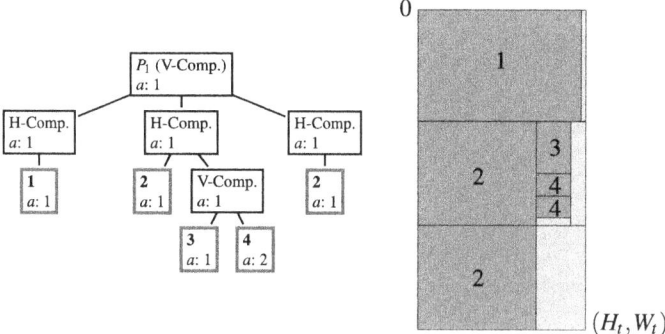

Fig. 1 A three-staged cutting tree (*left*) and the corresponding cutting pattern (*right*). The leaves represent actual elements of types 1 to 4 obtained after at most three stages of guillotine cuts

obtaining competitive solutions in this context. Lodi et al. [9] give a survey on the classical construction heuristics for cutting and packing problems. In contrast to the exponential runtime of the exact approaches, their runtime complexities range from $O(n \log n)$ to $O(n^3)$, where n is the total demand over all elements. Unfortunately, this advantage is still not enough when the total demand is very high. Moreover, these heuristics are rather inflexible and a strategy that does not prematurely fix the structure of the cutting tree and the sizes of the inserted subpatterns is needed. In a recent paper, Fleszar [6] proposed three more involved construction heuristics for the cutting stock problem with only a single sheet type achieving excellent results in short time. A promising approach to boost the quality of constructed solutions is sequential value-correction, an idea that has been successfully applied, among others, to the two-dimensional strip packing problem by Belov et al. [1] and the two-dimensional bin packing problem with a single sheet type by Cui et al. [4]. The basic idea is to associate each element type with a value signifying that type's priority. Multiple solutions are constructed based on these values, which are continuously adapted according to certain quality criteria, leading to a guided diversification process. The values are increased for types that led to patterns with high relative waste, s.t. those types are used earlier in the following iterations. Considering multiple sheet types, Hong et al. [7] embedded fast construction heuristics in a backtracking framework to address the problem of a meaningful sheet type selection. Solutions are constructed sheet by sheet, where backtracking is applied to revise the choice of a sheet type that led to a poor solution. In recent work, we followed a similar strategy by successfully employing a beam-search algorithm to the K2CSV [5], which serves as a basis for the sequential value-correction framework we propose here and is summarized in the following section.

3   A Congruency-Aware Construction Heuristic for the K2CSV

The main component of the algorithm is a construction-heuristic for computing the pattern for a single sheet, more precisely, the heuristic solves the K-staged two-dimensional knapsack problem. By operating on element types rather than the single elements separately, the heuristic is highly scalable w.r.t. element type demands avoiding the excessive runtimes of element-based construction heuristics. A solution is constructed by considering for each element type $i \in E$ the insertion of a completely filled grid of $a_i^{\text{vert}} \times a_i^{\text{hor}}$ instances of i. The best position and grid size in the sheet pattern are determined according to a heuristic fitness criterion. Note that in contrast to the original algorithm, we use here the more complex *sufficiency criterion* as proposed in [2]. Due to the compact solution representation in the cutting tree, where congruent subpatterns are stored only once with an associated quantity, the heuristic can then simultaneously apply this insertion to as many congruent subpatterns as possible. Another drawback of conventional construction heuristics we avoid is their inflexibility w.r.t. already placed elements. The heuristic considers the current pattern flexible in the sense that compounds are not fixed after their initialization, but can be resized, if necessary to accommodate additional subpatterns.

In order to make meaningful decisions on the choice of sheet types, this construction heuristic is used in a beam-search algorithm generating the solution sheet by sheet. Each node in the search tree corresponds to a (partial) solution, starting with the empty solution at the root. A branch from a node reflects the decision for one of the sheet types from T, the computation of a new pattern on a sheet of that type using the construction heuristic and adding this pattern to the solution with a quantity as high as possible considering the residual element type demands. At each level, all the nodes on that level are evaluated and all but the BW best ones are pruned, where BW is the chosen beam-width. The quality of a partial solution is determined by the average relative waste over all used sheets weighted by the cost factors of the respective types. The lower the relative waste, the better the solution. This procedure continues until all requested elements are used.

4   Sequential Value-Correction

The effectiveness of the beam-search algorithm is demonstrated in [5], where experimental results showed that the approach computes high-quality solutions in relatively short time. Nonetheless, being heuristic in nature, the algorithm is not guaranteed to always find good solutions. In order to compensate for this drawback, we embed it in a sequential value-correction framework.

We associate each element type $i \in E$ with a value v_i that is initially equal to its area, i.e. $v_i = h_i w_i$. The beam-search algorithm is then called for a certain number of iterations and each time a solution has been computed, these values are updated. Let elems$_i(P_j)$ be the number of elements of type i in pattern P_j and let further wr(P_j) denote the waste ratio of pattern P_j, i.e. the ratio of unused area to total area in P_j. For a given solution, the value v_i of each element type $i \in E$ is adapted for each of the sheet patterns $P_j \in \mathscr{P}$ according to the following formula:

$$v_i \leftarrow (1 - g) \cdot v_i + \frac{g \cdot (h_i w_i)^p}{1 - \text{wr}(P_j)}, \tag{2}$$

where p is a parameter slightly larger than 1 (e.g. 1.02) and g is defined as

$$g = \frac{\text{elems}_i(P_j) \cdot a_j}{d_i + d_i^r} \tag{3}$$

where d_i^r is the residual demand of element type i after computing pattern P_j.

The intuition behind this formula is the following:

- The deterministic weighting factor g ensures that the value of each element type i is updated to an extent that is proportional to the number of elements of type i used in P_j. In particular, we have $g = 0$ if elems$_i(P_j) \cdot a_j = 0$, i.e. v_i remains unchanged if no element of type i occurs in the pattern.
- As $p > 1$, the values are overproportional to the element types' areas. The intention is to prefer element types that have a relatively large area and are therefore harder to pack.
- Similarly, combinations of elements that yield a high waste ratio will lead to higher values for the respective element types and to their preference in the following iterations as they are hard to combine.

Since the decisions in both the beam-search algorithm as well as the fitness-criterion of the underlying construction heuristic are based on the element types' areas, they can easily be adapted to utilize the values v_i, for $i \in E$, instead.

5 Computational Results

Our algorithms were implemented in C++, compiled with GCC version 4.8.4, and executed on a single core of a 3.40 GHz Intel Core i7-3770. We tested the sequential value-correction approach for 30 iterations and BW $= 20$ (SVC) on the benchmark set by Hopper and Turton [8]. It comprises three instance categories of increasing complexity, each consisting of five randomly generated instances with $|T| = 6$, $2 \leq q_t \leq 4$ and c_t proportional to t's area, for all $t \in T$, and $d_i = 1$, for all $i \in E$. We compared SVC with the HHA algorithm by Hong et al. [7] and with applying the pure beam-search for BW $= 500$ (BS500) and, using the original fitness criterion from [5], for BW $= 5000$ (BS5000). As HHA does not use a stage limit, we set $K = 10$ for our algorithms. Table 1 reports for each category the average percentage of the used area on the sheets $\overline{a(\mathscr{P})}$, to be comparable with the results from [7], and the

Table 1 Comparison of area utilization for the three instance categories M1 to M3

Instance category	$|E|$	$|T|$	HHA		BS500		BS5000		SVC	
			$\overline{a(\mathscr{P})}$	$\bar{t}(s)$	$\overline{a(\mathscr{P})}$	$\bar{t}(s)$	$\overline{a(\mathscr{P})}$	$\bar{t}(s)$	$\overline{a(\mathscr{P})}$	$\bar{t}(s)$
M1	100	6	**98.4**	60	97.7	21.6	**98.4**	34.2	**98.4**	18.6
M2	100	6	95.6	60	**96.3**	12.2	96.1	32.0	**96.3**	13.0
M3	150	6	97.4	60	96.8	45.8	96.5	103.1	**97.6**	35.9

The best value for $\overline{a(\mathscr{P})}$ in each row is printed in bold

average runtime $\bar{t}$. Although $d_i = 1$, for all $i \in E$, i.e. we cannot exploit congruency, our algorithms yield competitive results in comparison to HHA. While the pure beam search variants achieve the same or slightly worse results, SVC yields the best results for all categories demonstrating the effectiveness of our value-correction strategy. Moreover, the runtimes of SVC are comparable to those of BS500 and better than both those of BS5000 and HHA, which always runs for 60 s.

6   Conclusions and Future Work

In this work, we extended a successful constructive algorithm for the $K2CSV$ by a sequential value-correction framework in order to improve the overall quality of the constructed solutions. The basic algorithm, which was presented in [5], is a beam-search constructing a solution sheet by sheet using a congruency-aware construction heuristic, which makes the approach highly scalable and allows it to solve large real-world instances within reasonable time. By computing multiple solutions while adapting values associated to the element types after each iteration, this approach can be seen as a guided diversification process. Experiments on benchmark instances document the effectiveness of this strategy.

In future work, we intend to develop a subsequent improvement heuristic for which the excellent results provided by this algorithm are used as initial solutions.

References

1. Belov, G., Scheithauer, G., Mukhacheva, E.A.: One-dimensional heuristics adapted for two-dimensional rectangular strip packing. J. Oper. Res. Soc. **59**(6), 823–832 (2008)
2. Charalambous, C., Fleszar, K.: A constructive bin-oriented heuristic for the two-dimensional bin packing problem with guillotine cuts. Comput. Oper. Res. **38**(10), 1443–1451 (2011)
3. Cintra, G., Miyazawa, F., Wakabayashi, Y., Xavier, E.: Algorithms for two-dimensional cutting stock and strip packing problems using dynamic programming and column generation. Eur. J. Oper. Res. **191**(1), 61–85 (2008)
4. Cui, Y., Yang, L., Zhao, Z., Tang, T., Yin, M.: Sequential grouping heuristic for the two-dimensional cutting stock problem with pattern reduction. Int. J. Prod. Econ. **144**(2), 432–439 (2013)

5. Dusberger, F., Raidl, G.R.: A Scalable Approach for the *K*-Staged Two-Dimensional Cutting Stock Problem with Variable Sheet Size. In: Computer Aided Systems Theory – EUROCAST 2015, LNCS. Springer (2015). *to appear*
6. Fleszar, K.: Three insertion heuristics and a justification improvement heuristic for two-dimensional bin packing with guillotine cuts. Comput. Oper. Res. **40**(1), 463–474 (2013)
7. Hong, S., Zhang, D., Lau, H.C., Zeng, X., Si, Y.: A hybrid heuristic algorithm for the 2D variable-sized bin packing problem. Eur. J. Oper. Res. **238**(1), 95–103 (2014)
8. Hopper, E., Turton, B.C.H.: An empirical study of meta-heuristics applied to 2D rectangular bin packing - part i. Stud. Inf. Universalis **2**(1), 77–92 (2002)
9. Lodi, A., Martello, S., Vigo, D.: Recent advances on two-dimensional bin packing problems. Discret. Appl. Math. **123**(13), 379–396 (2002)
10. Pisinger, D., Sigurd, M.: The two-dimensional bin packing problem with variable bin sizes and costs. Discret. Optim. **2**(2), 154–167 (2005)

A Logic-Based Benders Decomposition Approach for the 3-Staged Strip Packing Problem

Johannes Maschler and Günther R. Raidl

Abstract We consider the 3-staged Strip Packing Problem, in which rectangular items have to be arranged onto a rectangular strip of fixed width, such that the items can be obtained by three stages of guillotine cuts while the required strip height is to be minimized. We propose a new logic-based Benders decomposition with two kinds of Benders cuts and compare it with a compact integer linear programming formulation.

1 Introduction

In the 3-staged Strip Packing Problem (3SPP) we are given n rectangular items and a rectangular strip of width W_S and unlimited height. The aim is to pack the items into the strip using minimal height, s.t. all items can be received by at most three stages of guillotine cuts. In the first stage the strip is cut horizontally from one border to the opposite one and yields up to n levels. In the second stage the levels are cut vertically and at most n stacks are received. In the third stage the stacks are cut again horizontally and the resulting rectangles of the three consecutive stages of guillotine cuts correspond to the n items and the waste.

The general Strip Packing Problem (SPP) was proposed by Baker et al. [1] and has received a large amount of attention: on the one hand many real-world applications, such as glass, paper and steel cutting, can be modeled as SPPs; on the other hand it is strongly-NP hard and has turned out to be a demanding combinatorial problem. We study the special case of the SPP, where only guillotine cuts are allowed as already

The authors thank Christina Büsing from TU Wien for her support on this paper.

J. Maschler (✉) · G.R. Raidl
Institute of Computer Graphics and Algorithms, Tu Wien, Favoritenstraße 9–11, Vienna, Austria
e-mail: maschler@ac.tuwien.ac.at

G.R. Raidl
e-mail: raidl@ac.tuwien.ac.at

considered by Hifi [4] and Lodi et al. [6]. We restrict ourselves to three stages of guillotine cuts. This can be motivated from glass cutting [8].

One of the leading exact approaches for the SPP proposed by Côté et al. [2] is based on a Benders decomposition using combinatorial Benders cuts, which can be seen as an implementation of the logic-based Benders decomposition (LBBD) introduced by Hooker and Ottosson [5]. Their master problem cuts items into unit-width slices and solves a parallel processor problem that requires all slices belonging to the same item have to be packed next to each other. The subproblem consists of transforming a solution of the master problem into a solution of the SPP. However, this algorithm cannot be trivially extended to solve the 3SPP.

In this work we suggest a different form of LBBD specifically for the 3SPP, compare it to a compact integer linear programming (ILP) formulation and show that its performance is competitive. The proposed master problem assigns items to levels and defines the number of stacks in which the items can appear. The resulting subproblems pack items of the same width assigned to the same level into the given number of stacks. Two kinds of Benders cuts are provided, from which the first one are rather straightforward, while the second one are more general.

2   A Logic-Based Benders Decomposition for 3SPP

The proposed LBBD consists of a master problem which is a relaxation of the 3SPP. From an optimal solution of the master problem subproblems are derived that yield a complete solution for the 3SPP. If this solution is not yet optimal we improve the master problem with Benders optimality cuts and resolve the master problem.

2.1   Master Problem

The master problem considers n levels and aims to pack the items into these levels. For symmetry breaking item i is only allowed to be packed into level j if $i \geq j$ and only if item j is also packed into level j. This way of symmetry breaking has been already used by Puchinger and Raidl [7]. To better exploit scenarios where many items have the same widths and/or lengths, let us more precisely define the set of all appearing widths as $W = \{w_1, \ldots, w_p\}$, the set of all appearing heights as $H = \{h_1, \ldots, h_q\}$, and the dimensions of item $i \in I = \{1, \ldots, n\}$ as w_{ω_i} and h_{λ_i} with $\omega \in \{1, \ldots, p\}^n$ and $\lambda \in \{1, \ldots, q\}^n$. We further assume w.l.o.g. that the items and the widths in W are given in a non-decreasing order.

We consider in the master problem different variants of each item, in which depending on a parameter $e \in \{1, 2, \ldots\}$ its width is increased by $w_{\omega_i} * e$ while its height is decreased by h_{λ_i}/e. We denote the modified width of an item i by the parameter e with $w_{\omega_i e}$, and analogously, the modified height with $h_{\lambda_i e}$. We require that for each level all packed items of the same width have to be reshaped by the same factor e. The variant of the items models on how many stacks the items meant to be placed. The total height of the packed variant of items with the same width represents the height if the items can be partitioned between the stacks ideally. Therefore, the master problem is indeed a relaxation of the original problem since it models the first two stages but not the third.

The next consideration concerns the maximal number of variants that has to be provided for each item. For a given item, the parameter e can be restricted on the one hand by the width of the strip and on the other hand by the total number of items of that width that can be packed into the considered level. The maximal number of different variants for a level j and a width w_g is

$$e_{\max}(j, g) = \min\left(\left\lfloor \frac{W_S - w_{\omega_j}}{w_g} \right\rfloor + \begin{cases} 1 & \omega_j = g \\ 0 & \text{otherwise} \end{cases}, \ |\{i = j, \ldots, n \mid \omega_i = g\}|\right).$$

The first term of the minimum calculates the maximum number of items of width w_g that can be placed next to each other without exceeding W_S. The second term yields the number of items of width w_g that can be packed into level j.

The master problem is modeled by using three sets of variables: Binary variables x_{jie} which are set to 1 iff item i in variant e is assigned to level j. Binary variables y_{jge} which are set to 1 iff an item in variant e with original width w_g is assigned to level j. Integral variables z_j which are set to the height of the corresponding level j. To ease the upcoming notation we denote with $[i, n]$ the set $\{i, \ldots, n\}$ and with $E(j, g)$ the set $\{1, \ldots, e_{\max}(j, g)\}$. The master problem is defined as follows:

$$\min \sum_{j \in [1,n]} z_j \tag{1}$$

$$\text{s.t.} \sum_{j \in [1,i]} \sum_{e \in E(j, \omega_i)} x_{jie} = 1 \qquad \forall i \in [1, n] \tag{2}$$

$$x_{jie} \leq y_{j\omega_i e} \qquad \forall j \in [1, n], \ \forall i \in [j, n], \ \forall e \in E(j, \omega_i) \tag{3}$$

$$\sum_{e \in E(j, g)} y_{jge} \leq 1 \qquad \forall j \in [1, n], \ \forall g \in [\omega_j, p] \tag{4}$$

$$\sum_{i \in [j, n] \mid \omega_i = g} x_{jie} \geq e \, y_{jge} \qquad \forall j \in [1, n], \ \forall g \in [\omega_j, p], \ \forall e \in E(j, g) \tag{5}$$

$$\sum_{e_1 \in E(j, \omega_j)} x_{jje_1} \geq x_{jie_2} \qquad \forall j \in [1, n-1], \ \forall i \in [j+1, n], \ \forall e_2 \in E(j, \omega_i) \tag{6}$$

$$\sum_{g \in [\omega_j, p]} \sum_{e \in E(j, g)} w_{ge} y_{jge} \leq W_S \qquad \forall j \in [1, n] \tag{7}$$

$$h_{\lambda_i} x_{jie} \leq z_j \qquad \forall j \in [1, n], \ \forall i \in [j, n], \ \forall e \in E(j, \omega_i) \tag{8}$$

$$\sum_{i \in [j, n] \mid \omega_i = g} h_{\lambda_i e} x_{jie} \leq z_j \qquad \forall j \in [1, n], \ \forall g \in [\omega_j, p], \ \forall e \in E(j, g) \tag{9}$$

Inequalities (2) force that each item has to be packed in a single variant exactly once. Equations (3) link the variables x_{jie} and $y_{j\omega_i e}$. The restriction that items of the same width have to be packed in the same variant is guaranteed by (4). Inequalities (5) ensure that items are not meant to be placed on more stacks than the number of items. Constraints (6) impose that items can only be packed into a level j iff also item j is packed into level j. Constraints (7) disallows that the total width of the packed item variants exceed the strip width W_S. Constraints (8) and (9) ensure that the level is at least as high as every single item in it and at least as high as the total height of all packed items of the same width in their corresponding variant.

2.2 Subproblem

The master problem is a relaxation of the 3SPP, since it assumes that items of the same width can be partitioned, s.t. the resulting stacks are of equal height. The subproblems determine the actual packing of the items and with it the actual level height. The resulting subproblems consist of assigning items, of the same width packed by the master into the same level, into the number of stacks determined by their item variant. The aim is the minimize the height of the highest stack. This problem corresponds to the $P||C_{max}$ problem [3]. We use for the subproblem a straightforward ILP formulation which was solved in less than a second in all considered instances.

2.3 Benders Cuts

The aim of Benders cuts is to incorporate the knowledge obtained in the subproblems back into the master problem. In the simple case a Benders cut states that if a set of items is packed in a certain variant, then the height of the level is at least as high as the result of the corresponding subproblem.

Let $\bar{I}$ and $\bar{e}$ be the set of items and their variant that have defined a subproblem and let $\bar{z}$ be the objective value of a corresponding optimal solution. Moreover, the set J' contains those levels that allow an assignment of items, s.t. the Benders cut can get activated. The simple version of our Benders cuts is

$$\left(\sum_{i \in \bar{I}} x_{ji\bar{e}} - |\bar{I}| + 1 \right) \bar{z} \leq z_j \quad \forall j \in J'. \tag{10}$$

These Benders cuts have the disadvantage that they do not affect item assignments differing from I' only in that items are exchanged by congruent items, i.e., items having the same width and height. The extended Benders cuts aim to overcome this drawback. Let $\bar{H} \subseteq H$ be the set of heights of the items from a subproblem defined by $\bar{I}$ and $\bar{e}$. We introduce for each height $h \in \bar{H}$ a binary variable u_h which is set to true iff at least as many items of the same width, height and variant are packed into a level as it has been packed in the considered subproblem. The set $I' \subseteq I$ contains all items having the corresponding width and height. The constraints that set the u_h variables are given by

$$\sum_{i \in I'} x_{ji\bar{e}} - |\bar{I}| + 1 \leq u_h(|I'| - |\bar{I}| + 1) \quad \forall h \in \bar{H}, \forall j \in J'. \tag{11}$$

The extended Benders cuts impose that the height of a level is at least as high as in the subproblem if all corresponding u_h variables are set to 1 and is defined as

$$\left(\sum_{h \in \bar{H}} u_h - |\bar{H}| + 1 \right) \bar{z} \leq z_j \quad \forall j \in J'. \tag{12}$$

Moreover, we iteratively exclude the smallest item of $\bar{I}$ and resolve the subproblem as long as the objective of the optimal solution does not change. The resulting Benders cuts are in general stronger and reduce the number of master iterations.

3 Compact Formulation

The used compact formulation is straightforward, hence we omit an exact specification. The main idea is to pack items first into stacks and then pack stacks into levels. To model this, we use binary variables v_{ki} that are set to one if item i is packed into strip k and binary variables u_{jk} to express that stack k resides in level j. Each item has to be packed exactly once and each stack containing items is allowed to appear in exactly one level. Moreover, we have to ensure that the total width of all stacks belonging to the same level does not exceed W_S. For each of the potentially n levels an integer variable is used which has to be at least as high as the highest residing stack. Furthermore, we applied the symmetry breaking described in Sect. 2.1.

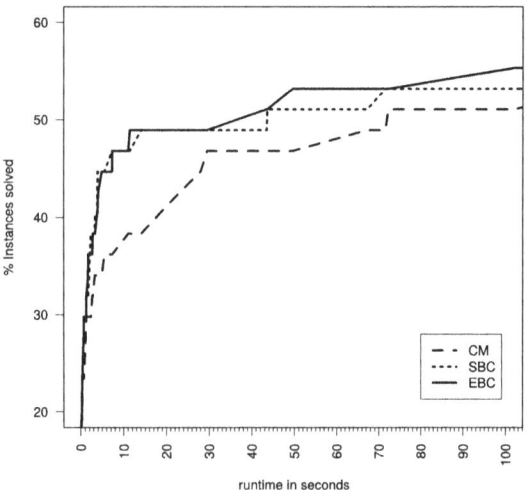

Fig. 1 Performance profile of the first 100 s for the compact formulation (CM) and for the presented LBBD with simple Benders cuts (SBC) and extended Benders cuts (EBC) on the instance sets beng, cgcut, gcut, ht and ngcut from [2]

4 Computational Results

The algorithms have been implemented in C++ and tested on an Intel Xeon E5-2630 v2, 2.60 GHz using Ubuntu 14.04. The ILP formulations have been solved with IBM Ilog Cplex 12.6.2 using the same parameter setting as in [2]. All algorithms had a time limit of 7200 s. For the benchmark we use the instance sets beng, cgcut, gcut, ht and ngcut from [2].

We compare the compact formulation against our LBBD with simple Benders cuts and with extended Benders cuts. The compact model could solve 31 out of 47 test instances to optimality, which is only marginally outperformed with 32 optimally solved test instances by the LBBD using either simple or extended Benders cuts. However, the LBBD can solve some instances considerably faster as Fig. 1 shows. For instance, after 10 s the LBBD with simple and with extended Benders cuts could solve 22 test instances to optimality, while the compact model could optimally solve 17 test instances.

5 Conclusion

We proposed a novel LBBD for the 3SPP and compared it with a compact formulation. The master problem relaxes the 3SPP s.t. only the first two stages of guillotine cuts are determined. The resulting subproblems are iteratively resolved to strengthen the generated Benders cuts. In addition, we proposed two kinds of Benders cuts. The experimental results have shown that the presented LBBD can solve substantially more test instances in the first 100 s. The LBBD can solve one test instance more than the compact model within the time limit. More testing is necessary to see under which conditions the proposed approach works especially well.

References

1. Baker, B.S., Coffman Jr., E.G., Rivest, R.L.: Orthogonal packings in two dimensions. SIAM J. Comput. **9**(4), 846–855 (1980)
2. Côté, J.F., Dell'Amico, M., Iori, M.: Combinatorial benders' cuts for the strip packing problem. Operations Research **62**(3), 643–661 (2014)
3. Graham, R.L., Lawler, E.L., Lenstra, J.K., Kan, A.R.: Optimization and approximation in deterministic sequencing and scheduling: a survey. Ann. Discret. Math. **5**, 287–326 (1979)
4. Hifi, M.: Exact algorithms for the guillotine strip cutting/packing problem. Computers & Operations Research **25**(11), 925–940 (1998)
5. Hooker, J.N., Ottosson, G.: Logic-based benders decomposition. Math. Program. **96**(1), 33–60 (2003)
6. Lodi, A., Martello, S., Vigo, D.: Models and bounds for two-dimensional level packing problems. J. Comb Optim. **8**(3), 363–379 (2004)
7. Puchinger, J., Raidl, G.R.: Models and algorithms for three-stage two-dimensional bin packing. Eur. J. Oper. Res. **183**(3), 1304–1327 (2007)
8. Puchinger, J., Raidl, G.R., Koller, G.: Solving a real-world glass cutting problem. Evolutionary Computation in Combinatorial Optimization, pp. 162–173. Springer, Berlin (2004)

Optimization Model for the Design of Levelling Patterns with Setup and Lot-Sizing Considerations

Mirco Boning, Heiko Breier and Dominik Berbig

Abstract Production levelling (Heijunka) is one of the key elements of the Toyota Production System and decouples customer demand from production orders. For the decoupling period a levelling pattern has to be designed. Existing approaches for the design of levelling patterns are majorly limited to large-scale production. Therefore, this article proposes a novel optimization model regarding the requirements of lot-size production. Relevant, sequence-dependent changeovers are considered. An integer, combined lot-sizing and scheduling model is formulated. The four target criteria changeover times, smoothness of daily workload, variance of lot-sizes and similarity of production sequences are aggregated into one optimization model. In a real case study of an existing production plan a clear improvement of changeover times, similarity and smoothness of workloads is realized.

1 Introduction to Production Levelling

One major problem of production planning is caused by the limited flexibility which exists in adapting the output of the production resources to a varying, fluctuating customer demand. In a globalized, highly-competitive market only limited rules for the timing of customer orders can be established. Therefore, a strict following of customer orders by production leads to undesired inefficiencies in production plans. One approach to tackle this issue is proposed by the well-known Toyota Production System with the concept of levelling (also production smoothing or Heijunka) [9]. Levelling decouples customer demand from production orders for a fixed period of time. For this levelling period, a levelling pattern needs to be designed. The pattern

M. Boning (✉) · H. Breier
Karlsruhe Institute of Technology (KIT), 76131 Karlsruhe, Germany
e-mail: mirco.boning@web.de

H. Breier
e-mail: heiko.breier@kit.edu

D. Berbig
Robert Bosch GmbH, 70771 Leinfelden-echterdingen, Germany
e-mail: Dominik.Berbig@de.bosch.com

© Springer International Publishing Switzerland 2017
K.F. Dœrner et al. (eds.), *Operations Research Proceedings 2015*,
Operations Research Proceedings, DOI 10.1007/978-3-319-42902-1_54

determines at which production day, which product, in which quantity (lot-size) and in which position (order) has to be produced. Levelling aims at patterns which are balanced in production volume as well as in production mix [4]. As a result, a reliable, balanced plan and a smoothened production rhythm can be communicated with all suppliers of the underlying supply chain. The impact of the bullwhip effect can be decreased and spare capacity or stocks to cope with demand peaks can be reduced [5].

2   Existing Approaches and Related Problems

The design of levelling patterns is nothing new and many approaches have been described in literature. Existing approaches can be classified into procedure models and optimization models [1]. Procedure models describe systematic approaches which contain a set of structured rules for the design of levelling patterns. Such approaches are presented in [11, 13, 14]. A good summary can be found in [2]. A major disadvantage of all procedure models is the lack of specific analytical descriptions, rules or algorithms for the design of levelling patterns. Therefore, the second class of optimization model tries to close this gap. For large-scale production a lot of research has been published on designing levelling patterns for mixed-model assembly lines. The underlying problem is referred to as Production Smoothing Problem (PSP) or level scheduling. The PSP aims at finding a production sequence which minimizes the deviation from ideal to actual objective values [1]. An excellent literature survey can be found in [3]. But due to the specific assumptions of the PSP (lot-size one and negligible changeover times) a generated production plan will not satisfy the requirements of traditional lot-size production. For the levelling of lot-sizes some existing research focuses on the Batch Production Smoothing Problem (BSP) which still ignores changeover times [7]. For lot-sizes and changeover times a promising approach is presented by [2]. The author uses the Traveling Salesmen Problem (TSP) for the generation of levelling patterns, but the smoothness of the production plan is not assured on a mathematical basis. Therefore, this article closes this research gap by capturing the levelling targets in an optimization model for lot-size production with relevant changeover times.

3   Modeling Approach

The basis of this model is the Distance-Constrained Vehicle Routing Problem (DCVRP), see [8] for an introduction. The DCVRP has been selected due to many analogies between routing and scheduling problems [12]. The following notation is introduced: In $k \in K$ workdays $i \in I$ products with a specific demand D_i must be produced. n denotes the total number of products and n_{WD} the number of workdays. A dummy product 0 is introduced to represent an idle state at the beginning and end of each day. PT_k denotes the available production time on day k and PTU_k models the used production time. $t_{CT,i}$ denotes the cycle time of i. $t_{CO,ij}$ denotes the changeover time from i to j. The binary decision variable y_{ijk} equals 1 if a changeover from i to j is conducted on day k. An integer decision variable x_{ik} models the production quantity of i on day k. For each product a specific $EPEI_i$ (Every Part Every Interval) has to be regarded: If $EPEI_A = 1$, the runner product A must be produced every day.

The following assumptions are drawn: The capacity of the production resources is limited. Planning is based on the final product stage (no levelling of subassemblies or components). Demand must be fulfilled and stock-outs are not permitted. A maximum of one lot per product can be produced per day. The changeover status at the end of one production day is not taken over to the next day. All input parameters are deterministic. Stochastic or dynamic influences are not considered. Changeover times are decision-relevant and lot-size one is impossible. The model can now be formulated as:

$$
\min \ \lambda_{Uti} \sum_{k=1}^{|K|-1} \left| \frac{PTU_k}{PT_k} - \frac{PTU_{k+1}}{PT_{k+1}} \right| + \lambda_{CO} \sum_{k \in K} \frac{\sum_{i \in I} \sum_{j \in V} y_{ijk} \cdot t_{CO,ij}}{PT_k}
$$

$$
- \lambda_{Sim} \frac{1}{|K|-1} \sum_{k=1}^{|K|-1} \frac{\sum_{i \in I} \sum_{j \in J} y_{ijk} \cdot y_{ijk+1}}{\sum_{i \in I} \sum_{j \in J} sign(y_{ijk} + y_{ijk+1})}
\tag{1}
$$

$s.t. :$

$$
\sum_{j \in I} \sum_{k \in K} y_{0jk} = n_{WD}
\tag{2}
$$

$$
\sum_{i \in I} \sum_{k \in K} y_{i0k} = n_{WD}
\tag{3}
$$

$$
\sum_{i \in I} y_{ihk} = \sum_{j \in I} y_{hjk} \qquad \forall h \in I, \forall k \in K
\tag{4}
$$

$$
\sum_{j \in I} y_{ijk} \leq 1 \qquad \forall i \in I, \forall k \in K
\tag{5}
$$

$$
\sum_{k \in K} x_{ik} = D_i \qquad \forall i \in I
\tag{6}
$$

$$
x_{ik} - M \sum_{j \in I} y_{ijk} \leq 0 \qquad \forall i \in I, \forall k \in K
\tag{7}
$$

$$
\sum_{j \in I} \sum_{k=\tilde{k}}^{\tilde{k}+\overline{EPEI}_i-1} y_{ijk} = 1 \qquad \forall i \in I, \qquad \forall \tilde{k} \in \{K : \tilde{k} \leq |K| - \overline{EPEI}_i + 1\}
\tag{8}
$$

$$
\sum_{i \in I} \sum_{j \in I} (y_{ijk} \cdot t_{CO,ij} + x_{ik} \cdot t_{C,i}) \leq PT_k \qquad \forall k \in K
\tag{9}
$$

$$PTU_k = \sum_{i \in I} \sum_{j \in J} (y_{ijk} \cdot t_{CO,ij} + x_{ik} \cdot t_{C,i}) \qquad \forall k \in K \quad (10)$$

$$x_{ik} \leq UB_i \qquad \forall i \in I, \forall k \in K \quad (11)$$

$$u_{0k} = 1 \qquad \forall k \in K \quad (12)$$

$$2 \leq u_{ik} \leq n + 1 \qquad \forall i \in I, \forall k \in K \quad (13)$$

$$u_{ik} - u_{jk} + 1 \leq n \cdot (1 - y_{ijk}) \qquad \forall i \in I, \forall j \in I, \forall k \in K \quad (14)$$

$$u_{ik} \in \{2, 3, \dots, n, n + 1\} \qquad \forall i \in I, \forall k \in K \quad (15)$$

$$y_{ijk} \in \{0, 1\}, x_{ik} \in \mathbb{Z}_+ \qquad \forall i \in I, \forall j \in I, \forall k \in K \quad (16)$$

The proposed target function (1) combines three levelling targets: The first part assures that the deviations of daily utilizations should be as smooth as possible. The second part minimizes the sum of changeover times relative to the production time. The third part assures that the order of runner-products should be as similar as possible to benefit from economies-of-repetition. A similarity measure for the VRP based on the Jaccard-Index has been proposed by [6] and is adapted in this article for levelling purposes. The $sign()$-function is taking the value 1 if either y_{ijk} or y_{ijk+1} equal 1 and can be modeled with a binary auxiliary variable. As similarity is maximized, the negative sign is used. All three components are weighted with the factors λ_{Uti}, λ_{CO} and λ_{Sim}.

Constraints (2) and (3) assure that the dummy state is reached at the beginning and end of each day. Equation (4) assures that if a changeover to a product h is planned, a changeover from h to another product must be conducted as well. Equation (5) assures that each product can only be produced once per day. Fulfilling the demand is assured by (6). Equation (7) constrains that product i can only be produced if a changeover from i is conducted (M presents a big number, e.g. the total demand for i). Equation (8) models the EPEI. Example if the EPEI is 3 for 3 production days, production must occur exactly once on either day 1, 2 or 3. Equation (9) is the capacity constraint which assures that the available daily production time is not exceeded. Equation (10) calculates the daily used production time which is necessary for the target function. One disadvantage of solutions of (1) is that lot-sizes on product-level can fluctuate at lot. From the viewpoint of the Lean-Philosophy only a small variance of lot-sizes should be reached to avoid demand peaks for part suppliers. Therefore, restriction (11) captures an upper bound UB_i for each x_i. For the calculation of UB_i the demand is spread evenly over the production occurrences: $UB_i = \left\lceil \frac{D_i}{\frac{n_{WD}}{EPEI_i}} \right\rceil$

Equations (12)–(15) represent the subtour elimination constraints to exclude impossible subcyles in the production plan according to the formulation of [10]. u_{ik} is an auxiliary variable which indicates the position of i in the production sequence on day k. Equation (16) restricts the range of the decision variables.

4   Results and Discussion

With the proposed target function four levelling targets are achieved: Production plans are smooth, repetitive, balanced in production mix and economic (long changeovers are avoided). The following example taken from a lot-size producer in the manufacturing industry demonstrates the desired properties.

For a production line with 17 products a levelling pattern needs to be designed for a fixation horizon of 10 days (2 weeks). 6 runner products with an EPEI of 1 are produced every day. Production orders are placed only in multiples of full pallet-sizes. All product specific input data is presented in Table 1. On each day 630 min are available for production and changeover times; all further OEE-losses are already considered. The weights for the target function λ_{CO}, λ_{Uti} and λ_{Sim} are all set to $\frac{1}{3}$. Initial and final setup times to the dummy product are set to 0. The optimization model has been implemented with Gurobi version 6.0.0 on a standard PC with 3.0 GHZ and 4 GB of RAM in multi thread mode with four cores. Optimization runtime is limited to 1 h. The optimized production plan is visualized in Fig. 1.

It can easily be seen that all four levelling targets are achieved. All runner products are produced in a repetitive sequence. Except for day 9 the daily production volume is almost perfectly smooth. The deviation on day 9 results due to the uneven demand of pallets which do not match the production days (e.g. demand for product B is 36 pallets, so 3,6 pallets is the ideal production rate per day, but only full pallets are allowed). Moreover, the results reveal that the planned available production time is too high and can be significantly reduced.

Compared to the previous production plan in Fig. 2 calculated by a myopic heuristic only considering changeover times, workload smoothness can be improved by 62 % (first component of (1)), changeover times by 37 % and similarity by 19 %. Due to the production in full pallet-sizes the fluctuation of lot-sizes can't be improved. For

Table 1 Product-specific input data

Product i	Demand D_i (pallets)	$EPEI_i$ (days)	Cycle time $T_{C,i}$ (min per pallet)	Product i	Demand D_i (pallets)	$EPEI_i$ (days)	Cycle time $T_{C,i}$ (min per pallet)
Product A	20	1	25, 8	Product J	4	5	25, 8
Product B	36	1	30, 6	Product K	20	1	25, 4
Product C	4	5	25, 8	Product L	8	3	28, 6
Product D	20	1	25, 8	Product M	4	5	28, 6
Product E	4	5	25, 8	Product N	20	1	25, 6
Product F	8	2	25, 8	Product O	4	5	29, 5
Product G	8	2	25, 8	Product P	4	5	25, 4
Product H	20	1	25, 8	Product Q	4	5	25, 4
Product I	8	2	25, 8				

Fig. 1 Levelled production plan

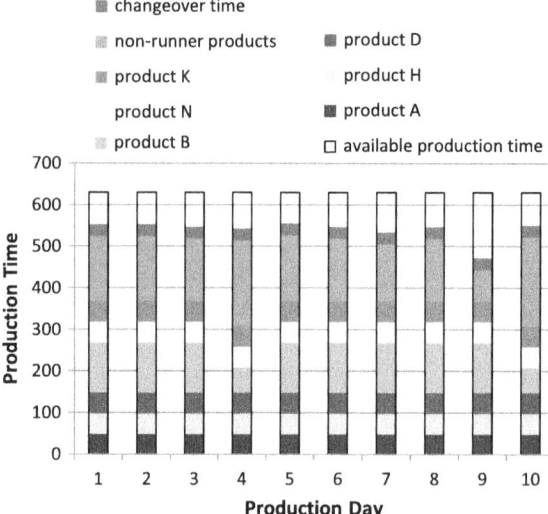

Fig. 2 Previous production plan created by myopic changeover heuristic

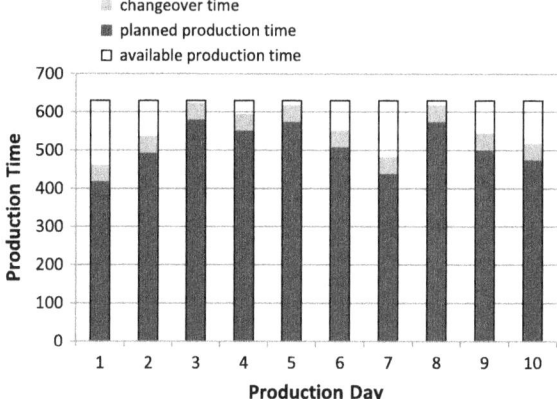

future research the proposed model offers many opportunities for either refinement or more efficient solution methods. Our experiments show that for models up to 100 products the solver can find acceptable solutions with an optimality gap below 10 %. However, due to the exponentially growing number of variables, solutions for bigger problems do not possess the desired properties any more. Therefore, the development of meta-heuristics such as multi-criteria genetic algorithms offers interesting potential for future research.

Appendix

References

1. Bohnen, F., Buhl, M., Deuse, J.: Systematic procedure for leveling of low volume and high mix production. CIRP J. Manuf. Sci. Technol. **6**(1), 53–58 (2013)
2. Bohnen F.M.: Eine Methodik zur Produktionsnivellierung auf der Basis von Fertigungsfamilien, Industrial Engineering, vol 12. Dissertation. Shaker Verlag, Aachen (2013)
3. Boysen, N., Fliedner, M., Scholl, A.: Sequencing mixed-model assembly lines: Survey, classification and model critique. European Journal of Operational Research **192**(2), 349–373 (2009)
4. Dennis, P.: Lean Production Simplified, 2nd edn. Productivity Press, New York (2007)
5. Furmans, K.: Models of heijunka-levelled kanban-systems. In: Papadopoulos, C. (ed.) 5th International Conference on Analysis of Manufacturing Systems - Production and Management, pp. 243–248. Publishing Ziti, Thessaloniki (2005)
6. Garcia, A., Bullinaria, J.A.: Bi-objective optimization for the vehicle routing problem with time windows: using route similarity to enhance performance. In: Ehrgott, M. (ed.) Evolutionary Multi-Criterion Optimization: 5th International Conference, EMO 2009, pp. 275–289. Springer, Berlin (2009)
7. Kubiak, W., Yavuz, M.: Just-in-time smoothing through batching. Manufacturing & Service Operations Management **10**(3), 506–518 (2008)
8. Laporte, G.: The vehicle routing problem: An overview of exact and approximate algorithms. Eur. J. Oper. Res. **59**(3), 345–358 (1992)
9. Liker, J.K.: The Toyota Way. McGraw-Hill, New York (2004)
10. Miller, C.E., Tucker, A.W., Zemlin, R.A.: Integer programming formulation of traveling salesman problems. J. ACM **7**(4), 326–329 (1960)
11. Monden, Y.: Toyota Production System: An Integrated Approach to Just-In-Time, 4th edn. CRC Press, Boca Raton (2012)
12. Selensky E. (2001) On mutual reformulation of shop scheduling and vehicle routing. In: Proceedings of the 20th UKPLANSIG, pp. 282–291
13. Smalley, A.: Creating level pull, 1st edn. Lean Toolkit Workbooks, The Lean Enterprise Institute, Brookline (2004)
14. Takeda, H.: The Synchronized Production System. Going Beyond Just-In-Time Through Kaizen. Kogan Page, London (2006)

Treating Scale-Efficiency Gaps
in Peer-Based DEA

Andreas Dellnitz

Abstract Data envelopment analysis (DEA) is a method for calculating relative efficiency as a ratio of weighted outputs to weighted inputs of decision making units (DMUs). It is well-known that DEA can be done under the assumption of constant returns to scale (CRS) or variable returns to scale (VRS). One major disadvantage of the classical approach is that each DMU optimizes its individual weighting scheme–often called self-appraisal. To overcome this flaw cross-efficiency evaluation has been developed as an alternative way of efficiency evaluation and ranking of DMUs. Here all individual weighting schemes–called price systems–are applied to the activities of all DMUs. The derived cross-efficiency matrix can form the basis for seeking a consensual price system–a peer–, and hence this price system can be used for a peer-based activity planning. The present contribution shows that a scale-efficiency gap can occur when peer-based activity planning under VRS is applied, i.e. there is no feasible point in which self-appraisal efficiency under CRS, VRS and peer-appraisal efficiency under VRS coincide. As a consequence, we propose a mixed integer linear problem to avoid this drawback.

1 Introduction

Data Envelopment analysis (DEA), in the essentials developed by [3], is a method for comparative efficiency analysis among profit and non-profit entities, so called Decision-Making-Units (DMUs). In this procedure the activity–observed inputs and outputs–of a DMU is compared with the activities of all other DMUs, which results in an efficiency ratio for the DMU under evaluation. Furthermore, the DMU gets to know even more about its economic position: Pure technical and scale-efficiency can be determined separately by consideration of different models–one under variable

A. Dellnitz (✉)
Department of Operations Research, University of Hagen,
Universitätsstraße 41, 58097 Hagen, Germany
e-mail: andreas.dellnitz@fernuni-hagen.de

© Springer International Publishing Switzerland 2017   409
K.F. Dœrner et al. (eds.), *Operations Research Proceedings 2015*,
Operations Research Proceedings, DOI 10.1007/978-3-319-42902-1_55

returns to scale (VRS) and the other one under constant returns to scale (CRS). If a DMU's activity is efficient under both models the DMU has a good economic position called most productive scale size (mpss), cf. [1]. Moreover, the multiplier form under VRS informs about the returns to scale (RTS) of the respective DMU, cf. [2, 6] have made a first attempt to utilize scale effects for activity changes, based on such RTS. This is what classical self-appraisal is about.

The main criticism of the self-appraisal approach is that each unit rates its own efficiency in the most favorable way. A supervising institution might oppose such opportunistic attitude and force the DMUs to accept the weight system of a so called peer-DMU as a common denominator for them all. Hence, there is an increasing interest in crosswise evaluation of DMUs by other DMUs in recent DEA literature, cf. [5]. The choice of a peer has far-reaching consequences, however. Inefficient DMUs now must improve their activities from the viewpoint of a peer rather than from their own position. For first approaches towards peer-based activity planning cf. [7]. Nevertheless, these authors focus on peer-selection and (cross-)efficiency improvements by considering *one* model–VRS or CRS–, only. A global perspective–VRS *and* CRS–on the technology and a corresponding mpss concept is still missing. In the present paper we pick up this issue. Even a mpss-DMU does not necessarily meet the peer's philosophy of (cross-) efficiency, we call this effect scale-efficiency gap. After presenting basics of DEA in Sect. 2, this new phenomenon is developed in Sect. 3.1. Section 3.2 then introduces a mixed integer program to avoid such gaps. Finally, Sect. 4 summarizes all findings and points to future research.

2   Preliminaries

This section is dedicated to the necessary theoretical concepts of DEA. Equation (1) presents the classical CCR-model in multiplier form and problem (2) the respective BCC-model. CCR and BCC are acronyms of their creators [2, 3]; whenever convenient we apply such acronyms. Activities $(\mathbf{x}_j, \mathbf{y}_j)$ are the inputs and outputs of DMU $j, j = 1 \ldots J$. $\mathbf{U}_k, \mathbf{V}_k, u_k$ and $\bar{\mathbf{U}}_k, \bar{\mathbf{V}}_k, \bar{u}_k$ are the price systems in the corresponding models. In the remainder of this contribution we focus on multiplier form and input orientation.

For DMU $k, k \in \{1, \ldots, J\}$ solve

$$\max g_k = \mathbf{U}_k^\mathsf{T} \mathbf{y}_k + u_k$$
$$s.t. \ \mathbf{V}_k^\mathsf{T} \mathbf{x}_k = 1$$
$$\mathbf{U}_k^\mathsf{T} \mathbf{y}_j + u_k - \mathbf{V}_k^\mathsf{T} \mathbf{x}_j \leqq 0 \quad \forall j$$
$$\mathbf{U}_k, \mathbf{V}_k \geq \mathbf{0} \text{ and } u_k = 0 \tag{1}$$

or

$$\max \bar{g}_k = \bar{\mathbf{U}}_k^\mathsf{T} \mathbf{y}_k + \bar{u}_k$$
$$s.t. \quad \bar{\mathbf{V}}_k^\mathsf{T} \mathbf{x}_k = 1$$
$$\bar{\mathbf{U}}_k^\mathsf{T} \mathbf{y}_j + \bar{u}_k - \bar{\mathbf{V}}_k^\mathsf{T} \mathbf{x}_j \leqq 0 \quad \forall j$$
$$\bar{\mathbf{U}}_k, \bar{\mathbf{V}}_k \geqq \mathbf{0} \text{ and } \bar{u}_k \text{ free} \tag{2}$$

Let $\mathbf{U}_k^*, \mathbf{V}_k^*, u_k^* = 0, g_k^*$ be the optimal solution of (1) and $\bar{\mathbf{U}}_k^*, \bar{\mathbf{V}}_k^*, \bar{u}_k^*, \bar{g}_k^*$ be the optimal solution of (2). The sign of $\bar{u}_k^* = / > / < 0$ indicates DMU k's RTS-characteristics–constant-/increasing-/decreasing-RTS, cf. [2]. Obviously, $u_k^* = 0$ in problem (1) makes DMU k work always under constant RTS.

Definition 1 If a DMU k is scale-efficient SE $= \frac{g_k^*}{\bar{g}_k^*} = 1$ and meets the condition $g_k^* = \bar{g}_k^* = 1$, then the DMU is called mpss-DMU.

If a DMU k is scale-efficient but not efficient, the input projection of DMU k has most productive scale size (mpss), cf. [1].

All these findings are valid in case of self-appraisal and widely unexplored for peer-appraisal, however. The main concept of peer-appraisal is cross-efficiency evaluation. Cross-efficiencies are the efficiencies of DMUs from the viewpoint of other DMUs. For an arbitrary DMU $l \in \{1, \ldots, J\}$ the CCR cross-efficiency from k's perspective is defined as

$$g_{kl}^* = \frac{\mathbf{U}_k^{*\mathsf{T}} \mathbf{y}_l}{\mathbf{V}_k^{*\mathsf{T}} \mathbf{x}_l} \tag{3}$$

and for BCC it is

$$\bar{g}_{kl}^* = \frac{\bar{\mathbf{U}}_k^{*\mathsf{T}} \mathbf{y}_l + \bar{u}_k^*}{\bar{\mathbf{V}}_k^{*\mathsf{T}} \mathbf{x}_l}. \tag{4}$$

These cross-efficiencies are typically ordered in a table called cross-efficiency matrix. For (3) each of its entries tells us about the relative input distance of l with respect to k's CCR-efficiency hyperplane at level 1 and for (4) the respective BCC-efficiency hyperplane. In general the price systems $\mathbf{U}_k^*, \mathbf{V}_k^*, u_k^* = 0$ and $\bar{\mathbf{U}}_k^*, \bar{\mathbf{V}}_k^*, \bar{u}_k^*$ are not unique and consequently the cross-efficiencies not either. Doyle and Green [5] as well as [7] give a two-stage approach to avoid such ambiguities. In the last contribution the authors propose decision rules how to determine a global price system–a peer. The peer prices now are obligatory weights for the remaining DMUs.

3   Treating Scale-Efficiency Gaps in Peer-Based Activity Planning

3.1   Scale-Efficiency Gap: A Peer-Based Problem

If a peer has been selected, all DMUs are invited to improve their cross-efficiencies from the viewpoint of the respective peer. So the cross-inefficient DMUs should strive in the direction of cross-efficient activities. In [7] the authors show that doing so each DMU can achieve a fully cross-efficient activity in either model. Unfortunately, the authors did not tackle the problem whether or not such activity changes make a DMU mpss; this is not always the case for (2)! Before we deepen this issue, we introduce a cross type of a mpss-DMU.

Definition 2   If a DMU l meets the condition $g_l^* = \bar{g}_l^* = \bar{g}_{kl}^* = 1$, then the DMU is called a cross mpss-DMU.

Now back to the problem. For the sake of transparency consider Fig. 1. Therein the line passing through the origin forms the boundary of the CCR-technology, the piecewise linear part that one of the BCC-technology. Table 1 contains the BCC cross-efficiencies for all possible peers k.

If we now for whatever reason nominate the price system of DMU 1 to be the peer in this particular case, the illustrated dashed line is the corresponding BCC-efficiency hyperplane.

We observe that for the activity $(\mathbf{x}_i, \mathbf{y}_i)$, which is BCC-efficient $\bar{g}_i^* = 1$, both efficiencies–the CCR-efficiency and the BCC cross-efficiency from the viewpoint of DMU 1–coincide and are less than one. Obviously, this is due to the input projection of i onto the intersection of both hyperplanes–point $(\tilde{\mathbf{x}}, \tilde{\mathbf{y}})$. If we move from activity i towards the activity of DMU 3, the activity then becomes a mpss but loses BCC cross-efficiency. This effect reverses when i strives in the direction of DMU 2. After

Fig. 1 Scale-efficiency gap

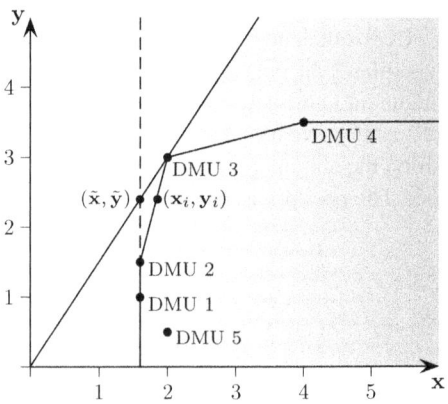

Table 1 BCC cross-efficiencies

DMU	$l = 1$	$l = 2$	$l = 3$	$l = 4$	$l = 5$
$k = 1$	1.00	1.00	0.80	0.40	0.80
$k = 2$	0.92	1.00	1.00	0.53	0.67
$k = 3$	0.92	1.00	1.00	0.53	0.67
$k = 4$	−3.75	−2.50	1.00	1.00	−4.00
$k = 5$	1.00	1.00	0.80	0.40	0.80

a first visual introduction of the problem we proceed to a more formal mathematical approach.

Let $(\tilde{\mathbf{x}}, \tilde{\mathbf{y}})$ be an element of the intersection of the boundary of CCR-technology and an arbitrary BCC-efficiency hyperplane. If the BCC-efficiency of $(\tilde{\mathbf{x}}, \tilde{\mathbf{y}})$

- is equal 1, $(\tilde{\mathbf{x}}, \tilde{\mathbf{y}})$ lies
- is greater than 1, $(\tilde{\mathbf{x}}, \tilde{\mathbf{y}})$ does not lie

in BCC-technology.

Proposition 1 Let $(\mathbf{x}_i, \mathbf{y}_i)$ be an activity with $g_i^* = \bar{g}_{ki}^* < 1$, $\bar{g}_i^* \leq 1$ and $\bar{g}_i^* \neq g_i^*$, then

- any improvement of its CCR-efficiency under constant BCC-efficiency worsens BCC cross-efficiency.
- any improvement of its BCC cross-efficiency under constant BCC-efficiency worsens CCR-efficiency.

The proof is beyond the scope of our paper; more details can be found in [4], see Theorem 5.2 on p. 145, Corollary 5.1 and Conclusion 5.2 on p. 146. An immediate consequence of Proposition 1 is the unreachability of a cross mpss-DMU in this case. The geometric location of $(\mathbf{x}_i, \mathbf{y}_i)$ implies in a tradeoff between the two objectives BCC cross-efficiency and scale-efficiency maximization. We call such a stuck scale-efficiency gap (SE-gap), and for $(\mathbf{x}_i, \mathbf{y}_i)$ it is measured by

$$\text{SE-gap}_i = \frac{g_i^*}{\bar{g}_i^*} = \frac{\bar{g}_{ki}^*}{\bar{g}_i^*} < 1. \tag{5}$$

A peer price system should always permit a cross mpss-DMU, we feel. In order to reach this goal a mixed integer problem is developed in the next section.

3.2 Avoid SE-Gap by Bridging CCR- and BCC-World

In Sect. 3.1 we discussed the possible problem of scale-efficiency gaps after a price system for activity planning purposes is fixed. To avoid such effects we have to ensure that prior to the peer seeking process each optimal BCC-solution contains at least one CCR-efficient activity. We achieve this goal through the following mixed integer problem called SE-gapless.

For DMU $k, k \in \{1, \ldots, J\}$ solve

$$\max \bar{g}_k = \bar{\mathbf{U}}_k^\top \mathbf{y}_k + \bar{u}_k$$

$$s.t. \quad \begin{array}{ll} \mathbf{V}_k^\top \mathbf{x}_k = 1 & \\ \mathbf{U}_k^\top \mathbf{y}_j - \mathbf{V}_k^\top \mathbf{x}_j \leq 0 & \forall j \\ \mathbf{U}_k^\top \mathbf{y}_j - \mathbf{V}_k^\top \mathbf{x}_j + M d_j \geq 0 & \forall j \end{array} \left. \begin{array}{l} \\ \\ \end{array} \right\} \text{CCR-World} \qquad (6)$$

$$\begin{array}{ll} \bar{\mathbf{V}}_k^\top \mathbf{x}_k = 1 & \\ \bar{\mathbf{U}}_k^\top \mathbf{y}_j + \bar{u}_k - \bar{\mathbf{V}}_k^\top \mathbf{x}_j \leq 0 & \forall j \\ \bar{\mathbf{U}}_k^\top \mathbf{y}_j + \bar{u}_k - \bar{\mathbf{V}}_k^\top \mathbf{x}_j + M d_j \geq 0 & \forall j \end{array} \left. \begin{array}{l} \\ \\ \end{array} \right\} \text{BCC-World} \qquad (7)$$

$$\sum_{j=1}^{J} d_j \leq J - 1 \qquad \qquad \left. \begin{array}{l} \\ \end{array} \right\} \text{Bridge} \qquad (8)$$

$$\mathbf{U}_k, \mathbf{V}_k, \bar{\mathbf{U}}_k, \bar{\mathbf{V}}_k \geq \mathbf{0}, \bar{u}_k \text{ free and } d_j \in \{0, 1\} \ \forall j$$

Let $\bar{g}_k^{**}$ be the solution of the above problem, then the relation $\bar{g}_k^{**} \leq \bar{g}_k^*$ holds because it is a combination of (1), (2) and thus it is more restrictive than (2). The bridging element–Eq. (8)–demands for at least one efficient activity in both worlds. The "sufficiently big" constant M refers to the well-known Big M method.

Proposition 2 *The SE-gapless problem is always feasible.*

The proof follows immediately from the fact that the BCC-technology contains at least one CCR-efficient activity, cf. [1].   □

Figure 2 indicates the allowable set of hyperplanes that meet the conditions of SE-gapless. Table 2 shows the impact of the new peer weights in accordance with

Fig. 2 Feasible hyperplanes

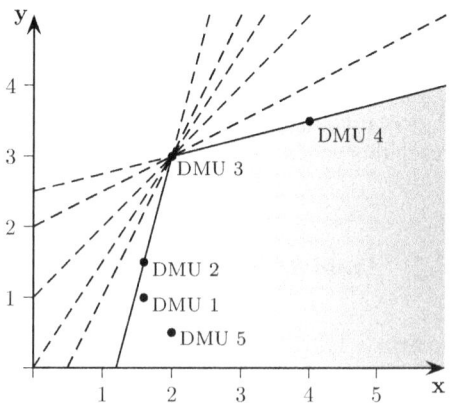

Table 2 SE-gapless BCC cross-efficiencies

DMU	$l = 1$	$l = 2$	$l = 3$	$l = 4$	$l = 5$
$k = 1$	0.92	1.00	1.00	0.53	0.67
$k = 2$	0.92	1.00	1.00	0.53	0.67
$k = 3$	0.92	1.00	1.00	0.53	0.67
$k = 4$	−3.75	−2.50	1.00	1.00	−4.00
$k = 5$	0.92	1.00	1.00	0.53	0.67

(6)–(8) on the BCC cross-efficiencies. From the perspective of the DMUs 2, 3 and 4 the BCC cross-efficiencies remain unchanged. However, the situation for the other two DMUs is quite different. In order to fulfill the second goal–integrating a feasible cross mpss-DMU–within the peer-selection process, the possible peer-DMUs 1, 5 have to revise their cross-efficiency assessment.

Finally, for treating multiple optimal solutions and selecting a peer one can grab the idea of benevolent cross-efficiency evaluation mentioned in [5, 7] and apply it to the SE-gapless problem.

4 Conclusion and the Road Ahead

In this contribution the phenomenon of scale-efficiency gaps within a peer-based DEA is introduced. To overcome this problem a mixed integer program is proposed. There is still one important open question: What is the most important goal in peer-based DEA–cross-efficiency maximization versus optimization of the individual evaluation? Generally speaking: What are the underlying economic key factors in peer-selection? And how can we incorporate such factors into this process?

References

1. Banker, R.D.: Estimating most productive scale size using data envelopment analysis. Eur. J. Oper. Res. (EJOR) **17**, 35–44 (1984)
2. Banker, R.D., Charnes, A., Cooper, W.W.: Some models for estimating technical and scale inefficiences in data envelopment analysis. Manag. Sci. **30**, 1078–1091 (1984)
3. Charnes, A., Cooper, W.W., Rhodes, E.: Measuring the efficiency of decision making units. Eur. J. Oper. Res. **2**, 429–444 (1978)
4. Dellnitz A (2015) Produktivitäts- und Effizienzverbesserungen in der DEA – Von der Selbst- zur Kreuzbewertung. Unpublished Dissertation, Hagen
5. Doyle, J., Green, R.: Efficiency and cross-efficiency in DEA: derivations, meanings and uses. J. Oper. Res. Soc. **45**, 567–578 (1994)
6. Kleine, A., Dellnitz, A., Rödder, W.: Sensitivity analysis of BCC efficiency in DEA with application to european health services. Operations Research Proceedings 2013, pp. 243–248. Springer, Berlin (2014)
7. Rödder, W., Reucher, E.: Advanced x-efficiencies for CCR- and BCC-models - towards peer-based DEA controlling. Eur. J. Oper. Res. **219**, 467–476 (2012)

On the Association Between Economic Cycles and Operational Disruptions

Kamil J. Mizgier, Stephan M. Wagner and Stylianos Papageorgiou

Abstract In this research we empirically verify the relationship between operational disruptions and economic cycles in the manufacturing industry of the United States. Contemporary and lagged correlation estimates are measured to demonstrate the degree of co-movement between the severity and the number of operational disruptions and several macroeconomic variables. Our findings suggest that the severity of operational disruptions follows the economic cycles with a lag of two years.

1 Motivation and Research Question

The management of operational disruptions has received increasing attention in operations and supply chain management research (e.g., [7, 8]). While researchers have mostly studied the classification, analysis and management of disruptions [4, 6], a pertinent knowledge gap exists with regard to economic cycles (or business cycles) and their relationship with operational disruptions. Therefore, in this research we focus on the identification of the relationship between operational disruptions (measured by operational risk losses) and economic cycles (measured by several key macroeconomic variables). Prior research dealing with the relationship between the state of the economy and operational risk is scarce and limited to operational disruptions in the financial services industry [1, 10]. We contribute to this existing

K.J. Mizgier (✉) · S.M. Wagner
Chair of Logistics Management, Department of Management,
Technology and Economics, Swiss Federal Institute of Technology Zurich,
Weinbergstrasse 56/58, 8092 Zurich, Switzerland
e-mail: kmizgier@ethz.ch

S.M. Wagner
e-mail: stwagner@ethz.ch

S. Papageorgiou
Chair of Macroeconomics: Innovation and Policy, Department
of Management, Technology and Economics, Swiss Federal Institute
of Technology Zurich, Zürichbergstrasse 18, 8092 Zurich, Switzerland
e-mail: spapageorgiou@ethz.ch

© Springer International Publishing Switzerland 2017
K.F. Dœrner et al. (eds.), *Operations Research Proceedings 2015*,
Operations Research Proceedings, DOI 10.1007/978-3-319-42902-1_56

knowledge by investigating whether relationships also exist in the manufacturing industry.

The remainder of this paper is organized as follows. In Sect. 2 we describe our data and methodology. Next, in Sect. 3 we present the results of the cross-correlation analysis. Finally, we conclude in Sect. 4.

2 Data and Methodology

In what follows, we discuss the data and methodology used in our research.

2.1 Data Sample

We obtained the number and severity of operational disruptions from the SAS OpRisk Global Data database. The database is composed of 29,374 observations from all over the world and covers all industry sectors. To filter out the country effects, we only examined operational disruptions which occurred in the United States (66.1 % of all events). Moreover, we selected operational disruptions that occurred between 1992 and 2013 in the manufacturing industry, which resulted in the final sample of 2,873 data points used for aggregation into yearly counts. In order to soften the impact of extreme events, the severity of losses has been transformed using the annual average of the natural logarithms of the losses, i.e., the sum of the logarithm of the losses in each year divided by the number of operational disruptions that occurred in the same year.

2.2 Methodology

The existence of a relationship between the cyclical components of the time series that describe the state of the economy and the operational disruptions is investigated by measuring their cross correlation [5]. More specifically, the co-movements of two variables X_{t+j} (representing the cycles of operational disruptions) and Y_t (representing the economic cycle) are described by the correlation coefficient $\rho(j)$ where j is an integer that takes both negative and positive values. In other words, we measure the correlation coefficient between the value of the variable Y at time t and the value of the variable X at time $t + j$, where $j = 0, \pm1, \pm2, \pm3$, that is, time points before, at and after time t. Following [5], the variable X_t is procyclical with respect to the variable Y_t when $\rho(0) > 0$ and countercyclical when $\rho(0) < 0$. In economic terms, a procyclical variable tends to increase with the expansion of the economy, whereas a countercyclical variable tends to increase when the economy is slowing down.

The procyclicality and the countercyclicality is considered as strong when the correlation coefficient is statistically significant at the level of 1 % and weak when it is statistically significant at the level between 1 and 5 %. Moreover, when the maximum absolute value of $\rho(j)$ occurs for a positive j, the variable X lags behind the variable Y, whereas when the maximum absolute value of $\rho(j)$ occurs for a negative j, the variable X leads variable Y.

The state of the economy is described by the concept of economic cycles that are defined as a type of fluctuation found in the aggregate economic activity of nations that organize their work mainly in business enterprises [2, p. 3]. Expansions and recessions that occur almost simultaneously in many industry sectors are the components of a cycle. Defining the term of economic cycles, [2] underline both the recurrent changes between expansions and recessions, and the absence of periodicity. To describe economic cycles, several macroeconomic variables can be found in the literature. We selected GDP, the annual volume of the United States Imports (Goods Imports), the Chicago Fed National Activity Index (CFNAI) and the Purchasing Manager Index (PMI) to operationalize the concept of the economic cycles. Please note that PMI and CFNAI are by definition cyclical variables, whereas GDP and Goods Imports first need to be decomposed into the trend and the cycles of the underlying time series. Therefore, we used the Hodrick-Prescott filter to detrend the GDP and calculated growth rates to detrend the Goods Imports.

3 Discussion of Results

To facilitate the discussion, we plot the time series of the average severity of operational disruptions and GDP in Fig. 1. It can be seen that during the period under investigation several economic cycles are captured, which are represented by the peaks and troughs of the GDP, followed by the peaks and troughs of the operational disruptions.

Next, we focus on the results of the cross correlation analysis as summarized in Table 1. It shows that the cycles of operational disruptions lag behind economic cycles. More specifically, there is a strong positive correlation between the second-order lags of economic cycles with the cycles of the operational disruptions. The strongest correlation is found between the severity of operational disruptions and GDP (0.677 at the 1 % significance level). Thus, the peaks and troughs of severity of operational disruptions occur two years after the peaks and troughs of the GDP cycles. As a test of robustness, Goods Imports, CFNAI and PMI all confirm this result. Furthermore, the severity of losses seems to be countercyclical with respect to economic cycles (negative coefficients for $j = 0$). However, that is not a statistically strong argument because only the coefficient $\rho(j = 0)$ between PMI and the time series is statistically significant and again only at the level of 5 %.

Beyond the relationship between economic cycles and the severity of operational disruptions, we investigate the relationship between economic cycles and the number of operational disruptions. For reason of comparison, the macroeconomic indicators

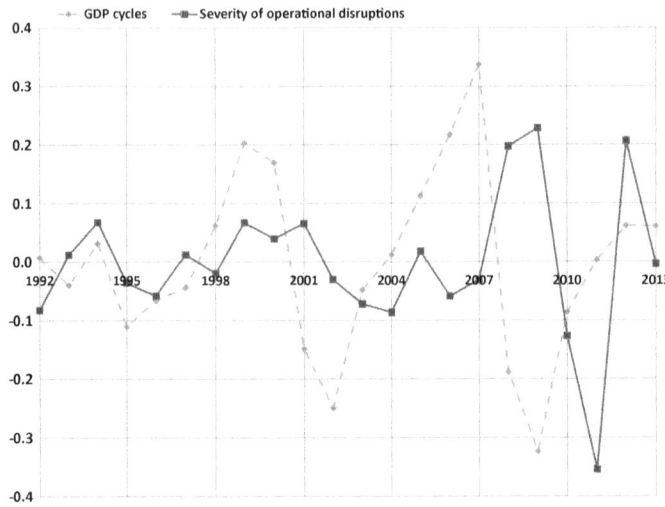

Fig. 1 Time series of the average severity of operational disruptions and GDP cycles, 1992–2013

Table 1 Correlation coefficients between the severity of operational disruptions and economic cycles, 1992–2013

Y_t	X_{t-3}	X_{t-2}	X_{t-1}	X_t	X_{t+1}	X_{t+2}	X_{t+3}
GDP	−0.162	−0.215	−0.308	−0.203	0.355	**0.677****	0.012
Goods imports	0.275	0.140	−0.226	−0.166	0.326	**0.544****	−0.634**
CFNAI	−0.029	−0.088	−0.057	−0.037	0.273	**0.535***	−0.148
PMI	0.187	0.001	0.097	−0.475*	0.153	**0.423***	0.207

**correlation significant at the 0.01 level
*correlation significant at the 0.05 level
Bold: Maximum value of the correlation coefficient

Table 2 Correlation coefficients between the number of operational disruptions and economic cycles, 1992–2013

Y_t	X_{t-3}	X_{t-2}	X_{t-1}	X_t	X_{t+1}	X_{t+2}	X_{t+3}
GDP	−0.588**	−0.269	0.279	0.311	**0.318**	0.010	−0.126
Goods imports	0.004	0.216	0.193	0.185	−0.078	**0.271**	−0.165
CFNAI	−0.030	0.156	0.325	0.261	0.184	**0.275**	0.269

**correlation significant at the 0.01 level
*correlation significant at the 0.05 level
Bold: Maximum value of the correlation coefficient

of the GDP, the Goods Imports, and the CFNAI were used in that case as well. The corresponding cross correlations are shown in Table 2. Also here, even though the maximum positive values of correlation are for $j = 1$ as measured by GDP and

$j = 2$ as measured by Goods Imports and CFNAI, none of the values are statistically significant. Therefore, we cannot make any statement about the association between the number of operational disruptions and the economic cycle.

The managerial implications of our study are summarized in the next section.

4 Conclusion and Implications

Our empirical analysis on the association between operational disruptions and economic cycles in the manufacturing industry of the United States reveals several important findings. We report that the severity of operational disruptions shows a weak countercyclical tendency. Moreover, we identify that the severity of operational disruptions follows the economic cycle with a lag of two years as measured by several macroeconomic variables. We do not find a statistically significant pattern concerning the frequency of operational disruptions and economic cycles. Given the complexity and interdependency of global supply chains [9, 11], these operational disruptions can lead to immense losses for firms in the supply chain. By taking into account our results, firms' operations and supply chain managers can incorporate the economic cycle effects into their capital planning activities to better manage the risk of operational disruptions. Moreover, since the severity of losses is substantial, the regulators shall impose more adequate policies on manufacturing firms to limit the possibility of such losses materializing in the future. One efficient mitigation strategy can be the transfer of residual risk to the (re-)insurance companies in form of business interruption insurance [3]. However, more theoretical and empirical research is needed to further explore this mechanism.

Acknowledgements We gratefully acknowledge the contribution of the SAS Institute (Switzerland) for granting access to the SAS OpRisk Global Data.

References

1. Allen, L., Bali, T.G.: Cyclicality in catastrophic and operational risk measurements. J. Bank. Financ. **31**(4), 1191–1235 (2007)
2. Burns, A.F., Mitchell, W.C.: Measuring Business Cycles. National Bureau of Economic Research, New York (1946)
3. Dong, L., Tomlin, B.: Managing disruption risk: the interplay between operations and insurance. Manag. Sci. **58**(10), 1898–1915 (2012)
4. Fahimnia, B., Tang, C.S., Davarzani, H., Sarkis, J.: Quantitative models for managing supply chain risks: a review. Eur. J. Oper. Res. **247**(1), 1–15 (2015)
5. Fiorito, R., Kollintzas, T.: Stylized facts of business cycles in the G7 from a real business cycles perspective. Eur. Econ. Rev. **38**(2), 235–269 (1994)
6. Heckmann, I., Comes, T., Nickel, S.: A critical review on supply chain risk-definition, measure and modeling. Omega **52**, 119–132 (2015)

7. Hendricks, K.B., Singhal, V.R.: An empirical analysis of the effect of supply chain disruptions on long-run stock price performance and equity risk of the firm. Prod. Oper. Manag. **14**(1), 35–52 (2005)
8. Mizgier, K.J., Hora, M., Wagner, S.M., Jüttner, M.P.: Managing operational disruptions through capital adequacy and process improvement. Eur. J. Oper. Res. **245**(1), 320–332 (2015)
9. Mizgier, K.J., Wagner, S.M., Jüttner, M.P.: Disentangling diversification in supply chain networks. Int. J. Prod. Econ. **162**, 115–124 (2015)
10. Moosa, I.: Operational risk as a function of the state of the economy. Econ. Model. **28**(5), 2137–2142 (2011)
11. Simchi-Levi, D., Schmidt, W., Wei, Y., Zhang, P.Y., Combs, K., Ge, Y., Gusikhin, O., Sanders, M., Zhang, D.: Identifying risks and mitigating disruptions in the automotive supply chain. Interfaces **45**(5), 375–390 (2015)

Simulated Annealing for Optimization of a Two-Stage Inventory System with Transshipments

Andreas Serin and Bernd Hillebrand

Abstract A two-level inventory system under a periodic review with lateral transshipments is considered. The supply chain is composed of the external manufacturer, the central warehouse and three identical retail outlets. By moving stock between retail outlets, the supply chain can maintain a service target while decreasing inventories. The aim is to optimize the order-up-to levels under a fill rate constraint. We combine a simulation with a barycentric interpolation at the Chebyshev points and a degree reduction technique to construct low-degree polynomial tensor product surfaces for the objective function and the constraint. The approximate optimization problem is solved by simulated annealing.

1 Introduction

Supply chains may reduce the operating costs by introducing lateral transshipments between stocking locations at the same level. Thus, stocking locations reduce their safety stocks while maintaining fill rates. A current review on related research is provided by [7].

The aim of this paper is to extend a single-level model according to [10] and to achieve an approximation for a two-level inventory system with transshipments under fill rate constraints, which can be easily treated by optimization techniques. The occurrence of transshipment flows needs to be approximated quite accurately in the state space to establish cost benefits sufficient to be exploited in practice. Though the accuracy of simulated data is in fact sufficient, using a time-consuming

A. Serin (✉)
Production and Supply Chain Management, Mercator School of Management,
Universität Duisburg-Essen, Lotharstraße 65, 47057 Duisburg, Germany
e-mail: andreas.serin@uni-due.de

B. Hillebrand
Production and Logistics, Junior Professorship Supply Chain Management,
Technische Universität Dortmund, Martin-Schmeißer-Weg 12,
44227 Dortmund, Germany
e-mail: bernd.hillebrand@tu-dortmund.de

© Springer International Publishing Switzerland 2017
K.F. Dœrner et al. (eds.), *Operations Research Proceedings 2015*,
Operations Research Proceedings, DOI 10.1007/978-3-319-42902-1_57

exhaustive simulation for optimization is not feasible. Therefore, our approach combines simulated data with a barycentric interpolation at the Chebyshev points to construct low-degree polynomial tensor product surfaces for the objective function and the constraint. An excellent paper on barycentric Lagrange interpolation is provided by [1].

Simulated annealing is a probabilistic meta-heuristic which can escape local optima by accepting uphill moves. Thus, simulated annealing can provide high quality solutions and short computation times. Survey articles are provided inter alia by [2, 6, 9].

2   The Model

We consider a single-product supply chain consisting of the external manufacturer, the central warehouse and three identical retail outlets under periodic review and static uncertainty. Warehouse shipments as well as transshipment flows are determined dynamically, but the order-up-to levels and the allocation policies are fixed in advance.

If the pre-transshipment stock on hand is sufficient, the retail outlet i fulfills the incoming demand completely. The remaining inventory can be offered to other retail outlets experiencing shortages. In other cases, the retail outlet i requests a lateral transshipment from the others. Transshipment lead times are negligible as opposed to deterministic replenishment lead times. If transshipment requests are feasible, then either two sources and one destination or one source and two destinations arise. Risk Balancing Policy (RBP) determines how much to transship from each source to each destination by balancing the next period stockout probabilities at the lower echelon respectively, cf. [10]. The excess demand, which can't be fulfilled in the same period even by means of lateral transshipments, is lost. At the end of each review period, retail outlets attempt to raise their inventory positions up to S_r. The central warehouse fills the orders as far as possible. If the warehouse experiences a stockout, the available inventories are rationed according to RBP. At the end of the period, the warehouse increases its inventory position up to S_c. Additionally, the stock on hand is forwarded to the next period, while the unsatisfied demand is lost.

The objective is to minimize the expected costs of the inventory system which are holding costs at each stocking location and transshipment costs. Let $\eta_r \geq \eta_c > \tau$ be the corresponding unit cost parameters in (1). β_i denotes the real end-customer fill rate at location i using transshipments where appropriate, while b_r is an aggregate fill rate target at the lower echelon.

$$\min_{S_r, S_c} f(S_r, S_c) = \tau \, ET + \eta_c \, EI_c^+ + \sum_{i \in M} \eta_r \, EI_i^+ \tag{1}$$

$$\text{s.t. } \beta_i(S_r, S_c) \geq b_r, \quad (S_r, S_c) \in \mathbb{N}^2, \quad i \in M, \quad M = \{1, 2, 3\}.$$

Fig. 1 State space

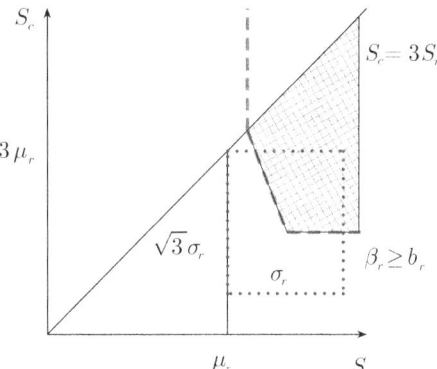

Considering a particular point of the solution space, there are two occasions for transshipments. First, stockouts at the retail outlets may occur despite inventory positions as high as S_r. In this case, high demand induces immediate transshipment flows. Second, stockout situations at the central warehouse cause time-delayed transshipment flows as a consequence of the fact that retail outlets are not able to raise their inventory positions up to S_r. If both reasons arise simultaneously, there is an interdependence between the warehouse and the retail outlets determining transshipment flows in the "close" neighborhood of $S_c = 3\mu_r$, if $S_c < 3S_r$ and $S_r \geq \mu_r$. For this reason, a fill rate constraint outside of this close neighborhood can be approximated from a single-stage model by ignoring the interdependencies between the stages. For the formulae, see [8].

The occurrence of transshipments is similar to a physical superposition of two or more waves resulting in a new wave pattern of greater or lower amplitude. Waves can travel in space, but there is an absorption. This analogy provides an intuition for why we can introduce boundaries making the state space finite. f and β_r are dependent on both S_r and S_c. For the desired end-customer fill rates $b_r \in [0.90, 0.95]$, we expect to solve the problem (1) setting $S_r \in [\mu_r, \mu_r + \sigma_r]$, $S_c \in [3\mu_r - \sqrt{3}\sigma_r, 3\mu_r]$, as shown in Fig. 1.

3 Proposed Approximations and Simulated Annealing

In order to achieve cost benefits sufficient for practical purposes, the occurrence of transshipment flows needs an accurate approximation. A simulation-based approach provides a sufficient accuracy. Since the simulation is a time-consuming process, an approach should require only a few values obtained that way. The use of Lagrangian interpolation is deplored for its numerical properties, but the second form of the barycentric formula is appropriate due to its numerical stability. The oscillation may be eliminated by choosing interpolation points at Chebyshev nodes. The corresponding barycentric weights w_j are usually contained in software packages. For a good introduction to barycentric Lagrange interpolation, we point to [1].

This numerical study considers a normal demand with parameters $\mu_r = \{200, 300, 400\}$ and $\sigma_r = \{45, 60, 75\}$.

In order to obtain a numerically stable interpolation, the state space defined above is first linearly transformed to $[-1, 1] \times [-1, 1]$, w.l.o.g. Then, Chebyshev points of the first kind are selected for the interpolation with respect to the S_c axis. For the demand parameters mentioned above, a sufficient number of points ranges between 10 and 13. Approximations are obtained by truncating high-order terms in the resulting Chebyshev series. For instance, quintic polynomials lead to a negligible maximum relative error $\varepsilon_{rel}(f) = \max |(\mathrm{appr}(f) - f)/f| = 0.6\,\%$. Lower sensitivities to changes in S_r justify an interpolation on four equally spaced points with respect to the S_r axis. By mapping uniform grid points into Chebyshev knots as described by [3], we avoid the Runge phenomenon and complete approximations.

The approximations of EI_r^+, EI_r^-, EI_c^+, EI_c^- and ET for different normal distributions differ only by a scaling factor σ_r. By eliminating σ_r, proposed approximations are valid for all normal distributions with an appropriate coefficient of variation. Thus, instead of approximating β_r directly, we consider an approximation of EI_r^-/σ_r. An approximation of β_r for a particular distribution can then be calculated according to $\mathrm{appr}(\beta_r) = 1 - \mathrm{appr}(EI_r^-/\sigma_r)\,\sigma_r/\mu_r$. For instance, quintic polynomials lead to a negligible maximum absolute error $\varepsilon_{abs}(\beta_r) = \max |\mathrm{appr}(\beta_r) - \beta_r| = 0.0003$.

Under arbitrary i.i.d. demand conditions, the use of the above approach is justified by Berry-Esseen theorem. By specifying a bound on the maximum error of approximation between the normal CDF and the CDF of the normalized sample mean of arbitrary i.i.d. values, we can assess the required number of demand periods to be simulated.

Different combinations of the variables lead to various local minima because the order-up-to levels are required to take integer values. Thus, the state space is discontinuous and disjointed. Such optimization problems are treated either by probabilistic or enumerative search algorithms.

Simulated annealing is a meta-heuristic algorithm used to address inter alia combinatorial optimization problems. An implementation requires four functional relationships to be determined. First, the neighborhood is specified by a probability density $g(\mathbf{x})$ determining the set of all feasible moves from the current solution $\mathbf{x}$ in the d-dimensional state space. $g(\mathbf{x})$ is required to generate integer-valued moves. Second, the acceptance function is defined by a probability density $h(\mathbf{x})$ for the acceptance of a new state given the current state. All approaches mentioned below accept an uphill move with a probability $p = \exp(-\Delta/T)$, $\Delta = f(\mathbf{x}_{new}) - f(\mathbf{x})$ being the difference between the current solution under consideration and the new one. Downhill moves are always accepted. Third, the annealing schedule is specified by a finite sequence of decreasing temperatures $T_0, T_1, \ldots, T_{fi}$, and a finite number of transitions at each temperature. Finally, the objective function $f(\mathbf{x})$ to be minimized is specified by (1). Thereby, the feasibility bounds holding the search within the space for which the approximation is valid are implemented explicitly, while the fill rate constraint is treated as a penalty in the objective function.

Table 1 Numerical results for $\mu_r = 400$, $\sigma_r = 75$, $\eta_r = \eta_c$ and $\tau = 0.9\eta_r$

Approach	Transship. policy	b_r	sugg. sol. (S_c, S_r)	obj. val.	comp. time, s.
Cauchy I	RBP	0.90	(1080, 455)	47.64	13.91
Cauchy II	RBP	0.90	(1080, 455)	47.64	16.07
VFSA	RBP	0.90	(1080, 455)	47.64	12.81
VFSA	No transshipment	0.90	(1082, 435)	73.85	14.88

A variety of simulated annealing approaches has been developed. A d-product of one-dimensional Cauchy distributions with the cooling schedule $T_k = T_0/k^{1/d}$ is referred herein as Cauchy I. A recent approach using d-dimensional Cauchy neighbors and the fast annealing schedule $T_k = T_0/k$ (see [5]) is referred as Cauchy II. A very fast simulated annealing according to [4] with the cooling schedule $T_k = T_0 \exp(-k^{1/d})$ is referred as VFSA. All computations are performed using an Intel Core i5-3210 CPU, 8 GB RAM, Windows 7 (64 bit) and R 3.2.1. VFSA achieves the fastest computation times in our test instances, even though all approaches lead to the same results. Cauchy II offers no advantages in a small two-dimensional state space. Some representative results are presented in Table 1.

4 Conclusions

This paper provides three main insights regarding the problem under consideration. First, lateral transshipments enable substantial cost benefits due to lower expected inventory levels at the end of the period. Thereby, the desired fill rate b_r uniquely determines the optimal inventory policy. As a consequence, only μ_r and σ_r influence the expected transshipments at the optimum. Furthermore, each test instance is computationally adjustable by fixing this optimum w.r.t. quantiles of normal distribution. In contrast, unit cost parameters influence only the nominal amount at the optimum, but not the optimal policy itself. For the instance mentioned in Table 1, transshipments reduce the expected costs of the inventory system by 35.49 % at the optimum, other things being equal. Second, approximations based on the polynomials of degree 5 lead to $\varepsilon_{abs}(\beta_r) = 0.0003$ and $\varepsilon_{rel}(f) = 0.6\%$. Approximations can be generalized for all normal distributions with a reasonable coefficient of variation. Under certain circumstances, the use is appropriate for general i.i.d. demands. Third, the approximate problem can be quickly solved. In our instances, VFSA obtains the fastest computation times while providing the same objective values at the optimum as other approaches.

The research can be easily extended to incorporate limited transport or handling capacities by introducing higher unit costs for the utilization of external capacities.

References

1. Berrut, J.-P., Trefethen, L.N.: Barycentric Lagrange interpolation. SIAM Rev. **46**(3), 501–517 (2004)
2. Eglese, R.W.: Simulated annealing: a tool for operational research. Eur. J. Oper. Res. **46**(3), 271–281 (1990)
3. Gaure, S.: Details of chebpol (2013). http://cran.r-project.org/web/packages/chebpol/vignettes/chebpol.pdf. Accessed 14 July 2015
4. Ingber, L., Rosen, B.: Genetic algorithms and very fast simulated reannealing: a comparison. Math. Comput. Model. **16**(11), 87–100 (1992)
5. Nam, D., Lee, J.-S., Park, C.H.: N-dimensional Cauchy neighbor generation for the fast simulated annealing. IEICE Trans. **87-D**(11), 2499–2502 (2004)
6. Nikolaev, A.G., Jacobson, S.H.: Simulated annealing. In: Gendreau, M., Potvin, J.-Y. (eds.) Handbook of Metaheuristics. International Series in Operations Research and Management Science, vol. 146, pp. 1–39. US, Springer (2010)
7. Paterson, C., Kiesmüller, G., Teunter, R., Glazebrook, K.: Inventory models with lateral transshipments: a review. Eur. J. Oper. Res. **210**(2), 125–136 (2011)
8. Serin, A., Hillebrand, B.: Inventory management with transshipments under fill rate constraints. In: Huisman, D., Louwerse, I., Wagelmans, A.P. (eds.) Operations Research Proceedings, pp. 437–442. Springer, Berlin (2014)
9. Suman, B., Kumar, P.: A survey of simulated annealing as a tool for single and multiobjective optimization. J. Oper. Res. Soc. **57**(18), 1143–1160 (2006)
10. Tagaras, G.: Pooling in multi-location periodic inventory distribution systems. Omega **27**(1), 39–59 (1999)

Optimized Modular Production Networks in the Process Industry

Dominik Wörsdörfer, Pascal Lutter, Stefan Lier and Brigitte Werners

1 Introduction

One innovative production concept is currently highly discussed in the process industry: transformable, modular plant designs implemented in standardized transportation iso-containers [3]. Demonstration plants, consisting of apparatus modules, in container design have already been developed and constructed within several research projects like the EU funded "F^3-Factory" project or the "CoPIRIDE" project [2, 3]. Motivations for such transformable, highly standardized plant designs in process industry are predominantly identified by market dynamics such as shortened product life cycles, an intense product differentiation and volatile product demands [3, 6, 8]. The provided high flexibility in capacity, product and location as well as short technology development times are attributes which lead to a high degree of attraction in case of facing the described market changes. The impact of the inherent uncertainty occurring in such markets can be reduced with the described innovative and flexible production system.

Alongside with the inherent mobility (using transportation container format) and scalability (by numbering up or down containers) new opportunities regarding supply chain and network structure are provided. Production locations can be placed directly in customers or resources proximity and containers can be relocated or easily adjusted in capacity over time in case of demand shifts. Identifying the most cost efficient

D. Wörsdörfer (✉) · S. Lier
Laboratory for Fluid Separations, Department of Mechanical Engineering,
Ruhr University Bochum, Bochum, Germany
e-mail: woersdoerfer@fluidvt.rub.de

P. Lutter · B. Werners
Chair of Operations Research and Accounting, Faculty of Management
and Economics, Ruhr University Bochum, Bochum, Germany
e-mail: pascal.lutter@rub.de

© Springer International Publishing Switzerland 2017
K.F. Dœrner et al. (eds.), *Operations Research Proceedings 2015*,
Operations Research Proceedings, DOI 10.1007/978-3-319-42902-1_58

production network is crucial in order to benefit of the given flexibilities. Due to the high complexity and the inherent interdependencies of the involved planning decisions, mathematical programming is used to calculate a production network at minimum cost level.

Network design models are well known in literature [5, 7] and have already been successfully applied to the chemical process industry [1]. While related problems focus on large scale production facilities, the scenario under consideration deals with small scale production facilities in iso-containers. These production containers can easily be shipped from one location to another within a few days and thus allow for completely modular production networks. Two different mixed-integer linear programming formulations are proposed to optimally support the production network design decision on a strategic and tactical level. Both model formulations are evaluated and compared on the basis of real-world test instances.

The remainder of this paper is structured as follows. Given opportunities and benefits regarding network design of transformable plants are briefly discussed in Sect. 2. Afterwards the mathematical optimization model to optimally design such innovative production networks is introduced in Sect. 3. Computational results are presented in Sect. 4 while Sect. 5 concludes with final remarks and perspectives.

2 Production Network Design of Transformable Plants

Transformable plant designs in small scale are currently in focus of the process industry. Such a plant design contains the five enablers for transformability (modularity, universality, scalability, compatibility, mobility), see [4]. Providing these enablers for transformability results in inherent flexibility in three dimensions. Production locations can be shifted as mobility of the plants is given by ISO-container surroundings. Product changes are feasible as modular apparatuses and standardized interfaces offer the possibility to reconfigure the production process. Finally, production capacity can be adapted by either numbering up or numbering down apparatuses (or entire processes) or replacing existing apparatuses by larger or smaller ones. As apparatuses are designed in standardized, predetermined formats and provide standardized interfaces, entire processes are quickly developed and constructed. A high degree of standardization in combination with the described simple scale-up process results in short process development times. Therefore, time to markets are drastically reduced compared to conventional customized large scale plants. Transformable plant designs are well appropriate for decentralized production due to the mentioned specifications. The next section describes optimization models for designing networks of transformable plants.

3   Optimizing Modular Production Networks

Decentralized modular production networks allow permanent relocations of production containers. This means that production modules can be shifted within a very short period of time allowing for flexible adjustments of location sites and their production volumes. As a consequence, the network structure can be modified in dependency of current product demands at each time step. In addition, the production in customer's proximity combined with customer specific lot sizes offers significant advantages in comparison to centralized world scale production with conventional large scale plant designs.

Modular production networks require a plurality of location and relocation decisions as well as the allocation and reallocation of customer demands to production facilities. At each decision step, container modules can either be added or removed from the network, relocated within the network, customer demands produced in (re-)assigned production container modules and new locations are set up or existing locations are closed. In summary the following decisions have to be made in every time step:

1. Insert or remove production containers from the network
2. (Re-)Assignment of customer demands to production containers
3. Open new locations or close old locations
4. Reposition/Relocation of production containers

Thus a variety of reacting options are offered and enabled by combining these mentioned decisions in order to provide a highly efficient network. Cost components comprise investment costs for each production container, fixed and variable costs for operating, mounting and dismounting of production containers, site development costs, transportation costs of products, containers and raw materials and purchasing costs for raw materials. In order to minimize overall costs for a given time horizon such that entire customer demands are fulfilled, two different mixed-integer programming approaches are proposed. The innovative production concept predominantly provides transformability, mobility and short times to market leading to a considerable reduction of the impact of the mentioned market uncertainties. Hence, the explicit treatment of uncertainty within the model formulation is neglected.

Given a set of M production containers, a set of J possible production locations and a set of K customers, the goal is to find an allocation of modules to production locations as well as an assignment of production outputs to customer demands such that total costs over the entire planning horizon are minimized and customer demands are fully satisfied. We restrict our model to the one product case. An extension to multiple products is straightforward.

An integrated mixed-integer programming model simultaneously deciding about all components (1.-4.) is proposed as shown in Fig. 1. Besides decisions concerning the optimal choice of flexible production locations being built and the assignment of production volumes to customers, the model considers the number, the location and the relocation of modules—the most crucial aspect. The location j of module m

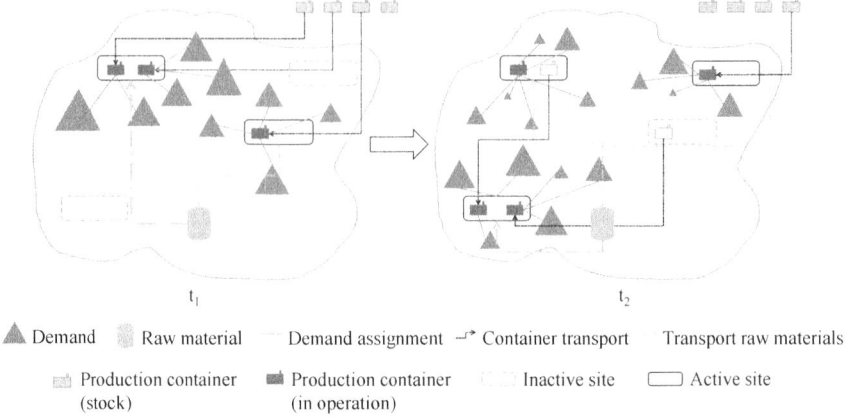

Fig. 1 Illustration of a production network for different points in time

at time t is given by the binary variable γ_{mjt}, where $\gamma_{mjt} = 1$ indicates that container m is placed at location j at time t. Location shifts of modules are denoted by the binary decision variables $\tau_{jj'mt}$, where $\tau_{jj'mt} = 1$ holds if module m is transported from location j to location j' ($j \neq j'$) in period t. Hence, if module m was located at j in period $t - 1$ and is located at site j in period t, this module requires a location shift at the beginning of period t. This leads to the following constraints:

$$\gamma_{mjt} + \gamma_{mj',t-1} - \tau_{jj'mt} \leq 1 \quad \forall j \in J_0, \forall j' \in J_0 \setminus \{j\}, \forall m \in M, \forall t \in T, \qquad (1)$$

where J_0 denotes all production locations as well as the module storage location. The general model formulation reads as

min   Cost function

s.t.   Selection of active sites/production locations

Location of production containers

Capacity supply

Assignment of customer demands to production containers

Relocation of modules (1).

The simultaneous consideration of facility locations, material and module flows over the entire time horizon involves a huge number of variables, which significantly increases solution times, especially in real world instances. Preliminary tests have shown that a real-world instance provided by one of our project partners could not be solved to a reasonable optimality gap after one week of computation time. In order to solve real-world problems, a two stage decomposition is proposed. The number of variables is reduced by hierarchically decomposing the decision process into two stages, where the first stage decides about the entire network structure and the second stage optimizes module and production flows within the given network. This leads to the second model formulation:

1. The first stage covers strategic and certain tactical decisions such as location planning, the number of modules at each location, raw material flows and the assignment of production output to customers.
2. The second stage optimizes module flows in the given network created by the first stage optimization model. Locations as well as the corresponding module demands at each location are given at this stage. The goal is to optimize the transportation routes of the modules such that costs–including investment costs for modules–are minimized and module demands are fulfilled.

The first stage of the optimization process is modeled as a network design problem with the objective to minimize the overall costs of the network. In this stage, transportation costs for modules are neglected, in order to reduce the number of variables. Hence, the initial location of facilities and the corresponding number of modules required at each location are part of the decision. Additionally, customer demands are assigned to facilities and the flow of raw materials is considered. Every relocation of modules induces mounting and dismounting costs as well as transportation costs. Hence, the resulting first stage locations may be sub-optimal with respect to the second stage problem structure including further cost components. Small changes in the given network structure can lead to significant cost savings due to decreased transportation effort. Further cost savings can be achieved by increasing the number of modules if additional investment costs are overcompensated by reduced transportation costs. In order to adept the previous decision, facilities can be merged together and new modules can be purchased, if total costs decrease. The second model has the advantage, that computation times considerably decrease and solutions of very high quality are generated. In the next section, both approaches are compared on the basis of real-world test instances.

4   Case Study and Results

The performance of the proposed model formulations were analyzed on the basis of modified real-world data provided by one of our industrial partners. All calculations were performed on a desktop PC with Intel Core i5 CPU (3.3 GHz) and 8GB RAM under Windows 7. Both approaches were implemented in Xpress-Mosel, a modeling and programming language included in the FICO Xpress Optimization Suite. The 64 bit version of FICO Xpress Version 7.6 was used as solver with default settings. A set of 20 test instances were generated on the basis of a large real-world data set. The planning horizon contains ten periods and customer demands are assumed to be known for each period. Test instances were systematically derived by varying the number of customers and potential facility location sites. Both models minimize the total costs. In the two step approach, the total costs are computed as the sum of

Table 1 Computational results for all 20 test instances

Instance	Integrated approach		2 step approach		
(#Loc./#Cust.)	gap (%)	time (s)	cost diff. (%)	gap (%)	time (s)
30/10	0.00	605	0.00	0.00	1
30/20	0.00	380	0.00	0.00	1
30/30	7.29	**tl**	−6.86	0.00	14
30/40	27.30	**tl**	−26.56	0.02	**tl**
30/50	25.67	**tl**	−25.03	0.11	**tl**
30/60	24.59	**tl**	−24.10	0.22	**tl**
30/70	24.98	**tl**	−23.18	0.17	**tl**
30/80	22.67	**tl**	−22.09	0.18	**tl**
30/90	21.31	**tl**	−20.86	0.24	**tl**
30/100	20.37	**tl**	−20.25	0.26	**tl**
50/10	0.85	**tl**	−0.82	0.00	1
50/20	0.00	1,092	0.00	0.00	1
50/30	10.34	**tl**	−9.49	0.00	23
50/40	3.27	**tl**	−2.17	0.00	150
50/50	2.70	**tl**	−1.89	0.04	**tl**
50/60	2.39	**tl**	−1.72	0.14	**tl**
50/70	24.52	**tl**	−23.85	0.11	**tl**
50/80	23.25	**tl**	−22.65	0.15	**tl**
50/90	22.43	**tl**	−21.98	0.14	**tl**
50/100	21.63	**tl**	−21.32	0.30	**tl**

Time limit (3,000 s) indicated by **tl**

both objectives and thus contain the same components as the integrated approach. Table 1 reports obtained results for each model after 3,000 s of running time. It turns out that the two stage formulation outperforms the integrated approach in terms of computation time and solution quality. As indicated by the column *cost difference*, in nearly all instances the two step approach found a considerably better solution in terms of total costs. The average relative cost improvement was about 13 %. In all instances, the two stage approach obtained the best objective values and exhibits the smallest optimality gaps.

5   Conclusion and Outlook

With the emersion of container based, transformable plant designs in chemical and pharmaceutical industry, new opportunities and potentials regarding a more customer orientated and more flexible production network design are provided. The mobility of production containers allows for frequent reconfigurations of the production network. In order to optimally decide about the network design for a given period of time, two mixed-integer linear optimization models are proposed. While the first formulation simultaneously considers all decisions, the two stage approach decomposes the entire problem into two smaller size optimization problems. Numerical experiments indicate that the latter decomposition shows clear advantages regarding computation times and solution quality. Future research should focus on the integration of demand uncertainty.

References

1. Berning, G., Brandenburg, M., Gürsoy, K., Mehta, V., Tölle, F.-J.: An integrated system solution for supply chain optimization in the chemical process industry. OR Spectr. **24**(4), 371–401 (2002)
2. Bieringer, T., Buchholz, S., Kockmann, N.: Future production concepts in the chemical industry: modular - small-scale continuous. Chem. Eng. Technol. **36**(6), 900–910 (2013)
3. Buchholz, S.: Future manufacturing approaches in the chemical and pharmaceutical industry. Chem. Eng. Process.: Process Intensif. **49**(10), 993–995 (2010)
4. Hernández, R., Wiendahl, H.-P.: Die wandlungsfähige Fabrik—Grundlagen und Planungsansätze. In: Kaluza, B., Blecker, T. (eds.), Erfolgsfaktor Flexibilität. Strategien und Konzepte für wandlungsfähige Unternehmen., pp. 203–227 (2005)
5. Hsu, C.-I., Li, H.-C.: An integrated plant capacity and production planning model for high-tech manufacturing firms with economies of scale. Int. J. Prod. Econ. **118**(2), 486–500 (2009)
6. Küppers, S., Ewers, C.: Supply chain event management in der Pharmaindustrie—Status und Möglichkeiten. Supply Chain Event Management, pp. 37–55 (2007)
7. Melo, M.T., Nickel, S., Da Gama, F.S.: Dynamic multi-commodity capacitated facility location: a mathematical modeling framework for strategic supply chain planning. Comput. Oper. Res. **33**(1), 181–208 (2006)
8. Shah, N.: Process industry supply chains: advances and challenges. Comput. Chem. Eng. **29**(6), 1225–1235 (2005)

Management Coordination for Multi-Participant Supply Chains Under Uncertainty

Kefah Hjaila, José M. Laínez-Aguirre, Luis Puigjaner and Antonio Espuña

Abstract A game decision support tool is developed to suggest the best conditions for the coordination contract between different stakeholders with conflictive objectives in a multi-participant Supply Chain (SC). On the base of dynamic games, the interaction between the involved stakeholders is modeled as a non-cooperative non-zero-sum Stackelberg's game under the leading role of one of the partners. The leader designs the first game move (price offered) based on its optimal conditions and taking into consideration the uncertain conditions of the follower. Consequently, the follower responds by designing the second move (quantity offered at this price) based on its best current/uncertain conditions, until the Stackelbergs payoff matrix is built. The expected follower payoffs are obtained taking into consideration the risks associated with the uncertain nature of the 3rd party suppliers. Results are verified on a case study consisting of different providers SC around a client SC in a global decentralized scenario. The results show improvements in the current/expected individual profits in the SCs of both leader and follower when compared with their standalone cases.

K. Hjaila (✉) · L. Puigjaner · A. Espuña
ETSEIB, Chemical Engineering Department, Universitat Politécnica de Catalunya,
Av. Diagonal 647, 08028 Barcelona, Spain
e-mail: kefah.hjaila@upc.edu

L. Puigjaner
e-mail: Luis.Puigjaner@upc.edu

A. Espuña
e-mail: antonio.espuna@upc.edu

J.M. Laínez-Aguirre
Department of Industrial and Systems Engineering, University at Buffalo, Buffalo, NY, USA
e-mail: jmlainez@gmail.com

© Springer International Publishing Switzerland 2017
K.F. Dœrner et al. (eds.), *Operations Research Proceedings 2015*,
Operations Research Proceedings, DOI 10.1007/978-3-319-42902-1_59

1 Introduction

Current tools for supporting SC planning decisions are based on the optimization of an overall target by assuming a centralized organization. Such approach disregards the complexity that may arise when considering the different objectives, possibly conflicting, of the different involved stakeholders, since in reality, each one usually seeks to optimize its own profits with no consideration of the uncertain reaction of the other players.

Some works have been carried out to solve these conflicting objectives through cooperative negotiations such as [1], who propose a cooperative multi-agent approach for the optimization of a Brazilian oil global SC. The objective in this work is to reach to an agreement to identify the oil products distribution plan. Zhao et al. [6] develop a bi-directional option contract (call option/put option) for a one manufacturer-one retailer decentralized SC. For the call option contracts, the manufacturer must buy a specific amount of products with a specific price, while for the put option, the retailer has to pay an allowance for cancelling or returning an order.

On the other hand, few works have been carried out to solve the conflicting objectives based on non-cooperative games. The work of [5] solve the interaction between different suppliers/retailers and one manufacturer through game theory. The competitiveness among the suppliers/retailers has been modeled as cooperative games through Nash Equilibrium (NE), while the interactions between the manufacturer and the suppliers/retailers have been modeled as non-cooperative Stackelberg games. Hjaila et al. [2] develop a scenario based dynamic-negotiation (SBDN) approach to solve the conflicts among the participating independent stakeholders within a decentralized SC. The authors consider the uncertain reaction of the followers SC as a probability of acceptance.

To the best of our knowledge, most of the decentralized SCs optimization models, based on either cooperative or non-cooperative games, focus on SC structures, where the interactions among the different stakeholders is hardly analyzed, leading to lose some practicality. Moreover, current methods based on game theory allow to provide individual decisions based on static cases, without considering the whole SC picture and how the other partners may react, thus giving a powerful position to one player (leader provider or client). This may lead to a bias representation of the decision-making process, particularly when the game players are subjected to risks due to the uncertainty in the expected response of their 3rd parties.

Accordingly, this work aims to suggest the optimal conditions for the coordination between different stakeholders, with different interests, within a decentralized SC superstructure. To illustrate the practicality of the proposed game approach, the developed models are implemented and solved for a superstructure of a manufacturing-distribution SC case study which is based on real data parameters. The decisions to be optimized are the resource flows and transfer prices between the participating SC stakeholders, production, inventory, and distribution levels.

2   Problem Statement

The Multi-Participant SC superstructure under study consists of several interacting manufacturing-distribution SCs (Providers and Clients SCs). The main players are the Provider and the Client, while the other stakeholders involved are considered as 3rd parties. The provider SC produces internal products which may be of interest to the Client SC and/or final products to external markets using resources from 3rd parties, so the Provider has the option to sell the same internal product to external markets, and the client can purchase the internal product from 3rd parties, giving more flexibility to both parties.

Based on dynamic (perfect information) games, in which each player information, strategies, 3rd parties, uncertain conditions, and benefits are known to each player, the interaction between the Provider and the Client is modeled as a non-cooperative non-zero-sum Stackelberg game under the leading role of the Client. The game takes into consideration the expected individual benefits of the Provider (Follower). The game items are the quantity and the transfer price of the internal product along the discrete planning time horizon. The reaction function of the follower is identified to be the quantity of the internal product at each planning time period.

Each player acts to optimize its individual benefits by taking into account that the other player is following the same goal. The leader player designs the game first move by offering the transfer price based on the available information, then the follower player reacts by providing the quantity. This is repeated until the Stackelberg payoff matrix is built, considering the follower current and uncertain conditions.

3   Mathematical Model

A set of SCs ($sc1$, $sc2 \ldots SC$) is considered to represent the game with their new subsets linking each SC to its game player (leader L or follower F). The game items are the inner product $\acute{r}$ flows (RG) and the transfer price (p). The objective function is to maximize the SC Payoff (Eq. 1),

$$Payoff_{sc} = SALE_{sc} - COST_{sc} \quad \forall sc \in SC \tag{1}$$

The SC revenue ($SALE$) (Eq. 2) is the summation of the sales to external markets m and to the leader SC (L); rp is the final product price, t is the discrete time period ($t1$, $t2 \ldots T$), RD is the final product flow each time period, r is the final product resource.

$$SALE_{sc} = \sum_{r \in R} \sum_{m \in M} \sum_{t \in T} rp_{r,sc}.RD_{r,sc,m,t} + \sum_{r' \in R} \sum_{t \in T} p_{r'}.RG_{r',sc' \in F,t} \quad \forall sc \in SC, sc' \in SC \tag{2}$$

The SC Cost is the summation of the external resources purchase, production, storage, distribution, and the internal product costs, respectively (Eq. 3). Here, it can be seen the conflicting objectives, as the game term is considered as a sale in the follower SC model (Eq. 2), while as a cost in the leader SC model (Eq. 3).

$$COST_{sc} = CRM_{sc} + CPR_{sc} + CST_{sc} + CTR_{sc} + \sum_{r' \in R}\sum_{t \in T} p_{r'}.RG_{r',sc' \in L,t} \quad \forall sc \in SC, sc' \in SC$$

(3)

Managing uncertainty

The expected payoff (Eq. 4) of the follower is obtained using a Monte Carlo Simulation method. A sample consisting of N risk scenarios is generated.

$$ExPayof_{sc'} = \sum_{n \in N} \frac{Payoff_{sc',n}}{N} \quad \forall sc' \in F$$

(4)

The mathematical model formulations result in Mixed Integer Non-Linear Programms (MINLP), for both leader and follower models. The complexity of the generic model stems from considering the policies of the third parties as part of the system. This is achieved by following the piecewise pricing model proposed by [3].

4   Results and Discussion

4.1   Case Study

The proposed approach has been implemented to solve a case study modified from [3]. The decentralized SC network (Fig. 1) consists of two main stakeholders: a polystyrene manufacturing-distribution SC stakeholder (leader) and an energy generation SC stakeholder (follower). The leader SC consists of 3 polystyrene manufacturing plants, 2 distribution centers (DC1, DC2), 3 markets (m1, m2, m3). The leader SC produces two products (A, B) using 4 raw materials (rm1, rm2, rm3, rm4) supplied from 4 vendors (sup1, sup2, sup3, sup4) and energy which is purchased from the local Grid. The follower SC consists of 6 renewable energy generation plants which are supplied by 4 biomass raw materials. The follower SC generates energy which is sold to final energy markets and the local Grid. The game is played to determine the optimal internal energy flows (economic/physical) between the follower and the leader SCs. To play the game, the leader offers energy prices (0.14–0.22/kWh), and consequentially, the follower responds by providing the internal energy amounts (3.0–24.71 GWh). The case study is modeled using the General Algebraic Modelling System (GAMS). The resulting MINLP tactical models are solved for 6 time periods, which consist of 1000 working hours each, using Global mixed-integer quadratic optimizer GloMIQO [4].

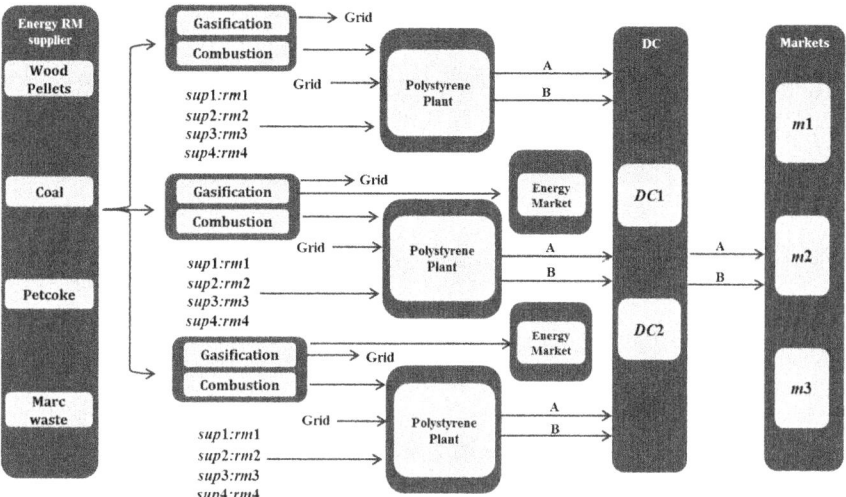

Fig. 1 The decentralized SC network [3]

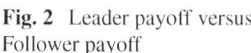

Fig. 2 Leader payoff versus Follower payoff

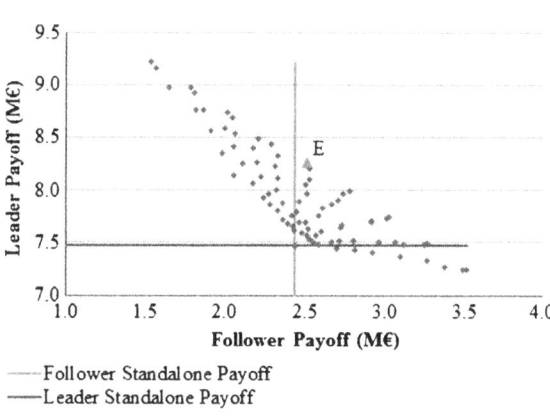

4.2 Results-Deterministic Conditions

The resulting Stackelberg's payoff matrix under the current energy prices around the follower SC has been projected on Fig. 2. It can be seen that the highest leader payoff value is 23 % higher than its Standalone payoff (7.47 M€), which corresponds to 59 % loss in the follower payoff compared with its Standalone payoff. The first win-win Stackelberg solution is the point E (Fig. 2), which guarantees 10.6 and 3 % profits improvements in the leader and follower payoffs, respectively in comparison with their corresponding standalone cases. Then, the resulting Stackelberg strategy would be to reach an agreement and signed a coordination contract that ensures a service of 24.71 GWh at a price of 0.18 €/kWh.

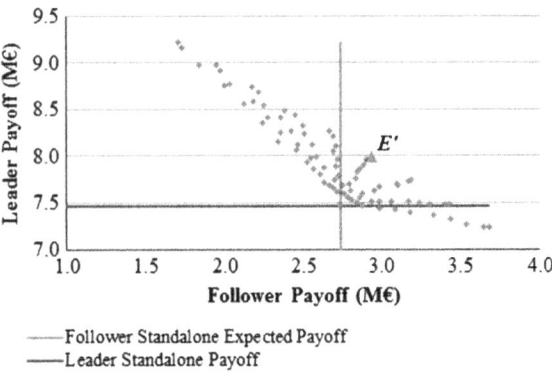

Fig. 3 Leader payoff versus Follower expected payoff

4.3 Results—Uncertain Conditions

The Stackelberg payoff matrix is built and projected on Fig. 3 based on the leader payoffs and the follower expected payoffs. This results are obtained from 500 generated scenarios using a Monte Carlo sampling based on the following parameters: energy prices mean = 0.22 €/kWh, standard deviation = 0.03 €/kWh. It is worth noticing that the Stackelberg solution has been shifted from E to É in order to mitigate the risks associated with the uncertain reaction of the follower. In this case, É represents the first win-expected-win solution. The coordination contract will be that one corresponding to the leader final strategy: 24.71 GWh at 0.19 €/kWh. Such a strategy results in 6.9 and 7.3 % profit improvements in the leader payoff and follower expected payoff, respectively in comparison with their Standalone cases.

Finally, the follower evaluates the game outcome based on its SC nominal expected payoff. To do so, the follower SC expected payoff (2.94 M€) according to the leader final strategy (É: 0.19 €/kWh) is compared with its expected nominal payoff at the leader strategy (E: 0.18 €/kWh). This is done by considering different 500 generated scenarios. The results show 6.6 % improvements in the follower expected payoff compared with its nominal expected payoff (2.76 M€).

5 Conclusions

A non-cooperative non-zero sum game approach is proposed for the optimization of decentralized SC. The methodological framework is based on determining the best coordination contract between the stakeholders of conflicting objectives (providers/clients) that guarantees win-win outcomes under the provider (the follower player) uncertain conditions. The Game approach results in different MINLP model implementations which have been solved to a real data case study that consists of different production-distribution providers SC (follower) around an industrial manufacturing-distribution SC (leader). The results show improvements in the stakeholders profit expectations when compared with their standalone situations. The uncertain behaviour of the follower affects the stackelberg outcome which induces the leader to change its strategy while keeping a win-win game outcome. The proposed approach provides a flexible decision-support tool that is able to mitigate the uncertain reaction of the providers, thus allowing to anticipate the mechanisms different manufacturers may use to modify their relationships with their providers during the decision-making process.

Acknowledgements Financial support received from the Agncia de Gesti d'Ajuts Universitaris i de Recerca AGAUR, the Spanish Ministry of Economy and Competitiveness and the European Regional Development Fund, both funding the Project SIGERA (DPI2012-37154-C02-01), and from the Generalitat de Catalunya (2014-SGR-1092-CEPEiMA), is fully appreciated.

References

1. Banaszewski, R.F., Arruda, L.V., Simão, J.M., Tacla, C.A., Barbosa-Póvoa, A.P., Relvas, S.: An application of a multi-agent auction-based protocol to the tactical planning of oil product transport in the Brazilian multimodal network. Comput. Chem. Eng. **59**, 17–32 (2013)
2. Hjaila, K., Puigjaner, L., Espuña, A.: Scenario-based price negotiations versus game theory in the optimization of coordinated supply chains. Comput. Aided Chem. Eng. **36**, 1853–1859 (2015)
3. Hjaila, K., Laínez-Aguirre, J.M., Zamarripa, M., Puigjaner, L., Espuña, A.: Optimal integration of third-parties in a coordinated supply chain management environment. Comput. Chem. Eng. **86**, 48–61 (2016)
4. Misener, R., Floudas, Ch.: GloMIQO: global mixed-integer quadratic optimizer. J. Glob. Optim. **57**, 350 (2013)
5. Yue, D., You, F.: Game-theoretic modeling and optimization of multi-echelon supply chain design and operation under stackelberg game and market equilibrium. Comput. Chem. Eng. **71**, 347–361 (2014)
6. Zhao, Y., Ma, L., Xie, G., Cheng, T.C.E.: Coordination of supply chains with bidirectional option contracts. Eur. J. Oper. Res. **229**, 375–381 (2013)

A Simulation Based Optimization Approach for Setting-Up CNC Machines

Jens Weber, André Mueß and Wilhelm Dangelmaier

Abstract The "Intelligent work preparation based on virtual tooling machines" research project presents an idea for pursuing an automatically optimized machine setup to obtain minimized tool paths and production time for CNC tooling machines. A simulation based optimization method was developed and will be combination with a virtual tooling machine to validate the setup parameters and configuration scenarios. The features of the machine simulation such as material removal and collisoon detection are associated with a sharp increase in the simulation complexity level which leads to a high effort for a simple simulation based optimization approach where a high number of iterations are typically necessary to evaluate the optimization results. This contribution focuses on the implementation of a machine setup optimization in a way that is practical as pre-processing estimation for workpiece positions. Therefore a simulation using a rastered workspace model, combined with an asynchronous PSO implementation will be introduced to avoid needless simulation runs.

1 Introduction

Today, for production processes CNC simulations by virtual tooling machines are established as a standard for validating complex cutting processes. The current research project (*Spitzencluster its OWL InVorMa*), supported by the *German Federal Ministry of Education and Research* contains the research field of improving the virtual tooling setup process of machines and production systems by using simulation models of tooling machines. The goal is to build a system, which automatically generates valid setup parameters for workpiece and workpiece clamp positions in the workspace.

In order to be able to successfully implement the setup optimizer, the idea arose to combine a virtual tooling machine with a simulation based optimization system using metaheuristics such as the Particle Swarm Optimization (PSO) algorithm. The

J. Weber (✉) · A. Mueß · W. Dangelmaier
Heinz Nixdorf Institute Paderborn, Business Computing, esp. CIM,
Fuerstenallee 11, 33102 Paderborn, Germany
e-mail: jens.weber@hni.upb.de

© Springer International Publishing Switzerland 2017
K.F. Dœrner et al. (eds.), *Operations Research Proceedings 2015*,
Operations Research Proceedings, DOI 10.1007/978-3-319-42902-1_60

PSO algorithm generates solution candidates for the workpiece position, evaluated by the simulation model (collision detection, production time, tool distance and circumstantial paths detection). The particle swarm will then threw the workspace until a stop criterion is reached and a solution is found. This combination generally leads to a high number of simulation runs and iterations in order to be able to provide useful setup parameters. For this reason, several pre-processing approaches have been developed to evaluate the workpiece positions. One of the current proofs of concept is presented in this contribution and will be further developed in future contributions.

2 Related Work About Parameter Optimization in SBO and Virtual Tooling

The literature shows that simulation based optimization (SBO) can be successfully used in the domains of logistics, layout planning, scheduling and production planning, esp. material flow systems [1]. The implementation of a SBO-system requires individual configurations and constraints [2, 3]. The contribution [4] directly shows that the material flow models are able to parameterize by using the (PSO) approach. This approach was compared with the 3-phases-PSO-algorithm, see [5] resulting in improved runtime. In order to identify useful metaheuristics for an SBO approach to improve the setup of virtual tooling, the contribution of [6] shows test results using defined benchmark functions of certain metaheuristic algorithms with regards to convergence behavior, runtime and the number of dimensions. The contribution of [7] contains a further method to accelerate the optimization systems of the SBO approach using metaheuristics as well as handling stochastic node failures and distributed and limited resources. The PSO-algorithm is then further developed and is reviewed as a synchronous, asynchronous and partially synchronous PSO.

3 Concept of the Pre-Processing Position Validator for Machine Setups

The first approach is developed in a 2D-Environment. Several position setups of workpiece and clamps can be defined and the contours of clamps and workpieces can be represented alongside the contours of the tool arm. The workpiece clamps and other areas in the unavailable spaces are defined as (unspecified) obstacles as well. The point of origin of the workspace is found in the top left corner. The workpiece point of origin, which is important for the NC-program, is also defined as the top left edge of the workpiece. The approach estimates only possible positions of workpieces, obstacles and clamps. The target geometry data of the workpiece as well as material removal will be ignored. The simulation evaluates whether the relationship of tool paths and workpiece based on the NC-program is feasible. The workspace of the machine is rasterized into discrete millimeter squares in which each occupied square of the total area is represented by the coordinate point in the top left corner. The developed system presents validated paths between workpiece and tool arm, eliminating collisions or non-reachable approach angles. For that, the NC program is read in by a NC-interpreter which recognizes movements of the tool in the workspace as well as movements between tool change point and application point at the workpiece. The tool arm will travel through the rasterized squares by a single shortest path search. In combination with the PSO algorithm [7], many position parameters are able to be investigated using the shortest path found by the abstracted simulation system as a fitness function. The standard PSO algorithm based on the contribution of [8] is used. The stop criteria of the PSO is reached when no significant improvement can be observed for 10 generations. The goal is to achieve short paths for the tool arm, as these will result in shorter work duration to evaluate setup configuration. In this way, the simulation time is negligible compared to the real 1:1 simulation model. The use cases are defined by an user interface to develop a 1:1 workspace area of the real tooling machine with consideration of real sizes. For each use case, 4 objects are defined. One object represents a workiece including workpiece clamps. The remaining objects are defined as obstacles.

To conclude the concept, the dimensions reduction from 3D to 2D is given by the restriction of the workspace by the coordinate $z = 0$. Because of the tool arm and the real world ability to drive through the 3D-workspace, there are two basic premises:

- If a tool path is found in 2D, there will be also a valid tool path in the 3D workspace
- If there is no tool path solution in 2D, no statement will be possible.

The workspace is rasterized by a transformation function $f : \mathbb{R}^2 \to \mathbb{Z}^2$ with an arbitrary rasterize-constant a such that for all $z \in \mathbb{Z}^2$ hold:

$$z \text{ is occupied } \leftrightarrow \exists r \in \mathbb{R}^2 : r \text{ is occupied AND } z_1 \leq r_1 < z_1 + a \text{ AND } z_2 \leq r_2 < z_2 + a$$

when a tool path is found in $\mathbb{Z}^2$, there will also be a tool path in $\mathbb{R}^2$ or if there is no tool path in $\mathbb{Z}^2$, no statement about a tool path in $\mathbb{R}^2$ will be possible. The use cases consist of a multi-position setup scenario where several raw workpieces are placed in the machine workplace. The workpiece from the current job is marked as workpiece while the remaining workpieces are marked as obstacles. This use case stands for the proof of concept of a successful pre-processing run to find a good or better setup position using a specified constant NC-program.

4 Results and Discussion of the Use Cases

Figure 1 gives an overview of the four use cases evaluated in this contribution. For an improved understanding of the optimization results, Fig. 2 illustrates the fitness landscape of the workpiece position for the four use cases. The x and y-axis represent the machine table coordinates of the workspace of the tooling machine. The fitness

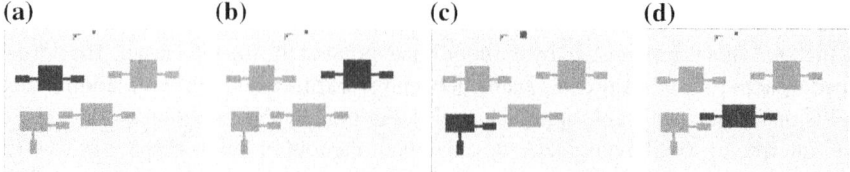

Fig. 1 Use cases. **a** Use case 1. **b** Use case 2. **c** Use case 3. **d** Use case 4

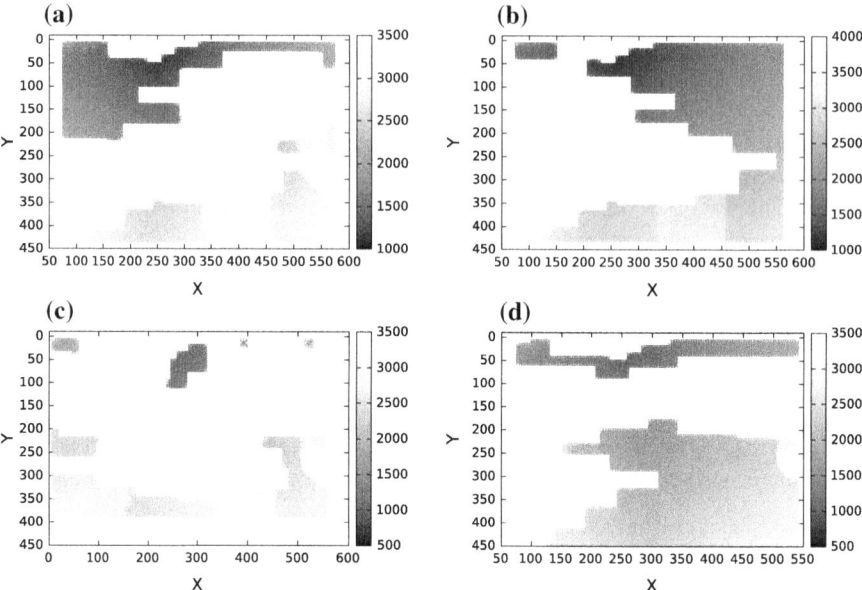

Fig. 2 Fitness landscapes of all four use cases from Fig. 1. **a** Use case 1. **b** Use case 2. **c** Use case 3. **d** Use case 4

value represents the number of steps needed during the simulation with the given setup.

The gray area represents valid setup positions for the workpiece, while the saturation represents the fitness value depending on obstacles. The best position of the workpieces can be analytically calculated using mathematical formulas, but this requires high effort and is impractical for real world problems. The analytical position coordinates which represent the best workpiece position and a valid simulation for all use cases are $x = 263$ and $y = 38$. The required number of raster steps during the simulation is given in Table 1.

The following Fig. 3 gives an overview about the achieved results of the treated approach. The x-axis represents the use cases and the y-axis differs depending on the metric. The bars represents the minimum and maximum results. The black line shows the Median and the rectangle the elements between the first and third quantile.

Table 1 Number of raster steps during simulation runs with analytical best position coordinates

Use case	Number of raster steps during simulation
1	1067
2	1051
3	929
4	929

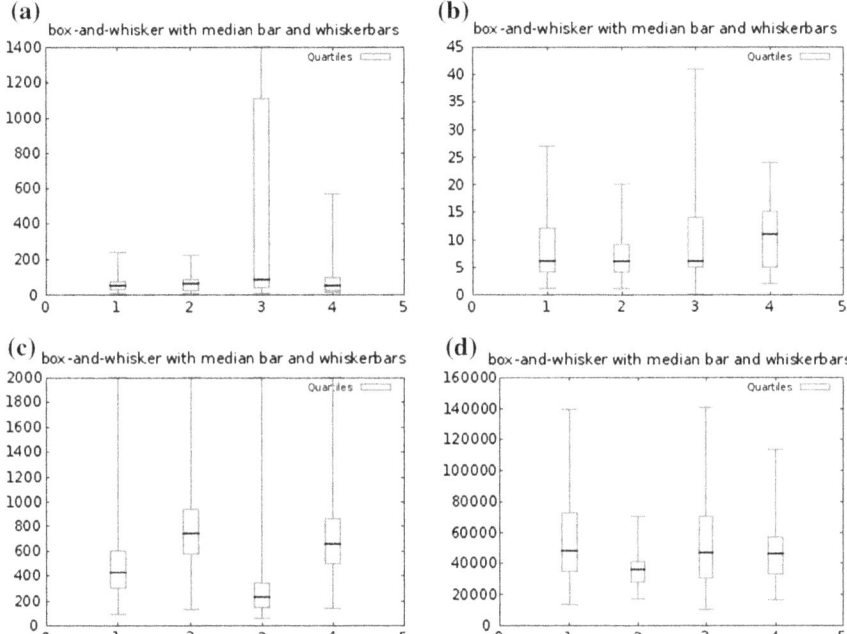

Fig. 3 Results comparing the four Use cases from Figs. 1 and 2. **a** Distance to analytical optimum. **b** Generations until PSO converges. **c** Simulation duration. **d** Optimization duration

In Fig. 3a, the differences in raster steps between the analytically calculated optimum and the solution given by the PSO are visualized (y-axis). In general the found solutions need about 50 steps more. It is notable that use case three shows a high deviation which is probably caused by the clamp position. The resulting fitness landscape (see Fig. 2c) indicates this finding. Furthermore, it determines an impractical search space for the PSO because of the huge white area surrounding the global optima. Use case one, two and four show similar results. Outliers are notable only in use case three and four. Figure 3b shows the number of PSO generations, needed to converge. On average more generations are needed for the use cases three and four. This is caused by the search space, having mostly invalid solutions near the analytical optimum.

Figure 3c illustrates the simulation duration of the use cases. On average a simulation run takes about 600 ms. A fast result is reached because the simulation focuses on the collision detection, considering 2 dimensions only. The experimental result of each use case determines low deviations, but there are also high outliers. The outliers are reflected by 2 % of the sample. These results are expected in cases where the direct path would run threw a concave obstacle, causing the shortest path algorithm to achieve worst case run times. Figure 3d shows the optimization duration of the use cases. The optimization duration seems to be independent from the chosen use case.

5   Outlook

The proposed contribution deals with a dimension (3D to 2D) reduced setup optimization process for the identification of a near optimal workpiece position. The pre-processing is necessary because of the high simulation durations using a full virtual tooling machine. To demonstrate the functionality, four use cases have been evaluated, each of them having different workpiece and obstacle positions. An optimization component improved the position, using an asynchronous PSO algorithm. The results differed depending on the obstacle, workpiece and workpiece clamp positions. The benchmark of the optimization process is shown by the average difference of raster steps, by the simulation run, using the analytical and the calculated optimal workpiece position. The simulation durations of the use cases are compared to each other and the convergence behavior is shown to be dependent on the distribution of valid solutions around the analytical optimum. For the future, the pre-processing approach is extendable for 3D-scenarios as well as more complex optimization problems with different tool sizes and NC-programs. Also a comparison between the shown approach and a virtual tooling machine will determine more results about the performance and accuracy of the systems as well as will offer a clearly evaluation of the model.

References

1. Krug, W., März, L., Rose, O., Weigert, G.: Simulation und Optimierung in Produktion und Logistik
2. Wenzel, S., Collisi-Böhmer, S., Rose, O.: Qualitäskriterien Für Die Simulation in Produktion und Logistik: Planung und Durchführung Von Simulationsstudien. Springer (2008)
3. Rabe, M., Spieckermann, S., Wenzel, S.: Verifikation und Validierung für die Simulation in Produktion und Logistik: Vorgehensmodelle und Techniken. Springer Science & Business Media (2008)
4. Laroque, C., Urban, B., Eberling, M.: Parameteroptimierung von Materialflusssimulationen durch Partikelschwarmalgorithmen. Multikonferenz Wirtschaftsinformatik **2010**, 449 (2010)
5. Angeline, P.J.: Using selection to improve particle swarm optimization. In: Proceedings of IEEE International Conference on Evolutionary Computation, vol. 89 (1998)

6. Weber, J., Boxnick, S., Dangelmaier, W.: Experiments using meta-heuristics to shape experimental design for a simulation-based optimization system: intelligent configuration and setup of virtual tooling. In: Asia-Pacific World Congress on Computer Science and Engineering (APWC on CSE), pp. 1–8. IEEE, New York (2014)
7. Reisch, R.-E., Weber, J., Laroque, C., Schröder, C.: Asynchronous optimization techniques for distributed computing applications. In: Proceedings of the 2015 Spring Simulation Multi Conference, 48th Annual Simulation Symposium, vol. 47, 2nd edn., pp. 49–57. IEEE, New York (2015)
8. Kennedy, J., Eberhart, R.: Particle swarm optimization. In: Proceedings of the 4th International Conference on Neural Networks, vol. 4, pp. 1942–1948. IEEE, New York (1995)

Simulation-Based Modeling and Analysis of Schedule Instability in Automotive Supply Networks

Tim Gruchmann and Thomas Gollmann

Abstract Within automotive supply chains, instability of order schedules of original equipment manufacturers (OEMs) creates inefficiencies in suppliers' production processes. Due to the market power of the OEM, first tier suppliers are not always able to influence the scheduling behavior of their customers. However, addressing the root causes of schedule instability, in particular the unreliability of suppliers' production processes, can help to curtail short-term demand variations and increase the overall supply chain efficiency. To this end, we introduce a stylised assembly supply chain model with two suppliers and a single OEM. This supply chain can be disrupted by a shortage occurring at one of the two suppliers due to random machine breakdowns, what consequently creates dependent requirements variations affecting both the buyer and the other supplier. Therefore the paper at hand contains two main sections. At first, a simulation model is developed containing the said mechanism causing schedule instability. Secondly, a simulation study is carried out to derive managerial and theoretical implications accordingly.

1 Introduction

Companies within automotive supply chains exchange and update demand information on a regular basis. Typically, suppliers receive from their customers daily updates of short-term delivery requirements and mid-term demand forecasts as a basis for production planning. It is however not uncommon that required shipment quantities and due dates become revised by the customer at short notice [13] and result in particular from the adjustment of the planned production schedule by the OEM. Such adjustments need not result exclusively from a change in the market

T. Gruchmann (✉)
ZNU - Centre for Sustainable Corporate Leadership, Witten/Herdecke University,
Alfred-Herrhausen-Str. 50, 58448 Witten, Germany
e-mail: tim.gruchmann@uni-wh.de

T. Gollmann
NTT DATA Deutschland, Hans-Döllgast-Str. 26, 80807 Munich, Germany
e-mail: Thomas.Gollmann@nttdata.com

© Springer International Publishing Switzerland 2017   453
K.F. Dœrner et al. (eds.), *Operations Research Proceedings 2015*,
Operations Research Proceedings, DOI 10.1007/978-3-319-42902-1_61

demand but may occur due to re-scheduling of production orders at the OEM. The German Association of the Automotive Industry (VDA) cites the complexity of the supply chain, technical and quality-related problems as well as other contingencies as sources of the said demand variations [16]; similar causes are mentioned in [1]. As a consequence, suppliers in automotive supply networks face an increased volatility within their production processes which lead to additional tool change-overs and inefficient lot sizes [6]. VDA estimates the resulting costs to be up to 5 % of the suppliers' turnover [16].

The described phenomenon is termed in the literature as *schedule instability*. Specifically, schedule instability is defined as incessant adjustment of OEM's production schedule which reacts to changed conditions [10]. Sivadasan et al. [13] describe the schedule instability as a mutual interaction between the adjustment of the buyer's orders and the production planning of the supplier, especially affecting the given lot sizing decisions [4]. In the past, a number of existing studies have addressed schedule instability and its effects. In particular, Liker and Wu [8] made a comparison of short-term schedule instability between US-American and Japanese OEMs. For the European automotive industry Childerhouse, Disney and Towill [1] discovered a significant correlation between demand variations at the OEM and schedule instability. In an experimental research design, Pujawan [9] as well as Herrera and Thomas [3] received similar results.

To the best of our knowledge, none of the existing studies addressed schedule instability as result of suppliers' vulnerability to disruptions in a model-based framework. Therefore, the main objective of the paper consists in developing a formal model and conducting a simulation-based analysis of the interdependencies within supply networks taking the unreliability into account. In detail, we study a model of an assembly network supply chain [17] and extend the above line of research by generating insights into the effect of supply unreliability on the extent of schedule instability in such a network.

2   Model Description

Generally, supply disruptions can occur on every supplier-buyer link in this network. Those links in which vendors operate at their capacity limit can be described as critical [15]. Short-term capacity shortages occurring at the supply side of one such link are likely to trigger demand variations across the supply network because of the supplier's inability to ship the required parts. This forces the buyer to re-schedule its production in order to avoid a standstill. For that purpose, the OEM increases the requirements for other suppliers' parts at short notice. At the same time, the requirements for the part in short supply are reduced. This creates horizontal ties between the suppliers; these ties are manifested in an indirect relationship between demand variations at the nodes which supply non-complementary products. This kind of relationships can be termed as *dependent requirements variations*. Consequently, dependent requirements variations are likely when disruptions occur, especially under single sourcing procurement. Note that if another critical link will, as a consequence of re-scheduling at the OEM, experience a demand surge and a resulting capacity shortage, the above effect is expected to propagate further. To focus on main effects, the present work refers to a simplified setting with two suppliers producing two different parts, for instance manual and automatic gearshifts, which are assembled on the same assembly line of the OEM into two different final products.

In particular, we model the operation of the supply chain under study as follows: The OEM faces a deterministic market demand per time period over an infinite horizon, and provides its suppliers a constant forecast based on its assembly line capacity. If a manufacturing disruption due to a random machine breakdown occurs at the unreliable supplier (supplier 1), the production of the disrupted batch halts; the items manufactured are shipped to the buyer while the remaining parts of the batch turn to a backorder. In contrast, the fully reliable supplier (supplier 2) is able to increase the capacity at short notice following the shortage of supplier 1 and the subsequent re-scheduling by the buyer. Specifically, while supplier 1's part is unavailable, the buyer dedicates a higher portion of its capacity to the assembly of supplier 2's part and increases accordingly its requirements, which supplier 2 is able to accommodate. By re-planning, the buyer avoids negative consequences of the disruption such as lost sales and standstill costs. After supplier 1 having recovered its backlog, the buyer will dedicate a higher portion of its capacity to the assembly of supplier 1's part and decreases accordingly its requirements for supplier 2's part.

3 Simulation Study

To conduct the simulation study, we built a System Dynamics simulation model in order to gain deeper insights into the nature of dependent requirements variations. In general, Systems Dynamics simulation is seen as a probate instrument to analyse problems with dynamic complexity [14]. Concentrating on dependent requirements variations, the re-scheduling interactions between the buyer and its supply network caused by random, discrete machine output at supplier 1 were analysed. In that context, Schönsleben [12] indicates that for discrete objects, such as a random machine output, the Poisson distribution provides a suitable representation. In the end, comparative simulation runs with different unreliability levels at supplier 1 were carried out. Figure 1 illustrates the simulation model as implemented in Anylogic. Ivanov et al. [5] state that Anylogic is a suitable tool when Systems Dynamics simulation is required. Table 1 describes the actors of the simulation model.

Following [7] at least three to five simulation replications for each setting should be performed. Using the confidence intervals method [11], three replications were identified as being sufficient. In the end, three replications for each unreliability level ($\lambda = 100$, 90 and 80) for 10,000 periods were conducted. Analysing the created data base, the quantities of the "AdditionalProduction" as indicator for dependent requirements variations took centre. In particular, the probability distribution of the quantities of "AdditionalProduction" was tested using the Anderson-Darling test

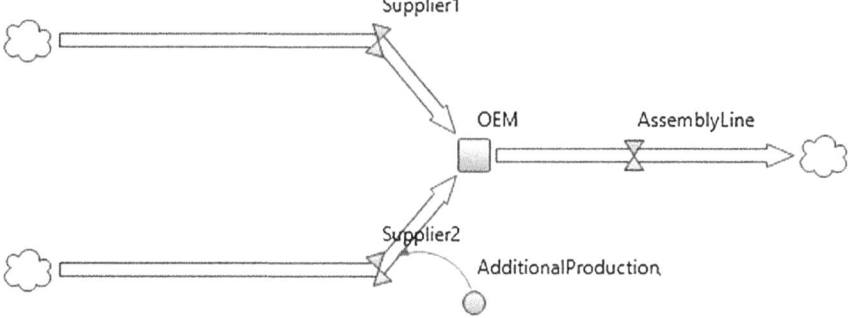

Fig. 1 Anylogic simulation model

Table 1 Actors of the simulation model

Actor	Source	Type	Setting	Character
Supplier 1	Yes	Flow	Production batch, Poisson distributed with $\lambda = 100$, 90 or 80 per period	Random
Supplier 2	Yes	Flow	Production batch, deterministic 100 units per period	Stable
AssemblyLine	OEM	Flow	Market demand, deterministic 200 units per period	Stable
AdditionalProduction	Yes	Dynamic	Indicator for dependent requirements variations	Dependent

Table 2 Simulation results for the actor "AdditionalProduction"

Replication	$\lambda = 100$			$\lambda = 90$			$\lambda = 80$		
	Ω	μ	σ	Ω	μ	σ	Ω	μ	σ
1	Normal	0.06	10.03	Normal	10.15	9.46	Normal	19.85	8.98
2	Normal	−0.03	9.88	Normal	9.97	9.48	Normal	20.04	8.95
3	Normal	0.05	10.01	Normal	9.98	9.5	Normal	19.82	8.94

[11]. Table 2 summarises the descriptive results of the simulation runs for each unreliability level.

The simulation results show clearly that due to the incessant adjustment of OEM's production schedule as reaction to supplier 1's unreliability, the schedule instability in the whole supply network increases. In particular, the mean of the quantities of the "AdditionalProduction" as indicator for dependent requirements variations shifts from an average of 0.09 (unreliability level $\lambda = 100$) to an average of 10.03 (unreliability level $\lambda = 90$) and to an average of 19.90 (unreliability level $\lambda = 80$). Additionally, it can be seen that the probability distribution of the quantities of the "AdditionalProduction" is estimated to be normally distributed and does not change with an increasing unreliability level. As expected, an approx. positive linear relation can be determined between the unreliability level and the mean of the quantities of the "AdditionalProduction" while the standard deviation decreases slightly with increasing unreliability level. Without loss of generality, the observed effects of indirect horizontal ties between suppliers will remain unchanged in a single-buyer-multiple-vendor-setting with in a single bill of material.

4 Conclusion and Outlook

Following [2] simulation is being used increasingly as a methodology for theory development. Particularly for longitudinal and nonlinear processes, simulation can help to build a more comprehensive and precise theory from so-called simple theory [2]. Applying the suggested roadmap by Davis et al. [2], the problem statement was first embedded in the general theoretical concept of schedule instability. As the theoretical concept of schedule instability could not answer the question as to how schedule instability spreads in more complex supply networks and to which extent unreliability of a certain member within that network affects schedule instability, the concept of dependent requirement variations was set up in a Systems Dynamics simulation model to investigate causal loops such as reinforcing or dampening feedback in a second step. As a result, it could be seen that for the stylised assembly supply chain consisting of one buyer and two suppliers, the schedule instability grew approx. linearly with an increasing unreliability level, not changing the probability distribution shape.

To summarise the results for practical use, OEMs should consider within their sourcing decisions not just the effects of disruptions for themselves, but also for their whole supply network, especially when the network is unable to compensate the unreliability of few. As the OEM got the market power to influence supplier reliability, an increase of the competitiveness of the supply network is possible by decreasing the number of disruptions. Also in the literature, the investment into the supplier reliability [18] is seen as a proven instrument to reduce schedule instability.

With respect to future research, two major directions can be pursued. First, an extension of the simulation-based model should be conducted, in particular including material flows and finite production capacities to gain further theoretical insights regarding dependent requirements variations. Second, the known instruments of decreasing the schedule instability (frozen zone, safety stocks, etc.) should be evaluated in the context of dependent requirements variations to decrease supply-chain-wide costs.

Acknowledgements We would like to express our sincere gratitude to JProf. Dr. Grigory Pishchulov for his comments and suggestions on the manuscript.

References

1. Childerhouse, P., Disney, S., Towill, D.: On the impact of order volatility in the European automotive sector. Int. J. Prod. Econ. **114**, 2–13 (2008)
2. Davis, J., Eisenhardt, K., Bingham, C.: Developing theory through simulation methods. Acad. Manag. Rev. **32**, 480–499 (2007)
3. Herrera, C., Thomas, A.: Formulation for less master production schedule instability under rolling horizon. In: Proceedings of the International Conference on Industrial Engineering and Systems Management (2009)
4. Ho, C.: Evaluating dampening efforts of alternative lot-sizing rules to reduce MRP system nervousness. Int. J. Prod. Res. **40**, 2633–2652 (2002)
5. Ivanov, D., Sokolov, B., Kaeschel, J.: A multi-structural framework for adaptive supply chain planning and operations control with structure dynamics considerations. Eur. J. Oper. Res. **200**, 409–420 (2010)
6. Jacobsen, A.: Messung und Bewertung der Nachfragedynamik und logistischer Agilität in der Automobilzulieferindustrie. PZH GmbH, Germany (2006)
7. Law, A., McComas, M.: Secrets of successful simulation studies. In: Proceedings of the 23rd Conference on Winter Simulation, pp. 21–27 (1990)
8. Liker, J., Wu, Y.: Japanese automakers, US suppliers and supply chain superiority. Sloan Manag. Rev. **42**, 81–93 (2000)
9. Pujawan, N.: Schedule instability in a supply chain: an experimental study. Int. J. Inventory Res. **1**, 53–66 (2008)
10. Pujawan, N., Smart, A.: Factors affecting schedule instability in manufacturing companies. Int. J. Prod. Res. **50**, 2252–2266 (2012)
11. Robinson, S.: Simulation: The Practice of Model Development and Use. Wiley, UK (2004)
12. Schönsleben, P.: Integrales Logistikmanagement. Springer, Germany (2007)
13. Sivadasan, S., Smart, J., Huatuco, L., Calinescu, A.: Reducing schedule instability by identifying and omitting complexity-adding information flows at the supplier-customer interface. Int. J. Prod. Econ. **145**, 253–262 (2013)
14. Sterman, J.: Business Dynamics: Systems Thinking and Modeling for a Complex World. McGraw-Hill, USA (2000)
15. Tempelmeier, H.: Bestandsmanagement in Supply Chains. Books on Demand, Germany (2006)
16. VDA (Hrsg.): VDA Empfehlung 5009: Forecast-Qualittskennzahl: Definition und Anwendung. Germany (2008)
17. Wang, Y., Gerchak, Y.: Capacity games in assembly systems with uncertain demand. Manufact. Serv. Oper. Manag. **5**, 252–267 (2003)
18. Wang, Y., Xiao, Y., Yang, N.: Improving reliability of a shared supplier with competition and spillovers. Eur. J. Oper. Res. **236**, 1–30 (2014)

Performance Evaluation of a Lost Sales, Push-Pull, Production-Inventory System Under Supply and Demand Uncertainty

Georgios Varlas and Michael Vidalis

Abstract A three stages, linear, push-pull, production-inventory system is investigated. The system consists of a production station, a finished goods buffer, and a retailer following continuous review (s, Q) policy. Exponentially distributed production and transportation times are assumed. External demand is modeled as a compound Poisson process, and a lost sales regime is assumed. The system is modeled as a continuous time—discreet space Markov process using Matrix Analytic methods. An algorithm is developed in MatLab to construct the transition matrix that describes the system for different parameters. The resulting system of linear equations provides the vector of the stationary probabilities, and then key performance measures such as customer service levels, average inventories etc. are computed. The proposed model can be used as a descriptive model to explore the dynamics of the system via different scenarios concerning structural characteristics. Also, it may be used as an optimization tool in the context of a prescriptive model.

1 Introduction

Depending on the timing of their execution relatively to end customer demand, processes in a supply chain can be categorized as push or pull [2]. Push processes are executed in anticipation of customer orders, while in pull processes execution is initiated in response to customer demand. In usual hybrid push/pull systems production at the upstream stations is push-type, while distribution at downstream stations is controlled by pull-type policies. Such systems have been found to perform better than pure push, or pure pull systems, while they are more flexible to address growing product variety, shorter product life cycles and the need of keeping inventory costs as low as possible [4–6]. However, their analysis is more complicated.

G. Varlas (✉) · M. Vidalis
Department of Business Administration, University of the Aegean,
Michalon 8, 82100 Chios, Greece
e-mail: g.varlas@aegean.gr

M. Vidalis
e-mail: m.vidalis@aegean.gr

© Springer International Publishing Switzerland 2017
K.F. Dœrner et al. (eds.), *Operations Research Proceedings 2015*,
Operations Research Proceedings, DOI 10.1007/978-3-319-42902-1_62

Cochran and Kim [3] study with Simulated Annealing a horizontally integrated hybrid production system with a movable junction point. Ghrayeb et al. [6] investigate a hybrid push/pull system of an assemble-to order manufacturing environment using discreet event simulation along with a genetic algorithm. Finally, Cuypere et al. [4] introduce a Markovian model for push-pull systems with backlogged orders, basing their analysis on quasi birth-and-death processes.

The main goal of our work is to provide an algorithm for the exact evaluation of a push-pull supply network with lost sales. The resulting descriptive model can be used for the optimization of the parameters of the system.

2   Description of the System

A single product, linear, push-pull supply chain is investigated (Fig. 1). A reliable and never starved station S_1 produces units at a rate μ_1 and exponentially distributed production times. Finished products are stored in a finished goods buffer (FGB) of finite capacity b. At any given time t, inventory level at buffer is denoted by the random variable B_t. In the case where S_1 completes processing, but on completion FGB is full, station S_1 blocks (blocking after processing). $0 \leq B_t \leq b + 1$, where the case $B_t = b + 1$ corresponds to blocking. Station S_1 consists the push section of the system. Downstream, the retailer R holds inventory and follows continuous review inventory control policy with parameters (s, Q). At time t, inventory level at the retailer is denoted by the random variable I_t. When inventory I_t reaches the reorder point s, a replenishment order of Q units is placed on the buffer ($0 \leq I_t \leq s + Q$). The actual level of the sent order depends on the available inventory at buffer. In the case where FGB is empty, dispatching is suspended until one unit finishes processing at S_1, upon which it is immediately forwarded for transportation to the retailer. Transportation is modelled as a virtual station T. Inventory in transit at time t is denoted by the random variable T_t, where $0 \leq T_t \leq Q$. Exponentially distributed transportation times are assumed with transfer rate μ_2. The retailer faces external demand with compound Poisson characteristics. Customers inter-arrival times are exponentially distributed with arrival rate λ and demand per customer is uniformly distributed in the space [1, n]. The following assumptions are also made:

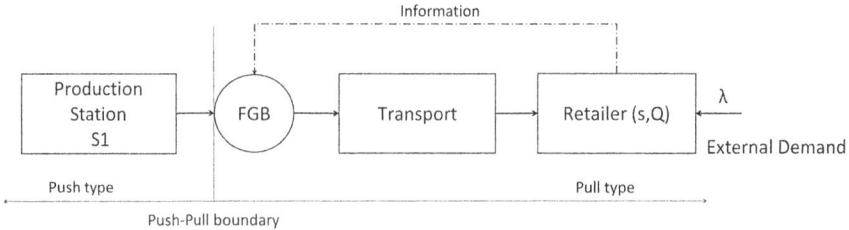

Fig. 1   System layout

1. There are no back-orders. Demand that cannot be met from inventory on hand is lost both at the retailer and the buffer. If there is insufficient inventory on hand, demand is partially met.
2. At any given time only one order can be in transit from FGB to the retailer. The one-outstanding-order assumption is a common assumption, necessary to maintain a tractable level of complexity [1].
3. When S_1 is blocked, the blocked unit is considered part of the buffer.

3   Description of the Model

The system is modelled as a continuous time, discreet space Markov Process using matrix analytic methods. Taking advantage of repeating structures, an algorithm is developed to generate the transition matrix for different parameters of the system.

The design variables that determine the dimension and structure of the transition matrix are the capacity of the finished goods buffer b, the reorder point at the retailer s, and the quantity of the orders requested by the retailer Q.

At any moment t, the state of the system can be defined by a three dimensional vector (B_t, T_t, I_t). The state space S of the Markov process is comprised of all the possible triplets (B_t, T_t, I_t). It can be easily proved that for any value of the given parameters, the dimension of the state space is given by

$$N_B^{s,Q} = (s + 1) + (s + 2) \cdot Q \cdot (B + 2) \tag{1}$$

We use the lexicographical ordering for the states [7]. We take as basic level the subset of all states corresponding to a fixed buffer inventory B_t. Within each level the states are grouped according to the inventory in transit T_t. For fixed buffer level and fixed inventory in transit, the states are ordered by inventory at retailer I_t.

The state of the system can be altered instantaneously by three kinds of events.

1. The completion of processing of one product unit at station S_1. In this case B_t increases by one unit.
2. The arrival of an outstanding order at the retailer. In this case the inventory on hand of the retailer I_t increases by T_t units. If the new value of I_t is not above the reorder point, a new transfer from FGB is initiated.
3. The occurrence of external demand. Each customer may ask for $d = 1, 2, 3, \ldots$ or n units. The inventory on hand of the retailer I_t decreases by the demanded quantity. If the updated I_t does not exceed s, a replenishment order is given to the FGB. Each value of d has equal probability of occurrence $1/n$.

Similar transitions or events, correspond to similar patterns in the transition matrix, so that sub-matrices with well defined and predictable characteristics can be defined. In general the transition matrix can be divided into three zones: the diagonal, the upper diagonal, and the below the diagonal (Fig. 2).

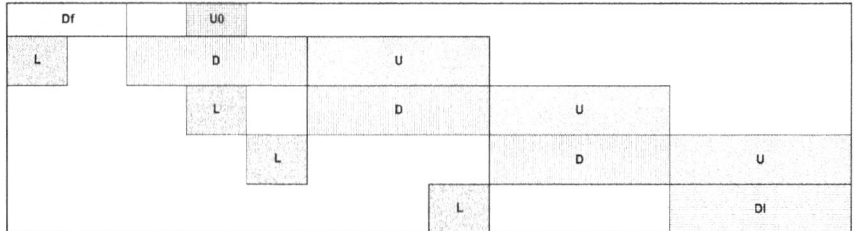

Fig. 2 Transition matrix structure

Sub-matrices on the diagonal correspond to transitions where no replenishment order is initiated. The first diagonal sub-matrix D_f corresponds to the boundary states where $T_t = 0$, $I_t < s$ and $B_t = 0$, while the last diagonal sub-matrix D_l corresponds to the boundary states where station S_1 is blocked ($B_t = b + 1$).

Upper diagonal sub-matrices correspond to arrivals from S_1 to buffer and are simple diagonal matrices of μ_1. Since S_1 processes one unit at a time, only transitions to adjacent levels occur. The first upper diagonal sub-matrix U_0 corresponds to arrivals at buffer, while the system is at the boundary states where $T_t = 0$, $I_t < s$ and $B_t = 0$.

The repeating block L below the diagonal describes transitions where there is triggering of replenishment order from the buffer to the retailer. L corresponds to transitions between non-adjacent levels and its exact positions in the transition matrix depend on the parameters B, s and Q.

From the transition matrix, the corresponding system of linear equations can be determined and the vector of stationary probabilities X can be computed. Using the stationary probabilities, key performance metrics for the system under consideration can be computed. Again, we take advantage of the structure of the transition matrix. For example, the probability of having inventory in transit equal to $(g + 1)$ can be computed as the double sum

$$T(g+1) = \sum_{j=0}^{b+1} \sum_{i=0}^{s} X(r+i) \qquad (2)$$

$$r = s + Q + 2 + (s+1) \cdot g + j \cdot ((s+1) \cdot Q + Q) \qquad (3)$$

and the average inventory in transit can be easily computed as the sum $\sum_{i=1}^{Q} i \cdot T(i)$.

In a similar way we can also calculate the rest performance measures of concern, including average inventory at the retailer (WIP retailer), average inventory at buffer (WIP buffer), the percentage of external customers whose demand is fully met by the inventory on hand at the retailer (Order Fill Rate, OFR), and the percentage of total external demand (in terms of product units) that is met from the inventory on hand at the retailer (service level, SL_2).

The validity of the algorithm was tested with simulation. 1360 different scenarios were tested for various combinations of b, s and Q, as well as for different μ_1, μ_2, and λ relations. Simulation results were consistent with the results from the analytic algorithm.

4 Results

We limit our investigation in systems where supply and demand are balanced. In our model this translates to $\mu_1 = \lambda \cdot d_{avg}$, where d_{avg} is the average demand per customer. We investigate the effect of the decision variables b, s and Q on the various performance measures.

With regard to buffer capacity (b), both order fill rate and service level increase with increasing b, but as expected, the effect decreases as OFR and SL_2 asymptotically approach their maximum value. The practical implication is that rising buffer capacity is not always an efficient way to increase customer satisfaction. The average inventory at the retailer exhibits a similar pattern, while the average inventory at buffer increases almost linearly with b. An interesting observation is that the evolution of OFR and SL_2 can be described by a logarithmic equation over a wide range of values with R^2 values above 0.94 (Fig. 3).

Since the maximum value of Q depends on b, we investigate the compound effect of b and Q on the performance measures. For fixed b, OFR and SL_2 increase with Q, with the performance measures closing asymptotically to a maximum value below 100 %. As expected, higher values of b, give better customer satisfaction. With regard to average inventories, WIP retailer increases with Q. Higher values of b also correspond to higher WIP Retailer, but the effect of b is less manifest than the effect of Q. WIP buffer decreases with increasing Q. For the performance measures under investigation the change of Q gives rise to repeating patterns across different classes of b (Fig. 4).

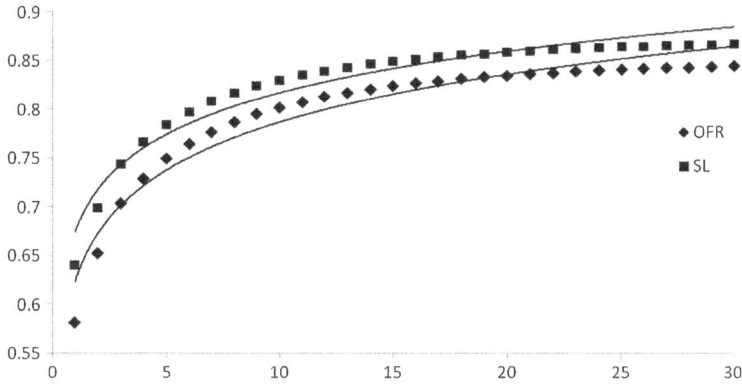

Fig. 3 Evolution of order fill rate and service level with b ($s = 2$, $Q = 4$, $n = 3$)

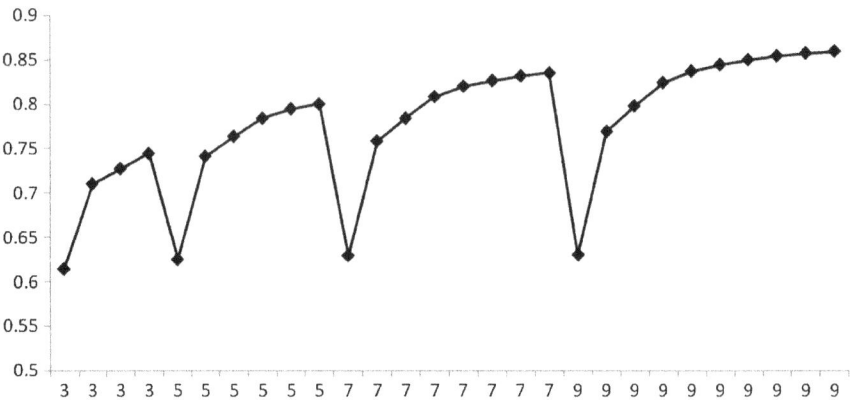

Fig. 4 Evolution of service level with Q for different classes of b (Q = 1 to b + 1)

Finally, both *OFR* and SL_2 show similar behaviour with changing s. By increasing s, better customer satisfaction can be obtained, but with decreasing efficiency. WIP retailer, increases with increasing s. For $Q > 1$, this dependence can be described linearly with good fitting. Average inventory at buffer decreases with s, suggesting that bigger s values can help in decreasing unnecessary stock at intermediate stages.

5   Conclusions

In our work we developed an exact algorithm for the analysis of a simple, serial, push-pull supply chain. The proposed descriptive model captures relationships between variables, offers insight on key features of the system at hand, and can be used for the evaluation of appropriate systems.

As indicated by the results for a balanced system, all parameters B, s and Q can have an impact on system performance. The relative importance of each depends on the specific range of its values. Despite the dynamic nature of the system under consideration, under certain circumstances performance measures could be described adequately with simple equations and this could have practical implications. Moreover, when the compound effect of parameters was investigated, performance measures exhibited repeating patterns suggesting symmetries that hold for different system configurations.

In a further step of our research, the algorithm can be expanded for phase type distributions (Erlang, Coxian) instead of exponential distribution. Different system architectures can also be an object of future investigation.

Acknowledgements The author Michael Vidalis has been co-financed by the European Union (European Social Fund ESF) and Greek national funds through the Operational Program "Education and Lifelong Learning" of the National Strategic Reference Framework (NSRF)—Research Funding Program: Thales: Investing in knowledge society through the European Social Fund.

References

1. Bijvank, M., Vis, I.: Lost-sales inventory theory: a review. Eur. J. Oper. Res. **215**, 1–13 (2011)
2. Chopra, S., Meindl, P.: Supply Chain Management, Strategy, Planning and Operations, 3rd edn, pp. 44–72. Pearson, New Jersey (2007)
3. Cochran, J.K., Kim, S.S.: Optimum junction point location and inventory levels in serial hybrid push/pull production systems. Int. J. Prod. Res. **36**(4), 1141–1155 (1998)
4. Cuypere, E., Turck, K., Fiems, D.: A queueing theoretic approach to decoupling inventory. analytical and stochastic modeling techniques and applications. In: Proceedings 19th International Conference ASMTA, Grenoble (2012)
5. Donner, R., Padberg, K., Hofener, J., Helbing, D.: Dynamics of supply chains under mixed production strategies. In: Proceedings Progress in Industrial Mathematics at ECMI, Mathematics in Industry, vol. 15. Springer, Berlin (2010)
6. Ghrayeb, O., Phojanamongkolkij, N., Tan, B.A.: A hybrid push/pull system in assemble-to-order manufacturing environment. J. Intell. Manuf. **20**(4), 379–387 (2009)
7. Latouche, G., Ramaswami V.: Introduction to Matrix Analytic Methods in Stochastic Modeling. ASA-SIAM Series on Statistics and Applied Probability, Philadelphia (1999)

Part VII
Analytics and Forecasting

Automatic Root Cause Analysis by Integrating Heterogeneous Data Sources

Felix Richter, Tetiana Aymelek and Dirk C. Mattfeld

Abstract This paper proposes a concept for automated root cause analysis, which integrates heterogeneous data sources and works in near real-time, in order to overcome the time-delay between failure occurrence and diagnosis. Such sources are (a) vehicle data, transmitted online to a backend and (b) customer service data comprising all historical diagnosed failures of a vehicle fleet and the performed repair actions. This approach focusses on the harmonization of the different granularity of the data sources, by abstracting them in a unified representation. The vehicle behavior is recorded by raw signal aggregations. These aggregations are representing the vehicle behavior in a respective time period. At discrete moments in time these aggregations are transmitted to a backend in order to build a history of the vehicle behavior. Each workshop session is used to link the historic vehicle behavior to the customer service data. The result is a root cause database. An automatic root cause analysis can be carried out by comparing the data collected for an ego-vehicle, the vehicle the failure situation occurred, with the root cause database. On the other hand, the customer service data can be analyzed by an occurred failure code and filtered by comparing the vehicle behavior. The most valid root cause is detected by weighting the patterns described above.

F. Richter (✉) · T. Aymelek
Volkswagen AG, 38440 Wolfsburg, Germany
e-mail: felix.richter@volkswagen.de

T. Aymelek
e-mail: tetiana.aymelek@volkswagen.de

D.C. Mattfeld
Institute on Business Information Systems, Mühlenpfordtstraße 23,
38106 Braunschweig, Germany
e-mail: d.mattfeld@tu-braunschweig.de

© Springer International Publishing Switzerland 2017   469
K.F. Dœrner et al. (eds.), *Operations Research Proceedings 2015*,
Operations Research Proceedings, DOI 10.1007/978-3-319-42902-1_63

1   Introduction

Failures that occur while using a product, e.g. complex products like vehicles, result in customer dissatisfaction and increasing after sales costs for the company. Thus, detecting the root cause of failures in a fast and accurate way is necessary to deal with these problems. Current failure detection has the main challenge to overcome the time-delay between failure occurrence and diagnosis. This work introduces a concept integrating decisions based on vehicle and customer service data in order to determine the root cause of failure situations with and without a failure code. A common data structure for the used data sources is introduced in this context. Section 2 gives an overview of the proposed concept. The following sections describe the analysis of failure situations without a failure code (Sect. 3) and with a given failure code (Sect. 4). Both sections focus on transforming the data sources in a unified representation.

2   Concept Integrating Heterogeneous Data Sources

A concept to detect the root cause online, as it occurs, has to deal with two different failure situations. In the automotive context there are well described situations where the vehicle reacts on an anomalous behavior with a Diagnostic Trouble Code (DTC). In order to detect the root cause of these failure situations the analytical focus is the historic customer service data. On the other side there are failure situations without a DTC. In these cases anomalies of the vehicle behavior has to be identified and described to detect the root cause.

This paper proposes an approach to handle both failure situations described before. Figure 1 gives an overview of the concept. The dotted lines highlight the process of

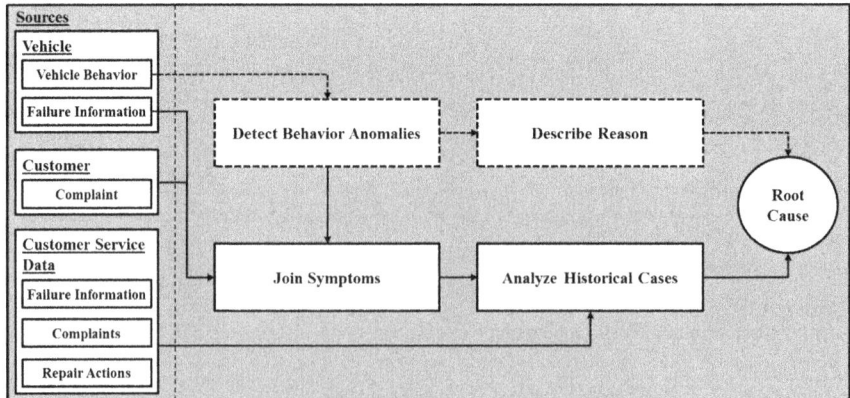

Fig. 1 Concept on root cause detection by analyzing heterogeneous data sources

analyzing failure situations without a DTC. In the following sections an outline of both analysis tasks, failure situations with and without a failure code, is given.

In terms of this concept each failure situation has to be represented in a unified structure, whether a failure code exists or not. This leads to the possibility of integrating all data sources. Thus, all failure situations are represented by Symptom-Action combinations, in the following named as failure cases. A failure case $FC_{v,t}$ of the vehicle v and the time interval t contains a set $S(v) = \{s_{e,1}, s_{e,2}, \ldots, s_{e,n}\}$ of symptoms with e as the symptom type and $R(v) = \{r_{a,1}, r_{a,2}, \ldots, r_{a,m}\}$ of performed repair actions with a as the type of the performed action. n denotes the number of symptoms and m the number of repair actions, though $|S(v)| = n$ and $|R(v)| = m$.

The shown data structure integrates three different data sources described in the following. The observed system itself, in this paper the vehicle, provides information on its own behavior and failure information, i.e. represented by DTCs, software and hardware versions or environmental parameters like the temperature. This paper considers the customer is a separate data source. Comparable to the sensor-based vehicle behavior the customer gives a description of the problem situation. The last data source is the customer service data warehouse, containing all historic failure information, recorded customer complaints (verbal description by the customer and description by the garage) and performed repair actions (detailed description in Sect. 4).

3 Analysis of Situations Without Failure Code

In terms of detecting the root cause of situations without a failure code the vehicle behavior while driving is analyzed. In such case the decision whether there are anomalies in the current vehicle dynamics based on its historical behavior. To increase the accuracy of this decision historical behavior of comparable vehicles in the fleet is used. This analysis has to take place in the backend to integrate all the historic data.

A challenge in analyzing vehicle data in a backend is the huge amount of data produced by vehicles. All control units in the vehicle are connected to a Controller Area Network bus (CAN-bus). Assuming an average data rate of 500 Kbps [4, p. 38] and the four buses mentioned in [5, p. 6ff.] a vehicle produces around 900 Mb of data per hour. In order to use the historical behavior of the ego-vehicle as well as the fleet, it has to be analyzed in a backend architecture. For that the data has to be transmitted in an aggregated form to overcome limitations by the huge amount of data.

The data within the vehicle is produced by sensors. These are continuous-time and continuous-value signals. Sensor values are read in defined intervals by the connected control units. Thus, the values are transformed into discrete-time. Further control units are limited in the storages value range, leading to the transformation in discrete values [5, p. 41ff.]. In a next step a control unit provides internal calculated values, based on the sensor inputs, and some raw sensor values to the CAN-bus.

Aggregation of data can be done in several ways. As the CAN-data represents a data stream, methods according to [6] are used. Table 1 gives a brief overview

Table 1 Aggregation methods and their applicability to represent the vehicle behavior

Method	Description	Applicable
Sum	Sum up all signal values (i.e. sum of the overall brake pressure per trip)	No
Mean/Median	One value representing a trip (i.e. mean brake pressure per trip)	No
Quartile	Representing the deviation of a trip	Partly
Count	Count the occurrence of values or the time a signal is active (i.e. profile of the brake pressure while driving)	Yes
Min/Max	Gives the extreme values of a trip (i.e. minimum brake pressure per trip)	No

of aggregation methods and highlights the applicability in representing the vehicle behavior.

The goal in analyzing vehicular data is to detect whether an aggregated trip divers form the vehicles historic behavior. If a new data instance is transmitted by the vehicle, a backend process calculates a score of representing an vehicle behavior anomaly. A survey of detection methods is given in [1]. An anomaly in a single trip can have different reasons which are not relevant for the analysis, i.e. traffic jams. To overcome this problem a scoring algorithm is used, in order to detect a subsequence of anomalous trips. The anomaly detection is not just based on a single recorded trip but on the anomaly frequency along the past $d \in \mathbb{N}$ trips. In terms of the failure cases FC introduced before, a detected anomaly can be seen as a symptom.

4   Analysis of Situations with DTC

The main focus in failure situations with a DTC is data stored by the customer service. Current research activities dealing with customer service data are building neural networks [2, 3] in order to determine the root cause. This paper describes an approach focusing on clustering failure cases in order to gain the possibility of iteratively improving the quality of proposed root cause solutions. The used customer service data consists of the following parts:

- **General vehicle data**: Information on the vehicle type (i.e. model, engine and gear box), the components and all production data (i.e. production date and plant)
- **Garage data**: This data stores verbal descriptions on the customer's complaint and a problem description of the garage
- **Diagnosis data**: In the diagnosis data all information on occurred DTCs is stored, including the date of occurrence and the affected control unit
- **Repair actions**: Stores all performed repair actions, spare parts, repair date and the reason of the failure.

In order to analyze customer service data a structure according to [3] is used. The customer service data is transformed to Symptom-Action combinations (failure cases). Thus, customer complaints and diagnosis data is handled as symptoms and actions are represented by garage descriptions and repair actions. This structure can be enhanced by anomalies detected in the vehicle data.

The data analysis of customer service data starts, related to the structure of failure cases, with a set of symptoms provided by the vehicle and customer. Using this set of data the customer data is searched. This results in all failure cases containing the given set of symptoms. In order to identify the root cause the most valid pattern has to be identified by grouping the data. Based on symptom specific similarity measures failure case patterns are detected. A formal definition of the similarity of two instances A and B is given in (1).

$$sim(A, B) = \frac{\sum_{n=1}^{N} w_n sim_{sym\_n}(A_{sym\_n}, B_{sym\_n})}{\sum_{n=1}^{N} w_n} \tag{1}$$

The value range is $0 <= sim(A, B) <= 1$ and N defines the number of compared symptoms. As not all attributes are similar important, i.e. older entries are less important, w_n defines a symptom specific weighting with $0 <= w_n <= 1$.

This grouping represents a clustering task. To retrieve the most valid pattern a threshold has to be defined. We propose the relative frequency of each cluster as a decision criterion, though the relative amount of instances in a cluster. As long as no cluster fulfills this threshold, further analysis has to be performed. This is shown in Fig. 2.

The first analysis deals with failure cases in a time interval of 14 days around the failure situation as a default. An example search criteria is a set of symptoms $S_{VN001} = \{s_{DTC,1}(\text{``}X1121\text{''}), s_{COMPLAINT,2}(\text{``}Engine\ gets\ too\ hot\text{''})\}$ for vehicle $VN001$.

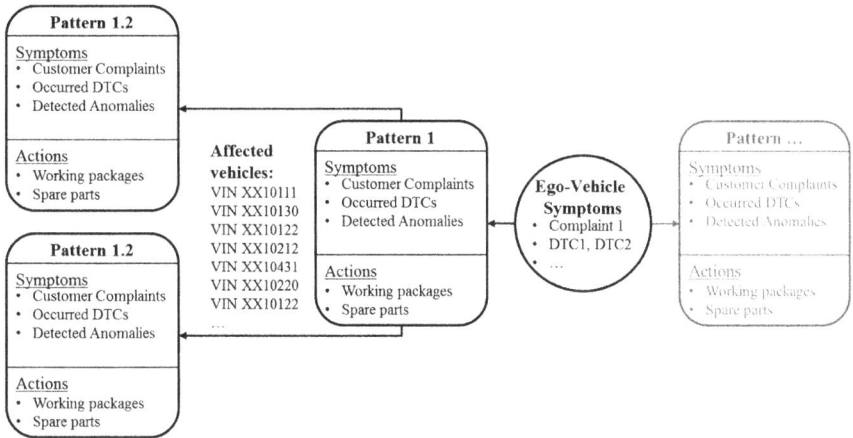

Fig. 2 Pattern detection with input symptoms by the ego-vehicle

The failure case database is filtered based on this set. The outcome is a set of failure cases containing the symptoms and actions in a time interval of 14 days around S_{VN001}. Within this data similar failure cases are going to be grouped based on the similarity defined in (1). The result is a set of failure patterns described by similar failure cases. In case of a frequency less then defined in the threshold a next iteration splitting the patterns is performed. In this case the time interval can be increased and/or anomaly symptoms can be used as a new filter criterion. This is done iteratively till the threshold is reached.

5 Conclusions and Further Research

This paper has introduced a common data structure to model failure cases as Symptom-Action combinations. By using this data structure all used data sources can be integrated in a unified representation. The proposed concept is able to use the data structure to provide the root cause of both possible failure situations, with and without DTC.

Further research has to investigate in evaluating the anomaly detection concept. Especially a threshold for the frequency of anomalies in the past trips has to be determined. In the area of analyzing customer service data the next research activities focus on the definition of symptom specific similarity functions.

Acknowledgements The authors would like to thank the team of VST/1 and NE-GQ/D from the Volkswagen AG for their professional support and revision of the proposed concept.

References

1. Chandola, V., Banerjee, A., Kumar, V.: Anomaly detection: a survey. ACM Comput. Surv. (CSUR) **41**(3), 15 (2009)
2. Hui, S., Jha, G.: Data mining for customer service support. Inf. Manag. **38**(1), 1–13 (2000)
3. Müller, T.C.: Neuronale modelle zur offboard-diagnostik in komplexen fahrzeugsystemen. Ph.D. thesis, Technische Universität Braunschweig (2011)
4. Reif, K.: Automotive Mechatronics: Automotive Networking, Driving Stability Systems. Electronics. Bosch Professional Automotive Information. Springer Fachmedien Wiesbaden, Heidelberg (2014)
5. Schäuffele, J., Zurawka, T.: Automotive Software Engineering. Springer Fachmedien Wiesbaden, Heidelberg (2013)
6. Shrivastava, N., Buragohain, C.: Aggregation and summarization in sensor networks. In: Gama, J., Gaber, M. (eds.) Learning from Data Streams, pp. 87–105. Springer, Heidelberg (2007)

Models and Methods for the Analysis of the Diffusion of Skills in Social Networks

Alberto Ceselli, Marco Cremonini and Simeone Cristofaro

We tackle the problem of analysing the diffusion of knowledge through social networks; indeed such an issue has already been highlighted in the literature [3]. We assume a set of individuals and a set of topics to be given. Each individual has a certain level of interest and skill on each topic, that change through interactions with other individuals. Links among individuals evolve according to these interactions. As shown in the literature such a phenomenon well represents the dynamics of opinions, relationships and trust. One of the main motivating applications of our work are e-learning forum analytics. In that case skills represent knowledge about a certain topic, individuals are students of a course on that topic and interactions are exchange of messages. We are therefore interested in means of understanding and improving student performances. The starting point of our research is a descriptive model detailed in [1] and used in an agent-based simulation tool. In this paper, we show that mathematical programming modelling allows at once to (a) handle large scale datasets more effectively and (b) naturally exploit the inner structure of the phenomenon, thereby combining computing efficiency with accuracy of results. Indeed, there is a renewed interest in modern optimization methods for data mining [2]. We first introduce a Linear Programming (LP) model for the network evolving over time, reproducing the results of [1] by exploiting the inner flow structure [4] of the diffusion of skills in such networks. Mainly, we show how to use our LP for predictive analytics (see Sect. 1). Then, we propose an Integer LP (ILP) prescriptive analytics model for a static scenario, that aims at finding most influential individuals in the network at a given point in time (see Sect. 2).

A. Ceselli (✉) · M. Cremonini · S. Cristofaro
Università Degli Studi di Milano – Dipartimento di Informatica, Via Bramante
65, 26013 Crema, Italy
e-mail: alberto.ceselli@unimi.it

M. Cremonini
e-mail: marco.cremonini@unimi.it

S. Cristofaro
e-mail: simeone.cristofaro@gmail.com

© Springer International Publishing Switzerland 2017   475
K.F. Dœrner et al. (eds.), *Operations Research Proceedings 2015*,
Operations Research Proceedings, DOI 10.1007/978-3-319-42902-1_64

1 Models for Predictive Analytics

Let I be the set of *nodes* in the network, M be a set of *topics* and K be a set of *time steps*. Let 1 and $|K|$ indicate the first and last time step. For each $i \in I$, $m \in M$ and $k \in K$, let $l_{im}^k \geq 0$ and $s_{im}^k \geq 0$ be the level of interest and skill, respectively, on topic m for node i at time k. For each $i, j \in I$ and $k \in K$, let f_{ij}^k be a measure of trust between nodes i and j at time k. For each $i, j \in I$, $m \in M$ and $k \in K$, let $x_{ijm}^k \geq 0$ be the amount of interaction on topic m between nodes i and j at time k, expressed in terms of number of messages sent.

Consider the following model parameters: let α be a coefficient linking skill improvement and amount of interactions, that is the units of skill improvement yielded by each unit of interactions on a certain topic; let β be a coefficient linking skill, interest and trust of nodes with maximum number of interactions of that node.

We assume the network to evolve according to the following two rules. First, the skill at time k can be obtained by increasing the skill at time $k - 1$ by a contribution given by the interactions at time k; we modelled such a rule as follows: for each $i \in I$, $m \in M$ and $k \in K$, $k \neq 1$

$$s_{im}^k = s_{im}^{k-1} + \alpha \sum_{j \in I} x_{jim}^k \qquad (1)$$

Second, the more a node is skilled or interested in a topic, the more interaction it can have with neighbours; furthermore, communication is more likely to occur between pairs of highly trusted nodes. We experimented on many formulations of that rule; after preliminary experiments we found the following to model more closely our systems: for each $i, j \in I$, $m \in M$ and $k \in K$, $k \neq 1$

$$x_{ijm}^k \leq \beta \cdot l_{im}^{k-1} \cdot s_{jm}^{k-1} \cdot f_{ij}^{k-1}. \qquad (2)$$

that is we assume the number of messages sent at time k from i to j concerning topic m is directly proportional to both the level of interest of i in m, the skill of j in m and the trust between i and j at time $k - 1$. Our prediction problem can be stated as follows: assuming that α and β are fixed, and that a subset of the $x_{jim}^k, l_{im}^k, s_{im}^k$ and f_{ij}^k terms are observable, thereby becoming data, find values for the remaining terms in order to satisfy constraints (1) and (2), maximizing some likelihood function.

Datasets. To test our models we created three sets of 10 instances each, corresponding to output of simulations of networks produced by the tool of [1]. Instances are respectively composed by 30 nodes and 7 topics (Dataset A), 40 nodes and 6 topics (Dataset B) and 50 nodes and 5 topics (Dataset C). This well reflects the size of our real world e-learning networks. Each simulation was stopped after 4000 steps, as we observed that networks reach a persistent state after that threshold, and predicting their behaviour becomes easy. In order to keep data of reasonable size we performed sampling, measuring the network state every 100 steps. Formally, each instance is described by a set of values $\tilde{s}_{im}^k$, $\tilde{l}_{im}^k$, $\tilde{x}_{ijm}^k$ and $\tilde{f}_{ij}^k$.

Training. Training our model consists in choosing values for α and β. We performed fitting to training data on Datasets A, B and C. Namely, we fixed each $s_{im}^k = \tilde{s}_{im}^k$, each $l_{im}^k = \tilde{l}_{im}^k$, each $f_{ij}^k = \tilde{f}_{ij}^k$ and each $x_{ijm}^k = \tilde{x}_{ijm}^k$. Then, we modified constraints (1) as

$$s_{im}^k = s_{im}^{k-1} + \alpha \sum_{j \in I} x_{jim}^k + \Delta_{im}^k \tag{3}$$

and constraints (2) as

$$x_{ijm}^k \leq \beta \cdot l_{im}^{k-1} \cdot s_{jm}^{k-1} \cdot f_{ij}^{k-1} + \Gamma_{ijm}^k \tag{4}$$

being Δ_{im}^k and $\Gamma_{ijm}^k \geq 0$ continuous support variables, and we minimized the following objective function

$$\sum_{i \in I, m \in M, k \in K} |\Delta_{im}^k| + \sum_{i \in I, j \in I, m \in M, k \in K} \Gamma_{ijm}^k ; \tag{5}$$

subject to constraints (3) and (4). That is, we allow the violation of constraints (1) and (2) at a price, and we search for values of α and β minimizing violations in norm one. Since (5) can be easily linearized with techniques from the literature, the resulting is a large scale LP problem. Overall, we found that for each experiment the relative violation of each constraint, that is either $|\Delta_{im}^k|/(s_{im}^{k-1} + \alpha \sum_{j \in I} x_{jim}^k)$ or $\Gamma_{tjm}^{k+1}/(\beta \cdot l_{im}^k \cdot s_{jm}^k \cdot f_{ij}^k)$ was low (about 4 % on average for constraints (3), less than 0.01 % for constraints (4)). Furthermore, we found that an optimal α was consistently about 3.0, while optimal β values ranged between 0.2 and 4.0, with a few outlier tests with much higher value. In each subsequent experiments we employed a cross-validation pattern: for each instance, α and β were fixed to the average values computed on the remaining instances of the same dataset, neglecting outliers.

Predicting the network evolution by observing skills. We consider the following task: assuming that skill and interest levels are known, predict the evolution of the network in terms of node connections. Formally, let $t_{ij}^k \geq 0$ be the amount of overall interaction between nodes i and j up to time k, that is

$$t_{ij}^k = \sum_{k' \in K : k' \leq k} \sum_{m \in M} x_{ijm}^{k'} . \tag{6}$$

Given a threshold constant γ, we say that nodes i and j are connected at time k if $t_{ij}^k \geq \gamma$, disconnected otherwise. In this task we assume that values $\tilde{s}_{im}^k$ and $\tilde{l}_{im}^k$, corresponding to skill and interest levels, are given while x_{ijm}^k and t_{ij}^k, together with Δ_{im}^k and Γ_{ijm}^k, are variables whose value has to be predicted. In a preliminary check, we found very high positive correlation between t_{ij}^k and $\tilde{f}_{ij}^k$; in fact, it turned out to be very easy to predict t_{ij}^k by including $\tilde{f}_{ij}^k$ as input: in our tests both accuracy, precision and recall values were always above 90 %. Therefore, we tackled a more complex prediction task: we assumed f_{ij}^k to be unobservable, and replaced constraints (4) with

$$x_{ijm}^{k+1} \leq \beta \cdot l_{im}^k \cdot s_{jm}^k \cdot t_{ij}^k + \Gamma_{ijm}^{k+1} \tag{7}$$

Then we minimized (5) subject to constraints (3), (6) and (7), obtaining again a large scale LP. We kept the experimental setting used for parameter fitting, defining $\tilde{t}_{ij}^k = \sum_{k' \in K : k' \leq k} \sum_{m \in M} \tilde{x}_{ijm}^{k'}$. We validated our method as follows: let True Positive (TP) be the number of node pairs having both $t_{ij}^{|K|} > \gamma$ and $\tilde{t}_{ij}^{|K|} > \gamma$, True Negative (TN) be the number of node pairs having both $t_{ij}^{|K|} \leq \gamma$ and $\tilde{t}_{ij}^{|K|} \leq \gamma$. Analogously, let False Positive (FP) be those with $t_{ij}^{|K|} > \gamma$ but $\tilde{t}_{ij}^{|K|} \leq \gamma$ and False Negative (FN) be those with $t_{ij}^{|K|} \leq \gamma$ but $\tilde{t}_{ij}^{|K|} > \gamma$. Charts of Fig. 1 report the average accuracy A = (TP+TN)/(TP+TN+FP+FN), precision P = TP/(TP+FP) and recall R = TP/(TP+FN) obtained by our method (y axis) on each Dataset for different values of γ (x axis). Even if A increases as γ increases, low values of R highlight underfitting. Still, in a scenario with a substantial lack of information, predictive power can be observed.

Predicting skills by observing the network structure. We first measured the correlation between the initial $\tilde{s}_{im}^1$ and final $\tilde{s}_{im}^{|K|}$ skill data levels in each simulation instance, finding correlation levels as low as 0.24 in Dataset A: skills evolve quickly, making it hard to estimate them without advanced analytics models. We therefore considered the following task: assuming that interactions and interest levels are known, predict the evolution of node skills. In this case, we consider $\tilde{x}_{ijm}^k$ to be given, together with initial skill $\tilde{s}_{im}^1$ and interest levels $\tilde{l}_{im}^k$, becoming data in our model. Any other term is kept as variable. As before, the resulting is still a large scale LP. Keeping the experimental setting described above, we compared the output s_{im}^k of our optimization with instance data $\tilde{s}_{im}^k$.

The chart of Fig. 2 (left) reports, for each $k \in K$ (x axis), the average correlation between the predicted skill levels and the simulated data (y axis) on each Dataset. Average correlation steadily keeps above 80 %, thus proving high accuracy. We performed a further experiment: for each $k \in K$ we measured the total predicted (resp. simulated) skill of each node $i \in I$, that is $\sum_{m \in M} s_{im}^k$ (resp. $\sum_{m \in M} \tilde{s}_{im}^k$). Then

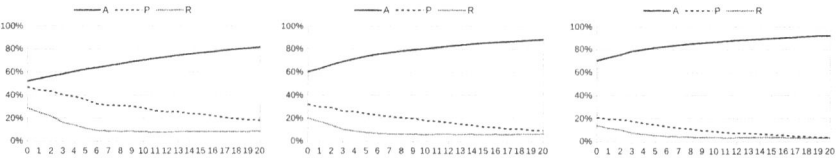

Fig. 1 Network structure prediction accuracy, precision and recall on Dataset A (*left*), B (*center*) and C (*right*)

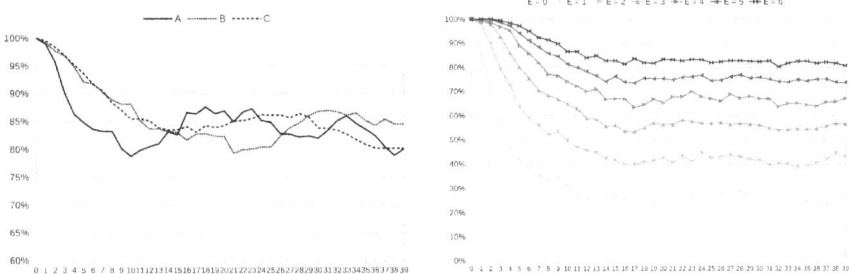

Fig. 2 Skill prediction accuracy: skills correlation on each Dataset (*left*), ranking accuracy on Dataset C (*right*)

we measured the rank of each node according to such a measure. In chart of Fig. 2 (right) results on Dataset C are reported. In particular, we depict the fraction of nodes whose ranking error is within a threshold E (y axis), for each $k \in K$ (x axis) and for values of E from 0 to 6 (grayscale data series). As can be seen, the fraction of nodes whose computed rank is exactly the simulated one ($E = 0$) is quite low, but allowing an error of 10 % in the ranking ($E = 5$) already yields an accuracy between 70 and 80 %.

Implementation. The agent-based simulation model of [1] was implemented in Netlogo 5.10. Our mathematical programming models were coded in AMPL; the corresponding large scale LPs were solved using the barrier algorithm of CPLEX 12.5. Each optimization required less than one minute of computation on a PC equipped with an Intel i7 2.8 GHz processor and 4GB of RAM. Actually, from a computational point of view, the bottleneck turned out to be our 32 bit AMPL interpreter, that could not handle larger data files. We also implemented a set of ad-hoc tools for data conversion and analysis using python 2.7.6 and plotting libraries.

2   Models for Prescriptive Analytics

Then, we elaborated on more complex analyses; these are mainly motivated by the e-learning application discussed in the introduction. We experimented on diverse models, trying to understand which features highlight nontrivial trends, and which approaches are computationally feasible. Hereafter detail one of them.

We assume that network evolution corresponds to students preparing exams in a certain semester. We focus on the final timeslot of such an evolution, that is a certain time frame just before the exams, and we consider each topic independently. Our aim is to elect a number of *tutors*, asking them to improve the skills of other students by sending messages during the final timeslot. We assume that all students having, either on their own or thanks to tutor messages, a skill level greater than or equal to a given threshold will obtain a positive evaluation in the final exam.

We suppose that Q tutors can be selected, and that each of them can send up to C messages in the final timeslot. We denote as L the skill threshold needed to obtain a positive evaluation. For the ease of notation, we indicate by s_i the skill of each node $i \in I$ at the beginning of the final timeslot, and we suppose that binary coefficients f_{ij} are 1 if node j trusts node i, 0 otherwise. The problem of finding the set of most influential tutors can be formulated as the following ILP:

$$\text{maximize} \sum_{j \in I} h_j \tag{8}$$

$$\text{s.t. } L \cdot h_j \leq s_j + \sum_{i \in I} z_{ij} \qquad \forall j \in I \tag{9}$$

$$\sum_{i \in I} y_i \leq Q \tag{10}$$

$$\sum_{j \in I} z_{ij} \leq C \cdot y_i \qquad \forall i \in I \tag{11}$$

$$0 \leq z_{ij} \leq C \cdot f_{ij} \qquad \forall i, j \in I \tag{12}$$

$$h_i, y_i \in \{0, 1\} \qquad \forall i \in I \tag{13}$$

where each binary variable h_j takes value 1 if node j may receive positive evaluation, each binary variable y_i takes value 1 if node i is selected as tutor, and each variable z_{ij} is the number of messages sent from tutor i to node j. Constraints (9) forbid to consider positive evaluation for nodes whose skill is below the threshold; constraint (10) limits the number of tutors; constraints (11) impose that no message is sent from nodes that are not tutors, and that at most C messages are sent by tutors; constraints (12) forbid nodes to accept messages from untrusted tutors. The objective (8) is to maximize the number of nodes above threshold. We coded this model in AMPL, and used the CPLEX 12.5 branch-and-cut to solve each ILP. We considered 30 networks status, corresponding to the last timeslot of instances in Datasets A, B and C. Each subproblem could be solved within a few seconds of CPU time. Charts of Fig. 3 report the average number of nodes above threshold in each dataset after the optimization process (y axis) as the number of tutors Q (grayscale data series) and the communication capacity C (x axis) change, fixing a threshold value $L = 0.9 \cdot \max_{i \in I} s_i$. From the modelling point of view, we verified that no feature acts as bottleneck: the number of nodes above threshold changes with both Q and C.

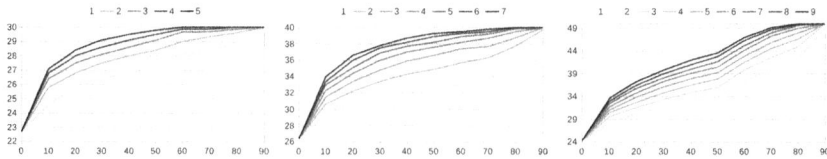

Fig. 3 Tutor optimization on Dataset A (*left*) B (*center*) and C (*right*)

Moreover, in our simulations, electing at least 4 tutors seems to allow a significant increase in number of nodes above threshold even for low C values.

Conclusion. We introduced a LP model for the diffusion of skills in networks that behaves as a linearized version of those from the literature. We proved that LP optimization can be used for effective predictive analytics on very high dimensional datasets. We also proved that parametric analyses with more complex prescriptive analytics ILP models are computationally viable. We are currently extracting data from real world repositories to validate our methods in practical settings.

Acknowledgements The work has been partially funded by Regione Lombardia/Fondazione Cariplo, Grant agreement no. 2015-0717, project "Towards Research on Decomposition Methods for Next Generation Analytics (REDNEAT)".

References

1. Allodi, L., Cremonini, M.: Self-organizing techniques for knowledge diffusion in dynamic social networks. Complex Network V, pp. 75–86. Springer, Heidelberg (2014)
2. Bertsimas, D.: Statistics and Machine Learning via a Modern Optimization Lens. P. Morse plenary lecture at INFORMS conference, San Francisco (2014)
3. Guille, A., Hacid, H., Favre, C., Zighed, D.A.: Information diffusion in online social networks: a survey. Sigmod Rec. **42**(2), 17–28 (2013)
4. Skutella, M.: An Introduction to Network Flows Over Time. Research Trends in Combinatorial Optimization, pp. 451–482. Springer, Heidelberg (2009)

Topological Data Analysis for Extracting Hidden Features of Client Data

Klaus B. Schebesch and Ralf W. Stecking

Abstract Computational Topological Data Analysis (TDA) is a collection of procedures which permits extracting certain robust features of high dimensional data, even when the number of data points is relatively small. Classical statistical data analysis is not very successful at or even cannot handle such situations altogether. Hidden features or structure in high dimensional data expresses some direct and indirect links between data points. Such may be the case when there are no explicit links between persons like clients in a database but there may still be important implicit links which characterize client populations and which also make different such populations more comparable. We explore the potential usefulness of applying TDA to different versions of credit scoring data, where clients are credit takers with a known defaulting behavior.

1 Introduction

When using powerful methods like neural networks in data modeling one is always concerned with the problem of *form versus function*. That is to say, many quite different looking shapes (forms) of models may lead to the same input-output (or functional) behavior. This in fact means that, conversely, when asserting some function one could reasonably infer a shape or subset of data producing this very function. The relatively new development of computational topology and especially TDA points towards a meaningful inference of shape induced by empirical data point clouds.

K.B. Schebesch (✉)
Faculty of Economics, Informatics and Engineering, Vasile Goldiş Western
University Arad, Campus UVVG, Liviu Rebreanu 86, 310414 Arad, Romania
e-mail: kbschebesch@uvvg.ro

R.W. Stecking
Department of Economics, Carl von Ossietzky University Oldenburg,
26111 Oldenburg, Germany
e-mail: ralf.w.stecking@uni-oldenburg.de

© Springer International Publishing Switzerland 2017   483
K.F. Dœrner et al. (eds.), *Operations Research Proceedings 2015*,
Operations Research Proceedings, DOI 10.1007/978-3-319-42902-1_65

TDA is indeed based on concepts of algebraic topology [1], a both venerable and super-active, but for the average computational scientist also difficult to access branch of mathematics. The main message of the TDA specialists seems to be that *shape conveys meaning* and can hereby help to advance the analysis of many data intensive, opaque practical modelling problems. Success of TDA in computational biology [6] is being popularized within certain research circles. Here some previously hidden classes of individuals and their atypical variants of disease were detected by identifying a tendril-like extension within a certain data projection accomplished by methods of TDA. All the methods used are computational and general in nature and they should be considered as candidates in the data analysis toolbox of other application domains like time series [7] or like politics and voting [3]. In the sequel we attempt to clarify in an exploratory way if, and to some extend how, TDA may bring additional information into applications concerned with modelling client data sets. To this end we use empirical credit client data. We use a state of the art but highly experimental software tool for the numerical computations.

2 How to Use TDA Methods in Practice?

A set of data points in an m-dimensional vector space contains N unconnected components, namely the data points themselves. We allow for a connection between two or more points to be established if they are within the reach of a distance (or limit of filtration). Then, increasing this distance from zero will encompass more and more points into such connections. The number and nature of these consecutive connections depends on the distribution (shape) of the data points. The resulting connections will be more or less persistent and will contain up to m-dimensional shapes. The construction of these connections is algorithmically realizable by simplicial complexes. The latter approximate the data by composing simplices of possibly different dimensions. As simplices are convex objects one may need a large number thereof in order to closely approximate more general (data) shapes. There are many ways of achieving such simplicial complexes but many of them are excessively expensive to compute. Vietoris–Rips complexes are to some extend computationally tractable. However, their computational load also sharply increases with the number of examples as the topological methods seek to infer possible higher dimensional features from the existing data.

In order to perform TDA in practice we use a state of the art but highly experimental software build around the C++ libraries *Dionysus* [5] and *Gudhi* [4]. This is an R-based numerical TDA package [2], which essentially allows for setting the parameters *limit of filtration* and *maximal dimension* of the topological objects to be discovered.

3 TDA on Some Artificial Data Instances

Applying the TDA software to 2-dimensional textbook examples proceeds as expected for sensible parameter values (see previous section). However, this restricts us to detecting 0-dimensional and 1-dimensional features. We prefer to be able to detect 2-dimensional features as well and henceforth we use a torus embedded in a cube of noise. We immediately remark that now the software starts to fail for quite low values of feature dimensions which we do not expect to improve with still higher dimensions (as it turns out this failure also depends on the shape of data).

Figure 1 depicts a case with textbook like behavior. The torus from the lhs plot, i.e. its 2-dimensional surface is embedded in 3-d. The points on the torus surface were generated randomly. The resulting diagram of the rhs plot is based on a Rips complex which attempts to connect neighboring points by a system of simplices. The different outputs (rhs plot) refer to 0-dim, 1-dim and 2-dim objects as described in Sect. 2. At least two 1-dim cyclic persistent components can be clearly evidenced (as are indeed present in a 2-d torus surface in 3-d).

4 Computational Results on Empirical Client Data

In order to explore the application of Rips complex computation on empirical examples, we use a credit client data set obtained from a German bank, containing $N = 139951$ cases and $m = 25$ features, i.e. with input matrix $x_{ij}, i = 1, \ldots, N, j = 1, \ldots, m$. We also have the corresponding $y \in \{-1, 1\}$ labels denoting eventual credit client behavior (i.e. $y_i = 1$ if the ith client was defaulting). The two sub-populations

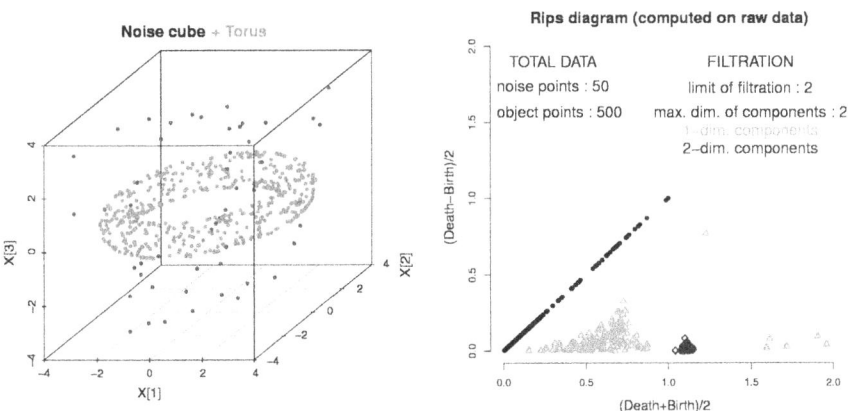

Fig. 1 A sparse 3-d noise cube superimposed by a 3-d torus. The corresponding Rips diagram with 0-d objects as dots on a diagonal and 1-d objects as small triangles (*rhs plot*)

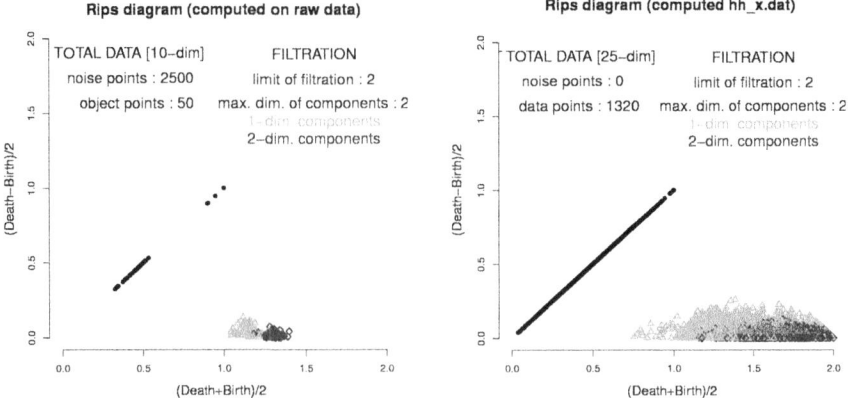

Fig. 2 Vietoris–Rips diagram using using a 25-dimensional noise-cube (*left plot*) and using the 25-dimensional credit client data for $N_{samp} = 1320$ (*right plot*), respectively

are called the **client classes**. These credit client data are also highly asymmetric in the sense that the entire population contains much less (around 3 %) defaulting clients.

Furthermore, in order to explore the difference between the original data and randomized versions thereof we compose different noisy data sets. The simplest variant used (A) are $N \times m$ matrices and smaller ones with entries drawn independently from a uniform distribution. The most restrictive (B) employs permutations over columns, i.e. they are feature-wise distribution conserving. The diagram of a run for variant (A) is depicted in the left plot of Fig. 2.

We then use the our credit client input data instead of the random test matrices. Here we first observe that we cannot compute Vietoris–Rips complexes with any meaningful parameterization on an average modern personal computer. We find that sub-sampling the $N \times m$ matrix (i.e. choosing some entries i at random) is necessary. In accordance with some past work [8] which analyzed various methods of sub-sampling in order to reduce the data model for client class forecasting we sub-sample the set of positives (defaulting clients) and that of negatives (non-defaulting clients) separately, with $N_{pos}^{samp} = N_{neg}^{samp}$, resulting in a total sample size of $N_{samp} = N_{pos}^{samp} + N_{neg}^{samp}$. A maximum sample size for our credit client data and the given computational endowment is $N_{samp} = 1320 \pm 10$. In order to monitor the growth of the Rips simplex we proceed by comparing the diagrams obtained for consecutive sample sizes, selecting $N_{samp} \in \{100, 150, 200, \ldots, 1320\}$. The resulting Rips complex sizes grow very fast and for bigger samples the software collapses. Figure 2, right plot, depicts the case of $N_{samp} = 1320$.

In order to obtain information about the shapes of the negative (nondefaulting) and the positive (defaulting) credit client populations, we essentially repeat the above computations for both populations separately, by using increased sample sizes N_{pos}^{samp} and N_{neg}^{samp}. Interestingly, at the display level of the V-R diagrams this partition of the data points does not show any remarkable structure. We refrain here from showing these results.

After observing that subdividing the data by client labels does not lead to expected differences in the Vietoris–Rips diagrams, we are turning to a subdivision (think of this as a coarse variant of slicing data) based on support vectors.

The labeled credit data have an associated optimization problem, namely finding a robust and well performing separation function between the two classes of clients [8]. This optimization problem subdivides the (positively and negatively labeled) data points into three sets: (1) those which clearly belong to a class (non support data vectors), those which belong to a region which cannot be clearly classified (bounded support vector data) and finally those which define the margins of a hypothetical, high dimensional box between the non support data vectors of opposite class (essential

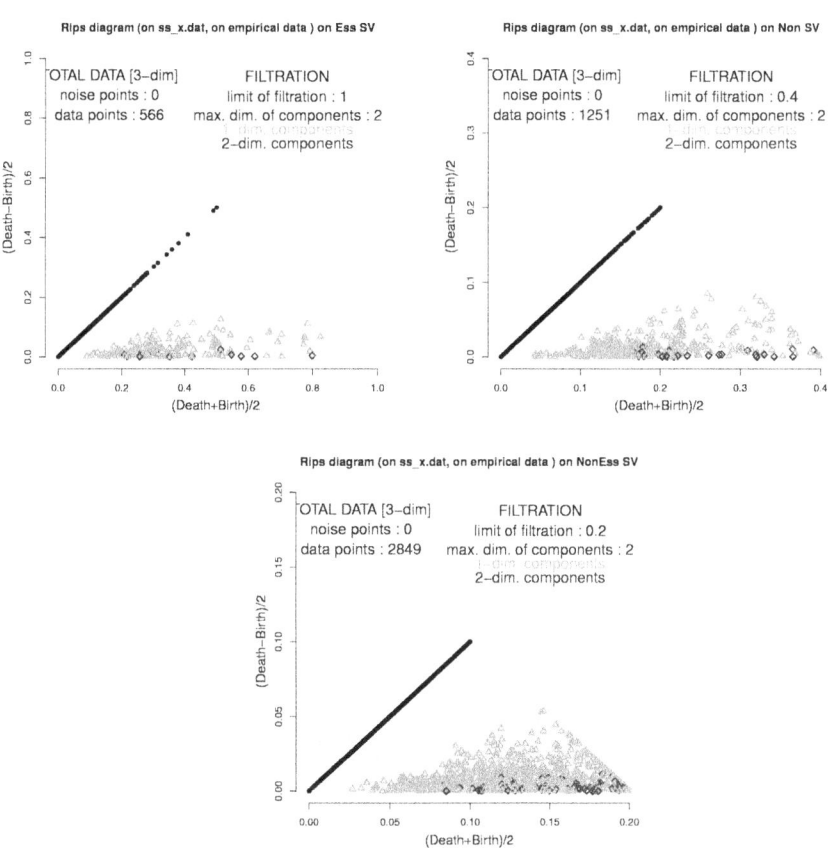

Fig. 3 Vietoris–Rips diagram using a 3-dimensional projection of the credit client data sample with $N = 4666$ persons. The *left upper plot* depicts the diagram for the set of bounded support vectors, the right h.s. plot for the essential support vectors and *left lower plot* that for the non-support vectors, respectively (see main text)

support vectors). In Fig. 3 we depict the Vietoris–Rips diagrams for the three subsets with a 90 % confidence interval computed by bootstrap within the TDA R-package [2]. Note the different scales (ranges) of both axes (i.e. limit of filtration and persistence). Attempts to compute beyond these very scales will result in very strong growth of complexes and in failure. The diagrams of the non-support vector data and those of the bounded support vectors depict more similar features then that for the essential support vectors. The latter appears to contain more persistent structure on the level of both 0-dimensional and 1-dimensional topological features.

5   Conclusion and Outlook

We applied a state of the art software library in order to elicit if any information may be extracted from credit client data sets by means of topological data analysis. Applying this TDA-software directly to the high dimensional data sets readily leads to computational overload. Therefore, not much information may be gained by naive exploration. The high dimensional sparse credit client data are in part real valued and entail some categorical features. Hence, in order to arrive at a slice-by-slice low dimensional characterization of the data, one cannot easily *cut slices* through the data as diverse continuous valued examples may suggest. Instead of geometrically motivated slices, we here propose to use the subsets of data points generated by the SVM-solutions of an associated client classification problem. For those three subsets we can observe certain distinct features of the resulting diagrams of the Vietoris–Rips complexes constructed on the data. Future analysis, including a planar graph-mapper for the topological features, should help reveal some (lower dimensional) data shapes like fingers, which may supplement the information extracted from clustering clients with more traditional statistical methods.

References

1. Carlsson, G.: Topology and data. Bull. (New Series) Am. Math. Soc. **46**(2), 255–308 (2009)
2. Fasy, B.T., Kim, J., Lecci, F., Maria, C.: Introduction to the R package TDA (with the CMU TopStat Group). https://cran.r-project.org/web/packages/TDA/vignettes/article.pdf
3. Lumm, P.Y., Carlsson, G., Vejdemo-Johansson, M.: The topology of politics: voting connectivity in the US House of Representatives. http://www.cs.cmu.edu/~sbalakri/Topology_final_versions/politics.pdf
4. Maria C.: GUDHI, Simplicial Complexes and Persistent Homology Packages (2014). http://project.inria.fr/gudhi/software/
5. Morozov, D.: Dionysus, a C++ library for computing persistent homology (2007). http://www.mrzv.org/software/dionysus/
6. Nicolau, M., Levine, A., Carlsson, G.: Topology based data analysis identifies a subgroup of breast cancers with a unique mutational profile and excellent survival. Proc. Nat. Acad. Sci. **108**, 7265–7270 (2011)
7. Perea, J., Harer, J.: Sliding windows and persistence: an application of topological methods to signal analysis (2014). arXiv:1307.6188v1

8. Schebesch, K.B., Stecking, R.: Clustering for data privacy and classification tasks. In: Huisman, D., Louwerse, I., Wagelmans, A.P.M. (eds.) Selected Papers of the International Conference on Operations Research OR2013, Rotterdam, 3–6 September 2013, Operations Research Proceedings, pp. 397–403. Springer (2014)

Supporting Product Optimization by Customer Data Analysis

Tatiana Deriyenko, Oliver Hartkopp and Dirk C. Mattfeld

Abstract This paper introduces a concept for product optimization support based on the integration of customer data sources. The motivation is a common misunderstanding gap between the manufacturer and the customer. While the customer has certain needs, the manufacturer aims at embedding them into the product design. However, due to imprecise understanding of the needs and subsequent development mistakes, the product can vary from what the customer actually requires. The concept combines two different data sources in order to reveal the gap between the product and the customer needs. The first source is represented by a customer-product interaction log file. The second source is social media delivering customer feedback regarding the product.

1 Introduction

Designing a product that satisfies the customer is a major goal in any manufacturing field. However, due to inaccurate identification and interpretation of the customer needs, as well as subsequent development mistakes, the product delivered to the customer can vary from what he requires. We can distinguish the following categories of the product non-compliance to the customer needs:

- The customer cannot reach his goal with the product. The required functionality is either not implemented or unavailable due to malfunctioning or other reasons;
- The customer can reach his goals with the product. However, some other aspect of the product does not meet the customer needs. For example, the customer finds the

T. Deriyenko (✉) · D.C. Mattfeld
Technische Universität Braunschweig, Muehlenpfordtstrasse
23, 38106 Braunschweig, Germany
e-mail: t.deriyenko@tu-braunschweig.de

D.C. Mattfeld
e-mail: d.mattfeld@tu-braunschweig.de

O. Hartkopp
Volkswagen AG, 38440 Wolfsburg, Germany
e-mail: oliver.hartkopp@volkswagen.de

© Springer International Publishing Switzerland 2017   491
K.F. Dœrner et al. (eds.), *Operations Research Proceedings 2015*,
Operations Research Proceedings, DOI 10.1007/978-3-319-42902-1_66

product too slow in operation or not visually attractive. Some of the functionality provided by the product can be not needed.

The task of product optimization is to minimize the gap between the product and the customer needs. Corresponding product optimization steps can include modification of the product appearance, elimination of non-demanded functions, introduction of new functions, etc.

In this paper we suggest a concept to support product optimization. The concept aims at discovering the gap between the product and the customer needs by integrating several data sources. The first source is represented by a structured customer-product interaction *log file*. The second source is *social media*, providing customer feedback about the product in a natural language form.

The remainder of this paper is structured as follows. Section 2 describes used data sources. The proposed concept is presented in Sect. 3. Section 4 summarizes the paper and provides an outlook to future work.

2 Customer Data Sources

We consider the system consisting of the product and the customer interacting with each other. The interaction is guided by the customer goal. People naturally communicate in multimodal nature: by means of spoken language in combination with gestures, mimics, and nonlinguistic sounds [2]. The same holds for the customer interaction with some products [7], so the interaction can involve different modalities.

We assume that the product non-conformity to the customer needs influences the way the customer and the product interact. Therefore, customer-product interaction analysis can be used for the product optimization. Related concepts include human performance analysis [3, 6], descriptive statistical analysis, sequential pattern mining, profile analysis [8], quality management [1] and others. Within the concept we consider the structured *log file* as a source that represents a detailed tracking of the actions taken by the customer and the product.

The knowledge about the state of the customer's mind is important to understand the customer interaction with the product and detect the product non-conformities. The human mind can be considered as a system, consisting of the following subsystems: perceptual, cognitive and motor one [4]. The perceptual subsystem carries sensations of the physical world. The cognitive subsystem connects perceptual and motor subsystems. Finally, the motor subsystem translates thoughts into actions. We consider voluntary customer feedback, expressing concerns regarding the product, as a source of knowledge about customer perception and intentions. *Social media* is selected as one of the most prominent sources of the customer feedback. The source comprises customer text messages from Internet blogs, discussion boards, social networks, etc. Related work include topic detection, sentiment analysis, summarization [5] and others.

3 Concept for Product Optimization Support

We suggest the concept integrating *social media* and *log file* to reveal the gap between the product and the customer. The concept facilitates detection of the product aspect non-complying with the customer needs.

3.1 Concept Overview

The concept overview is presented in Fig. 1. The customer and the product interact with each other. The interaction can be influenced by a wide range of factors, like the weather, geographical location, the customer health condition, etc. The interaction is tracked and recorded in the structured *log file*. At the same time, the customer voluntarily provides his feedback about the product in *social media*.

Data acquisition and preprocessing are left outside the scope of this paper. The analysis of the data sources implies completion of the following interconnected tasks:

- Detect common deviations of customer-product interaction from the manufacturer anticipations. Such deviations can be considered as possible indicators to the product non-compliance to the customer needs;
- Reveal common concerns expressed by the customer with regards to the product.

The output of the above mentioned tasks can be integrated in order to generate the hypothesis regarding the product aspect non-complying with the customer needs. The generated hypothesis can be confirmed or rejected by performing customer surveys

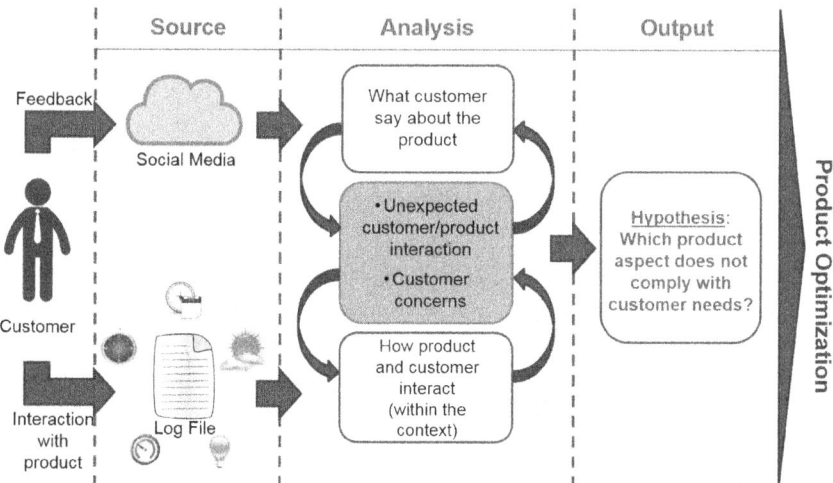

Fig. 1 The concept overview

or by consulting domain experts. If the hypothesis is confirmed, it can be used to support the product optimization.

3.2   Log File and Social Media Analysis

This section describes how the tasks can be completed. The *log file* is used to detect customer-product interaction deviating from the manufacturer expectation. Here the interaction is considered from the manufacturer perspective. Thus, the definition of the deviation is dependent on the business context.

The customers can be clustered according to their behavior. Each of the clusters can be analyzed independently. The interaction can be also considered within the relevant context. Simple examples of metrics used to detect unexpected interaction are function or interface element usage frequency or duration. The baseline for such metrics can be either predefined by the manufacturer or generated based on the results of different periods, functions, customers, etc. Advanced analysis can include sequential or frequent pattern mining, revealing unexpected combinations of steps frequently taken by the customer.

Social media analysis considers customer messages and therefore provides the customer view. The analysis can be supported by the derivation of structured information like words and relations between them. This could be achieved by applying associated rule mining against a set of importance metrics, as well as by other methods. Here we base on an assumption that frequent usage of the same words in one message indicates customer concerns. Based on the results of the analysis, the knowledge about common customer concerns can be gained. Social media can also describe the relevant context.

3.3   Integration Schema

In order to combine the unexpected customer-product interaction and customer concerns, a common integration structure is required. We suggest the following common set of dimensions:

- *Criteria* defining the product aspect deviating from the customer needs. The criteria can include functionality, reliability, performance, usability, security, understandability, design aesthetics, etc.;
- *Function* defining the logical block of product functionality that does not meet the customer needs against some of the criteria;
- *Modality* defining the way of customer-product interaction (by voice operation, gestures, etc.). The modality can deviate from the customer needs against some criteria as well;
- *Context* of interaction defining factors influencing customer-product interaction.

Analysis of both the *log file* and *social media* can be performed in accordance to the suggested schema. The mapping can be supported by introducing a common vocabulary. This way, the concepts from each of the sources get assigned to the common function, modality, etc.

3.4   Concept Illustration

This section provides some examples of generating the hypothesis regarding the product aspect non-complying with the customer needs. The hypothesis is created based on the integration of unexpected customer-product interaction and customer concerns.

We first consider the situation when the customer cannot achieve his goals with the product. In this case *social media* analysis is expected to reveal corresponding customer concerns. The *log file* analysis can find corresponding deviations in the customer-product interaction. This could be multiple attempts of the customer to find or launch the required functionality, as well as a search for possible workarounds. Product optimization would include implementation of the lacking functionality.

The second category covers the situations when the customer can reach his goals with the product, but some other product aspect does not fit his needs. For the first example let us assume that the *log file* analysis shows that the product function is operating more slowly than it is expected by the manufacturer. This finding can indicate that the function performance is non-complying with the customer needs. If corresponding customer concern regarding the function speed is present in *social media*, the product performance should be considered as a subject for optimization.

Another example is unexpectedly rare usage of some function detected by the *log file* analysis. If the *social media* analysis detects no corresponding customer concerns, the functionality can be considered as redundant. Alternatively, the concerns from *social media* can show that the customer finds the product not attractive or hard to understand. Corresponding product optimization steps can remove the non-needed functionality or modify the product interface and appearance.

Finally, if *social media* reveals that the customer does not reach his goals with the product and *log file* analysis shows that corresponding functionality is not used, the customer can be unaware of it. In that case the product understandability should be improved.

4   Summary and Outlook

In this paper we presented the concept to discover the knowledge about unexpected customer-product interaction and customer concern regarding the product. Integration of the findings helps to generate the hypothesis about the product aspect non-complying with the customer needs. This facilitates exploration of the gap between the product and customer needs and supports product optimization.

For perspective development of the concept methods of semi-automatic implementation will be explored in details. Methods of mapping the findings from each of the sources to the common integration schema, as well as ways of hypothesis generation support will be explored. Sentiment analysis can be implemented in order to incorporate the emotional component of the customer feedback.

References

1. Boehm, B., Brown, J.R., Lipow, M.: Quantitative evaluation of software quality. In: Proceedings of the 2nd International Conference on Software Engineering, San Francisco, California (1976)
2. Bunt, H.: Issues in multimodal human-computer communication. In: Proceeding of Multimodal Human-Computer Communication, Systems, Techniques, and Experiments (1998)
3. Card, S.K., Moran, T.P., Newell, A.: The keystroke-level model for user performance time with interactive systems. Commun. ACM CACM **23**(7), 396–410 (1980)
4. Card, S.K., Moran, T.P., Newell, A.: The Psychology of Human-Computer Interaction, 1st edn. Lawrence Erlbaum Associates, Mahwah (1983)
5. Fan, W., Wallace, L., Rich, S., Zhang, Z.: Tapping the power of text mining. Commun. ACM **49**, 76–82 (2006)
6. Geng, R.: Improving web navigation usability by comparing actual and anticipated usage. IEEE Trans. Hum. Mach. Syst. **45**(1), 84–94 (2015)
7. Quek, F., McNeill, D., Bryll, R., Duncan, S., Ma, X., Kirbas, C., McCullough, K.E., Ansari, R.: Multimodal human discourse: gesture and speech. ACM Trans. Comput. Hum. Interact. **9**(3), 171–193 (2002)
8. Srivastava, J., Cooley, R., Deshpande, M., Tan, P.: Web usage mining: discovery and applications of usage patterns from web data. ACM SIGKDD Explor. **1**(2), 12–23 (2000)

Inspection of the Validity in the Frequent Shoppers Program by Using Particle Filter

Shinsuke Suzuki and Kei Takahashi

1 Introduction

This paper discusses the validity of the frequent shoppers program (FSP) based on a particle filter estimation model. The FSP is one of the marketing methods for promoting sales at the individual consumer level and thus increasing profitability. Specifically, stores issue discount coupons to important customers through the FSP. From the viewpoint of customer relationship management, purchasing behavior has two characteristics: consumer heterogeneity and dynamic preference changes at the individual level. Therefore, in order to achieve appropriate revenue management, it is important for stores to accurately measure the effects of the FSP.

We construct a category purchase decision model in order to analyse the effects of the FSP based on scanner panel data of a retailer in Japan. In this paper, the category purchasing behavior is represented by the Nested Logit (NL) model. In the NL model, consumers decide whether or not to buy in an upper level and which brand to buy in a lower level. The parameters in our model are changed dynamically, and we adopt a particle filter that enables us to estimate parameters sequentially. In our model, explanatory variables include purchasing interval, household inventory, purchase price, internal reference price, brand loyalty, and capacity of each brand. Internal reference price (IRP) is a standard in evaluating a purchase price. Purchasing interval is the number of days from the present purchase date to the last purchase date. Household inventory is the amount that consumers keep at home. Brand loyalty is a scale of consumers' brand preference. With our constructed model, we inspect the validity of the FSP utilized by retailers pragmatically by using actual FSP data. In addition, we propose the appropriate content for the sales promotion via FSP.

S. Suzuki (✉)
Graduate School of Waseda University, 3-4-1 Okubo, Shinjuku-ku,
Tokyo 169-8555, Japan
e-mail: istorytale0904@gmail.com

K. Takahashi
The Institute of Statistical Mathematics, 10-3 Midoricho, Tachikawa,
Tokyo 190-8562, Japan
e-mail: k-taka@ism.ac.jp

© Springer International Publishing Switzerland 2017
K.F. Dœrner et al. (eds.), *Operations Research Proceedings 2015*,
Operations Research Proceedings, DOI 10.1007/978-3-319-42902-1_67

2   Method

2.1   The Model

We construct a model that represents purchasing behavior in order to measure the effects of FSP sales promotion. To achieve this aim, the model needs to represent the category purchase decision and the brand choice decision. Therefore, we model category purchase decisions based on the NL framework because the NL model can represent hierarchical decision making. In our model, the category purchase decision and the brand choice decision are represented in the upper level and lower levels, respectively.

The consumer h's purchase probability for brand j belonging to category c at time t can be expressed as:

$$P_t^h(j) = P_t^h(j|c) P_t^h(c),\qquad (1)$$

$$P_t^h(c) = 1/\left(1 + \exp(-V_{ct}^h)\right),\qquad (2)$$

$$P_t^h(j|c) = \frac{\exp V_{jt}^h}{\sum_{j'} \exp V_{j't}^h},\qquad (3)$$

where $P_t^h(c)$ is the consumer h's category incidence probability of category c and $P_t^h(j|c)$ denotes the consumer h's choice probability of brand j given category purchase incidence at time t. V_{jt}^h is the utility for each brand j. We represent V_{jt}^h as follows:

$$V_{jt}^h = \alpha_{1t}^h Y_{1jt}^h + \alpha_{2t}^h Y_{2jt}^h + \alpha_{3t}^h Y_{3jt}^h + \alpha_{4t}^h Y_{4jt}^h,\qquad (4)$$

where Y_{1jt}^h is the capacity of brand j, Y_{2jt}^h is the effect of the benefit, and Y_{3jt}^h is the effect of the loss. Both Y_{2jt}^h and Y_{3jt}^h depend on the difference between the IRP R_{jt}^h and the retail price M_{jt}^h. Dynamics of the IRP R_{jt}^h refer to a model of Briesch et al [2]. Y_{2jt}^h and Y_{3jt}^h are computed as:

$$Y_{2jt}^h = \begin{cases} R_{jt}^h - M_{jt}^h & \text{if } R_{jt}^h > M_{jt}^h \\ 0 & \text{otherwise} \end{cases}, \quad Y_{3jt}^h = \begin{cases} M_{jt}^h - R_{jt}^h & \text{if } R_{jt}^h \le M_{jt}^h \\ 0 & \text{otherwise} \end{cases}.\qquad (5)$$

Y_{4jt}^h is the loyalty toward brand j. $\alpha_{1t}^h, \alpha_{2t}^h, \alpha_{3t}^h$, and α_{4t}^h are utility function parameters and changed everyday. The changes reflect the dynamic consumer changes with regards to the brand choice decision in our model.

In terms of the category incidence probability $P_t^h(c)$, V_{ct}^h is the utility in category c. We represent V_{ct}^h as follows:

$$V_{ct}^h = \beta_{1t}^h X_{1t}^h + \beta_{2t}^h X_{2t}^h + \beta_{3t}^h X_{3t}^h,\qquad (6)$$

where X_{1t}^h is the category value; X_{2t}^h is the category inventory; X_{3t}^h is the purchasing interval; and β_{1t}^h, β_{2t}^h, and β_{3t}^h are utility function parameters and changed everyday. The changes reflect the dynamic consumer changes with regards to the category purchasing decision in our model. X_{1t}^h is expressed by the logarithm of the denominator of the brand choice model:

$$X_{1t}^h = \log \sum_j \exp V_{jt}^h. \tag{7}$$

Thus, the category purchasing decision is influenced by the utility of each brand V_{jt}^h. We assume that the dynamic parameters α_{1t}^h, α_{2t}^h, α_{3t}^h, α_{4t}^h, β_{1t}^h, β_{2t}^h, and β_{3t}^h conform smoothing prior distributions based on a model of Sato and Higuchi [3]. Moreover, we suppose that regulations of the parameters' changes are expressed as:

$$\begin{aligned} \boldsymbol{a}_t &= \boldsymbol{I}\boldsymbol{a}_{t-1} + \boldsymbol{I}\boldsymbol{v}_t, \\ \boldsymbol{a}_t &= \left[\alpha_{1t}^h, \alpha_{2t}^h, \alpha_{3t}^h, \alpha_{4t}^h, \log\left(\beta_{1t}^h / \left(1 - \beta_{1t}^h\right)\right), \beta_{2t}^h, \beta_{3t}^h\right], \\ \boldsymbol{v}_t &= \left[e_{\alpha_{1t}^h}, \dots, e_{\alpha_{4t}^h}, e_{\beta_{1t}^h}, \dots, e_{\beta_{3t}^h}\right], \end{aligned} \tag{8}$$

where $\boldsymbol{I}$ is the 7×7 unit matrix, and $\boldsymbol{v}_t$ is the vector of stochastic terms that follows:

$$\boldsymbol{v}_t \sim N\left(0, \boldsymbol{\Sigma}_h\right), \quad \boldsymbol{\Sigma}_h = \operatorname{diag}\left[\tau_{\alpha_{1t}^h}^2, \dots, \tau_{\alpha_{4t}^h}^2, \tau_{\beta_{1t}^h}^2, \dots, \tau_{\beta_{3t}^h}^2\right].$$

$\tau_{\alpha_{1t}^h}^2, \dots, \tau_{\alpha_{4t}^h}^2$ and $\tau_{\beta_{1t}^h}^2, \dots, \tau_{\beta_{3t}^h}^2$ are variants of each stochastic term.

2.2   Estimation

In the parameters estimation: there are three procedures, initializing by holdout data, the particle filter method and the Nelder–Mead simplex method. We estimate the dynamic parameters by particle filter on the basis of given hyper parameters. These hyper parameters are estimated outside of the dynamic parameters estimation via the Nelder–Mead simplex method. The whole estimation procedure is shown in Fig. 1.

We adopt a particle filter for estimating dynamic parameters that include α_{1t}^h, α_{2t}^h, α_{3t}^h, α_{4t}^h, β_{1t}^h, β_{2t}^h, and β_{3t}^h. The particle filter is one of the Bayesian estimation methods that allows for the sequential estimation of parameters. With the particle filter, a finite set of N particles is regarded as the distribution of parameters. Each particle is a vector of dynamic parameters and changes dynamically in accordance with formula (8).

Hyper parameters $\tau_{\alpha_{1t}^h}^2, \dots, \tau_{\alpha_{4t}^h}^2$ and $\tau_{\beta_{1t}^h}^2, \dots, \tau_{\beta_{3t}^h}^2$ are estimated via maximum likelihood estimation following the Nelder–Mead simplex method. The reason for using the Nelder–Mead simplex method is that it is impossible to use the gradient method because of the Monte Carlo error.

Fig. 1 Empirical method using particle filter and the Nelder–Mead simplex method. L is represented as $(L_{max} - L_{min})/L_{max}$

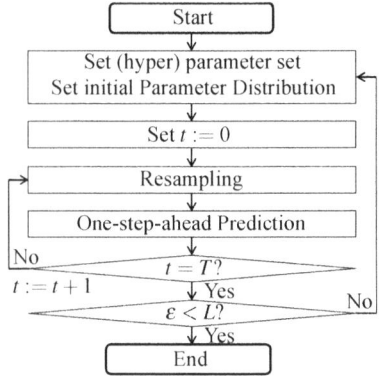

3 Application

3.1 Data

We use scanner-panel data from a Japanese retailer on the category of chicken eggs because it is a category that consumers buy frequently. We treat six brands: the five brands purchased most frequently in this category and "others", which is the set of other brands. The total share of the five brands reaches to about 90 % in this category. The data covers a period of 52 weeks. The first 26 weeks are for initializing, the next 22 weeks are for estimating our model, and the last 4 weeks are for forecasting. We chose to use the data of consumers who had visited the retailer more than 40 times during the period of the model estimation. We select 35 consumers randomly out of about 2800 consumers.

3.2 Result

To inspect our model, we compare it with that of Bell and Bucklin [1] through two hitting rates of forecasting: whether or not consumers buy (category purchase decision) and which brand they buy (brand decision). Table 1 shows the results of the comparison. We find that the two hitting rates of our model are better than those of Bell and Bucklin. Our model is also better model than that of Bell and Bucklin with regards to the estimation sample.

We carry out two analyses in our model. First, we use actual FSP data to inspect the validity of the FSP sales promotion used by retailers. Second, we propose the appropriate target brand and discount rate for sales promotion using the FSP.

Figure 2 shows utility $V_{jt}^{h'}$ of some brands for the specific consumer h'. Brand 1 is the target brand of the sales promotion using the FSP. Brand 1 was discounted during Period 1 and not in Period 2. Brands 2 and 3 are the brands that consumer h' purchases

Table 1 Comparison of hitting rates on forecasting

Model	Bell and Bucklin (%)	Our model (%)
Category purchase decision	46.29	58.86
Brand decision	64.18	65.67

Fig. 2 The transition of consumer h''s brand utilities $V_{jt}^{h'}$

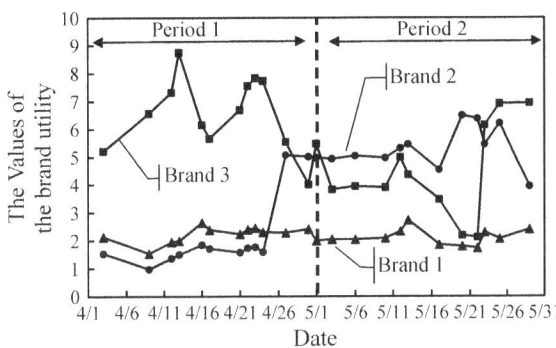

frequently. Periods 1 and 2 refer to the period of April 1st to 31st and May 1st to 30th, respectively. Figure 2 shows that utility $V_{jt}^{h'}$ of Brand 1 is lower than that of Brands 2 and 3. In addition, the difference in utility $V_{1t}^{h'}$ between Periods 1 and 2 is small. These results show that the sales promotion of Brand 1 can be considered ineffective. As a result of simulation via our model, about 80 % is needed for consumer h' to change his brand decision to Brand 1. However, 80 % discounting has a significant negative impact on sales. Therefore, it can be concluded that the sales promotion of Brand 1 to consumer h' is not effective. Similarly, Brand 1 is the target of the sales promotion using the FSP for the other 34 consumers. We analyse the other consumers in the same way that we did consumer h' and estimate that most consumers need a high discount rate in order to purchase Brand 1. In conclusion, Brand 1 is not suitable for the target of the sales promotion using the FSP.

Table 2 shows how the sales influenced the simulation in order to estimate the appropriate target brand and discount rate for an FSP sales promotion. In this simulation, we use the data from between April 1st and 31st. We estimate the total sales of 35 consumers under a sales promotion using the FSP by changing rules determining the discount rate or the target brand. The discount rate is determined by the brand in the rules of the actual FSP. Moreover, the target brand of the sales promotion is only

Table 2 The total sales of the simulation result

	Actual	Simulation A	Simulation B
The total (expected) sales	¥21, 456	¥33, 825	¥34, 361

Brand 1 of the chicken egg category. Therefore, in the first simulation (Simulation A), the sales promotion using the FSP is performed, the discount rate is determined by the brand, and the target brand is not only Brand 1. On the other hand, in the second simulation (Simulation B), the discount rate is determined by each consumer, and the target brand is not only Brand 1. Table 2 indicates that the changes in the rules determining the discount rate and the target brand increase the total sales. The difference between the sales on the Simulations A and B is small. If the cost of changing the rule for Simulation B is bigger than the difference in the total sales, the retailer should not adopt the rule for the Simulation B.

4   Conclusion

In this study, we inspect the validity of the FSP used by retailers from a practical perspective by examining actual FSP data. Our results indicate that sales promotion using the FSP is ineffective because of the inappropriate target brand. In addition, we propose the appropriate target brand and discount rate for sales promotion using the FSP. We change the rules that determine the discount rate and the target brand in the simulation. As a result, the total estimated sales increase.

There remains a need to apply this estimation for consumers who are engaging the retailer for the first time or who rarely purchase. The data for these consumers is limited. Therefore, the particle filter estimation could not be applied to such consumers.

Acknowledgements This work was supported by JSPS Number 28570821. The data set used for the work is offered in the 2014 data analysis competition by JASMAC.

References

1. Bell, D.R., Bucklin, R.E.: The role of internal reference points in the category purchase decision. J. Consum. Res. **26**, 128–143 (1999)
2. Briesch, R.A., Krishnamurthi, L., Mazumdar, T., Raj, S.P.: A comparative analysis of reference price models. J. Consum. Res. **24**, 202–213 (1997)
3. Sato, T., Higuchi, T.: An analysis of purchase incidence behavior using dynamic individual model. J. Mark. Sci. **16**(1,2), 49–73 (2009) (in Japanese)

Value-at-Risk Forecasts Based on Decomposed Return Series: The Short Run Matters

Theo Berger

Abstract We apply wavelet decomposition to decompose financial return series into a time frequency domain and assess the relevant frequencies for adequate daily Value-at-Risk (VaR) forecasts. Our results indicate that the frequencies that describe the short-run information of the underlying time series comprise the necessary information for daily VaR forecasts.

1 Introduction

Forecasting conditional volatility of financial returns represents a key aspect for risk measurement and management. As financial volatility is the major factor causing variability of expected portfolio returns, accurate volatility forecasts are crucial to assess financial risk.

In this vein, for the practitioner as well as for the researcher, Value-at-Risk (VaR) has become a popular standard to translate financial market risk into a single monetary risk figure (see [4]).

By applying wavelet methodology to the underlying return series, we indicate a novel approach to VaR forecasts. That is, we apply wavelet filter to decompose the underlying return series into different frequencies and assess daily VaR forecasts based on particular frequencies. In doing so, we are able to identify the relevant frequencies for adequate daily VaR forecast.

The remainder of this paper is structured as follows: Sect. 2 presents the a brief overview of the applied methodology. In Sect. 3, we indicate empirical results and Sect. 4 concludes.

T. Berger (✉)
University of Bremen, Wilhelm-Herbst-Str. 5, 28359 Bremen, Germany
e-mail: theoberger@uni-bremen.de

© Springer International Publishing Switzerland 2017   503
K.F. Dœrner et al. (eds.), *Operations Research Proceedings 2015*,
Operations Research Proceedings, DOI 10.1007/978-3-319-42902-1_68

2   Methodology

2.1   *Wavelet Decomposition*

Although wavelet decomposition is widely explored, its application to financial time series is relatively new (see [3, 5] amongst others).

As wavelet decomposition allows for the decomposition of a time series into several components, a financial return series can be decomposed into different seasonalities. That is, short-run fluctuations can be separated from long-term trends by decomposing the underlying financial return series into different frequency bands. Currently, wavelet decomposition is mainly applied to assess either dependence between filtered financial return series (see [1, 2, 6]) or to filter short-run noise from intra-day data (see [3]). In the remainder we assess Out-of-Sample VaR forecasts based on decomposed return series.

2.1.1   Maximum Overlap Discrete Wavelet Decomposition

We apply the Maximum Overlap Discrete Wavelet Transformation (MODWT) to decompose financial return series and this section provides an intuitive introduction of the applied wavelet transform and we refer to [6] for a thorough introduction to wavelet filtering.

Figure 1 presents a simulated noisy doppler signal. Let this signal be decomposed via wavelet decomposition, then (as presented in Fig. 2), the signal gets decomposed into four into four series comprising the fluctuations around the trend and a smooth trend. It is to note, that the first decomposition describes the short-run fluctuations

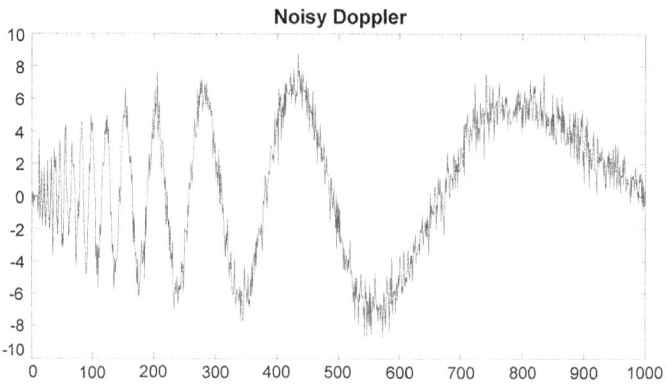

Fig. 1 Simulated noisy doppler signal

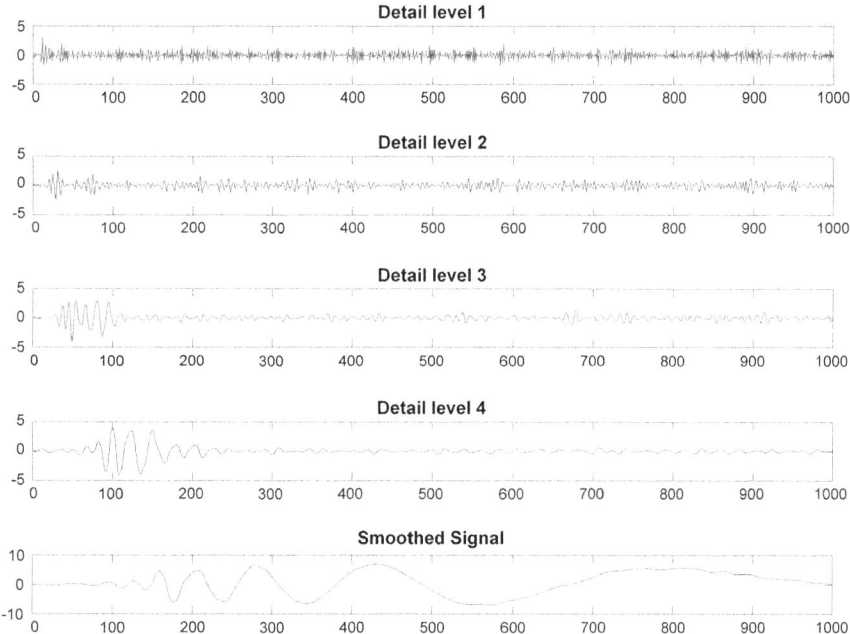

Fig. 2 Wavelet decomposition of the noisy doppler signal

around the trend, whereas the horizon increases with each decomposition. That is, the first scale comprises the short-run trend and the fourth scale, the long-run trends of the underlying signal. Furthermore, the original can be reconstructed by simply summing up the details and the smooth trend.

Now, in terms of financial time series, let the signal be described by a financial return series r, then the return series can be decomposed into a set of details and a smoothed version of the filtered data. The corresponding MODWT leads to the following decomposition of the return series:

$$r = \sum_{i=1}^{j} S_i + C_j. \tag{1}$$

In this context, S_i present the details of decomposition i, with $i = 1, \ldots, j$, and C_j presents the smoothed version of the signal after the jth decomposition.

2.2 Value-at-Risk

Based on the decomposed return series, we model VaR estimates for the original return series. Generally, as presented by [4] VaR defines a maximum loss limit which will not be exceeded with a given probability and describes the quantile of the return distribution. Due to the fact, that VaR forecast are described by quantiles, its accuracy can be tracked by the accuracy of the quantile forecasts.

3 Empirical Results

Now, in order to assess the quality of VaR forecasts based on decomposed return series, we apply an extensive out of sample study. Specifically, we decompose the underlying return series of 29 assets into eight scales and assess the accuracy of the achieved VaR forecasts for each scale via rolling window approach. That is, we take 1000 days, and forecast the 1001st day, and then carry on by shifting the window by one day to forecast the 1002nd day, and so on.

The data set comprises 29 daily return series of stocks that are listed in the Dow Jones Industrial Average from January 1st 2000 till September 30th 2013.[1]

Table 1 presents the accuracy of VaR forecasts based on different decomposed scales. As the 95 % VaR forecasts represent the 5 % quantile of the return distribution, we expect an empirical amount of VaR breaches that is close to 5 %. By successively adding the information of larger scales to the first scale, we are able to assess the relevant information that is needed to achieve adequate forecasts. More concretely, starting from the short-run noise that is described by the first decomposition (S1), we successively add decomposed scales until we achieve acceptable VaR backtesting performance.

Obviously, scales 1, 2 and 3 (S1-S3) lead to a failure rate that is characterized by the highest precision regarding individual 95 % VaR forecasts. Despite of Chevron (CHEVR), General Electric (GE) and Walt Disney (WD), adequate 95 % VaR for 26 assets are achieved. Adding more scales, does not improve the overall performance of VaR forecasts. Therefore, our results indicate that the first three scales comprise the necessary information for adequate VaR forecast. This means, that the relevant information for daily volatility forecasts is described by the first three scales. Thus long-run seasonalities could be discarded from the decomposed return series without any impact on the quality of VaR forecasts.[2]

---

[1]Due to its late listing we excluded Visa.

[2]For the interested reader, we are happy to provide a detailed methodological description of our study upon request.

Table 1 Out-of-sample performance of 95 % VaR Forecasts

Scales	S1	S1-S2	S1-S3	S1-S4	S1-S5	S1-S6	S1-S7	S1-S8
3M	10.54%**	6.50%**	5.08%	4.08%*	3.75%**	3.54%**	3.38%**	3.29%**
AT&T	11.04%**	6.25%*	5.00%	4.21%	3.71%	3.58%**	3.50%**	3.42%**
AM EXP	11.08%**	6.75%**	5.13%	4.58%	4.13%	3.83%*	3.42%**	3.38%**
BOEING	11.17%**	6.71%**	4.92%	4.33%	4.08%	3.75%**	3.67%**	3.58%**
CATER	11.54%**	6.96%**	4.83%	4.21%	3.71%	3.50%	3.38%**	3.08%**
CHEVR	11.88%**	7.58%**	6.29%**	5.54%	5.13%	4.83%	4.50%	4.63%
CISCO	9.58%**	5.79%	4.29%	3.79%*	3.42%**	3.33%**	3.17%**	3.04%**
COCAC	10.50%**	6.50%**	5.00%	4.38%	3.88%	3.75%	3.46%*	3.54%
E I DU	12.13%**	7.04%**	5.04%	4.50%	3.75%	3.58%*	3.46%*	3.29%**
EXXON	10.17%**	7.38%**	5.96%	5.25%	4.79%	4.63%	4.50%	4.46%
GE	11.00%**	7.21%**	5.50%*	4.50%*	4.13%**	3.88%**	3.79%**	3.63%**
GOLDS	12.04%**	7.00%**	4.71%	4.29%	4.04%	3.67%*	3.54%**	3.38%**
HOME	11.50%**	6.92%**	4.88%	4.25%	3.96%	3.67%**	3.50%**	3.42%**
INTEL	10.58%**	6.63%**	4.71%	3.83%	3.63%**	3.25%**	3.04%**	3.00%**
INTER	10.58%**	6.38%**	4.67%	4.08%	3.88%*	3.83%**	3.38%**	3.29%**
JP M	10.58%**	6.50%**	5.04%	4.33%	3.83%*	3.50%**	3.33%**	3.29%**
JJ	9.71%*	5.46%	4.42%	3.58%**	3.38%**	3.00%**	2.96%**	2.83%**
MCD	10.04%**	5.21%	4.04%	3.25%**	3.08%**	2.96%**	2.88%**	2.88%**
MERCK	11.17%**	6.04%**	4.08%	3.46%*	3.13%**	2.83%**	2.75%**	2.75%**
MS	11.04%**	6.96%**	4.50%	3.75%	3.46%*	3.33%**	3.21%**	3.25%**
NIKE	10.88%**	6.25%**	4.17%	3.71%	3.33%**	3.00%**	3.00%**	2.96%**
PFIZER	10.75%**	6.63%**	4.71%	4.08%	3.92%*	3.75%**	3.46%*	3.29%**

(continued)

Table 1 (continued)

	Scales							
	S1	S1–S2	S1–S3	S1–S4	S1–S5	S1–S6	S1–S7	S1–S8
P&G	10,42 %**	6,21 %**	5,00 %	4,04 %	3,75 %**	3,54 %**	3,54 %**	3,38 %**
TRAV	9,88 %**	6,25 %**	4,71 %	4,17 %	3,79 %	3,71 %*	3,54 %*	3,54 %**
UT	10,71 %**	6,50 %**	5,25 %	4,38 %	3,96 %**	3,58 %**	3,54 %**	3,50 %**
UNH	11,00 %**	6,83 %**	5,17 %	4,58 %	3,96 %*	3,63 %*	3,54 %*	3,33 %*
VERI	11,08 %**	6,63 %**	4,96 %	4,38 %	4,00 %*	3,63 %*	3,54 %*	3,38 %*
WAL	11,21 %**	6,54 %**	4,67 %*	4,08 %**	4,00 %**	3,88 %**	3,88 %**	3,83 %**
WD	10,63 %**	6,88 %**	5,29 %	4,67 %**	4,13 %**	3,58 %**	3,42 %**	3,21 %**

The percentages represent the relative amount of VaR breaches for the investigated period. return distribution and data (either returns or scales). */** indicate a rejection of the quality of the VaR forecasts with 95/99 % significance level

4 Conclusion

By this study, we illustrate that the relevant information for adequate daily VaR forecasts is stored in the short-run frequencies. Based on decomposed time series, adequate VaR forecasts can be achieved by exclusively taking the short-run and middle-run frequencies into account. Thus our results indicate that long-run seasonalities and long-run memory of financial time series do not impact the precision of daily VaR forecasts.

References

1. Berger, T.: A wavelet based approach to measure and manage contagion at different time scales. Physica A **436**, 338–350 (2015)
2. Gallegati, M.: A wavelet-based approach to test for financial market contagion. Comput. Stat. Data Anal. **56**, 3491–3497 (2012)
3. Gençay, R., Selçuk, F., Whitcher, B.: Scaling properties of foreign exchange volatility. Physica A **289**, 249–266 (2001)
4. Jorion, P.: Value at Risk: The New Benchmark for Controlling Derivatives Risk. McGraw, New York (2007)
5. Ramsey, J.B., Uskinov, D., Zaslavsky, G.M.: An analysis of U.S. stock price behavior using wavelets. Fractals **3**, 377–389 (1995)
6. Reboredo, J., Riveiro-Castro, M.: Wavelet-based evidence of the impact of oil prices on stock returns. Int. Rev. Econ. Financ. **29**, 145–176 (2014)

Feed-In Forecasts for Photovoltaic Systems and Economic Implications of Enhanced Forecast Accuracy

Oliver Ruhnau and Reinhard Madlener

Abstract The combination of governmental incentives and falling module prices has led to a rapid increase of globally installed solar photovoltaic (PV) capacity. Consequently, solar power becomes more and more important for the electricity system. One main challenge is the volatility of solar irradiance and variable renewable energy sources in general. In this context, accurate and reliable forecasts of power generation are required for both electricity trading and grid operation. This study builds and evaluates models for day-ahead forecasting of PV electricity feed-in. Different state-of-the-art forecasting models are implemented and applied to a portfolio of ten PV systems. More specifically, a linear model and an autoregressive model with exogenous input are used. Both models include inputs from numerical weather prediction and are combined with a statistical clear sky model using the method of weighted quantile regression. Forecasting-related economic implications are analyzed by means of a two-dimensional mean-variance approach. It is shown that enhanced forecast accuracy does not necessarily imply an economic gain.

1 Introduction

Several recent studies present feed-in forecasting models using fundamental equations, statistical methods, artificial intelligence techniques or a combination of these approaches (e.g. [2]). Those studies usually evaluate and compare forecast accuracy based on statistical error measures. Further studies analyze the economic implications of such forecasts with a focus on different renewable energy sources and market conditions (e.g. [5]), confirming the clearly positive value of forecasting.

O. Ruhnau (✉)
RWTH Aachen University, Templergraben 55, 52056 Aachen, Germany
e-mail: oliver.ruhnau@rwth-aachen.de

R. Madlener
Institute for Future Energy Consumer Needs and Behavior (FCN), School of Business and Economics/E.ON Energy Research Center, RWTH Aachen University, Mathieustrasse 10, 52074 Aachen, Germany
e-mail: RMadlener@eonerc.rwth-aachen.de

© Springer International Publishing Switzerland 2017
K.F. Dœrner et al. (eds.), *Operations Research Proceedings 2015*,
Operations Research Proceedings, DOI 10.1007/978-3-319-42902-1_69

Our work elaborates on the relationship between forecast accuracy and its economic implications by comparing errors and balancing costs involved when using different forecasting models. To this end, we combine theoretical knowledge from the fields of forecasting science and economics with practical insights from renewable electricity marketing. This interdisciplinary approach creates novel and thrilling results which question the business-as-usual unidimensional accuracy-based forecast assessment. A more detailed documentation of our research can be found in [6].

2   Methodology

2.1   Subject, Data and Context of Investigation

Our analysis is based on a portfolio of ten large-scale PV systems[1] installed all over Germany with a total capacity of 156.7 MW. Feed-in time series from 2014 are available for those systems in a quarter-hourly resolution. For each quarter-hour of the year, there is a value indicating how much electricity has been fed into the grid.

Our forecasting models use numerical weather prediction (NWP) for the parameters "solar irradiance" and "clear sky irradiance" as input. For testing, historical weather forecasts are retrieved from the ERA-Interim model of ECMWF.[2] Operating at the global scale, this model generates data at relatively low spatial (about 28×17 km) and temporal resolution (6 h). Hence, we use interpolation to generate site-specific time series at a quarter-hourly resolution.

The economic impacts of different feed-in forecasts are analyzed in the context of German electricity markets, more specifically the day-ahead and the intra-day market. In the day-ahead auction, electricity producers submit their bids for each delivery hour of the next day. In the case of solar electricity, those bids are based on day-ahead feed-in forecasts. After the gate closure at 12 pm, hourly market clearing prices and respective delivery schedules are derived from the merit order curve. Intra-day trading is possible from 3 pm (day-ahead) until 45 min before delivery and allows market participants to take corrective actions on these schedules, e.g. if updated solar feed-in forecasts deviate from day-ahead forecasts. Note that errors occuring after intra-day gate closure are balanced in the imbalance market. However, as we consider day-ahead forecasts only, this is beyond the scope of our study.

---

[1]Ranging from 0.8 to 82 MW of installed peak capacity.
[2]European Centre for Medium-Range Weather Forecasts.

2.2   Forecasting Models

The electricity output of a solar PV system is a function of (1) deterministic factors such as the sun's position and the collector's orientation, and of (2) stochastic weather impacts such as cloud cover, aerosols, fog, snow and temperature. In this study, deterministic and stochastic factors are modeled separately by means of a clear sky model. To this end, the actual solar irradiance I_t^{DA} is written as

$$I_t^{DA} = f_t^{CSI} I_t^{CS}, \tag{1}$$

where f_t^{CSI} is the (stochastic and unit-less) clear sky irradiance factor and I_t^{CS} is the (deterministic) irradiance under clear sky conditions. Similarly, the actual electricity output $E_t^{(DA)}$ is calculated as

$$E_t^{DA} = f_t^{CSE} E_t^{CS}, \tag{2}$$

where f_t^{CSE} is the clear sky electricity factor and E_t^{CS} is the clear sky electricity output.

Calculation of the *clear sky electricity* can follow either the fundamental or a statistical approach. In this study, an advanced statistical clear sky model based on [1] is implemented. Calculations use the actual feed-in time series E_t as presented in Sect. 2.1. This time series is rearranged such that

$$E_t = E(x_t, y_t), \tag{3}$$

where $x = 1, \ldots, 365$ is the day of the year and $y = 1, \ldots, 96$ is the quarter-hour of the day. The clear sky electricity time series can be seen as the upper surface of the resulting point cloud and can be estimated by statistical methods. In this study, we apply weighted quantile regression as described in the aforementioned study except for two refinements that we developed in order to treat two shortcomings of the original approach. On the one hand, the "day of year" distance function is modified in order to address scarcity of clear sky winter days. On the other hand, a simple correction tool deals with systematic overestimation of the clear sky electricity feed-in around the start and end of the daily feed-in (cf. [6]).

We use two different models to forecast the *clear sky electricity factor*, namely a linear model (LM) and an autoregressive model with exogenous input (ARX). The LM can be described by

$$f_t^{CSE} = \beta_0 + \beta_1 f_t^{CSI} + \varepsilon_t, \tag{4}$$

where f_t^{CSI} is obtained from NWP, β_0 and β_1 are regression coefficients and ε_t is the residual term. For each forecasting day, coefficients are estimated by ordinary least squares regression based on quarter-hours from a fixed number of preceding days, where the actual feed-in exceeded a given threshold. Thus, the model features seasonality and very unproductive quarter-hours are excluded from the regression, because f_t^{CSE} and f_t^{CSI} do not take reasonable values in that case. The ARX can be characterized by

$$f_t^{CSE} = \beta_0 + \beta_1 f_t^{CSI} + \beta_2 f_{t-d}^{CSE} + \varepsilon_t, \tag{5}$$

where f_{t-d}^{CSE} is the most recent observation of the clear sky electricity factor at the same time of the day which is available for forecasting. Regression coefficients are computed based on recursive least squares with exponential forgetting, as presented in [1]. Again, a threshold is applied in connection with actual feed-in.

2.3 Error Measure

Following [4], we find the mean absolute error (MAE) to be an appropriate error measure for our forecasting purpose. To allow for a comparison among PV systems of different size, the MAE is calculated with respect to actual electricity generation by

$$MAE = \sum |E_t^{DA} - E_t| / \sum E_t, \tag{6}$$

where the sums include values from all quarter-hours of a given period.

2.4 Evaluation of Economic Impacts

Balancing costs are calculated according to the following equation

$$BC_t = (E_t^{DA} - E_t)(p_t^{ID} - p_t^{DA}) \tag{7}$$

where p_t^{DA} and p_t^{ID} are the electricity prices at the day-ahead and intra-day market. Depending on the signs of the forecast error and the price spread between day-ahead and intra-day market, balancing efforts can generate costs ($BC_t > 0$) as well as earnings ($BC_t < 0$). Note that additional balancing costs occur in the imbalance market, but those depend on intra-day forecasting which goes beyond the scope of this study.

Intraday prices are highly time-dependent and intra-day balancing costs can be optimized through strategic balancing behavior (e.g. [3]). To exclude such influences from our analysis, the intra-day reference price is used, which is the volume-weighted average price of all deals that are related to a given quarter-hour and that were realised during the last 15 min before the respective intra-day gate closure. Thus, we assume that deviations are balanced right before gate closure at an average price which is in line with the reality of the German intra-day electricity market.

According to decision theory we assume forecast-users to aim at minimizing their expected costs and risk. In order to estimate those two dimensions for a given forecast, we calculate the mean and the standard deviation of daily balancing costs. To allow for comparison among PV systems of different size, both values are normalized using the installed capacity and then referred to as relative mean of daily costs (rMDC) and relative standard deviation of daily costs (rSDC).

3 Results

Figure 1 shows the MAE for all considered PV systems and the portfolio. Apparently, the ARX outperforms the LM for all PV systems, whereas the differences are relatively small. As expected, due to spatial averaging of forecasting errors, the MAE of the portfolio is remarkably lower than for single systems.

The left plot in Fig. 2 summarizes the rMDC and the MAE for both forecasting models. Surprisingly, for the majority of PV systems (seven out of ten), balancing costs increase when using the ARX instead of the LM, even though the ARX features higher accuracy. For the portfolio, this increase amounts to 5 % of the rMDC.

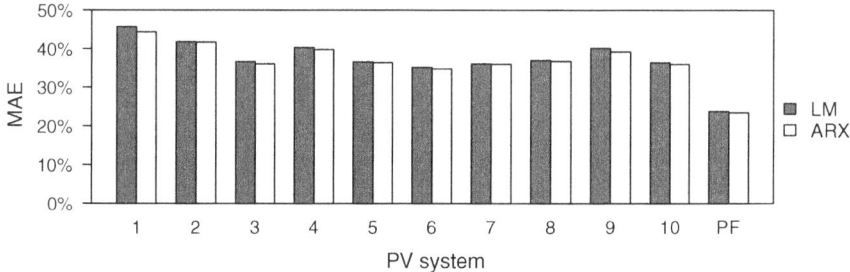

Fig. 1 Forecast accuracy for the ten considered PV systems and the respective portfolio (PF)

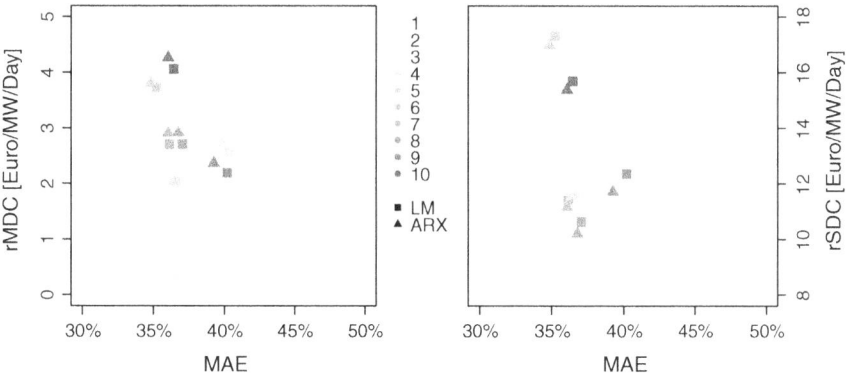

Fig. 2 Forecast accuracy and economic implications for ten different PV systems (*color*) and forecasting models (*shape*)

The right plot in Fig. 2 compares the rSDC and the MAE. Apparently, the volatility of the balancing costs decreases as forecast accuracy improves for each PV system. Thus, better forecasts clearly reduce the risk related to balancing costs.

4 Conclusion

Our investigation leads to the conclusion that enhanced forecast accuracy clearly reduces forecast-related risk while it tends to increase forecast-related costs. The latter can be explained by a greater correlation between forecast errors and market prices. Correlation of the enhanced forecasting model might be higher due to the fact that it is more similar to the forecasting models that are used by the majority of market participants. If so, there is evidence that forecasting science and practice should not only focus on forecast accuracy but also on forecast diversity.

References

1. Bacher, P., Madsen, H., Nielsen, H.A.: Online short-term solar power forecasting. Sol. Energy **83**(10), 1772–1783 (2009)
2. Fernandez-Jimenez, L.A., Muoz-Jimenez, A., Falces, A., Mendoza-Villena, M., Garcia-Garrido, E., Lara-Santillan, P.M., et al.: Short-term power forecasting system for photovoltaic plants. Renew. Energy **44**, 311–317 (2012)
3. Garnier, E., Madlener, R.: Balancing forecast errors in continuous-trade intraday markets. Energy Syst. **6**(3), 361–388 (2015)
4. Kostylev, V., Pavlovski, A.: Solar power forecasting performance-towards industry standards. In: 1st International Workshop on the Integration of Solar Power into Power Systems, 24 October 2011, Aarhus, Denmark (2011)
5. Parkes, J., Wasey, J., Tindal, J., Munoz, L.: Wind energy trading benefits through short term forecasting. In: Proceedings of the European Wind Energy Conference, 27 February–2 March 2006, Athens, Greece (2006)
6. Ruhnau, O., Hennig, P., Madlener, R.: Economic implications of enhanced forecast accuracy: the case of photovoltaic feed-in forecasts. FCN Working Paper No. 6/2015, Institute for Future Energy Consumer Needs and Behavior (FCN), RWTH Aachen University, June (2015)

Part VIII
Financial Modeling, Accounting and Game Theory

Transfer Pricing—Heterogeneous Agents and Learning Effects

Arno Karrer

Abstract In this paper we analyze the impact of heterogeneous agents and learning effects on negotiated transfer prices and the consolidated profit resulting at firm level. An agent-based simulation is employed to show potential results implied by learning and interaction effects between negotiating profit centers. In particular, intra-company profit centers can choose to trade with each other or with independent parties on an external market. Since the profit centers have incomplete and heterogeneous information about this external market, they are involved in a bargaining process with outside options. To achieve a maximized comprehensive income it may be favourable on profit center level or even on firm level to choose outside options. In the long run the intracompany option should be favourable on all levels, as it excludes the profit orientated external market. We investigate our agents' behaviour under different parameter settings regarding the incentive system set by the company-wide management. Results show how learning effects and different incentive systems affect the decision making process with respect to the firm's overall objective.

1 Introduction

To determine a profit center's success of a decentralized corporate organization, transfer prices are needed to account for delivery or service relationships between profit centers. Supplying profit centers account generated transfer prices as income, while receiving profit centers determine transfer prices of transferred goods or services as cost. Besides income recognition transfer prices also coordinate, motivate and steer intracompany profit centers. Literature recognizes numerous ways to determine transfer prices such as market-orientated, cost-orientated and negotiated transfer prices [1, 2].

Today's literature on transfer pricing can be divided into model analytical [3] and empirical [4] work. Hirshleifer's model [3] forms an important starting point,

A. Karrer (✉)
Alpen-Adria Universität Klagenfurt, Abteilung Für Controlling und Strategische Unternehmensführung, Universitätsstraße 65-67, 9020 Klagenfurt, Austria
e-mail: arno.karrer@aau.at

© Springer International Publishing Switzerland 2017
K.F. Dœrner et al. (eds.), *Operations Research Proceedings 2015*,
Operations Research Proceedings, DOI 10.1007/978-3-319-42902-1_70

which was basis for further investigations. Parts of further research were focused on tax optimization [5], which will not be topic of this paper. As summarized by Eccles [6] transfer prices cause internal discussions between profit centers as well as between profit centers and headquarter. The fundamental question how to determine the transfer price is influenced by numerous conditions. The presence of an external market, transaction freedom, purchase and supply constraints, technological constraints and the distribution of bargaining power are essential factors whether one chooses a cost-orientated, market-orientated or negotiated transfer price. Numerous models have been developed based on these options [4, 7, 8].

Although negotiated transfer prices create positive incentives for profit center managers by giving them autonomy, they also cause problems. Different information layers and unevenly distributed bargaining power cause situations where negotiated transfer prices have suboptimal effects on the comprehensive income [9, 10]. However there is almost no literature to answer the question how different information layers regarding market prices and cost structures affect the comprehensive income and profit center's profit [11] as well as how learning effects and incentive systems play into account.

2 Model

The existing literature is limited to closed mathematical models, which work with numerous simplifying model assumptions. In terms of a further approximation to reality, those assumptions are weakened or repealed by using an agent based simulation [12–14]. Heterogeneous actors interact dynamically over several transfer periods. To obtain useful and plausible conclusions, we developed a model which contains restricted market information, learning effects and influence by an incentive system.

Our model contains of a headquarter, two profit centers (A and B) and an external market. Profit center A supplies goods which can be sold to the external market or within the company to profit center B. B, which is in need of the goods provided by A, has the choice to accept the offer by A or to buy from the external market. The traded goods are homogeneous and we assume technological and demand independence. As the external market is profit orientated, it is on a company wide perspective suboptimal to choose the option of the external market in the long run. Nevertheless the profit centers may do business with the external market. The negotiation process between the profit centers is simulated with an ultimatum game, where B has the choice between the offer of A and the external market [15]. Due to restricted market information, both profit centers estimate the market price independently and uncorrelated. This may lead to suboptimal decisions where A may offer a too high price and B may have a too low price limit to purchase the goods within the company (Fig. 1).

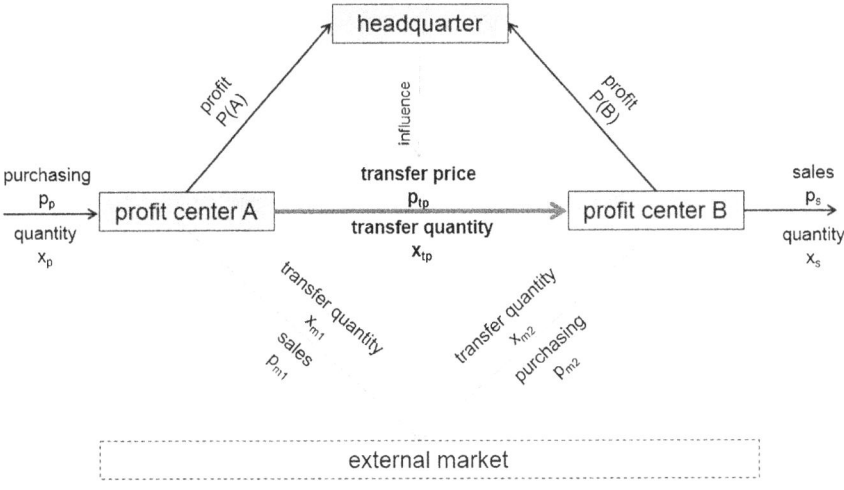

Fig. 1 Overview of the transfer pricing model

2.1 Basic Scenario

In our model we use autoregressive time series values (p) for different simulated prices $(p = c + \beta \cdot p_{(t-1)} + \varepsilon_t)$. Production prices (p_p), retail prices (p_s), estimated market prices and actual market prices (p_m) are simulated with different parameter settings. The comprehensive profit $P(C)$ and the profit center profits $P(A)$ and $P(B)$ are calculated as follows:

$$P(C) = P(A) + P(B) \tag{1}$$

$$P(A) = p_{tp} \cdot x_{tp} + p_{m1} \cdot x_{m1} - p_p \cdot x_p \tag{2}$$

$$P(B) = p_s \cdot x_s - p_{tp} \cdot x_{tp} - p_{m2} \cdot x_{m2} \tag{3}$$

In this model we assume that the two profit centers keep no goods on stock. Therefore we assume that x_p and x_s have a fixed value and the following restrictions:

$$x_p = x_{tp} + x_{m1} \tag{4}$$

$$x_s = x_{tp} + x_{m2} \tag{5}$$

As mentioned the crucial part of the model is the determination of the offered transfer price by A and the price limit by B. Both profit centers know the market price of the last period $t - 1$ and consequently estimate the market price for the current period t. In consequence to the estimated market conditions $(p_{meA}$ and $p_{meB})$ both profit centers have strategy options $(\phi_A$ and $\phi_B)$. An anticipated profit will let the profit center choose the offensive option, where the offered transfer price or the price limit is reasonable but profit orientated. Otherwise they choose the defensive option, where they try to reach an intracompany agreement and focus their offered transfer price or price limit in this direction. This leads to the following offered transfer prices and price limits in our first scenario without learning effects:

$$p_{tp(offer)At} = p_{meAt} + \phi_{At} \tag{6}$$

$$p_{tp(limit)Bt} = p_{meBt} + \phi_{Bt} \tag{7}$$

This first basic scenario has no influence by the decisions made in the past.

2.2 Scenario with Learning Effects

Based on the basic scenario we expand our model by the influence of decisions made last period $(t - 1)$. An incentive and penalty system (θ_t), based on the comprehensive profit, the profit center profit or a combination of both, is implemented.

$$p_{tp(offer)At} = p_{meAt} + \phi_{At} + \theta_{At} \tag{8}$$

$$p_{tp(limit)Bt} = p_{meBt} + \phi_{Bt} + \theta_{Bt} \tag{9}$$

The incentive and penalty system is mainly influenced by the difference between the actual comprehensive profit and the potential comprehensive profit of the last period (s_{t-1}), caused by suboptimal decisions. The variable u describes how much influence the incentive system takes on the decisions made by the profit centers. $u = 1$ describes an incentive and penalty system with 100% influence by the comprehensive profit and $u = 0$ describes a system with 100% influence by the profit center profit itself. The latter causes a scenario where profit centers act like they were not part of the same company.

$$\theta_t = s_{t-1} \cdot u \tag{10}$$

A penalty due to suboptimal decisions influences the offered transfer price or price limit. When profit centers receive no penalty their decisions were optimal and they get rewarded. This results in a consistent strategy based on last period's decisions.

3 Parameter Settings and Results

In this simulation we use autoregressive time series values $(p = c + \beta \cdot p_{(t-1)} + \varepsilon_t)$ where ε_t is a white noise process with zero mean and constant variance σ. We set different parameter settings for the production prices (p_p), retail prices (p_s), estimated

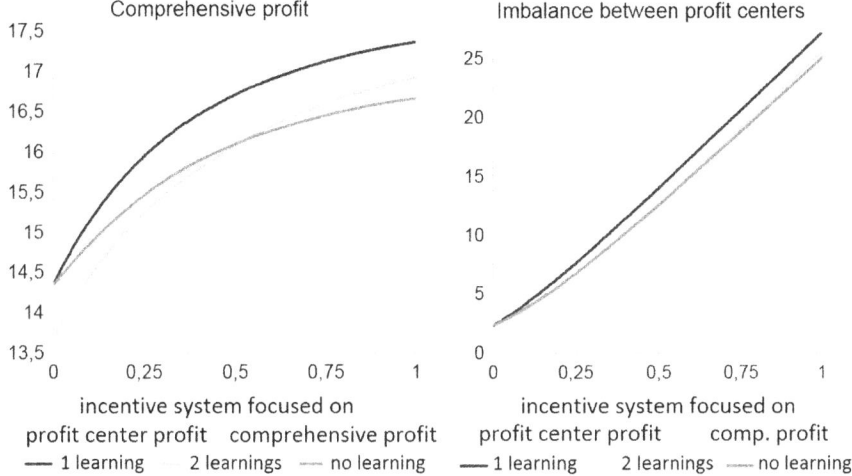

Fig. 2 Comprehensive profit and profit center profit imbalance with and without learning in different incentive settings

market prices and actual market prices (p_m):

$$c = 50(p_m), \ 45(p_p), \ 55(p_s), \ \beta = 0, 5, \ \sigma = 10 \tag{11}$$

Consequently the fixed averages of the production price, retail price and estimated and actual market prices are 90, 110 and 100. Thus, the potential profit is 20 for every item on average.

Regarding the influence of the incentive and penalty system we assume $u = [0, 1]$.

Our results represent average values of 5000 simulation runs. Each simulation run had 105 time periods, in which the first 5 were not considered as each time series started with a fixed value of 90, 100 and 110.

Based on a 99 % confidence level our results for the comprehensive income show a statistically significant difference between scenarios with learning effects over one period and scenarios without learning, as soon as the incentive systems contains proportions of the comprehensive profit. The scenario with a learning effect over one period performs statistically significant better than the scenario with learning effects over two time periods. Scenarios with learning effects over one period also have significantly higher and more sustainable internal transaction rates. The results on profit center profit level also show a statistically significant difference between the scenarios. As the scenario without learning has already an imbalance between the two profit centers, this imbalance increases when we add learning effects over one period. In scenarios where the incentive and penalty system is proportionally higher influenced by the comprehensive profit, A will act very careful, as the profit center profit itself is less important than the comprehensive profit. This causes high profits for B and also a high comprehensive profit (Fig. 2).

4   Conclusion and Further Research

Results show that short term learning is likely to increase the consolidated profit. Learning over more than one period still may have a better result than the basic scenario depending on the incentive system, but performs constantly worse than the scenario with only one period learning effects. Higher proportions of the comprehensive profit within the incentive system cause an significant increase in consolidated profit. Profit center B has an advantage in the setup as it can choose between accepting profit center A's offer or the external market. This advantage also pays off in the results, as its profit is significantly higher than profit center A's. The imbalance also significantly increases when applying learning effects to the model and further increases with an increasing proportion of the comprehensive profit within the incentive system.

In further research, we will introduce a more complex negotiation process moving away from the ultimatum game and test different correlation settings between market price estimations by the profit centers. Furthermore, we will investigate sustainability in learning effects and implement controlled shocks as well as further forms of volatility to move to moving-average time series data. Another point of future research will be the introduction of different layers of market knowledge to investigate if more market information pays off.

References

1. Ewert, R., Wagenhofer, A.: Interne Unternehmensrechnung. Springer, Heidelberg (2008)
2. Küpper, H.U., Friedl, G., Hofmann, C., Hofmann, Y., Pedell, B.: Controlling: Konzeption. Instrumente, Schäffer-Poeschel Verlag für Wirtschaft Steuern Recht GmbH, Aufgaben (2013)
3. Hirshleifer, J.: J. Bus. 29(3), 172 (1956)
4. Wolff, M., Staubach, S., Lindstädt, H.: Zeitschrift für Manag. 3(1), 27 (2008)
5. Horst, T.: J. Polit. Econ. 79, 1059–1072 (1971)
6. Eccles, R.G.: Harv. Bus. Rev. 61(6), 149 (1983)
7. Baldenius, T., Reichelstein, S.J.: (2004)
8. Vaysman, I.: J. Account. Econ. 25(3), 349 (1998)
9. Baldenius, T., Reichelstein, S., Sahay, S.A.: Rev. Account. Stud. 4(2), 67 (1999)
10. Dürr, O.M., Göx, R.F.: Schmalenbach Bus. Rev. 65, 27 (2013)
11. Pfeiffer, T., Schiller, U., Wagner, J.: Rev. Account. Stud. 16(2), 219 (2011)
12. Bonabeau, E.: Proc. National Acad. Sci. 99(suppl 3), 7280 (2002)
13. Davis, J.P., Eisenhardt, K.M., Bingham, C.B.: Acad. Manag. Rev. 32(2), 480 (2007)
14. Lorscheid, I., Heine, B.O., Meyer, M.: Comput. Math. Organ. Theory 18(1), 22 (2012)
15. Hayashida, T., Nishizaki, I., Saiki, K.: In: 2014 IEEE International Conference on in Systems, Man and Cybernetics (SMC), pp. 507–512. IEEE (2014)

Modeling of Dependent Credit Rating Transitions Governed by Industry-Specific Markovian Matrices

Dmitri V. Boreiko, Yuri M. Kaniovski and Georg Ch. Pflug

Abstract Two coupling schemes where probabilities of credit rating migrations vary across industry sectors are introduced. Favorable and adverse macroeconomic factors, encoded as values 1 and 0, of credit class- and industry-specific unobserved tendency variables, modify the transition probabilities rendering individual evolutions dependent. Unlike in the known coupling schemes, expansion in some industry sectors and credit classes coexists with shrinkage in the rest. The schemes are tested on Standard and Poor's data. Maximum likelihood estimators and MATLAB optimization software were used.

1 Motivation

Within the CreditMetrics approach, the study of changes in the credit quality of debtors through time is a corner stone, see Gupton et al. [3]. While the credit rating of each of them evolves as a time-homogeneous Markov chain, in order to model the joint distribution of a pool of debtors, a coupling scheme can be suggested. Then, introducing dependence among the migrations, the evolution of every debtor in the pool can be represented as a randomization of an idiosyncratic move and a common component. In particular, the model by Kaniovski and Pflug [4] considers a single common component for all debtors belonging to a credit class, while in the modification by Wozabal and Hochreiter [6] common components are debtor-specific. An intermediate situation, where the same common component affects all debtors characterized by a combination of a credit class and an industry sector, was introduced in Boreiko et al. [2]. In all three cases, the distribution of a common component depends on an unobserved binary tendency variable. It indicates whether

D.V. Boreiko · Y.M. Kaniovski (✉)
Faculty of Economics and Management, Free University of Bozen-Bolzano,
Piazza Università 1, 39100 Bolzano, Italy
e-mail: YKaniovskyi@unibz.it

G.Ch. Pflug
Department of Statistics and Decision Support Systems, University of Vienna,
Universitätstraße 5, 1090 Vienna, Austria

© Springer International Publishing Switzerland 2017
K.F. Dœrner et al. (eds.), *Operations Research Proceedings 2015*,
Operations Research Proceedings, DOI 10.1007/978-3-319-42902-1_71

the overall state of the economy is favorable or not for debtors belonging to the credit class in question. In other words, it is assumed that, being credit class-specific, the microeconomic factors affect all industry sectors in the same way. Let us label this pattern of tendency variables as *synchronous evolution of industries*.

In this paper, *asynchronously moving industries* are analyzed. That is, tendency variables are not synchronized across industry sectors: favorable conditions in some of them can coexist with adversities in the rest. Incorporating the known coupling schemes as particular cases, the settings introduced here could account better for the actually observed variability of the strength and of the direction of the macroeconomic factors across industry sectors. More importantly, industry-specific tendency variables are necessary for implementing the main new element of this paper— industry-specific Markovian matrices. This departure from the standard CreditMetrics approach allows, in particular, for default frequencies varying across industry sectors. See, for example, Nagpal and Bahar [5] who report industry-specific default frequencies for American firms.

A multidimensional Markov credit rating transition process with dependent coordinates is specified in Sect. 2. A list of parameters of the model and the corresponding estimators are given in Sect. 3.

In order to test the suggested coupling schemes, a Standard and Poor's (S&P's) data set covering OECD (Organization for Economic Co-operation and Development) countries was used. A pool of debtors mimicking the portfolio generating the Dow Jones iTraxx EUR index was considered. Input data and the corresponding estimates are reported and discussed in Sect. 4.

2   Main Assumptions

There is a portfolio consisting of credit contracts. Debtors are non-homogeneous in their credit ratings and they belong to different industry sectors. There are $M \geq 2$ non-default credit classes. Numbering them in a descending order, 1 is assigned to the most secure assets, while the next-to-default credit class is indexed by M. Defaulted debtors receive the index $M + 1$. There are $S \geq 1$ industry sectors. Departing from the CreditMetrics approach, see Gupton et al. [3], where the same Markovian transition matrix applies to all debtors, let us assume that (annual) credit rating migrations in industry sector s are governed by an $M \times (M + 1)$ Markovian matrix $P^{(s)}$ with entries $p_{i,j}^{(s)}$. Then $p_{i,j}^{(s)}$ is the probability of a migration of a debtor belonging to industry sector s within one year from ith credit rating to jth. The credit rating migrations take place at times $t = 1, 2, \ldots$. (Since a defaulted debtor never returns to business, $p_{M+1,M+1}^{(s)} = 1$ and $p_{M+1,j}^{(s)} = 0$ for all s and $j \neq M + 1$. That is, $M + 1$ is an absorption state.) At the beginning there are $\mathcal{N}(1)$ debtors in the portfolio numbered by $n = 1, 2, \ldots, \mathcal{N}(1)$. Set $X_n(t)$ for the credit rating at time $t \geq 1$ of the debtor indexed by n. Then $X_n(t)$ is a discrete-time Markov chain with $M + 1$ states. Its transient states are $1, 2, \ldots, M$, while $M + 1$ is an absorption state. The evolution of the whole portfolio is captured by a multi-dimensional random process $\vec{X}(t) = (X_1(t), X_2(t), \dots, X_{\mathcal{N}(1)}(t))$. Denote by $s(n)$ the industry sector of debtor n. The rating randomly evolves in time, becoming $X_n(2)$ at time $t = 2$, while the assignment to sector $s(n)$ does not change. Since the distributions in question are time-homogeneous, it is enough to consider a transition from time $t = 1$ to time $t = 2$.

Introduce $\mathcal{N}(1)$ independent in n random variables ξ_n such that $\mathbb{P}\{\xi_n = j\} = p_{X_n(1),j}^{(s(n))}$. Conceptually, ξ_n represents an *idiosyncratic component* of the transition from $X_n(1)$ to $X_n(2)$. Its contribution to the resulting move is determined by a Bernoulli random variable δ_n according to the formula:

$$X_n(2) = \delta_n \xi_n + (1 - \delta_n)\eta_n. \tag{1}$$

Here η_n stands for a *common component* in the transition from $X_n(1)$ to $X_n(2)$. Random variables δ_n are independent in n and $\mathbb{P}\{\delta_n = 1\} = q_{X_n(1),s(n)}$. The common components introduce dependence among individual migrations. The probabilities of success $q_{i,s}$ determine the strength of this dependence. The common components are dependent as well. The corresponding mechanism is described next.

Let $\{0, 1\}^{MS}$ be the set of all vectors with MS coordinates, each 0 or 1, and let $\pi(\cdot) = \{\pi(\vec{\chi}), \quad \vec{\chi} \in \{0, 1\}^{MS}\}$ be a probability distribution. A *tendency vector* $\vec{\Pi} = (\Pi_1, \dots, \Pi_{MS})$ is a random vector with distribution $\pi(\cdot)$. The coordinates Π_i are not observable. They are termed as *tendency, hidden* or *latent* variables. Let $\vec{\chi} = (\chi_1, \dots, \chi_{MS})$ be a realization of a tendency vector. Its coordinate $\chi_{M(s-1)+i}$ affects the evolution of debtors from credit class i and industry sector s. If $\chi_{M(s-1)+i} = 1$, all of the random variables η_n, such that $X_n(1) = i$ and $s(n) = s$, cannot assume the values larger than i. This is referred to as a *non-deteriorating* tendency for the corresponding debtors. If $\chi_{M(s-1)+i} = 0$, then all random variables η_n, such that $X_n(1) = i$ and $s(n) = s$, take on exclusively the values exceeding i. It can be characterized as a *deteriorating* tendency. The corresponding conditional probabilities $p_{i,j}^{(s)}(\cdot)$ read:

$$p_{i,j}^{(s)}(1) = \begin{cases} p_{i,j}^{(s)}/p_i^{(s)} & \text{if } j \le i, \\ 0 & \text{if } j > i; \end{cases} \quad \text{and} \quad p_{i,j}^{(s)}(0) = \begin{cases} p_{i,j}^{(s)}/[1 - p_i^{(s)}] & \text{if } j > i, \\ 0 & \text{if } j \le i. \end{cases}$$

Here $p_i^{(s)} = p_{i,1}^{(s)} + p_{i,2}^{(s)} + \cdots + p_{i,i}^{(s)}$ and it is assumed that $p_i^{(s)} \in (0, 1)$. Then $\pi(\cdot)$ and $P^{(s)}$ are related:

$$p_i^{(s)} = \sum_{\vec{\chi} \in \{0,1\}^{MS}: \chi_{M(s-1)+i}=1} \pi(\vec{\chi}). \tag{2}$$

These relations guarantee that the unconditional distribution of η_n and, consequently, the distribution of the mixture (1), coincides with the X_nth row of $P^{(s(n))}$. Conceptually, they mean that the coordinate $M(s - 1) + i$ of a tendency vector takes on value 1 with probability $p_i^{(s)}$.

The families of random variables $\{\delta_n\}, \{\xi_n\}$ and $\{\eta_n\}$ are independent. There are two possibilities for $\{\eta_n\}$, referred to as *coupling schemes* 1 and 2 in what follows next.

The first one assumes conditional on $\overrightarrow{\Pi}$ independence in n of common components. Given $\overrightarrow{\Pi}$, the second scheme requires stochastic independence of random variables η_n and η_l, if $s(n) \neq s(l)$, or $X_n(1) \neq X_l(1)$, while these variables have to coincide, $\eta_n = \eta_l$, if $s(n) = s(l)$ and $X_n(1) = X_l(1)$. In sum, common components are debtor-specific for the first scheme, whereas for the second scheme, the same common component applies to all debtors characterized by a combination of an industry sector and a credit class.

3 Parameters and Estimators

For both coupling schemes, the following inputs are required in order to run the model:

- $M \times (M + 1)$ Markovian transition matrices $P^{(s)}$, $s = 1, 2, \ldots, S$;
- a probability distribution $\pi(\cdot)$ on $\{0, 1\}^{MS}$;
- an $M \times S$ matrix Q formed by probabilities of success $q_{i,s}$ of Bernoulli random variables in (1).

Matrices $P^{(s)}$ are supposed to be given, while Q and $\pi(\cdot)$ have to be estimated. The support of $\pi(\cdot)$ consists of 2^{MS} sample points, whereas Q contains MS entries. Since the sum of the probabilities is 1, there are $MS + 2^{MS} - 1$ independent unknowns.

Parameters for coupling scheme l are estimated by maximizing $\ln L_l(\pi(\cdot), Q)$ subject to the corresponding constraints. Here

$$L_1(\pi(\cdot), Q) = \prod_{t=1}^{T} \sum_{\overrightarrow{\chi} \in \{0,1\}^{MS}} \pi(\overrightarrow{\chi}) \prod_{s=1}^{S} \prod_{m_1=1}^{M} \prod_{m_2=1}^{M+1} f(s, \overrightarrow{\chi}, m_1, m_2, Q)^{I^t(s, m_1, m_2)};$$

$$L_2(\pi(\cdot), Q) = \prod_{t=1}^{T} \sum_{\overrightarrow{\chi} \in \{0,1\}^{MS}} \pi(\overrightarrow{\chi}) \prod_{s=1}^{S} \prod_{m_1=1}^{M} g(t, s, \overrightarrow{\chi}, m_1, Q);$$

$$f(s, \overrightarrow{\chi}, m_1, m_2, Q) = \begin{cases} \dfrac{1 - q_{m_1, s}(1 - p_{m_1}^{(s)})}{p_{m_1}^{(s)}} & \text{if } m_1 \geq m_2, \ \chi_{M(s-1)+m_1} = 1, \\[2ex] \dfrac{1 - q_{m_1, s} \cdot p_{m_1}^{(s)}}{1 - p_{m_1}^{(s)}} & \text{if } m_1 < m_2, \ \chi_{M(s-1)+m_1} = 0, \\[2ex] q_{m_1, s} & \text{otherwise}; \end{cases}$$

$$g(t, s, \overrightarrow{\chi}, m_1, Q) = \sum_{m_2=1}^{M+1} p_{m_1, m_2}^{(s)} (\chi_{M(s-1)+m_1})(q_{m_1, s} + \frac{1 - q_{m_1, s}}{p_{m_1, m_2}^{(s)}})^{I^t(s, m_1, m_2)} \prod_{j=1, j \neq m_2}^{M+1} q_{m_1, s}^{I^t(s, m_1, j)}.$$

Time instants from $t = 1$ through $t = T$ correspond to the period of observation. $I^t(s, m_1, m_2)$ denotes the number of debtors in industry sector s that migrated from credit class m_1 to credit class m_2 in period t. There are linear equality constraints:

$$\sum_{\vec{\chi} \in \{0,1\}^{MS}} \pi(\vec{\chi}) = 1,$$

$$\sum_{\vec{\chi} \in \{0,1\}^{MS}, \; \chi_{M(s-1)+i}=1} \pi(\vec{\chi}) = p_i^{(s)}, \quad i = 1, 2, \ldots, M, \quad s = 1, 2, \ldots, S. \quad (3)$$

Constraints (3) correspond to relations (2). Linear inequality constraints require that elements of Q and probabilities $\pi(\cdot)$ belong to $[0, 1]$.

All industries will move synchronously, as in the known coupling schemes, if the following additional constraint is satisfied:

$$\sum_{\vec{\chi} = (\vec{\chi}^*, \vec{\chi}^*, \ldots, \vec{\chi}^*), \; \vec{\chi}^* \in \{0,1\}^M} \pi(\vec{\chi}) = 1. \quad (4)$$

It implies that the support of an admissible distribution $\pi(\cdot)$ consists of binary vectors formed by S identical blocks $\vec{\chi}^*$ of dimension M. Given relations (3), $p_i^{(s)}$ cannot depend on s if (4) takes place. Consequently, in the known settings Markovian matrices cannot be industry-specific. If the Markovian matrices are identical for all industries and constraint (4) holds true, then the setting of Wozabal and Hochreiter [6] corresponds to $l = 1$, while for $l = 2$ the coupling scheme introduced in Boreiko et al. [2] is obtained.

4 Input Data, Estimates and Their Conceptual Interpretation

A S&P's data set covering companies from 30 OECD countries for the period from 1991 through 2013 was used. Hence, $t = 1$ and $T = 23$ correspond to years 1991 and 2013 in what is reported next.

Five industry sectors involved in the portfolio generating the Dow Jones iTraxx EUR market index are considered: 1—auto and industrial; 2—consumer; 3—energy and utilities; 4—finance and insurance; 5—telecommunications, media and technology. There are two non-default credit classes: investment grade and non-investment grade debtors. Then $S = 5$ and $M = 2$ implying that the total number of independent unknowns is $10 + 2^{10} - 1 = 1033$. The investment grade debtors are characterized by S&P's ratings from AAA to BBB, while the non-investment grade ones occupy the ratings from BB and downward.

With this choice of parameters, using finite-difference approximations for derivatives, the suitable MATLAB software, the Interior Point Algorithm and the Sequential Quadratic Programming Method, required typically 4–8 h in order to find a solution.

Estimating parameters of a coupled Markov chain model from real data, one deals with mixtures of multinomial distributions. This class of statistical problems is known, see among others Allman et al. [1], to imply multiple solutions. Given this possibility, a variety of initial approximations have been tried, including the use of a solution obtained by one of the methods as a starting point for the other one. In all cases the results reported here were identical for both MATLAB algorithms.

The following Markovian transition matrices were used as inputs:

$$P^{(1)} = \begin{pmatrix} 0.9663 & 0.0334 & 0.0003 \\ 0.0378 & 0.9281 & 0.0341 \end{pmatrix}, \quad P^{(2)} = \begin{pmatrix} 0.9691 & 0.0300 & 0.0009 \\ 0.0283 & 0.9450 & 0.0267 \end{pmatrix},$$

$$P^{(3)} = \begin{pmatrix} 0.9808 & 0.0179 & 0.0013 \\ 0.0627 & 0.9209 & 0.0164 \end{pmatrix}, \quad P^{(4)} = \begin{pmatrix} 0.9709 & 0.0267 & 0.0025 \\ 0.0290 & 0.9276 & 0.0434 \end{pmatrix},$$

$$P^{(5)} = \begin{pmatrix} 0.9735 & 0.0257 & 0.0008 \\ 0.1621 & 0.8207 & 0.0172 \end{pmatrix}, \quad P = \begin{pmatrix} 0.9732 & 0.0257 & 0.0011 \\ 0.0792 & 0.8944 & 0.0264 \end{pmatrix}.$$

Here $P^{(s)}$ is formed by frequencies of credit rating migrations in industry sector s. Matrix P contains frequencies characterizing the whole pool of debtors.

In order to analyze distributions of hidden variables, consider a 5×5 matrix C containing below (above) the main diagonal the coefficients of correlation

$$c_{i,j} = Corr(\Pi_{2(i-1)+1}, \Pi_{2(j-1)+1})(Corr(\Pi_{2(i-1)+2}, \Pi_{2(j-1)+2}))$$

between non-deteriorating tendencies affecting investment (non-investment) grade debtors from industry sectors i and j. Since $Corr(\Pi_{2(i-1)+1}, \Pi_{2(i-1)+1}) = Corr(\Pi_{2(j-1)+2}, \Pi_{2(j-1)+2}) = 1$, all entries $c_{i,i}$ of the main diagonal are equal to 1. In the case of synchronously moving industries, the support of a tendency vector consists of the following four sample points:

$$(0000000000), (0101010101), (1010101010), (1111111111),$$

formed by the blocks 00, 01, 10 and 11, respectively. Hence, every element of C equals 1 in this case.

A conclusion concerning the synchronicity of motion can be reached by analyzing the support of $\pi(\cdot)$: every elementary outcome different from the above four ones is an argument against the synchronicity. The number of off-diagonal entries of C that differ from one and the amplitudes of these differences are measures of synchronicity as well. The larger these values are, the stronger is the evidence that the moves are asynchronous.

If transition matrices are industry-specific, for the first (second) coupling scheme, there are 7 (8) realizations of the tendency vector whose probabilities exceed 0.005. One of them occurs with probability 0.9768 (0.9818). For the first/second scheme matrix C reads:

$$\begin{pmatrix} 1.00/1.00 & 0.77/0.68 & 0.42/0.36 & 0.88/0.84 & 0.30/0.70 \\ 0.75/0.54 & 1.00/1.00 & 0.31/0.57 & 0.78/0.78 & 0.51/0.40 \\ 0.51/0.75 & 0.19/0.24 & 1.00/1.00 & 0.61/0.55 & 0.39/-0.01 \\ 0.52/0.56 & 0.55/0.96 & 0.12/0.25 & 1.00/1.00 & 0.39/0.62 \\ -0.03/-0.03 & -0.03/-0.03 & -0.02/-0.02 & -0.03/-0.03 & 1.00/1.00 \end{pmatrix}.$$

Since for both schemes the support of $\pi(\cdot)$ differs from the above four sample points and since all off-diagonal entries of C differ from 1, it follows that the industries migrate asynchronously through credit classes.

An asynchronicity pattern remains in the particular case when all industry sectors were governed by the same Markovian matrix P. Counterparts of the values reported above are as follows: there are 10 (7) realizations of the tendency vector whose probabilities exceed 0.005, one of them occurs with probability 0.9998 (0.9910); matrix C is

$$\begin{pmatrix} 1.00/1.00 & 1.00/-0.03 & 0.39/-0.03 & 1.00/1.00 & 0.58/0.48 \\ 0.59/0.48 & 1.00/1.00 & 0.39/1.00 & 1.00/-0.03 & 0.58/0.49 \\ 0.19/0.15 & 0.19/0.14 & 1.00/1.00 & 0.39/-0.03 & 0.18/0.49 \\ 0.59/0.48 & 1.00/1.00 & 0.19/0.14 & 1.00/1.00 & 0.58/0.48 \\ -0.03/-0.03 & -0.03/-0.03 & -0.03/-0.03 & -0.03/-0.03 & 1.00/1.00 \end{pmatrix}.$$

5 Summary and Outlook

The real data analyzed here indicate that, allowing for industry specific Markovian matrices and asynchronously evolving industries, more realistic models of dependent credit rating transitions can be developed.

Acknowledgements Financial support from the Free University of Bozen-Bolzano for the project "Coupled Markov chains models for evaluating credit and systemic risk" is gratefully acknowledged.

References

1. Allman, E.L., Matias, C., Rhodes, J.A.: Identifiability of parameters in latent structure with many observed variables. Ann. Stat. **37**, 3099–3132 (2009)
2. Boreiko, D.V., Kaniovski, Y.M., Pflug, G.Ch.: Modeling dependent credit rating transitions - a comparison of coupling schemes and empirical evidence. Central Eur. J. Oper. Res. (2015). doi:10.1007/s10100-015-0415-6
3. Gupton, G.M., Finger, Ch.C., Bhatia, M.: Credit Metrics – Technical Document. Technical report, J.P. Morgan Inc. (1997)
4. Kaniovski, Y.M., Pflug, G.Ch.: Risk assessment for credit portfolios: a coupled Markov chain model. J. Bank. Financ. **31**, 2303–2323 (2007)
5. Nagpal, K., Bahar, R.: Measuring default correlation. Risk **14**, 129–132 (2001)
6. Wozabal, D., Hochreiter, R.: A coupled Markov chain approach to credit risk modeling. J. Econ. Dyn. Control **36**, 403–415 (2012)

Replicating Portfolios: $\mathscr{L}^1$ Versus $\mathscr{L}^2$ Optimization

Jan Natolski and Ralf Werner

Abstract Currently, the major challenge in the life insurance sector is to find a numerically efficient and precise method for the estimation of the fair value of future liability cash flows. Besides least square Monte Carlo algorithms, the construction of replicating portfolios is very popular. However, there has been a debate as to how diversions between future discounted cash flows of the replicating portfolio and liabilities ought to be penalized. A frequently used argument against squared error penalization is that a few scenarios with abnormally high interest rates will cause big discrepancies between future cash flows. These scenarios will therefore dominate in the minimization with the consequence that the replicating portfolio badly approximates liabilities in the average scenario. In this article we undermine this argument by showing that the described observation will not take place when discounting with the appropriate numéraire.

1 Introduction

At present, life insurance companies struggle every quarter to estimate the market consistent value of their reserves and liabilities under predetermined shock scenarios. Due to the influence of nonhedgeable risk factors such as longevity and due to the complexity of crediting given surrender options, bonus schemes and penalties, they are facing a tough challenge. Practitioners commonly resort to Monte Carlo methods. The two methods currently dominating are least square Monte Carlo techniques as investigated in e.g. [1, 2] and the construction of replicating portfolios (see e.g. [3, 4] or [5]). A thorough mathematical foundation of replicating portfolios is yet missing and there is a wide range of potential approaches (c.f. [6]). One discussion which has been led for some time now is the question whether mismatches between liability

J. Natolski (✉) · R. Werner
Institut für Mathematik, Universität Augsburg, Universitätsstraße 14,
86159 Augsburg, Germany
e-mail: jan.natolski@math.uni-augsburg.de

R. Werner
e-mail: ralf.werner@math.uni-augsburg.de

© Springer International Publishing Switzerland 2017
K.F. Dœrner et al. (eds.), *Operations Research Proceedings 2015*,
Operations Research Proceedings, DOI 10.1007/978-3-319-42902-1_72

cash flows and portfolio cash flows should be measured by the $\mathscr{L}^1$ or $\mathscr{L}^2$ distance. In practice, the former has become more popular. The reason is an argument often applied to discredit the use of $\mathscr{L}^2$ distance. According to that argument under $\mathscr{L}^2$ measurement too much weight is put on few scenarios with extremely high interest rates so that standard scenarios with moderate interest rates are neglected in the minimization problem. The result is a replicating portfolio whose cash flows match liability cash flows inaccurately, which has different sensitivities and is thus in any way unsatisfactory. A second but less frequently addressed argument is that due to the diverse scales of cash flows of different financial instruments the minimization under $\mathscr{L}^2$ is badly conditioned and in consequence the optimal solution unstable. In this article, we show that with appropriate discounting of cash flows scenarios with high interest rates do not obtain a disproportionate weight in neither minimization problem. Furthermore, we show by example that one can avoid the problem of ill conditioned minimization problems by scaling cash flows.

2   Outline and Comparison

Throughout the article we restrict ourselves to matching of discounted terminal values since the choice of bucketing of cash flows or optimization under constraints does not play a role in our context. Alternative matching problems can be found in [6] or [4].

Denote by

- $N = (N_t)_{t=1,...,T}$ the cash account (with initial value $N_0 = 1$, paying no intermediate cash-flows),
- $\mathbb{Q}$ the risk neutral measure with numéraire N
- by $P(0, t)$ today's price of a zero bond with maturity t,
- by $A^F = (A_i^F)_{i=1,...,m}$ the terminal values of the m financial assets used for replication and
- by A^L the terminal value of the liability cash flow.

We consider the following four objective functions

$$\mathbb{E}^{\mathbb{Q}}\left(\left|N_T^{-1}\left(A^L - \alpha^T A^F\right)\right|\right) \tag{CA_1}$$

$$\left[\mathbb{E}^{\mathbb{Q}}\left(\left[N_T^{-1}\left(A^L - \alpha^T A^F\right)\right]^2\right)\right]^{\frac{1}{2}} \tag{CA_2}$$

$$\mathbb{E}^{\mathbb{Q}}\left(\left|P(0, T)\left(A^L - \alpha^T A^F\right)\right|\right) \tag{B_1}$$

$$\left[\mathbb{E}^{\mathbb{Q}}\left(\left[P(0, T)\left(A^L - \alpha^T A^F\right)\right]^2\right)\right]^{\frac{1}{2}} \tag{B_2}$$

Occasionally, practitioners use zero coupon bonds for discounting cash flows. However, the appropriate way to discount is using the cash account $(N_t)_{t=1,\ldots,T}$. The reason is that the bond $(P(0, t))_{t=1,\ldots,T}$ is not a numéraire for $\mathbb{Q}$ in general (see [6]). This implies that expected terminal values discounted with the bond are not equal to the fair value. This is contrary to our objective of matching fair values. Moreover, we will show that this is a source for the problem of overfitting scenarios with large interest rates as mentioned before. This will be illustrated by comparing cash flows of replicating portfolios generated by the four optimization problems.

3 Numerical Example

For a numerical example we assume there are three financial assets (N, S, Z) available for replication, the cash account, a stock and a portfolio with returns equal to a constant proportion of the returns of the former two. We model the assets as follows.

$$\ln(r_t) = \theta - a\ln(r_{t-1}) + \sigma_1\eta_t, \quad N_t = \exp\left(\sum_{s=1}^{t} r_s\right),$$

$$S_t = S_0 \exp\left(\sum_{s=1}^{t}\left(r_s - \frac{1}{2}\sigma_2^2 + \sigma_2\tilde{\eta}_s\right)\right),$$

$$Z_t = Z_0 \exp\left(\sum_{s=1}^{t}\left(r_s - \frac{1}{2}\gamma^2\sigma_2^2 + \gamma\sigma_2\tilde{\eta}_s\right)\right),$$

for $t = 1, \ldots, T$ and where $\theta, r_0, a, \sigma_1, \sigma_2$ and $\gamma \in (0, 1)$ are constants and $(\eta_t, \tilde{\eta}_t)_{t=1,\ldots,T}$ are i.i.d. with bivariate normal distribution and correlation $\rho \in (-1, 1)$.

The process Z therefore arises from an investment strategy in the cash account and stock maintaining constant yield proportions.

Here, the risk free rate is described by a discrete approximation of the Black Karasinski model (see [8]) and the stock by a geometric Brownian motion in discrete time.

On the liability side, we assume that there are no surrenders and mortality risk. The policy holder account $A^P = A^L$ is modelled as in [7]. Denote by A_t the value of the insurance company's asset side and by A_t^P the absolute value of total policy accounts at time t. Then

$$A_t = Z_t$$

$$A_t^P = A_{t-1}^P\left(1 + \max\left[g, \alpha\left(\frac{A_{t-1} - A_{t-1}^P}{A_{t-1}^P} - \beta\right)\right]\right),$$

for $t = 1, \dots, T$. Here, g is the constant rate guaranteed to the policy holder, α the participation ratio and β the target buffer ratio.

We generated 100000 sample paths of N, S and Z and computed the corresponding policy account A^P. The bond price $P(0, T)$ was approximated by taking the sample mean of $1/N_T$. Finally we obtained four replicating portfolios by minimizing the objective functions $CA1$, $CA2$, $B1$ and $B2$.

3.1 Appropriate Discounting

Figure 1 plots relative absolute differences between discounted terminal values of the replicating portfolio and the policy account in each scenario against the cash account on a logarithmic scale. The circles represent errors when discounting with the bond and the plus signs when discounting with the cash account.

In both, $\mathscr{L}^1$ and $\mathscr{L}^2$ optimization we observe that scenarios with large interest rates are strongly correlated negatively with large errors when discounting with the cash-account. The reason is that cash-flows of assets except the cash account itself discounted with the cash account become negligible in such scenarios. Consequently absolute errors also become small. The opposite occurs when discounting with the bond, where huge interest rates cause errors to grow rapidly. This explains why practitioners observe overfitting in the scenarios with large interest rates.

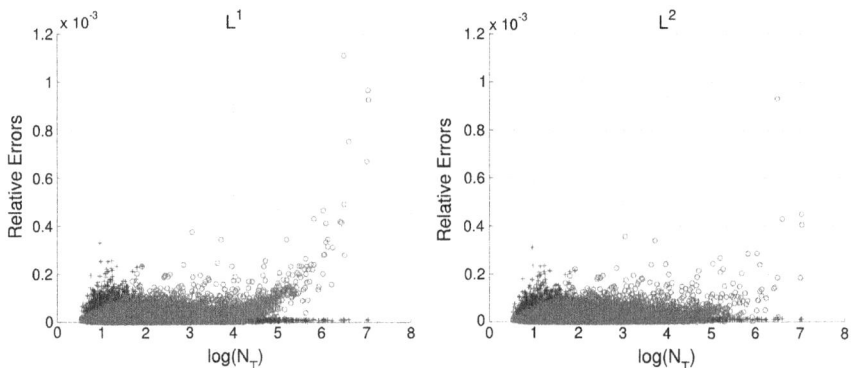

Fig. 1 Relative absolute errors in each scenario between terminal values of liabilities and the replicating portfolios discounted with the bond (o) and with the cash account (+) depending on the size of the cash account N_T. Parameters used: $A_0 = 1.1$, $S_0 = 1$, $A_0^P = 1$, $r_0 = 0.05$, $g = 0.02$, $\alpha = 0.95$, $\beta = 0.1$, $a = 0.08$, $\theta = -0.3$, $\sigma_1 = 0.23$, $\sigma_2 = 0.25$, $\rho = -0.3$, $\gamma = 0.5$

3.2 Conditionedness of $\mathscr{L}^2$

For the computation of portfolios minimizing the mean squared error we scale all cash-flows such that the initial market value of all assets is 1. We thus avoid ill-conditioning due to scales. In order to quantify the level of ill-conditionedness of the problem we need some indicator. To that end, we first need to look at a representation of portfolios minimizing $CA2$ and $B2$.

Given samples $(Z_T^k, S_T^k, A_T^{P,k}, N_T^k)_{k=1,\dots,n}$ and sample mean $\hat{P}(0,T) = \frac{1}{n} \sum_{k=1}^{n} \frac{1}{N_T^k}$ we define the matrices

$$\hat{Q}^N = \frac{1}{n} \sum_{k=1}^{n} \frac{1}{N_T^k} (N_T^k, S_T^k, Z_T^k) \frac{1}{N_T^k} (N_T^k, S_T^k, Z_T^k)^\top$$

$$\hat{Q}^P = \hat{P}(0,T)^2 \frac{1}{n} \sum_{k=1}^{n} (N_T^k, S_T^k, Z_T^k)(N_T^k, S_T^k, Z_T^k)^\top.$$

The portfolios minimizing $CA2$ and $B2$ respectively are given by

$$\hat{\alpha}_{\mathrm{opt}}^N = \left(\hat{Q}^N\right)^{-1} \frac{1}{n} \sum_{k=1}^{n} \frac{A_T^{P,k}}{N_T^k} \frac{1}{N_T^k} (N_T^k, S_T^k, Z_T^k)^\top,$$

$$\hat{\alpha}_{\mathrm{opt}}^P = \left(\hat{Q}^P\right)^{-1} \frac{1}{n} \sum_{k=1}^{n} \frac{A_T^{P,k}}{N_T^k} \frac{1}{N_T^k} (N_T^k, S_T^k, Z_T^k)^\top.$$

As a measure of conditionedness we therefore consult the condition numbers κ^N, κ^P of the matrices $\hat{Q}^N$ and $\hat{Q}^P$. We use the same samples as in Fig. 1 and compute the condition numbers once with original parameters from Fig. 1 and once by scaling S_0, A_0 and P by factor 1000. The condition numbers are given in Table 1.

Although a condition number of 90 is not ideal it is sufficient for precise and fast computation of the optimal replicating portfolio. Of course, as the number of replicating instruments grows, the condition number becomes worse and this has to be investigated in more detail for more complex cases. However, keeping in mind that the processes S, N and Z are highly correlated since Z is a process obtained from trading N and S dynamically, a condition number of this size does not appear to be excessively large.

Table 1 Condition numbers of the matrices $\hat{Q}^N$ and $\hat{Q}^P$

	κ^N	κ^P
Original	90	20
Scaled	8×10^6	6×10^7

However with the size of cash flows from assets other than the cash account scaled by the factor 1000 the condition number increases dramatically. This was to be expected and hence comes as no surprise since entries of the matrices $\hat{Q}^N$ and $\hat{Q}^P$ have very different sizes. This explains why ill-conditioning is usually observed by practitioners.

4 Conclusion

In this article we have introduced a simple toy model for the assets and liabilities of an insurer. By a simple numerical example we have demonstrated that it is essential to use the correct numéraire for discounting. The frequently addressed problem that least squares optimization leads to a disproportionate weighting of scenarios with exploding interest rates is therefore undermined. The reason is to be found in the fact that zero bonds are used for discounting instead of the cash account. Furthermore, we investigated the issue of ill-conditioning in least squares optimization. On the basis of the condition number, the numerical example does not show any significant evidence for systematic problems with matrix inversion provided cash flows are scaled to a common size. This scaling is occasionally omitted by practitioners which explains observed problems with conditionedness.

References

1. Bacinello, A.R., Biffis, E., Millossovich, P.: Regression based algorithms for life insurance contracts with surrender guarantees. Quant. Financ. **10**, 1077–1090 (2010)
2. Beutner E., Pelsser, A., Schweizer J.: Fast Convergence of Regress-Later Estimates in Least Squares Monte Carlo (2014). arXiv:1309.5274
3. Pelsser, A.: Pricing and hedging guaranteed annuity options via static option. Insur. Math. Econ. **33**, 283–296 (2003)
4. Chen, W., Skoglund, J.: Cash flow replication with mismatch constraints. J. Risk **14**, 115–128 (2012)
5. Daul, S., Vidal E.G.: Replication of Insurance Liabilities. Risk Metrics (2009)
6. Natolski, J., Werner, R.: Mathematical analysis of different approaches for replicating portfolios. Eur. Actuar. J. **4**, 411–435 (2014)
7. Hieber, P., Korn, R., Scherer, M.: Analyzing the effect of low interest rates on the surplus participation of life insurance policies with different annual interest rate guarantees. Eur. Actuar. J. **5**, 11–28 (2015)
8. Brigo, D., Mercurio, F.: Interest Rate Models - Theory and Practice, 2nd edn, pp. 82–84. Springer, Berlin (2006)

The Existence of Equilibria in the Leader-Follower Hub Location and Pricing Problem

Dimitrije D. Čvokić, Yury A. Kochetov and Aleksandr V. Plyasunov

Abstract We propose a model where two competitors, a Leader and a Follower, are sequentially creating their hub and spoke networks and setting prices. The existence of the unique Stackelberg and Nash pricing equilibria is shown. On the basis of these results we give the conclusion about existence of the profit maximising solution for the Leader.

1 Introduction

The Hub Location Problem consists of finding the optimal locations for one or more hubs with respect to some given objective. Because markets are usually oligopolies, the profit of a company is not only affected by the decision of its management, but also by the moves and responses of the competitors. Competition between firms that use hub and spoke networks has been studied mainly from a sequential location approach. An existing firm, the Leader, serves the demand in some region, and a new firm, the Follower, wants to enter. One thing that can strongly affect the competition is the price. In the Facility Location Theory, the pricing has been studied for some time now (some more recent works are [1, 2]), but that is not the case with the hub location problems. Recently, Lüer-Villagra and Marianov in [3] analysed a competitive case of hub location problem where the pricing is taken into account. They argued that a location, or route opening decisions, or even the entrance into a market can be very

D.D. Čvokić (✉)
University of Banja Luka, Mladena Stojanovića 2, 78000 Banja Luka,
Republika Srpska, Bosnia and Herzegovina
e-mail: dimitriye.chwokitch@yahoo.com

Y.A. Kochetov · A.V. Plyasunov
Sobolev Institute of Mathematics, Pr. Akademika Koptyuga 4, 630090
Novosibirsk, Russia
e-mail: jkochet@math.nsc.ru

Y.A. Kochetov · A.V. Plyasunov
Novosibirsk State University, St. Pirogova 2, 630090 Novosibirsk, Russia
e-mail: apljas@math.nsc.ru

© Springer International Publishing Switzerland 2017
K.F. Dœrner et al. (eds.), *Operations Research Proceedings 2015*,
Operations Research Proceedings, DOI 10.1007/978-3-319-42902-1_73

dependent on the revenues that a company can obtain by operating these locations and routes. In turn, revenues depend on the pricing structure and competitive context.

Here, we consider a sequential hub location and pricing problem in which two competitors, a Leader and a Follower, compete to attract clients in a given market. Each player tends to maximize his own profit rather than a market share. Customers choose which company and route to patronize by price. It is expected that the demand is split according to the logit model. The location of hubs, allocation of spokes, and pricing are to be determined so as to maximize the profit of the Leader. For this Stackelberg competition we show that there are Stackelberg and Nash pricing equilibria, if the networks of the competitors are already set. Besides their existence and uniqueness, transcendental equations for finding both pricing equilibria are provided. On the basis of these results we give the conclusion about existence of the profit maximising solution for the Leader.

2 A Leader-Follower Hub Location and Pricing Problem

The problem is defined over a directed multi-graph $G = G(N, A)$, where N is the non-empty set of nodes and A is the set of arcs. For every arc $(i, j) \in A$, there is an opposite arc $(j, i) \in A$. If a competitor wants to locate a hub at node $i \in N$, that would cost him some fixed amount f_i. Also, hubs can be shared and there are no capacity constraints. For every arc $(i, j) \in A$ there is a fixed (positive) cost g_{ij} for allocating it as a spoke, and a (positive) transport cost per unit of flow c_{ij}. The cost itself is a non-decreasing function of distance. On the inter-hub transfer there is a known fixed discount factor $\alpha \in (0, 1)$. At most two hubs are allowed to be on a single route. The transportation cost $c_{ij/kl}$ over a route $i \to k \to l \to j$ is given by the following expression $c_{ij/kl} = c_{ik} + \alpha c_{kl} + c_{lj}$. Demand w_{ij} for every OD pair $(i, j) \in A$ is assumed to be non-elastic and positive. Every customer is served either by the Leader or by the Follower. The logit model is used as a discrete choice model. There are no budget constraints. Following the work [4, 5], we address the setting where both players are forced to serve all nodes. Now, we introduce the decision variables for the players:

- $x_k = 1$ if the Leader locates a hub at node $k \in N$ and 0 otherwise
- $\lambda_{ij} = 1$ if the Leader establishes a direct connection between nodes $i, j \in N$, where $(i, j) \in A$, and 0 otherwise
- $p_{ij/kl}$ is the price charged by the Leader for the flows between nodes $i \in N$ and $j \in N$, using the intermediate hubs $k, l \in N$.
- $y_k = 1$ if the Follower locates a hub at node $k \in N$ and 0 otherwise
- $\zeta_{ij} = 1$ if the Follower establishes a direct connection between nodes $i, j \in N$, where $(i, j) \in A$, and 0 otherwise
- $q_{ij/kl}$ is the price charged by the Follower for the flows between nodes $i \in N$ and $j \in N$, using the intermediate hubs $k, l \in N$.

The Leader wishes to maximize his profit, anticipating that the Follower will react to his decision by creating own hub and spoke network and own pricing structure. This Stackelberg game can be presented as the following nonlinear mix-integer bilevel optimization problem. The model for the Leader is

$$\max \sum_{i,j,k,l \in N} (p_{ij/kl} - c_{ij/kl}) w_{ij} u_{ij/kl} - \sum_{i \in N} f_i x_i - \sum_{(i,j) \in A} g_{ij} \lambda_{ij} \tag{1}$$

$$\sum_{s,t \in N} x_s x_t \lambda_{is} \lambda_{st} \lambda_{tj} \geq 1, \quad \forall i,j \in N \tag{2}$$

$$u_{ij/kl} = \frac{x_k x_l \lambda_{ik} \lambda_{kl} \lambda_{lj} e^{-(\cdot)p_{ij/kl}}}{\sum_{s,t \in N} x_s x_t \lambda_{is} \lambda_{st} \lambda_{tj} e^{-(\cdot)p_{ij/st}} + \gamma_{ij}^*}, \quad \forall i,j,k,l \in N \tag{3}$$

$$\gamma_{ij}^* = \sum_{s,t \in N} y_s^* y_t^* \zeta_{is}^* \zeta_{st}^* \zeta_{tj}^* e^{-(\cdot)q_{ij/st}^*}, \quad \forall i,j \in N \tag{4}$$

$$((y_i^*), (\zeta_{ij}^*), (v_{ij/kl}^*), (q_{ij/st}^*)) \in F^*((x_i), (\lambda_{ij}), (u_{ij/kl}), (p_{ij/kl})) \tag{5}$$

$$p_{ij/kl} \geq 0, \quad \forall i,j,k,l \in N \tag{6}$$

$$x_i \in \{0,1\}, \quad \forall i \in N \tag{7}$$

$$\lambda_{ij} \in \{0,1\}, \quad \forall (i,j) \in A \tag{8}$$

Feasible solutions are tuples $((x_i), (\lambda_{ij}), (p_{ij/kl}), (y_i^*), (\zeta_{ij}^*), (q_{ij/st}^*))$ satisfying constraints (1)–(8), where (5) indicates that the Follower's problem has the optimal solution $F^*((x_i), (\lambda_{ij}), (u_{ij/kl}), (p_{ij/kl}))$ for a particular Leader's solution $((x_i), (\lambda_{ij}), (u_{ij/kl}), (p_{ij/kl}))$. The model for the Follower is

$$\max \sum_{i,j,k,l \in N} (q_{ij/kl} - c_{ij/kl}) w_{ij} v_{ij/kl} - \sum_{i \in N} f_i y_i - \sum_{(i,j) \in A} g_{ij} \zeta_{ij} \tag{9}$$

$$\sum_{s,t \in N} y_s y_t \zeta_{is} \zeta_{st} \zeta_{tj} \geq 1, \quad \forall i,j \in N \tag{10}$$

$$v_{ij/kl} = \frac{y_k y_l \zeta_{ik} \zeta_{kl} \zeta_{lj} e^{-(\cdot)q_{ij/kl}}}{\sum_{s,t \in N} y_s y_t \zeta_{is} \zeta_{st} \zeta_{tj} e^{-(\cdot)q_{ij/st}} + \eta_{ij}}, \quad \forall i,j,k,l \in N \tag{11}$$

$$\eta_{ij} = \sum_{s,t \in N} x_s x_t \lambda_{is} \lambda_{st} \lambda_{tj} e^{-(\cdot)p_{ij/st}} \quad \forall i,j \in N \tag{12}$$

$$q_{ij/kl} \geq 0, \quad \forall i,j,k,l \in N \tag{13}$$

$$y_i \in \{0,1\}, \quad \forall i \in N \tag{14}$$

$$\zeta_{ij} \in \{0,1\}, \quad \forall (i,j) \in A \tag{15}$$

The objective functions (1) and (9) are representing the profits of the competitors. Constraints (2) and (10) are assuring that all OD pairs are going to be served. Equations (3) and (11) are representing the Leader's and the Follower's market shares, respectively. Next, (4) is characterizing the effect of the Follower's optimal solution on the Leader's market share. Equation (12) characterizes the Leader's effect on the

Follower's market share. In addition, we are distinguishing two extreme cases for the Follower's behaviour: altruistic and selfish.

3   Stackelberg Pricing Equilibrium

We have the optimal pricing expression for the Follower, provided in [3]. Let H_{ij}^L denotes the Leader's set of inter-hub arcs which are connecting OD pair (i, j). Following that, let H_{ij}^{F*} represents the Follower's set of inter-hub arcs that are connecting OD pair (i, j), based on his optimal solution (a hub and spoke topology).

Theorem 1 ([3]) *The Follower's optimal price for every route $i \to k \to l \to j$ is given by $q_{ij/kl}^* = c_{ij/kl} + \frac{1}{\Theta}\left(1 + W_0\left(\frac{1}{\eta_{ij}}\sum_{(s,t)\in H_{ij}^{F*}} e^{-\Theta c_{ij/st}-1}\right)\right)$, where W_0 is the principal branch of the Lambert W function.*

One could ask if a similar result holds for the Leader? How many equilibria are there? Are they finite? But before we give some answers, we are going to prove one small lemma.

Lemma 1 *If the Follower uses a fixed margin in his best response on all OD pairs, then the Leader should also use a fixed margin in his best response.*

Proof It is enough to prove that First Order Conditions (FOC) are satisfied only for a fixed margin. Objective function is decomposable, so we can focus our attention to some particular OD pair $(i, j) \in N^2$, thus neglecting the OD indices. This reduces analysis to the objective

$$\max w \frac{\sum\limits_{(k,l)\in H^L} (p_{kl} - c_{kl})e^{-\Theta p_{kl}}}{\sum\limits_{(k,l)\in H^L} e^{-\Theta p_{kl}} + \gamma^*}$$

In next few lines, we give a sketch for the essentially straightforward proof. For particular hubs s and t, we can compute $\frac{\partial \gamma^*}{\partial p_{st}}$ from the FOC expression. The derivative can also be computed from (4), using the Theorem 1. These two expression are constructing an equation, hard-wired to the hubs s and t. We can derive the similar equation choosing some other hubs, e.g. $(m, g) \in H^L$, and from there to obtain that $p_{st} - c_{st} = p_{mg} - c_{mg}$.   □

Now, we present the theorem about the Stackelberg pricing equilibrium.

Theorem 2 *In LFHLPP, where hub and spoke networks are already given, there is a unique finite Stackelberg equilibrium in terms of pricing.*

Proof Like in the preceding lemma, we focus our attention to OD pairs. Using the definition of the Lambert W function, the subject of our analysis becomes

$$z(r) = \frac{wr}{1 + W_0(Qe^{\Theta r-1})}$$

where $r = p_{kl} - c_{kl}$, and $Q = \sum_{(k,l) \in H^{F*}} e^{-\Theta c_{kl}} / \sum_{(k,l) \in H^L} e^{-\Theta c_{kl}}$ (which is always greater than 0). Now, it is easy to see that $z(r)$ has a unique maximizer r^*. FOC can be written as a system

$$x^2 + (2 - \Theta r)x + 1 = 0 \tag{16}$$

$$x = W_0(Qe^{\Theta r - 1}) \tag{17}$$

The quadratic equation (16) gives us that the feasible solution exists only when $r \geq \frac{4}{\Theta}$. From the straightforward examination of the slopes for the left and right hand sides of (17), we can conclude that $z(r)$ has only one maximum. □

The number of possible hub and spoke networks for both players is finite. For each pair of them there is a Stackelberg pricing equilibrium. So, for both types of the Follower's behaviour, there exists a finite optimal solution for the Leader.

Theorem 3 *A Stackelberg equilibrium exists in the LFHLPP.*

4 Nash Pricing Equilibrium

One can think about relaxing the pre-commitment in terms of pricing, a.k.a "the price war" from [6]. Thus, we need to show if the Nash equilibrium exists.

Theorem 4 *For already given hub and spoke networks there is a unique finite Nash equilibrium in terms of pricing.*

Proof Again, we can focus our attention to the OD pairs, and neglect the corresponding indices. For both competitors we have the expressions for their best responses, that is $r_L(r_F) = \frac{1}{\Theta}\left(1 + W_0\left(Qe^{\Theta r_F - 1}\right)\right)$ and $r_F(r_L) = \frac{1}{\Theta}\left(1 + W_0\left(\frac{e^{\Theta r_L - 1}}{Q}\right)\right)$ for the Leader and the Follower, respectively. Here, $Q = \frac{\sum_{(k,l) \in H^L} e^{-\Theta c_{kl}}}{\sum_{(k,l) \in H^F} e^{-\Theta c_{kl}}}$. Now, the equation $r_L^* = r_L(r_F^*) = r_L(r_F(r_L^*))$ needs to be solved, which is equivalent to

$$t = W_0\left(Qe^{W_0\left(\frac{e^t}{Q}\right)}\right) \tag{18}$$

$$r_L^* = \frac{t + 1}{\Theta} \tag{19}$$

Taking into account that $W_0(Qe^{W_0(\frac{e^t}{Q})}) = Qe^{W_0(\frac{e^t}{Q})}$, we can transform (18)–(19) into

$$W_0(Qe^\xi) = \frac{1}{\xi} \wedge \xi > 0 \tag{20}$$

$$\xi = W_0\left(\frac{e^t}{Q}\right) \tag{21}$$

$$r_L^* = \frac{t + 1}{\Theta} \wedge r_L^* \geq 0 \tag{22}$$

The first equation always has a solution on $(0, \infty)$. Now, we check the feasibility of the solution, that is if $r_L^* \geq 0$. The last two equations result in

$$e^{t} = Q\xi e^{\xi} \wedge \xi > 0 \wedge t \geq -1 \quad \Leftrightarrow \quad \xi \geq W_0\left(\frac{1}{Qe}\right)$$

What is left to be shown is that $W_0(Qe^{W_0((Qe)^{-1})}) \leq 1/W_0((Qe)^{-1})$, for all $Q > 0$. For that one could just analyse a function $f(Q) = W_0((Qe)^{-1})W_0(Qe^{W_0((Qe)^{-1})})$ on the corresponding interval. □

When it comes to the profit, Stackelberg pricing equilibria is the best one-shot move, by its concept. But these two scenarios, could lead to different outcomes from the hub location point of view. It is not clear that the scenario with the pre-commitment in terms of pricing (LFHLPP) will bring more profit. Nevertheless, we have the following conclusion.

Theorem 5 *In the leader-follower hub location and pricing competition, where competitors are allowed to change their prices, there is a profit maximising solution for the Leader.*

5 Conclusion and Future Work

We have analysed the Leader-Follower setting for hub location and pricing problem, extending the results of Lüer-Villagra and Marianov [3]. It is shown that, when it comes to the pricing, there is a unique solution for the Leader to minimize the damage that can be done by the Follower. This result implied the existence of the solution for this problem.

In future, we plan to address this problem from the computational point of view, and to compare the solutions for LFHLPP and the "pricing war" version. Another line of the research is oriented to a setting where the demand is elastic.

Acknowledgements This research was partially supported by the RFBR grant 15-07-01141.

References

1. Diakova, Z., Kochetov, Yu.: A double VNS heuristic for the facility location and pricing problem. Electron. Notes Discret. Math. (2012). doi:10.1016/j.endm.2012.10.005
2. Panin, A., Pashchenko, M.: Plyasunov A (2014) Bilevel competitive facility location and pricing problems. Autom. Remote Control (2014). doi:10.1134/S0005117914040110
3. Lüer-Villagra, A., Marianov, V.: A competitive hub location and pricing problem. Eur. J. Oper. Res. **231**(3), 734–744 (2013). doi:10.1016/j.ejor.2013.06.006
4. Sasaki, M.: Hub network design model in a competitive environment with flow threshold. J. Oper. Res. Soc. Jpn. **48**, 158–171 (2005)
5. Sasaki M, Campbell JF, Ernst AT, Krishnamoorthy M (2009) Hub Arc Location with Competition. Technical Report of the Nanzan Academic Society Information Sciences and Engineering
6. Serra, D., ReVelle, Ch.: Competitive Location and Pricing on Networks. Geogr. Anal. (1999). doi:10.1111/j.1538-4632.1999.tb00972.x

Part IX
Continuous and Stochastic Optimization, Control Theory

Adaptive Representation of Large 3D Point Clouds for Shape Optimization

Milan Ćurković and Damir Vučina

Abstract A numerical procedure for adaptive parameterization of changing 3D objects for knowledge representation, analysis and optimization is developed. The object is not a full CAD model since it involves many shape parameters and excessive details. Instead, optical 3D scanning of the actual object is used (stereo-photogrammetry, triangulation) which leads to the big-data territory with point clouds of size 10^8 and beyond. The total number of inherent surface parameters corresponds to the dimensionality of the shape optimization space. Parameterization must be highly compact and efficient while capable of representing sufficiently generic 3D shapes. The procedure must handle dynamically changing shapes in optimization quasi-time iterations. It must be flexible and autonomously adaptable as edges and peaks may disappear and new ones may arise. Adaptive re-allocation of the control points is based on feature recognition procedures (edges, peaks) operating on eigen-value ratios and slope/ curvature estimators. The procedure involves identification of areas with significant change in geometry and formation of partitions.

1 Introduction

Contemporary trends of globalization impose very demanding challenges on the product design and development procedures as new products need to deliver excellent performance and reliability at low investment and short time-to-market. These mutually opposed aspirations can only be provided for by involving optimization at all stages. Such a motivation consequently formulates the scope of this paper which encompasses enhanced inverse re-engineering based on 3D shape acquisition [1, 2, 8, 13], adaptive parameterization and numerical optimization [7, 15]. It is likely that starting from the previous design, incremental shape optimization will deliver faster and better results. Nevertheless, using the CAD model of our previous design may

M. Ćurković (✉) · D. Vučina
FESB Department, University of Split, Split, Croatia
e-mail: milan.curkovic@fesb.hr

D. Vučina
e-mail: vucina@fesb.hr

© Springer International Publishing Switzerland 2017
K.F. Dœrner et al. (eds.), *Operations Research Proceedings 2015*,
Operations Research Proceedings, DOI 10.1007/978-3-319-42902-1_74

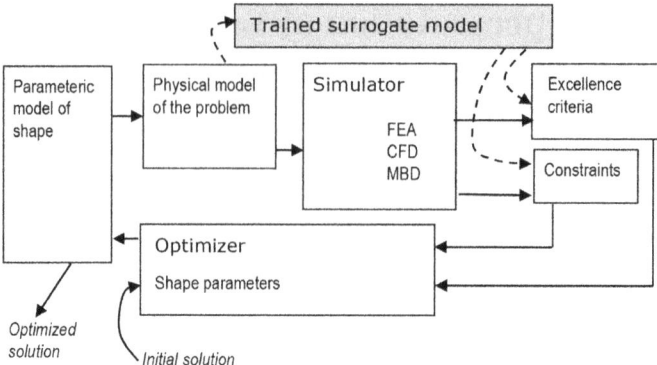

Fig. 1 Changing the initial shape towards excellence by numerical optimization (FEA = finite element analysis, CFD = computational fluid dynamics, MBD = multi-body dynamics)

frequently be inadequate due to the excessive degree of design detail. Therefore a rather general procedure is developed in this paper where a sparing parametric shape model can be derived for any object following the respective 3D scanning. Optical 3D scanning typically provides a huge amount of data in the respective point cloud (commonly in the order of 10^8 points) which needs to be transformed into compact (from the data-set size point of view) parametric surfaces without loss of essential information contained in the large point cloud. The resulting set of parameters needs to be compact such that the dimensionality of the subsequent optimization space is not too excessive, while still able to model sufficiently generic shapes. This ambition is provided for by developing an intelligent, feature-aware partitioning and parameterization approach for the overall shape.

The paper therefore makes reference to parametric shape optimization related to 3D shape acquisition, stereo-photogrammetry, and triangulation as the acquisition pipeline for the raw big-data set, while applying parametric models of shape [3], and fitting models to data as the data representation framework. Enhanced parameterizations are hereby proposed based on feature detection in 3D point clouds and feature-aware adaptive 3D shape parameterizations which can efficiently handle complex shapes and be adaptive in dynamic shape transitions occurring in shape optimization.

This is in line with the concept of enhanced re-engineering, Fig. 1.

2   Modeling Shape

Different approaches have been used to represent general shapes, [4]. Integral CAD models, [3], can be used directly in some cases. Associated sets of shape variables can involve the properties of individual primitives in the model, control points of individual curves or surfaces, operators evolving 2D contours into 3D shapes and other.

The question arises whether integral or partitioned shape parameterizations of the overall objects should be used instead. For point clouds obtained by 3D shape acquisition, [12], parameterizations are obviously necessary. Nevertheless, even shapes available as CAD models are sometimes inadequate for shape optimization since: excessive level of detail may be unnecessary for early-stage shape optimization, many shape variables inevitably lead to high dimensionality of the design space in corresponding optimization, many constraint relationships defined amongst the individual CAD primitives may lead to discontinuities or numerical problems, etc. Integral shape parameterization, [5, 10, 14], may therefore frequently be a better option using the existing design as the initial solution. Different parametric entities, [3] may be engaged, some of which are applied here.

B-splines automatically join low-order basis functions to provide a piecewise representation approach. A B-spline curve of degree d with a given set of $(n + 1)$ control points $\mathbf{Q}_i$ is defined as $\mathbf{P}(t) = \sum_{i=0}^{n} N_{i,d}(t) \cdot \mathbf{Q}_i$ with $t \in [0, 1]$ as the position parameter. The basis functions N use a non-decreasing sequence of scalars-knots t_i such that $0 \leq i \leq n + d + 1$ and can be evaluated recursively

$$N_{i,0}(t) = \begin{cases} 1, & t_i \leq t \leq t_{i+1} \\ 0, & \text{otherwise} \end{cases}, \quad 0 \leq i \leq n + d \tag{1}$$

$$N_{i,j}(t) = \frac{t - t_i}{t_{i+j} - t_i} N_{i,j-1}(t) + \frac{t_{i+j+1} - t}{t_{i+j+1} - t_{i+1}} N_{i+1,j-1}(t), \ 1 \leq j \leq d, \ 0 \leq i \leq n + d - j.$$

Shape variation represented in terms of a B-spline curve in one direction (for example x) can be combined with a B-spline based shape variation in another independent direction (e.g. y-direction). A B-spline surface is accordingly formulated for a 2D array $(n_0 + 1) \times (n_1 + 1)$ of control points $\mathbf{Q}$ by

$$\mathbf{P}(u, v) = \sum_{i_0=0}^{n_0} \sum_{i_1=0}^{n_1} N_{i_0,d_0}(u) \cdot N_{i_1,d_1}(v) \cdot \mathbf{Q}_{i_0 i_1}, \quad u, v \in [0, 1] \tag{2}$$

with u and v as the position parameters, d_0 and d_1 as the degrees of the surface and N the basis functions for the two directions.

3 Fitting Models to Data

Fitting a B-spline to a given $(m + 1)$ points data-set $\mathbf{P}$ can be based on a linear procedure, [6, 9], where the given points are ordered with increasing sample 'times' u, v, whereby the sample 'times' of the given points evaluate to parameter values according to

Fig. 2 Fitting a B-spline
surface to a 3D point cloud
of a boat hull

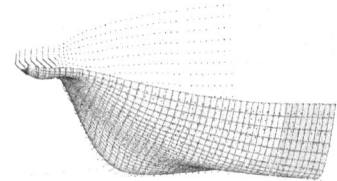

$$u_{j_0} = \frac{r_{j_0} - r_0}{r_{m_0} - r_0}, \quad v_{j_1} = \frac{s_{j_1} - s_0}{s_{m_1} - s_0} \tag{3}$$

$$E(\mathbf{Q}) = \frac{1}{2} \sum_{j_0=0}^{m_0} \sum_{j_1=0}^{m_1} \left(\sum_{i_0=0}^{n_0} \sum_{i_1=0}^{n_1} N_{i_0,d_0}(u_{j_0}) \cdot N_{i_1,d_1}(v_{j_1}) \cdot \mathbf{Q}_{i_0 i_1} - \mathbf{P}_{j_0 j_1} \right)^2 \tag{4}$$

which can be brought [9] into a linear system in $\mathbf{Q}$.

As an example, Fig. 2 presents the example where the front part of the hull of a boat was scanned into a point cloud, which was subsequently used to fit a B-spline surface.

4 Enhanced Parameterizations

As already mentioned, shape optimization may be computationally very demanding as it takes place in a high-dimensional design (search) space and involves complex performance simulations. It is therefore critically important to keep the number of shape parameters at its necessary minimum. It can be experimentally observed that the same multitude of control points can provide better adherence to the original geometry of the selected engineering object if they are distributed more densely around the edges or zones with intensive change of shape [16].

Therefore, the strategy applied in this paper is based on detecting features (e.g. edges, peaks, tips), [11, 16], in the 3D point cloud and subsequently use feature-aware parameterizations. Such shape features are important for: shape partitioning in the overall parameterization, dynamic adaptive parameterizations associated with topological changes and major shape changes during optimization (e.g. edges disappearing or new ones arising) and sparing shape parameterizations with reduced number of shape parameters without sacrificing shape representation capacity. There may be different approaches to feature detection in point clouds. When a local subset of the overall 3D point cloud at a selected point is selected, then different indicators can be applied. A point cloud is here assumed to be a large geometric data-set where generic statistical analysis methods can be applied. Finite-difference based approximations can be applied to estimate the local slope or to estimate local curvatures, for example (here in 1D) (Fig. 3),

$$f'(x) = (f_{i+1} - f_{i-1})/(2\Delta x), \quad f''(x) = (f_{i+1} - 2f_i + f_{i-1})/\Delta x^2 \tag{5}$$

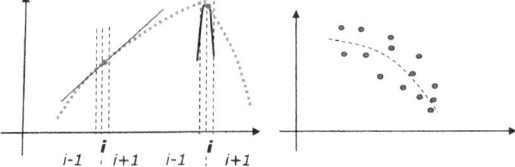

Fig. 3 Subsets of the point cloud, 1D illustration of locally fitting geometric entities, evaluating finite-differences and covariance towards detecting features

Fig. 4 Integral 3D shape parameterization, portion of small cylinder head

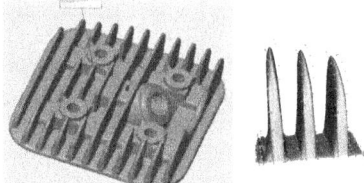

For a local sub-set of the point cloud, the variance measures the spread of data along a single dimension in the data set, and the covariance measures the mutual dependence of any two dimensions of the data-set. Principal component analysis may assist in identifying features contained in the data set by transforming correlated variables into new variables with less correlation. Edge detect formula based on eigenvalues (λ_i) of covariance matrix as a measure of correlation (for example an edge satisfies the relation $\lambda_i \ll \lambda_j \ll \lambda_k$)

$$\frac{\min\{\lambda_1, \lambda_2, \lambda_3\}}{\lambda_1 + \lambda_2 + \lambda_3} \qquad (6)$$

Different procedures are possible in enhancing the overall integral parameterization of a point cloud which contains edges. One approach is to assign more relative weight to the points in the vicinity of the edges during best-fitting. The same effect may be achieved if more points are extracted for best-fitting from the original cloud in the neighborhood of the edges, which again assigns those zones more impact on the overall best-fitting procedure. If the procedure is sufficiently efficient, then even integral (non-partitioned) parametric surfaces may prove satisfactory for objects which contain edges, peaks or generally elements of the surface with very small radii of curvature. Figures 4 and 5 demonstrate such cases.

5   Conclusions

By employing enhanced feature-aware 3D shape parameterizations, the search space in optimization becomes lower-dimensional without sacrificing geometric modeling capacity. Such parameterizations can be implemented dynamically such that even

Fig. 5 Integral 3D shape parameterization, part of formula-student body shell

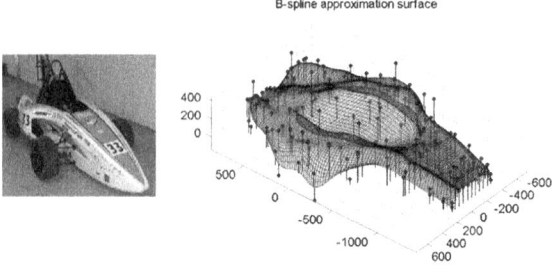

topological changes of shape are accounted for. Moreover, edges may dissolve and new ones may arise during optimization, and the parameterization becomes sufficiently adaptive to such changes.

Acknowledgements This paper presents introductory elements of the research supported by the Croatian science foundation, project number IP-2014-09-6130.

References

1. Barbero, B.R., Ureta, E.S.: Comparative study of different digitization techniques and their accuracy. Comput. Aided Des. **43**(2), 188–206 (2010)
2. Bernardini, F., Rushmeier, H.: The 3D model acquisition pipeline. Comput. Graph **21**(2), 149–172 (2002)
3. Bohm, W., Farin, G., Kahmann, J.: A survey of curve and surface methods in CAGD. Comput. Aided Geom. Des. **1**, 1–60 (1984)
4. Campbell, R.J., Flynn, P.J.: A survey of free-form object representation and recognition techniques. Comput. Vis. Image Underst. **81**, 166–210 (2001)
5. Ćurković, M., Vučina, D.: 3D shape acquisition and integral compact representation using optical scanning and enhanced shape parameterization. Adv. Eng. Inf. **28**(2), 111–126 (2014)
6. Cyganek, B., Siebert, J.P.: An Introduction to 3D Computer Vision Techniques and Algorithms. Wiley, Chichester (2009)
7. Deb K, Goel T (2000) Multi-objective evolutionary algorithms for engineering shape design. In: KanGAL report 200003, Indian Institute of Technology
8. DePiero, F.W., Trivedi, M.M.: 3-D Computer vision using structured light: design, calibration and implementation issues. Adv. Comput. doi:10.1016/S0065-2458(08)60646-4 (1996)
9. Eberly D (2010) Least-squares fitting of data with b-spline surfaces. http://www.geometrictools.com. Accessed on 30 July 2015
10. Eck, M., Hoppe, H.: Automatic reconstruction of b-spline surfaces of arbitrary topological type. In: IACM SIGGRAPH Conference. http://research.microsoft.com/en-us/um/people/hoppe/proj (1996). Accessed on 30 July 2015
11. Fathi, H., Brilakis, I.: Automated sparse 3D point cloud generation of infrastructure using its distinctive visual features. Adv. Eng. Inf. **25**, 760–770 (2011)
12. GOM GmbH. http://www.gom.com/metrology-systems/system-overview/atos.html (2012). Accessed 30 Jul 2015
13. Peng, T., Gupta, S.K.: Model and algorithms for point cloud construction using digital projection patterns. ASME J. Comput. Inf. Sci. Eng. **7**(4), 372–381 (2007)
14. Piegl, L.A., Tiller, W.: Parameterization for surface fitting in reverse engineering. Comput. Aided Des. **33**, 593–603 (2001)

15. Rao, S.S.: Engineering Optimization. Wiley Interscience, New York (1996)
16. Vučina, D., Lozina, Ž., Pehnec, I.: Computational procedure for optimum shape design based on chained Bezier surfaces parameterization. Eng. Appl. Artif. Intell. **25**(3), 648–667 (2012)

The Distortionary Effect of Petroleum Production Sharing Contract: A Theoretical Assessment

Fazel M. Farimani, Xiaoyi Mu and Ali Taherifard

Abstract The distortionary effect of upstream petroleum taxation has been discussed extensively by economists. The literature however, has largly neglected the Production Sharing Contract (PSC) which is widely used by the internationl petroleum industry. We examine how a PSC can distort the optimal time path of production from an oil reservoir. To do that, we use optimal control theory and solve the problem with Hamiltonian function. We show that, regardless of the contract parameters, a PSC always distort the time path of production unless the oil price changes at the rate of interest rate.

1 Introduction

Originally adopted by Indonesia in the 1960s, production sharing contract (PSC) is widely used by the international petroleum industry, particularly in Asia, Africa and Latin America. Under a PSC, the state government as the resource owner grants the foreign investor, normally an international oil company (hereinafter the "operator"), the right to explore and produce oil. If oil is found and produced, the operator recovers its costs from the produced oil.[1] The remaining oil is then divided between the host

---

[1] There is often a limit as to the proportion of oil available each year for the operator to recover its costs. This is called cost recovery limit. Uncovered costs are usually carried forward to future years.

F.M. Farimani (✉) · X. Mu
Center for Energy, Petroleum and Mineral Law and Policy, University of Dundee,
Nethergate, Dundee DD1 4HN, Scotland
e-mail: f.moridifarimani@dundee.ac.uk

X. Mu
e-mail: X.Mu@dundee.ac.uk

A. Taherifard
Faculty of Economics, Imam Sadiq University, 14655-159, Tehran, Iran
e-mail: Taherifard@isu.ac.ir

© Springer International Publishing Switzerland 2017
K.F. Dœrner et al. (eds.), *Operations Research Proceedings 2015*,
Operations Research Proceedings, DOI 10.1007/978-3-319-42902-1_75

government and the operator based on an agreed share. In essence, the PSC functions as a tax to the operator for the right to exploit the resource.[2]

This paper studies the effect of a PSC on the optimal time paths of production from an oil reservoir. Using dynamic optimization method, one can find a time path of production which maximizes the net present value (NPV) of the reservoir in a no-tax, no-contract environment. We call this the neutral path, which could occur when the resource owner is also the operator. However, under a PSC, there is a divide of benefit between the owner and operator. For a given set of contract and tax parameters, the operator will maximize its NPV of production, yielding its optimal time path of production. Will the OP's optimal path match the neutral path? How does it differ? These are the questions this paper seeks to analyze.

The natural resource economics literature has demonstrated that different taxes have varying degree of distortions on the time path of resource production.[3] However, to our knowledge, no previous papers have investigated analytically how a PSC distorts the optimal time path of oil production from the neutral case. Some recent studies have examined the optimal production profile of a specific oil field using numerical methods. For example, Leighty and Lin (2012) [2] and Ghandi and Lin (2012) [1] analyzed the effect of the Iranian buy back contracts and tax-royalty system respectively on the production profiles of specific oil fields. Zhao et al. (2012) [5] model the optimal production rate of an example field under a PSC. Smith (2014) [4] examines the effect of different fiscal regimes including PSC for a generic field. However, the results from these studies are necessarily confined to the specific field(s) and fiscal regimes and did not analyze the possible distortion of a PSC on optimal production paths.

We consider an oil reservoir that is to be developed under a stylized PSC. Using optimal control methods, we solve the optimization problems analytically for the neutral case and the operator under the general form of the PSC respectively. We find that the optimal time paths of production of the operator rarely match that of the neutral case except when the oil price changes at the rate of interest rate. The magnitude of the distortion is determined by the contract elements.

2   The Model

2.1   Contract Parameters and Assumptions

Without loss of generality, we consider a stylized PSC with the following features. For every barrel of oil produced, the operator (hereinafter "OP") first pays a flat rate of royalty α which is a percentage of output to the government. The proportion of oil that can be used by OP to recover its cost in any year is limited to β. The remaining

---

[2]Usually the operator also pays the normal business taxes such as the corporate income tax.
[3]See Smith (2012) [3] for a detailed literature review.

oil (i.e. the profit oil) is then shared between the host government (hereinafter "HC") and the OP, with γ being the share due to the HC and $1 - \gamma$ to OP. In addition, OP pays the corporate income tax at a rate of τ. $0 < \alpha, \beta, \gamma, \tau < 1$.

We assume both objective and constraint functions are concave to satisfy the sufficiency conditions of the optimization. Both the state and control variables are bounded. They cannot be negative or less than the agreed contractual parameters. Furthermore, the initial and terminal values of the state variables are fixed and the salvage value is assumed to be zero. The problem is an autonomous problem as time, the independent variable, does not appear explicitly as an argument in the integrand of the objective function or differential constraint. The cost of oil production is: $c = c(R(t), y(t))$ with $c_y > 0$, $c_R < 0$ where $R(t)$ and $y(t)$ are respectively the amount of reserve and production at year t. Reserve will be repleted by additional drilling activity $v(t)$ resulting in additional reserve $A(v(t))$. The cost of drilling $J(v(t))$ is increasing in $v(t)$: $Jv > 0$. The oil price is exogenous.

2.2   Model Structure

In a no-tax, no-PSC environment, the net cash flow to the HC in any year is the revenue from production, less the cost of oil production, less the drilling expense. So, in the neutral case, the optimization problem for the HC is to maximize the discounted net cash flow over the life of the PSC period, which can be written as:

$$Max \int_0^T \Pi_t^N e^{-rt} = \int_0^T \{p \cdot y(t) - c(R(t), y(t)) - J(v(t))\}e^{-rt}dt \qquad (1)$$

subject to the law of motion $\dot{R} = A(v(t)) - y(t)$ and the depletion rate $\frac{y(t)}{R(t)} = \rho_N$[4] where Π_t^N is the net cash flow in the neutral case in year t, T is the last period of the PSC contract, p is the oil price, $y(t), c(.), J(.), R(t)$ and $A(v)$ are as previously defined, $\dot{R}$ is the change of reserve and ρ is the field specific Maximum Efficient Rate, MER.[5]

Under the PSC regime, the OP maximizes its discounted after-tax cash flow over the contract period and solves the following objective function:

---

[4]In this simplified study, this constraint is a unique ratio of yearly depletion, while in reality it should be at least 3 different constraints depending on the life phases of the field, i.e. build up, plateau and declining phase.

[5]There is a controversy over the nature of Maximum Efficient rate (MER). Some authors consider it as a pure engineering factor while the others believe it is an economic engineering element.

$$Max \int_0^T \Pi_t^{OP} e^{-rt} =$$

$$\int_0^T \{py\,(1-\alpha)\left[1-(1-\beta)\,(\gamma+\tau\,(1-\gamma))\right]-c(.)-J(.)\}e^{-rt}dt$$

$$\text{subject to } \frac{y\,(t)}{R\,(t)} = \rho_{OP} \text{ and } \dot{R} = A\,(v\,(t))-y\,(t) \qquad (2)$$

The net cash flow to the host government in any year of the PSC is the sum of the royalty revenue, the profit oil and the income tax paid by the OP. In theory, HC would also have a preferred time path that maximizes its discounted revenue subject to the condition that the OP receives non-negative NPV. Thus, the HC may have the following optimization problem:

$$Max \int_0^T \Pi_t^{HC} e^{-rt} =$$

$$\int_0^T \{\alpha py + py\left[(1-\alpha)\,(1-\beta)\,(\gamma+\tau\,(1-\gamma))\right]\}e^{-rt}dt$$

$$\text{subject to } \frac{y\,(t)}{R\,(t)} = \rho_{HC} \text{ and } \dot{R} = A\,(.)-y\,(t) \text{ and } \int_0^T \Pi_t^{OP} e^{-rt} >= 0 \qquad (3)$$

Technically, as the main concern of the OP is maximizing the profit rather than keeping the reservoir for a long period of time as an intergenerational asset, its depletion rate should be greater than that of HC and the neutral case. The same argument applies to the discount rate of the HC and OP, where the discount rate of the OP is higher than that of the HC. However, in order to evaluate the distortion which stems from the contract per se, we freeze the effects of different depletion and discount rates. Accordingly we have: $\rho_N = \rho_{HC} = \rho_{OP} = \rho$ and $r_N = r_{HC} = r_{OP} = r$.

For the ease of calculation, in both cases we assume that the OP's costs are fully recovered in the same year as they are incurred, so there is no cost carried forward.

2.3   Solving the Model and Results

To solve all above mentioned problems analytically, we choose Optimal Control Theory and Hamiltonian function among the other methods. The result is as follows; we just present here the problem of OP:

$$H = \{py\,(1-\alpha)\left[1-(1-\beta)(\gamma+\tau\,(1-\gamma))\right] - c\,(.)$$

$$-J(.)\}e^{-rt}dt + \mu\,(t)\,(A\,(.) - y\,(t)) + \lambda\,(t)\,(\frac{y\,(t)}{R\,(t)} - \rho) \tag{4}$$

There is one control, y and one state variables R. First Order Condition is as follows:

$$H_y : -\frac{\lambda\,(t)}{\rho} - \mu\,(t) + e^{-rt}\,(p\,(1-\alpha))\left[1-((1-\beta)(\gamma+\tau\,(1-\gamma)\,\tau) - c^{(1,0)}(.)\right] = 0 \tag{5}$$

$$H_R : \lambda\,(t) - e^{-rt}c^{(0,1)}\,(.) = \mu'\,(t) \tag{6}$$

$$H_\mu : A(v\,(t)) - y\,(t) = R'\,(t) \tag{7}$$

$$H_\lambda : -\frac{y\,(t)}{\rho} + R\,(t) \tag{8}$$

To find the optimal path we take the first derivative of Eq. (5) respect to t:

$$-\frac{\lambda'}{\rho} - \mu' - e^{-tr}r\,(p\,(1-\alpha))$$

$$\left[1-((1-\beta)(\gamma+(1-\gamma)\,\tau) - c^{(1,0)}(.)\right] + e^{-rt}$$

$$((1-\alpha)\left[1-(1-\beta)(\gamma+(1-\gamma)\,\tau)\right]p' - R'c^{(1,1)}(.) - y'c^{(2,0)}(.)) \tag{9}$$

Considering Eq. (9), replacing $R'(t)$ and $\mu'(t)$ with their equivalent from Eqs. (6) and (7), assuming $y' = \theta y$,[6] solving the system of equations and applying further simplifications[7] result in:

$$y^*_{OP}\,(t) = \frac{PG-K}{\theta c^{(2,0)} - c^{(1,1)}} \tag{10}$$

---

[6] To avoid solving a complicated differential equation, we assumed that the production function has an exponential form $y(t) = e^{\theta t}$. This form is fairly compatible with the realities of the industry, in which, in the first period, buidup, the production increases exponentially and then remains stable in the peak for a period of time and finally decreases exponentially. Accordingly, θ would have 3 different values during the life cycle of the field production: For $0 \le t \le t_1$, build up period, $\theta > 0$, For $t_1 \le t \le t_2$, plateau period, $\theta = 0$, For $t_2 \le t \le t_3$, declining period, $\theta < 0$.

[7] We assume that $\lambda'(t) = \varepsilon\lambda(t)$ and $e^{rt}\lambda(t) = \phi$. The first assumption is true since from Hamiltonian function we know $\lambda(t) = \frac{\partial H}{\partial P}$. In fact, the Lambda represents the shadow price of the depletion rate. We know that the shadow price of the depletion rate decreases as the reserve decreases over the time. We can consider an exponential form for the Lambda as follows: $\lambda(t) = e^{\varepsilon t}$. If we take derivative from both side we get $\lambda'(t) = \varepsilon\lambda(t)$. Furthermore, we assume that: $(p' - rp) = P$, $\left(A(v(t)) - c^{(1,1)} + \phi\left(\frac{\varepsilon}{\rho} - 1\right) + rc^{(1,0)} - c^{(0,1)}\right) = K$, $((1-\alpha)\left[1-(1-\beta)(\gamma+(1-\gamma)\,\tau)\right]) = G$.

Following the same procedure we can derive:

$$y_N^* (t) = \frac{P - K}{\theta c^{(2,0)} - c^{(1,1)}}.$$ (11)

3 Analysis

As we discussed earlier, the main purpose of the paper is to show to what extent the neutral optimal path could be different from that of OP. To do that, we compare $y_N^* (t)$ and $y_{OP}^* (t)$. We find $\frac{y_N^* (t)}{y_{OP}^* (t)}$ as follows:

$$\frac{y_N^* (t)}{y_{OP}^* (t)} = \frac{P - K}{PG - K}$$ (12)

Theorem 1 *For any range of contract parameters i.e. α, β, γ and τ, we have*
if $P = 0 \Rightarrow y_N^* (t) = y_{OP}^* (t)$;
if $P > 0 \Rightarrow y_N^* (t) > y_{OP}^*$;
if $P < 0 \Rightarrow y_N^* (t) < y_{OP}^* (t)$.

Proof The first part is obvious. To prove the next two parts, first we show that $G < 1$. Since $0 < \alpha, \beta, \gamma, \tau < 1$, it is obvious that $0 < (\gamma + (1 - \gamma) \tau) < 1$. It also follows: $0 < (1 - \alpha) \left[1 - (1 - \beta) (\gamma + (1 - \gamma) \tau) \right]$ (or G) < 1. If $P > 0$, we can derive that $P > PG$ and $P - K > PG - K$. In other words $\frac{P-K}{PG-K} > 1$ or $\frac{y_N^* (t)}{y*_{OP}(t)} > 1$ or $y_N^* (t) > y_{OP}^* (t)$. If $P < 0$, we can derive that $P < PG$ and $P - K < PG - K$. In other words $\frac{P-K}{PG-K} < 1$ or $\frac{y_N^* (t)}{y*_{OP}(t)} < 1$ or $y_N^* (t) < y_{OP}^* (t)$.

The above results show that the optimal production path of the operator under a PSC always deviate from that of the neutral case unless the oil price changes at the rate of interest rate. If price grows faster than the interest rate, within an inter-temporal preferences framework, the operator prefers future production so the operator's path would be lower than the neutral case. Conversely, when the oil price grows slower than the interest rate, the operator prefers producing more in order to invest the revenue in the alternative (capital) market.

4 Conclusion

This paper studies the distortionary effect of a PSC on the optimal production path of an oil reservoir. We show that distortion is inevitable unless the oil price changes at the rate of interest rate. The results does not have any provision about the relationship between the optimal path of the OP and HC which we plan to analyze in future research. The study have two important assumptions about the production function and the shadow price of the depletion rate, which could be relaxed in the future works.

The novelty of the paper lies in the fact that the different optimal path of both parties have been modelled in a dynamic optimization context and the engineering aspects of the field are also considered in the form of optimization constraint.

References

1. Ghandi, A., Lin, Cynthian: Do Iran's buy-back service contracts lead to optimal production? The case of Soroosh and Nowrooz. Energy Policy **42**, 181–190 (2012)
2. Leighty, W., Lin, C.Y.C.: Tax policy can change the production path: A model of optimal oil extraction in Alaska. Energy Policy **41**, 759–774 (2012)
3. Smith, J.: Issues in extractive resource taxation: a review of research methods and models, No 12/287, IMF Working Papers, International Monetary Fund (2012)
4. Smith, J.L.: A parsimonious model of tax avoidance and distortions in petroleum exploration and development. Energy Econom. **43**, 140–157 (2014)
5. Zhao, X., Luo, D., Xia, L.: Modelling optimal production rate with contract effects for international oil development projects. Energy **45**, 662–668 (2012)

Together We Are Strong—Divided Still Stronger? Strategic Aspects of a Fiscal Union

D. Blueschke and R. Neck

Abstract In this paper we present an application of dynamic tracking games to a monetary union. We use a small stylized nonlinear two-country macroeconomic model of a monetary union for analysing the interactions between two fiscal (governments) and one monetary (common central bank) policy makers. We introduce a negative asymmetric demand side shock describing the macroeconomic dynamics within a monetary union similar to the economic crisis (2007–2010) and the sovereign debt crisis (since 2010) in Europe. We investigate the welfare consequences of three scenarios: fiscal policies by independent governments (the present situation), centralized fiscal policy (a fiscal union) with an independent central bank, and a fully centralized fiscal and monetary union. For the latter two scenarios, we investigate the effects of different assumptions about the weights for the two governments in the cooperative agreement.

1 Introduction

The recent financial and economic crisis has hit the Euro Area (EA) hard, especially because it was followed by a sovereign debt crisis in some member states. This revealed an asymmetry in the EA between a core of financially sound countries and a periphery lacking fiscal sustainability. One possible solution for these problems may be the creation of a fiscal union in addition to the monetary union. In this paper we examine macroeconomic effects of such a fiscal union with a view towards shocks like the recent ones, emphasizing strategic aspects of stabilization policies. We use a dynamic game approach for this purpose.

D. Blueschke (✉) · R. Neck
Alpen-Adria-Universität Klagenfurt, Klagenfurt, Austria
e-mail: dmitri.blueschke@aau.at

R. Neck
e-mail: Reinhard.Neck@aau.at

© Springer International Publishing Switzerland 2017
K.F. Dœrner et al. (eds.), *Operations Research Proceedings 2015*,
Operations Research Proceedings, DOI 10.1007/978-3-319-42902-1_76

2   The Dynamic Game Problem

We consider nonlinear dynamic games in discrete time given in tracking form. The
players aim at minimizing quadratic deviations of the equilibrium values from given
desired values. Each player minimizes an objective function (loss function) J^i:

$$\min_{u_1^i,\ldots,u_T^i} J^i = \min_{u_1^i,\ldots,u_T^i} \sum_{t=1}^{T} L_t^i(x_t, u_t^1, \ldots, u_t^N), \quad i = 1, \ldots, N, \tag{1}$$

with

$$L_t^i(x_t, u_t^1, \ldots, u_t^N) = \frac{1}{2}[X_t - \tilde{X}_t^i]'\Omega_t^i[X_t - \tilde{X}_t^i]. \tag{2}$$

The parameter N denotes the number of players (decision makers). T is the terminal period of the planning horizon. X_t is an aggregated vector

$$X_t = [x_t \ u_t^1 \ u_t^2 \ \ldots \ u_t^N]', \tag{3}$$

consisting of an ($n_x \times 1$) vector of state variables and N ($n_i \times 1$) vectors of control
variables. The desired levels of the state and the control variables enter (1)–(2) via
the terms

$$\tilde{X}_t^i = [\tilde{x}_t^i \ \tilde{u}_t^{i1} \ \tilde{u}_t^{i2} \ \ldots \ \tilde{u}_t^{iN}]'. \tag{4}$$

Finally, (2) contains a penalty matrix Ω_t^i weighting the deviations of states and
controls from their desired levels at any period t.

The dynamic system constraining the choices of the decision makers is given in
state-space form by a first-order system of nonlinear difference equations:

$$x_t = f(x_{t-1}, x_t, u_t^1, \ldots, u_t^N, z_t), \quad x_0 = \bar{x}_0. \tag{5}$$

$\bar{x}_0$ contains the initial values of the states, z_t contains non-controlled exogenous variables. Equations (1), (2) and (5) define a nonlinear dynamic tracking game problem,
which can be solved for different solution concepts. In order to solve this game we
use the OPTGAME algorithm as described in [1].

3   Set-Up of the Games

In this study we use a dynamic macroeconomic model of a monetary union consisting
of two countries (or two blocs of countries) with a common central bank called
MUMOD1. A description of the model is given in [2]. The model is calibrated to
deal with the problem of sovereign debt in a situation that resembles the one currently
prevailing in the European Union. Mainly based on the public finance situation, the
EA is divided into two blocs: a "core" and a "periphery". The first bloc includes ten

EA countries (Austria, Belgium, Estonia, Finland, France, Germany, Luxembourg, Malta, Netherlands, and Slovakia) with a more solid fiscal situation. The second bloc consists of seven countries with higher public debt and/or deficits and higher interest and inflation rates on average (Cyprus, Greece, Ireland, Italy, Portugal, Slovenia, and Spain).

MUMOD1 is formulated in terms of deviations from a long-run growth path and contains the following state variables: output (y), real interest rate (r), nominal interest rate (I), inflation (π), union-wide inflation and output (π_E, y_E), public debt (D) and interest rate on government bonds (BI). The model includes three decision-makers: the common central bank decides on the prime rate R_{Et} (a nominal rate of interest); the national governments decide on fiscal policy: g_{it} denotes country i's ($i = 1, 2$) real fiscal surplus (or, if negative, deficit), measured in relation to real GDP.

It is largely agreed upon that the recent sovereign debt crisis in Europe is to a certain extent due to the asymmetry between the core and the periphery in the EA. Several solutions have been proposed for this problem, in particular with respect to the difficulties of Greek governments to achieve sustainable public debt. One such alleged remedy, a "haircut" (partial debt relief) for the periphery, was examined in the context of the MUMOD1 model in [2]. We found that such a "haircut" can be disadvantageous not only for the lending countries (in our model, the government of the core) but also for the indebted country (the periphery). Here we investigate another possible solution often proposed in the political debate, namely the creation of a fiscal union. As in the earlier study, we assume the dichotomy between core and periphery to be strict and immutable and do not differentiate between various core or periphery countries.

We also emphasise that the model of the fiscal union is rather strict in assuming that there is one common policy enacted by the core and the periphery, which are treated together as one player only; hence we consider a coalition (in the sense of cooperative game theory) between the governments of the two countries (or blocs). A pure fiscal union (denoted by PFU) is modelled as a noncooperative (feedback Nash) game with two players, the central bank and the joint fiscal policy maker (the "EA minister of finance" or the "EA government"). The latter has two instruments, the budgetary surpluses of the core and the periphery, which are determined jointly. Alternatively, we also consider Pareto optimal solutions for the game, which are obtained as solutions to optimum control problems with three instruments. They can be interpreted as a centralized monetary cum fiscal union (denoted by MFU), where all decisions about monetary and fiscal policies are made jointly, and may be considered as a benchmark for the advantages of cooperation on the level of the EA.

4   Results of the Games

In the following, we report the results of some numerical game simulations. In all cases, the central bank gets a weight of 1/3 and the two governments get a joint weight of 2/3 in the joint objective function. For the fiscal union, we vary the weights of the core and the periphery from 0.9 and 0.1 respectively (denoted by 09-01 in the figures) to 0.1 and 0.9 respectively (01-09 in the figures). These weights express the relative importance of each government in the fiscal union and can be interpreted as indicators of the relative power of the core and the periphery respectively. We run a baseline game of three independent policy makers for the noncooperative solution, which is the same as in [2]. The baseline Pareto solution from [2] is the 05-05 MFU. In the following figures, we only show the results for the baseline solution, the extreme cases of 09-01 (a "Schäuble fiscal union") and 01-09 (a "Varoufakis fiscal union"), and the intermediate case of 05-05 (equal weights for both countries).

Figures 1, 2 and 3 display the feedback Nash equilibrium solution trajectories of the fiscal policy instrument variables (fiscal balance) and the two most important target variables, the output gap and public debt. In the baseline, three players play Nash while in the other scenarios (for the PFU), the fiscal union and the central bank are the two Nash players. For lack of space, we do not show the trajectories of the overall Pareto solutions (MFU), as they are qualitatively similar to those of the PFU.

It turns out that the weights of the governments in the fiscal union are important for the distribution of the "burden" of policy making in the fiscal union, while there are no strong differences with respect to monetary policy (not shown here). The larger the weight of the respective player, the more he can shift the task of stabilizing output or debt to the other one. In particular, the 09-01 scenario leads to a very

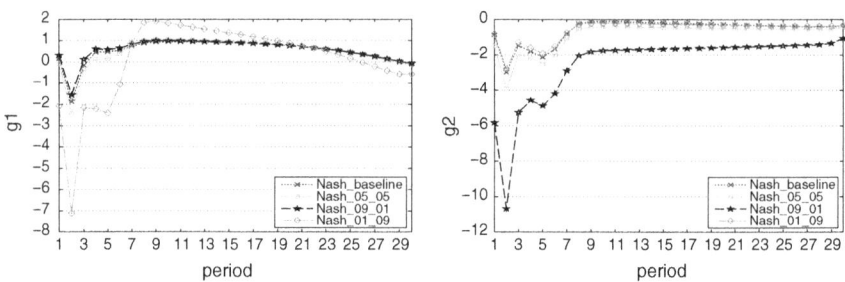

Fig. 1 Country i's fiscal surplus g_{it} (control variable) for $i = 1$ (core; *left*) and $i = 2$ (periphery; *right*); baseline scenario

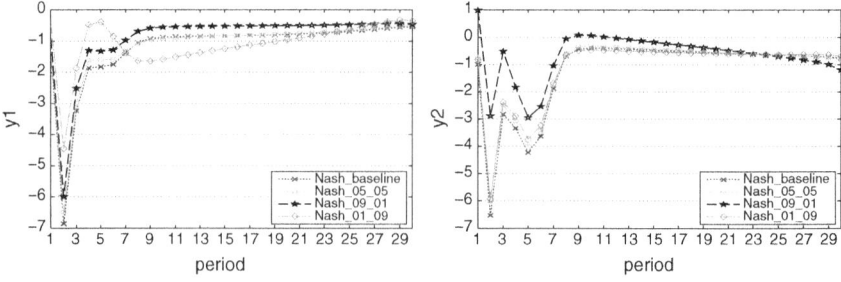

Fig. 2 Country i's output y_{it} for $i = 1$ (core; *left*) and $i = 2$ (periphery; *right*)

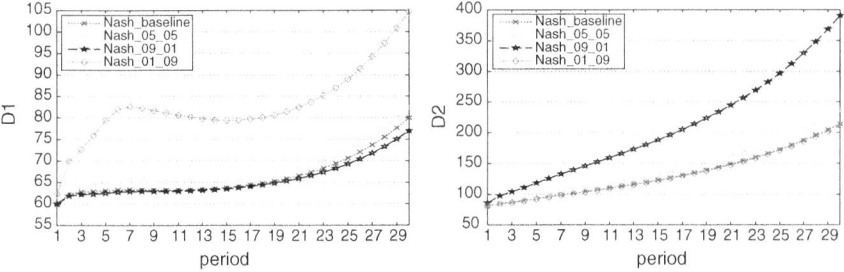

Fig. 3 Country i's debt level D_{it} for $i = 1$ (core; *left*) and $i = 2$ (periphery; *right*)

cautious use of fiscal policy in the core and a very expansionary one in the periphery, which has to "do all the work" (Fig. 1). In contrast, in the 01-09 solution, the core has to fulfil a locomotive function during the crisis while the periphery may use its instrument in a way close to the noncooperative baseline solution. Note that this enforces an expansionary fiscal policy upon the core in spite of its preference for a more restrictive use of this instrument. Resistance against such a type of fiscal union by the core decision maker is understandable, as is the quest in favour of it by the periphery decision maker.

The different policy mixes result in relatively close developments of the output gap among different scenarios (Fig. 2). However, the development of the government debt is strongly dependent on the weights in the fiscal union (Fig. 3). The weaker country in the fiscal union obtains an enormous increase of its public debt, which leads to a clearly unsustainable level of nearly 400 % of GDP in the case of the 09-01 PFU. The intermediate 05-05 fiscal union avoids these extremes and leads to trajectories similar to those in the baseline scenario.

Table 1 shows the values of the objective functions of the players, the central bank (CB), the core (C1) and the periphery (C2) governments, the fiscal union (C1 + C2) and the overall loss (CB + C1 + C2). We can see that the MFU for given weights within the fiscal union always dominates the PFU for the fiscal policy makers, as does the baseline Pareto solution (the 05-05 MFU) over the baseline Nash solution. The same is true for the joint loss of the fiscal union, which is always lower in the MFU than in the PFU. On the other hand, the central bank loses by being part of a full MFU relative to being excluded from an agreement with the fiscal union (PFU); however, this may change when its weight relative to the fiscal union (1/3) increases. This does not imply, however, that joint actions are always better, as can be seen from comparing the Nash baseline solution with the 05-05 PFU scenario. Here the fiscal union is better for the periphery and the central bank but worse for the core and the joint fiscal union. Although the differences are small, this shows that even an intermediate fiscal union with an equal distribution of weights among core and periphery need not be in everybody's interest and may hence be unstable politically.

Table 1 Objective function values of dynamic games for MUMOD1

Scenario	CB	Core (C1)	Periphery (C2)	C1 + C2	CB + C1 + C2
Nash_baseline	41.41	52.55	66.62	119.17	160.58
PFU_0.9-0.1	13.05	36.45	171.27	207.72	220.76
MFU_0.9-0.1	24.75	20.93	146.66	167.59	192.35
PFU_0.8-0.2	23.95	44.43	95.81	140.25	164.19
MFU_0.8-0.2	35.22	24.97	83.24	108.20	143.43
PFU_0.7-0.3	29.53	47.86	78.22	126.08	155.61
MFU_0.7-0.3	40.82	27.25	64.41	91.66	132.48
PFU_0.6-0.4	32.61	49.97	71.61	121.58	154.19
MFU_0.6-0.4	44.06	29.08	56.16	85.24	129.30
PFU_0.5-0.5	34.30	51.76	68.28	120.04	154.34
MFU_0.5-0.5	45.88	31.08	51.61	82.68	128.56
PFU_0.4-0.6	35.00	53.96	66.04	120.03	155.03
MFU_0.4-0.6	46.62	33.98	48.52	82.50	129.12
PFU_0.3-0.7	34.71	57.93	64.00	121.93	156.64
MFU_0.3-0.7	46.23	39.45	45.89	85.34	131.57
PFU_0.2-0.8	32.88	68.57	61.16	129.73	162.60
MFU_0.2-0.8	44.09	53.34	42.82	96.16	140.26
PFU_0.1-0.9	27.08	121.00	54.74	175.75	202.82
MFU_0.1-0.9	37.66	111.07	37.36	148.43	186.09

5   Conclusion

In this paper we used a dynamic game approach to analyse macroeconomic effects of a fiscal union in a model of a monetary union with two countries, the core and the periphery. We showed that a fiscal union in most but by no means all cases gives better results for the fiscal policy makers than noncooperative behaviour. However, the desirability of a fiscal union depends strongly on the weight of the core versus the periphery in the joint objective function of the union, reflecting institutional rules about the decision processes in the union.

References

1. Blueschke, D., Neck, R., Behrens, D.A.: OPTGAME3: A dynamic game solver and an economic example. In: Krivan, V., Zaccour, G., (eds.) Advances in Dynamic Games. Theory, Applications, and Numerical Methods, pp. 29–51. Birkhäuser (2013)
2. Neck, R., Blueschke, D.: "Haircuts" for the EMU periphery: virtue or vice? Empirica **41**(2), 153–175 (2014)

Adaptive Simulated Annealing with Homogenization for Aircraft Trajectory Optimization

C. Bouttier, O. Babando, S. Gadat, S. Gerchinovitz,
S. Laporte and F. Nicol

Abstract In air traffic management, most optimization procedures are commonly based on deterministic modeling and do not take into account the uncertainties on environmental conditions (e.g., wind) and on air traffic control operations. However, aircraft performances in a real-world context are highly sensitive to these uncertainties. The aim of this work is twofold. First, we provide some numerical evidence of the sensitivity of fuel consumption and flight duration with respect to random fluctuations of the wind and the air traffic control operations. Second, we develop a global stochastic optimization procedure for general aircraft performance criteria. Since we consider general (black-box) cost functions, we develop a derivative-free optimization procedure: noisy simulated annealing (NSA).

C. Bouttier (✉) · O. Babando · S. Laporte
Airbus Operations SAS, 316 route de Bayonne, 31060 Toulouse Cedex 9, France
e-mail: clement.bouttier@airbus.com

O. Babando
e-mail: olivier.babando@airbus.com

S. Laporte
e-mail: serge.laporte@airbus.com

S. Gadat
TSE, Toulouse School of Economics, 21 allée de Brienne,
31015 Toulouse Cedex 6, France
e-mail: sebastien.gadat@tse-fr.eu

S. Gerchinovitz
Institut de Mathématiques de Toulouse, Université Toulouse III,
118 route de Narbonne, 31062 Toulouse Cedex 9, France
e-mail: sebastien.gerchinovitz@math.univ-toulouse.fr

F. Nicol
ENAC, 7 avenue douard Belin, 31015 Toulouse Cedex 4, France
e-mail: nicol@recherche.enac.fr

© Springer International Publishing Switzerland 2017
K.F. Dœrner et al. (eds.), *Operations Research Proceedings 2015*,
Operations Research Proceedings, DOI 10.1007/978-3-319-42902-1_77

1 Introduction

The aircraft trajectory optimization problem (ATOP) aims to find a four-dimensional path connecting two cities, which minimizes the sum of costs imputable to the flight called Direct Operating Costs (DOC) such as fuel consumption, salary and maintenance costs or environmental taxes [2]. Since airlines face a continuously growing competition, inaccurate DOC calculations are not affordable anymore. To ensure the robustness of the DOC computation, it is necessary to consider the uncertainty on its influence parameters. Moreover an accurate aircraft performance model should be used.

This work aims to investigate uncertainty into aircraft dynamics and to give some evidence of the sensitivity of aircraft performance with respect to a combination of random inputs such as the weather or the air traffic control (ATC) operations. It relies on previous mono-source uncertainty impact investigations on simplified models such as [9]. We then provide an optimization procedure adapted to accurate stochastic aircraft performance models.

2 Mathematical Models of Uncertainty Sources and NSA Optimization Procedure

The wind information available on-board of aircrafts as well as in the airline operation centres is far from today's meteorological centre maximal capabilities. Meteorological data transmission to the aeronautical community has not evolved since the standardization by the ICAO[1] in [6]. The wind data are only available at discrete locations in time and space. Assuming this resolution unchanged, we provide a method for generating a physically acceptable error field associated to the gridded wind data. This approach was already introduced in [10] or [4].

Wind Model: We divide the wind error field into two components, the current wind estimation error $\Delta W_{e,t}$ at time t and the forecasting error $\Delta W_{f,t}(\tau)$ at time t for time $t + \tau$, so that $\Delta W_{f,t}(0) = \Delta W_{e,t}$. We assume that the forecasting error is a Gaussian field, centered with stationary and isotropic correlations. The correlation in space and time has the following form:

$$Corr\left(\Delta W_{f,t}(x_i, \tau_i), \Delta W_{f,t}(x_k, \tau_k)\right) = \rho_x(x_i, x_k)\, \rho_t(\tau_i, \tau_k)$$

$$= exp\left(-\frac{\|x_i - x_k\|}{d^*}\right) exp\left(-\frac{|\tau_i - \tau_k|}{t^*}\right),$$

where x_i and x_k are two geographical points, d^* is a characteristic distance, τ_i and τ_k are two times and t^* is a characteristic time. We assume the randomness of space and time to be independent. The forecasting error field is obtained at future discrete

---

[1] International Civil Aviation Organization.

times by propagating the wind error with the following propagation Formula 1. We assume the variance to be increasing with the time horizon ($\sigma_{\tau_0}^2 < \sigma_{\tau_1}^2$ if $\tau_0 < \tau_1$).

$$\Delta W_{f,t}(x, \tau_1) = \sigma_{\tau_1}\sigma_{\tau_0}^{-1}\left(\rho_t(\tau_0, \tau_1)\Delta W_{f,t}(x, \tau_0) + (1 - \rho_t(\tau_0, \tau_1))Z_{\tau_1}(x)\right), \quad (1)$$

where Z_{τ_1} is a centered Gaussian random field independent of $\Delta W_{f,t}(\cdot, \tau_0)$ and with covariance $Cov(Z_{\tau_1}(x), Z_{\tau_1}(x')) = \frac{1+\rho_t(\tau_0,\tau_1)}{1-\rho_t(\tau_0,\tau_1)} Cov(\Delta W_{f,t}(x, \tau_0), \Delta W_{f,t}(x', \tau_0))$.

Estimation of the wind model parameters: The variance of $\Delta W_{e,t}$ at each point of the gridded wind data is estimated using a classical maximum likelihood estimator. The variance estimation is key for the representativeness of this model. As proposed in [10] we assume that some geographic and seasonal clusters of wind predictions can be defined. This is reasonable as data from completely different periods or regions should not be compared. The value of the wind estimation error field $\Delta W_{e,t}$ at any point is estimated through a local linear regression.

ATC disturbance model: ATC mainly interacts with the aircraft through negotiations requested by the pilot. This interaction is modelled as a sequence of Bernoulli trials. We assume that each pilot request is accepted with probability p, independently from the past. In fact, the ATC may propose some alternative solutions to the pilot. This will be addressed in future work. Figure 1 provides a good representation of the impact of the negotiation procedure on the aircraft vertical profile. It should be noticed that DOC grows rapidly with airspace congestion.

If one negotiation fails, the pilot waits until he enters a new ATC area and then starts a new negotiation. We set δ to be the characteristic distance between two ATC negotiations. In the sequel, we denote by u_{ATC} the uncertainty induced by the ATC negotiation process, and by u_{Wind} the uncertainty induced by the wind error field.

Optimization procedure: Noisy Simulated Annealing (NSA): Since we consider a stochastic environment (wind and ATC), the DOC associated to a flight are in turn stochastic. Classical methods for optimizing aircraft trajectories are thus no longer applicable. We choose to minimize the expected cost:

$$\min_{path} \mathbb{E}(DOC(path, u_{ATC}, u_{Wind})).$$

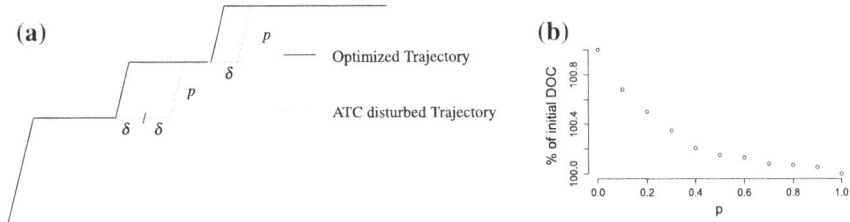

Fig. 1 Evidence of ATC impact on aircraft performance. **a** Theoretical profile disturbed by ATC operations. **b** Influence of p on the DOC

We propose an alternative optimization approach to handle this case since the expected DOC computation can only be performed approximatively. Simulated Annealing (SA) is a global optimization procedure able to deal with non convex problems [1]. We have chosen to use the Monte Carlo Markov Chain procedure based on the Metropolis–Hastings algorithm. It is particularly well suited to optimization problems for which the constraints on the state variables are only known through an oracle answer [7], which is the case of the ATOP. However, it was not designed for optimizing noisy cost functions but, as proved in [5], it remains efficient under the assumption of a sufficiently decreasing noise ($\sigma_{noise} = \mathcal{O}\,(k^{-\gamma})$ with $\gamma > 1$, where k is the iteration number of the algorithm). Following [3], we propose an alternative simulated annealing process, which allows us to relax this noise condition.

Algorithm 1 NSA

---

procedure NSA (**Input:** Neighborhoods structure $(S_i)_{i \in S}$)
Choose one initial path: i
  for k from 1 to N **do**
    Draw one new path : $j \in S_i$ uniformly at random
    $\hat{E}_k(i) = \frac{1}{N_k} \sum\limits_{l=1}^{N_k} (DOC(i, ATC_l, Wind_l))$ and $\hat{E}_k(j) = \frac{1}{N_k} \sum\limits_{l=1}^{N_k} (DOC(j, ATC_l, Wind_l))$
    with $N_k \propto \log(2 + \lfloor \frac{k}{L} \rfloor)^2$
    if $\hat{E}_k(j) < \hat{E}_k(i)$ **then** $i := j$
  end for
end procedure

Considering a Gaussian centered estimation error with variance σ_k^2 on the cost function difference, we obtain the transition probability $P_{i \to j}^k(T)$ of the NSA at step k from i to j of the solution space S:

$$\forall i, j \in S, \quad P_{i \to j}^k = \begin{cases} \dfrac{\mathbb{1}_{j \in S_i} \Phi\left(-\frac{E(j)-E(i)}{\sigma_k}\right)}{|S_i|} & \text{if } j \neq i \\[2ex] 1 - \sum\limits_{l \in S_i/\{i\}} P_{i \to l}^k & \text{if } j = i, \end{cases}$$

where Φ is the cumulative distribution function of the standard normal distribution and $E(i) = \mathbb{E}\,(DOC(i, u_{ATC}, u_{Wind}))$. By analogy with the Glauber transition mechanism as it is done in [3], we obtain the heuristic $\sigma_k = T_k \sqrt{\frac{\pi}{8}}$, where T_k is the classical temperature parameter of the SA, for assessing the accuracy σ_k at step k.

Considering a classical cooling scheme [1], we deduce the corresponding accuracy. Using for example a Monte Carlo expected DOC estimation procedure, we obtain the number $N_k = \left\lfloor \frac{\pi}{8} \frac{\log(2+\lfloor \frac{k}{L} \rfloor)^2}{\Delta^2 (L+1)^2} \right\rfloor$ of Monte Carlo shootings per path at step k, where L and Δ are two constants of the classical cooling scheme.

3 Simulation Results

Using an accurate aircraft performance model, we numerically investigate the impact of wind variations and ATC interventions on the aircraft cruise performance. The wind and ATC scenarios are drawn independently at random from the models of Sect. 2. Then we provide some evidence of the potential reduction in the expected cost when we use our NSA algorithm to optimize the trajectory.

Uncertainty Impact on Aircraft Performance

We considered a single aisle aircraft flying a standard continental mission. In the first part of Table 1, all DOC values (means and standard deviations) are calculated on the same trajectory obtained via a standard deterministic optimization tool [8]. As we can see in Table 1, the wind uncertainty source mainly impacts the variance of the DOC whereas the ATC uncertainty implies some bias. This is therefore important

Table 1 Uncertainty impact on DOC

Uncertainty source	Deterministic optimized trajectory	NSA optimized trajectory
No uncertainty	14 000 (0)[a]	–
Wind	14 000 (200)	14 000 (10)
ATC	14 100 (10)	14 050 (10)
Wind + ATC	14 200 (200)	14 100 (10)

[a] mean DOC (standard deviation)

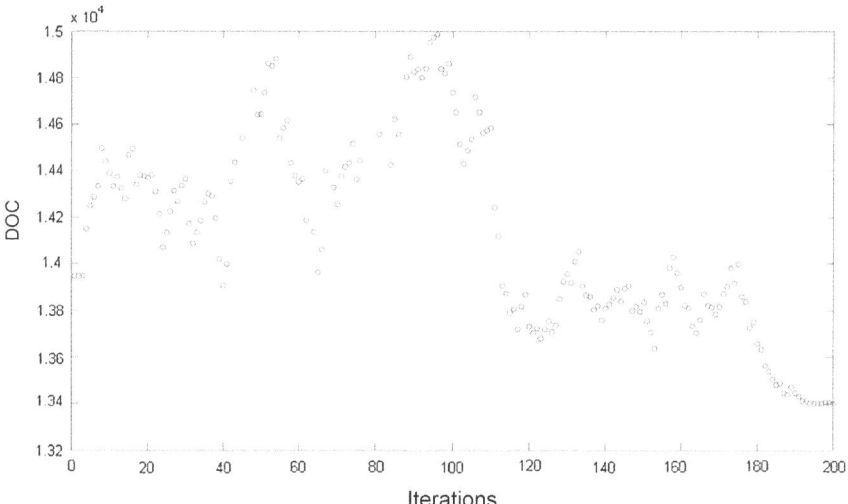

Fig. 2 Example of a NSA descent. Evaluation of the DOC at each iteration

to consider a combination of both effects. The magnitude of the variation is quite important with respect to the usual variation of the DOC w.r.t the vertical flight plan.

Noisy Simulated Annealing: Numerical Alidation

In the second part of Table 1, we consider the same setting as above, but we now use our NSA method to optimize the trajectory and compute the expected DOC in each of the three uncertainty scenarios. As shown in Table 1, both DOC variance and bias can be significantly reduced. The magnitude of the reduction is non negligible compared to the today's airlines targeted improvements. As can be seen on Fig. 2, the optimal solution is attained after only few iterations.

4   Conclusion

We showed that the ATC and wind uncertainties should be taken into account simultaneously in order to optimize the DOC. We also provided a methodology to quickly optimize the expected DOC. As mentioned in Sect. 2, the ATC uncertainty model can be enriched. The probability of success of the negotiation with ATC is set here as a constant value but it is in fact a function of the congestion of the airspace. Therefore, it is both a function of the geographic position of the aircraft and of the time. Future works will focus on providing an estimation procedure for this quantity. Concerning the NSA procedure, a work for providing some theoretical assessment to validate the numerical results is under progress.

References

1. Aarts, E., Korst, J.: Simulated annealing and Boltzmann machines. Wiley, New York (1988)
2. Chakravarty, A.: Four-dimensional fuel-optimal guidance in the presence of winds. J. Guid. Control Dyn. **8**(1), 16–22 (1985)
3. Fink, T.MA.: Inverse protein folding, hierarchical optimisation and tie knots. PhD thesis, University of Cambridge (1998)
4. Glover, W., Lygeros, J.: A multi-aircraft model for conflict detection and resolution algorithm evaluation. HYBRIDGE Deliverable D **1**, (2004)
5. Gutjahr, W.J., Pflug, GCh.: Simulated annealing for noisy cost functions. J. Glob. Optim. **8**(1), 1–13 (1996)
6. ICAO. Annex 3 to the convention on international civil aviation: Meteorological service for international air navigation (2007)
7. Kalai, A.T., Vempala, S.: Simulated annealing for convex optimization. Math. Oper. Res. **31**(2), 253–266 (2006)
8. Liden, S.P.: Apparatus and method for computing wind-sensitive optimum altitude steps in a flight management system, November 12 1996. US Patent 5,574,647
9. Vazquez, Rafael, Rivas, Damián: Propagation of initial mass uncertainty in aircraft cruise flight. J. Guid. Control Dyn. **36**(2), 415–429 (2013)
10. Zheng, Q.M., Zhao, J.Y.: Modeling wind uncertainties for stochastic trajectory synthesis. In: Proceedings of the 11th AIAA Aviation Technology, Integration, and Operations (ATIO) Conference, pp 20–22 (2011)

Optimization Under Uncertainty Based on Multiparametric Kriging Metamodels

Ahmed Shokry and Antonio Espuña

Abstract Different reasons can hinder the application of multiparametric programming formulations to solve optimization problems under uncertainty, as the high nonlinearity of the optimization model, and/or its complicated structure. This work presents a complementary method that can assist in such situations. The proposed tool uses kriging metamodels to provide global *multiparametric metamodels* that approximate the optimal solutions as functions of the problem uncertain parameters. The method has been tested with two benchmark problems of different characteristics, and applied to a case study. The results show the high accuracy of the methodology to predict the multiparametric behavior of the optimal solution, high robustness to deal with different problem types using small number of data, and significant reduction in the solution procedure complexity in comparison with classical multiparametric programming approaches.

1 Introduction

The MultiParametric Programming (MPP) approach is an efficient tool widely used to manage the uncertainty in some of the process model parameters, when this model is used for optimization [3]. In this problem (1), an objective function $Z(x_i)$ is to be minimized satisfying a set of constraints $g_l(x)$, while some Uncertain Parameters (UPs) θ_j within specific bounds are affecting the problem [2].

$$\underset{x_i}{\text{Max}} \quad Z = f(x_i, \theta_j)$$

$$S.T. \quad g_l(x_i, \theta_j) \leq 0, \qquad l = 1, 2, \ldots L \tag{1}$$

A. Shokry (✉) · A. Espuña
Department of Chemical Engineering, Universitat Politecnica de Catalunya,
Av. Diagonal, 647, 020280 Barcelona, Spain
e-mail: ahmed.shokry@upc.edu

A. Espuña
e-mail: antonio.espuna@upc.edu

© Springer International Publishing Switzerland 2017
K.F. Dœrner et al. (eds.), *Operations Research Proceedings 2015*,
Operations Research Proceedings, DOI 10.1007/978-3-319-42902-1_78

$$lb_{x_i} \leq x_i \leq ub_{x_i}, \quad lb_{\theta_j} \leq \theta_j \leq ub_{\theta_j}, \quad x_i \in R^K, \quad \theta_j \in R^k$$

The MPP solution provides a set of P simple mathematical expressions (2) describing the optimal solutions (variables and objective) as a function of the UPs; each expression p is valid for a certain partition of the UPs space which is called a *critical region* [2]. So as the UPs vary, the optimal solution is calculated by these simple expressions, without the need to solve the optimization problem every time [1, 3].

$$Z_p^* = f_{0p}(\theta_j), \quad x_{ip} = f_{ip}(\theta_j) \quad p = 1, 2, \ldots P \tag{2}$$

Different MPP algorithms have been developed [2, 3] depending on the optimization problem nature (linear MPP, quadratic MPP, and many other MPP algorithms most of them based on different iterative approximation procedures), which imply a deep mathematical and programming knowledge to develop these techniques. Additionally, many reasons can hinder the MPP applications, as the difficulties to get a clear mathematical model (e.g. black box, and sequential simulation based models) and/or the mathematical complexity of the resulting model (high nonlinearity), that frequently stems from the embedded highly nonlinear relations in the model, as thermodynamics and reaction kinetics ... etc. [3, 4].

This work proposes a data-based multiparametric analysis method that can be used generally, based on Ordinary Kriging (OK) metamodels [6], which are trained using input–output training data (UPs-optimal variables and objective) to find black box multiparametric relations that accurately approximate the optimal solution as a function of the UPs. The input–output data are generated through the optimization of the original process model several times, using different values of the UPs. The method has been applied to two examples from the MPP literature and a case study, showing high accuracy and robustness compared to the classical MPP solutions.

2 Methodology

The first part of the method is the sampling over the UPs space and data generation: To obtain accurate metamodel prediction, the training data should include—as much as possible—information about the optimal solution behavior in every sub-region of the total input (UPs) space. Consequently, a sampling plan $[\theta]_n$ ($\theta \in R^k$) should be designed to uniformly cover/span the whole input (UPs) domain of the metamodels, where n is the number of sample points (inputs combinations) and k is the number of UPs (number of input variables). To achieve this goal, a hybrid sampling design technique of Hammersley sequence [5, 6] and fractional factorial design is used, because it requires low computational cost. The number of sample points n proportionality depends on the complexity of the multiparametric behavior of the optimal solution, and also on the inputs dimensionality (k). After designing a sampling plan $[\theta]_n$, the optimization problem is solved n times, each with a different combination of the UPs values (a row of $[\theta]_n$), to obtain the outputs $[Z^*, x_i^*]_n, x_i \in R^K$.

The second part is the multiparametric metamodel fitting: Given a set of input–output training data $[\theta_i, y_i]$, $i = 1, 2, \ldots n$, $\theta \in R^k$, $y \in R$, the OK is used to obtain the functional mapping $\hat{y} = f(\theta)$; more details can be found in [5, 6]. The OK is adopted here due to its ability to precisely approximate highly nonlinear systems using relatively small number of training data. To validate the fitted metamodel, a different set of input–output data $[\theta_v]_{nv}$, $[y_v]_{nv}$ is generated; the metamodels prediction of the output of the validation set $[\hat{y}]_{nv}$ is compared by the real output $[y]_{nv}$, and the Normalized Root Mean Square Error (NRMSE %) is calculated to assess the estimation accuracy, $\text{NRMSE} = 100 * (\frac{1}{n_v} \sum_{i=1}^{nv} (y_{vi} - \hat{y}_i)^2)^{0.5} / (y_{v_{max}} - y_{v_{min}})$.

3 Applications

3.1 Mathematical Examples

The first example (3) is taken from [2], and consists of a bilinear objective subjected to two linear constraints, both affected by an UP θ. The method is applied as follows: Over the domain $[0 : 1]$ of the UP (input), a sampling plan $[\theta]_{60}$ is designed; then for each UP value in the plan, the optimization problem is solved to obtain the corresponding optimal variables and objective (outputs) $[x_1^*, x_2^*, Z^*]_{60}$.

$$Min \ \ Z = x_1 x_2, \ \ \ S.T. \begin{bmatrix} 2 & 1 \\ 1 & 3 \end{bmatrix} \begin{bmatrix} x_1 \\ x_2 \end{bmatrix} \geq \begin{bmatrix} 1 \\ 0.5 \end{bmatrix} \theta, \ \ -1 \leq x_1, x_2 \leq 1, \ \ 0 \leq \theta \leq 1$$

(3)

Using these input–output training data, three metamodels ($\hat{z}^* = f_0(\theta)$, $\hat{x}_1^* = f_1(\theta)$ and $\hat{x}_2^* = f_2(\theta)$) are fitted, one for each of the optimal objective and variables. A different validation data set is generated ($[\theta_v]_{150}$, $[x_{1v}^*, x_{2v}^*, Z_v^*]_{150}$), and the inputs $[\theta_v]_{150}$, are used to predict the outputs $[\hat{x}_1^*, \hat{x}_2^*, \hat{Z}^*]_{150}$ using the three metamodels, and the NRMSE (Table 1) of the prediction is calculated for each. Additionally, the same validation set $[\theta_v]_{150}$ is used to calculate the optimal objective and variables using the deterministic MPP solution provided in [2] (see Fig. 1). The results in Table 1 and Fig. 1 show that the method is able to approximate the multiparametric solution with very high accuracy via simple interpolation using the metamodels with low time requirements (0.2 s), saving a huge time quantity of the real optimization (16.2 s), see Table 2.

Table 1 The NRMSE of the Metamodels

Metamodels	Example 1			Example 2			Case study		
	$\hat{Z}$	$\hat{x}_1$	$\hat{x}_2$	$\hat{Z}$	$\hat{x}_1$	$\hat{x}_2$	$\hat{C}$	$\hat{S}_F$	$\hat{S}_T$
NRMSE (%)	0.01	0.8	0.4	0.84	0.84	0.11	0.61	0.77	0.31

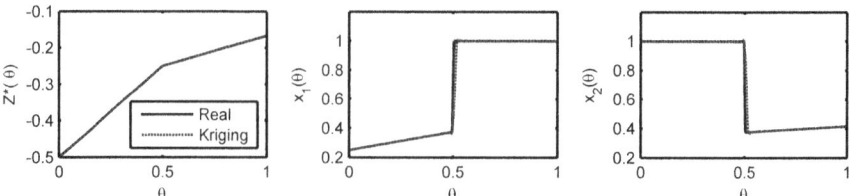

Fig. 1 Validation of the multiparametric metamodels (*solid line*) versus the deterministic MPP solutions (*dashed line*)

Table 2 Time of training data generation, fitting, validation data generation, and OK prediction

	Time (s) $HP-dc$ 7900, *Intel core* $2-duo$ 3.16 *GHz*. *RAM* 6 *GB*			
	Training		Validation	
	Optimization	Fitting	Optimization	Prediction
Example 1	6.9	3.01	16.2	0.2
Example 2	0.8	12.8	4.5	1.5
Case study	48.4	3.67	300.8	0.2

In the second example (4), a quadratic objective to be minimized, subjected to six linear constraints and two UPs, more details can be found in [1].

$$Min \quad Z = c^T \begin{bmatrix} x_1 & x_2 \end{bmatrix}^T + 0.5 \begin{bmatrix} x_1 & x_2 \end{bmatrix} Q \begin{bmatrix} x_1 & x_2 \end{bmatrix}^T$$

$$S.T. \quad A \begin{bmatrix} x_1 & x_2 \end{bmatrix}^T \leq b + F \begin{bmatrix} \theta_1 & \theta_2 \end{bmatrix}^T, \quad -1 \leq x_1, x_2 \leq 1, \quad 0 \leq \theta_1, \theta_2 \leq 1 \quad (4)$$

A sampling plan $[\theta_1, \theta_2]_{80}$ is designed over the UPs domain. The problem is solved 80 times to obtain the outputs $[x_1^*, x_2^*, Z^*]_{80}$, then three metamodels are fitted: $\hat{z}^* = f_0(\theta_1, \theta_2)$, $\hat{x}_1^* = f_1(\theta_1, \theta_2)$ and $\hat{x}_2^* = f_2(\theta_1, \theta_2)$. The validation is done using another data set $([\theta_{1v}, \theta_{2v}]_{400}, [x_{1v}^*, x_{2v}^*, Z_v^*]_{400})$, then the inputs $[\theta_{1v}, \theta_{2v}]_{400}$, are used to predict the outputs $[\hat{x}_{1v}^*, \hat{x}_{2v}^*, \hat{Z}_v^*]_{400}$ using the metamodels, and the NRMSE is calculated (Table 1). The same validation set $[\theta_{1v}, \theta_{2v}]_{400}$ is used to calculate the optimal objective and variables form the deterministic MPP solution provided in [1] (see Fig. 2). The results in Table 1 and Fig. 2 show the high accuracy of the method compared to the individual optimization results, and the MPP solutions found in [1].

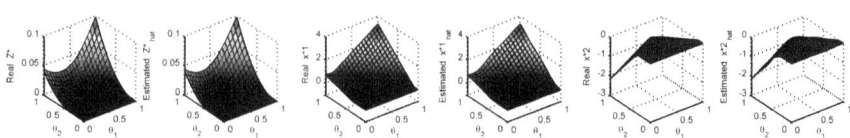

Fig. 2 MPP deterministic solutions versus metamodels estimated solutions: objective, x_1^* and x_2^*

3.2 Case Study

This case study includes a black box simulation model (Fig. 3) of a utility system, which supplies mechanical energy (Qw) to an industrial process. The system is composed of a boiler (E-1) that receives water and supplies high pressure steam to a steam turbine (T1); its outlet steam is condensed (E-3) and the water is pumped (P-1) back to the boiler inlet. The problem (5) is to minimize the system operational cost (C), which includes the costs of energies (Q-2, Q-1, Q-5) consumed by the boiler (E-1) and pumps (P-1, P-2) respectively and the cooling water (stream no. 6) cost. The operational cost is modeled as a function of the boiler outlet steam flowrate (S_F) and temperature (S_T). However, two UPs [θ_1, θ_2] affect the system, which are: the power demand that must be satisfied by the turbine work (Qw) and varies between [53000 kW: 57000 kW], and the turbine efficiency that varies in the range [75 % : 95 %]. This case represents a difficulty for the classical MPP approaches, as the simulation model is a black box that includes different embedded complex functions, which are used to calculate thermodynamic properties and simulate the behavior of different process units (e.g. boiler, cooler). The model can be used to obtain optimized values for the decision variables (S_F, S_T), once the uncertainty [θ_1, θ_2] is unveiled.

$$\operatorname*{Min}_{S_F, S_T} \quad C = f(S_F, S_T, \theta_1, \theta_2)$$

$$S.T. \quad Qw(S_F, S_T, \theta_2) \geq \theta_1 \tag{5}$$

$$36000 < S_F < 79200 \text{ kgMole/hr}, \quad 162 < S_T < 360 \,^\circ C$$

A sampling plan [θ_1, θ_2]$_{70}$ is designed over the domain [53000 : 57000, 75 : 95]. The black box simulation based optimization problem is then solved 70 times (fmincon Matlab optimizer is used) to get the optimal variables and objective [S_F^*, S_T^*, C^*]$_{70}$, then three metamodels are fitted: $\hat{C}^* = f_0(\theta_1, \theta_2)$, $\hat{S}_F^* = f_1(\theta_1, \theta_2)$ and $\hat{S}_T^* = f_2(\theta_1, \theta_2)$. Another data set is generated [$\theta_{v1}\theta_{v2}$]$_{400}$, [$S_{Fv}^*, S_{Tv}^*, C_v^*$]$_{400}$, the inputs [$\theta_{1v}, \theta_{2v}$]$_{400}$ are used to predict the outputs using the metamodels to obtain [$\hat{S}_{Fv}^*, \hat{S}_{Tv}^*, \hat{C}_v^*$]$_{400}$ (see Fig. 4), and the NRMSE is calculated. Tables 1, 2 and Fig. 4 show the high potentials of the method: the optimal decisions are accurately predicted via simple interpolations using the metamodels in almost no time (0.2 s), saving a

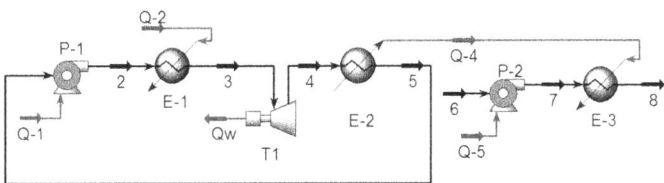

Fig. 3 Black box model of the utility plant modeled by ASPEN HYSYS modeling and simulation environment

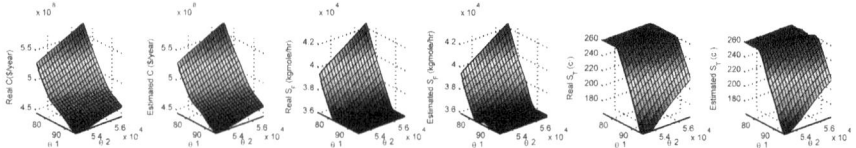

Fig. 4 Real versus estimated C^* (*left*), S_F^* (*middle*), S_T^* (*right*)

huge time amount needed by the real simulation based optimization (300.8 s), which represents a strong tool to immediately manage the uncertain parameters variations during the process online operations. Additionally, Table 2 shows that the method advantage increases as the optimization problem complexity increases: in example 2, the optimization problem is quite easy (single global optima), thus the percentage of the time saved using the method was $66\%(100 \times (4.5 - 1.5)/4.5)$. However, as the problem complexity increased in example 1 (bilinear function includes a saddle behavior) the percentage of the time saved increased to 98.7%. Finally, when a complex nonlinear black box optimization problem is used (case study) the amount of the saved time reached to 99.9%.

4 Conclusions

A data based multiparametric analysis and optimization method is presented, which includes sampling design for computer experiments, and machine learning (OK) techniques. The method has been applied to benchmark examples and a case study. The results show that the method can approximate the multiparametric solutions using relatively small number of training data, with very high accuracy. More importantly, significant differences with the results of the standard MPP appear; (1) in all the tested cases, a single relation was enough to correctly reproduce the optimal solution multiparametric behavior over the whole UPs domain, instead of several mathematical relations each is applicable to a certain partition of the UPs space, (2) a systematic robust method is able to solve different problem natures (bilinear, quadratic, nonlinear black box) instead of many classical MPP algorithms each for different problem nature. The method achieves two goals: First, A simple and robust way for MPP applications, without going through complex mathematical formulations. Second, it can assist in situations where it is difficult to apply traditional MPP algorithms.

Acknowledgements Financial support received from the Spanish Ministry of Economy and Competitiveness and the European Regional Development Fund (both funding the research Project SIGERA, DPI2012-37154-C02-01), and from the Generalitat de Catalunya (2014-SGR-1092-CEPEiMA) is fully appreciated.

References

1. Dua, V., Bozinis, N.A., et al.: A multiparametric programming approach for mixed-integer quadratic engineering problems. Comput. Chem. Eng. **26**, 715–733 (2002)
2. Pistikopoulos E.N., Georgias M.C et al.: Multiparametric programming: Theory, Algorithim, and Application. Wiley, New York (2007)
3. Pistikopoulos, E.N., et al.: Theoretical and algorithmic advances in multi-parametric programming and control. Comput. Manage. Sci. **9**, 183–203 (2012)
4. Rivotti, P., et al.: Combined model approximation techniques and multiparametric programming for explicit nonlinear model predictive control. Comput. Chem. Eng. **42**, 277–287 (2012)
5. Shokry, A., Bojarski, A.D., Espuna, A.: Using surrogate models for process design and optimization. Uncertainty Modeling in Knowledge Engineering and Decision Making, 483–488 (2012)
6. Shokry, A., Espuna, A.: Applying Metamodels and Sequential Sampling for Constrained Optimization of Process Operations. LNCS. **8468**, 396–407 (2014)

Part X
Scheduling, Project Management and Health Services

Influence of Appointment Times on Interday Scheduling

Matthias Schacht, Lara Wiesche, Brigitte Werners
and Birgitta Weltermann

Abstract In primary care mainly two types of patient requests occur: walk-ins without an appointment and patients with a prescheduled appointment. The number and position of such prescheduled appointments influence waiting times for patients, capacity for treatment and the utilization of physicians. An integer linear model is developed, where the minimum number of appointments prescheduled for a weekly profile is determined. Since the number of patient requests differs significantly between seasons, weekdays and daytime, efficient appointment scheduling has to take different scenarios into account. Using an intensive monte-carlo simulation, we compare appointment strategies with respect to their performance for different scenarios.

1 Introduction

Primary Care Physicians (PCPs) face a complex task when planning clinic appointment schedules in order to take two types of patients into account: *urgent patients*, who have to see their PCP as soon as possible at the time of request and *non-urgent patients*, who seek a prescheduled appointment in order to have less waiting time in clinic. For these patients, capacity has to be blocked in terms of prescheduled appointment slots on certain days and times. The willingness to wait for non-urgent patients not only depends on medical reasons but also on their individual preferences. Visiting the consultation-hour or taking an appointment is decided by patients and strongly depends on the actual status of the queuing-system for appointments.

An optimal appointment schedule considers patients as well as PCP preferences [2]: *patients prefer* a schedule with a reasonable amount of capacity for urgent

M. Schacht (✉) · L. Wiesche · B. Werners
Chair for Management, esp. Operations Research and Accounting, Faculty
of Management and Economics, Ruhr University Bochum, Bochum, Germany
e-mail: or@rub.de

B. Weltermann
Essen University Hospital, Institute for General Medicine, Essen, Germany
e-mail: Birgitta.Weltermann@uk-essen.de

© Springer International Publishing Switzerland 2017
K.F. Dœrner et al. (eds.), *Operations Research Proceedings 2015*,
Operations Research Proceedings, DOI 10.1007/978-3-319-42902-1_79

requests who need same-day treatment. It should be avoided that urgent patients are not treated by the PCP on the day of request. Those patients are named urgent overflow patients. *PCPs seek* for a schedule in which clinic's utilization is high and balanced throughout the week with a minimum amount of overtime for clinic's staff. These preferences are influenced by the number of scheduled appointments offered and must be taken into account when generating the appointment schedule and deciding on the number and position of scheduled appointment slots throughout the week. Existing literature has broadly discussed optimal intraday appointment scheduling where optimal time slots on a given day are generated [3, 4]. Decisions as to the optimal amount of appointment slots per day on a tactical decision level have gained little attention.

As primary care clinics face variations in demand for treatments, this paper analyzes how varying demand affects the quality of an appointment schedule which has been determined on a tactical level with respect to the patient preferences. An integer linear model is developed, where the minimum number of appointments prescheduled for a weekly profile is determined. It is described how short-term adjustments of PCPs workload can be incorporated in order to increase patients' satisfaction by allowing an acceptable amount of overtime.

2 Reservation of Prescheduled Appointments in Primary Care

While working time of an idealized physician is constant for each working day, demand for treatments varies [1]. Figure 1a shows that the maximum time for treatments c_i on workday i is equal over weeks whereas demand per day (filled boxes) varies. Some patients have to be treated on the next day if patient demand exceeds the capacity. Such overflow patients occur on Monday and have to be shifted to Tuesday. Because of the overflow patients from Monday, there remains less capacity for Tuesday requests and some requests have to be shifted to Wednesday.

The idea of appointment scheduling is to offer as many scheduled appointments as needed in order to match capacity with demand on each workday. Figure 1b shows the consequences of scheduling some patients requesting on Monday to appointments on Wednesday. By this, we satisfy capacity restriction on Monday, increase

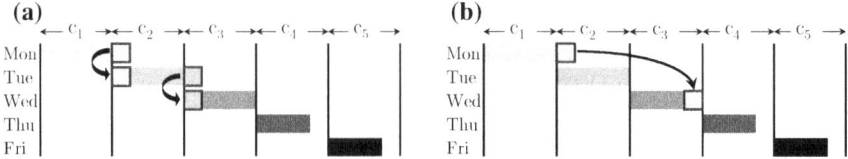

Fig. 1 Matching capacity with demand: **a** urgent overflow patients without prescheduled appointments and **b** solution after shifting patients

utilization on Wednesday and no overflow patients occur. Nevertheless, an appointment slot might be booked by a patient requesting an appointment on another day or there might be urgent patients who need same day care. Therefore, Sect. 3 presents an innovative model featuring these considerations and determining an optimal appointment schedule taking patient preferences into account.

3 Optimal Interday Appointment Scheduling

In this section, an integer linear model for **O**ptimal **R**eservation of **C**apacity for **A**ppointments (*ORCA*) is presented. The *ORCA* model supports the decision maker deploying a tactical plan for the appointment schedule as a weekly profile which is valid for several weeks or months as long as the demand remains comparable. Taking into account varying numbers of patient requests during the day, each working day $i \in I := \{1, \ldots, 5\}$ is divided into a morning and a post-lunch session denoted by $j \in J := \{1, 2\}$. The decision variable $x_{i,j} \in \mathbb{N}_0$ denotes the number of appointment slots planned for day i and session j. The number of patient requests in the morning who are treated in the following post-lunch session is denoted by variable $v_i \in \mathbb{N}_0$. Each day $i \in I$ and session $j \in J$ is associated with a corresponding number of patient requests $d_{i,j}$, with $d_{i,j} = d_{i,j}^r + d_{i,j}^u$. The number of same-day regular patients is denoted by $d_{i,j}^r$. The number of urgent patients who seek treatment on the same day is denoted by $d_{i,j}^u$ (urgent). Depending on their preferences and current queuing status, regular patients prefer an appointment or visit the consultation-hour. The total demand of regular patients per week is denoted by $D^r := \sum_{i \in I} \sum_{j \in J} d_{i,j}^r$. Parameter b denotes the treatment duration for one patient. Parameter $c_{i,j}$ describes the capacity on day i in session j. A varying number of patient requests during the week results in varying chances of a slot being booked by patients with requests from different days. This probability is defined as the *appointment request ratio* $p_{i,j} = d_{i,j}^r/D^r$. Then, $p_{i,1} \sum_{\ell \in I} \sum_{j \in J} x_{\ell,j}$ gives back the average number of patients with request on day i in the morning ($j = 1$) who will take a prescheduled appointment. The integer linear optimization model *ORCA* can be formulated as follows:

$$\min \quad \sum_{i \in I} \sum_{j \in J} x_{i,j} + \sum_{i \in I} v_i \tag{1}$$

$$\text{s.t.} \quad (d_{i,1}^r + d_{i,1}^u)\, b + x_{i,1}\, b - v_i\, b - p_{i,1} \sum_{\ell \in I} \sum_{j \in J} x_{\ell,j}\, b \leq c_{i,1} \quad \forall i \in I \tag{2}$$

$$(d_{i,2}^r + d_{i,2}^u)\, b + x_{i,2}\, b + v_i\, b - p_{i,2} \sum_{\ell \in I} \sum_{j \in J} x_{\ell,j}\, b \leq c_{i,2} \quad \forall i \in I \tag{3}$$

$$v_i \leq d_{i,1} \qquad\qquad\qquad\qquad\qquad \forall i \in I \tag{4}$$

$$x_{i,j} \in \mathbb{N}_0 \qquad\qquad\qquad\qquad\qquad \forall i \in I, \forall j \in J \tag{5}$$

$$v_i \in \mathbb{N}_0 \qquad\qquad\qquad\qquad\qquad \forall i \in I \tag{6}$$

The objective (1) is to minimize the number of appointment slots—to ensure reasonable capacity for urgent requests—and the number of patients who have to be shifted from morning to post-lunch session. We formulate the capacity constraints for a given morning session on day i in Eq. (2): The left hand side of the constraints denotes the duration of patients being treated in the morning session on day i. Total treatment time for patients requesting a morning session is expressed by $(d^r_{i,1} + d^u_{i,1})\, b$. To consider additional treatment duration for appointments in this session, $x_{i,1}\, b$ is added. Treatment time for patients who book an appointment is subtracted by the appointment request ratio as well as time for patients being shifted to the post-lunch session on the same day ($v_i\, b$). Treatment time of all patients in the slot has to be less or equal to the available time $c_{i,1}$. Capacity constraints for post-lunch sessions differ slightly to morning sessions (3). By this, we avoid systematic overtime. Note that in a steady state system with specified parameters no overtime occurs. Constraints (4) prevent shifting more requests from the morning to the post-lunch session than there are requests in the morning. Decision variables are integer and non-negative ((5) and (6)). We test our model with respect to sensitivity in an exemplary case study in Sect. 4 by incorporating an extensive simulation model.

4   Optimal Appointment Schedules and Sensitivity Analysis

The advantage of *ORCA* is a straightforward way of modeling an average week of a PCP, in which only little data for input are required. Results were tested in a discrete-event simulation by considering a single server PCP with a five (three) hour working time in the morning (post-lunch) session over one year. For all results, simulation runs were repeated twenty times and expected values were calculated. Treatment time for same-day patients and schedules as well as interarrival times and maximum willingness to wait for a scheduled appointment are modeled with uncertainty.

In the first analysis, no overtime is allowed (NO). All urgent patients who cannot be treated on the same day become urgent overflow patients. Figure 2 compares the average waiting time of the optimized appointment schedule with a half morning policy (HM) which is widely applied. In HM, each morning session consists of one half out of consultation-hour and the other half out of scheduled appointments. The varying average waiting times show that the *ORCA* model provides a substantially better appointment schedule with respect to the patient preferences. This goes

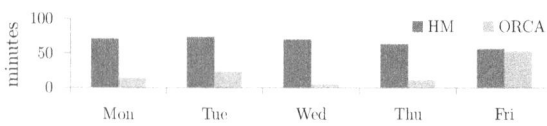

Fig. 2 Comparison of waiting times during consultation-hour optimized (*ORCA*) strategy and a regular half morning (HM) strategy

along with a significantly lower number of overflow patients (20% less overflow patients). Therefore, the optimized appointment profile allows a significantly higher number of treatments for urgent patients. Nevertheless, the decision maker cannot predict weekly or monthly variations in patient requests for treatments. Therefore, this section analyzes the sensitivity of the generated results using *ORCA* for a PCP clinic.

The optimal solution generated by *ORCA* with an average of 255 patients per week is generated and retained. The optimal solution is evaluated in different scenarios in which patient demand is increased or decreased by 10%. The influence of varying patient occurrence on the mean number of overflow patients per week and the mean waiting time during consultation hour is presented in Fig. 3. The gray bars represent the results of a simulation based on an optimal appointment schedule for an expected number of 255 patients. Both performance criteria are very sensitive to varying patient demand. Especially, an increase of the number of patients leads to a much higher number of overflow patients and waiting time. If the optimization model is adjusted to varying patient occurrences, the scenario-optimal appointment profile reaches slightly better solutions, which are marked by the black dots. But even the scenario-optimal solution with 280 patients per week cannot improve the results for patients substantially. This means that strategic planning of scheduled appointment slots is not capable of anticipating this high amount of patients if capacity is not increased. Therefore, the solution for the *ORCA* with 255 anticipated patients is presented given that PCP allows for a maximum overtime of 15 min for each session (O). This increases capacity and allows more same-day treatments. Figure 4 shows that allowing overtime by only 30 min per day (increase 6.25% of regular capacity) reduces the number of urgent overflow patients by 75%. Thus, a small capacity increase leads to a substantially better performance for patients. Summarized we can show that increasing capacity by allowing overtime on an operational level in busy seasons is a significantly better strategy than changing the appointment schedule.

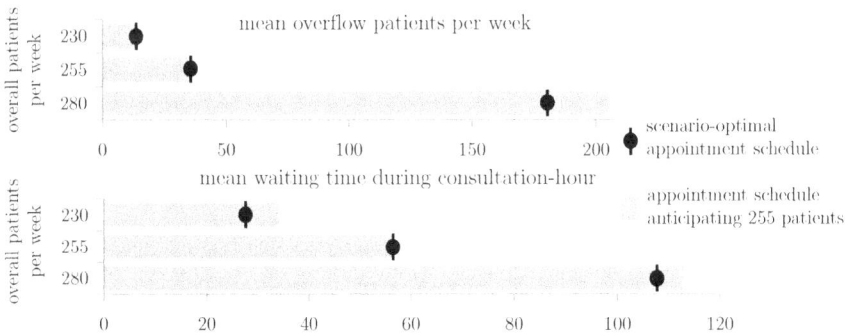

Fig. 3 Evaluation of determined appointment schedules with varying number of patient requests per week

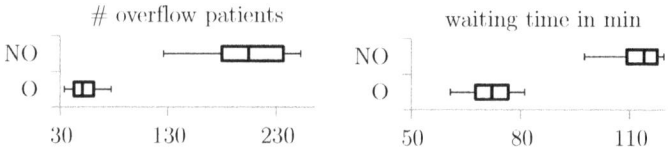

Fig. 4 Comparison of patient oriented criteria without overtime (NO) and with overtime (O)

5   Conclusion and Outlook

This contribution analyzed the influence of varying numbers of the patient requests on a determined appointment schedule. With the presented *ORCA* model, an optimal appointment schedule was generated, which is substantially better with respect to the patient preferences than basic appointment schedules. The solution has been tested with varying parameters for patient demand. Results from analyses show that criteria for patient satisfaction sensitively reacts on such variations. Higher patient demand leads to significantly longer waiting times and more overflow patients which cannot be compensated by an adjusted appointment schedule.

In addition to the objective of ORCA it is possible to also take into account a balanced utilization during a week. Weighting those two criteria allows the decision maker to adjust the model with respect to her personal preferences. In order to improve strategies for interday appointment scheduling, a robust appointment schedule which anticipates changing patient demands as well as reactions of the PCP could be deployed, e.g. in [5].

References

1. Cao, W., Wan, Y., Tu, H., Shang, F., Liu, D., Tan, Z., Sun, C., Ye, Q., Xu, Y.: A web-based appointment system to reduce waiting for outpatients: A retrospective study. BMC Health Serv. Res. **11**, 318 (2011)
2. Cayirli, T., Veral, E.: Outpatient scheduling in health care: a review of literature. Prod Oper Manag **12**(4), 519–549 (2003)
3. De Vuyst, S., Bruneel, H., Fiems, D.: Computationally efficient evaluation of appointment schedules in health care. Eur J Oper Res **237**(3), 1142–1154 (2014)
4. Gupta, D., Denton, B.: Appointment scheduling in health care: challenges and opportunities. IIE Trans **40**(9), 800–819 (2008)
5. Rachuba, S., Werners, B.: Robust approach for scheduling in hospitals using multiple objectives. J Oper Res Soc **65**(4), 546–556 (2014)

Time-Dependent Ambulance Deployment and Shift Scheduling of Crews

Lara Wiesche

Abstract For patients requesting emergency medical services (EMS) in a life-threatening emergency, the probability of survival is strongly related to the rapidness of assistance. An especially challenging task for planners is to allocate limited resources while managing increasing demand for services. The provision of sufficient staff resources for the ambulances has great impact on the initial treatment of patients and thus on the quality of emergency services. Data-driven empirically required ambulance location planning as well as the allocation of staff for these vehicles are successively optimized in the proposed approach to support emergency medical service decision makers. According to the identified problem structure, an integer linear programming model is proposed. An exemplary case study based on real-world data demonstrates how this approach can be used within the emergency medical service planning process.

1 Introduction

The efficient use of resources for EMS is crucial for an overall efficient health care system. Emergency medical vehicles operate in an environment where requests have to be served as soon as possible. Allocating limited resources is a complex task since allocation of ambulances to stations and decisions on dispatching ambulances are difficult optimization problems, particularly with regard to an expected increase in future ambulance demand. While the ambulance location and relocation have been extensively studied in the literature (see [1]), the particular ambulance crew scheduling has received only little attention so far. Since staff costs in EMS account for up to 90% of the total costs, optimal staff planning is inevitable for an efficient EMS. In the German EMS system, most ambulance services require at least two crew members for every ambulance with different levels of qualification. The emergency medical care is given by a parametric and an emergency medical technician, in

---

L. Wiesche (✉)
Chair for Management, Faculty of Management and Economics, esp. Operations
Research and Accounting, Ruhr University Bochum, Bochum, Germany
e-mail: lara.wiesche@rub.de

© Springer International Publishing Switzerland 2017
K.F. Dœrner et al. (eds.), *Operations Research Proceedings 2015*,
Operations Research Proceedings, DOI 10.1007/978-3-319-42902-1_80

life-threatening situations supported on scene by an emergency physician getting there in a separate emergency physicians vehicle. In this paper an optimization model for shift scheduling of ambulance staff is presented to support EMS decision makers. Based on the time-dependent optimization model for ambulance location and relocation by Degel et al. [2], special attention is given to the ambulance shift schedule. In particular the time-dependent empirically coverage requires a specific pattern of ambulances and therefore staff shifts and specific requirements will be taken into account. The remainder of this paper is structured as follows. Section 2 describes the decision making process for emergency medical services. In Sect. 3 a two-stage approach is introduced to include ambulance crew scheduling in the decision process. Section 4 presents a case study for evaluating the proposed approach. A conclusion and topics for further research are given in Sect. 5.

2 Decision Making for Emergency Medical Services

Due to the complexity of the entire emergency service process, capacity planning can be divided into the long-term strategic, the medium-term tactical and short-term operational planning. Following the classification of Hulshof et al. [5] Table 1 summarizes different decisions within the EMS. There is no clear separation between the hierarchical levels and the different EMS decision types. The typical process of scheduling staff to satisfy demand is to forecast demand and determine the number of staff needed. Afterwards, optimal staff shift schedules have to be identified and on the operational decision making level the roster of staff, meaning the allocation of individual staff members to shifts, has to be taken into account (see [4]). It is important to notice that these staff scheduling decisions are highly interdependent and cannot be taken into account separately. Empirical studies considering the EMS process provide important insights into dynamic changes during the day [7]. In particular,

Table 1 Decision making in the emergency medical service dependent on the time horizon

Strategic (Resource dimension)	Tactical (Resource allocation)	Operational (Resource assignment)
Location of EMS departments	Use of flexible locations	Dispatching of requests (ambulance, crew, hospital choice)
Division of city into districts	Time-subdivisions (time-periods)	Dynamic ambulance management
Number of ambulances	Allocation of ambulances to departments	Time-dependent relocation
Number of staff members	Staff shift scheduling and crew pairing	Staff to shift assignment task assignment

tactical planning has a great impact on identifying and implementing time-dependent adjustments and will be the main focus of this paper.

Degel et al. [2] introduced an optimization model for ambulance location and relocation on a tactical decision level called *Empirically Required Coverage Problem* (ERCP). With the objective of maximizing the empirically required coverage taking into account relocations and flexible locations, the presented model integrates the variation of demand for emergencies and ambulance travel time, dynamic adjustments and flexible ambulance locations. Existing coverage models in the literature commonly use a fixed (double) coverage level. In case of more than one (two) parallel operation(s), the fixed objective functions do not ensure sufficient available vehicles. In contrast in the ERCP an empirically determined coverage is used, taking parallel operations into account. Analyses regarding parallel operations show that an empirical coverage can be determined which takes temporal and spatial variations into account and guarantees an empirically required coverage. The concept of ERCP leads to an optimization model which simultaneously minimizes undersupply and prevents wasting resources. Since specific staff requirements are not considered in the ERCP, the influence of integrating these requirements in the ambulance planning is analyzed. Using the minimum number of ambulances and therefore the minimum number of crews required for each period generated from the ERCP as input, an approach to minimize the number of crews considering specific crew requirements is presented.

3 Time-Dependent Shift Scheduling of Ambulance Crews

Most of the work in the area of EMS services focuses on decisions on departments and ambulance locations. Whereas the task of shift scheduling has been studied extensively in literature (see [4]), ambulance shift scheduling received very little attention [3, 6]. In [3, 6] a two-stage integrated approach for ambulance scheduling is developed using the output from a coverage model as an input in a second integer programming model allocating ambulance crews to shifts. In contrast to the existing literature, time-dependent variations during the day are captured in this paper.

To estimate the optimal number and location of ambulances, the method proposed by Degel et al. [2] is used: they determine the required coverage of demand nodes empirically considering the number of emergency requests occurring simultaneously. To calculate the necessary coverage degree of each demand node i, the probability that an emergency call could be served is greater than β, or in other words that:

$$\text{Probability} \left\{ \begin{array}{c} \text{\# of ambulances covering} \\ \text{demand node } i \end{array} \geq \begin{array}{c} \text{\# of parallel emergencies in} \\ \text{the area around } i \end{array} \right\} \geq \beta$$

Thereby, the empirically required coverage ensuring the coverage of demand nodes with a sufficient number of ambulances is determined for example for $\beta = 95\%$ with a probability greater than 95 %. An optimal allocation of ambulances is obtained by

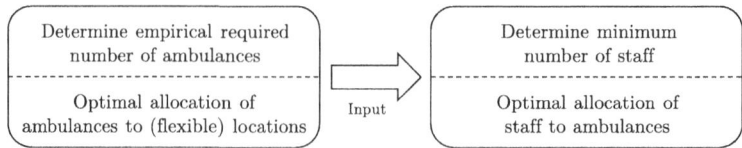

Fig. 1 Decision making process of time-dependent shift scheduling of crews

the ERCP model taking these probabilities into account. Specific staff requirements like the assignment of crews to stations, shift lengths, workload for crews etc. are not taken into account in this consideration. The question is how an optimal ambulance allocation solution performs when specific staff requirements are considered afterwards. Therefore an integer linear programming model is developed that minimizes the number of staff shifts during one week. Using the optimal allocation of ambulances from the ERCP as an input, the optimal allocation of staff to ambulances is modeled (see Fig. 1). According to the ERCP, each day $d \in D$ is divided in different time periods $t \in T$. The entire shift schedule consists of $\bar{S}$ different predefined shifts which are summarized in the set $S := \{1, \ldots, \bar{S}\}$. The decision variable $x_{d,s} \in \mathbb{N}_0$ denotes the number of crews working shift s on day d. The output of the ERCP is considered in the model as the parameter $A_{d,t}$ denoting the optimal number of ambulances on day d in time period t. $p_{t,s}$ denotes predefined shifts and captures different shift types and lengths. Note that one shift represents one crew and implies usually a parametric and an emergency medical technician.

$$\sum_{s \in S} x_{d,s} \, p_{t,s} \geq A_{d,t} \qquad \forall d \in D, \, t \in T \qquad (1)$$

Using the inputs from the ERCP, the model ensures that there are at least as many ambulance crews available as required by the ERCP (1). The objective of the crew-ambulance matching model is to minimize the total number of shifts during one week assuming a fixed allocation of each crew to a specific ambulance. By performing What-if analyses and predefined shift schedule variations, improved insights and thus an ideal ambulance staff schedule can be obtained.

4 Application and Results

To illustrate the application of the model, a real world case study from the city of Bochum (Germany) is analyzed. Bochum has more than 30,000 annual operations and an infrastructure of 22 potential ambulance locations serving 163 demand nodes (each $1 \times 1 \text{ km}^2$). In Bochum, a typical demand profile for EMS and therefore a typical distribution of the empirically required coverage with few requests during night periods and rush-hours between 8am–2pm is observable. Following the intensive

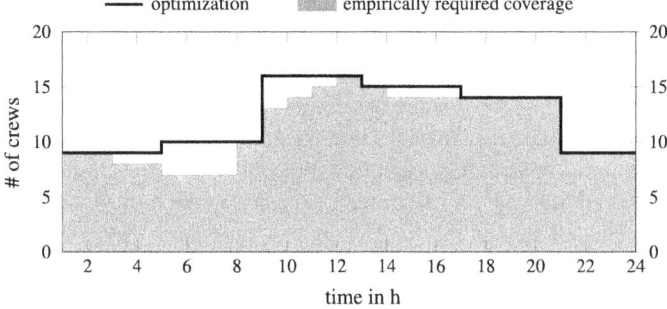

Fig. 2 Number of crews: empirically required and solution of the optimization model

analyses of [2], for Bochum an equidistant division of the total planning horizon of 24 h into $T = 6$ time periods is most practical. Shifts lengths of 8, 12 and 24 h are considered starting at the beginning of each time period taking minimum/maximum shift lengths, possible starting points and weekly working hours into account. Based on the empirically required number of ambulances from the ERCP, an optimal shift schedule is determined and depicted for one day in Fig. 2.

Figure 2 illustrates the number of crews determined by the optimization model (black solid line) in relation to the empirically required coverage (gray area). The figure clearly visualizes that at any time at least ambulance crews are available as required by the ERCP. Taking staff requirements into account more staff is disposable than ambulances needed in the ERCP as can be seen for example in hours 4 to 8, where crews are scheduled 4 h earlier than needed in order to adhere predefined shift lengths. Note that each crew is assigned to one specific ambulance. Relaxing this assumption and looking over the whole city the minimum number of shifts needed can be determined. Figure 3 compares the number of crews and shifts for

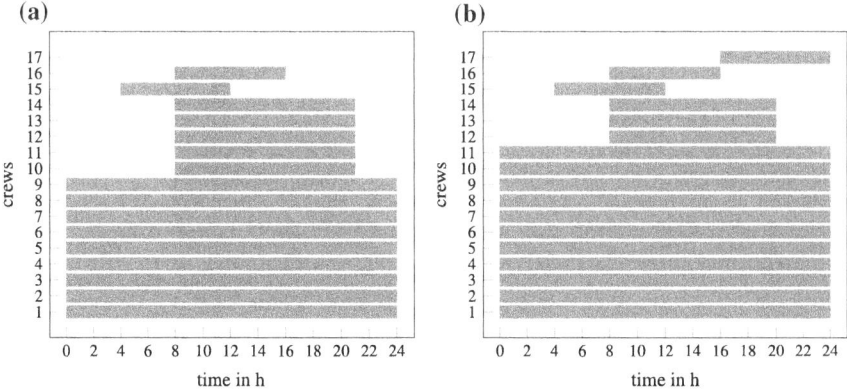

Fig. 3 Time-dependent shift scheduling: minimum number of shifts (**a**), optimal solution (**b**)

minimizing the number of shifts (a) and the crew-ambulance matching model (b). Taking the ambulance-crew matching into account not only leads to one additional crew but also to 32 more staff hours (10%). The first analysis shows that the solution of the ERCP model underestimates the number of required ambulances. Since the availability of ambulances is strongly influenced by specific shift requirements, an integrated planning approach is essential. Not only the fixed input of ambulances for the staff scheduling but an integrated solution approach to solve the ambulance deployment and staff scheduling for EMS decision makers is recommended.

5   Conclusion and Outlook

In this paper the influence of staff requirements on ambulance allocation models is analyzed. The considered approach focuses on the time-dependent amount of staff and their schedule based on an empirically required coverage. Results indicate that an ambulance planning neglecting staff requirements underestimates the number of necessary ambulances. The solution strongly suggests to combine ambulance and shift scheduling to capture interdependences between vehicle and staff planning. The evaluation of first results indicates the potential of this approach and justifies more intensive research in this area. Future research has to consider constraints like legal (rest days, break times), managerial (flexible locations) and staffing requirements. Particularly a fairly allocation of staff to shifts as well as a balanced workload for crews during the shifts offer potential for further research.

References

1. Brotcorne, L., Laporte, G., Semet, F.: Ambulance location and relocation models. Eur. J. Oper. Res. **147**(3), 451–463 (2003)
2. Degel, D., Wiesche, L., Rachuba, S., Werners, B.: Time-dependent ambulance allocation considering data-driven empirically required coverage. Health Care Manag. Sci. **18**(4), 444–458 (2015)
3. Erdogan, G., Erkut, E., Ingolfsson, A., Laporte, G.: Scheduling ambulance crews for maximum coverage. J. Oper. Res. Soc. **61**(4), 543–550 (2010)
4. Ernst, A.T., Jiang, H., Krishnamoorthy, M., Sier, D.: Staff scheduling and rostering: a review of applications, methods and models. Eur. J. Oper. Res. **153**(1), 3–27 (2004)
5. Hulshof, P., Kortbeek, N., Boucherie, R.J., Hans, E., Bakker, P.: Taxonomic classification of planning decisions in health care: a structured review of the state of the art in or/ms. Health Syst. **1**(2), 129–175 (2012)
6. Rajagopalan, H., Saydam, C., Setzler, H., Sharer, E.: Ambulance deployment and shift scheduling: an integrated approach. J. Serv. Sci. Manag. **4**(1), 66–78 (2011)
7. Wiesche, L.: Time-dependent dynamic location and relocation of ambulances. In: Huisman, D., Louwerse, I., Wagelmans, A.P.M. (eds.) Operations Research Proceedings 2013, pp. 481–486. Operations Research Proceedings (2014)

Personnel Planning with Multi-tasking and Structured Qualifications

Tobias Kreiter, Ulrich Pferschy and Joachim Schauer

Abstract We present a fairly involved ILP-model for a complex personnel planning problem arising from a real-world application. Two main aspects distinguish the problem from standard models: (i) Several tasks may be executed by the same person simultaneously. However, this multi-tasking is subject to certain complicated conditions. (ii) Qualification of personnel is complex and totally inhomogeneous. We introduce a representation for both issues that is at the same time fairly general and still easy enough to operate for the personnel manager.

1 Introduction

In a standard personnel assignment problem, every person can fulfill every given task, therefore the planning problem boils down to covering the required time periods by a feasible roster. If each person is assigned a certain skill level on an ordinal scale and each task requires a minimal skill level (a frequent scenario e.g. in hospitals), a feasible plan has to make sure that each person assigned to a task has at least the required skill level. Due to space restrictions we refer the reader to the surveys [1, 2, 4], instead of giving a detailed literature review.

In this contribution we consider a much more complicated setting: each person has a certain portfolio of unordered, categorical skills and can only fulfill tasks it is qualified for. Hence, each person has to be considered individually and the personnel structure is very inhomogeneous. The main innovation of our problem is the feature that a person might fulfill several tasks at the same time. This is a major deviation from classical personnel planning literature although complex skill levels were considered

T. Kreiter (✉) · U. Pferschy · J. Schauer
Department of Statistics and Operations Research, University of Graz,
Universitaetsstrasse 15, E3, 8010 Graz, Austria
e-mail: tobias.kreiter@uni-graz.at

U. Pferschy
e-mail: ulrich.pferschy@uni-graz.at

J. Schauer
e-mail: joachim.schauer@uni-graz.at

© Springer International Publishing Switzerland 2017   597
K.F. Dœrner et al. (eds.), *Operations Research Proceedings 2015*,
Operations Research Proceedings, DOI 10.1007/978-3-319-42902-1_81

e.g. in [3]. Moreover, the possibility of combining tasks is again highly non-uniform and depends both on the individual task and the considered person.

2 Representation of Task Combinations

The possibility of combining tasks, i.e. of carrying out more than one task at the same time by the same person, requires compatibility of tasks and ability of persons to perform the corresponding work by multi-tasking. Each of these two aspects has to be characterized and represented in a suitable way.

2.1 Representation of Task Compatibility

The simple exclusion of one person performing two tasks with overlapping time windows can be represented by a *conflict matrix* with $c_{jk} = 1$ indicating that tasks j and k require different persons. However, the practical planning problem this work originates from also contains a number of tasks which require little operative work. Instead, they mainly consist of monitoring activities. This means that a person qualified for such an activity has to be present during its duration (such an activity is modelled as a task). Since there are no (or little) actual operations involved the person assigned can very well carry out one or more additional tasks at the same time.

This leads to the question of possible task combinations a person can perform at the same time. Producing a full list of all feasible combinations would be impractical because of the exponential length of such a list and the necessity for the personnel planner to consider highly improbable combinations. Hence, we propose the following task classification scheme which is easy to use for the personnel planner and leaves a high degree of freedom from the modelling point of view.

class A: These tasks require the full concentration of the assigned person—hence they can not be combined with any other task.

class B: These tasks could be combined with at most one task of class C and arbitrary many tasks of class D, but with no task of classes A or B.

class C: These tasks can be added to at most one task of class B or C and arbitrary many tasks of class D.

class D: Arbitrary many tasks of this class (which contains mainly functions with no operative aspects) can be combined with tasks of classes B, C or D.

In the mathematical model we will use two conflict matrices for representing the possible simultaneous execution of any pair of tasks. As introduced above $c_{jk} = 1$ prevents tasks j and k being carried out by the same person. To capture also the feasibility of combining two tasks, but not three, we use a second conflict matrix S where $s_{jk} = 1$ indicates that there may be situations where tasks j and k can not be carried out by the same person. This applies e.g. for tasks of class C: two tasks of class C can be carried out by the same person, but only if no third task of classes B or C is assigned to the person in the same time window. An entry of $s_{jk} = 0$ indicates that tasks j and k can be combined independently from the remaining task assignments. In our application this will apply only for tasks for class D. With the help of this second conflict matrix S we can avoid using a three dimensional matrix. Its entries are generated according to the table at the end of this section. Their usage in a mathematical programming formulation will be described in Sect. 3.

2.2 Structured Qualifications of Personnel

Usually, in standard personnel planning problems a list of qualifications is given. Each task requires one particular qualification and each person provides one particular qualification. However, in our planning scenario most of the qualifications even have different levels: level 2 indicates a higher level of ability or more experience compared to level 1. Each task requires a certain qualification at either level 1 or level 2. In the former case any person with this qualification could perform the task while in the latter case only a person with level 2 can be assigned.

It is a common observation that the multi-tasking ability required for the combination of tasks is not the same for each person. Moreover, the ability to carry out several tasks at the same time (provided that the tasks fulfill the necessary conditions described in Sect. 2.1) does not only depend on the person but also on how familiar the person is with the tasks in question. Defining the multi-tasking ability for each staff member and each pair or triple of compatible tasks would clearly overwhelm any human resource manager in charge of such a massive task. Hence we introduce the following simple classification for each qualification of a person:

+: The person can combine a task with the respective qualification with other tasks according to the class scheme of tasks.

−: For the required qualification, the person can carry out at most one task of classes A and B, while two tasks of class C and arbitrary many tasks of class D could be performed in parallel.

Note that this $+/-$ scheme is not only person dependent but also qualification dependent. Thus, a person with long experience in activity a could easily combine a with a minor activity, e.g. of class C. However, the same person might have little experience or prior knowledge on another activity b of class B. Thus, the person has to devote all its concentration on b and cannot add another task of class C.

This $+/-$ classification of a person for a certain qualification may also differ depending on the associated level 1 or 2: a person could be able to multi-task while performing some activity a_1 with level 1, but the more demanding activity a_2 with level 2 does not allow this.

Clearly, this notion of representing qualification dependent multi-tasking abilities does not capture every possibility of task combinations. For example for a person in training one might also want to rule out the combination of two tasks of class C. However, one should keep in mind that the qualification profiles also have to be defined and kept updated by a human resource manager. In our application scenario the above solution seemed to be a reasonable compromise.

For the mathematical model we have to personalize the task dependent conflict matrix C by introducing c_{ijk} for each person i. The entries marked by x are 0 or 1 and represent the actual multi-tasking ability of a person. As an example, a person assigned a—for the qualification required for a task of class B, can not carry out any additional task of class C at the same time and thus x would be set to 1.

entries of c_{ijk} for person i				
task j/k	A	B	C	D
A	1	1	1	1
B	1	1	x	0
C	1	x	0	0
D	1	0	0	0

entries of s_{jk}				
task j/k	A	B	C	D
A	1	1	1	1
B	1	1	1	0
C	1	1	1	0
D	1	0	0	0

3 General Mathematical Model

In the following we will describe the core of a mixed integer linear programming formulation (MILP). This comprises only a fraction of the actual model used to solve the real-world problem. We concentrate in our presentation on the aspects of task combinations and omit the tedious time-planning part of the model.

We are given a list of m persons $i = 1, \ldots, m$ and n tasks $i = 1, \ldots, n$. The competencies of all personnel are listed in a matrix W where $w_{iq} = 1$ states that person i has qualification q. Each task j requires exactly one person of qualification q_j. The main binary decision variable with $x_{ij} = 1$ states that person i performs task j.

$$\sum_i x_{ij} = 1 \quad \ldots \forall j \tag{1}$$

$$x_{ij} \leq w_{iq_j} \quad \ldots \forall i, j \tag{2}$$

$$x_{ij} + x_{ik} \leq 2 - c_{ijk} \quad \ldots \forall i, j, k \neq j \tag{3}$$

$$x_{ij} + x_{ik} + x_{i\ell} \leq 5 - s_{jk} - s_{j\ell} - s_{k\ell} \quad \ldots \forall i, j, k \neq j, \ell \notin \{j, k\} \tag{4}$$

Constraints (3) enforce that at most one of two conflicting tasks can be performed by the same person i. Constraints (4) are implemented only for triples of tasks j, k, ℓ where all three right-hand side entries of s are 1 and enforce that at most two of the three tasks can be combined which rules out combinations of classes B, C, C and C, C, C.

Often tasks do not occur independently from each other, but belong to a certain project, customer or other logical bracket. For organizational efficiency it is preferred by the company that a minimum number of different persons are engaged with the tasks of a project. We assign to each task j the number of the project p_j it belongs to. A binary variable z_{ip} states in (5) if person i is involved in project p.

From the workers perspective, a major focus is put on days off, i.e. days without any task at all. This number is also subject to numerous legal and contractual constraints. Often employees prefer uninterrupted sequences of free days. Therefore, we require as input for each task j the day d_j it is to be executed (days are numbered from 1 to D) and keep a binary variable y_{id} stating through (6) whether person i performs any task on day d. If $y_{id} = 0$ then d is a free day for i. It is easy to add by (7) an upper bound d_{max} on the number of consecutive working days. To put emphasis on the concatenation of free days versus single free days, we introduce a binary bonus variable p_{id} which is set to 1 if days d and $d+1$ are both free for person i, see (8) and (9). Clearly, $p_{iD} = 0$.

$$z_{ip} \geq x_{ij} \quad \dots \forall i, j \text{ with } p_j = p \tag{5}$$

$$y_{id} \geq x_{ij} \quad \dots \forall i = 1 \dots n, j \text{ with } d_j = d \tag{6}$$

$$\sum_{d=s}^{s+d_{max}} y_{id} \leq d_{max} \quad \dots \forall i, s = 1 \dots D - d_{max} \tag{7}$$

$$p_{id} \leq 1 - y_{id} \quad \dots \forall i = 1 \dots n, d = 1 \dots D - 1 \tag{8}$$

$$p_{id} \leq 1 - y_{id+1} \quad \dots \forall i = 1 \dots n, d = 1 \dots D - 1 \tag{9}$$

In practical applications the details of working times and shift time restrictions tend to be complicated and company specific. A frequent time restriction which is based e.g. on the Austrian Arbeitsruhegesetz (ARG) Sect. 4, and is part of many collective agreements (Kollektivvertrag), requires for each person a weekly period of rest lasting for 36 h covering a full day (e.g. a weekend). The treatment of this condition can serve as a blueprint for other kinds of time restrictions. Therefore, we introduce a binary variable t_{id} stating whether day d is fully included in a rest period of at least 36 h for person i. We also use integer variables r_{id}^s and r_{id}^e to describe the starting time of the earliest task and finishing time of the last task performed by i on day d. For each task j let h_j be its length (in hours) and b_j its starting time. The variable wt_{id} denotes the actual working time of employee i on day d. The following constraints establish the connections between these variables.

$$r_{id}^s \leq b_j + 24(1 - x_{ij}) \quad \dots \forall i, j \text{ with } d_j = d \tag{10}$$

$$r_{id}^e \geq (b_j + h_j) x_{ij} \quad \dots \forall i, j \text{ with } d_j = d \tag{11}$$

$$r_{id}^e - r_{id}^s \leq wt_{id} \quad \dots \forall i, d \tag{12}$$

$$r_{id}^s \leq r_{id}^e \quad \dots \forall i, d \tag{13}$$

$$\sum_{7(w-1)<d\leq 7w} t_{id} + p_{id} \geq 1 \quad \dots \forall i, \forall \text{ weeks } w = 1, 2, \dots \tag{14}$$

$$36 + r_{i.d+1}^s - r_{i.d-1}^e \geq 24 \cdot t_{id} \quad \dots \forall i, d = 2 \dots D - 1 \tag{15}$$

$$t_{id} \leq (1 - y_{id}) \ \ldots \forall i, d \tag{16}$$

$$t_{i1} = t_{iD} = 0 \ \ldots \forall i \tag{17}$$

Note that the quality of a *good* working plan is not easily captured by an objective function focusing only on monetary aspects. In particular, worker satisfaction can not be easily quantified. Usually, one resigns to a linear combination of different parameters with appropriate weight factors. In our setting, we combine the personnel cost based on an hourly cost l_i for each person i, the number of different persons involved in each project and the bonus for consecutive free days (which should be maximized) as follows:

$$\min \ \alpha_1 \sum_i l_i w t_{id} + \alpha_2 \sum_p \sum_i z_{ip} - \alpha_3 \sum_d \sum_i p_{id} \tag{18}$$

4 Aspects of the Real-World Personnel Planning Problem

This work originates from the personnel planning task encountered at *Grazer Spielstätten* (GS), a major culture and event organization company in charge of several venues and responsible for providing technical personnel and support staff to all kinds of performances. The highly diversified technical qualifications and requirements were the origin for Sect. 2.2. Tasks that may be combined as discussed in Sect. 2.1 could be e.g. a sound engineer (task B) who also acts as the contact person for the artist (task C) and the fire safety engineer (task D). Special demands of individual artists or performing companies also play a role in the planning.

While the task and qualification settings of our application follows mostly the description of Sect. 2, the aspect of time windows is much more complicated than the model in Sect. 3. In the application the unit of time is 30 min, working shifts can be split (which is unpopular), legal and union regulations require different kinds of breaks, night's rest and free time. Overtime hours are subject to complicated accounting rules and—to make matters worse—personnel is divided into three groups with quite different legal situations. Special rules apply for tasks that end after midnight and thus intersect with two calender days. Employees prefer to have their shift-start as uniform as possible over consecutive days. Part of the project involved going through the 36-page Kollektivvertrag, the Austrian Arbeitszeitgesetz (AZG) and Arbeitsruhegesetz (ARG). Further discussions with the HR-manager led to the actual implementation of these sometimes fuzzy rules.

A massively extended version of the model from Sect. 3 was implemented in *Python* 3.3 using *PuLP*. The resulting ILP-model was solved to optimality by the non-commercial *COIN-OR CBC MILP Solver*. The test instances provided by GS could be solved to optimality for a three-week planning horizon with 25 persons and ≈ 100 tasks stemming from 15 events (=projects) within one hour of running time on a standard Laptop. Clearly, using a commercial MILP solver and more powerful hardware one should be able to solve also larger instances within a few minutes.

Acknowledgements We gratefully acknowledge the helpful cooperation with practitioners Michael Tassis, Kurt Schulz and Heike Herrgesell from GS. Ulrich Pferschy and Joachim Schauer were supported by the Austrian Science Fund (FWF): [P 23829-N13].

References

1. De Bruecker, P., Van den Bergh, J., Belin, J., Demeulemeester, E.: Workforce planning incorporating skills: state of the art. Eur. J. Oper. Res. **243**, 1–16 (2015)
2. De Causmaecker, P., Berghe, G.V.: A categorisation of nurse rostering problems. J. Sched. **14**, 3–16 (2011)
3. Golalikhani, M., Karwan, M.H.: A hierarchical procedure for multi-skilled sales forces patial planning. Comput. Oper. Res. **40**, 1467–1480 (2013)
4. Van den Bergh, J., Belin, J., De Bruecker, P., Demeulemeester, E., De Boeck, L.: Personnel scheduling: a literature review. Eur. J. Oper. Res. **226**, 367–385 (2013)

Algorithmic System Design Using Scaling and Affinity Laws

Lena C. Altherr, Thorsten Ederer, Christian Schänzle, Ulf Lorenz and Peter F. Pelz

Abstract Energy-efficient components do not automatically lead to energy-efficient systems. Technical Operations Research (TOR) shifts the focus from the single component to the system as a whole and finds its optimal topology and operating strategy simultaneously. In previous works, we provided a preselected construction kit of suitable components for the algorithm. This approach may give rise to a combinatorial explosion if the preselection cannot be cut down to a reasonable number by human intuition. To reduce the number of discrete decisions, we integrate laws derived from similarity theory into the optimization model. Since the physical characteristics of a production series are similar, it can be described by affinity and scaling laws. Making use of these laws, our construction kit can be modeled more efficiently: Instead of a preselection of components, it now encompasses whole model ranges. This allows us to significantly increase the number of possible set-ups in our model. In this paper, we present how to embed this new formulation into a mixed-integer program and assess the run time via benchmarks. We present our approach on the example of a ventilation system design problem.

L.C. Altherr (✉) · C. Schänzle · P.F. Pelz
Chair of Fluid Systems, TU Darmstadt, Darmstadt, Germany
e-mail: lena.altherr@fst.tu-darmstadt.de

C. Schänzle
e-mail: christian.schaenzle@fst.tu-darmstadt.de

P.F. Pelz
e-mail: peter.pelz@fst.tu-darmstadt.de

T. Ederer
Discrete Optimization, TU Darmstadt, Darmstadt, Germany
e-mail: ederer@mathematik.tu-darmstadt.de

U. Lorenz
Chair of Technology Management, Universität Siegen, Siegen, Germany
e-mail: ulf.lorenz@uni-siegen.de

© Springer International Publishing Switzerland 2017
K.F. Dœrner et al. (eds.), *Operations Research Proceedings 2015*,
Operations Research Proceedings, DOI 10.1007/978-3-319-42902-1_82

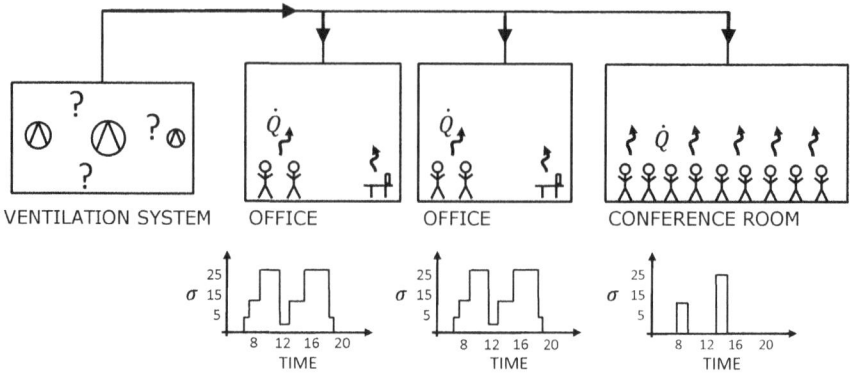

Fig. 1 We illustrate our approach by planning the energy optimal ventilation system for an office floor. Office staff, lights and computers produce heat. A time-dependent occupation density depicts how many people are working on average in each room at a given daytime

Table 1 Load scenarios considered in the optimization problem

Scenario	Volume flow in m³/s	Pressure in Pa	Time fraction (%)
1	6.88	200	15
2	5.16	175	30
3	3.44	150	55

1 Technical Application

The ventilation system design problem will be introduced very briefly. For more information please refer to [5]. We consider a ventilation system for a building with several offices and a conference room, cf. Fig. 1. The function of the ventilation system is to provide fresh air. Following guideline VDI 2078, each person corresponds to a heat source of $\dot{Q} = 120\,\text{W}$ and each technical device to a heat source of $\dot{Q} = 30\,\text{W}$. For a comfort cooling system, a temperature difference of $\Delta T = 2\,\text{K}$ between the supplied air and the room temperature is recommended [3]. For every room an occupation density is assumed that is used to derive three different load scenarios with specific pressure and volume flow demands, cf. Table 1.

2 Mathematical Model

Our objective is to design the ventilation system that is able to fulfill the given load scenarios in the most energy efficient way. We compare all possible systems, i.e. combinations of fans that fulfill the load and minimize the expected energy costs.

According to [1] we model all possible systems by the complete graph $G = (V, E)$ with edges E corresponding to possible components, and vertices V representing connection points inbetween them. The set of edges consists of the set of fans F and the set of interconnections C, i.e. pipes. We introduce a binary variable $b_{i,j}$ for each optional component $(i, j) \in E$ which indicates whether a component is bought ($b_{i,j} = 1$) or not ($b_{i,j} = 0$). For each loading scenario sc in the set of loading scenarios Sc we copy the graph G once. Binary variables $a_{i,j,sc}$ for each edge (i, j) of the graph G_{sc} corresponding to scenario sc indicate whether a component is operating in scenario sc ($a_{i,j,sc} = 1$) or is switched off ($a_{i,j,sc} = 0$).

If we assume incompressibility of the air conveyed by the fans, the conservation of the volume flow $\dot{V}_{i,v,sc}$ is given by

$$\forall sc \in Sc, \, v \in V: \qquad \sum_{(i,v) \in E_{sc}} \dot{V}_{i,v,sc} = \sum_{(v,j) \in E_{sc}} \dot{V}_{v,j,sc} . \qquad (1)$$

An additional condition with an adequate upper limit $\dot{V}_{max}$ makes sure that only active components convey air:

$$\forall \, sc \in Sc, \, (i, j) \in E: \qquad \dot{V}_{i,j,sc} \leq \dot{V}_{max} \cdot a_{i,j,sc} . \qquad (2)$$

Each fan increases the pressure according to

$$\forall \, sc \in Sc, \, (i, j) \in E: \qquad p_{j,sc} \leq p_{i,sc} + \Delta p_{i,j,sc} + M \cdot a_{i,j,sc} \qquad (3)$$
$$p_{j,sc} \geq p_{i,sc} + \Delta p_{i,j,sc} - M \cdot a_{i,j,sc} \qquad (4)$$

if it is active ($a_{i,j,sc} = 1$). Otherwise, the pressure at fan inlet $p_{i,sc}$ and the pressure at fan outlet $p_{j,sc}$ are uncoupled by means of a big-M formulation.

The resulting increase of pressure Δp depends on the fan's rotational speed n and on the volume flow $\dot{V}$ the fan conveys. This dependency is described by characteristic head curves for each fan, cf. Fig. 2. For each rotational speed, a different characteristic curve for volume flow and pressure increase is given. The fan's efficiency η and thus its power consumption P also depends on n and $\dot{V}$. Our objective is to minimize the total power consumption

$$P_{\text{system}} = \sum_{sc \in Sc} \sum_{(i,j) \in F_{sc}} P_{i,j,sc} \qquad (5)$$

of the fans F. The next section describes in detail how to integrate the fan characteristic curves into the mathematical model.

3 Modeling the Construction Kit of Fans

In previous works, we included characteristic diagrams like those shown in Fig. 2 for each single component into our optimization model by piecewise linearization [6]. In this case, for each optional fan with diameter d one has to provide data points on these characteristic diagrams for different values of the rotational speed n and the volume flow $\dot{V}$. A fan slot can be occupied by one of $|D|$ different fans, if D is the set of possible discrete diameters d. This results in $2 \cdot |D|$ two-dimensional piecewise linearizations: For each diameter one has to linearize the pressure increase Δp and the efficiency η over n and $\dot{V}$.

Making use of laws derived from similarity theory, only data points on two dimensionless characteristic curves have to be provided for each fan. We are able to describe a whole model range by only three bivariate piecewise linearizations. A model range consists of geometrically similar fans, that are constructed in the same way, but have different diameters d, cf. Fig. 3.

The characteristic diagrams for all fans of a model range can be described by just two univariate dimensionless curves: The coefficient of pressure ψ and the efficiency η as functions of the flow coefficient φ, cf. Fig. 3. By using affinity [2] and scaling laws [4], the specific head curve and the power consumption of a fan with diameter d and rotational speed n can be derived:

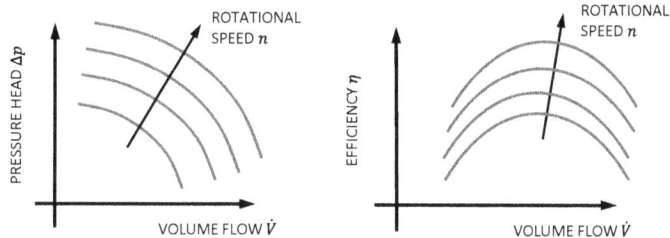

Fig. 2 Characteristic curves for a fan with specific diameter d and variable speed n

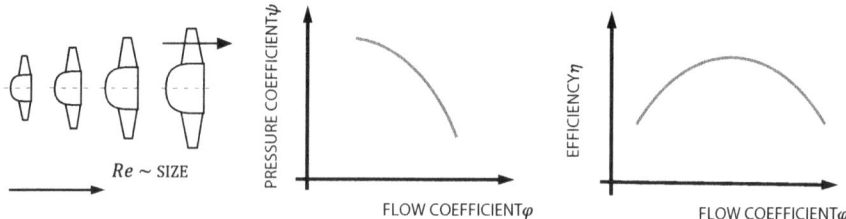

Fig. 3 Dimensionless curves representing a whole model range, i.e. geometrically similar fans with different diameters d

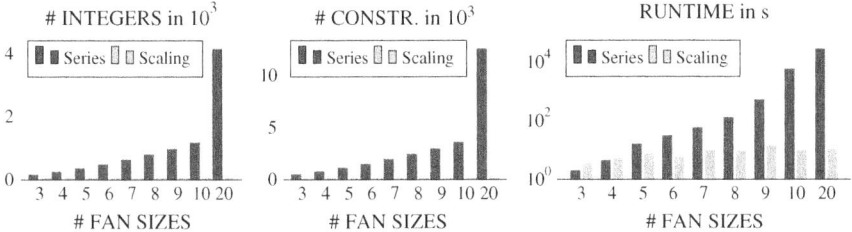

Fig. 4 Benchmarks comparing the model series formulation to the new similarity-based approach. *One* load scenario and *three* available fan slots are fixed. Number of available diameters varies

$$\dot{V} = \frac{\pi^2}{4}\,\varphi\,n\,d^3, \qquad \Delta p = \frac{\pi^2}{2}\psi\rho n^2 d^2, \qquad P = \frac{\pi^4}{8}\,\frac{\varphi\psi}{\eta}\,\rho\,n^3 d^5, \qquad (6)$$

where ρ is the assumed constant density of air. Nine data points on each of the two dimensionless curves were chosen and Eq. 6 were integrated into the model by piecewise linearization using a logarithmic formulation [6].

4   Benchmarks of the Two Modeling Approaches

We compare the new modeling approach based on similarity theory to the one with piecewise linearized characteristic curves for each single fan with benchmarks using cplex 12.6, an Intel Core i7 with 3.8 GHz and one thread. The size of our models depends on three parameters: the number of fan slots (i.e. places where an optional fan could possibly be built in), the number of fan sizes (i.e. the number of discrete diameters), and the number of load scenarios.

By modeling a whole model range, the number of integer variables and constraints decreases significantly, compared to our traditional approach. If we consider one load scenario and three possible fan slots and vary the number of possible fan sizes, the scaling approach provides a speed up in run time if five and more sizes are available in the construction kit, cf. Fig. 4. Doubling the number of load scenarios increases the run time significantly, both for the scaling approach as well as for the approach with single components, cf. Fig. 5. The speed up in run time for the scaling approach now holds for six fan sizes and above.

If we consider one load scenario and five fan sizes and vary the number of fan slots, our new approach is consistently faster, cf. Fig. 6. If we fix the number of available fan sizes and available fan slots and vary the number of scenarios, the scaling approach is not always the best choice, cf. Fig. 7, as the run time is strongly dependent on the number of available fan sizes.

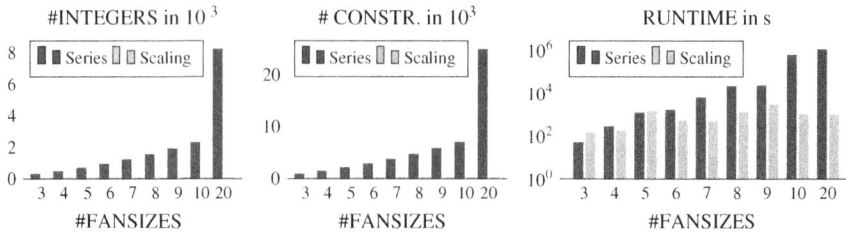

Fig. 5 Benchmarks comparing the model series formulation to the new similarity-based approach. *Two* load scenarios and *three* available fan slots are fixed. Number of available diameters varies

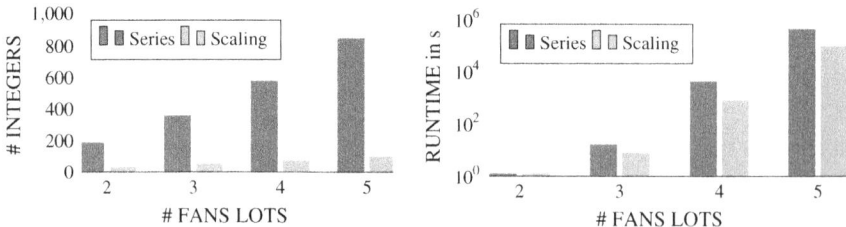

Fig. 6 Benchmarks comparing the model series formulation to the new similarity-based approach. *One* load scenarios and *five* available diameters are fixed. Number of available slots varies

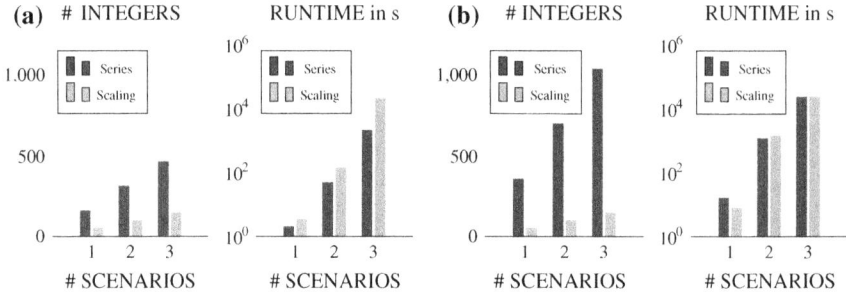

Fig. 7 Benchmarks comparing the model series formulation to the new similarity-based approach. *Three/Five* available diameters and *three* available slots are fixed. Number of load scenarios varies

5   Conclusion and Outlook

In this work, we integrated an efficient formulation based on similarity theory into a MILP for the optimal design of a technical system. By this approach, it is possible to model whole model ranges of technical components in an efficient way and to reduce the number of integer variables and linear constraints significantly. For the example of a ventilation system, benchmarks showed a speed up in run time compared to the traditional approach if one considers many different fan sizes in the construction kit. For few available fan sizes and if we consider more load scenarios, the new approach is however inferior to the traditional one. Our formulation based on similarity theory can be applied to other technical systems, e.g. the energy efficient pump system we planned and validated in [1].

Acknowledgements The authors would like to the German Research Foundation DFG for funding this research within the Collaborative Research Center 805 "Control of Uncertainties in Load-Carrying Structures in Mechanical Engineering" and in the framework of the Excellence Initiative, Darmstadt Graduate School of Excellence Energy Science and Engineering (GSC 1070).

References

1. Altherr, L.C., Ederer, T., Pöttgen, P., Lorenz, U., Pelz, P.F.: Experimental validation of an enhanced system synthesis approach. In: OR2014, Aachen, Germany (2014)
2. Carolus, T.: Ventilatoren: Aerodynamischer Entwurf, Schallvorhersage, Konstruktion. Springer, Heidelberg (2013)
3. Ihle, C., Bader, R., Golla, M.: Tabellenbuch Sanitär, Heizung, Lüftung, 2. Auflage. Verlag Dr. Max Gehlen (1998)
4. Pelz, P.F., Stonjek, S.S.: The influence of reynolds number and roughness on the efficiency of axial and centrifugal fans—a physically based scaling method. J. Eng. Gas Turbines power **135**(5), 052601 (2013)
5. Schänzle, C., Altherr, L.C., Ederer, T., Lorenz, U., Pelz, P.F.: As good as it can be - ventilation system design by a combined scaling and discrete optimization method. In: FAN 2015 - Proceedings of the International Conference on Fan Noise, Fan Technology and Numerical Methods (2015)
6. Vielma, J.P., Ahmed, S., Nemhauser, G.: Mixed-integer models for nonseparable piecewise-linear optimization: unifying framework and extensions. Oper. Res. **58**(2), 303–315 (2010)

A New Hierarchical Approach for Optimized Train Path Assignment with Traffic Days

Daniel Pöhle and Matthias Feil

1 Introduction

In today's German timetabling process, train paths are only planned when operators apply for specific train services. The train path application includes specific train characteristics such as train length, braking power, etc. Each train path is planned manually, which is time-consuming and results in an inefficient use of the infrastructure capacity. German Railways Infrastructure division DB Netz has started to introduce a more efficient timetabling process for rail freight timetabling (see Feil and Pöhle [2]). This process contains two main stages: a pre-planning of standardized train paths (called slots) and the assignment of train path applications to pre-planned slots. For train path assignment, an optimization model has been introduced by Nachtigall and Opitz [4]. Their approach has been tested for long-term timetable scenarios, shows promising results and is also able to yield benefit for detection of bottlenecks in the infrastructure (see Pöhle and Feil [5]). Long-term timetable scenarios have a model scope of one traffic day. However, train path applications for the year-to-year network timetable have diverse traffic days (e.g. Mon–Fri). Hence, an extended train path assignment model with traffic days is needed. In this work, a heuristic approach for solving the train path assignment problem with traffic days is presented. This paper is organized as follows: The Sect. 2 compares customer needs for a train path assignment with the infrastructure managers point of view and analyzes the existing patterns in traffic days of train path applications. Section 3 presents the developed approach based on traffic day patterns and Sect. 4 discusses the

D. Pöhle (✉) · M. Feil
DB Netz AG, Theodor-Heuss-Allee 7, 60489 Frankfurt, Germany
e-mail: daniel.poehle@deutschebahn.com

M. Feil
e-mail: matthias.feil@deutschebahn.com

© Springer International Publishing Switzerland 2017
K.F. Dœrner et al. (eds.), *Operations Research Proceedings 2015*,
Operations Research Proceedings, DOI 10.1007/978-3-319-42902-1_83

computational results for an extended long-term timetable with traffic days. Additionally, the results are compared to the recently developed optimization model by Nachtigall [3] addressing the same problem.

2   Traffic Days Analysis and Customer Needs

Railway freight companies specify in their train path applications on which days the train will be operated. On the one hand, each train path application has a traffic day which determines the days of the week the train is running, e.g. Monday till Friday. Additionally, there are restrictions, for instance before and after holidays. Butzbach [1] gives a detailed overview about the German traffic day system. On the other hand, each train path application has an operation period which defines the time span within the timetable year, e.g. April 1st till September 30th. Traffic days and operation periods are a characteristic for the slots, too. Due to a different plan of operation for passenger trains on different days of the week or construction works, certain slots can only be planned on some days. Hence, there is an additional dimension of complexity for the train path application with traffic days.

When a railway company plans a freight train with multiple traffic days, it is desired that this train will get the same train path for each operation day. This is due to needs for a coherent communication of departure and arrival times, simpler rolling stock scheduling and staff planning. On the other hand, the infrastructure manager wants to optimize its network utilization and maximize its sales. Only allowing train path assignments which consist of the same train paths for each traffic day can induce a reject of train path applications, especially when the traffic day patterns are very heterogeneous. This is due to fragmented train paths which cannot be used for an identical train path assignment for all traffic days. Hence, there is a tradeoff between identical train paths on all traffic days and a high number of fulfilled train path applications. If it is not possible to get the same train path every day, the number of different train paths for an application shall be minimized. For example, if a daily train application cannot be fulfilled by the same train path for every day, two different train paths (e.g. one for Monday till Friday and another for the weekend) are much better than a different train path for every single day of the week. To detect if some patterns occur more often than others an analysis of real train path applications of the year 2013 has been worked out. The main results are:

- There are 13 groups of traffic day patterns which cover around 91 % of the total demand
- The most frequent patterns are Monday till Friday, daily, Saturday and Sunday
- The remaining 9 % of the demand is within patterns covering less than 1 % of the total demand each and cannot be combined appropriately to a bigger cluster

Figure 1 shows the distribution of coverage for the 13 biggest groups of traffic day patterns. If a group is flagged with an asterisk it is a combined pattern. For example, the combined pattern Mon + Wed + Fri* includes the additional patterns of Mon

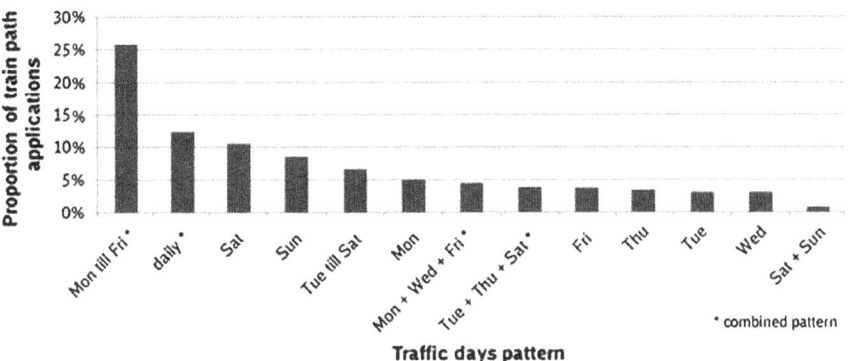

Fig. 1 Distribution of traffic days

+ Wed, Mon + Fri and Wed + Fri. If a train path application has a traffic day of Mon + Fri it will be treated as if it would also run on Wednesday. The operation periods of train path applications are much more scattered and there are hardly any frequent patterns. The biggest group with 39 % of all train path applications has the operation period of the whole timetable year. The second biggest group has only a proportion of 2 % and contains the thirteen last weeks of the timetable year. Due to the absence of useful clusters and to limit complexity for the first approach of train path assignment with traffic days, the heuristic algorithm and the optimization model is focused on traffic days for a week only. In the next section the heuristic approach will be explained.

3 Hierarchical Approach for Train Path Assignment

The linear optimization model for train path assignment for one single traffic day was introduced by Nachtigall and Opitz [4] and yielded good solutions for several long-term scenarios. The objective function maximizes the total quality of all assigned train paths. For each train path application (or request) r a route from origin to destination, called itinerary p, is selected from a set of possible itineraries.

$$\sum_{p,r} (\rho \cdot \tau(p_r^*) - \tau(p)) \cdot x_{p,r} \rightarrow max$$

$$\forall C_s \in \mathscr{C} : \sum_{r,p \in C_s} x_{p,r} \leq 1 \tag{1}$$

$$\forall H \in \mathscr{H} : \sum_{r,p \in P(H)} x_{p,r} \leq c_H \tag{2}$$

$$x_{p,r} \in \{0, 1\}$$

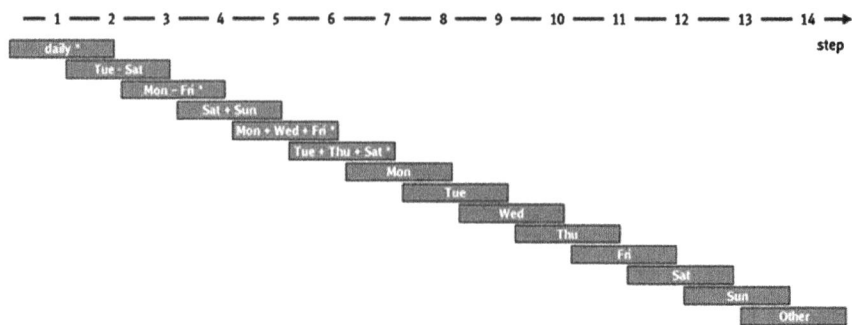

Fig. 2 Optimization sequence of the heuristic approach

The binary decision variable $x_{p,r}$ indicates whether request r uses itinerary p. ρ is a detour factor and $\tau(p)$ is the actual travel time for itinerary p. Constraint (1) ensures that every slot can only be used by one request (C_s is the set of all itineraries using slot s). Constraint (2) limits the node capacity, so that at most c_H trains can dwell in a node simultaneously for each time interval H ($P(H)$ is the set of all itineraries within H). If slot or node capacity is not sufficient, requests can be rejected.

Core idea of the heuristic approach for train path assignment with traffic days is to split the optimization into a 14 step sequence, where each step contains only requests of one single traffic day pattern (see Fig. 2). In the first step all requests with the traffic day pattern daily* are optimized simultaneously and all rejected requests are separated and put into subsequent traffic day patterns. The train path assignments of all fulfilled requests are now fixed and in the second step all requests with traffic day pattern Tuesday–Saturday are now optimized with the remaining capacity. For each traffic day pattern a single rule for separation is defined, if a request cannot be fulfilled identically for all days of the traffic days. For example, the traffic day pattern daily* will be divided into the pattern Monday–Friday and Saturday + Sunday whatever actual day(s) cause the need for separation. This sequential process of descending traffic days and the separation rules produce identical train path assignments for many request and separates the requests only for the case of infeasibility. The next section describes the computational results of this heuristic approach and compares them with the direct train path assignment with traffic days by Nachtigall [3].

4   Computational Results

A long-term timetable test scenario for Germany including the most important freight train lines is used to evaluate both the heuristic and the optimization approach for one master week. In total, 2727 requests with different traffic days are used. On average, each request has 3.5 traffic days. To ensure a sufficient optimization potential for both approaches, the scenario has a network overload compared with the predicted demand. Hence, no inference to the actual network capacity of the network can be drawn from this scenario.

Fulfillment and need for separation of requests Using the heuristic approach, approximately 70 % of all requests are fulfilled with an identical train path for all traffic days. Less than 1 % is also fully fulfilled but needs a separation for the traffic days. 289 requests cannot get a train path for all their traffic days and they are only partly fulfilled. Approximately 20 % of all requests have to be rejected due to a lack of capacity (see Fig. 3). In comparison, the optimization approach can fulfill 93 % of all requests identically for all traffic days. This is an increase of about 34 %. The fulfillment is even higher than the sum of fully and partly fulfilled requests of the heuristic. The optimization approach is up to now not able to separate requests if they cannot be fulfilled with an identical train path for every day. Hence, the total number of rejected requests is assumed to decrease if this feature will be added.

Travel time The average travel time of all assigned train paths, quantified as BFQ (actual travel time divided by minimal travel time), for the heuristic approach is 1.53. In contrast, the optimized assignments have an 11 % lower average BFQ of 1.36 (see dashed lines in Fig. 3). Additionally, the travel times' variance using the optimization is lower, which means that the infrastructure manager offers more homogeneous train paths.

Rejected train path applications Figure 4 shows how many requests have been rejected dependent on the number of traffic days for the heuristic approach and for the time interval Monday from 12 am (noon) till 6 pm the tracks with high capacity utilization. The proportion for the rejected requests for the optimization is very similar and is for this reason not presented. Most rejected train path applications have either one or five traffic days and the biggest rejection group is traffic day pattern Monday–Friday. This can be explained by the network overload, because most train path requests have this traffic day pattern (see Fig. 1). Second most rejections are on Saturdays, third most on Mondays. Half of the rejections on Saturday are between 12 am (noon) and 6 pm. On Monday, around 40 % of the rejects are in the same time interval and 24 % are between 6 am and 12 am. Consequently, during the identified time intervals there is the greatest lack of capacity in the network. It is also possible to locate the lack of capacity in the network (see Fig. 4 for Monday).

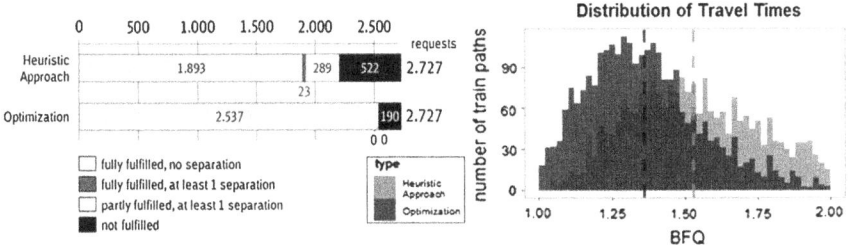

Fig. 3 Comparison of the optimization results

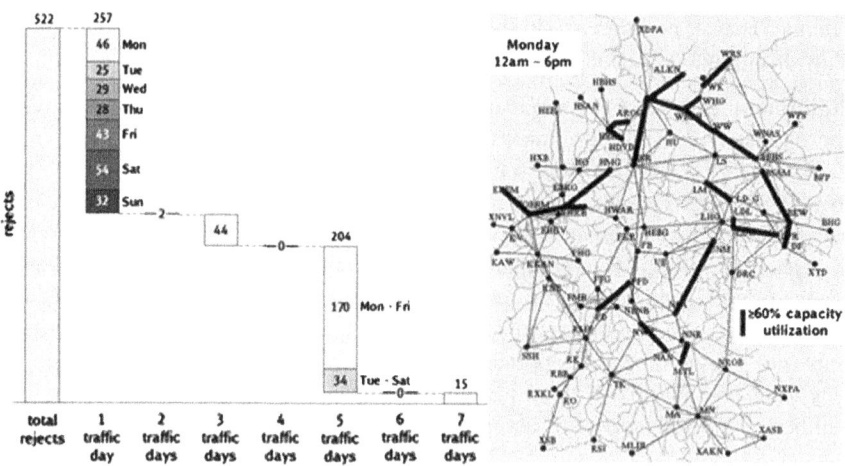

Fig. 4 Heuristic approach: analysis of rejects and highly utilized lines in the network

5 Conclusion

German Railways Infrastructure division DB Netz has started to introduce a more efficient timetabling process for freight trains. For train path assignment in the network timetable there is the need to extend the existing optimization algorithm by Nachtigall and Opitz. The first heuristic approach splits the requests into 14 groups of traffic day patterns and uses the existing optimization model in a hierarchical sequence. Rejected requests are separated and put into subsequent groups. Using his approach gives plausible results and achieves for about 70 % of the demand an identical train path for all traffic days in a long-term timetable scenario. Around 11 % of all requests have to get different train paths during their desired traffic days. However, compared to the second optimization approach of Nachtigall [3] the results appear less favorable. The optimization approach can significantly fulfill more requests, provides a lower average travel time and does not need to split the requests into smaller traffic day patterns than desired by the customers. In future work the presented approaches will be extended from one week to the whole timetable year. This will be a challenge due to the extraordinary growth of the planning horizon. It is, however, aspired to use the train path assignment model for the network timetable and in daily business within the next three to four years.

References

1. Butzbach, V.: Der Verkehrstagesschlüssel im elektronischen Zeitalter. Deine Bahn **2**, 44–47 (2012)
2. Feil, M., Pöhle, D.: Why does a railway infrastructure company need an optimized train path assignment for industrialized timetabling? In: Proceedings of International Conference on Operations Research, Aachen (2014), to appear
3. Nachtigall, K.: Modelling and solving a train path assignment model with traffic day restriction. In: Proceedings of International Conference on Operations Research, Vienna (2015)
4. Nachtigall, K., Opitz, J.: Modelling and solving a train path assignment model. In: Proceedings of International Conference on Operations Research, Aachen (2014), to appear
5. Pöhle, D., Feil, M.: What are new insights using optimized train path assignment for the development of railway infrastructure? In: Proceedings of International Conference on Operations Research, Aachen (2014), to appear

Modelling and Solving a Train Path Assignment Model with Traffic Day Restriction

Karl Nachtigall

Abstract The German Railway Company (DB Netz) schedules freight trains by connecting pre-constructed slots to a full train path. We consider this problem with special attention to traffic day restrictions and model it by a binary linear decision model. For each train request a train path has to be constructed from a set of pre-defined path parts within a time-space network. Those train requests should be realized only at certain days of the week. Each customer request has a specific traffic day pattern, which is a difficult challenge for the allocation process. Infrastructure capacity managers intend to achieve an efficient utilization of the capacity, whereas customers are interested in homogeneous train paths, i.e. they want the same traffic path connection for all requested traffic days. We discuss those partly contradictory requirements within the context of our binary linear decision model. The problem is solved by using column generation within a branch and price approach. We give some modeling and implementation details and present computational results from real world instances.

1 Introduction and Motivation

Freight train planning often suffers from the fact that passenger trains are scheduled first and the freight trains may only use the remaining capacity. As a result, those schedules are often of bad quality. The German Railway Company (DB Netz) changes the planning process as follows:

1. Passenger Trains and freight train slots between construction nodes are scheduled simultaneously.
2. Train assignment for freight train demand by connecting the slots to a full train path. Ad hoc requests will be solved by a greedy approach. Long term requests shall be handled by optimization.

K. Nachtigall (✉)
Chair of Traffic Flow Science, Faculty of Transportation and Traffic Sciences,
Dresden University of Technology, 01062 Dresden, Germany
e-mail: karl.nachtigall@tu-dresden.de

© Springer International Publishing Switzerland 2017
K.F. Dœrner et al. (eds.), *Operations Research Proceedings 2015*,
Operations Research Proceedings, DOI 10.1007/978-3-319-42902-1_84

Long term requests contain complicated pattern of the requested traffic days distributed over the year. The annual planning can be carried out on sample weeks as a rolling scheduling process. In this paper we focus on the problem to find a good plan for one master week, i.e. train slots and freight train demand are restricted or required for a certain subset $D \subseteq \{Mon, Tue, Wed, Thu, Fri, Sat, Sun\}$. This paper substantially extends previous results (see [2, 4, 5]) for the train path assignment problem. A request is said to be homogeneous satisfiable, if for all requested traffic days the same traffic path connection will be assigned.

The treatment of not homogeneous satisfiable demands can be done differently:

(a) Homogeneous satisfiable requests have top priority. Not homogeneous satisfiable requests are split and, if possible, satisfied heuristically.
(b) Homogeneous and non-homogeneous satisfiable demands are treated simultaneously in the optimization model.

In this paper we present a general mathematical model for the case (b), which generalizes the special case (a). A more detailed discussion of assigning different train paths for different traffic days is given by Feil and Pöhle (see [6]).

2   Basic Model

A train request r is defined by

- An origin node O_r with preferred departure time d_r
- A destination node D_r with preferred arrival time a_r
- A set of intermediates stopping nodes X_i with preferred arrival and departure times a_i and d_i, which are indexed by $i = 1, \ldots, r_n$.
- a set of traffic days $D_r \subseteq W := \{Mon, Tue, Wed, Thu, Fri, Sat, Sun\}$, for which a train path is required. The assignment should be as homogeneous as possible, i.e. for each operating day the train path should preferably be the same.

In order to fulfill this demand a system of freight train slots is used. Each slot is some kind of placeholder, which might be used to run a freight train. The slots are constructed between pre-defined construction nodes and planned simultaneously with the passenger trains. The freight train slots are only operating at special days,i.e. for each slot s_i we are given a set $D(s_i) \subseteq W$ of traffic days for which this slot is available. A train path for a request r can be modeled by a path within a time-space network, the so-called slot network. To keep the model general, we will allow inhomogeneous solutions. This means, that the assignments at different days of operation is achieved by different train paths or cannot be fulfilled for all requested days. With each of those potential train paths p we associate a binary decision variable $x_{p,r,D}$, to indicate whether train request r uses path p:

$$x_{p,r,D} = \begin{cases} 1, & \text{if } r \text{ takes path } p \text{ for all days } d \in D \subseteq W \\ 0, & \text{otherwise} \end{cases}$$

By P we denote the set of all paths and P_r is the set of all paths for request r.

Some of the potential train paths cannot be carried out simultaneously, because both paths are using one common slot or a pair of conflicting slots. For each slot $s \in \mathscr{S}$ we define by C_s the set of all train paths, which either contain s or a conflicting slot s', which cannot be used simultaneously with s. Then a feasible solution fulfills the slot conflict constraint $\forall d \in D, C_s \in \mathscr{C} : \sum_{p:p \in C_s, r, D:d \in D} x_{p,r,D} \leq 1$.

For the stop of the train there must be enough halting place capacity, which can be modelled by halting constraints (see Fig. 1).

For a halting position h we denote by $P(h)$ the set of all paths stopping at h and define by κ_h its capacity, i.e. the maximum number trains which are allowed to wait at the same time. Hence, for each time interval $T \in \mathscr{T}(h,d)$ we define a halting constraint $H = (h, T, \kappa_h)$ with

$$\sum_{p:p \in P(h), D:d \in D} x_{p,r,D} \leq \kappa_h$$

In general, not all requests can be fulfilled simultaneously. Minimizing the number of rejected requests performs bad, because a lot of the generated solutions have too much running and waiting time. The most promising approach will be to maximize total quality. This quality of a train path p for the request r is measured by the travel time $\tau_r(p) := a_p - d_r$ with respect to the arrival time a_p of the train path and the preferred departure time d_r. We use a detour factor ρ to define the quality of the train path by the objective coefficient $\omega_{p,r} := \rho \cdot \tau_r(p_r^*) - \tau_r(p)$.

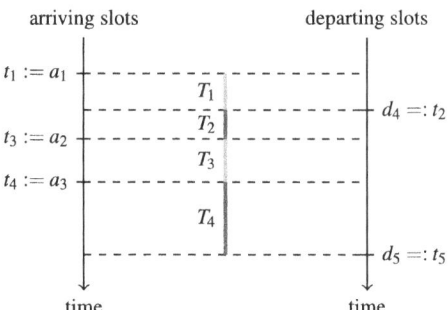

Fig. 1 Sort the sequence of all arrival and departure times of all slots of one traffic day d with access to the halting position h by $t_1 < t_2 < t_3 < \cdots < t_k$. Then, the time interval system $T_j := |t_j, t_{j+1}|$ has the property, that the configuration of halting trains cannot be changed during each of those intervals. The system of all those AD time intervals is denoted by $\mathscr{T}(h,d)$

Fig. 2 Factorization of non-homogenous solutions

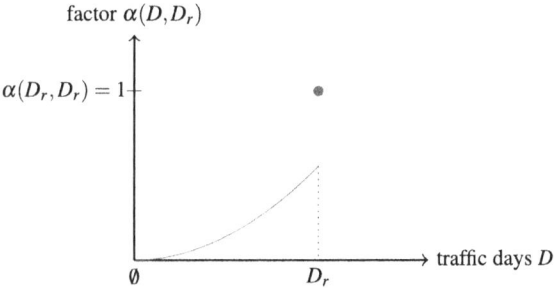

We have to extend this concept to assess the quality of a train path operating only for a traffic day subset $D \subseteq D_r$. We will do this by the approach

$$\omega_{p,r,D} := \alpha(D, D_r)\left(\rho \cdot \tau_r(p_r^*) - \tau_r(p)\right)$$

$\alpha(D, D_r)$ defines a factor, which evaluates the number of days operated by the same train path. $\alpha(D, D_r)$ should be a monotone increasing function, i.e. for $D \subseteq D' \subseteq D_r$ there holds $\alpha(D, D_r) \leq \alpha(D', D_r)$ (Fig. 2).

Hence, total quality is maximized by the model

$$\sum_{p,r,D} \omega_{p,r,D} \cdot x_{p,r,D} \to \max$$

$$\forall d \in D, \ C_s \in \mathcal{C}: \quad \sum_{p:p\in C_s, \ r, \ D:d\in D} x_{p,r,D} \leq 1$$

$$\forall d \in D, \ H \in \mathcal{H}: \quad \sum_{p:p\in P(h), \ r, \ D:d\in D} x_{p,r,D} \leq \kappa_h \qquad (1)$$

$$x_{p,r,D} \in \{0, 1\}$$

For the customer it will be important to recieve a train path for each of the requested traffic days. This can only be modeled by using additional variables to indicate that the request is fulfilled for all days of the request. $y_r \in \{0, 1\}$ can be set to 1, if and only if for each required traffic day a path is available.

$$\sum_{p,r,D} \omega_{p,r,D} \cdot x_{p,r,D} \sum_{r\in\mathcal{R}} +\beta_r y_r \to \max$$

$$\forall d \in D, \ C_s \in \mathcal{C}: \quad \sum_{p:p\in C_s, \ r, \ D:d\in D} x_{p,r,D} \leq 1$$

$$\forall H \in \mathcal{H}: \quad \sum_{p:p\in P(h), \ r, \ D:d\in D} x_{p,r,D} \leq \kappa_h \qquad (2)$$

$$\forall r,\ d \in D_r : \sum_{p:p \in P,\, p,D:d \in D} x_{p,r,D} \geq y_r \tag{3}$$

$$x_{p,r,D} \in \{0, 1\}$$

$$y_r \in \{0, 1\}$$

3 A Branch Cut and Price Approach

We use a branch-and-cut-and-price approach, which can be extended from the problem without traffic days (see [4]). The process of dynamically generating variables is called column generation (see [1]) and done by computing the reduced cost of the non-active column, which will be added to the model if it has negative reduced cost. Lower bounds may either be calculated directly from a feasible integral solution, or, for the case that only a fractional solution of the LP-Relaxation is available, by applying a problem specific rounding heuristic. We use a rounding method by applying a greedy approach for the set of train requests with non-zero, fractional decision variables. Adding the most promising candidates to the model, will improve the actual best solution. Finding those variables is called the pricing problem and done by identifying those variables with minimum negative reduced cost. Searching for new columns with minimum negative cost can be formulated as a modification of a shortest path problem in the underlying slot network, the so-called regular language constrained shortest path problem (see [3]).

The initial solution can be calculated by a greedy approach: All requests are sorted with respect to departure time. According to this sequence for each request the best possible train path will be assigned. We obtain an upper bound by using the well known concept of Lagrange relaxation.

4 Computational Results

We implemented the method to maximize the quality of homogeneous satisfied requests, i.e. we used model (2) and only variables $x_{p,r,D}$ with $D = D_r$ are generated. The method had been validated by a real world instance from DB Netz. The instance has 18541 slots and a demand of 2727 freight trains requests (see Fig. 3).

Table 1 reports the results in detail. Each train path p is qualified by the detour coefficient $\rho(p) = \frac{\tau_r(p)}{\tau_r(p_r^*)}$. The tables report the average value of the detour coefficient of each solution. During the optimization iteration, the number of rejected requests

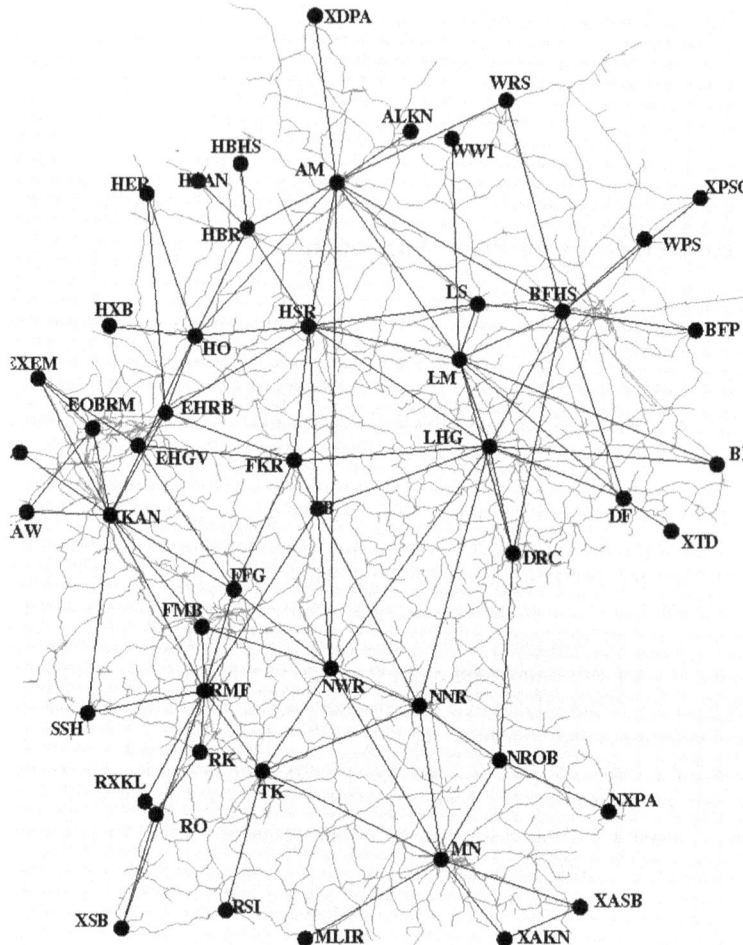

Fig. 3 The figure represents the space network for all 18541 slots between the construction nodes

Table 1 Computational results

Iter.	CPU(s)	Upper bound	Objective	$\overline{\rho}$	Rejected
Start heuristic	1343	2227.9559	956.88	1.4331	1029
1	2435	2120.3143	1389.88	1.3580	552
2	3572	1978.1653	1496.14	1.3562	393
3	5074	1887.3424	1550.59	1.3571	305
4	6346	1814.1011	1581.79	1.3586	251
5	7715	1755.3105	1603.93	1.3595	213
6	9289	1720.7749	1621.33	1.3602	183
7	10998	1688.8051	1623.00	1.3603	180
8	12876	1658.2678	1632.10	1.3607	164

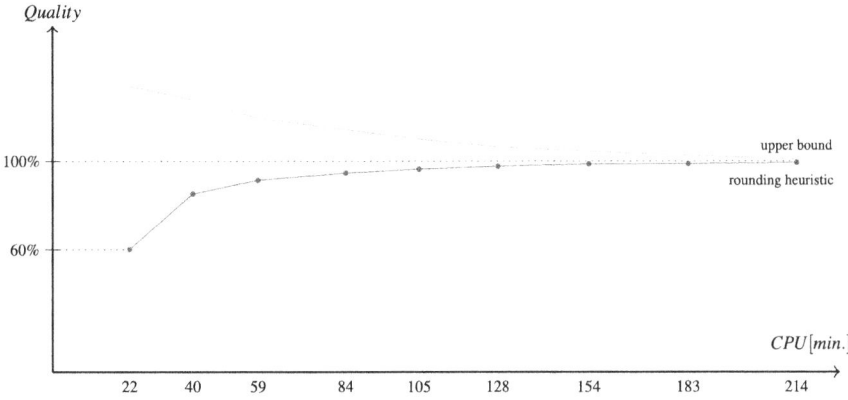

Fig. 4 Duality gap during column generation

can be considerably reduced (see column 'rejected'). Performance and results of the method are principally satisfying; especially the small duality gap indicates, that we have found a good solution near to the optimal one (Fig. 4).

References

1. Barnhart, C., Johnson, E.L., Nemhauser, G.L., Savelbergh, M.W.P., Vance, P.H.: Branch-and-Price: Column Generation for Solving Huge Integer Programs. Operations Research, pp. 316–329 (1998)
2. Feil, M., Poehle, D.: Why does a railway infrastructure company need an optimized train path assignment for industrialized timetabling? In: Proceedings of International Conference on Operations Research, Aachen (2014), to appear
3. Jacob, R., Barrett, C., Marathe, M.: Formal language constrained path problems. SIAM J. Comput. **30**(3), 809–837 (2001)
4. Nachtigall, K., Opitz, J.: Modelling and solving a train path assignment model. In: Proceedings of the International Conference on Operations Research, Aachen (2014)
5. Poehle, D., Feil, M.: What are new insights using optimized train paths assignment for the development of railway infrastructure? In: Proceedings of the International Conference on Operations Research, Aachen (2014)
6. Poehle, D., Feil, M.: A new hierarchical approach for optimized train path assignment with traffic days. In: Proceedings of the International Conference on Operations Research, Wien (2015)

Part XI
Energy

Portfolio Management and Stochastic Optimization in Discrete Time: An Application to Intraday Electricity Trading and Water Values for Hydroassets

Simone Farinelli and Luisa Tibiletti

Abstract Hydro storage system optimization is becoming one of the most challenging task in Energy Finance. Following the Blomvall and Lindberg (2002) interior point model, we set up a stochastic multiperiod optimization procedure by means of a "bushy" recombining tree that provides fast computational results. Inequality constraints are packed into the objective function by the logarithmic barrier approach and the utility function is approximated by its second order Taylor polynomial. The optimal solution for the original problem is obtained as a diagonal sequence where the first diagonal dimension is the parameter controlling the logarithmic penalty and the second is the parameter for the Newton step in the construction of the approximated solution. Optimimal intraday electricity trading and water values for hydroassets are computed. The algorithm is implemented in Mathematica.

JEL classification C06 · G13 · G17 · G11

1 Introduction

The liberalised electricity market poses new challenges to power generating companies for the electrical grid. A key driver to set up economically efficient grids is the capacity to store electricity through hydro storage systems and thereby decouple electricity generation from electricity consumption. So, the hydro storage system optimization is becoming one of the most challenging task in Energy Finance, as highlighted in [6] and in [5].

S. Farinelli · L. Tibiletti (✉)
Department of Management, University of Torino, Corso Unione Sovietica
218 bis, 10134 Torino, Italy
e-mail: luisa.tibiletti@unito.it

S. Farinelli
e-mail: simone@coredynamics.ch

S. Farinelli · L. Tibiletti
Core Dynamics GmbH, Scheuchzerstrasse 43, 8006 Zurich, Switzerland

© Springer International Publishing Switzerland 2017
K.F. Dœrner et al. (eds.), *Operations Research Proceedings 2015*,
Operations Research Proceedings, DOI 10.1007/978-3-319-42902-1_85

The aim of the present paper is to set up a stochastic dynamic programming to optimize intraday electricity trading and model at the same time water values for hydroassets.

Starting from the seminal Blomvall and Lindberg model (see [1–4]), a stochastic multiperiod optimization problem in discrete time for a generic utility function is discretized in the space dimensions by means of a "bushy" recombining tree (i.e. a $k \gg 1$-lattice), so that we do not have deal with the dimensionality curse nor are we annoyed by heuristic arguments concerning the choice of representative branches in a non recombining tree, as Blomvall and Lindberg implicitly have to deal with in their original work. Inequality constraints are packed into the objective function by the logarithmic barrier approach and the utility function is approximated by its second order Taylor polynomial. The optimal solution for the original problem is obtained as a diagonal sequence where the first diagonal dimension is the parameter controlling the logarithmic penalty and the second is the parameter for the Newton step in the construction of the approximated solution. The obtained algorithm is implemented in Mathematica and applied to optimize intraday electricity trading and model at the same time water values for hydroassets.

This paper is structured as follows. Section 2 we illustrate the differences between the classical Blomvall and Lindberg model and ours. In Sect. 3 the model is applied to the intraday electricity trading to find an optimal strategy and to determine water values of hydroelectric infrastructures to be used for market bids. Section 4 concludes the note.

2   The Blomvall and Lindberg Model and Behind

The implemented model differs from the Blomvall–Lindberg original one for the following formal and substantial points:

- Blomvall–Lindberg formulate directly the optimization problem on the nodes of a non-recombining tree. We formulate it for a general filtration. This is a rather a formal distinction, because the formulae are essentially the same.
- The system states in the Blomvall–Lindberg approach are *external states*, that is, they depend on the control rules, e.g. portfolio values at different times. The system states in our approach are *internal states*, that is, they do not depend on the control rules, e.g. base assets' values at different times. Our choice leads to simplified formulae for the terms appearing in Riccati's equation, when establishing the inputs for the optimal control rules' formula.
- The objective function in the Blomvall–Lindberg approach at time t is a function of the risk factors realizations at time t. The objective function in our approach at time t can be seen as the expectation at time t of the discounted sums of Blomvall–Lindberg's objective functions at times $s = t + 1, \ldots, T$.
- The implementation of the algorithm occurs on a lattice. We see a lattice with many branches as a totally recombining tree. Therefore, being the number of nodes in a

time layer a linear function of time, the full fledged model is implementable even on standard computers. Of course, the main challenge is to fill the nodes with state realizations in such a way that these are compatible with their dynamics on one hand, and with the full recombining property of the graph, on the other. To our knowledge this method is new, an is a generalization of binomial and trinomial trees' construction utilized for option pricing. This choice has the advantage of being extendible to the case where no (semi-)closed solution on the nodes exists, and is thus a viable implementation method for a numerical solution of Bellman's backward recursion.

3 Application: Water Values and Intraday Electricity Trading

The new version of the Blomvall–Lindberg model has been utilized to optimize intraday electricity trading and model at the same time water values for hydroassets. Everyday by 11:00 CET all the participants to the Swiss electricity spot market have to submit to the energy exchange their aggregated bids for the day-ahead both demand and supply. These, in the "ask"–case specify for every hour of the following day, from 00:00 till 24:00$^-$ CET the quantity of energy E_t^{Ask} in MWh that one participant is willing to deliver during that hour $t = 0, \ldots, 23$ if the electricity price S_t then is greater than or equal to a certain value GP_t^{Ask}, called *generation water value*. In the "bid"–case they specify for every hour of the following day the quantity of energy E_t^{Bid} in MWh that the participant is willing to buy during that hour $t = 0, \ldots, 23$ if the electricity price S_t then is smaller than or equal to a certain value GP_t^{Bid}, called *delivery water value*. For every hour the energy exchange aggregates all asks and all bids two monotone step functions, the ask curve and the bid curve, representing the quantity of energy deliverable (ask) or requested (bid) as a function of the price. The intersection point of the two curves, i.e. the market clearing price at time t is the spot price which will hold for the hour t of the next day. The 24 spot prices for the day-ahead are published at around 11:15 CET of the current day. Note that none of the market participants is due to deliver or to buy the quantity of energies specified during the bidding process. That information was formally only utilized by the energy exchange to establish the day-ahead spot prices.

As soon as the new day starts, the intraday market tries to generate a profit by looking at the given fixed water values GP_t^{Ask} and comparing them with the current energy price X_t, according to a simple strategy, which for the trader of an hydroasset reads:

- If $X_t \geq GP_t^{Ask}$ turbine water to produce E_t^{Ask} energy and deliver it.
- If $X_t < GP_t^{Ask}$ do not turbine water, buy E_t^{Ask} energy at price X_t in the intraday market and deliver it.

At the same time all the trades for the day ahead settled between 11:15 and 23:59 CET, where energy quantities $E_t^{\text{Spot, Sell}}$ and $E_t^{\text{Spot, Buy}}$ will be sold and respectively bought at hour t of the next day at price S_t have to be taken into account by the trading strategy. We analyze now the objective function of the optimization problem

$$\max_{(u_t)_{t=0,\ldots,T-1}} \mathbb{E}_0\left[\sum_{t=1}^{T} \beta_t V_t(X_t, u_{t-1})\right] \tag{1}$$

For the implementation we chose the representation

$$\max_{\substack{(u_t^j)_{t=0,\ldots,T-1} \\ \text{Restrictions for } u_t^j \\ j=1,\ldots,N}} \sum_{t=1}^{T} \beta_t \mathbb{E}_0[f_t(u_{t-1}X_t)], \tag{2}$$

where f_t is a function representing the risk-reward trade-off at time t. Since the dynamics of the hydro-infrastructure can be very complex, for the ease of implementation, it is substituted by the following simple equality restriction, imposing a maximal energy $E_{\text{tot}}^{\text{Ask}}$ to be produced during the 24 h:

$$\sum_{t=0}^{23} \sum_{j=1}^{M} u_t^j \leq E_{\text{tot}}^{\text{Ask}}. \tag{3}$$

Note that such a constraint has an intertemporal nature and can not be covered by the formulae developed so far, which need the appropriate extension.

After 11:15 CET, when the spot prices for the following day are published, the day ahead intraday strategy is determined and we do not need to solve the optimization (2), because we already know the solution. But before 11:00 CET we can utilize (2) to determine the generation water values GP_t^{Ask} for $t = 0, \ldots, 23$ for the hydro-infrastructure whose bids we want to aggregate in our bid for the energy exchange. That is, we extend the set of stochastic optimization variables to

- $(u_t^j)_{t=0,\ldots,T-1}$ and
- $(\text{gp}_t^{\text{Ask}})_{t=0,\ldots,T-1}$,

and we add following restrictions to the optimization problem:

- If $X_t \geq \text{gp}_t^{\text{Ask}}$, then $u_t^j > 0$ (i.e. turbine water),
- If $X_t < \text{gp}_t^{\text{Ask}}$, then $u_t^j \equiv 0$ (i.e. do not turbine water).

Therefore, we obtain

$$
\begin{array}{c}
\max_{\substack{(u_t^j)_{t=0,\ldots,T-1} \\ (gp_t^{Ask})_{t=0,\ldots,T-1} \\ u_t^j(\{X_t \geq gp_t^{Ask}\})>0 \\ u_t^j(\{X_t < gp_t^{Ask}\})=0 \\ \text{Restrictions for } u_t^j \\ j=1,\ldots,N}} \quad \sum_{t=1}^{T} \beta_t \mathbb{E}_0[f_t(u_{t-1}X_t)].
\end{array}
\tag{4}
$$

The solution of the optimization problem (4) *before* 11:00 CET delivers a stochastic process $(gp_t^{Ask})_{t=0,\ldots,T-1}$ which we can use to find production water values for the bid, by taking

$$
GP_t^{Ask} := \mathbb{E}_0\left[gp_t^{Ask}\right].
\tag{5}
$$

Remark 1 The model proposed is intrinsically *balance-energy neutral* for the balance group which the hydro-infrastructure belongs to. A *balance group* is a set of electricity meters measuring 15 min consumption and production for net users. The *transmission system operator* makes sure that every balance group is in an equilibrium state, by adding or subtracting electric energy in such a way that the total sum of energies vanishes for every quarter of an hour. Of course this comes at a certain expensive price with which the TSO charges the balance group owner, which can be (but not necessarily is) the hydro-infrastructure owner as well. Thus, there is an incentive not to generate or at least to reduce balance energy, in order to minimize costs.

4 Conclusion

A stochastic multiperiod portfolio optimization problem in discrete time for a generic utility function is discretized in the space dimensions by means of a lattice. Inequality constraints are packed into the objective function by means of a logarithmic penalty and the utility function is approximated by its second order Taylor polynomial. A converging sequence of solutions of the approximated problem converging to the optimal solution of the original problem is implemented in an algorithm coded in Mathematica. As an application we model optimal intraday electricity trading and time water values for hydroassets.

Acknowledgements We would like to thank Sai Anand and Rémi Janner for their hints and their very valuable feedbacks.

References

1. Blomvall, J., Lindberg, P.O.: A riccati-based primal interior point solver for multistage stochastic programming. Eur. J. Oper. Res. **143**(2), 452–461 (2002)
2. Blomvall, J., Lindberg, P.O.: A riccati-based primal interior point solver for multistage stochastic programming—extensions. Optim. Methods Softw. **17**, 383–407 (2002)
3. Blomvall, J., Lindberg, P.O.: A multistage stochastic programming algorithm suitable for parallel computing. J. Parallel Comput. Spec. Issue Parallel Comput. Numer. Optim. **29**(4), 431–445 (2003)
4. Blomvall, J., Lindberg, P.O.: Back-testing the performance of an actively managed option portfolio at the Swedish stock market 1990–1999. J. Econ. Dyn. Control **27**(6), 1099–1112 (2003)
5. Egerer, J., Scharff, R., Söder, L.: A description of the operative decision-making process of a power generating company on the Nordic electricity market. Energy Syst. **5**(2), 349–369 (2014)
6. Löhndorf, N., Minner, S., Wozabal, D.: Optimizing trading decisions for hydro storage systems using approximate dual dynamic programming. Oper. Res. **61**(4), 810–823 (2013)

Investments in Flexibility Measures for Gas-Fired Power Plants: A Real Options Approach

Barbara Glensk and Reinhard Madlener

Abstract The promotion of electricity from renewable energy in Germany by means of guaranteed feed-in tariffs and preferential dispatch leads to difficulties in the profitable operation of many modern conventional power plants. Nevertheless, conventional power generation technologies with enhanced flexibility in their operational characteristics can contribute to balancing electricity supply and demand. For this reason, the operational flexibility of conventional power plants becomes important for the system and has an inherent economic value. The focus of this research is on high efficiency gas-fired power plants; we tackle the following three research questions from a plant owner's perspective: (1) How can already existing conventional power plants be operated more flexibly and thus be made more profitable? (2) Which flexibility measures can be taken under consideration? (3) What is the optimal timing to invest in flexibility measures? To answer these questions we propose an optimization model for the flexible operation of existing gas-fired power plants that is based on real options analysis (ROA). In the model, the economic and technical aspects of the power plant operation are explicitly taken into account. Moreover, the spark spread, which is an important source of uncertainty, is used for the definition of the flexible plant operation in terms of different load levels and corresponding efficiency factors. The usefulness of the proposed model is illustrated with a case study mimicking the retrofitting decision process.

1 Introduction

Markets with high shares of fluctuating renewable energy sources require that conventional power plants are able to react to these fluctuations in a more flexible manner. Besides grid expansion, storage capacity, and demand-side management, the balance

B. Glensk (✉) · R. Madlener
School of Business and Economics/E.ON Energy Research Center,
Institute for Future Energy Consumer Needs and Behavior (FCN),
RWTH Aachen University, Mathieustrasse 10, 52074 Aachen, Germany
e-mail: BGlensk@eonerc.rwth-aachen.de

R. Madlener
e-mail: RMadlener@eonerc.rwth-aachen.de

© Springer International Publishing Switzerland 2017
K.F. Dœrner et al. (eds.), *Operations Research Proceedings 2015*,
Operations Research Proceedings, DOI 10.1007/978-3-319-42902-1_86

between the renewables-based and conventional power generation technologies can be achieved by better exploiting modern, highly-efficient power plants. Their flexible operation can be enhanced by using additional components such as power electronic converters, storage systems, or upgrades of existing components to the best available technology. Being specific to a certain power plant, and given the lack of a liquid market, the investment in upgrading equipment can be regarded as irreversible, justifying the real options approach, which becomes more relevant in an uncertain energy market environment.

ROA offers the consideration of different types of options (e.g. to invest, abandon, expand, contract, shut down, choose etc.), regarding project specification and action to be undertaken. A useful overview of ROA applied to different industries can be found in [8, 12]. A comprehensive review of the use of ROA in the energy sector is provided by Fernandez et al. [4]. From a historical perspective, these authors present various applications of ROA to the oil industry, power generation, energy markets, as well as emission mitigation policy. Moreover, they point out different solution methods, including partial differential equation modeling, binomial option valuation, Monte Carlo simulation, and dynamic programming. Regarding gas-fired power plants, Näsäkkälä and Fleten [9] study investments in base-load or peak-load gas-fired power plants, using a two-factor model for price processes, considering both the short-term mean revision and long-term uncertainty. They use the spark spread, defined as the difference between the electricity price and the cost of gas necessary for power generation, and they find that the increase in the variability of the spark spread, on the one hand, increases the value of a peak-load power plant (i.e. this investment is more attractive); on the other hand, it delays investment. Later on in [5], they investigate the investment and technology upgrade of a gas-fired power plant, using the already mentioned two-factor model for price processes [9]. From the analysis conducted they find that the possibility to ramp the power plant up and down becomes significant. As shown in the valuation model proposed by Deng and Oren [2] during the operation process of the power plants, the characteristics, such as start-up and shut-down costs, ramp-up constraints, as well as the operating-level-dependent heat rate, are important parameters. In the numerical analysis they show that the operational characteristics affect the valuation of the power plant, but also depend on the operating efficiency and the assumptions made regarding the development of electricity and fuel prices (choice between mean-reversion or Brownian motion process).

In our study, we consider one of several technical options for enhancing the flexibility of gas-fired power plants and try to find the optimal time to invest. Applying ROA, based on the approach proposed by Deng and Oren [2], the investor compares the present value of the existing power plant with the value after the retrofitting.

2   Model Specification

The model proposed is based on the real options approach and consists of a three-step procedure.

In the first step, the so-called operation strategy (operation regime) of the power plant is defined for each hour. Here, the spark spread is used as a profitability indicator and source of uncertainty. For a gas-fired power plant the spark spread is the difference between the electricity price and the cost of gas necessary for power generation. In the typical definition of the spark spread the gas price is multiplied by the heat rate (see [9]) or divided by the net efficiency of the power plant. Nevertheless, the net efficiency of the power plant changes with the load level (output). Regarding this dependency we define the spark spread ($Spread_t$) as follows:

$$Spread_t = P_{elec,t} - \left(\beta \cdot \frac{P_{gas,t}}{\eta(load\ max)} + (1 - \beta) \cdot \frac{P_{gas,t}}{\eta(load\ min)} \right) \quad (1)$$

where $P_{elec,t}$ and $P_{gas,t}$ denote the electricity and gas price, respectively, $\eta(\cdot)$ is the load-level-dependent net efficiency of the power plant, and $\beta \in [0, 1]$. For the long-term evolution of the spark spread, an arithmetic Brownian motion (ABM) process is used,[1] and defined as:

$$Spread_t = Spread_{t-1} + \alpha dt + \sigma dZ_t \quad (2)$$

where α (drift) and σ (volatility) are constants, and Z_t is a standard Brownian Motion process. Comparing the spark spread from the ABM process ($Spread_{t,abm}$) with the sum of the variable operation and maintenance costs (OM_{var}) and CO_2 costs, the operation load levels for each hour can be defined.[2]

In the second step of the procedure, the project's (power plant's) cash flows are calculated regarding operation strategy, CO_2 price development (accordingly ABM processes), variable OM costs, as well as start-up, shut-down and marginal ramping costs ($c_{start-up}$, $c_{shut-down}$, $c_{ramp-up}$).[3] Furthermore, considering the assumption in terms of possible operational output levels, and following the model proposed by Deng and Oren [2], the three operational stages (S1—the power plant is off, S2—the power plant is at min load operation, and S3—the power plant is at max load operation), as well as three possible actions (A1—the power plant runs at full-capacity level, A2—the power plat runs at low-capacity level, and A3—the power plant is turned off) are possible. Regarding all these assumptions, the operating cash flow CF_t of the power plant for each hour with respect to operational stage and undertaken action can be represented as follows:

---

[1] The arithmetic Brownian motion process offers an alternative to the standard geometric Brownian motion often used in ROA, especially when negative values are also expected [1].

[2] For simplicity reasons, we assume that the power plant has only three possible operational load (output) levels: maximum, minimum, and zero. For more information see [6].

[3] For more information see [6].

$$CF_t(P_t, A_t, S1) = \begin{cases} \text{for} & A_t = A1 : -c_{start\text{-}up} - c_{ramp\text{-}up}(load\ max_t) \\ \text{for} & A_t = A2 : -c_{start\text{-}up} - c_{ramp\text{-}up}(load\ min_t) \\ \text{for} & A_t = A3 : 0 \end{cases} \quad (3)$$

$$CF_t(P_t, A_t, S2) = \begin{cases} \text{for} & A_t = A1 : -c_{ramp\text{-}up}(load\ max_t) \\ \text{for} & A_t = A2 : P_t(load\ min_t) \\ \text{for} & A_t = A3 : -c_{shut\text{-}down} \end{cases} \quad (4)$$

$$CF_t(P_t, A_t, S3) = \begin{cases} \text{for} & A_t = A1 : P_t(load\ max_t) \\ \text{for} & A_t = A2 : P_t(load\ min_t) \\ \text{for} & A_t = A3 : -c_{shut\text{-}down} \end{cases} \quad (5)$$

where

$$P_t(load_t) = \left(Spread_t - \frac{P_{CO_2,t} \cdot e_{spec}}{\eta(load_t)} - OM_{var}\right) \cdot load_t \quad (6)$$

and $load_t \in \{load\ max_t, load\ min_t\}$, $P_{CO_2,t}$ denotes the CO_2 price, and e_{spec} is the specific emission factor for gas.

The specific functional form for the ramp-up costs ($c_{ramp\text{-}up}$), following [11], is decomposed into the ramping fuel requirement (rf), and the decreased depreciation due to ramping (d), represented as:

$$c_{ramp\text{-}up} = rf \cdot \left(P_{gas} + P_{CO_2} \cdot e_{spec}\right) + d. \quad (7)$$

In the last step, applying ROA, the optimal timing for the decision: to continue power plant operation or to invest in the retrofit measure can be determined. The decision to retrofit the power plant with the new additional technical components, or to upgrade already existing elements, can be seen as an option to expand. The option to expand implies that an expansion should not be made unless the present value of the future expected cash flows exceeds the value of the option to expand [12]. The decision process can be defined as follows:

$$\begin{aligned} \text{if} \quad & PV_t < Option\ to\ expand & \text{invest in considered flexibility measure} \\ \text{if} \quad & PV_t \geq Option\ to\ expand & \text{wait with the retrofitting} \end{aligned} \quad (8)$$

where

$$PV_t = \sum_{t=0}^{T} \frac{CF_t - OM_{fixed,t} - Dep_t}{(1 + WACC)^t} \quad (9)$$

and

$$Option\ to\ expand = max\left(RPV_{t+\Delta t} - retrofit_{investment}, 0\right) \quad (10)$$

OM_{fixed} represents fixed operation and maintenance costs, Dep_t denotes depreciation, RPV_t defines the project value after retrofitting (calculated using Eq. (9)), Δt

describes the time needed for retrofitting, and *WACC* (Weighted Average Cost of Capital) is used as the discount rate. The application of *WACC* can be justified with two arguments: (1) it is the most popular discount rate among corporate investors, as it takes debt and equity capital into account (thus reflecting a company's capital structure well); (2) its value is firm-specific, and can thus serve as a useful investment indicator.

3　Case Study

In the case study, we apply the proposed model to a highly energy-efficient and recently built gas-fired combined cycle power plant in Germany. The power plant analyzed was commissioned in 2010, with a net installed capacity of 845 MW. Its net thermal efficiency is 59.7 %, and the total investment volume 400 million Euros [3, 7]. Increases in the operational flexibility of combined cycle power plants can be achieved by the optimization of the minimum load as well as its part-load efficiency, load ramps and reserve capacity, or start-up times. To improve the minimum load and the part-load efficiency, the power plant can be retrofitted, for instance, with variable-pitch guide vanes at the gas turbine compressor or an air preheater [10]. The implementation of such a component takes some time when the power plant should be totally turned off, and leads to additional investment costs.[4] More detailed information regarding technical and economic parameters as well as start-up and shut-down times and costs can be found in [6].

In the first step of the proposed model, the operation strategy of the analyzed power plant was determined and used in the secod module of our procedure for the cash flow calculations (Eqs. (3)–(5)). Furthermore, the project value (Eq. (9)) for each time period of plant operation (i.e. over 28 years remaining after the commissioning in 2010) was calculated for the situation without and with retrofit measure (which decreases the minimum load for the operation from 40 to 30 % of total installed power).

The results obtained are then used to find out the optimal investment time for the retrofitting measure. Assuming different realization times, during the investment phase where the power plant will have to be shut down (for about 5 till 8 months), the decision about the investment in the technical improvements should be made immediately, i.e. in the first period analyzed (see Table 1). This is due to the use of the real options approach described in Eqs. (8)–(10). Furthermore, we observe the positive values of the project after retrofitting (the second column in Table 1), in comparison to the project value before retrofitting ($PV_{01.2013} = -77, 477, 904$), which also indicates that the enhancement of the plant is viable.

---

[4]In our case study, it amounts to approximately 5 % of the price of a new power plant (information based on the discussion with experts from industry).

Table 1 Present value of the retrofitted power plant, option value and decision

Realization time (in months)	RPV of power plant (in €)	Option value (in €) according Eq. (10)	Decision for $PV_{01,2013} = -77,477,904$
5	1,064,357	−18,935,642	Invest
6	1,449,620	−18,550,379	Invest
7	732,366	−19,267,633	Invest
8	1,918,045	−18,081,954	Invest

4   Conclusions

The analysis conducted shows that the real options approach developed can be useful in the decision-making process regarding investments for the more flexible operation of conventional power plants. Furthermore, we demonstrate that the proposed model can simply take advantage from market uncertainties incorporated into the model's structure, as well as consider important technical and economic characteristics such as start-up times and costs, or ramp-up costs. These parameters, and also the operation strategy of the power plant, are crucial for power plant owners' profitability calculation. In this respect, there is still scope for the further enhancement of the proposed procedure.

References

1. Alexander, R.D., Mo, M., Stent, A.F.: Arithmetic Brownian motion and real options. Eur. J. Oper. Res. **219**(1), 114–122 (2012)
2. Deng, S.-J., Oren, S.S.: Incorporating operational characteristics and start-up costs in option-based valuation of power generation capacity. Probab. Eng. Inf. Sci. **17**, 155–181 (2003)
3. E.ON. Kraftwerk Irsching (2014). http://www.eon.com/de/ueber-uns/struktur/asset-finder/irsching.html. Accessed 10 July 2015
4. Fernandez, B., Cunha, J., Ferreira, P.: The use of real options approach in energy sector investments. Renew. Sustain. Energy Rev. **15**, 4491–4497 (2011)
5. Fleten, S.-E., Näsäkkälä, E.: Gas-fired power plants: investment timing, operating flexibility and CO$_2$ capture. Energy Econ. **32**, 805–816 (2010)
6. Glensk B., Madlener R.: Real options analysis of the flexibile operation of an enhanced gas-fired power plant. FCN Working Paper No. 11/2015, RWTH Aachen University, August (2015)
7. Mainova. Hocheffizientes Gas- und Dampfturbinenkraftwerke setzt neue Maßstäbe (2014). http://www.mainova.de/unternehmen/press/2835.html. Accessed 28 April 2014
8. Mun, J.: Real options analysis: tools and techniques for valuing strategic investment and decisions. Wiley, Hoboken (2006)
9. Näsäkkälä, E., Fleten, S.-E.: Flexibility and technology choice in gas fired power plant investments. Rev. Financ. Econ. **14**, 371–393 (2005)

10. Pickard, A., Meinecke, G.: The Future Role of Fossil Power Generation. Siemens AG, Erlangen (2011)
11. Traber, T., Kemfert, C.: Gone with the wind? Electricity market prices and incentives to invest in thermal power plants under increasing wind energy supply. Energy Econ. **33**, 249–256 (2011)
12. Trigeorgis, L.: Real options: managerial flexibility and strategy in resource allocation, 5th edn. The MIT Press, Cambridge (2000)

Bidding in German Electricity Markets—Opportunities for CHP Plants

Nadine Kumbartzky, Matthias Schacht, Katrin Schulz and Brigitte Werners

Abstract Since the liberalisation of the German energy markets, supply companies strive to gain additional revenues by trading in different electricity markets. Thus, trading strategies have to be adjusted to lower but more volatile energy prices caused by a rising share of renewable energy feed-in. Due to the increasing importance of combined heat and power (CHP) plants, an energy supply company is considered that operates a CHP plant with heat storage. A new modelling approach is presented to support generation companies scheduling their participation in two sequential electricity markets, the balancing market and the day-ahead spot market. In both markets, specific bidding procedures for trading exist. To derive optimal bidding strategies for CHP plants, we formulate the bidding problem as an innovative and detailed multi-stage stochastic programming model taking into account the sequencing of market clearing. The optimisation model simultaneously determines the operation of the CHP plant and heat storage device. An exemplary case study illustrates the benefit from coordinated bidding in sequential electricity markets.

1 Introduction

The liberalisation of the German electricity markets let to fundamental changes in the German energy system. The emergence of new markets offers additional revenue potential for energy supply companies operating flexible conventional power plants. However, trading in electricity markets is accompanied by an increased competition. Moreover, a high installed capacity of renewable energy leads to a volatile power feed-in from renewables. As a consequence, the general price level has decreased whereas market price volatility has increased. Hence, power supply companies are forced to develop new trading strategies for conventional power plants.

Bidding in multiple electricity markets has already been discussed in literature, for example in [4, 6]. In this paper, special attention is paid to highly efficient CHP plants.

N. Kumbartzky (✉) · M. Schacht · K. Schulz · B. Werners
Faculty of Management and Economics, Chair of Operations Research
and Accounting, Ruhr University Bochum, Bochum, Germany
e-mail: nadine.kumbartzky@rub.de

© Springer International Publishing Switzerland 2017   645
K.F. Dœrner et al. (eds.), *Operations Research Proceedings 2015*,
Operations Research Proceedings, DOI 10.1007/978-3-319-42902-1_87

These conventional power plants simultaneously generate heat and power resulting in a high utilisation ratio of the fuel used. However, the participation of a CHP plant in electricity markets is restricted as costumers' power and heat demand have to be fulfilled in any case. The usage of heat storage enables a more flexible operation, but requires an integration of storage policies into the optimisation procedure. A combined optimisation of trading in German electricity markets and the operation of a complex cogeneration plant has not been analysed in literature so far. Therefore, our approach extends current work and combines relevant aspects of cogeneration with bidding in different electricity markets.

2 Bidding in German Electricity Markets

Our approach considers two different electricity markets, the balancing market and the day-ahead spot market. The amount of electricity traded in both markets is allocated based on a specified bidding process which is explained below.

Since electricity is practically non-storable in any major quantities, power supply and demand have to match exactly at all times. To ensure system stability, the Transmission System Operators (TSOs) conduct an online auction at regelleistung.net for the provision of balancing power. In Germany, three different qualities of balancing power are traded: primary control reserve, secondary control reserve and minute reserve. Primary and secondary control reserve have to be delivered within 30 seconds respectively 5 min. In contrast, minute reserve is electronically activated within 15 min. In the following, we consider a flexible gas-fired CHP plant with a start-up time less than 15 min. Consequently, a provision of minute reserve is guaranteed even if the plant is currently shut down. To avoid a must-run condition, we focus on trading of minute reserve.

Minute reserve is divided into positive (up-regulating) and negative (down-regulating) reserve power. The auction takes place on weekdays one day ahead of delivery and is closed at 10 a.m. Trading of minute reserve is split up into six 4-h time slices per day and is organised as a pay-as-bid auction. Thus, prices paid to suppliers are based on their individual bids. Every bid consists of an offered volume and two prices, a capacity price and an energy price. The capacity price is used for the remuneration of preserving capacity and is paid if the bid is accepted. If minute reserve is actually delivered, suppliers receive a reimbursement based on the offered energy price. All submitted bids are aggregated according to capacity prices and the marginal capacity price is determined as the last bid accepted to fulfil the reserve capacity. Since minute reserve is rarely called, the profit gained from trading in the balancing market is primarily driven by the capacity price.

In the German day-ahead spot market, which is operated by the European Power Exchange (EPEX SPOT SE), day-ahead power is traded for the following day. In contrast to minute reserve, trading in the spot market is executed as a uniform-price, double-sided call auction. This means that buyers and sellers simultaneously submit price-volume-bids and, if successful, are paid the market clearing price. Trading takes place 7 days a week for every hour of the following operation day. The market is closed at 12 p.m. All received supply and demand bids are aggregated and the uniform market clearing price is determined as the intercept of the supply and demand curve. After market clearing, equilibrium prices for every hour of the following operation day are announced and participants are informed whether their bids are accepted or rejected.

3   Multistage Stochastic Optimisation Model

To support generation companies in the bidding process in both electricity markets, we developed a multistage stochastic optimisation model. The objective is to minimise total net acquisition costs which are defined as generation costs (including start-up costs) plus purchase costs subtracted by revenues from power trading. The model captures the detailed bidding procedures as well as the current supply situation of the generation company. This includes the customers' power and heat demand and the operating status of the CHP plant and heat storage device. These factors determine whether and to which amount the generation company is able to participate in the considered electricity markets. The heat storage device provides further flexibility by decoupling heat supply and demand. Regardless of the dispatched volumes in both markets, a stable power and heat supply for customers has to be guaranteed. Thus, the sum of power generation and volumes purchased in the spot market subtracted by volumes sold has to be equal to power demand at all times. Heat demand can be fulfilled by own generation plus heat discharged minus heat charged from the storage. The CHP plant's characteristic operating region is modelled as a convex combination of extreme points as introduced in [3]. In addition, the operation of the storage is optimised in combination with the CHP plant according to [5].

Bidding in sequential electricity markets is an optimisation problem under uncertainty since market prices are unknown when bids are submitted. Uncertain prices are represented by different scenarios and expected net acquisition costs are minimised. In the optimisation model, the information structure and bidding procedures of the two markets are explicitly depicted. Figure 1 illustrates the decision-making process

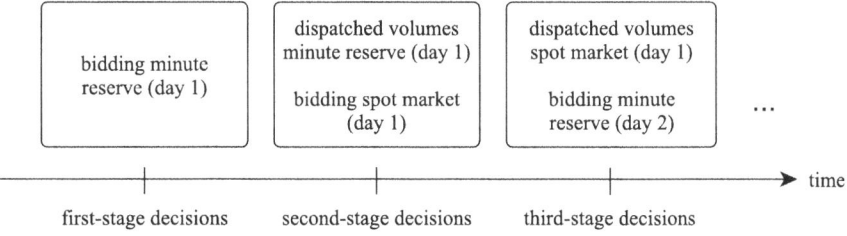

Fig. 1 Decision-making process for bidding in sequential electricity markets

considering the timing of bidding and clearing for both markets in sequential order. We consider a time horizon of D days. At day 0, decisions regarding power generation and trading for day 1 have to be made. Since the auction for minute reserve ends at 10 a.m, first-stage decisions are the quantities bid in the market for positive and negative minute reserve for day 1. After market clearing, actual prices of the six time slices are announced and determine which bids are accepted. Therefore, second-stage decisions are the corresponding minute reserve volumes dispatched as well as the volumes bid in the spot market for day 1. We assume that spot market participants either submit a supply or demand curve for every hour. After market closure at noon, market clearing prices for the 24 h of the following operation day are revealed. This information determines the volumes dispatched which are modelled as third-stage decisions. At the same stage, volumes bid in the minute reserve market for day 2 need to be assigned. This succession of decisions continues until day D. Thus, the sequential bidding problem forms a multistage stochastic optimisation model. Note that decisions made at a certain stage depend only on the information available in this stage and not on future realisations.

The bidding process in detail is modelled similar to [1] using step-wise bidding curves. We exemplarily describe the bidding formulation for selling power to the spot market. Let $\mathscr{J}$ denote the set of price indices, $\mathscr{T}$ the set of time periods and $\mathscr{S}$ the set of scenarios. To formulate the bidding problem as a linear program, bid volumes w_{jts} as well as dispatched volumes z_{ts} are decision variables whereas bid prices PZ_{jt} are assumed to be fixed parameters. With λ_t^s denoting the uncertain spot price, the bidding problem is modelled as follows:

$$z_{ts} = w_{jts} \quad \text{if } PZ_{jt} \leq \lambda_t^s < PZ_{j+1,t} \qquad \forall j \in \mathscr{J}, t \in \mathscr{T}, s \in \mathscr{S} \tag{1}$$

$$w_{jts} \leq w_{j+1,ts} \qquad \forall j \in \mathscr{J} \setminus \{J\}, t \in \mathscr{T}, s \in \mathscr{S}. \tag{2}$$

A bid is accepted if the bid price is less or equal to the uncertain spot price as in (1). Since the supply curve is assumed to be monotonically increasing, constraints (2) ensures that the bid volume at price index $j + 1$ is greater or equal to the bid volume at price index j. Note that bid volumes in the spot market are defined as the accumulated quantities bid at a certain price.

4   Case Study

The developed multistage MILP was applied to an exemplary generation company operating a CHP plant with heat storage specified according to [5]. The volumes offered in the minute reserve and spot market as well as the operation of the CHP plant and storage are simultaneously optimised considering a time horizon of 2 weeks. Similar to [1, 2], uncertain day-ahead spot and minute reserve prices are fitted to seasonal ARIMA models. To approximate the underlying probability distribution, a finite set of scenarios is generated.

In the following, computational results of the expected value approach are shown exemplarily for one scenario. Figure 2 depicts the relationship between forecasted prices and volumes traded in (**a**) the day-ahead spot market and (**b**) the balancing market. If power generated by the CHP plant does not equal costumers' demand, e.g. due to technical restrictions, missing or additional power is compensated by trading in the spot market. This can be seen, for example, in Fig. 2a at $t = 28$ and $t = 40$, where the plant is shut down since it is profitable to satisfy total demand by purchasing power in the day-ahead market. Regarding minute reserve, it is possible to offer either positive or negative reserve. Figure 2b shows that if the plant is operating, negative minute reserve is dispatched at all times. This is caused by a higher price level of negative in contrast to positive minute reserve and the fact that revenues can be generated from the spot and negative minute reserve market at the same time. If the CHP plant is currently shut down, positive minute reserve can still be offered due to a start-up time of less than 15 min, see hours 40–47. However, in hours 28–31, the price for positive reserve is too low to make a start-up worthwhile. Figure 2c

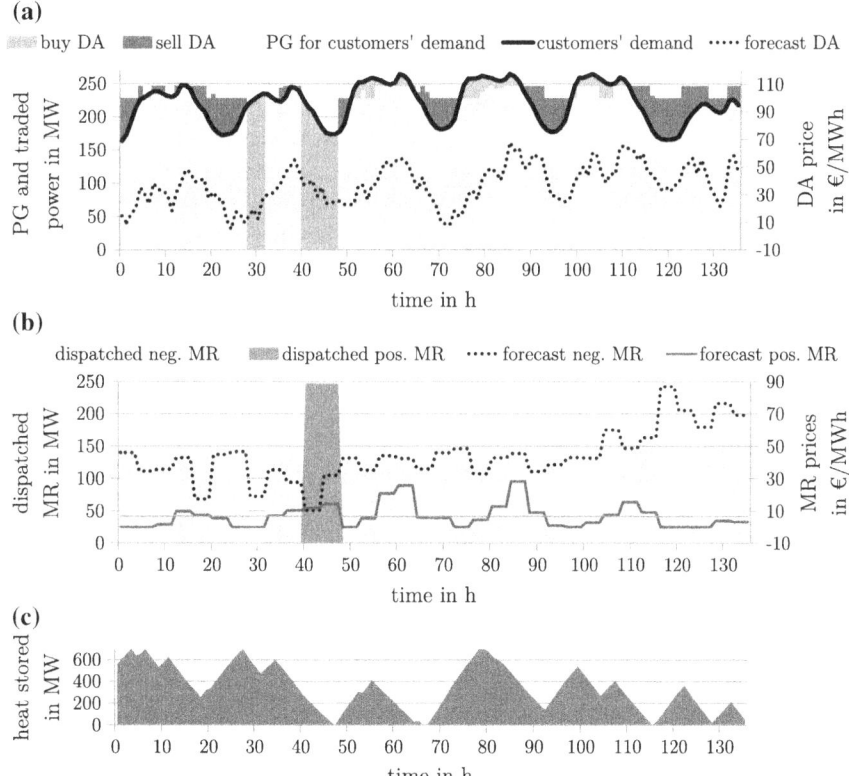

Fig. 2 (**a**) Power generation (PG) attuned to trading in the day-ahead (DA) market. (**b**) Dispatched volumes and forecasted prices for minute reserve (MR). (**c**) Operation of heat storage device

depicts the corresponding storage level. By decoupling heat supply and demand, heat storage enables a flexible operation of the CHP plant. Heat demand can be fulfilled by discharging the storage in times of low market prices, e.g. in hours 28–31, since it is not profitable to operate the CHP plant. Alternatively, the rising storage level in hours 67–78 implies that heat generation exceeds demand to charge the storage enabling additional revenues in the day-ahead spot and balancing market. This shows that the storage policy is optimised according to the trading opportunities of the CHP plant in both electricity markets.

5   Conclusion

We have presented a multistage stochastic optimisation model for trading in German electricity markets utilising a CHP plant with heat storage, which has not been examined in literature so far. Special focus lies on accurately modelling the specific structure of the German day-ahead spot and balancing market. A widely accepted approach for optimal operating CHP plants has been adapted to an energy supply company to support their participation in two electricity markets. Due to the CHP plant with heat storage, trading in sequential markets is a highly complex task. The case study shows that optimal bidding policies as well as plant operation react sensitively to changes in current market price levels. We conclude that an adequate representation of uncertain parameters by different scenarios is crucial. Due to computational tractability, the number of scenarios considered in the optimisation model is limited. In further studies, scenario reduction techniques are applied.

References

1. Boomsma, T.K., Juul, N., Fleten, S.: Bidding in sequential electricity markets: the Nordic case. Eur. J. Oper. Res. **238**(3), 797–809 (2014)
2. Fleten, S., Kristoffersen, T.K.: Stochastic programming for optimizing bidding strategies of a Nordic hydropower producer. Eur. J. Oper. Res. **181**(2), 916–928 (2007)
3. Lahdelma, R., Hakonen, H.: An efficient linear programming algorithm for combined heat and power production. Eur. J. Oper. Res. **148**(1), 141–151 (2003)
4. Plazas, M.A., Conejo, A.J., Prieto, F.J.: Multimarket optimal bidding for a power producer. IEEE Trans. Power Syst. **20**(4), 2041–2050 (2005)
5. Schulz K, Schacht M, Werners B (2014) Influence of fluctuating electricity prices due to renewable energies on heat storage investments. In: Huisman D et al. (eds.) Operations Research Proceedings, pp 421–428 (2013)
6. Swider, D.J.: Simultaneous bidding in day-ahead auctions for spot energy and power systems reserve. Int. J. Electric. Power Energy Syst. **29**(6), 470–479 (2007)

Well Drainage Optimization in Abandoned Mines Under Electricity Price Uncertainty

Reinhard Madlener and Mathias Lohaus

Abstract In this pit drainage study, we investigate how pump control optimized with respect to the prevailing electricity prices impacts the operating costs. The optimization of the well operation takes the dependency of the electrical power on the water volume lifted and the changing water levels into account. The nonlinear dependency is transformed into a linear optimization problem in multiple stages. First, a superstructure optimization is used. Second, the characteristic pump profiles are linearized piecewise, resulting in a simplified problem where only the multiplication of a binary and a positive real variable remains. The multiplication of the two variables is replaced by a new variable, transforming the optimization problem into a mixed integer linear optimization (MILP) problem. The results of the superstructure optimization yield the optimal pump size and the minimal costs incurred, which are then used to optimize the maintenance strategy. We find that well costs can be reduced markedly by the optimization proposed.

1 Introduction

The German coal mining industry in the Ruhr area will be terminated in 2018, altering the conditions for mine water drainage, which needs to be pursued also after the phaseout of coal production for environmental protection reasons [8]. Therefore, cost-effective continuous operation is of utmost importance, which is why a switch from conventional drainage to well drainage is currently under way ([4]:127). Furthermore, well drainage enables the usage of underground retention space and thus drainage optimization based on the electricity price level [6].

R. Madlener (✉)
Institute for Future Energy Consumer Needs and Behavior (FCN),
School of Business and Economics/E.ON Energy Research Center, RWTH Aachen
University, Mathieustrasse 10, 52074 Aachen, Germany
e-mail: RMadlener@eonerc.rwth-aachen.de

M. Lohaus
RWTH Aachen University, Templergraben 55, 52056 Aachen, Germany
e-mail: mathias.lohaus@rwth-aachen.de

© Springer International Publishing Switzerland 2017
K.F. Dœrner et al. (eds.), *Operations Research Proceedings 2015*,
Operations Research Proceedings, DOI 10.1007/978-3-319-42902-1_88

The aim of this study is to determine the cost-saving potential associated with the construction and operation of a well drainage. Here, an optimum well drainage is sized based on the uncertain electricity price development of the entire modeling horizon. The results indicate the "best practice case" for the studied time period.

For achieving the research goals described above, a mathematical model is developed that is set up as an MILP problem in GAMS and optimized in terms of profit maximization by means of the CPLEX solver. Instead of computing the total cost, however, the differences in costs and revenues compared to continuous operation are calculated. This operating point is used as a reference point because the drainage has to be operated in any case. Additional costs, e.g. for personnel, maintenance, and monitoring, are therefore neglected.

The remainder of this paper is organized as follows. Section 2 describes the optimization model for the well drainage. In Sect. 3, the optimization results are presented. Data for the years 2013 and 2014 are used for analyzing the impact of the electricity price. Section 4 discusses the findings and concludes.

2   Model Description and Mode of Operation

The time horizon for the analysis of the mine drainage site analyzed is one year. The granularity of the optimization model is one hour. To assess the impact of the electricity price, the results of two years are compared. EEX electricity spot market price data for the years 2013 and 2014 are used (day-ahead market).

For reasons of simplification, we assume that there are no starting operations (e.g. of the well pumps). Based on these framework conditions, three variants are examined, which result from the different modes of well drainage: (1) *Reference case.* Operation of a single well and three pumps, of which two are in permanent operation. The water level is treated as being independent from the electricity price, and allowed to be as high as possible. (2) *Pump operation.* The operation of the wells depends on the electricity price. (3) *Pump and turbine operation.* The drainage site is operated as a small pumped storage hydro power plant (PSHPP). For the economic calculations, we assume a depreciation of the investment in pumps and sloped pipelines over 20 years and an interest rate of 5 %.

In the following, the optimization model of the well drainage with a pump and a riser pipe is described. For submersible borehole pumps, which are designed specifically for the mining industry, a positive correlation between the generated pressure and the conveyed flow can be observed. In turn, the pressure is proportional to the height difference between the water levels of the upper and the lower basin. Thus, for a chosen pump, the pumped volume flow depends on the height (and thus pressure) difference.

The efficiency of a pump also depends on the flow rate and thus on the height difference. The pumps used for the mine water drainage have several stages. The model for the performance of each stage contains the flow rate $\dot{V}$, the height to be overcome, h, and the efficiency, η_P (Eq. (1)). For the total energy needs, the number of stages, η, is multiplied by the stage output, per one stage (Eq. (2)). The pressure can be seen as being equal to the performance. Per pump stage, a certain pressure can be generated depending on the pump type and the volume flow. The pressure curve is a function of the volume flow. Since the pressure is set by the actual height difference between the upper and the lower water level, the volume flow of the pump can be determined. The pressure loss Δp_{loss} shown in Eq. (3) is caused by the pipe friction; ρ and g denote the density of water and the gravity constant.

$$P_{el}^{onestage} \eta_P(V) - \dot{V} \cdot p(h) \tag{1}$$
$$P_{el} = \quad P_{el}^{onestage} \cdot \eta \tag{2}$$
$$p = \rho \cdot g \cdot h + \Delta p_{loss} \tag{3}$$

As a lower limit for the dimensioning of the pumps, two pumps must be sufficient to pump up the mine water in the case of failure of one pump, or when replacing a pump. This way, an adequate security level for the groundwater and drinking water can be achieved. For the investment costs of the pump, we assume a linear increase with power at 630 k€/MW [1].

To optimize well drainage, the use of a superstructure is suitable. To this end, the superstructure of the model must be specified. In this particular case of well drainage, it is pre-determined which pump types and how many pump stages can be selected. The superstructure is represented mathematically and converted into a mixed integer nonlinear optimization problem (MINLP) ([9]:378ff). For well drainage this is illustrated generally as follows. The functions of the stage performance and the steps to create pressure are presented as a function of $\dot{V}$, and can generally by determined by Eqs. (1)–(3).

$$max(Z) = \quad f(\delta, P_{nom}, P(t)) \tag{4}$$
$$g_i(\delta, P_{nom}, P(t), p(t)) = \quad 0 \tag{5}$$
$$h_j(\delta, P_{nom}, P(t), p(t)) = \quad 0 \tag{6}$$
$$\delta = 0, 1; P_{nom}, P(t); p(t) \in \mathbb{R} \tag{7}$$

Since for the well drainage primarily costs occur, minimizing the cost of lifting the mine water is indirectly connected to this. Investment and operating costs of the well drainage, which are specific to the variant considered, are both taken care of in the objective function Z. The integer binary variables (δ) describe the selected pump type and the selected pump stage. Volume flows, pump capacity, pressure, and other properties are implemented as floating point numbers in the model. The non-linear correlation between flow rate and pump power, pressure level, and efficiency thus poses a non-linear optimization problem. Likewise, the multiplication of binary variables by the floating point number variables also represents a non-linear system of equations.

A mixed-integer nonlinear system is problematic with regard to finding solutions, because in nonlinear optimization problems it cannot clearly be determined whether the optimal solution found describes a local or a global optimum ([2]:1115ff.). In order to avoid such problems, the optimization problem is transformed from an MINLP into an MILP, which leads to additional equations and additional binary variables for the optimization problem (for details see [7]). Firstly, the power and pressure curve functions with the dependency of the volume flow get piecewise linearized. The more supporting points, the merrier the calculation can become. At the same time the calculation costs rise ([2]:1119). This way, the optimization problem is still an MINLP, because it consists of the multiplication of a binary and a flow variable (for details see [7]). This non-linear function can be transformed exactly into a linear one, by using the method described by Yokoyama et al. ([10]:777). The method is applied several times for replacing the nonlinear equations.

3   Results of the Well Drainage Superstructure Optimization

As part of the superstructure, three different pump models were selected, each with two different numbers of pump stages (Table 1). Table 2 shows the results of the superstructure optimization for the three alternatives considered, using the current prices of the EEX day-ahead market for the years 2013 and 2014 [3]. As can be seen, the objective function value between the two alternative operation cases compared to the reference case shows a significant cost reduction (of approx. 200 k€ p.a.). When comparing the objective function values between the two alternative modes of operation, we can see that the additional turbine operation improves the objective function value only marginally compared to the pure pump operation. Due to the additional degrees of freedom, the objective function of the alternative "pump and turbine operation" achieved a higher objective function value. Note that in these calculations, the commodity price is simply considered as the electricity price traded on the EEX exchange.

The choice of the mode of operation changes when using the electricity price data in the objective function, but not the result of the chosen superstructure (Table 2). When comparing these variations, it can be noted that the potential of a pumped storage power plant offers no significant economic gain, and results mainly from the activities on the day-ahead spot market.

Table 1 Information on the three available pump models (cf. [5])

	Pump model 1	Pump model 2	Pump model 3
P_N per stage [kW]	99.4	225.9	158.4
$\dot{V}_N$ [m³/h]	850	1620	2400
Stages	19 or 20	16 or 17	34 or 35

Table 2 Results for the superstructure optimization

	Reference case		Pump operation only		Pump and turbine operation	
Electricity data of year	2013	2014	2013	2014	2013	2014
P_N [MW]	5.665	5.665	7.391	5.665	7.391	5.665
Selected pump	3 × mod. 1[a]		2 × mod. 1[a]; 1 × mod. 2[b]	3 × mod. 1[a]	2 × mod. 1[a]; 1 × mod. 2[b]	3 × mod. 1[a]
Pump head [m]	650	650	655.28	655.14	656.97	655.41
Investment cost [€]	−4,726,891	−4,824,391	−5,880,832	−4,834,414	−5,884,127	−4,834,941
Electricity purchase/ production [MWh]	32,817	32,817	33,269	33,101	35,978	33,459
	−	−	−	−	1464	195
Av. el. price [€/MWh]	37.36	32.76	25.96	26.11	27.00	26.40
Electricity costs [€]	−1,226,043	−1,075,085	−863,836	−864,418	−971,238	−883,301
Relative error [%]	0	0	1.86	0.30	2.05	0.5
Objective function [€]	−1,719,082	−1,577,517	−1,503,473	−1,386,636	−1,493,151	−1,385,885

Note [a] 19 stages, [b] 16 stages

The results in Table 2 for the superstructure optimization clearly show that the data used are highly dependent on the year, thus leading to significant differences. For instance, much better objective function values in general are achieved for 2014 compared to 2013. In comparison, although the average electricity price for 2013 is higher than for 2014, the average purchase price for 2013 is lower for the pure pump operation due to the usage of large pumps with a higher throughput. These can consume more (less) energy in times of low (high) electricity prices. This greater difference enables the superstructure optimization to tolerate higher investment costs for pumps with a higher throughput.

4 Discussion and Conclusion

One aim of this study was the economic optimization of the well drainage of an abandoned coal mine as part of the "eternal costs" by minimizing the operating costs for pumping. Formally, this was implemented by superstructure optimization and several distinct steps to transform the problem at hand into an MILP problem. The model created allows the mapping of pump characteristics and, therefore, the mathematical describing of the relationships between flow, pressure, and power.

Using the model, the optimal pump dimensioning can be computed for systems that are operated with widely varying water levels. Both the optimal type of pump and the optimum number of steps are calculated by means of superstructure optimization. For optimal results by means of superstructure optimization, a narrowing of the problem is crucial.

For the optimization of the drainage site considered, the minimum number of stages has achieved the best results because the available pump strokes have not been fully exploited. This is due to the ratio between the number of pumps to the existing base area available per meter of rising water.

In our study, the effect that the pumps are not only working in pump operation but can also be used as a turbine, was also investigated. In this case, the well drainage can be considered as a small pumped storage hydro power plant. In order to increase the cost-effectiveness when operating as a PSHPP, other electricity markets need to be considered as well.

The aim of our study was to find the "best practice case" for the well drainage previously calculated in the superstructure optimization, in the presence of electricity price uncertainty, given that future well drainage can indeed be operated cost-efficiently. While the current price development for the entire observation period is known in the superstructure optimization, this is not the case in practice. For this reason, an optimization of well drainage operation was carried out with a day-by-day 168-hour prediction horizon for the operation of the well, from which we can conclude the following (details see [7]): (1) There are marked economic potentials for the optimized operation calculated by means of a superstructure optimization; (2) Even with the minimum installed pump size, the electricity price-dependent operation of the wells can significantly improve the cost effectiveness. The use of smaller pumps can be advantageous for the stability of the mining area [6], given that the pit water level can only be lowered slowly.

References

1. Andritz Ritz GmbH. Bestätigung ber Investitionskostenschätzung für Tauchmotorpumpen. Telephone interview, 3 March 2015 (2015)
2. Borghetti, A., D'Ambrosio, C., Lodi, A., Martello, S.: An MILP approach for short-term hydro scheduling and unit commitment with head-dependent reservoir. IEEE Trans. Power Syst. **23**(3), 1115–1124 (2008)
3. EEX: EEX.com (2015-03-09) (2015). http://www.eex.com/de/marktdaten/strom/spotmarkt/. Accessed 9 March 2015
4. Fischer, P., Frankenhoff, H.: Die Entwicklung der Wasserhaltung an der Ruhr. In: Sroka, A. (Hrsg.). 12. Geokinematischer Tag des Institutes für Markscheidewesen und Geodäsie: Am 5. und 6. Mai 2011 in Freiberg. Schriftenreihe des Institutes für Markscheidewesen und Geodäsie an der Technischen Universität Bergakademie Freiberg 2011, 1. Essen: VGE, Verl. Glückauf, pp. 120–128 (2011)
5. Flowserve Hamburg GmbH, Pleuger Pumps: Unterwasserpumpen Submersible Pumps (2008). https://www.maschinensucher.de/dokumente/906581.pdf. Accessed 20 March 2015
6. Frankenhoff, H.: Personal interview M.L., 22 December 2014 (2014)

7. Madlener R., Lohaus, M.: Well Drainage Management in Abandoned Mines: Optimizing Energy Costs and Heat Use under Uncertainty, FCN Working Paper No. 12/2015, Institute for Future Energy Consumer Needs and Behavior, RWTH Aachen University, June (2015)
8. RAG. Konzept zur langfristigen Optimierung der Grubenwasserhaltung der RAG Aktiengesellschaft für das Saarland: gemäß §4 Erblastenvertrag zur Bewältigung der Ewigkeitslasten des Steinkohlenbergbaus der RAG AG im Rahmen der sozialverträglichen Beendigung des subventionierten Steinkohlenbergbaus in Deutschland vom 14.08.2007. Herne (2014)
9. Voll, P., Klaffke, C., Hennen, M., Bardow, A.: Automated superstructure-based synthesis and optimization of distributed energy supply systems. Energy **50**, 374–388 (2013)
10. Yokoyama, R., Hasegawa, Y., Ito, K.: A MILP decomposition approach to large scale optimization in structural design of energy supply systems. Energy Convers. Manag. **43**(6), 771–790 (2002)

The Future Expansion of HVDC Power Transmission in Brazil: A Scenario-Based Economic Evaluation

Christian Köhnke Mendonça, Christian A. Oberst and Reinhard Madlener

Abstract This paper investigates the future need of the Brazilian electric grid for High-Voltage-Direct-Current (HVDC) transmission lines using expansion scenarios. Currently, electricity is produced mainly in large hydropower plants with enormous reservoirs, but the model is approaching a limit because the potential for hydropower near load centers is almost depleted. To preserve a low-carbon electrical energy mix, renewable energy sources in far more distant locations need to be exploited. This paper focuses on the expansion potential of wind, solar and hydropower and its spatial mismatch with the expected future electricity demand. Linear optimization is used to determine the best HVDC connections. The results show clear differences in the costs for transmission lines in different scenarios, confirming that the transmission capacity is a critical factor in the process of expanding the renewable energy generation capacity in Brazil. This study provides insights for policy makers and industry.

1 Introduction

Renewable energies (RES) account for 83 % of installed power generation in Brazil 2013, of which large hydropower plants account for 69 % [7]. However, the current hydrothermal energy system is approaching its limits and probably it will be unable to meet the growing future demand (see [13]). The hydroelectric potential available near load centers (Southeast and South) is essentially depleted [4]. To exploit promising alternative renewable energy sources (in particular wind and hydropower in the

C.K. Mendonça (✉)
RWTH Aachen University, Templergraben 55, 52062 Aachen, Germany
e-mail: christian.koehnke@rwth-aachen.de

C.A. Oberst · R. Madlener
Institute for Future Energy Consumer Needs and Behavior (FCN), School
of Business and Economics/E.ON Energy Research Center, RWTH Aachen
University, Mathieustrasse 10, 52074 Aachen, Germany
e-mail: COberst@eonerc.rwth-aachen.de

R. Madlener
e-mail: RMadlener@eonerc.rwth-aachen.de

© Springer International Publishing Switzerland 2017
K.F. Dœrner et al. (eds.), *Operations Research Proceedings 2015*,
Operations Research Proceedings, DOI 10.1007/978-3-319-42902-1_89

Amazon rain forest), long-distance transport (2000 km and more) of large amounts of electricity is needed [10]. HVDC is technically and economically a superior transmission technology for such long distances (over 700 to 900 km), but expensive DC-AC converter stations constitute a crucial cost factor for planning the future grid and generation sites [16].

This paper investigates to which extent certain expansion scenarios of the RES capacity in the period 2015–2035 would entail the construction of additional HVDC transmission lines. The focus is on modeling scenarios for a weak and a strong expansion of hydropower, wind and/or solar power according to their unused potentials (hereafter referred to as "potential").

2 The Electricity Sector in Brazil

There is a spatial mismatch between the RES power potential and the expected increase in electricity demand by 2035 (see Fig. 1). The highest potentials for wind power and hydropower are in the Northeast and in the North/West-Central (Amazon rainforest) regions, respectively [1, 14]. However, the highest increase in electricity demand is expected in the Southeast. The RES potential in the rainforest is restrained

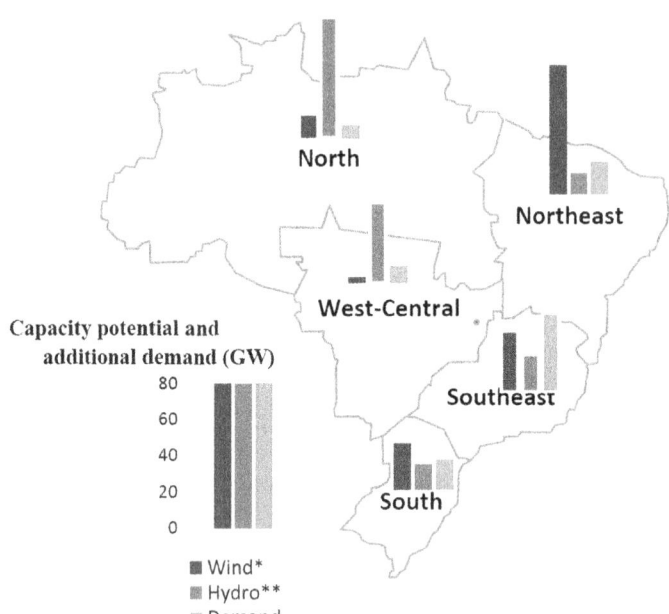

Fig. 1 Geographical distribution of wind and hydropower potential (nominal) and expected increase in electricity demand in 2035 *Source* Based on [1, 3–5]

by environmental concerns and difficult accessibility from which we abstract in our analysis [12]. Solar potential is ubiquitously available.

With or without the further expansion of hydropower capacity, the immense storage capacity of the hydrothermal energy system, which is large enough to absorb the volatile availability and limited predictability of RES [6, 15, 17], supports the integration of high shares of alternative RES in the power matrix. Regional seasonal complementarity of wind and hydro-power can increase storage capacity further [2].

3 Methodology

The optimal grid solution is sought by grid designs that enable the cost-efficient provision of power originated from RES in each scenario for the future electric mix in 2035. The linear optimization of installed transmission capacity (HVDC grid designs) for each scenario is similar to a procedure applied in [19] for a pan-European powers system in 2050 and [18] for a 2030 US grid expansion plan that includes 20 % of wind energy. Mathematically, it can be formulated as:

$$min \sum_{i \in N_{GEN}, j \in N_{LOAD}} y_{(i,j)} l_{(i,j)} + C \sum_{i \in N_{LOAD}} p_i^{NR} \tag{1}$$

$$\text{s.t.} \quad f_{(i,j)} \leq y_{(i,j)} \qquad \forall i \in N, j \in N \tag{2}$$

$$\sum_{j \in N} f_{(i,j)} = -p_i \qquad \forall i \in N_{GEN} \tag{3}$$

$$\sum_{j \in N} f_{(i,j)} = d_i \qquad \forall i \in N_{LOAD} \tag{4}$$

$$f_{(i,j)} = -f_{(j,i)} \qquad \forall i \in N, j \in N \tag{5}$$

$$|f_{(i,j)}| \geq t_{(i,j)} \bigvee f_{(i,j)} = 0 \qquad \forall i \in N, j \in N \tag{6}$$

$$d_j = d_j^C - p_j^{NR} \qquad \forall j \in N_{LOAD} \tag{7}$$

$$p_i = p_i^R - s_i \qquad \forall i \in N_{GEN} \tag{8}$$

$$s_i, p_j^{NR} \geq 0 \qquad \forall i \in N_{GEN}, j \in N_{LOAD} \tag{9}$$

The decision variable $y_{(i,j)}$ represents the amount of installed transmission capacity between regions i and j, while $l_{(i,j)}$ represents the length as an approximation of the costs. N represents all nodes in the grid, composed of the regions with surplus generating capacity N_{GEN} and the regions with a power deficit N_{LOAD}. The possibility for power generation with non-renewable sources within the regions is also considered, but this is only done when there is not enough renewable capacity given to fulfill the demand. To take this into account, the non-renewable generation capacity given by p_i^{NR} is included in the objective function (1) and multiplied with

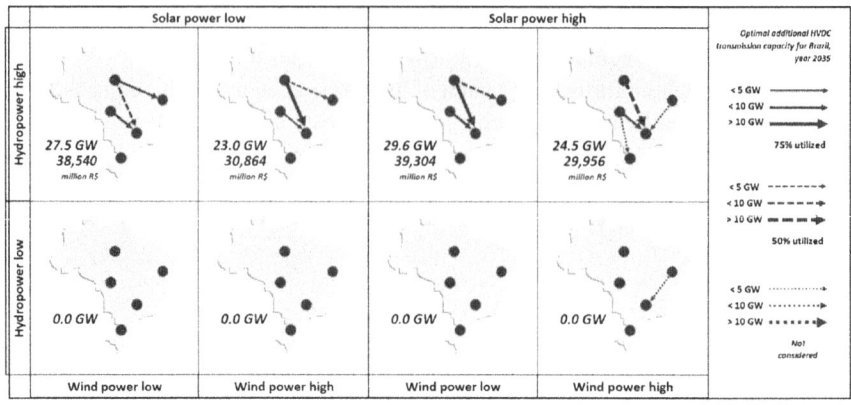

Fig. 2 Summary of results by scenario

$C \gg y_{(i,j)}l_{(i,j)}, \forall i, j \in N$. The power flow $f_{(i,j)}$ between i and j needs to be smaller or equal to the transmission capacity $y_{(i,j)}$ (2); the flow leaving regions with surplus needs to be equal the surplus renewable power p_i (3); and the flow arriving at a region with power deficit needs to be equal to the demand d_i (4). The regional demand d_j is defined as the actual regional demand d_j^C minus the non-renewable generation capacity p_j^{NR} installed in the region (7). The regionally available power p_i is defined as the difference between the power from renewable sources p_i^R and the surplus s_i (8). The non-renewable capacity and the surplus are both positive (9) and are only different from zero if $p_i^R \neq d_j^C$, i.e. when the renewable generation capacity is different than the actual demand. The threshold $t_{(i,j)}$ is defined as the minimum threshold a power needs to surpass.

The optimization for different expansion scenarios was performed using a computer program developed specifically for this task (MS Visual Basic .NET). The optimization problem was solved using an iterative simplex algorithm for parameters within the range of possible combinations of additional installed RES. For additional information see [11]. In total 16 scenarios were examined, considering:

- Low (0–75 GW) and high (75–150 GW) additional hydropower capacity.
- Low (0–65 GW) and high (65–130 GW) additional wind power capacity.
- Low (0 GW) and high (100 GW) additional solar power capacity.

4 Results

The further expansion of hydropower (distant from demand foci) is identified as a main driver for the grid expansion with HVDC (see Fig. 2). Increased wind power capacity reduces the need for long distance transmission, whereas the assumed expansion of solar power has no relevant impact.

The need for HVDC is particularly high in the scenario of high hydropower expansion and low wind and solar power expansion, which would represent the continuation of the current model and sustain Brazil's high share of RES. Southeast and Northeast have a deficit in RES power supply and could draw green electricity from the West-Central region and the North. This means that hydropower plants in the Amazon rainforest (North) substantially supply electricity to the Northeast. The connection would have an average utilization rate of above 75 % for a transmission capacity of up to 11 GW (requiring multiple HVDC connections).

The Southeast region does have an even larger deficit in RES supply, and thus needs to import power from the West-Central region. A transmission capacity of up to 10.8 GW would be used at a 50 % rate (mainly in the rainy part of the year), or with a capacity of 5 GW at a 75 % rate. The deficit in the Southeast could be further met with hydropower from the North, but the added HVDC connections would mainly be used during the rainy season and implying a 50 % usage rate (by a capacity of 11 GW). HVDC lines connecting the North and the West-Central region to the Northeast and the Southeast have to traverse dense tropical forest and savanna areas [9]. Their implementation would certainly require the clearing of vegetation which adds to the environmental impact of expanding hydropower generation that is distant to load centers in Brazil.

Scenarios with expanded wind power mainly benefits the Northeast, erasing the power deficit during the windy second half of the year. Moreover, the reduced power flow to the Northeast enables the transport of more RES electricity to the Southeast (increasing the utilization rate to 75 % at 12 GW capacity). Furthermore a possibility exists to build HVDC transmission lines between the Northeast and the Southeast, in order to enable the export of RES power from the Northeast to the South. Although this would be a good complement to the seasonality of hydropower, the simulation showed only low power flows between the regions.

Further, if a HVDC line would be built to connect the Northeast to the Southeast, the main challenge would be to surpass several mountain chains in the state of Minas Gerais in the Southeast [8]. Considering the high unit energy costs as well as the geographical challenges, all-in-all a connection between the Northeast and the Southeast seems disadvantageous.

5 Conclusion

In light of increasing electricity demand and imminent limits of the present electricity market with predominantly large hydropower plants near load centers, Brazil has to expand and diversify its electricity mix. This paper investigates the amount of additional HVDC transmission capacity needed in several expansion scenarios of RES over the period 2015–2035. Scenarios of weak and strong expansion of hydro, wind, and solar power are used as parameters of an HVDC transmission capacity optimization, and the actual need for transmission lines is assessed using both technical and economic criteria.

The results from this study confirm the need for large investments in HVDC transmission lines for certain scenarios, in particular with the development of large hydroelectric power plants in the North and West-Central region. Wind power expansion reduces the power deficit in the Northeast, Southeast and South, thus alleviating the potential needs to import electricity through HVDC lines. However, seasonality patterns limit the overall wind power potential. The expansion of solar power shows mixed results, in general decreasing the need to transport, but this is less important for the HVDC compared to hydro and wind power expansion. Considering all the aspects mentioned, the present results call for a stronger utilization of alternative RES besides hydropower. An expansion of the hydropower capacity would most likely entail high transmission costs (apart from the environmental impact), whereas an expansion of wind and solar power (and likely biomass and small hydropower capacities) is potentially less expensive in a holistic approach for the energy system.

References

1. Amarante O., Zack M., de Sá A (2001) Atlas do Potencial Eólico Brasileiro, Brasília
2. Amarante O., Schultz D., Bittencourt R., Rocha N.: Wind/Hydro Complementary Seasonal Regimes in Brazil, DEWI Magazin Nr. 19, Rio de Janeiro RJ, Brazil (2001)
3. Agencia Nacional de Águas (2005). Caderno de Recurso Hídricos: Disponibilidade e Demandas de Recursos Hídricos no Brasil, Brasília
4. ANEEL: Atlas de Energia Elétrica do Brasil, 3rd edn. Agência Nacional de Energia Elétrica, Brasília (2008)
5. Banco de Informaões de Geraão (BIG) (2015), ANEEL (Brazilian Electricity Regulatory Agency) http://www.aneel.gov.br/aplicacoes/capacidadebrasil.cfm (accessed July 24, 2015)
6. Dester, M., Andrade, M.T.O., Bajay, S.V.: New renewable energy sources for electric power generation in Brazil. Energy Sources, Part B: Econ. Plan. Policy 7(4), 390–397 (2012)
7. EPE (2013) Plano Decenal de Expansão de Energia (PDE) 2022, Rio de Janeiro, Brazil
8. IBGE (2006) Mapa de Unidades de Relevo do Brasil, Rio de Janeiro, Brazil
9. IBGE (2006). Mapa de Vegetao do Brasil, Rio de Janeiro, Brazil
10. Jardini, J., dos Santos, M., Casolari, R., Vasquez-Arnez, R., Saiki, G., Sousa, T., Nicola, G.: Power transmission over long distances: economic comparison between HVDC and half-wavelength line. IEEE Trans. Power Deliv. 29(2), 502–509 (2014)
11. Köhnke Mendonça C., Oberst C.A., Madlener R.: The future expansion of HVDC power transmission in Brazil: a scenario-based economic evaluation. FCN Working Paper No. 14/2015, Institute for Future Energy Consumer Needs and Behavior, RWTH Aachen University, October (2015)
12. Mesarović: Theory of Hierarchical, Multilevel Systems. Academic Press, New York (1970)
13. Ministério de Minas e Energia, Plano Nacional de Energia 2030, Brasília, Brazil (2007)
14. Pereira, M., Camacho, C., Freitas, M., da Silvam, N.: The renewable energy market in Brazil: current status and potential. Renew. Sustain. Energy Rev. 16(6), 3786–3802 (2012)
15. Steinberger.: Integrao em larga escala de gerao eólica em sistemas hidrotérmicos. (Ph.D. Thesis), UFRJ/COPPE, Rio de Janeiro, Brazil (2012)
16. Sousa T., dos Santos M.L., Jardini J.A., Casalori R.P., Nicola G.: An Evaluation of the HVDC and HVAC Transmission Economic, Transmission and Distribution: Latin America Conference and Exposition (T&D-LA), 2012 Sixth IEEE/PES, Montevideo, September 3-5, 2012
17. The Electricity Supply Reliability from the Perspective of the Expansion Planning and the Power Systems Operation Regarding the Integration of Intermittent Sources in the Brazilian Energy Matrix. IAEE, New York, USA (2014)

18. U.S. Department of Energy 20% Wind Energy by 2030: Increasing Wind Energy's Contribuition to U.S. Electricity Supply, Washington DC (2008)
19. Wiernes P.E., Schabram J., Linnemann C., Kraemer C., Moser A., Mercado P.E. (2013) Joint Strategic Planning of Generation and Transmission Network, Energytech 2013, p. 1-5, IEEE

Two-Stage Heuristic Approach for Solving the Long-Term Unit Commitment Problem with Hydro-Thermal Coordination

Alexander Franz, Julia Rieck and Jürgen Zimmermann

Abstract In the last decade the share of renewable feed-in increased sharply. More than 25 % of the gross electricity consumption in Germany is typically covered by renewable sources. Despite the positive effects, e.g., low marginal costs and a sustainable electricity supply with less emissions, a generation portfolio should also contain thermal power plants as well as energy storages, mainly hydro storages, to guarantee a long-term security of electricity supply. The problem of flexible coordination and dispatch of thermal power plants and hydro storages in order to meet the highly volatile residual demand (energy demand minus the volatile feed-in of renewables) becomes even more challenging for generating companies. Therefore, we present a new two-stage heuristic approach for solving the resulting long-term unit commitment problem with hydro-thermal coordination (UCP-HT), where system operating costs have to be minimized. Within a comprehensive performance analysis, the results of the approach are compared to near-optimal solutions obtained by CPLEX on the basis of a tight mixed-integer linear formulation for the UCP-HT. In particular, well-known instances from the literature with real-world energy demands, a planning horizon of one year, and hourly time steps are solved within minutes ensuring a solution gap of less than 1 %.

1 Introduction and Problem Specification

The increasing share of prioritized renewable energies leads to challenging problems for generating companies and system operators. Due to the volatile and stochastic characteristic of renewable feed-ins and residual power demands, the need for highly flexible, but cost-efficient, power plant operations arises. Energy storages, mainly hydro storages, are a promising mean to support these flexible requirements and to smooth the residual demand for thermal power plants. In hydro storages, excess

A. Franz (✉) · J. Rieck · J. Zimmermann
Operations Research Group, Institute of Management and Economics,
Clausthal University of Technology, 38678 Clausthal-Zellerfeld, Germany
e-mail: alexander.franz@tu-clausthal.de

© Springer International Publishing Switzerland 2017   667
K.F. Dœrner et al. (eds.), *Operations Research Proceedings 2015*,
Operations Research Proceedings, DOI 10.1007/978-3-319-42902-1_90

power is used to pump water into a higher reservoir to be used later in reverse operation providing electricity.

The dispatching and coordinating of power plants and energy storages result in the unit commitment problem with hydro-thermal coordination (UCP-HT) [1, 7]. A solution of the treated problem is a production schedule for a set of power plants and energy storages that meet the system residual demand (energy demand minus the volatile feed-in of renewables) as well as techno-economic plant specific and system-oriented constraints. In particular, we consider a long-term planning horizon (i.e., one year with hourly time steps) and a cost-based objective function, where operating costs have to be minimized.

The UCP and obviously the UCP-HT can be formulated as a mixed-integer linear program (MILP) [7]. Thereby, binary decision variables (e.g., u_{it} that indicate if power plant i is active in period t or not) have to be included. Since the numbers of binary decision variables as well as the number of constraints in the model depend on the length of the planning horizon, only instances with a short-term planning horizon (i.e., one day or one week) can be solved to optimality within reasonable time. For long-term instances, heuristic approaches must be used. Heuristics for the UCP (or the UCP-HT) can be classified into conventional techniques and metaheuristic algorithms (see, e.g., survey in [5]). Conventional techniques include simple priority list-based methods, Dynamic Programming and decomposition techniques like Lagrangian Relaxation (e.g., [7]), which tends to be an industrial standard. Metaheuristics are often based on local search, genetic algorithms or simulated annealing (e.g., [3, 6]).

In what follows, we present a new two-stage heuristic approach for the UCP-HT which enhances a priority list-based approach. Promising results are already presented by e.g., Delarue et al. [2], who focus on repairing an infeasible start solution (without any consideration of energy storages). Among a similar local repair step, we introduce a multiple local improvement step in the first stage of our algorithm. In the second stage, the total operating costs are reduced by successively decommitting power plants and replacing them by energy storages. Moreover, e.g., even negative residual energy demands are considered using a preprocessing mechanism in order to fix necessary energy storage activities. Additionally, a random generation of (infeasible) start solutions allows us to embed the approach in a multi-start scheme. Computational experiments using well-known instances from the literature as well as real-world data show very convincing results in terms of solution gap and computational time.

2   Model

In order to describe the proposed cost-based UCP with energy storages in more detail, the consideration of a mathematical model formulation is beneficial. Therefore, we present an overview of a tight MILP for the UCP-HT. The basic idea of the model comes from Morales-España et al. [4] and Carrión et al. [1], who formulated the problem without energy storages. We extended the existing model in order to manage the coordination between thermal and storage units. Furthermore, a reformulation of constraints is performed so that the resulting model is able to find near-optimal solutions significantly faster than [4].

The objective of the UCP-HT under consideration is to minimize the total system operating costs in order to obtain the production schedule for each plant $i \in I$ and storage $j \in J$ over all time intervals $t \in T$. Hence, a solution can be represented by (binary) on/off status variables u_{it} and (continuous) power generation levels P_{it} of power plants i as well as generation and pumping levels of (hydro) energy storages j. The total system operating costs covers relevant production-related cost components, consisting of thermal production costs FC_{it}, thermal start-up costs SC_{it}, (hydro) storage operating costs HC_{jt} as well as non-served energy costs $c^{non} N_t$. Thermal production costs arise mainly from fuel consumption as well as emission certificates and can typically be considered as the major cost component. Storage operating costs can be neglected, since no (direct) fuel costs occur for pumping or generating and the remaining costs for e.g., maintenance are quite low. Non-served energy costs penalize a (negative) deviation from the given energy demand in the amount of N_t and help to decrease computational time effort. Taking these aspects into account, the general objective function (1), which is to be minimized, has the following form:

$$\text{minimize} \quad \sum_{t=1}^{|T|} \sum_{i=1}^{|I|} (FC_{it} + SC_{it}) + \sum_{t=1}^{|T|} \sum_{j=1}^{|J|} HC_{jt} + \sum_{t=1}^{|T|} c^{non} N_t \qquad (1)$$

A feasible solution for the UCP-HT must satisfy several thermal and storage system restrictions as well as technical and logical constraints that can be formulated as linear constraints. In particular, each unit (power plant or energy storage) has to be operated within its technical limits. This means that the power generation level can vary from a minimum to a maximum level (a specific parameter of each individual unit to ensure an optimal and secure usage). In case of a thermal power plant, it should be guaranteed that, once a thermal generator is started (shutdown), it has to be online (offline) for at least its minimum up time (minimum down time). For (hydro) storages, energy balance and energy flow conservation equations are to be considered to calculate the amount of energy retained in each storage at each time t. Furthermore, logical constraints are necessary to determine the binary status of a shutdown or start-up of a thermal plant as well as to differentiate between e.g., a cold and hot start-up of a plant. Moreover, the fulfillment of the energy demand and the provision of a sufficient spinning reserve (for the event of an unpredictable outage) over all active plants and storages have to be ensured. Finally, constraints define the domain of each decision variable (e.g., $u_{it} \in \{0, 1\}$, $P_{it} \geq 0$, $N_t \geq 0$ etc.). A standard constraint formulation can be found in e.g., [1, 4] or [7]. The resulting MILP model can be given to a solver (e.g., CPLEX). Since the run times are quite large for long-term instances (several hours), heuristic solution approaches are required.

3 Two-Stage Heuristic Approach

In this section, a two-stage heuristic approach for the UCP-HT is presented that is able to solve instances from the literature with real-world energy demands, a planning horizon of one year, and hourly time steps with a gap of less than 1 % within minutes. The basic concept of our considered heuristic is illustrated in Fig. 1. In the first stage, a thermal plant optimization is performed and in the second stage, the hydro-thermal coordination is considered. In addition, a preprocessing is included (prior to the first stage) to ensure feasibility in case of missing total thermal capacity, negative residual demands or high gradients of the residual demand.

The *Thermal Plant Optimization* preselects certain plants to fulfill the fluctuating residual demand and spinning reserve requirements without using any energy storages. Moreover, several techno-economic parameters like power output specifications, minimum up times (MUT) and minimum down times (MDT) or time-dependent startup costs are considered. In particular, three steps are determined:

Create start solution An initial (typical) infeasible solution is obtained by using a greedy algorithm. Thereby, all plants are ranked and sorted according to their marginal costs (i.e., their merit order). Then, the plants are committed in this order as long as the demand and reserve requirements are fulfilled. A random perturbation of marginal costs can be provided in order to generate many different start solutions building a multi-start scheme for the two-stage heuristic.

Deduce feasible solution A rule-based repair mechanism (similar to the approach of [2]) creates a feasible solution which satisfies all constraints described in Sect. 2. Mainly, a correction of the MUT and MDT is performed in the repair mechanism by enlarging an operating or idle phase. Quite short operating phases of base load power plants must be prevented. Only peak-load power plants can be operated with a short duration in order to avoid high production costs.

Improve feasible solution Further improvements can be achieved by local optimization. Thereby, each unit operating phase is assessed and, if applicable, replaced by an operating phase of a set of other plants providing the highest amount of cost savings. Thus, the solution is locally changed, if it is economically reasonable.

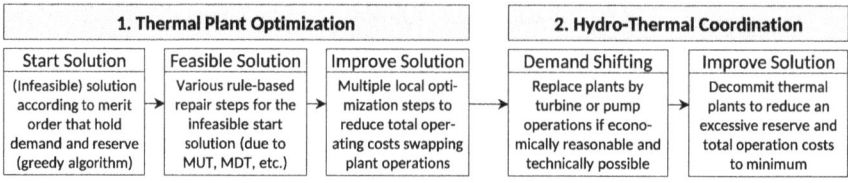

Fig. 1 Principal concept of the considered two-stage heuristic approach for the UCP-HT

The second stage improves the solution determined in the first stage by implementing a *Storage-Thermal* or *Hydro-Thermal Coordination* for pumped storages. Two steps have to be executed:

Shift demand using energy storages In the demand shifting, the best allocation of energy storages is found by replacing committed plants by storage operations to achieve further cost reductions. A gradient analysis identifies suitable time intervals for generating (high energy demand) or retaining energy to a storage (low energy demand). For each possible generating time, a decommitment of each active thermal plant is evaluated iteratively, starting with the plant that offers the maximum saving potential. Consequently, suitable and cost efficient storing phases (which are identified before) are allocated to balance the energy withdrawal due to the generation. If this particular storage operation causes a reduction of the total system operating costs, the solution schedule is updated. Moreover, in each iteration a self-repairment and deallocation step is performed to maintain a feasible solution (e.g., in terms of energy balance or upper and lower levels of the stored energy).

Improve feasible solution A last improvement step is performed to locally replace or decommit (redundant) operating phases of thermal plants if cost savings are achievable or if excessive spinning reserve is existent. Finally the solution schedule consists of an hourly dispatch of each thermal and storage unit for, e.g., a yearly time-horizon.

4   Computational Study

Within our performance analysis, we compared the results of the two-stage heuristic approach to the results obtained by CPLEX using our MILP formulation. Here, we present only an extract of our numerical study consisting of 100 problem instances classified in two test sets. Test set T_1 was first introduced by [3] and is commonly used for analyzing UCP models (e.g., in [2, 4] or [6]). All instances consist of ten basic thermal units which are exactly replicated in order to get a 20, 30, ..., 100 plant system. Thereby, the basic 24 h demand is adjusted relatively to the total capacity of the replicated system. For testing long-term planning horizons the 24 h planning horizon is copied to get instances of 1, 4, 12, and 20 weeks as well as 1 year. In test set T_2, the recurrent demand pattern is replaced by the actual residual demand in Germany in 2012 (again, adjusted to the total thermal capacity). Furthermore, in each instance of T_1 and T_2 hydro storages are added according to the total thermal capacity (about 7 %, which is the current situation in Germany).

The average results of the 10–100 plant system are presented for each time horizon in Table 1. Additionally, worst-case gap-values are written in brackets. Columns 3–6 refer to T_1 and columns 7–10 to T_2. It can be observed that the presented two-stage heuristic approach delivers the best results for long-term instances, whereas the solution time stands out and is always substantially lower compared to the MILP. Using the multi-start scheme with 100 iterations further gap-improvements of 0.1 %

Table 1 Computational results for T_1- and T_2-instances (avg. gap (max. gap) [%], avg. t^{cpu} [s])

Time horizon	#plants	MILP (T_1)		Heuristic (T_1)		MILP (T_2)		Heuristic (T_2)	
		gap	t^{cpu}	gap	t^{cpu}	gap	t^{cpu}	gap	t^{cpu}
1 week	10–100	0.2 (0.8)	11	0.8 (0.9)	0.1	0.3 (0.6)	31	0.9 (1.0)	0.1
4 weeks	10–100	0.5 (0.9)	177	0.8 (0.9)	0.5	0.6 (1.0)	198	0.6 (0.8)	0.3
12 weeks	10–100	0.7 (0.9)	558	0.9 (1.0)	4.2	0.5 (0.9)	3,406	0.7 (0.9)	2.4
20 weeks	10–100	0.6 (0.9)	1,103	0.9 (1.0)	11.0	7.4 (68.9)	3,586	0.8 (1.0)	5.8
1 year	10–100	6.3 (48.4)	14,245	**0.9 (1.0)**	**72.1**	62.8 (99.8)	14,167	**0.9 (1.0)**	**58.9**

Performed on an Intel Core i7-2760QM CPU with 2.7 GHz and 8 GB of RAM; GAMS 24.0 and CPLEX 12.4 used for MILP solutions (duality gap: 1%, time-limit: 5 h)

can be achieved. All in all, our two-stage heuristic approach can be classified as a fast method for the UCP-HT which solves problems to near-optimality for all tested instances (according to the worst-case values no solution exceeds the 1% gap).

5 Conclusion

After introducing the UCP-HT problem and its MILP formulation, we proposed a short introduction to our two-stage heuristic approach to solve it. This method significantly outperforms the MILP as well as other tested decomposition techniques (e.g., Benders Decomposition) for mid-term and long-term planning horizons. In detail, the solution time is substantially less than 1% of the original CPLEX-solution time needed to solve the UCP-HT and the solution quality is comparable or even better. Therefore, it can be concluded that the presented construction and improvement heuristic is well suited for solving unit commitment problems with storages in energy applications.

In future investigations, we focus on applying the considered heuristic approach for real-world applications (e.g., a case study of the German electricity market) and on embedding our work into a hierarchical structure for a more strategic perspective to support establishment, retrofit or retirement decisions of thermal plants.

References

1. Carrión, M., Arroyo, J.M.: A computationally efficient mixed-integer linear formulation for the thermal unit commitment problem. IEEE Trans. Power Syst. **21**, 1371–1378 (2006)
2. Delarue, E., Cattrysse, D., D'Haeseleer, W.: Enhanced priority list unit commitment method for power systems with a high share of renewables. Electric Power Syst. Res. **105**, 115–123 (2013)
3. Kazarlis, S.A., Bakirtzis, J.M., Petridis, V.: A genetic algorithm solution to the unit commitment problem. IEEE Trans. Power Syst. **11**, 83–92 (1996)

4. Morales-España, G., Latorre, J.M., Ramos, A.: Tight and compact MILP formulation for the thermal unit commitment problem. IEEE Trans. Power Syst. **28**, 4897–4908 (2013)
5. Saravanan, B., Das, S., Sikri, S., Kothari, D.P.: A solution to the unit commitment problem - a review. Front. Energy **7**, 223–236 (2013)
6. Viana, A., Pinho de Sousa, J., Matos, M.A.: Fast solutions for UC problems by a new meta-heuristic approach. Electric Power Syst. Res. **78**, 1385–1395 (2008)
7. Wood, A.J., Wollenberg, B.F., Sheblé, G.B.: Power Generation, Operation, and Control. Wiley, Hoboken (2014)

Bilevel Model for Retail Electricity Pricing

Georgia E. Asimakopoulou, Andreas G. Vlachos
and Nikos D. Hatziargyriou

Abstract The advent of smart electricity meters is bound to affect the electricity distribution sector in more ways than one can imagine. Thus, companies involved in the retail electricity market will need to adjust to the new conditions, especially with regard to the electricity pricing, which constitutes an important part of the activities of a retailer. In this work, a bilevel model is proposed as a tool for facilitating the retailer's task in optimally selecting price levels to be announced to various local resources composing his portfolio. Application of proper linearizations allows obtaining and equivalent one-level mixed integer linear programming problem (MILP) which is solved using CPLEX solver under GAMS. Examination of several scenarios regarding the form of the pricing scheme shows that the number of time zones as well as the grouping/assignment of each hour of the day to one time zone or another affects significantly the profitability of the retailer's business.

1 Introduction

The electricity distribution is currently undergoing significant changes both in its mode of operation and in its form. New types of loads (shiftable, curtailable), as well as the existence of distributed generation units pose new challenges, while advances in the Information and Communication Technologies provide new opportunities for the network and its users (Operator, retailers, prosumers). Smart meters are seen as a means for enabling customers having a better control over their electricity bills and for helping the retailer in the process of data gathering. More importantly, they

G.E. Asimakopoulou (✉) · N.D. Hatziargyriou
School of Electrical and Computer Engineering, National Technical
University of Athens, 9, Iroon Polytechniou Str., 15773 Athens, Greece
e-mail: gasimako@power.ece.ntua.gr

N.D. Hatziargyriou
e-mail: nh@power.ece.ntua.gr

A.G. Vlachos
Regulatory Authority for Energy, 132, Pireos Str., 11854 Athens, Greece
e-mail: agvlachos@rae.gr

© Springer International Publishing Switzerland 2017
K.F. Dœrner et al. (eds.), *Operations Research Proceedings 2015*,
Operations Research Proceedings, DOI 10.1007/978-3-319-42902-1_91

constitute an invaluable ally for facilitating the pricing task of the retailer, reflecting more efficiently the true cost of electricity.

Price signals made available through in-home displays are expected to affect the behavior of the distribution network users in a way difficult to predict using existing tools (e.g. load forecasting). In attempting to participate in the retail market cost-efficiently, the electricity retailer needs to incorporate such a complex environment in its operations. The question that arises is: *at what level should the retail price be set and what form the pricing scheme should have in order to induce the consumers in a behavior optimal both for them and for the retailer?* In order to answer this question we first identify the hierarchical decision-making process underlying the aforementioned interaction between electricity retailer and consumer. We, then, formulate a bilevel programming problem (BLPP)—the mathematical equivalent of a Stackelberg game—that enables testing a variety of pricing schemes. The results are compared in terms of profitability for the retailer and cost for the customer.

An extensive review of relevant applications can be found in [1].

2 Model Description and Mathematical Formulation

Management and integration of local resources to the operation of the retail electricity market are considered to be a responsibility of the retailer. His role as a mediator between the local resources and the wholesale market encompasses several relevant tasks: keeping the energy balance of his portfolio, selling to/buying from the wholesale market eventual energy surplus/deficit, determining the price levels to be announced to the various categories of local resources. The local resources, in turn, based on the given price signals and individual preferences select the respective energy volume (to be consumed, produced or curtailed), thus forming the BLPP[1,2]:

$$
\min_{\substack{e^{p,t},RP^{p,z},CP^{p,z},\\ PP^{p,z},pm^{p,z}}} \sum_{p,z} \sum_{t \in T_z} w(p) \Bigg[SMP^{p,t} e^{p,t} + \sum_{pb,s} PP^{p,z} x_{pb}^{p,t,s}
$$
$$
+ \sum_{dc,s} (CP^{p,z} x_{dc}^{p,t,s} - \overline{RP}\, \overline{Q}_{dc}^{p,t}) - \sum_{dp} RP^{p,z} x_{dp} Q_{dp}^{p,t} - \sum_{db,s} RP^{p,z} x_{db}^{p,t,s} \Bigg] \quad (1)
$$

---

[1]The model presented here is based on the one appearing in [2]. However, for ensuring the integrity of the paper, the equations describing the model are reproduced here in order to emphasize the individual characteristics that differentiate the current implementation from the previous one.

[2]Term $\overline{RP}\, \overline{Q}_{dc}^{p,t}$ represents the payment made by demand entity dc to the retailer for baseline energy consumption $\overline{Q}_{dc}^{p,t}$ charged at the predefined price $\overline{RP}$ and is constant. Thus, the optimization results are not affected by this term; it is only given here for completeness and is subsequently ignored.

Table 1 Nomenclature

Indices and sets	
$dr \in DR$	Set of distributed resources, where dr includes $db \in DB$ for consumers with load bids, $dc \in DC$ for consumers with curtailable loads, $pb \in PB$ for producers with production bids
$dp \in DP$	Set of consumers with demand profiles
$s \in S$	Set of steps of the submitted bids of the distributed resources
$p \in P$	Set of characteristic periods
$z \in Z$	Set of dispatch time zones
$t \in T$	Set of dispatch intervals
$t_z \in T_z$	Set of dispatch intervals in time zone z of period p
Parameters	
N_z	Number of dispatch time zones
$w(p)$	Scenario weight of each characteristic period
$P_{dr}^{p,t,s}, Q_{dr}^{p,t,s}$	Price-quantity pair at period p, dispatch interval t and block s for the hourly priced bids submitted by distributed resource dr (in €/MWh and MWh, respectively)
$P_{dp}, Q_{dp}^{p,t}$	Price-quantity pair of profile consumer dp at period p and dispatch interval t (in €/MWh and MWh, respectively)
$\overline{Q}_{dc}^{p,t}$	Demand volume without curtailment at period p and dispatch interval t of consumer dc (in MWh)
$\overline{RP}$	Unit charge for the baseline load of consumers with curtailable loads (in €/MWh)
$SMP^{p,t}$	Forecasted wholesale market price (system marginal price) at period p and dispatch interval t (in €/MWh)
Variables	
$RP^{p,z}$	Retail price for consumers with load bids and demand profiles in time zone z of period p (in €/MWh)
$CP^{p,z}$	Curtailment price for consumers with curtailable loads, in time zone z of period p (in €/MWh)
$PP^{p,z}$	Production price for producers with production bids, in time zone z of period p (in €/MWh)
$e^{p,t}$	Energy from wholesale market at period p and dispatch interval t (in MWh)
$x_{dr}^{p,t,s}$	Cleared quantity of distributed resource dr at period p, dispatch interval t and block s (in MWh)
x_{dp}	Commitment of profile consumer dp

subject to the energy balance constraint (2) (Table 1).

$$e^{p,t} + \sum_{pb,s} x_{pb}^{p,t,s} + \sum_{dc,s}(x_{dc}^{p,t,s} - \overline{Q}_{dc}^{p,t}) - \sum_{dp} x_{dp}Q_{dp}^{p,t} - \sum_{db,s} x_{db}^{p,t,s} = 0, \forall p, t \quad (2)$$

$$x_{db}^{p,t,s} = \arg\left\{ \min_{x_{db}^{p,t,s}} \sum_s (RP^{p,z} - P_{db}^{p,t,s})x_{db}^{p,t,s}, \text{ s.t. } x_{db}^{p,t,s} \le Q_{db}^{p,t,s}, \forall p, t, s \right\}, \forall db \quad (3)$$

$$x_{pb}^{p,t,s} = \arg\left\{\min_{x_{pb}^{p,t,s}}\sum_s (P_{pb}^{p,t,s} - PP^{p,z})x_{pb}^{p,t,s}, \text{ s.t. } x_{pb}^{p,t,s} \leq Q_{pb}^{p,t,s}, \forall p, t, s\right\}, \forall pb \quad (4)$$

$$x_{dc}^{p,t,s} = \arg\left\{\min_{x_{dc}^{p,t,s}}\sum_s (P_{dc}^{p,t,s} - CP^{p,z})x_{dc}^{p,t,s}, \text{ s.t. } x_{dc}^{p,t,s} \leq Q_{dc}^{p,t,s}, \forall p, t, s\right\}, \forall dc \quad (5)$$

$$x_{dp} = \arg\left\{x_{dp} \leq 1 \perp \tau_{dp} \geq 0, -P_{dp} + \frac{\sum_z RP^z}{N_z} + \tau_{dp} \geq 0 \perp x_{dp} \geq 0\right\}, \forall dp \quad (6)$$

Objective function (1) represents the expenses (for the energy procured from the wholesale market, for the energy provided by the local producers and for the curtailed consumption of the local consumers) minus the revenues (from selling energy to consumers with load bids and load profiles) of the retailer. The lower level entities seek to maximize the notional benefit (consumer's or producer's surplus) described by (3) for consumers with demand bids, (4) for producers with production bids, and (5) for consumers with curtailment bids, while consumers with load profiles are described by the complementarity conditions (6). The BLPP is equivalently transformed into a MILP problem by appending to the upper level problem the Karush-Kuhn-Tucker conditions of the lower level problems and applying proper linearizations (Strong Duality Theorem [3], big-M formulation [4]), and is solved using CPLEX solver under GAMS.

3 Implementation and Results

The simulation is performed over a time horizon of 24 h (one characteristic period; scenario weight $w(p)$ equal to 1; 24 hourly dispatch intervals $T = \{t_1, \ldots, t_{24}\}$), with 24 time zones $Z = \{z_1, \ldots, z_{24}\}$. This means that each hour constitutes a different zone and, thus, the variables representing price levels are free to vary from hour to hour. Each customer category comprises 20 entities with different price-quantity bids: consumers with consumption bids (9–73 €/MWh, 8–40 MWh); consumers with curtailment offers (13.8–78.4 €/MWh, 0.9–4.9 MWh); producers with production bids (10–15 €/MWh, 14.9–59.3 MWh); consumers with load profiles (15–42 €/MWh, 11–40 MWh). Wholesale market prices are in the range of 7.15–15 €/MWh. These characteristics of the various distributed resources are available to the retailer through contractual statements or consumers'/producers' declarations, thus ensuring that proper consent on behalf of the customer has been given for the use of this data by the retailer. Furthermore, the retailer is considered to have knowledge of the external characteristics (i.e. wholesale electricity prices) through observation of historical data or by means of forecasting tools.

Two different cases, in terms of consumption and curtailment bids, are examined:

Case 1   The price-quantity pairs comprising the bids are identical throughout the 24-h horizon (but are different for each customer).

Case 2   The bid of each customer is identical throughout the 24-h horizon in terms of quantity, but the respective prices belong to one of two zones (on-peak, off-peak).

Case 2, as opposed to Case 1, takes into account the differentiation in customer preferences between hours of the same day in forming the price-quantity bids. As more appliances are operating during the hours of higher load levels, thus allowing greater margins for load management, consumers present a more flexible behavior and price responsiveness is higher. This rational behavior is reflected in the construction of the bids in Case 2, while it is completely ignored in Case 1. For each case mentioned above, and in order to test different forms of pricing on the side of the retailer, we consider eight scenarios regarding the number of time zones, into which the 24-h period is divided. These are: 24 zones of 1 h each, 12 zones of 2 h each, 8 zones of 3 h each, 6 zones of 4 h each, 4 zones of 6 h each, 3 zones of 8 h each, 2

Table 2 Price zones per scenario

Hour of day	SMP	2 zones	3 zones	4 zones	6 zones	8 zones	12 zones	24 zones
1	7.9	2	3	3	5	6	9	18
2	7.6	2	3	4	5	7	10	19
3	7.4	2	3	4	6	7	11	21
4	7.3	2	3	4	6	8	11	22
5	7.2	2	3	4	6	8	12	24
6	7.2	2	3	4	6	8	12	23
7	7.5	2	3	4	5	7	10	20
8	12.9	1	2	2	3	4	5	10
9	14.5	1	1	1	1	2	2	4
10	14.9	1	1	1	1	1	2	3
11	14.3	1	1	1	2	2	3	6
12	13.0	1	2	2	3	3	5	9
13	12.0	1	2	2	3	4	6	11
14	10.6	2	2	3	4	5	7	14
15	10.4	2	3	3	5	6	9	17
16	10.5	2	2	3	4	6	8	16
17	10.5	2	2	3	4	5	8	15
18	13.2	1	1	2	2	3	4	8
19	13.3	1	1	2	2	3	4	7
20	15.0	1	1	1	1	1	1	1
21	15.0	1	1	1	1	1	1	2
22	14.4	1	1	1	2	2	3	5
23	11.9	1	2	2	3	4	6	12
24	10.7	2	2	3	4	5	7	13

Table 3 Optimization results per scenario for Case 1 and Case 2

Zones per day	Retailer's revenues		Aggregated notional benefit			
			$db \in DB$		$dc \in DC$	
	Case 1	Case 2	Case 1	Case 2	Case 1	Case 2
24	75,449	121,413	1,415	2,396	236	343
12	75,222	117,611	1,415	2,534	250	332
8	75,020	115,092	1,354	2,290	226	329
6	74,318	113,167	1,346	2,050	226	362
4	73,771	114,799	1,362	2,196	235	325
3	71,616	110,319	1,150	2,232	181	309
2	69,946	104,501	2,064	2,413	201	307
1	61,199	100,348	2,036	2,037	142	401

zones of 12 h each, 1 zone of 24 h. In the current implementation this grouping of the hours of the day to the various zones is predefined. In particular, it is performed by grouping together the hours with similar wholesale market prices as presented in Table 2. In Table 3 the results of the optimization for the retailer, for consumers with demand bids and for consumers with curtailment bids are presented. Evidently the retailer's revenues deteriorate as the number of zones decreases. This is attributed to the fact that with greater number of zones, the retailer has more degrees of freedom in determining price levels that in line with the variations in the wholesale electricity market. Concerning consumers with consumption and curtailment bids, regardless of the number of zones, the notional benefit is higher in Case 2 than in Case 1. This result indicates that it is favourable for these types of consumers to define their bids by incorporating the on-peak/off-peak characteristics of their behavior.

4 Conclusions and Further Research

The definition of time zones for pricing the energy produced, consumed, or curtailed locally by entities managed and represented in the market transactions by a retailer is a task that affects the profitability of the retailer. On the other hand the definition of the entities' bids affect the perceivable benefit of the local entities. The model presented in this paper could prove useful for a retailer deciding upon the form of the pricing scheme to offer to the local resources, taking into account not only external factors such as the wholesale market prices but also the customer side and its characteristics.

Future research could focus on testing various types of pricing schemes that differ not only in terms of number of zones, but could also include different types of charges, e.g. regarding peak consumption. On the customer side, investigating different forms of bids could also assist both the retailer in promoting and the customer in accepting a new type of interaction facilitated by the infrastructure inherent in the smart grid.

References

1. Saad, W., Han, Zhu, Poor, H.V., Basar, T.: Game-theoretic methods for the smart grid: an overview of microgrid systems, demand-side management, and smart grid communications. IEEE Signal Process. Mag. **29**(5), 86–105 (2012)
2. Asimakopoulou, G.E., Vlachos, A.G., Hatziargyriou, N.D.: Hierarchical decision making for aggregated energy management of distributed resources. IEEE Trans. Power Syst. **30**(6), 3255–3264 (2015)
3. Luenberger, D.G., Ye, Y.: Linear and Nonlinear Programming, 3rd edn. Springer, New York (2008)
4. Fortuny-Amat, J., McCarl, B.: A representation and economic interpretation of a two-level programming problem. J. Oper. Res. Soc. **32**(9), 783–792 (1981)

Risk Analysis of Energy Performance Contracting Projects in Russia: An Analytic Hierarchy Process Approach

Maria Garbuzova-Schlifter and Reinhard Madlener

Abstract The market for Energy Performance Contracting (EPC) in Russia is just emerging, but progress has so far been slow despite of promising forecasts. The successful realization of EPC projects requires a sound understanding of the main project risks. It highly depends on effective risk analysis and management, which should be essential parts of daily business activities of Energy Service Companies (ESCOs) and Energy Service Providing Companies (ESPCs) engaged in EPC projects in Russia. This study provides a first risk analysis framework for EPC projects executed in the industrial sector in Russia. General risks associated with EPC projects identified from the international literature were validated by Russian EPC practitioners in expert interviews. An Analytic Hierarchy Process approach was used to rank the identified risk factors and causes of risk in terms of their contribution to the riskiness of EPC projects. The data were obtained from a web-based survey among the Russian ESCOs and ESPCs. A widely usable risk management framework is proposed that is potentially very useful for EPC practitioners and can support the development of proactive risk analysis in EPC projects required both by third-party capital lenders and the further development of EPC market regulation. We also identify causes of risk related to the financial and regulatory aspects contributing the most to the riskiness of industrial sector EPC projects in Russia.

1 Introduction

Energy Performance Contracting (EPC) projects are considered to be one of the key vehicles for the aspired energy- and carbon-efficient modernization of the Russian economy. Worldwide, most EPC projects are executed by Energy Service Companies

M. Garbuzova-Schlifter (✉) · R. Madlener
School of Business and Economics/E.ON Energy Research Center, Institute
for Future Energy Consumer Needs and Behavior (FCN), RWTH Aachen University,
Mathieustrasse 10, 52074 Aachen, Germany
e-mail: MGarbuzova@eonerc.rwth-aachen.de

R. Madlener
e-mail: RMadlener@eonerc.rwth-aachen.de

© Springer International Publishing Switzerland 2017
K.F. Dœrner et al. (eds.), *Operations Research Proceedings 2015*,
Operations Research Proceedings, DOI 10.1007/978-3-319-42902-1_92

(ESCOs). In Russia, by contrast, other types of energy service providing companies (ESPCs) may also accomplish EPC projects[1] [2]. The Russian market for energy services is just forming and in spite of promising forecasts, the progress of EPC projects has so far been rather slow. Most EPC projects are ranked as a "risky" investment, and ESCOs suffer from the limited access to funds at reasonable rates.

EPC projects are complex, their realization bears technical and performance risks for ESCOs, as their project investments are refinanced through a guaranteed amount of energy savings that results from the implemented energy conservation measures at the client's site. Moreover, ESCOs may also assume investment and financial risks. In order to guarantee the anticipated energy savings, ESCOs are obliged to know the main EPC project risks and, if certain risks cannot be eliminated, to manage and mitigate these [4].

Practical experience shows, however, that EPC projects often underperform, as most energy performance contractors use a "rule of thumb" to analyze project risks [5]. Some international researchers are providing the first attempts to identify and structure some of the risks that ESCOs face while executing EPC projects (e.g. [4, 9]). However, it remains unclear as to which risks Russian ESCOs and ESPCs face while executing EPC projects.

The objective of this study is to identify, classify and quantify (i.e. rank) the main risk factors and causes of risk in terms of their contribution to the riskiness of EPC projects that ESCOs and ESPCs face in the Russian industrial sector. In doing so, we conduct a qualitative risk analysis (risk identification and assessment). Such risk analysis is one of the most important phases of the project risk management process [1].[2] For this purpose, the Analytic Hierarchy Process (AHP) methodology [12], a non-statistical methodology, is applied.

The remainder of this paper is structured as follows. Section 2 provides a review on the data acquisition procedure (risk identification and the web-based questionnaire survey) and the AHP methodology. The latter is applied to rank the identified risks associated with EPC projects executed in the Russian industrial sector. Section 3 discusses the main results of the survey conducted among the Russian ESCOs and ESPCs by introducing a ranked order of the identified risk factors and causes of risk in terms of their contribution to the riskiness of EPC projects. Section 4 concludes. More details on this study can be found in [3].

---

[1]When considering risks associated with EPC projects in Russia, we refer only to ESCOs, implying ESCOs and ESPCs, unless we explicitly differentiate between those types of company.

[2]Risk factors are triggered by causes of risk, affect timely achievement for the guaranteed amount of energy savings and impede the refinancing of an EPC project on time and in full. A cause of risk is defined as events or sets of circumstances " […] which exist in the project or its environment, and which give rise to uncertainty" [6].

2   Data and Methodology

The main general risks that ESCOs may face while executing EPC projects are extracted from the relevant academic and business/governmental literature (e.g. [4, 10]). These general risks were then validated for the Russian market by EPC practitioners in the six semi-structured interviews that were conducted in Moscow in May 2013. The six interviewees were selected in view of their direct experience and expertise in realization of EPC projects in Russia; all of them are in upper management positions (deputy general directors, executive and general directors). They represented four ESCOs, one non-commercial self-regulatory organization (NP SRO), and one scientific center with a focus on energy efficiency technologies. The identified specific risks that Russian ESCOs and ESPCs may face were classified into risk factors and causes of risk.

To rank these risk factors in terms of their contribution to the riskiness of an EPC project and the causes of risk in terms of their contribution to the related parent risk factor, a web-based questionnaire survey based on the AHP methodology was compiled and disseminated among 162 ESCOs and ESPCs operating in Russia in 2014. The target companies, which are presumed to execute EPC projects in Russia, were identified by screening the companies' web pages as well as by scientific and business conference proceedings and interviews (using key words related to EPC projects and their implementation processes). The participants were able to select one of the three suggested sectors (industrial; housing and communal services; or public), where they presume that they have the most expertise and experience to assess risks arising in EPC projects.

The application of the AHP method in our study can be divided into four major phases. In phase 1, the identified risk factors and causes of risk were structured into a risk hierarchy. In phase 2, the expert's subjective judgements on the relative importance of the identified risk factors and causes of risk were elicited by means of pairwise comparisons in line with the questionnaire-based surveys. In phase 3, the resulting priories (relative weights of the assessed risk factors and causes of risk) were composed into comparison matrices and the weights of the normalized priority vector were then derived based on the eigenvalue method (Eq. (1)):

$$W = \lim_{k \to \infty} \frac{A^k \cdot e}{e^T \cdot A^k \cdot e}, \tag{1}$$

where $W = (w_1, \ldots, w_n)$ is a priority vector, A defines a comparison matrix, k is a sufficiently large power of A, and e is a unit vector. In phase 4, the subjective judgments were controlled for consistency. In the case that they do not satisfy a previously defined consistency level, there are three possibilities to overcome this issue: inconsistent judgments should be (1) reviewed by survey participants, (2) mathematically corrected by analysts, or, if neither is possible, (3) excluded from further calculations. The first method could not be applied in this study because the questionnaire survey was conducted in anonymized form. The application of the third method would denote the loss of a substantial part of information gained in the survey. To improve the consistency of the AHP results, two mathematical methods were adopted: (1) the Maximum Deviation Approach (MDA) proposed by [12, 13] and (2) the Induced Bias Matrix Model (IBBM), recently developed by [8]. These two methods are executed in parallel. For each matrix order, each adjusted result was compared with the original one to determine which of the two methods causes the least changes to the original results and, therefore, preserves as much original information as possible until the AHP consistency condition has been satisfied.

3 Results

For the Russian industrial sector, 12 participants performed 48 pairwise comparisons while assessing the contribution of each risk factor to the riskiness of an EPC project and of each cause of risk to the corresponding parent risk factor. Most of these companies are located in the central part of Russia. Among 12 respondents, ESCOs constitute 41.7 % of the respondents. The remainder encompasses: 16.7 % of equipment suppliers; energy audit companies, non-commercial self-regulatory organizations (NP SRO), energy service providers represented 8.3 % each, and 16.7 % of the respondents selected an option "others" (an energy retail company and a study center for energy audits). These remaining companies are classified as ESPCs. 72.7 % of these companies organize their business regionally and 27.3 % nationally. The number of employees working for these companies ranges from 4 to 700 (full-time equivalents). 58.3 % of these companies employ neither a risk analyst, a risk manager, or a risk consultant to perform risk analysis in EPC projects. 25 % of the respondents state that their companies apply a formal, systematic approach for risk assessment.

For the AHP calculations, 108 pairwise comparison matrices were generated. The AHP results were then checked for inconsistencies and consistency improvements occurred either with the MDA or the IBBM method. Our results showed that the MDA method requires fewer changes of the elements in the upper triangle in the inconsistent matrices of a larger order until the CR condition ($CR \leq 0.1$) has been fulfilled. In contrast, IBBM emerges as being more effective for the inconsistent matrices of a smaller order. Therefore, we applied the MDA method for the consistency improvements of 8×8 matrices and the IBBM for 3×3 and 4×4 matrices.

The analysis of the pairwise comparisons of risk factors and causes of risk associated with EPC projects at the aggregated level results in scores of their relative importance. The calculated local and global priorities of risk factors and causes of risk (expressed in terms of adjusted and normalized geometric means) as well as the rank order of 22 causes of risk are presented in Table 1. In this context, the higher the value of the global normalized and adjusted priority of a cause of risk, the greater its contribution to the corresponding risk factor and, hence, to the EPC project's riskiness.

Table 1 Local and global priorities of risk factors and causes of risk associated with an EPC project executed in the Russian industrial sector

	Risk factors	Local priorities	Structural adjustment	Causes of risk	Local priorities	Global priorities adjusted and normalized	Rank
A	Risks of project preparation and execution phases	0.110	3/22	1. Project tendering exclusively price-based	0.291	0.035	16
				2. Lack of reliable data for baseline estimation of energy consumption of a client	0.194	0.023	21
				3. Unreliable energy certification provided by an external energy audit company	0.515	0.051	5
B	Contractual risks	0.114	2/22	1. No explicit risk pricing in EPC	0.536	0.044	10
				2. Poor prior risk division between an ESCO and a client	0.464	0.038	13
C	Technical and operational risks	0.084	3/22	1. Improper operation of the installed equipment by a client	0.292	0.027	19
				2. Improper verification of energy savings (approach/instruments)	0.482	0.044	11
				3. Energy supply disruptions	0.226	0.021	22
D	Financial risks	0.159	3/22	1. Poor investment capacity of an ESCO	0.271	0.047	8
				2. No long-term funding without a governmental or third-party guarantee for a loan	0.309	0.053	7
				3. Delayed energy saving payments from a client	0.420	0.072	3

(continued)

Table 1 (continued)

	Risk factors	Local priorities	Structural adjustment	Causes of risk	Local priorities	Global priorities adjusted and normalized	Rank
E	Client's risks	0.109	3/22	1. Client's bankruptcy risk	0.210	0.025	20
				2. Fluctuation in client's energy consumption due to undisclosed changes in productive capacity	0.318	0.037	14
				3. Difficulty of an ESCO to prove energy savings have been achieved for a client	0.473	0.056	6
F	Human and behavioral risks	0.107	2/22	1. Lack of management and technical expertise	0.466	0.036	15
				2. Client's mistrust of an ESCO	0.534	0.041	12
G	Political and regulatory risks	0.165	3/22	1. Poor and unstable legislation base for EPC projects	0.390	0.069	4
				2. Lack of tax exemptions for EPC or an ESCO	0.419	0.074	2
				3. Cross subsidization	0.192	0.034	17
H	Market risks	0.151	3/22	1. Unpredictably fluctuating energy prices	0.202	0.033	18
				2. Poor market demand and lack of incentives to invest in energy efficiency	0.277	0.045	9
				3. High interest rates for bank or third-party lending	0.521	0.085	1

The *high interest rates for bank or third-party lending* is ranked highest (0.085) among the potential causes of risk. This result is not surprising, since for most banks and other lenders, EPC is still a relatively new concept in Russia. Most lenders lack the technical expertise to evaluate and verify the return on investment of an EPC project that equals to the actual amount of energy savings achieved. In this context, the interest rates set for EPC projects can reach 15–20% in Russia. The *lack of tax exemptions for EPC or an ESCO* was ranked second highest (0.074), as the existent tax exemptions do not provide a systematic tax framework for the ECMs and are related rather to the clients of ESCOs than to ESCOs themselves. The *delayed energy saving payments from a client* was ranked third highest (0.072) among the potential causes of risk as some clients complete the payments for other obligations before ESCOs get paid.

The participants considered the *energy supply disruptions* to contribute least to the riskiness of an EPC project (0.021). This result is surprising as around 40% of the Russian population is affected by frequent electricity supply disruptions [11]. Also unexpected is the result that the *lack of reliable data for baseline estimation of energy consumption of a client* is the second lowest ranked cause of risk. The supposition that a lack of energy consumption accounting practices and installed metering systems in Russia would increase the relative significance of this cause of risk is not supported by this result.

4 Conclusion

This paper presents the main results of a first systematic analysis of risks associated with EPC projects faced by ESCOs and ESPCs in Russia. These results support the conclusion that EPC projects require systematic and effective analysis and management of project risks (PRM), which should be integrated into daily business activities of ESCOs and ESPCs in Russia. Furthermore, it presents a ranking of the main risk factors and causes of risk associated with EPC projects by taking into account the complexity and interdisciplinarity of their execution in the Russian industrial sector. The risk factors and causes of risk related to financial issues contribute most to the riskiness of EPC projects. In addition, issues related to political and regulatory risks were also ranked highly, confirming that the actual regulatory framework for the EPC projects exhibits numerous loopholes. Although the identified risks are assumed to cover most of the general risks associated with such projects, the produced list is not exhaustive. Our study provides some new insights into the methodological aspects of the consistency improvements of the AHP results by means of the MDA and the IBBM methods. As neither method controls for a number of elements in an inconsistent matrix to be adjusted, we assumed that only 50% of the elements of the upper triangle needed to be changed to retain as many as possible of the original preference judgments (information) provided by the survey participants.

References

1. Chapman, R.J.: The controlling influences on effective risk identification and assessment for construction design management. Int. J. Proj. Manag. **19**, 147–160 (2001)
2. Garbuzova-Schlifter, M., Madlener, R.: Prospects and barriers for Russiaś emerging ESCO market. Int. J. Energy Sect. Manage. **7**, 113–150 (2013)
3. Garbuzova-Schlifter, M., Madlener, R.: Risk analysis of energy performance contracting projects in Russia: an analytic hierarchy process approach. FCN Working Paper No. 10/2014. RWTH Aachen University, Aachen (2014)
4. Hansen, S.J.: Performance Contracting: Expanding Horizons, 2nd edn. Fairmont Press, Lilburn (2006)
5. Heo, Y., Augenbroe, G., Choudhary, R.: Risk analysis of energy-efficiency projects based on bayesian calibration of building energy models. In: Soebarto, V., Bennetts, H., Bannister, P., Thomas, P.C., Leach, D. (eds.) Proceedings of the Building Simulation 2011: 12th Conference of International Building Performance Simulation Association, pp. 2579–2586. IBPSA Australasia and AIRAH, Sydney (2011)
6. Hillson, D.: When is a risk not a risk? Prof. Mag. Int. Proj. Manag. Assoc. (IPMA) **1**, 6–7 (2006)
7. Jackson, J.: Energy Budgets at Risk (EBaR): A Risk Management Approach to Energy Purchase and Efficiency Choices. Wiley, Hoboken (2008)
8. Kou, G., Ergu, D., Peng, Y., Shi, Y.: Data Processing for the AHP/ANP (Quantitative Management), vol. 1. Springer, Heidelberg (2013)
9. Lee, P., Lam, P.T.I., Lee, W.L.: Risks in energy performance contracting (EPC) projects. Energy Build. **92**, 116–127 (2015)
10. Mills, E., Kromer, S., Weiss, G., Mathew, P.A.: From volatility to value: analysing and managing financial and performance risk in energy savings projects. Energy Policy **34**, 188–199 (2006)
11. Public Opinion Fund & Schneider Electric (2014). Survey conducted by the Public Opinion Fund & Schneider Electric: Almost 40% of the Russian population is confronted with electricity supply disruptions, 8:58–64 (in Russian)
12. Saaty, T.L.: The Analytic Hierarchy Process: Planning, Priority Setting, Resource Allocation. McGraw-Hill, New York (1980)
13. Saaty, T.L.: Decision-making with the AHP: why is the rincipal eigenvector necessary. Eur. J. Oper. Res. **145**, 85–91 (2003)

Risk-Adapted Capacity Determination of Flexibility Investments for Distributed Energy Resource Systems

Katrin Schulz

Abstract Distributed energy resource (DER) systems, composed of different small to medium-sized renewable and conventional energy generators, allow to balance the volatile supply from renewable energies. Considering a DER system with power and district heat, small combined heat and power (CHP) plants and their flexible operation play a central role. Flexibility investments such as heat storage devices provide further flexibility for the DER system. In this contribution, the profitability of an investment in a heat storage device and its optimal capacity are examined. An innovative decision support approach for capacity determination of such flexibility investments is applied considering the investment's benefit throughout its entire lifetime. System dependency as well as the decision maker's risk attitude are taken into account. The developed two-stage stochastic programming model simultaneously optimizes the operation of the DER system and the investment's capacity. In a simulation study, the calculated capacities are evaluated. An exemplary case study illustrates the advantages of the proposed approach.

1 Introduction

As a result of today's energy and environmental challenges, DER systems have attracted much attention in recent years. A DER system is composed of different small to medium-sized energy generators sited near customers [3]. Here, a DER system with power and district heat is considered for which different technologies can be applied: conventional CHP plants, renewable energies like solar photovoltaic (PV) and wind generators. Moreover, a DER system can comprise responsive loads and energy storages such as a heat storage device [1]. Compared to centralized generation, DER systems offer considerable advantages. Due to the position of the generation units, transmission and distribution losses are reduced and the voltage profile is improved. Efficient conventional technologies (e.g., CHP) and renewable

K. Schulz (✉)
Chair of Operations Research and Accounting, Faculty of Management
and Economics, Ruhr University Bochum, 44780 Bochum, Germany
e-mail: katrin.schulz@rub.de

© Springer International Publishing Switzerland 2017
K.F. Dœrner et al. (eds.), *Operations Research Proceedings 2015*,
Operations Research Proceedings, DOI 10.1007/978-3-319-42902-1_93

energy resources decrease primary energy consumption. The integration of renewable energies requires a flexible operation of the whole DER system because flexible conventional generation is needed in times of low power feed-in from renewable resources. Such balancing power is also needed outside the system so that excess energy can be sold to the market. The higher the system's flexibility, the more revenues can be gained from trading. Additional flexibility for the DER system can be provided using heat storage to decouple heat generation from demand [5]. According to the findings of Katulić et al., energy companies prefer the heat storage capacity that provides maximum flexibility [2]. With regard to an investment decision, the optimal storage capacity has to be determined considering an adequate amortization of the resulting investment expenditures. The profitability of a certain capacity is derived from the optimal operation of the storage device within the DER system. Thus, the optimal capacity and the optimal operation have to be determined simultaneously. Here, an innovative decision support approach for capacity determination [6] is suggested for flexibility investments in DER systems.

2 Integrated Approach for Flexibility Investments in DER Systems

The investment decision and especially the capacity determination constitutes a challenge. On the one hand, the flexibility investment and the DER system are interdependent. This means that the chosen capacity results in a certain profitability which, in turn, has to ensure the amortization of the corresponding investment expenditures. On the other hand, the future development of the DER system influences the possible operation of the flexibility investment. Thus, the planning horizon for the investment decision is determined as the time period that has continuity with the DER system. This continuity is defined as minimum of the economic lifetime of other system components or by contractual obligations. The storage device may have a remaining lifetime beyond this planning horizon. However, the investment's benefit in its remaining lifetime is typically not considered for the investment decision. Since a higher capacity is accompanied by an increasing profitability ceteris paribus, the determination of the optimal capacity only for the planning horizon may lead to an underestimation of needed capacity in the remaining lifetime.

In this complex investment decision, a two-stage approach for capacity determination of flexibility investments is applied [6]. First, the optimal capacity for the planning horizon is determined which results from the optimal operation of the storage within the system. To consider the system's uncertain supply situation, different patterns of power prices and heat and power demand are taken into account as uncertain parameters. Therefore, a mixed integer linear programming model depicting the complex supply situation of a DER system [3] is converted into a two-stage stochastic optimization model to determine the optimal heat storage capacity for different scenarios. The heat storage's capacity and its operation within the DER system are optimized simultaneously considering the amortization of the resulting investment expenditures. It is considered reasonable not to equate the amortization period completely with the planning horizon. With regard to the storage's benefit in its remaining lifetime, part of the payback can result from the remaining lifetime.

In a second step, the possible operation of the heat storage device in its remaining lifetime is examined. For this purpose, frequent modes of operation are analyzed and the needed storage capacity is calculated. Based on this calculation, the decision maker may adapt the capacity determined in the first step. To which extent this is done depends on the decision maker's risk attitude as explained in detail below.

3   Risk-Adapted Approach for Capacity and Cost Determination

The operator of a DER system seeks to ensure the continued supply of heat and power at minimum overall net acquisition costs. For each scenarios $s \in S$, net acquisition costs z_s are defined according to [4] as

$$z_s = generation\ costs_s + power\ purchase\ costs_s - power\ sales\ revenues_s. \quad (1)$$

Since the storage influences the optimal operation of the DER system and the resulting trading opportunities, net acquisition costs differ in a supply situation with z_s and without z_s^* storage. This difference is named added value and reflects the storage's profitability. For a certain scenario s, the scenario-specific added-value av_s of the storage capacity is determined according to (2).

$$av_s = z_s^* - z_s \qquad\qquad \forall s \in S \qquad\qquad (2)$$
$$av_s \geq aic \qquad\qquad \forall s \in S \qquad\qquad (3)$$

The investment expenditures themselves depend on the chosen optimal heat storage capacity which, in turn, influences the added value. Thus, the investment expenditures and the adapted amortization costs aic are calculated in the optimization model for the same time horizon (using a corresponding adequate effective annual percentage rate). The added value av_s must exceed or at least equal aic in all scenarios (3) to ensure the amortization of the investment within the planning horizon. With regard to the investment's benefit in its remaining lifetime, it is reasonable for a decision maker to chose a slightly higher capacity. Therefore, the chosen approach for the two-stage stochastic optimization model to represent the decision maker's risk attitude is also used to adjust the strict formulation in constraint (3).

A risk-neutral decision maker tends to decide based on expected values ev and strives to minimize expected net acquisition costs (4). Accordingly, the expected added value has to be always greater than or equal to the amortized investment costs *aic* (5). Using the expected value criterion, the probability of occurrence for each scenario has to be known and influences the optimal storage capacity.

$$\min \quad \sum_{s \in S} p_s \cdot z_s^{ev} \tag{4}$$

$$\text{s. t.} \quad \sum_{s \in S} p_s \cdot av_s^{ev} \geq aic \tag{5}$$

Energy companies seek to determine a heat storage capacity with a high added value independent of the scenario that will occur. Therefore, a solution is sought that is robust with regard to the result of the optimization model. This is modeled by the minimax regret criterion which evaluates a solution according to the maximum absolute or relative regret and is especially preferred by risk-averse decision makers. The absolute regret *ar* for all scenarios is defined as the difference of the scenario-specific net acquisition costs z_s^{ar} and the net acquisition costs with the scenario optimal storage capacity $z_s^{hsc^*}$. Thus, for each scenario the optimal heat storage capacity hsc_s^* and the resulting net acquisition costs $z_s^{hsc^*}$ are calculated.

Regarding the amortization of the investment expenditures, the idea of the absolute regret is picked up. The scenarios-specific added value is determined as difference of the net acquisition costs without heat storage z_s^* and the net acquisition costs in the absolute regret approach z_s^{ar} (8). An absolute regret is determined as maximum difference between the scenario-specific added value av_s and the amortized investment costs *aic* over all scenarios. As this absolute regret *aric* is also to be minimized, it is included in the objective function using a parameter $\alpha \in [0, 1]$ to weight the two objectives. The adapted part of the optimization model is:

$$\min \quad \alpha \cdot ar + (1 - \alpha) \cdot aric \tag{6}$$

$$\text{s. t.} \quad ar \geq z_s^{ar} - z_s^{hsc^*} \qquad \forall s \in S \tag{7}$$

$$av_s = z_s^* - z_s^{ar} \qquad \forall s \in S \tag{8}$$

$$aric \geq aic - av_s \qquad \forall s \in S \tag{9}$$

Analogously, instead of an absolute regret, a relative regret can be applied to determine an optimal investment capacity. This has the advantage that more importance is given to losing net acquisition costs in unfavorable scenarios.

4   Case Study

In this section, the presented two-stage stochastic optimization model is applied exemplarily to a DER system The three different solution approaches, i.e. expected value and absolute and relative regret (for the latter two with $\alpha = 0.8$) are calculated to consider different risk attitudes. For each approach, the resulting optimal heat

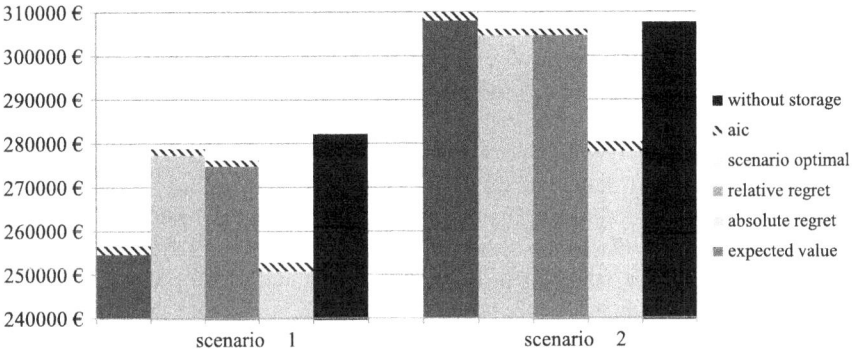

Fig. 1 Net acquisition costs and adapted amortized investment expenditures *aic* considering the optimal heat storage capacity for different solution approaches

storage capacities and amortized investment costs are determined. Different scenarios depicting the uncertain supply situation are considered with equal probability of occurrence. Figure 1 shows the results for each approach for two exemplary scenarios. The net acquisition costs are depicted as grey bars and the amortized investment costs *aic* are illustrated as the hatched area above the net acquisition costs.

With regard to the investment decision in a heat storage device, net acquisition costs plus amortized investment costs shall be less than the net acquisition costs in a supply situation without storage. This is the case for scenario 1: the net acquisition costs with storage plus *aic* are below the net acquisition costs without storage (black bar), independent of the risk-approach taken. Thus, the amortization of the investment expenditures within the planning horizon is ensured for the considered scenario. In contrast, the investment's expenditures for the storage capacity determined with the expected value approach are not amortized in scenario 2. The reason is the equal weighting of all scenarios considered. The net acquisition costs in scenario 1 are considerably lower with this storage capacity.

The heat storage capacity derived from the expected value approach is approximately 15 % higher than the optimal capacity with a strict amortization schedule. This higher capacity is seen as an investment proposal for a risk-neutral decision maker. Whether this proposal is accepted depends on the estimated added value of such a higher capacity in its remaining lifetime. Therefore, the calculated capacity is evaluated regarding the future development of the DER system. It is analyzed for different modes of operation for the heat storage in the uncertain future. A distinction is made between long-term uncertainty due to fundamental trends and the resulting short-term consequences for the DER system like an increasing volatility of power prices. With respect to the investment decision considered here, it has to be examined which short-term uncertainty prevails regarding the heat demand and which modes of operation result.

One possible future scenario is the occurrence of high power prices for several hours. During this period heat demand should be fulfilled using heat from the storage.

Thus, a simulation study is conducted to analyze whether the proposed capacities are adequate to provide heat from the storage for 4 to 8 h in winter for the whole DER system. In the expected value approach, the heat storage capacity calculated with the strict amortization schedule and the adapted capacity bridge this period with a probability of 42.6 and 64.8 %, respectively. This analysis shows that the increased capacity has a clearly higher benefit in the investment's remaining lifetime. Taking into account this future benefit, the decision maker has to determine the final heat storage capacity. Similar analyses have to be conducted for other frequent modes of operation to examine the benefit of the proposed capacities for the DER system throughout the investment's entire lifetime.

5   Conclusion

In this contribution, an innovative approach for capacity determination has been applied to heat storage investments in DER systems. A widely recognized mixed-integer linear programming model depicting the complex supply situation of a DER system has been expanded for the investment decision. The developed two-stage stochastic optimization model has been solved with an expected value as well as an absolute and relative regret approach. Special emphasis is given to the consequences of a risk-adapted amortization schedule. With regard to the investment's benefit in its remaining lifetime, the amortization period is not strictly limited to the planning horizon but varies according to the decision maker's risk attitude.

Results show that the storage's mode of operation varies between the scenarios resulting in different scenario optimal capacities. In the two-stage optimization model, the optimal capacity for all scenarios depends on the decision maker's risk attitude. Due to the interdependencies between the interest rate and the length of the planning horizon, findings depend on the assumptions made. A profound analysis and quantification of these resulting effects is the focus of further work.

The evaluation of the optimal capacity derived from the expected value approach exemplarily shows that adapted investment capacities can be justified by the investment's expected benefit in the remaining lifetime. It is evident that the uncertain future development is crucial for investment decisions. Thus, future research will focus on the question of how consequences of fundamentally uncertain developments in the energy sector can be incorporated into strategic investment decisions.

References

1. Alarcon-Rodriguez, A., Ault, G., Galloway, S.: Multi-objective planning of distributed energy resources: a review of the state-of-the-art. Renew. Sustain. Energy Rev. **14**(5), 1353–1366 (2010)
2. Katulić, S., Čehil, M., Bogdan, Ž.: A novel method for finding the optimal heat storage tank capacity for a cogeneration power plant. Appl. Therm. Eng. **65**(1–2), 530–538 (2014)

3. Omu, A., Choudhary, R., Boies, A.: Distributed energy resource system optimisation using mixed integer linear programming. Energy Policy **61**, 249–266 (2013)
4. Rong, A., Lahdelma, R.: An efficient envelope-based Branch and Bound algorithm for non-convex combined heat and power production plants. Eur. J. Oper. Res. **183**(1), 412–431 (2007)
5. Schulz, K., Schacht, M., Werners, B.: Influence of fluctuating electricity prices due to renewable energies on heat storage investments. In: Huisman, D., et al. (eds.) Operations Research Proceedings 2013. Springer, New York (2014)
6. Schulz, K., Werners, B.: Capacity determination of ultra-long flexibility investments for district heating systems. J. Bus. Econ. **85**(6), 663–692 (2015)

Decision Support System for Intermodal Freight Transportation Planning: An Integrated View on Transport Emissions, Cost and Time Sensitivity

Andreas Rudi, Magnus Froehling, Konrad Zimmer and Frank Schultmann

Abstract The evaluation and selection of intermodal routes with regard to the key objectives, i.e., transit time, transport emissions and cost, is the main challenge in the design of intermodal networks. The aim of this paper is to present a decision support system for intermodal freight transportation planning, which offers methodological contributions to the research on transport mode, route and carrier selection as well as results for industrial practitioners for the assessment of emission abatement potentials. Core of this approach is a capacitated multi-commodity network flow model considering three minimization objectives, i.e. costs, time and CO_2-equivalents. In this contribution a tri-objective mixed-integer linear model formulation minimizes the number of transported and transshipped full truck loads taking into account tied in-transit capital and the distance travelled. The decision support system is validated in an exemplary case study application analyzing the sensitivity of objectives on optimal route and carrier choice. By applying the augmented ε-constraint method, a Pareto-efficient frontier is determined to investigate the tradeoff between economic and ecological objectives in intermodal freight transportation planning.

1 Introduction

With the growing demand for freight transportation, the amount of released air emissions from transport increases [6]. To mitigate climate relevant air emissions from freight transportation, policy-makers stimulate the application of intermodal freight transport chains [9]. The common flexible but environmentally less favorable road transport by truck can be combined with the more environmentally friendly transportation by rail and sea [4]. The evaluation and selection of intermodal routes con-

A. Rudi (✉) · M. Froehling · K. Zimmer · F. Schultmann
Karlsruhe Institute of Technology (KIT), Institute for Industrial
Production (IIP), Hertzstr. 16, 76187 Karlsruhe, Germany
e-mail: Andreas.Rudi@kit.edu

© Springer International Publishing Switzerland 2017   699
K.F. Dörner et al. (eds.), *Operations Research Proceedings 2015*,
Operations Research Proceedings, DOI 10.1007/978-3-319-42902-1_94

sidering the often conflicting key objectives, i.e., greenhouse gas (GHG) emission, transportation cost and time, is crucial in the design of intermodal networks [2]. Therefore, the aim of this paper is to provide support for decision makers in industry concerning route and carrier choice in transport service design and the assessment of emission abatement potentials with respect to economic and ecological objectives.

2 Tri-Objective Model Formulation

The following generic tri-objective model formulation is based on the capacitated multi-commodity network flow model (CMCNF) formulated by Rudi et al. [8]. The authors frame a single objective mixed-integer linear problem considering multiple criteria, i.e., CO_2-equivalents, cost and time, as well as in-transit holding costs. By introducing criteria weightings and applying the weighted sum scalarization method, a linear scale transformation of the criteria is enabled. As a result, the model optimizes for a single solution with certain properties according to the decision maker's criteria preferences. Despite the fact that the weighted sum scalarization method is commonly used in multi-criteria decision making [3], the detection of the Pareto-efficient frontier is challenging due to the increased number of model runs to determine possible solution alternatives [7]. To provide support in the planning of intermodal freight transportation networks, the identification of Pareto-efficient solutions (Pareto set) is valuable. Therefore, a single objective CMCNF model is transformed into a tri-objective model formulation and sketched in the following. The three objectives are represented by linear utility functions and each one minimizing the transport costs in monetary units (MU), emissions in tCO_2e or times in hours of transported and transshipped full truckloads (FTL).

The network is described by the physical movement of a product $p \in P$ with the intermodal carrier $s \in S$ on a link, which is in accordance to the node-to-node demand defined as a doublet (i, j) with $i, j \in V$ between origins $V_a \subset V \backslash (V_h \wedge V_b)$ and destinations $V_b \subset V \backslash (V_h \wedge V_a)$, through intermediate stages $(h, i), (i, j) \in E$. The transport flow at capacitated intermediate nodes $V_h \subset V \backslash (V_a \wedge V_b)$ is represented by either terminal nodes $V_t \subset V_h \backslash V_d$ and the corresponding dummy nodes $V_d \subset V_h \backslash V_t$, or carrier transshipment nodes. Those transshipment nodes are defined as terminal nodes for carrier transshipments $V_{t^{TS}} \subset V_t$ and their corresponding carrier transshipment dummy nodes $V_{d^{TS}} \subset V_d$, or transfer nodes $V_h \backslash (V_t \wedge V_d)$. Two decision variables are defined representing the number of FTLs transported ($x_{psij}^{TR} \in \mathbb{N}$) or transshipped ($x_{psij}^{TS} \in \mathbb{N}$) carrying one product p with carrier s between two locations i and j. The specific assessment factors for the transport ($f_{psij}^{TR^k}$), transshipment ($f_s^{TS^k}$) and in-transit (f_p^{INV}) process are defined individually to evaluate the objectives $k \in \{Cost, CO_2e, Time\}$.

Objective Functions

The first objective function (1) minimizes the transport, transshipment and in-transit holding costs in accordance to the cost assessment factors for the transport and transshipment as well as the in-transit process. Whereas the first term of the objective function expresses the transport and transshipment costs, the second term integrates the time dimension to calculate the in-transit holding cost while the product is processed taking into account the distance (d_{ij}) and the carrier payload (l_s). The second objective function (2) minimizes the emissions and the third objective function (3) the times in accordance to the emission/time assessment factors. The objective functions (1)–(3) summarize the challenge of finding the optimal route through the intermodal network design by minimizing the number of transported and transshipped FTLs in accordance to minimal costs, emissions and times.

$$Min_x^{Cost} \sum_{p \in P} \sum_{s \in S} \sum_{(i,j) \in E} \left[x_{psij}^{TR} \cdot f_{psij}^{TR^{Cost}} + x_{psij}^{TS} \cdot f_s^{TS^{Cost}} \right.$$

$$\left. + \left(x_{psij}^{TR} \cdot f_{psij}^{TR^{Time}} + x_{psij}^{TS} \cdot f_s^{TS^{Time}} \right) \cdot \frac{f_p^{INV}}{l_s} \right] \cdot d_{ij} \tag{1}$$

$$Min_x^{CO_2e} \sum_{p \in P} \sum_{s \in S} \sum_{(i,j) \in E} \left[x_{psij}^{TR} \cdot f_{psij}^{TR^{CO_2e}} + x_{psij}^{TS} \cdot f_s^{TS^{CO_2e}} \right] \cdot d_{ij} \tag{2}$$

$$Min_x^{Time} \sum_{p \in P} \sum_{s \in S} \sum_{(i,j) \in E} \left[x_{psij}^{TR} \cdot f_{psij}^{TR^{Time}} + x_{psij}^{TS} \cdot f_s^{TS^{Time}} \right] \cdot \frac{d_{ij}}{l_s} \tag{3}$$

Constraints

The model is described by the distribution of complete transport units (FTL) carrying one category of product, starting from origins, crossing several transfer and/or terminal nodes and ending at destinations. The origins provide a certain amount of product supply (A_{pi}) (4), whereas, destinations demand the equivalent (B_{pj}) (5). The mass conservation constraint (6) models the FTL transports between terminals and dummy terminals for terminal process evaluation. The transshipment flow constraint (7) allows the transshipment of transports on different carriers and ensures that transshipped input and output are in balance as well. The transfer flow constraint (8) guarantees the equilibrium between the input and the output of FTLs at each transfer point per product-carrier combination. The carrier replacement constraint (9) enables the transshipment of FTLs at predefined terminals onto new carriers. Thereby, x_{pshi}^{TR} is the entering, x_{psij}^{TR} the leaving and x_{psij}^{TS} the transshipped number of FTLs onto the new carrier. In contrast, variable v_{pshi}^{TS} refers to the replaced carrier. To ensure the accurate replacement of the former carrier by the new one, the carrier's capacities have to correspond with the payload flow constraint (10). This constraint becomes valid when the truckload is transshipped during transport and ensures that the truckload is permitted to be transshipped by the new carrier. The consideration of terminal capacities (U_{ij}) of processed FTLs per planning period is important for the realistic implementation of the model (11). Thus, since commodity networks share terminals, the terminal capacity is the critical parameter that transforms multiple single-commodity network flow problems into one capacitated multi-commodity network flow problem.

$$\sum_{s \in S} \sum_{(i,j) \in E} \left[x_{psij}^{TR} \cdot l_s - x_{psij}^{TR} \cdot l_s \right] \leq A_{pi} \qquad \forall p \in P;\ i \in V_a;\ j \in V \tag{4}$$

$$\sum_{s \in S} \sum_{(i,j) \in E} \left[x_{psij}^{TR} \cdot l_s - x_{psij}^{TR} \cdot l_s \right] \geq B_{pj} \qquad \forall p \in P;\ i \in V;\ j \in V_b \tag{5}$$

$$\sum_{(h,i) \in E} x_{pshi}^{TR} = x_{psij}^{TR} \qquad \forall p \in P;\ s \in S;\ h \in V;\ i \in V_t;\ j \in V_d \tag{6}$$

$$\sum_{s \in S} \sum_{(h,i) \in E} x_{pshi}^{TR} = \sum_{s \in S} \sum_{(i,j) \in E} x_{psij}^{TR} \qquad \forall p \in P;\ h \in V_{tTS};\ i \in V_{dTS};\ j \in V \tag{7}$$

$$\sum_{(i,j) \in E} x_{psij}^{TR} = \sum_{(i,j) \in E} x_{psji}^{TR} \qquad \forall p \in P;\ s \in S;\ i \in V;\ j \in V_h \backslash V_{dTS} \tag{8}$$

$$x_{pshi}^{TR} - \sum_{(i,j) \in E} x_{psij}^{TR} = v_{pshi}^{TS} - x_{psij}^{TS} \qquad \forall p \in P;\ s \in S;\ h \in V_{tTS};\ i \in V_{dTS};\ j \in V \tag{9}$$

$$\sum_{s \in S} x_{pshi}^{TS} \cdot l_s = \sum_{s \in S} x_{psij}^{TS} \cdot l_s \qquad \forall p \in P;\ h \in V_{tTS};\ i \in V_{dTS};\ j \in V \tag{10}$$

$$\sum_{p \in P} \sum_{s \in S} x_{psij}^{TR} \leq U_{ij} \qquad \forall i \in V_t;\ j \in V_d \tag{11}$$

Augmented ε-constraint

In order to determine the Pareto-efficient frontier of the tri-objective intermodal freight transportation model, the augmented ε-constraint method, according to Mavrotas [7], is applied. Therefore, the set of Pareto optimal solutions is detected by using lexicographic optimization to generate the payoff table. The augmented ε-constraint method formulation with three minimization objectives $f_1(x), f_2(x)$ and $f_3(x)$ is as follows (12) and (13). $\mathbf{x}$ represents the vector of decision variables, S the solution space, s_2 and s_3 the slack variables, r_2 and r_3 the according objective function ranges, eps a sufficiently small number as well as e_2 and e_3 the variation parameters:

$$Min f_1(\mathbf{x}) + eps \cdot \left(\frac{s_2}{r_2} + \frac{s_3}{r_3} \right) \tag{12}$$

$$s.t.\ f_2(\mathbf{x}) - s_2 = e_2;\ f_3(\mathbf{x}) - s_3 = e_3;\qquad \forall s_2, s_3, r_2, r_3 \in \mathbb{R}^+ \tag{13}$$

3   Model Application

For the review of the detailed input data in the following example (e.g., carrier, GHG emission calculations and cost and time figures) the reader is referred to the publications by Froehling et al. [5] and Rudi et al. [8]. The application of the tri-objective model uses the case study presented by Froehling et al. [5], which describes a transport network structure of a multinational supplier for the automotive industry.

The 25,473-km long transport network is characterized by 57 nodes linked with 90 edges encompassing Great Britain, the Netherlands, Belgium and France, defining two product-dependent transport lanes. Each product transport lane starts at one origin (Dundee; Ballymena), a production facility, and ends at the same destination (Rouvignies), a warehouse. 27 terminals act as transshipment points enabling the modal shift and forming an overlapping multi-stage network. The distance between the terminal and the dummy node is set to 10 km to assess the transshipment process. The planning period is one year for the distribution of supply ($A_{11} = 41,521\ t$ and $A_{22} = 35,871\ t$) to satisfy the according demand (B_{11}, B_{21}) restricted by a terminal processing capacity ($U_{ij} = 2,500\ FTL$). In summary, the network design involves, starting from the origins, the in-haulage to the transshipment points by road, the subsequent main run by road, rail and/or sea transport, and the out-haulage to the final destination via truck. Within the spanning network twelve carriers compete with each other in terms of costs, times and GHG emissions. To deliver decision support regarding route and carrier selection under predefined requirements is the key objective of this example application.

4 Results and Conclusion

Whereas Rudi et al. [8] focus on the tradeoff sensitivity of transport cost, emissions and time by applying the weighted-sum method and analyzing different weighting scenarios, the present approach finds the Pareto-efficient frontier using the augmented ε-constraint method. This contribution is based on scenario 4 of the aforementioned publication. The applied CPLEX solver computes the Pareto-efficient frontier with 13,154 non-zero elements and 1,650 integer variables in a B&C tree with 350,265 nodes in 581,859 iterations and 12 s. The calculations are carried out using a 2.6 GHz i5 CPU with 8 GB RAM memory. The Pareto-efficient frontier of the tri-objective intermodal freight transportation model is shown in Fig. 1. The figure evidently shows a tradeoff between the cost and CO_2e objectives for a time objective range from approx. 130–155 h of average transportation time summing up both transport lanes. Furthermore, the correlation between cost and emission appears to be linear with a significant negative correlations coefficient of −0.95 shaping a direct dependency of emissions as a function of costs in Mio. MU. Hence, a one-one relationship between transport cost and emissions can be derived such that an increase of costs by 100 MU reduces emissions by 1 tCO_2e on average. In the following, specific model solutions are presented and discussed. Taking into account the time range from 110–130 h, the Pareto-efficient frontier shows irregularities, especially in the blue segment of the illustration depicted by Circle 1. This segment is characterized by a lower average transportation time and the lowest CO_2e emissions, but by the highest costs. This is the result of both expensive but environmentally friendly transport mode rail and fast transportation by road. The domination of rail transport due to its emission-efficient nature is emphasized; however, due to the high economic impact its application is unrealistic. A practical application with stable objectives is provided by Circle 2. This solution results in equally shared utilization of the

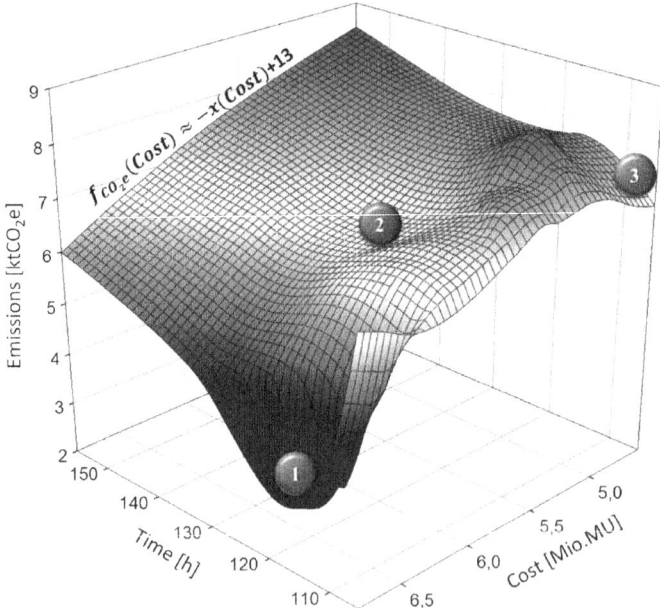

Fig. 1 Pareto-efficient frontier of the tri-objective model formulation

different transport modes. The most applicable solution is presented by Circle 3 describing a balanced situation of objectives with lowest costs and times, and comparable low emissions. This situation is characterized by an equal share of transport by road and sea. The illustration of the Pareto-efficient frontier enables the decision maker to understand the interdependent relations between the objectives and their efficient solutions to select the most preferred among them. The competitive equilibrium corresponds to Pareto-viable network flow designs, that state a precise Pareto tradeoff between the contending objectives [1]. By identifying potential solution scenarios the decision maker is guided towards the final solution. In summary, the findings of the model application highlight an existing interrelation between costs and emissions, and confirm the hypothesis that intermodal long-haul freight transportation is a key idea towards the design of low-emission transportation chains and sustainable logistics.

References

1. Bektas, T., Laporte, G.: The pollution-routing problem. Transp. Res. Part B **45**, 1232–1250 (2011)
2. Crainic, T.G., Kim, K.H.: Intermodal transportation. Handb. Oper. Res. Manag. Sci. **14**, 467–537 (2007)

3. Ehrgott, M.: A discussion of scalarization techniques for multiple objective integer programming. Ann. Oper. Res. **147**, 343–360 (2006)
4. EU transport in figures: Statistical Pocketbook 2014. European Union, Luxembourg (2014)
5. Froehling, M., Zimmer, K., Schultmann, F.: A case study on route and haulier choice considering carbon emissions. Working paper, Karlsruhe Institute of Technology (2013)
6. Helmreich, S., Keller, H.: Freightvision: Sustainable European Freight Transport 2050. Springer, Heidelberg (2011)
7. Mavrotas, G.: Effective implementation of the ε-constraint method in multi-objective mathematical programming problems. Appl. Math. Comp. **213**, 455–465 (2009)
8. Rudi, A., Froehling, M., Zimmer, K., Schultmann, F.: Freight transportation planning considering carbon emissions and in-transit holding costs: a capacitated multi-commodity network flow model. EURO J. Transp. Logist. (2014). doi:10.1007/s13676-014-0062-4
9. SteadieSeifi, M., Dellaert, N.P., Nuijten, W., Van Woensel, T., Raoufi, R.: Multimodal freight transportation planning: a literature review. Eur. J. Oper. Res. **233**, 1–15 (2014)

A Multi-objective Time Segmentation Approach for Power Generation and Transmission Models

Viktor Slednev, Valentin Bertsch and Wolf Fichtner

Abstract The complexity of large-scale power system models often necessitates the choice of a suitable temporal resolution. Nowadays, mainly simple heuristic approaches are used. An adequate decision support related to power generation and transmission optimisation in systems with a high RES share, however, requires preserving the complex intra-period and intra-regional links within and between the volatile electricity demand and supply profiles. Focussing on power systems operation, we are able to show that even an amount of less than 300 time segments may be sufficient for the modelling of a whole year, if chosen carefully.

1 Introduction

The rapid expansion of renewable energy sources (RES) necessitates an extensive structural rearrangement of the power system. The power grid plays a key role in this context. In order to provide valuable decision support for power systems operation and analyse grid utilisation, especially during times of peak load or generation, models are needed which are able to consider a high regional and temporal input data granularity. This requirement inevitably leads to a target conflict between model complexity and computational intensity on the one hand and model accuracy on the other hand. When simplifying the representation of time, it is therefore crucial to minimise the loss of relevant information. In particular, in case of a combined consideration of power generation and transmission in systems with a high RES share, it is important to preserve the complex intra-period and intra-regional links within and between the volatile electricity demand and supply profiles. So far, however, mainly simple heuristic approaches are used (see e.g., [2–5]).

V. Slednev (✉) · V. Bertsch · W. Fichtner
Chair of Energy Economics, Institute for Industrial Production (IIP),
Karlsruhe, Germany
e-mail: viktor.slednev@kit.edu

V. Slednev · V. Bertsch · W. Fichtner
Karlsruhe Institute of Technology (KIT), Karlsruhe, Germany

© Springer International Publishing Switzerland 2017
K.F. Dœrner et al. (eds.), *Operations Research Proceedings 2015*,
Operations Research Proceedings, DOI 10.1007/978-3-319-42902-1_95

We therefore propose a multi-objective optimisation approach for time segment selection in Sect. 2 allowing for an explicit analysis of the sensitivity of the temporal structure. Focussing on operating decisions in this paper, we present selected results in Sect. 3 and show that even an amount of less than 300 time segments may be sufficient for the modelling of a whole year. In Sect. 4, we conclude and indicate needs for future research.

2 The Multi-objective Approach for Time Segment Selection

The developed approach is aimed at representing the nodal profiles of electricity demand and supply-dependent volatile RES generation of a specific year through a subset of the initial hourly time structure. Preserving the complex intra-period and intra-regional links within and between the volatile electricity demand and supply profiles is a major challenge. The time segment selection is therefore based on the solution of a two-step multi-objective integer optimisation problem.

An adequate generation and transmission optimisation within load flow constrained power system models is basically determined through local or global bottlenecks. In consequence, the trade-off between selecting the critical hours for the power grid usage and the typical periods for the unit dispatch constitutes a serious challenge, especially as the critical hours are not known in advance. In the following, we therefore choose extreme combinations of load, wind and solar photovoltaic energy as constraints for the time segment selection. This approach is based on [1], where the eight possible combinations of high and low electricity demand and feed-in from wind and photovoltaic energy are used to define critical situations for grid utilisation. For the following selection of time slices, the "lwp"-(**l**oad-**w**ind-**p**v) constraint requires that at least one time segment in the reduced set is an element of the critical "lwp" sets (S^{lwp}).

As mentioned above, our overall target is to provide a time segment selection which accounts for both, typical and extreme demand and supply profiles on a nodal and global level and thus minimises the hourly deviations between the original and reduced profiles. In Sect. 2.1, we therefore introduce our grid impact based error measure. In Sect. 2.2, we subsequently describe the first step of our two-step approach, i.e. the selection of 'typical' days by solving a multi-objective binary clustering problem. In Sect. 2.3, we describe the second step of our two-step approach, i.e. the intraday time slice reduction by means of constraint programming.

The implementation of the developed time segment selection approach is based on MATLAB and GAMS. MATLAB is used for the data handling and preprocessing, such as the calculation of the clustering distances and the critical sets as well as for the calculation of the initial MIP start solutions, based on a k-Means clustering. The multi-objective binary clustering problem is implemented in GAMS and solved with CPLEX.

2.1 A Grid Impact Based Error Measure

We evaluate the reduction of a time series to its characteristic values based on the resulting error's impact on the solution space. In direct current optimal power flow approaches (DC-OPF), the solution space is determined by the load flow equations:

$$P_f = \Phi \cdot P_{inj} \,, \tag{1}$$

where the relation between the bus injection P_{inj} and the branch flow P_f is determined through the power transfer distribution factor (PTDF) matrix Φ. Splitting the bus injection into a variable and fix part by defining the electricity demand and RES-E feed-in as exogenous parameters, and clustering the original right hand side injection vector along the temporal dimension, the impact of the resulting deviation E of the underlying values to their representative within the cluster on the original solution can be expressed by:

$$P_f - \Phi \cdot P_{inj}^{var} = \Phi \cdot P_{inj}^{fix} + \Phi \cdot E \,. \tag{2}$$

The impact of a clustering policy on the solution space of the load flow equations is therefore given by the product of the cluster distance of the right hand side parameters of a specific hour with the PTDF matrix. Reducing the resulting $(L \times 1)$ vector, where L corresponds to the number of branches l ($1 \leq l \leq L$), by the L_2 norm, we can define the single hour distance c for a deviation from the exogenous bus injection:

$$c = \left(\sum_{l \in L} (\Phi \cdot E)^2 \right)^{1/2} \,. \tag{3}$$

2.2 Selection of the 'Typical' Days (Step 1)

In the first step, the 'typical' days are selected based on the clustering of the 365 daily 24-h vectors of a year subject to a minimisation of a distance function. In order to include constraints for the time segment selection, a multi-objective combinatorial optimisation is chosen. Restricting the definition of typical days to elements of the underlying vector set, an optimal clustering for a given cluster number (CL^{lim}) may be obtained based on Eqs. (4)–(6)

$$\sum_{\tilde{d} \in D^{SS}(d)} x_{d,\tilde{d}} = 1 \,, \tag{4}$$

$$\sum_{\tilde{d} \in D} x_{d,\tilde{d}} \leq M \cdot x_{\tilde{d}} \,, \tag{5}$$

$$\sum_{\tilde{d} \in D} x_{\tilde{d}} \leq CL^{lim} \,, \tag{6}$$

where $x_{\tilde{d}} \in \{0, 1\}, x_{d,\tilde{d}} \in \{0, 1\}$ and $x_{d,\tilde{d}}$ denotes the binary clustering decision of representing day d through the profile of $\tilde{d}$ under a certain mapping policy.[1] Demanding that each day is assigned to exactly one typical day (Eq. 4), the selection of the typical $x_{\tilde{d}}$ is defined by the Big-M method (Eq. 5). The implementation of the preliminary discussed constraints of the time segment selection is achieved by demanding the critical sets S^{lwp} to be nonempty under the current clustering decision:

$$\sum_{\tilde{d} \in S^{lwp}(\tilde{d})} x_{\tilde{d}} \geq 1 . \tag{7}$$

For handling the multiple objectives of selecting 'typical' time slices for every energy conversion technology and energy consumption profile on a nodal and global basis, the PTDF based error measure defined above is used. Basically, this approach allows reducing the multiple objectives to a single distance which captures the impact of clustering the residual load bus injection vector from a flow based point of view. In order to avoid a balancing between the different bus injection types (load, wind and photovoltaic energy), which may be undesirable in the further model-based processing, we define the 24 h clustering distance for all target dimensions $\tau \in \{load, wind, pv, residual\ load\}$ as follows:

$$c_{\tau,d,\tilde{d}} = \sqrt{\sum_{h \in \{1,\dots,24\}} \sum_{l \in L} \left(\sum_{n \in N} \phi_{n,l} \cdot \left(v_{n,\tau,d,h} - v'_{n,\tau,\tilde{d},h} \right) \right)^2} , \tag{8}$$

where v and v' are the hourly nodal profile vectors defined over the set of 24-h vectors of an underlying year and $\phi_{n,l}$ is the power transfer distribution factor, determining the impact of an injection at bus n on the flow over branch l. Based on the reduced coefficient matrix of the clustering distance and the binary decision variable $x_{d,\tilde{d}}$, we obtain the following clustering costs:

$$\gamma_\tau = \sum_{d,\tilde{d}} c_{\tau,d,\tilde{d}} \cdot x_{d,\tilde{d}} . \tag{9}$$

For an optimisation of the remaining multiple dimensions of the clustering cost a general goal programming formulation is chosen:

$$min_{z^{L_1},z^{L_\infty}} z = \alpha \cdot z^{L_1} + (1 - \alpha) \cdot z^{L_\infty} , \tag{10}$$

$$\text{with} \quad z^{L_1} = \sum_{\tau \in \{load, wind, pv, residual\ load\}} \omega_\tau \cdot \frac{p_\tau}{\gamma_\tau^{min}} , \tag{11}$$

---

[1] Restricting the mapping to a subset of days $D^{SS}(d)$ may be desirable, e.g., to avoid a mapping of profiles from working days to weekends.

$$z^{L_\infty} \geq \omega_\tau \cdot \frac{p_\tau}{\gamma_\tau^{min}} \, , \qquad (12)$$

$$\gamma_\tau^{min} = \gamma_\tau + p_\tau - n_\tau \, , \qquad (13)$$

where z^{L_1} and z^{L_∞} represent the weighted positive percentage deviations from the minimal cost targets γ_τ^{min} based on the L_1 and L_∞ norm for a multi-objective optimisation with a focus on an efficient or balanced solution, respectively.

2.3   Intraday Time Slice Reduction (Step 2)

A further reduction of the time slice number on an intraday basis is similarly modelled to the previous clustering of the daily 24-h vectors. Restricting the definition of the optimal time slices of the previously selected typical days to the underlying 24 h of a day and the set of possible aggregations of subsequent hours on a two hour level or three hour level (during the night), an optimal aggregation of time slices for a given upper limit (TS^{lim}) may be obtained by:

$$\sum_{ts \in TS(ts,h)} x_{\tilde{d},ts} = 1 \quad \forall h \, , \qquad (14)$$

$$\sum_{ts \in TS(ts,h)} \sum_{h \in TS(ts,h)} x_{\tilde{d},ts} = 24 \quad \forall h \, , \qquad (15)$$

$$\sum_{\tilde{d} \in \tilde{D}} \sum_{ts \in TS(ts,h)} x_{\tilde{d},ts} \leq TS^{lim} \, , \qquad (16)$$

with $x_{\tilde{d},ts} \in \{0, 1\}$, $TS = \{ts_1, \ldots, ts_{54}\}$ and

$$\begin{cases} ts_1 = \{h_1\}, & \ldots \, ts_{24} = \{h_{24}\} & \text{for } 1-\text{hour intervals} \, , \\ ts_{25} = \{h_1, h_2\}, & \ldots \, ts_{47} = \{h_{23}, h_{24}\} & \text{for } 2-\text{hour intervals} \, , \\ ts_{48} = \{h_1, h_2, h_3\}, & \ldots \, ts_{54} = \{h_{22}, h_{23}, h_{24}\} & \text{for } 3-\text{hour intervals} \, , \end{cases} \qquad (17)$$

where $x_{\tilde{d},ts}$ denotes the binary decision of clustering the corresponding hours of the previously selected optimal typical day $\tilde{d} \in \tilde{D}$. Similar to the clustering in step 1, Eq. (14) defines that each underlying value is assigned to exactly one cluster, while Eq. (15) demands that the combinations of the time slices need to represent the 24 h of a day. Analogously to step 1, the constraints of the time segment selection are defined by requiring that the critical sets S^{lwp} should be nonempty under the current clustering decision:

$$\sum_{(\tilde{d},ts) \in S^{lwp}(\tilde{d},ts)} x_{\tilde{d},ts} \geq 1 \, . \qquad (18)$$

Given a known clustering policy of 24-h vectors $\tilde{D}(d, \tilde{d})$, the distance function for the clustering of time slices is defined as follows:

$$c_{\tau,\tilde{d},ts} = \sqrt{\sum_{d\in\tilde{D}(d,\tilde{d})} \sum_{h\in TS(ts,h)} \sum_{l\in L} \left(\sum_{n\in N} \phi_{n,l} \cdot \left(v_{n,\tau,d,h} - v'_{n,\tau,\tilde{d},h}\right)\right)^2}. \qquad (19)$$

Based on the reduced coefficient matrix of the clustering distance and the binary decision variable $x_{\tilde{d},ts}$, we obtain the following clustering costs:

$$\gamma_\tau = \sum_{\tilde{d}\in\tilde{D},ts\in TS} c_{\tau,\tilde{d},ts} \cdot x_{\tilde{d},ts}. \qquad (20)$$

For an optimisation of the remaining multiple dimensions of clustering costs, a constraint programming formulation is chosen, with the goal of finding the efficient supported solutions. Due to computational efficiency, the four target dimensions (load, wind energy, pv, residual load) are reduced to two target dimensions based on the L_1 norm (efficiency objective) and L_∞ norm (balancing objective). The algorithmic implementation of the first phase of the two phase method corresponds to [6]. A detailed explanation is therefore omitted.

3   Selected Results

A time segment selection based on simulated transmission grid injection data of 2012 shows that the right hand side error of the load flow restriction (Eq. 2) becomes rather insensitive above a certain temporal resolution. An intraday reduction of the time structure showed no significant increase of the error in the range of 14 to 21 typical days illustrating that the marginal gain of additional time segments decreases after a number of approximately 300 time segments (see Fig. 1).

Obviously, an even lower potential load flow error could be achieved with the same temporal resolution, in the event that the multi-objective nature of the problem or the need of including extreme situations could be ignored. In this case, a less advanced and quicker clustering technique, such as a k-Means clustering approach could be utilised. In our application, however, the multi-objective nature and extreme situations cannot be ignored. Nevertheless, Fig. 1 visualises the impact of the proposed new PTDF-based distance measure. Despite the higher degree of freedom in optimisation, a k-Means clustering is not able to outperform the proposed multi-dimensional approach for time slice selection if the proposed PTDF-based distance measure is not applied.

Fig. 1 Development of the difference in cumulative load flow over all lines for different numbers of typical days: comparison of our approach ('Multidim. constrained') with two k-Means variants

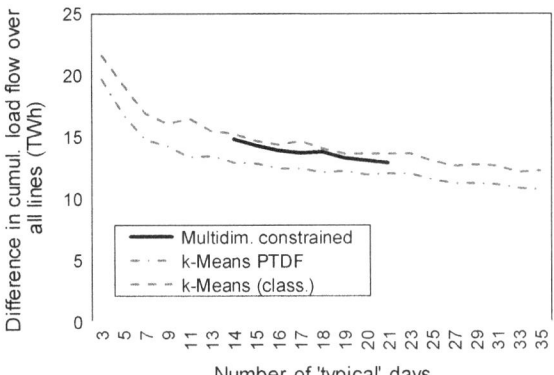

4 Conclusions and Outlook

We proposed a structured, multi-objective approach for time segment selection to handle the complexity of large-scale power system models. In this paper, we focussed on time segment selection for power systems operation optimisation with a special emphasis on power grid utilisation. Our approach is therefore aimed at preserving both, intra-period and intra-regional links within and between the volatile electricity supply and demand profiles. We could show that even an amount of less than 300 time segments may be sufficient for the modelling of a whole year, if chosen carefully. Future enhancements of the approach should include the extension to power generation and transmission expansion planning problems, simulation-based sensitivity analyses, accounting for variations between the importance of load, wind and solar energy profiles respectively, as well as further validation studies for a series of realistic energy economic problems.

References

1. Consentec and IAEW. Regionalisierung eines nationalen energiewirtschaftlichen Szenariorahmens zur Entwicklung eines Netzmodells (NEMO)-Survey on behalf of the Bundesnetzagentur, Bonn, Germany. (Consentec GmbH and Institut für Elektrische Anlagen und Energiewirtschaft (IAEW), 02-04-2012), 2012
2. Fehrenbach, D., Merkel, E., McKenna, R., Karl, U., Fichtner, W.: On the economic potential for electric load management in the german residential heating sector-an optimising energy system model approach. Energy **71**, 263–276 (2014)
3. Hawkes, A., Leach, M.: Impacts of temporal precision in optimisation modelling of micro-combined heat and power. Energy **30**(10), 1759–1779 (2005)

4. Haydt, G., Leal, V., Pina, A., Silva, C.A.: The relevance of the energy resource dynamics in the mid/long-term energy planning models. Renew. Energy **36**(11), 3068–3074 (2011)
5. Sandberg, J., Larsson, M., Wang, C., Dahl, J., Lundgren, J.: A new optimal solution space based method for increased resolution in energy system optimisation. Appl. Energy **92**, 583–592 (2012)
6. Stidsen, T., Andersen, K.A., Dammann, B.: A branch and bound algorithm for a class of biobjective mixed integer programs. Manage. Sci. **60**(4), 1009–1032 (2014)

Practical Application of a Worldwide Gas Market Model at Stadtwerke München

Maik Günther

Abstract In this paper the worldwide gas market model WEGA and the base case scenario of Stadtwerke München (SWM) are described. The potential of WEGA in practical application is shown through three sensitivity analyses. Firstly, it is demonstrated that shale gas in Europe has only a small impact on European gas prices in future. Nevertheless, shale gas is very important for security of supply in Europe. The second sensitivity analysis is calculated with modified U.S. liquefied natural gas (LNG) export volumes. A surge of LNG would significantly decrease the gas price in Europe. On the other hand, Europe will be oversupplied till the end of the twenties and consequently, the prices will be relatively immune to few U.S. LNG export volumes in this period. At the third sensitivity analysis, gas demand is modified. The results reveal that the gas prices are very sensitive to changes of demand.

1 Introduction

SWM has invested in all stages of the value chain of natural gas. It ranges from exploration and production in North Sea to distribution and downstream. SWM also owns gas-fired power plants and heating plants. Thus, it is important for SWM to have a detailed knowledge of the gas market and gas prices in the next 30 years. Additionally, the knowledge of price sensitivity to modifications of parameters such as gas demand or geopolitical situations is a competitive advantage. SWM has been applying the worldwide gas market model WEGA since 2013 to calculate the gas prices in long-term horizon as a basis for investment decisions.

WEGA is a mathematical model based on Linear Programming (LP). The model optimizes daily gas flows till 2040 to cover the demand of each country/trading point with the cheapest possible solution. Results are detailed gas flows and hub prices, e.g. for NetConnect Germany (NCG). This contribution answers to the following questions using WEGA: How the gas prices react to establishment of European moratorium on fracking, changes of U.S. LNG export volumes and increase or decrease

M. Günther (✉)
Stadtwerke München GmbH, 80287 Munich, Germany
e-mail: guenther.maik@swm.de

© Springer International Publishing Switzerland 2017
K.F. Dœrner et al. (eds.), *Operations Research Proceedings 2015*,
Operations Research Proceedings, DOI 10.1007/978-3-319-42902-1_96

of gas demand. To answer these questions the following research methods are used. Data about the worldwide gas market are collected by analysis of public and commercial literature and databases. In this context forecasting techniques are applied to prepare data. Thereafter experiments are done with WEGA. The results are analyzed and discussed in expert interviews, to check their plausibility.

The paper is structured as follows. In the next section, the gas market model WEGA will be described. Subsequently, the SWM base case scenario of the gas market is introduced. The sensitivity analyses are performed on the base case in Sect. 4. This paper ends with a conclusion.

2   Gas Market Model WEGA

Gas market models can be categorized into models with oligopolistic competition and with perfect competition. WEGA dispatches the sources through a perfect competition model under various constraints. Market power of producers as well as investment decisions are fixed. However, these values can be modified through introduction of scenarios. The modeling concept of WEGA can be compared with the TIGER model from EWI [3]. For an overview of gas market models reference is made to Chyong and Hobbs [1] and Holz et al. [4].

WEGA is a worldwide gas market model with a focus on Europe. Europe is therefore, modeled much more in detail than the rest of the world. Modeled assets in WEGA consist of LNG liquefaction terminals, LNG regasification plants, pipelines, interconnections, storages and gas fields. WEGA also contains a database of contracts with volumes, take or pay clauses and price formulas. Due to the large amount of data, business intelligence software is integrated in the graphical user interface of WEGA to analyze input data and results.

WEGA is a LP based model developed under FICO Xpress Optimization Suite. It is based on the commercial gas market model Pegasus from Pöyry Management Consulting (UK) Ltd [2]. SWM purchased the model source code and the dataset from Pöyry. However, SWM has own assumptions and forecasts of parameters, e.g. the gas demand for power sector in Europe comes from an own fundamental power model [5]. Thus, the dataset from Pöyry was changed by SWM and all scenarios and results from WEGA in this paper are an own view of SWM. To modify the original dataset SWM uses public sources as well as commercial data services from PIRA Energy Group, Wood Mackenzie, IHS Cera and Bloomberg New Energy Finance. For plausibility checks, NCG, NBP and TTF future prices are frequently calculated to make sure that WEGA is generating accurate market prices of the next years. In addition, results are discussed with experts of SWM and other companies.

The worldwide gas market in WEGA is modeled as a network consisting of a set of nodes and edges. Each edge is a connection of two nodes. Figure 1 shows an example of two demand zones with interconnections, storages, demand shedding (DS), indigenous production and delivery points (DP). A delivery point is a pipeline

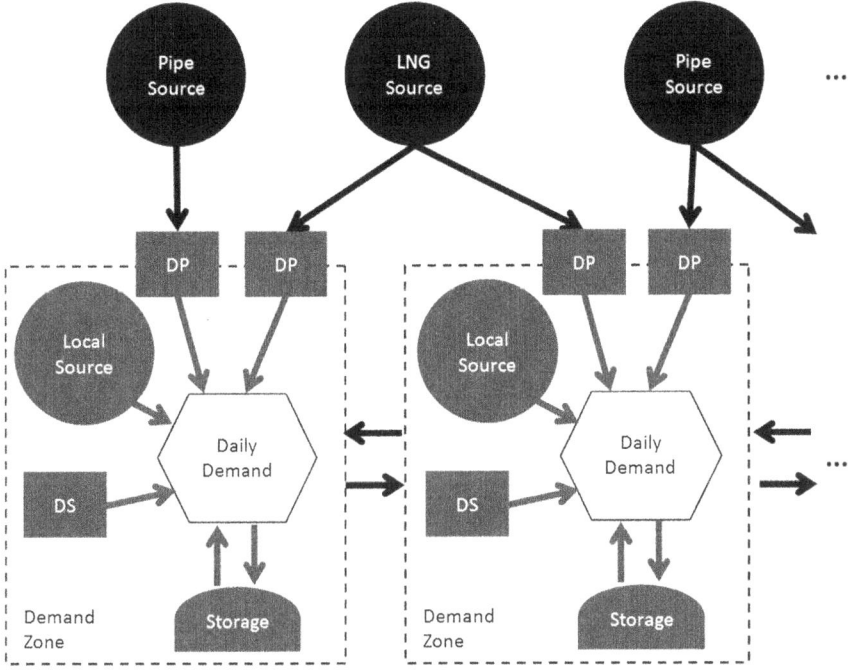

Fig. 1 Example of two demand zones

landing point or a LNG regasification plant. A demand zone is a trading point, a region or a country.

The planning horizon is always a gas year. Gas years are optimized separately under perfect foresight. The fitness function can be described as follows: The overall costs to cover the demand in each demand zone have to be minimized. These costs include costs of production, transport and processing besides market entry costs, costs of storages and costs for demand shedding (if necessary). It is a hard constraint to cover the demand of each demand zone on a daily basis. Further hard constraints are e.g. capacity restrictions of assets, take or pay clauses and volume restrictions of contracts.

3 Base Case Scenario

The base case scenario of the worldwide gas market is an own view of SWM. SWM considers the calculated gas prices as confidential data and restricts publication of them. However, important parameters will be described here to have a good understanding of the base case.

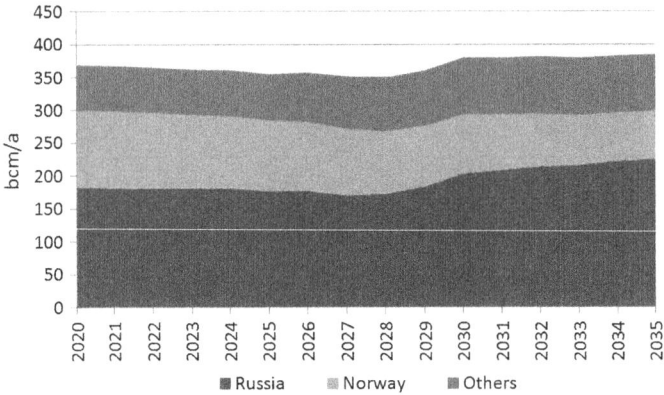

Fig. 2 Production capacities for pipeline supply from outside Europe

The gas demand in Europe is expected to reach 570 billion cubic meters of gas per annum (bcm/a) in 2035, implying a slight increase compared with today. An important factor in European supply is LNG. Worldwide LNG export capacity will be nearly doubled in the next 20 years reaching to approx. 630 bcm/a in 2035. The additional volumes come mainly from Australia, U.S., South East Africa and Russia. Indigenous production in Europe is important for security of supply, but will decline from 120 bcm/a in 2020 to 62 bcm/a in 2035. The potential of shale gas in Europe with 8 bcm/a in 2035 is relatively low. The volumes are mainly expected from GB, Poland and Spain. Production capacities for pipeline supply to Europe are depicted in Fig. 2. While the production volumes in Norway shrink, it grows in Russia. Some of these Russian gas fields will be connected to China and Europe in future.

4 Experiments and Results

Due to the internal restrictions on publishing the prices, the results of the scenarios are presented by the percentage of deviation from base case prices. Only results for NCG are published to avoid overloaded diagrams. NCG prices are relatively close to TTF or NBP prices under normal conditions.

At first, the shale gas assumptions are modified in base case. A scenario in which a complete ban on fracking in Europe occurs is compared with another scenario which expects a shale boom in Europe. It is assumed that the level of indigenous production in 2020 is not declining and is stabilized with the shale gas production. It has to be noted that a shale boom in Europe, especially in such a short time, is unrealistic. This modification is only calculated to show the maximal effect on NCG prices (dotted line in Fig. 3). This price effect of is relatively small, because the European gas market is oversupplied till the end of the twenties. There are enough cheap sources to cover the demand and shale gas production in Europe is on average relatively expensive

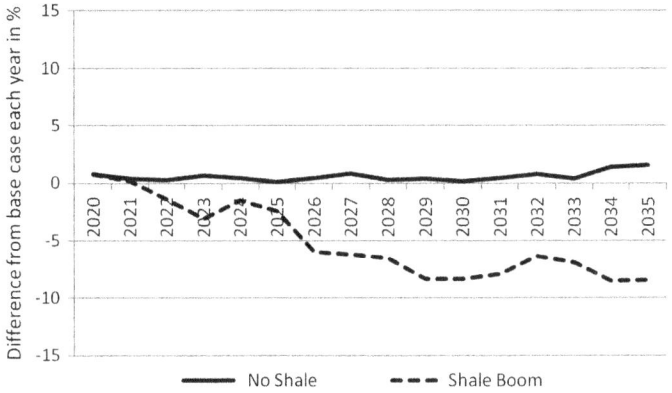

Fig. 3 Price effect at NCG without shale gas or with a shale gas boom in Europe

compared to the shale gas produced in North America. The black line shows a ban on fracking. This would have only a small effect on NCG prices, because expected shale gas volumes in base case are relatively small and expensive. It should be noted that a strong indigenous production is important for security of supply.

The second sensitivity analysis was done with U.S. LNG exports. Here, the annual export capacity of each year is doubled or halved. Figure 4 reveals that the U.S. LNG exports are an influential factor in price development at NCG. Halving the U.S. export volumes in the beginning of the period has a low impact on prices mainly due to small U.S. export volumes and a well supplied European market. However, the price difference will soar as the time pass and remain between 10 to 15 %. On the other hand, doubling the U.S. LNG export volumes of each year reduces the prices by more than 10 %.

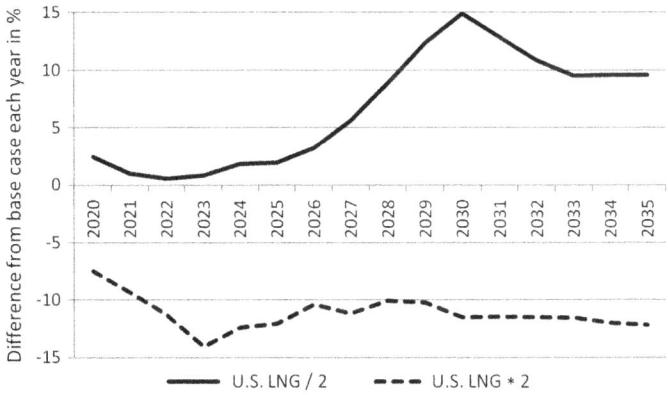

Fig. 4 Price effect at NCG with halved or doubled U.S. LNG export capacity each year

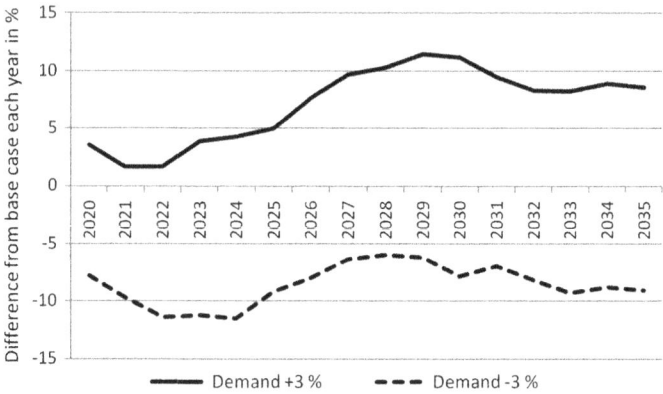

Fig. 5 Price effect at NCG with ±3 % gas demand each year

At the last sensitivity analysis, gas demand in Europe is modified by increasing or decreasing gas demand in each year by 3 %. Figure 5 demonstrates the results of the analysis. The impact of increasing gas demand is not harsh in the beginning of the period; however, will fluctuate around 10 % in the rest of the period. With a reduction in European gas demand, price differences range between −6 and −12 %.

5   Conclusion

In this paper the gas market model WEGA and the base case scenario of SWM were described. Three sensitivity analyses were presented. It was demonstrated that shale gas in Europe has only a small impact on European gas prices in future. Nevertheless, shale gas is very important for security of supply in Europe. A second sensitivity analysis was calculated with modified U.S. LNG export volumes. A surge of LNG would significantly decrease the gas price in Europe. On the other hand, Europe will be oversupplied till the end of the twenties and consequently, the prices will be relatively immune to few U.S. LNG export volumes in this period. At the third sensitivity analysis, gas demand was modified. The results reveal that the gas prices are very sensitive to changes of demand.

Only one parameter in each scenario was modified. A combination of scenarios with the same price direction (e.g. no shale gas, less U.S. LNG exports and higher demand) may result in much higher variations.

References

1. Chyong, C.K., Hobbs, B.F.: Strategic eurasian natural gas market model for energy security and policy. Analysis **44**, 198–211 (2014)
2. Davies, G., Sarsfield-Hall, R.: Gas SCR—Cost Benefit Analysis for a Demand-Side Response Mechanism. A report to Ofgem, Pöyry (2004)
3. Hecking, H., John, C., Weiser, F.: An Embargo of Russian Gas and Security of Supply in Europe. EWI (2014)
4. Holz, F., von Hirschhausen, C., Kemfert, C.: A strategic model of european gas supply (GAS-MOD). Energy Econ. **30**, 766–788 (2008)
5. Schaber, K., Roth, H., Fallahnejad, M.: Can the gas sector provide the flexibility to the power sector for the integration of renewables? In: 13th WIW, pp. 46–51 (2014)

9 783319 429014